Liquid Metals, 1976

Liquid Metals, 1976

Invited and contributed papers from the Third International Conference on Liquid Metals held at the University of Bristol, 12–16 July 1976

Edited by R Evans and D A Greenwood

Conference Series Number 30

The Institute of Physics
Bristol and London

CODEN IPHSAC 30 1–672 (1977)

British Library Cataloguing in Publication Data

International Conference on Liquid Metals, 3rd, University of Bristol, 1976
Liquid metals, 1976. (Institute of Physics. Conference Series; No. 30)
ISBN 0 85498 120 9
ISSN 0305 2346
1. Title 2. Evans, R 3. Greenwood, D A 4. Series
546′.3 TA463
Liquid metals – Congresses

The Third International Conference on Liquid Metals was organized by The Institute of Physics

Honorary Editors

R Evans and D A Greenwood

Published by The Institute of Physics, Techno House, Redcliffe Way, Bristol BS1 6NX, and 47 Belgrave Square, London SW1X 8QX, in association with the American Institute of Physics, 335 East 45th Street, New York, NY 10017, USA.

Set in 10/12 Press Roman and printed in Great Britain by Adlard and Son Ltd, Dorking, Surrey.

Preface

The Third International Conference on Liquid Metals was held in the H H Wills Physics Laboratory of the University of Bristol from 12–16 July 1976. Since the Second International Conference at Tokyo in 1972 there has been an economic crisis, and we sometimes wondered, as we made preparations for the Third, if it would indeed take place. In fact scientific budgets proved sufficiently resilient and interest in liquid metals sufficiently keen to bring some 140 delegates from overseas and 40 British participants to Bristol.

The very full programme of the conference reflected the shifts and growths of interest. There were, for example, crowded sessions on the transition and rare earth metals and these sessions also included discussions on these metals in the amorphous state. More coherent patterns are beginning to emerge in the study of liquid semiconductors and this session attracted many experimental papers. The considerable theoretical progress in understanding the thermodynamics and structure of simple liquid metals was well represented. In planning the programme, attempts were made to give some emphasis to the links between the physics and chemistry of liquid metals and this resulted in an interesting session, though the number of chemists we attracted to the conference was rather small.

We must thank both the International Advisory Board: N W Ashcroft (USA), L E Ballentine (Canada), V Bortolani (Italy), F Cyrot-Lackman (France), P A Egelstaff (Canada), H Endo (Japan), H-J Güntherodt (Switzerland), F Hensel (West Germany), J Jortner (Israel), A Lodding (Sweden), N H Nachtrieb (USA) and M Watabe (Japan), and the British Committee: C C Addison, N E Cusack, J E Enderby, T E Faber, D A Greenwood (Conference Secretary), J S L Leach, N H March, E F W Seymour and J M Ziman (Conference Chairman), for their help and advice at various stages in the planning of the conference. We are grateful to the University of Bristol for providing facilities for the conference, to The Institute of Physics for their logistic and financial support and to the City of Bristol for hospitality.

We are grateful to the various referees who helped us with the difficult task of selecting papers for publication. The Proceedings would not have appeared so promptly without the efforts of Ken Hall, Staff Editor at The Institute of Physics and we thank him for his invaluable assistance.

At an informal meeting of the International Advisory Board it was agreed there was scope for a Fourth International Conference. We may look forward to this being held in Grenoble, France, in 1980.

August 1976

R Evans
D A Greenwood

Contents

Part 2: Electronic properties

Inst. Phys. Conf. Ser. No. 30 © 1977: Chapter 1, Part 1

Interatomic forces and thermodynamic properties of liquid metals: a hard-sphere description

W H Young
School of Mathematics and Physics, University of East Anglia, Norwich, UK

Abstract. Good theoretical hard-sphere structure factors and entropies are now available both for one- and two-component systems and can be used in conjunction with the corresponding observed quantities to gain insight into the broad features of real interatomic forces.

On the basis of structure factor evidence, most liquid metals behave quite like hard spheres. This is consistent with the usual view of the interatomic potential with a hard core and a minimum near which nearest neighbours are found. It is shown that for such systems the measured structure factors can be used to obtain lower bounds to the absolute entropies and upper bounds to the heat capacities.

Loss of core hardness is signalled by damping and a phase shift inwards at high momentum transfers. Such behaviour is particularly noticeable in the alkalis; their differences from the good hard-core metals is only one of degree. The qualitatively different cases (e.g. Sn, Bi) exhibit a low-lying shoulder on the first peak which suggests that the interatomic potential has a destabilizing effect on nearest neighbours.

In the hard-core cases, entropy observations on simple metals can confirm the structure factor evidence as well as provide a measure of hardness; in transition metals the two types of observations together lead to the isolation of the electronic entropy contributions and therefore to the density of states parameters.

On mixing, a binary hard-sphere model is quite successful in explaining the excess entropies, provided the *observed* volumes are used and the *total* hard-core volume for both species is conserved.

Ab initio pseudopotential calculations embodying many of the above features have had some degree of success. For pure metals, entropies and heat capacities of about the right size and with roughly the right trend from metal to metal usually are obtained, while the variation of the latter over a wide temperature range (the initial fall and the subsequent rise) can be described. Furthermore, provided the pseudopotentials are carefully chosen, the same data which serve for the pure metals can be used to predict entropies of mixing.

1. Introduction

Our understanding of the electronic properties of liquid metals and alloys has been influenced crucially by pseudopotential theory. In particular we now have a qualitative understanding of interatomic forces (how they arise, what they look like and what their effects are) and, in favourable cases, can even make semi-quantitative calculations.

While the above work was being developed, successes were being achieved in classical liquid theory where the aim was to predict the thermodynamic properties for a system with prescribed forces (e.g. Lennard-Jones fluids). The thrust of some recent work is to see to what degree the classical liquid techniques can be successfully applied to liquid metals.

Progress on the classical problem hinged on two developments which, in brief, were (i) the ability to describe with high precision the thermodynamic and related properties of hard spheres and (ii) the advent of perturbation techniques that allow the properties of real fluids to be expressed in terms of those appropriate to hard spheres. In this way, the properties of a Lennard-Jones fluid have been successfully described by Mansoori and Canfield (1969), Weeks *et al* (1971) (the WCA theory) and Anderson *et al* (1971).

The approach of Mansoori and Canfield was called the Gibbs–Bogoliubov (GB) method by Isihara (1968). In its simplest form (that in which all calculations have been performed so far), the perturbation series for the Helmholtz free energy (Lukes and Jones 1968) is terminated and the resulting upper bound provides the basis for a variational method. This technique, first applied to metals by Jones (1971), is the subject of the present review.

The more sophisticated WCA approach has also been used in liquid metal theory (e.g. by Wehling *et al* 1972, Hasegawa and Watabe 1974 and Kumaravadivel and Evans 1977). It will not be reviewed here partly because the picture for the method is less complete and partly because the work of Kumaravadivel and Evans follows in these proceedings. Its relevance to the present paper will, however, emerge.

In §2, the GB method is introduced and applied to pure metals. Only the more general inferences (independent of such computational features as pseudopotential shapes and screening techniques) are considered at this stage. In §3 a similar discussion for alloys takes place. Then, in §4 *ab initio* calculations are discussed and finally, in §5 there is a summary.

2. Pure metals

2.1. *Hard spheres*

Consider a one-component hard-sphere fluid at temperature T. Suppose that the mass of a sphere is M and its diameter is σ so that its volume is $\omega = \pi\sigma^3/6$. Thus, if Ω is the container volume per sphere, the packing fraction is $\eta = \omega/\Omega$. At densities relevant to liquid metals, the thermodynamic and related properties of such a system are now known virtually exactly as a result of computer studies.

In figure 1 are shown typical results for $g_{hs}(r)$, the radial distribution function, and $a_{hs}(q)$, the corresponding structure factor. Of the simple approximate theories, that of Percus and Yevick (PY) has had the greatest empirical success and results for this case are shown also.

Carnahan and Starling (1969) have proposed a semi-empirical formula for the equation of state of a hard-sphere fluid. This leads to an entropy per ion

$$S_{hs} = S_{gas} + S_\eta \tag{1}$$

where

$$S_{gas}/k = 2{\cdot}5 + \ln\,[\Omega\,(MkT/2\pi\hbar^2)^{3/2}] \tag{2}$$

is a perfect gas contribution and

$$S_\eta/k = -(\zeta - 1)(\zeta + 3) \tag{3}$$

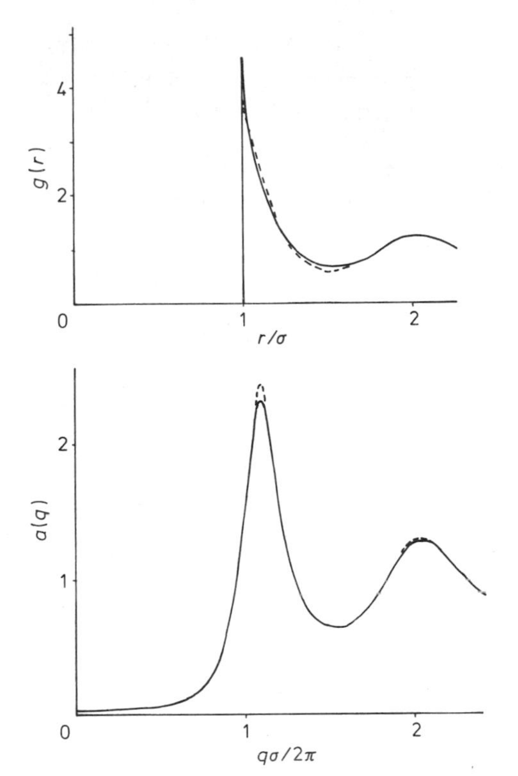

Figure 1. Hard-sphere radial distribution function $g_{hs}(r)$ and structure factor $a_{hs}(q)$ for packing fraction 0·445. The full line $g_{hs}(r)$ corresponds to essentially exact machine calculations (Barker and Henderson 1971); the corresponding transform $a_{hs}(q)$ is due to B N Perry and M Silbert (private communication) and is based on the real space parametrization of Verlet and Weis (1972). The broken lines correspond to the PY approximation (Throop and Bearman 1965 in real space, Thiele 1963 and Wertheim 1963 in inverse space).

(where $\zeta = (1 - \eta)^{-1}$) is the modification arising from finite packing. Equation (1) is compatible with the PY approximation but comparison with computer studies (Alder and Wainwright 1957) shows it to be essentially exact at packings of interest.

Ideally, equation (1) should be used in applications together with exact structure factors. In practice though, the PY versions of the latter have been used very often. This is because (i) the latter are available in convenient closed form (Wertheim 1963, Thiele 1963) and continue to be so even for binary mixtures, and (ii) errors in other parts of an application (e.g. in the calculation of interatomic forces in metals; see §2.3 below) are often much more serious.

2.2. *Thermodynamic variational method*

Starting from the work of Zwanzig (1954), various forms of perturbation theory have been developed for real fluids. They have in common the underlying assumption that in the zero order, such fluids can be described by hard spheres. The basic theorem for the first-order variational GB method to be discussed in this paper now follows.

Let $H = K + V$ be the Hamiltonian of a system of interest and $H_{hs} = K + V_{hs}$ be that for a reference system of hard spheres having the same masses and occupying the same volume as the actual particles. Then for any chosen sphere diameter σ, an upper bound to the Helmholtz free energy per ion, at fixed temperature and volume, is

$$\begin{aligned} F &= F_{hs} + \langle H - H_{hs} \rangle_{hs} \\ &= F_{hs} + \langle V \rangle_{hs} . \end{aligned} \qquad (4)$$

Here angular brackets denote an appropriate expectation value per atom and $F_{hs}=\frac{3}{2}kT-TS_{hs}$ with S_{hs} given by equation (1). If, at most, V contains pairwise interactions, $\langle V\rangle_{hs}$ can be calculated using the g_{hs} (or a_{hs}) discussed in §2.1; many-ion interactions would require knowledge of higher-order hard-sphere correlation functions.

One must always minimize the free energy with respect to σ and so obtain the 'best' hard-sphere reference system. Accordingly, it is now to be understood that

$$(\partial F/\partial\sigma)_{\Omega,T}=0 \tag{5}$$

and that the optimizing σ is used in all expressions. In particular, this σ will define a hard-sphere structure factor a_{hs} which is meant to approximate that of the real system.

Another immediate and important consequence is that the entropy estimate for the real system contains no contribution from the thermal variation of σ (Edwards and Jarzynski 1972). For

$$S = -\left(\frac{\partial F}{\partial T}\right)_{\Omega} = -\left(\frac{\partial F}{\partial T}\right)_{\Omega,\sigma} - \left(\frac{\partial F}{\partial\sigma}\right)_{\Omega}\left(\frac{\partial\sigma}{\partial T}\right)_{\Omega}$$

the final term vanishes via equation (5). Thus, by equation (4),

$$S = S_{hs} - \left(\frac{\partial}{\partial T}\langle V\rangle_{hs}\right)_{\Omega,\sigma}. \tag{6}$$

Now g_{hs} and a_{hs} have no explicit T-dependence. Thus for insulators, when V is independent of T, the final term vanishes. For metals it is non-vanishing to the extent that the electron gas is non-degenerate; to first order in T (all that is needed in practice) it may be written (Silbert *et al* 1975, Meyer *et al* 1976)

$$S = S_{hs} + S_{elec} \tag{7}$$

where

$$S_{elec} = \frac{\pi^2}{3}N(E_F)k^2T. \tag{8}$$

This result is valid even when multi-ion forces are present.

2.3. *Interatomic potentials*

I will now write

$$V = V_0 + \tfrac{1}{2}\sum_{i\neq j} v(\mathbf{r}_i - \mathbf{r}_j) + \dots. \tag{9}$$

For metals, pseudopotential theory tells us how to describe the various terms of this series. In particular, many calculations of pairwise interactions have been made. The technical problem (Shaw and Heine 1972) is to define a normalized energy wavenumber characteristic $\Phi_N(q)$ and effective valency z^* in terms of the pseudopotential and the properties of the electron gas and then transform to obtain

$$v(r) = \frac{z^{*2}}{\pi}\int_0^\infty [1-\Phi_N(q)]\frac{\sin qr}{qr}\,dq. \tag{10}$$

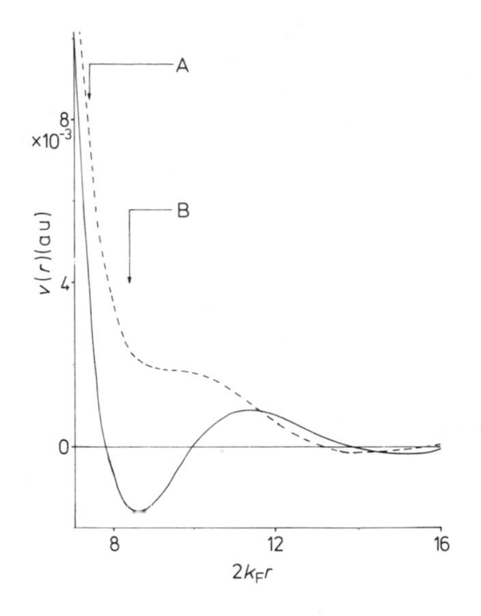

Figure 2. Calculated interionic potentials for Mg and Zn (Appapillai and Heine 1972). Point A corresponds to the hard-core diameter assuming $\eta = 0{\cdot}45$ and B is the nearest-neighbour distance in solid.

Examples of calculated v are shown in figure 2 (taken from Appapillai and Heine 1972) for Mg and Zn. That shown for Mg is, qualitatively, the v that is generally held to be characteristic of simple metals. It is quite like a Lennard-Jones potential at near-neighbour distances (Jones 1973a) so the GB and similar methods should be applicable.

That shown for Zn has a core but there is also something of a ledge. A hard-sphere description will be deficient to some extent in any system for which such a potential is appropriate (and it is not being suggested that this curve *is* appropriate to Zn). On the other hand there is still a hard core so one might expect a simple hard-sphere reference system to continue to operate at a more limited level of usefulness.

Despite much effort, the v are still quite poorly known in general and, for a given metal, a search of the literature will often reveal calculated curves of each of the types discussed above. The primary cause of this is the sensitivity of the results to the pseudopotential and screening details (Shaw and Heine 1972) but even thermally induced volume changes can produce significant variations.

Finally, it is appropriate to bear in mind that almost all work to date has been to second order in the pseudopotential. Recently, however, Hasegawa (1976) has calculated quite large third-order contributions to v for Na and K. Fortunately, the situation is possibly retrievable; Rasolt and Taylor (1975) and Dagens *et al* (1975) suggest that a suitable parametrization of the pseudopotential takes higher-order terms into account.

2.4. Structure factors

Observed structure factors do indeed resemble those for hard spheres. A common way of demonstrating this is to compare them with hard-sphere forms characterized by packing fractions chosen to match the observed heights of the principal peaks. It has been known since the work of Ashcroft and Lekner (1966) that $\eta \sim 0{\cdot}45$ is about right at the melting temperature.

To illustrate this point, figure 3 shows results for Na and Al. In Na, significant phase and amplitude differences develop in the structure factors at high scattering angles due

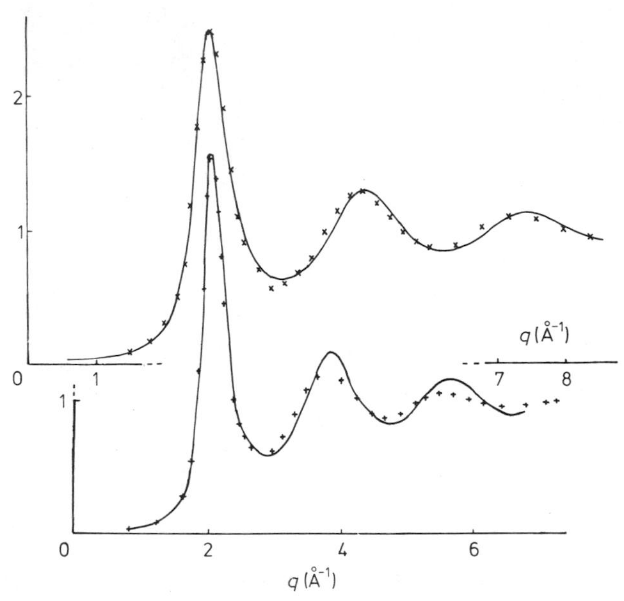

Figure 3. Structure factors for Na at 373 K (plusses) and Al at 943 K (crosses). Also drawn (full lines) are the corresponding PY hard-sphere forms with η = 0·475 and 0·45 respectively, chosen to match the observed first peak heights.

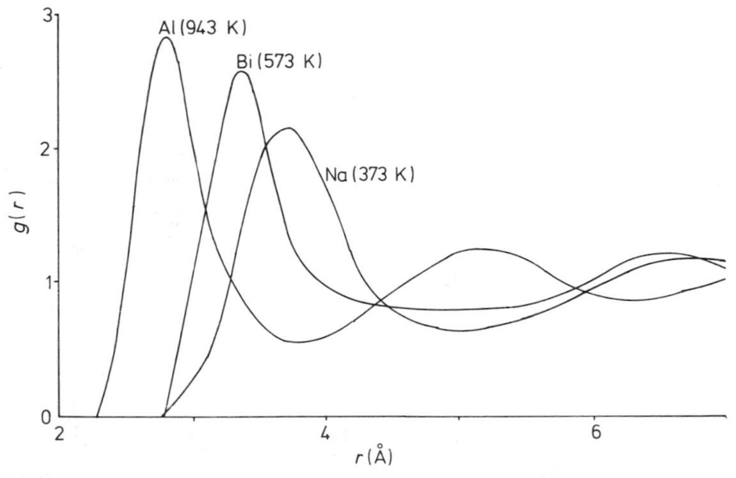

Figure 4. Radial distribution functions for Na, Al and Bi near their respective melting temperatures. They correspond to the observed data shown in figures 3 and 5.

to the relatively soft core in this metal (Page *et al* 1969, Hansen and Schiff 1973) and similar behaviour occurs in the other alkalis. The second case is more typical; quite good hard-sphere behaviour occurs in Mg, Al, In, Pb (Waseda and Suzuki 1973), the alkaline earths (Waseda *et al* 1974), the noble and transition metals (Waseda and Ohtani 1974, 1975a, Waseda and Tamaki 1975) and in a number of rare earths (Y Waseda private communication). A slight skewness of the first peaks occurs in Zn and Cd (Waseda and

Suzuki 1973) but otherwise these metals are also quite well described by hard spheres.

The same point can be demonstrated alternatively by calculating the $g(r)$ corresponding to the measured $a(q)$. Results, once again for Na and Al, are shown in figure 4 and it is evident from the behaviour at low values of r that Al has the harder core. For sufficiently small $g(r)$, this can be quantified (e.g. see Rice and Gray 1965) since there, $g(r) \sim \exp(-v/kT)$.

However, in Ga, Si, Ge, Sn, Sb, Bi and (to a very modest extent) in Tl there is a shoulder on the high-angle side of the principal peak which cannot be described by hard spheres alone. For the most part, these are the cases which do not form close-packed solids. Heine and Weaire (1970) have shown that the Coulomb forces favour close packing and that other cases can be explained by special circumstances. For example, if the nearest-neighbour distance for close packing occurs near the energetically unfavourable first maximum, it pays for some such neighbours to move inwards and the others to move out. Heine and Weaire suggest that similar considerations have a role to play in liquid-state structure theory. Certainly the implication that the bumps and dips of $g(r)$ will be less pronounced is borne out on Fourier transformation of the observed $a(q)$ for such cases (Waseda and Suzuki 1973). The completely typical result for Bi may be contrasted with those for Na and Al in figure 4.

A calculation by Silbert and Young (1976) is also relevant. These writers have taken at their face value, curves of the type shown for Zn in figure 2, and have performed model calculations for hard spheres with repulsive tails. Their results are of the correct

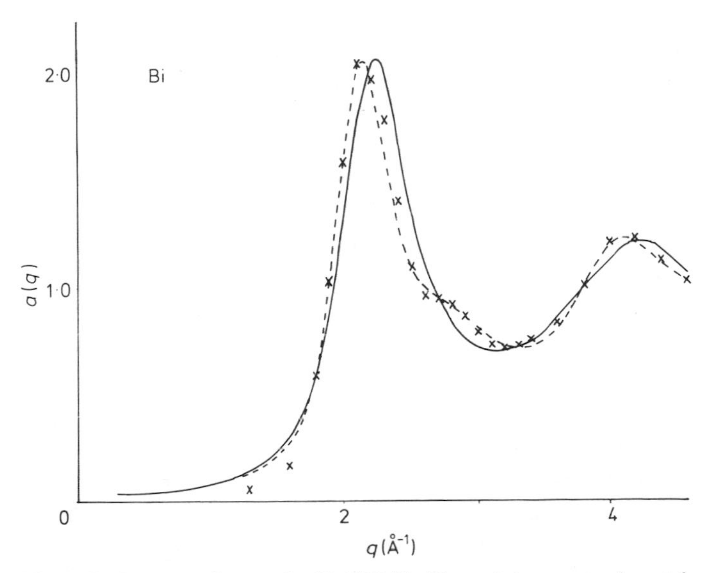

Figure 5. Structure factors for Bi (573 K). The points are experimental values (Waseda and Suzuki 1973), the full line is the PY curve for $\sigma = 5{\cdot}65$ a.u. and the broken line is the RPA corrected version (Silbert and Young 1976) resulting from the interaction:

$$v = \begin{cases} \infty & (r < \sigma) \\ \epsilon k T & (\sigma < r < \lambda\sigma) \\ 0 & (\lambda\sigma < r) \end{cases}$$

where $\sigma = 5{\cdot}65$ a.u., $\lambda = 2{\cdot}1$ and $\epsilon = 0{\cdot}44$.

Table 1. Observed data at melting temperature (from Faber 1972). These data should be compared with the typical theoretical value $a_{hs}(0) = 0{\cdot}028$ corresponding to $\eta = 0{\cdot}45$ (see equation (11)). The discrepancies between theory and experiment arise because volume terms contribute to the compressibilities, especially of polyvalent metals.

Li	Na	K	Rb	Cs	Cu	Ag
0·031	0·024	0·023	0·022	0·028	0·021	0·019
Zn	Cd	Hg	Al	In	Tl	Pb
0·014	0·012	0·005	0·018	0·007	0·011	0·009

type, as their curve for Bi (figure 5) shows. Clearly, more work needs to be done on such systems but from the structure factor evidence it can already be inferred that the hard-sphere model is deficient though still retaining semi-quantitative meaning†.

Turning finally to the low-angle region we have the very general result (Watabe and Hasegawa 1973, Chihara 1973, Gray 1973, March and Tosi 1973, Jones 1973b)

$$a(0) = kT\beta_T/\Omega \tag{11}$$

where β_T is the isothermal compressibility. The right side of this equation may be evaluated using observed data for the melting point and for hard spheres with $\eta = 0{\cdot}45$. The agreement is poor for polyvalent metals (table 1) and in fact there is good reason for this. Being density-dependent, V_0 contributes to β_T and therefore to the very long wavelength region. Direct calculation (Hasegawa and Watabe 1972) shows that as a fraction of the total, this contribution is small for the alkalis and large for polyvalents thus resolving the apparent problem presented by table 1.

Fortunately, for most purposes the successful application of techniques of the present kind does not hinge on a good description of this region. To see this, note (by the use of equations (4), (9) and (10)) that within a pairwise description, the method requires us to minimize

$$\frac{2z^{*2}}{\pi}\int_0^\infty [1-\Phi_N(q)]\,[a_{hs}(q)-1]\,\mathrm{d}q - TS_{hs} \tag{12}$$

with respect to σ. Now $1-\Phi_N(q) = O(q^2)$ at small q and is very near to unity around and beyond the position of the principal peak in the structure factor. It follows that variation of the integral with σ is decided by the latter and not the former region of q.

2.5. Entropies

A general prediction of the GB method is (§2.2) that the diameter which characterizes the structure factor also describes the entropy. This correlation can be tested for any metal possessing a good hard-sphere structure factor. By fitting the radiation data at the first peak (a reasonable procedure according to §2.4), a packing fraction can be extracted and used in equation (1) to find S_{hs}. Then, using the free-electron form of S_{elec}, the total entropy may be found from equation (7) and compared with that measured by using the usual thermodynamic route.

The results of such an exercise for Pb (figure 6) would appear to confirm this correla-

† Mon, Chester and Ashcroft have proposed a third explanation! (See reference in Ashcroft's paper in this volume.)

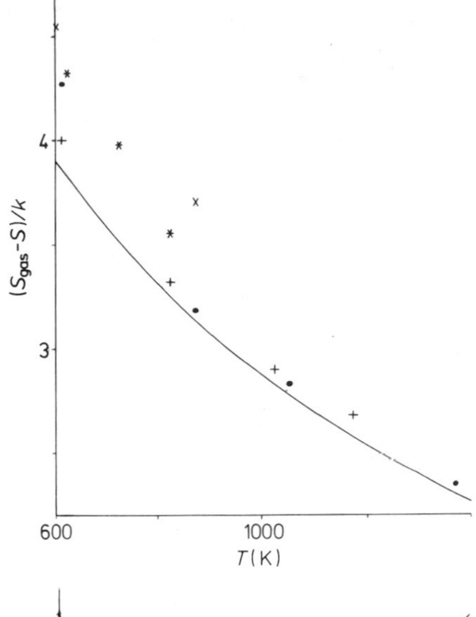

Figure 6. Excess entropies (relative to a corresponding ideal gas with $\eta = 0$) for Pb. The full line corresponds to the data of Hultgren *et al* (1973); with • North *et al* (1968) (neutrons); × Kaplow *et al* (1965) (x-rays); + Waseda and Suzuki (1972, 1973) (x-rays); ∗ Steffen (1976) (x-rays). The differences between the points and the line define S_s in equation (7*a*).

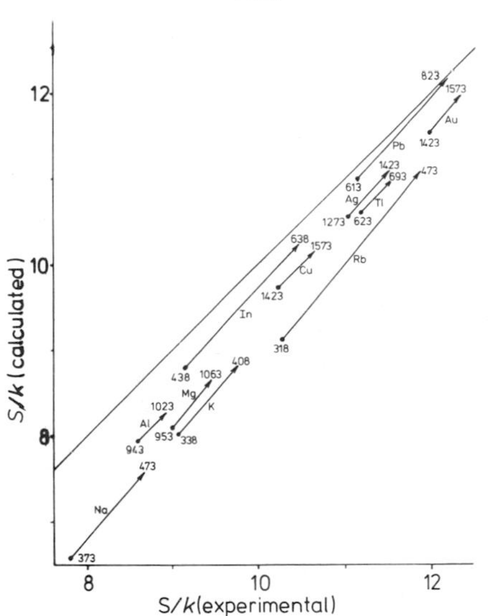

Figure 7. Entropies per atom; observed (Hultgren *et al* 1973) against calculated. In the latter, the following radiation data are used: Na, K, Greenfield *et al* (1971); Rb, Wingfield and Enderby (from Howells 1973); Cu, Breuil and Tourand (1970); Ag, Au, Waseda and Ohtani (1974); Mg, Waseda and Ohtani (1975b); Al, Fessler *et al* (1966); In, Ruppersberg and Wintersberg (1971); Tl, Halder and Wagner (1966); Pb, Waseda and Suzuki (1972, 1973). The full circles correspond to the lowest temperatures (K) at which the radiation experiments were performed and the arrowheads to the second lowest.

tion. The figure suggests, however, that better than equation (7) is the empirical equation:

$$S = S_{hs} + S_{elec} + S_s \qquad (7a)$$

with S_s (the subscript s specifies soft, for reasons which will become clear shortly) decreasing monotonically towards zero as T increases. A detailed study of other systems that are now to be discussed, confirms this.

Shown in figure 7 is a plot of S using thermodynamic tables, against $S_{hs} + S_{elec}$ calculated as above, in each case† for the lowest pair of temperatures for which radiation data exist. Always $S_s \geq 0$ and in every case the trajectory moves towards the 45° axis

† Except for In where the second lowest result of a set of six was ignored because it seemed out of line with the others.

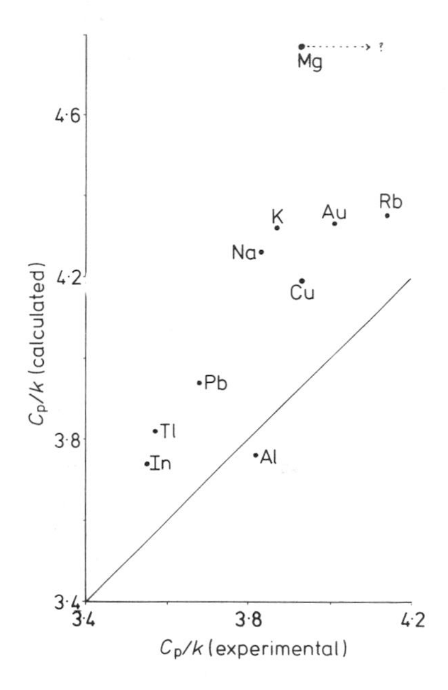

Figure 8. Heat capacities per atom; observed (Hultgren *et al* 1973) against calculated. These results follow from the data of figure 7. The possible revision indicated for Mg corresponds to an alternative value quoted by Hultgren *et al* (1973).

when T becomes larger. The corresponding average heat capacities, consistent with this graph, are shown in figure 8. As indicated above, the radiation data provide over-estimates and there is some degree of correlation between the two sets of results.

There is a simple explanation of the above behaviour. We can expect a hard-sphere system to become more applicable as T becomes higher so that $S_s(T) \to 0$ as $T \to \infty$ is understandable. Furthermore, $S_s > 0$ since the blurring of the best hard-sphere potential to create a more physical one will increase the entropy. From this point of view the relative softness of the alkali cores is apparent from figure 7 and confirms the conclusion of §2.4.

In the WCA theory, the departure of the actual potential from that of hard spheres modifies the hard-sphere structure factor and gives rise to an explicit term in the entropy. The latter might be roughly identifiable with S_s though it has to be remembered that the packing fraction of WCA is not identical with that of GB. This particular problem will be discussed by Kumaravadivel and Evans (1977).

So far, S_{elec} has played a marginally useful role (cf figure 6 and table 2). For transition metals, however, its appearance in the formalism is vital since it is usually an order of magnitude bigger than for simple metals. This is partly because the observations are

Table 2. Contributions to the entropy for two metals. S_{gas} and S_η are calculated from measured data, the latter by fitting the observed structure factor peak height to that for hard spheres. $S_{elec} + S_s$ is then deduced by subtraction from the measured S_{obs} (cf equation (7*a*)). A free-electron calculation for Pb gives $S_{elec}/k = 0{\cdot}28$.

	T(K)	S_{obs}/k	S_{gas}/k	S_η/k	$(S_{elec} + S_s)/k$
Pb	1373	14·03	16·33	−2·62	0·33
Fe	1560	12·07	13·81	−3·94	2·20

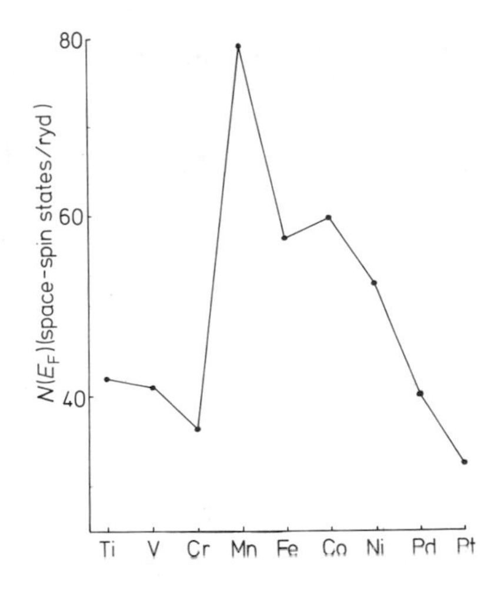

Figure 9. $N(E_F)$ for liquid transition metals, near the melting temperatures, obtained from the x-ray data of Waseda. The results shown here are based on (essentially) exact hard-sphere structure factors (B N Perry and M Silbert private communication) and the experimental data of Hultgren *et al* (1973). The earlier results of Meyer *et al* (not much different from those given here) were based on the PY theory (the larger cause of difference) and the experimental data of Hultgren *et al* (1963). All calculations assume $S_s = 0$ in equation (7*a*) and will, if anything, produce overestimates for this reason.

invariably for higher temperatures (a consequence of the higher melting temperatures) and partly because the $N(E_F)$ are larger (cf equation (8)).

As a consequence (table 2), it becomes a practical proposition to isolate $S_{elec} + S_s$ (cf equation (7*a*)) by evaluating $S - S_{hs}$ using experimental information and the technique described above. If as expected, S_s is small for transition metals (a few tenths of k in every case), an estimate of $N(E_F)$ may be made. Proceeding in this way, the results (overestimates) shown in figure 9 may be established.

Knapp and Jones (1972) have performed a somewhat similar analysis for a group of transition metal solids. Only Pd and Pt are common to both the present analysis and their analysis; in each of these cases the density of states in the liquid is higher than that in the hot solid and comparable with the expected low-temperature value (excluding phonon enhancement). For further discussion see Meyer *et al* (1976) and Young (1976).

3. Alloys

In this section the generalization of the above to alloys is considered.

3.1. Hard spheres

Consider a binary mixture of hard spheres, atomic fractions c_1 and $c_2 = 1 - c_1$, masses M_1 and M_2 and diameters σ_1 and σ_2. The mean sphere volume is $\omega = (\pi/6)(c_1\sigma_1^3 + c_2\sigma_2^3)$ so that if, as before, Ω is the container volume per sphere, the packing fraction is once more $\eta = \omega/\Omega$.

Work on finding very accurate partial structure factors for hard-sphere mixtures is proceeding (Grundke and Henderson 1972) but for the present the only practical proposition appears to be to use the PY forms. These are available as closed expressions (Ashcroft and Langreth 1967a, Enderby and North 1968) and are invoked where necessary below.

The work of Carnahan and Starling has also been generalized in a natural way by Mansoori *et al* (1971) and their results, when tested against computer studies, prove to be very accurate. Accordingly, their description is adopted below.

The entropy corresponding to the description of Mansoori *et al* is (Umar *et al* 1976)

$$S_{hs} = S_{gas} + S_c + S_\eta + S_\sigma \tag{13}$$

where S_{gas} is as in equation (2) with $M = M_1^{c_1} M_2^{c_2}$, the total packing contribution S_η is exactly as in equation (3) and S_c is the ideal mixing term defined by

$$S_c/k = -(c_1 \ln c_1 + c_2 \ln c_2).$$

The final diameter mismatch term is given by

$$S_\sigma/k = A c_1 c_2 (\sigma_1 - \sigma_2)^2 \tag{14}$$

where

$$A = (\zeta - 1)\,[(\zeta + 3)\,Y_1 + \zeta Y_2] - (Y_1 + Y_2) \ln \zeta \tag{15}$$

with

$$Y_1 = (\sigma_1 + \sigma_2)/(c_1 \sigma_1^3 + c_2 \sigma_2^3) \tag{16}$$

$$Y_2 = \sigma_1 \sigma_2 (c_1 \sigma_1^2 + c_2 \sigma_2^2)/(c_1 \sigma_1^3 + c_2 \sigma_2^3)^2. \tag{17}$$

All this of course is for the additive case, when $\frac{1}{2}(\sigma_1 + \sigma_2)$ is the collision diameter for spheres of different types. Some work has begun on the more general problem (Adams and McDonald 1975, Melnyk and Sawford 1975) but it is not complete enough to be incorporated into the present formalism.

3.2. *Thermodynamic perturbation theory*

The generalization of §2.2 is trivial; the equations continue to apply provided one reads σ_i, $i = 1, 2$ for σ and, of course, S_{hs} is now given by the considerations of §3.1.

3.3. *Interatomic potentials*

Electronic behaviour in alloys is inevitably much more complex than in pure metals. Compound formation and charge transfer for example, are not very well understood but are, nevertheless, important effects which have sometimes to be reckoned with. The method developed in recent years by Bhatia, March and co-workers (Bhatia 1977) comes into its own in this regard and in this sense their approach and the present one are somewhat complementary.

However, for some alloys we can hope that the nearly free electron model is adequate. For these, the one-component pseudopotential formalism generalizes readily. Ashcroft and Langreth (1967b) first calculated interatomic potentials in alloys in this way. Recent calculations have been made by Takeda and Watabe (1977) and typical results are shown in figure 10. Evidence of this kind points to the conclusion that

$$v_{12} \sim \tfrac{1}{2}(v_{11} + v_{22}) \tag{18}$$

in such cases.

Of course one has to bear in mind that equation (18) is largely predetermined by the zero-order free-electron theory on which the perturbation theory is based. Nevertheless, it provides some motivation for using the hard-sphere model with additive diameters

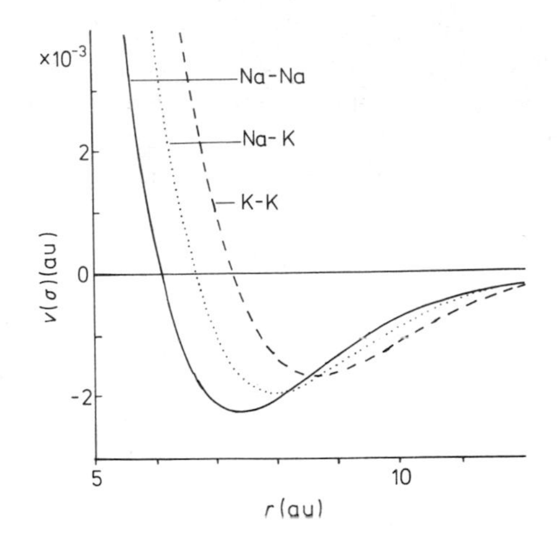

Figure 10. Pair potentials in NaK at 373 K (Takeda and Watabe 1977). Geldart–Vosko (1966) screening has been used in this calculation. In the Kleinman (1967) case (not shown), minima are about twice as deep though in much the same positions. Nevertheless, the qualitative result given by equation (18) continues to hold.

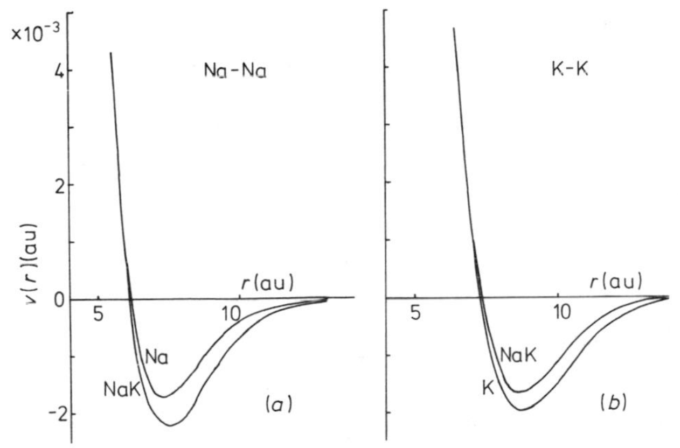

Figure 11. (*a*) Na–Na pair potential at 373 K in pure Na and in KNa alloy. (*b*) K–K pair potential at 373 K in pure K and in KNa alloy. Calculations due to Takeda and Watabe (1977).

(§3.1). On the other hand it is perhaps also worth bearing in mind that the model may be applicable in some cases, even though pseudopotential theory in its standard form is not.

A further effect which also needs to be noted is that the interaction between a pair of ions in the pure metal will be different from the same pair in an alloy, even at the same temperature. This point is illustrated in figure 11 where the changed density in the alloy affects the screening.

3.4. Structure factors

All the complexities discussed in §2.4 are also of course present for alloys. Nevertheless, Ashcroft and Langreth (1967a) and Enderby and North (1968) have shown that the x-ray scattering, at least in the important region of intermediate momentum transfer, can sometimes be interpreted in terms of the PY theory of §3.1.

In other cases, for example when double-headed peaks appear (Faber 1972, §6.11) or when compound formation clearly occurs (Bhatia and Ratti 1976), such a description is deficient. Nevertheless, just as with the pure metals, structure factors may be rather sensitive indicators and hard-sphere inadequacies here might not preclude a tolerable description of the entropy. At any rate, I proceed on this assumption occasionally below.

3.5. *Entropies*

The possibility of extracting hard-sphere diameters from the radiation data for alloys and proceeding much as in §2.5 has not yet been explored. It is clear however that very good data are needed before one can hope to detect changes in hard-core diameters (cf figure 11) on mixing.

Yet some progress seems possible without this information. Figure 11 suggests that on alloying, core diameter changes might be small. If they are assumed to be zero, then the excess entropies of mixing, defined generally in terms of the pure metal entropies (S_1 and S_2) by

$$\Delta S_E \equiv S_{alloy} - c_1 S_1 - c_2 S_2 - S_c, \qquad (19)$$

can be predicted as follows.

The pure metal diameters are obtained from the observed entropies and volumes using equation (7) (i.e. ignoring S_s in equation (7*a*)). The *same* values, together with the

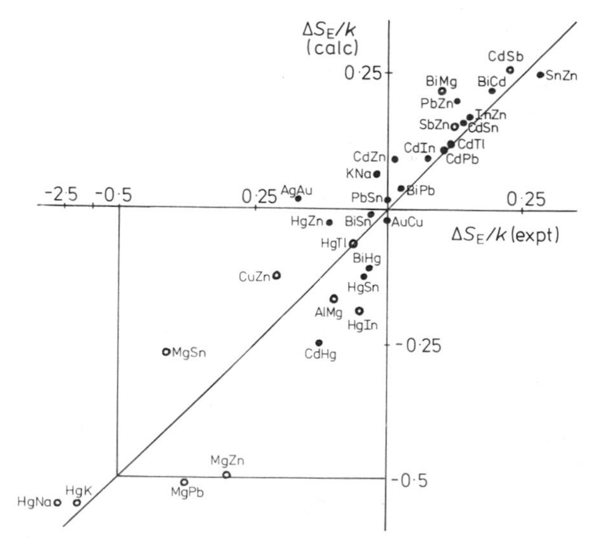

Figure 12. Excess entropies, ΔS_E. Full circles correspond to apparently simple systems; open circles to possibly more complex ones. The calculated values are based on the assumption that core diameters, found by using equation (7) for fitting the observed pure liquid entropies, do not change on alloying. Each amalgam (with the exceptions of HgK and HgNa) is at a temperature below that of the melting temperatures of the second component. There are, as a consequence, greater uncertainties in the experimental values quoted (Umar *et al* 1976) and these are almost certainly responsible for the poorer correlation for this group.

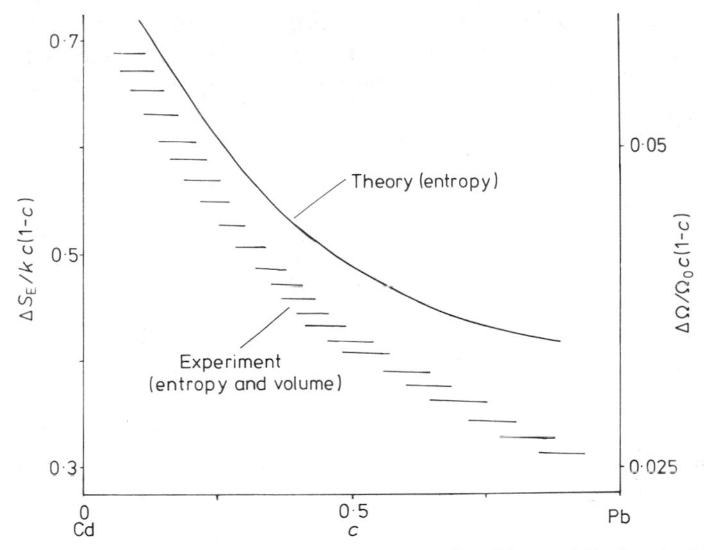

Figure 13. Entropy data for Cd_cPb_{1-c} at 773 K. Observed S_E (hatched line); calculated S_E (full line). The volumes were measured at 623 K and the usual assumption is made that these may be used as input information at 773 K. There is a certain amount of scatter but, roughly, $\Delta S_E(\text{expt})/k \propto \Delta\Omega/\Omega_0$ so that the hatched line serves for both cases.

observed volume per ion of the alloy can then be used in equation (19), S_{elec} once again taking its free-electron form for simple metals. Figure 12 shows the results so obtained by Umar *et al* (1976) for all 50–50 alloys of simple metals for which data were available.

The variation of excess entropy with composition can also be explained in this way (Umar *et al* 1976). As an example, the results for Cd_cPb_{1-c} are shown in figure 13. The correlation of these results with the excess volumes of mixing, $\Delta\Omega$, is no accident. Indeed the latter is of fundamental significance in considerations of the above type as I now indicate.

In first order (an excellent approximation for variations of a few per cent), Umar *et al* (1974a) showed that

$$\Delta S_E \sim \Delta S_{E_0} + (k - \eta_0 S_{\eta_0}{}') \frac{\Delta\Omega}{\Omega_0} + \eta_0 S_{\eta_0}{}' \frac{\Delta\omega}{\omega_0} . \tag{20}$$

Here, quantities carrying a zero subscript apply to the two pure metals before mixing *considered as a single system.* Such small terms as S_{elec} contribute negligibly to equation (20) and one might expect this also to be true for S_s in equation (7*a*) (which is not of course strictly within this formalism).

In the present discussion, the fractional increase in hard-sphere volume $\Delta\omega/\omega_0$ was by hypothesis taken to be zero. But the formula demonstrates that the assumption on which the results (e.g. figures 12, 13) are based, is $\Delta\omega = 0$ rather than $\Delta\sigma_i = 0$. This is a weaker and thus more plausible condition (see figure 11).

From equation (3) one calculates that characteristic values for $-\eta_0 S_{\eta_0}{}'/k$ are 8·4 (for $\eta_0 = 0{\cdot}45$) and 5·9 (for $\eta_0 = 0{\cdot}40$). Therefore this effect always dominates the contribution from the perfect gas term and it will be seen that a 1% excess volume corresponds to an entropy change per ion of about $0{\cdot}09\,k - 0{\cdot}07\,k$.

Umar *et al* (1976) have gone on to discuss heat capacities and departures from Kopp's law. This involves an additional assumption (that $\Delta\Omega$ is temperature-independent) and the experimental data are not very accurate. Within these limitations, the theory seems to continue to work.

4. *Ab initio* calculations

The next question is what happens when detailed pseudopotentials and screening functions are fed into the formalism. A full prescription for doing this for simple metals and their alloys is given by Umar *et al* (1974b) (see also Stroud 1973). Note, however, that the present entropy expressions are slightly better than those used by these authors.

4.1. Pure metals

On minimizing the free energy (cf equations (4), (5)) and calculating the diameters and thus the entropies, one might hope to obtain agreement with experiment. A number of such calculations have been performed (Edwards and Jarzynski 1972, Stroud and Ashcroft 1972, Jones 1973a, Silbert *et al* 1975 and papers in this volume). Entropies of about the right size are almost always obtained and the same is true for heat capacities for low valencies ($z \lesssim 3$). For higher valencies however, theory tends to underestimate heat capacities. Some quite typical calculations of the latter are given in the work of Yokoyama *et al*(1977).

The poorer results they find for Sn and Bi might be attributable to the inadequacy of hard-sphere reference systems in such cases (§2.4). But if the deficiency for Pb is real (and further calculations using a variety of pseudopotentials and screening functions are, perhaps, called for before this conclusion is accepted), then it means the calculated σ does not decrease quickly enough as T increases. Since figure 6 points to the existence of a fairly satisfactory $\sigma(T)$ curve, all that would appear to be lacking is an ability to calculate it and the incorporation of nonlinearity in one form or another (Hasegawa 1976, Rasolt and Taylor 1975) seems to be called for.

At this point it is appropriate to mention a successful calculation by Jones (1973a) and by Umar (1974) of the dependence of heat capacity on T for liquid Na. The *ab initio* calculation of the shape of this curve (figure 12) is a notable achievement of the

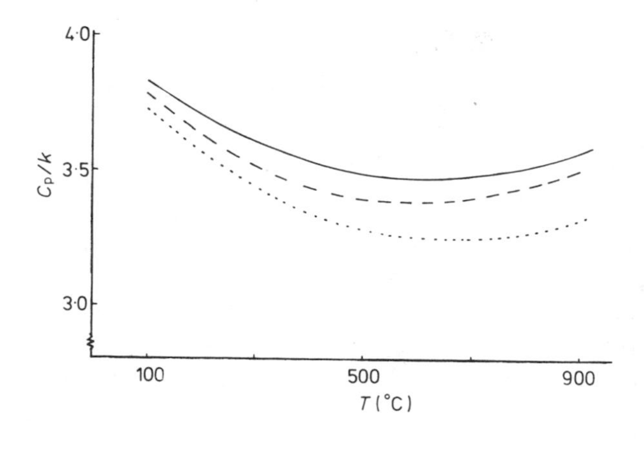

Figure 14. Heat capacity of liquid sodium. The full line is the experimental curve, the dotted line was calculated without an electronic contribution and the broken line with an electronic term (Umar 1974).

theory. From equation (7), the constant volume heat capacity is $(C_{hs})_\Omega + S_{elec}$ where

$$\frac{(C_{hs})_\Omega}{k} = \frac{3}{2} + TS_\eta' \left(\frac{\partial \eta}{\partial T}\right)_\Omega . \tag{21}$$

As T increases, the ions become harder and so the final (positive) term in equation (21) decreases. This explains the initial fall in C_p shown in figure 14; it is a reflection of the increasing steepness of $v(r)$ as r becomes smaller. If equation (21) were the whole story, the heat capacity would continue to decrease. However, at higher T the role of thermal expansion becomes more significant. The larger volume implies an increased entropy and therefore an enhanced heat capacity. As figure 14 shows, this causes C_p to rise and this trend is accentuated by the electronic term.

4.2. *Alloys*

Calculated excess entropies of mixing are very sensitive to the pseudopotentials used; also perturbation theory will become deficient when the constituent ions have large valence differences. Accordingly, Yokoyama *et al* (1977b) took all 50–50 alloys of which they were aware, for which the valence difference was not more than unity and where the volume and entropy data were known. Then they asked if suitably parametrized empty-core pseudopotentials could be found so that first, the calculated excess entropies should agree with experiment very closely and second, the pure metal entropies should differ from observation by at most 5%.

A report on the above work will be given at this conference (Yokoyama *et al* 1977a) but the main points are worth summarizing here:

(i) Satisfactory parameters exist and correlate quite well with those known (Cohen and Heine 1970) to satisfy observed Fermi level data. There was no *a priori* guarantee that a solution could be found since some metals occur in a number of the alloys, and thus there are, in almost all cases, more conditions than parameters.
(ii) Most of the calculated pure metal entropies are low, in qualitative agreement with the present work and that of Kumaravadivel and Evans (1977).
(iii) The proposition suggested by figure 11 and discussed in §3.5 is confirmed; that is, on alloying one diameter shrinks while the other expands. However, in contrast with the philosophy adopted in establishing figure 12, the fractional variation in packing,

$$\Delta\eta/\eta_0 \sim \Delta\omega/\omega_0 - \Delta\Omega/\Omega_0$$

is, on average, as strongly affected by the first as the second term. But as was remarked in connection with equation (18), such results might be unduly influenced by the uniform electron distribution of zero order and therefore be exaggerated. If so, then core size variation in reality plays a secondary but non-negligible role.

5. Summary and conclusions

It would seem that, for present purposes at least, liquid metals really are a lot of balls. Whether they are hard balls or soft balls can be inferred most directly from observed structure factors (figures 3, 4 and 5). In the hard case and for simple metals, such radiation data lead to realistic entropy (under)estimates and heat capacity (over)estimates.

For transition metals, the experimental structure factors and entropies together lead to (over)estimates of $N(E_F)$ (figure 9). Soft cases possibly fall into two types: there are the alkalis which differ only in degree (figures 3, 4 and 7) from those I have chosen to call hard. Then there are those like Bi where the characteristic shoulder (figure 5) might signal a ledge-type feature in the pair potential (figure 2) or a standard potential with a hypothetical hard-sphere first coordination shell occurring in an energetically unfavourable position.

On mixing, the excess entropies can be interpreted in a rather simple manner (by equation (20)) and in a fairly satisfactory way (figures 12 and 13) using pure liquid observations and the assumption that the swelling in one species compensates for the shrinkage in the other ($\Delta\omega = 0$). Sufficiently accurate radiation experiments might also lead to estimates of $\Delta\omega$ but this has not so far been attempted.

Ab initio pseudopotential calculations have been quite successful (Yokoyama *et al* 1977a,b), particularly for lower valence cases, and seem to suggest that $\Delta\omega$ may have some role to play in tidying up figures 12 and 13. Alloying properties in particular, however, are sensitive to the pseudopotential parametrizations which should therefore be chosen with care.

Lack of space has prevented any discussion of the volume term V_0 of equation (9) and such related properties as internal energies, pressures and compressibilities. A formalism exists for discussing this problem (Heine and Weaire 1970, Sommer 1973) but has not, at least within the liquid state context, been as systematically explored as that for the pairwise contributions.

Acknowledgments

I am grateful to the SRC for support of this work and to R Evans, A Meyer, B Perry, M Silbert, M Stott, K Takeda, M Watabe and I Yokoyama for permission to draw upon their results in advance of publication.

References

Adams D J and McDonald I R 1975 *J. Chem. Phys.* **63** 1900
Alder B J and Wainwright T E 1957 *J. Chem. Phys.* **27** 1208
Anderson H C, Weeks J D and Chandler D 1971 *Phys. Rev.* **A4** 1597
Appapillai M and Heine V 1972 *Solid State Theory Group, Cavendish Laboratory, Cambridge, Technical Report* No 5
Ashcroft N W and Langreth D C 1967a *Phys. Rev.* **156** 685 (see also erratum 1968 *Phys. Rev.* **166** 934)
—— 1967b *Phys. Rev.* **159** 500
Ashcroft N W and Lekner J 1966 *Phys. Rev.* **145** 83
Barker J A and Henderson D 1971 *Molec. Phys.* **21** 187
Bhatia A B 1977 this volume
Bhatia A B and Ratti V K 1976 *J. Phys. F: Metal Phys.* **6** 927
Breuil M and Tourand G 1970 *J. Phys. Chem. Solids* **31** 549
Carnahan N F and Starling K E 1969 *J. Chem. Phys.* **51** 635
Chihara J 1973 *The Properties of Liquid Metals* ed S Takeuchi (London: Taylor and Francis) p 137
Cohen M L and Heine V 1970 *Solid St. Phys.* **24** 37 (New York: Academic Press)
Dagens L, Rasolt M and Taylor R 1975 *Phys. Rev.* **11** 2726
Edwards D J and Jarzynski J 1972 *J. Phys. C: Solid St. Phys.* **5** 1745

Enderby J E and North D M 1968 *Phys. Chem. Liquids* **1** 1
Faber T E 1972 *Introduction to the Theory of Liquid Metals* (London: Cambridge UP)
Fessler R R, Kaplow R and Averbach B L 1966 *Phys. Rev.* **150** 34
Geldart D J W and Vosko S H 1966 *Can. J. Phys.* **44** 2137
Gray P 1973 *J. Phys. F: Metal Phys.* **3** L43
Greenfield A J, Wellendorf J and Wiser N 1971 *Phys. Rev.* **A4** 1607
Grundke E W and Henderson D 1972 *Molec. Phys.* **24** 269
Halder N C and Wagner C N J 1966 *J. Chem. Phys.* **45** 482
Hansen J P and Schiff D 1973 *Molec. Phys.* **25** 1281
Hasegawa M 1976 *J. Phys. F: Metal Phys.* **6** 649
Hasegawa M and Watabe M 1972 *J. Phys. Soc. Japan* **32** 14
—— 1974 *J. Phys. Soc. Japan* **36** 1510
Heine and Weaire D 1970 *Solid St. Phys.* **24** 249 (New York: Academic Press)
Howells W S 1973 *The Properties of Liquid Metals* ed S Takeuchi (London: Taylor and Francis) p 43
Hultgren R, Desai P D, Hawkins D T, Glesier M and Kelley K K 1973 *Selected Values of the Thermodynamic Properties of the Elements* (Metals Park, Ohio: Am. Soc. Metals)
Hultgren R, Orr R L, Anderson P D and Kelley K K 1963 *Selected Values of Thermodynamic Properties of Metals and Alloys* (New York: Wiley)
Isihara A 1968 *J. Phys. A: Gen. Phys.* **1** 539
Jones H 1971 *J. Chem. Phys.* **55** 2640
—— 1973a *Phys. Rev.* **A8** 3215
Jones W 1973b *J. Phys. C: Solid St. Phys.* **6** 2833
Kaplow R, Strong S L and Averbach B L 1965 *Phys. Rev.* **A138** 1336
Kleinman L 1967 *Phys. Rev.* **160** 585
Knapp G S and Jones R W 1972 *Phys. Rev.* **B6** 1761
Kumaravadivel R and Evans R 1977 this volume
Lukes T and Jones R 1968 *J. Phys. A: Gen. Phys.* **1** 29
Mansoori G A and Canfield F B 1969 *J. Chem. Phys.* **51** 4958
Mansoori G A, Carnahan N F, Starling K E and Leland T W 1971 *J. Chem. Phys.* **54** 1523
March N H and Tosi M P 1973 *Ann. Phys., NY* **81** 414
Melnyk T W and Sawford B L 1975 *Molec. Phys.* **29** 891
Meyer A, Stott M J and Young W H 1976 *Phil. Mag.* **33** 381
North D M, Enderby J E and Egelstaff P A 1968 *J. Phys. C: Solid St. Phys.* **1** 1075
Page D I, Egelstaff P A, Enderby J E and Wingfield B R 1969 *Phys. Lett.* **29A** 296
Rasolt M and Taylor R 1975 *Phys. Rev.* **B11** 2717
Rice S A and Gray P 1965 *Statistical Mechanics of Simple Liquids* (New York: Wiley) §2.5
Ruppersberg H and Wintersberg K H 1971 *Phys. Lett.* **A34** 11
Shaw R W and Heine V 1972 *Phys. Rev.* **B5** 1646
Silbert M, Umar I H, Watabe M and Young W H 1975 *J. Phys. F: Metal Phys.* **5** 1262
Silbert M and Young W H 1976 *Phys. Lett.* to be published
Sommer F 1973 *Acta Metall.* **21** 1289
Steffen B 1976 *Phys. Rev.* **B13** 3227
Stroud D 1973 *Phys. Rev.* **B7** 4405
Stroud D and Ashcroft N W 1972 *Phys. Rev.* **B5** 371
Takeda K and Watabe M 1977 to be published
Thiele E 1963 *J. Chem. Phys.* **39** 474
Throop G J and Bearman P J 1965 *J. Chem. Phys.* **42** 2408
Umar I H 1974 *PhD Thesis* University of East Anglia
Umar I H, Meyer A, Watabe M and Young W H 1974b *J. Phys. F: Metal Phys.* **4** 1691
Umar I H, Watabe M and Young W H 1974a *Phil. Mag.* **30** 957
Umar I H, Yokoyama I and Young W H 1976 *Phil. Mag.* **34** 535
Verlet L and Weis J J 1972 *Phys. Rev.* **A5** 939
Waseda Y and Ohtani M 1974 *Phys. Stat. Solidi* **B62** 535
—— 1975a *Z. Phys.* **B21** 229
—— 1975b *Z. Naturf.* **A30** 801

Waseda Y and Suzuki K 1972 *Phys. Stat. Solidi* **B49** 339
—— 1973 *Sci. Rep. Res. Inst., Tohoku Univ.,* **A24** 139
Waseda Y and Tamaki S 1975 *Phil. Mag.* **32** 273
Waseda Y, Yokoyama K and Suzuki K 1974 *Phil. Mag.* **30** 1195
Watabe M and Hasegawa M 1973 *The Properties of Liquid Metals* ed S Takeuchi (London: Taylor and Francis) p 133
Weeks J D, Chandler D and Anderson H C 1971 *J. Chem. Phys.* **54** 5237
Wehling J H, Shyu W M and Gaspari G D 1972 *Phys. Lett.* **39A** 59
Wertheim M S 1963 *Phys. Rev. Lett.* **10** 321
Yokoyama I, Meyer A, Stott M J and Young W H 1977a to be published
Yokoyama I, Stott M J, Umar I H and Young W H 1977b this volume
Young W H 1976 *Ber. Bunsenges. Phys. Chem.* **80** 749
Zwanzig R W 1954 *J. Chem. Phys.* **22** 1420

Concentration fluctuations and structure factors in binary alloys †

A B Bhatia

Department of Physics, University of Alberta, Edmonton, Alberta, Canada

Abstract. The dependence of the various structure factors in binary mixtures on deviations from ideality is examined. In the long wavelength ($q \to 0$) limit the effects are conveniently displayed by the concentration fluctuation structure factor $S_{CC}(q)$. The variation of $S_{CC}(0)$ with concentration, and hence of the Faber–Ziman structure factors $a_{ij}(0)$, is discussed for different types of mixtures (regular, athermal, compound forming etc). For $q \neq 0$, the usual hard-sphere calculations are such that $S_{CC}(0)$ does not agree with experiment even for simple (weakly interacting) systems such as Na–K. Some recent work on the calculations of $S_{CC}(q)$ etc, which attempts to rectify approximately this difficulty, for (i) weakly interacting and (ii) compound-forming mixtures, is also described.

1. Introduction

The long wavelength limit of the various structure factors of a binary mixture is determined, apart from the compressibility and a volume dilatation factor, by the mean square fluctuations in the composition of the alloy (Bhatia and Thornton 1970). The present paper, intended as a review of some work by the author and his collaborators, first describes the typical behaviour of the concentration fluctuations $S_{CC}(0)$ in different types of mixtures (ideal, conformal, compound-forming etc) using appropriate phenomenological models. This knowledge is then used to discuss for some alloys the concentration dependence of the long wavelength limit of the Faber–Ziman partial structure factors. Finally, some results on the calculation of $S_{CC}(q)$ and other structure factors, at arbitrary wavenumber q, are briefly reported for (i) weakly interacting (like conformal) solutions and (ii) strongly interacting (compound-forming) systems.

2. The structure factors

The partial structure factors $a_{ij}(q)$ are defined as (Faber and Ziman 1965):

$$a_{ij}(q) = 1 + \frac{N}{V} \int (g_{ij}(r) - 1) \exp(\mathrm{i}\mathbf{q}\cdot\mathbf{r}) \, \mathrm{d}^3 r, \tag{2.1}$$

where $i, j = 1, 2$, refer to the two types of atoms in the alloy, V is the volume, $N = N_1 + N_2$ the total number of atoms in the alloy and $g_{ij}(r)$ $(= g_{ji}(r))$ is the pair distribution function. Note that these are different from the structure factors $S_{ij}(q)$ (Ashcroft and Langreth 1967, Enderby and North 1968), the relationship being

$$S_{ij}(q) = \delta_{ij} + (c_i c_j)^{1/2} (a_{ij}(q) - 1), \tag{2.2}$$

† Work supported in part by the National Research Council of Canada.

where $c_i = N_i/N$ is the concentration of ith-type atoms. In the following, it will sometimes be convenient to write the concentrations c_1 and c_2 also as $c_1 \equiv c$, $c_2 = 1 - c$.

The long wavelength limit of the partial structure factors is conveniently discussed in terms of those of the number–concentration structure factors $S_{NN}(q)$, $S_{NC}(q)$ and $S_{CC}(q)$ introduced by Bhatia and Thornton (1970). These are, respectively, associated with (number) density–density, density–concentration and concentration–concentration correlations and are related to $a_{ij}(q)$ by (Bhatia and Thornton 1970)

$$\begin{aligned} c_1^2 a_{11}(q) &= c_1^2 S_{NN}(q) + 2c_1 S_{NC}(q) + S_{CC}(q) - c_1 c_2 \\ c_2^2 a_{22}(q) &= c_2^2 S_{NN}(q) - 2c_2 S_{NC}(q) + S_{CC}(q) - c_1 c_2 \\ c_1 c_2 a_{12}(q) &= c_1 c_2 S_{NN}(q) + (c_2 - c_1) S_{NC}(q) - S_{CC}(q) + c_1 c_2. \end{aligned} \tag{2.3}$$

We note, for future reference, that as $q \to \infty$, $S_{NN}(q) \to 1$, $S_{NC}(q) \to 0$ and $S_{CC}(q) \to c_1 c_2$, while all the three $a_{ij}(q) \to 1$. For a completely uncorrelated system, these limiting values, of course, hold at all q.

In the long wavelength ($q \to 0$) limit, the number–concentration structure factors have simple physical meaning for fluid mixtures, namely

$$S_{NN}(0) = \langle(\Delta N)^2\rangle/N, \qquad S_{CC}(0) = N\langle(\Delta c)^2\rangle \quad \text{and} \quad S_{NC}(0) = \langle \Delta N \Delta c\rangle \tag{2.4}$$

where $\langle(\Delta N)^2\rangle$ is the mean square fluctuation in the number of particles in a subvolume, $\langle(\Delta c)^2\rangle$, the mean square fluctuation in concentration and $\langle \Delta N \Delta c\rangle$ correlation between the two fluctuations. One may show from statistical mechanics that† (Bhatia and Thornton 1970)

$$S_{NN}(0) = \theta + \delta^2 S_{CC}(0), \qquad \theta = (N/V) k_B T \kappa_T \tag{2.5}$$

$$S_{NC}(0) = -\delta S_{CC}(0) \tag{2.6}$$

$$S_{CC}(0) = \frac{N k_B T}{(\partial^2 G/\partial c^2)_{T,P,N}} \tag{2.7}$$

where $c \equiv c_1$, κ_T is the isothermal compressibility at constant composition, P, T are pressure and temperature respectively, G is the Gibbs free energy and δ is a dilatation factor defined by

$$\delta = \frac{1}{V}\left(\frac{\partial V}{\partial c}\right)_{P,T,N} = \frac{v_1 - v_2}{c v_1 + (1-c) v_2} \tag{2.8}$$

where v_1 and v_2 are the partial molar volumes of the two species.

If the partial molar volumes of the two species are the same, then $\delta = 0$ and the fluctuations in number density are independent of those in concentration ($S_{NC}(0) = 0$), as might be expected intuitively. Further $S_{NN}(0)$ is then θ, like the expression for the structure factor of a pure liquid.

To calculate $S_{CC}(0)$ we need to know G or the free energy of mixing G_M which is defined by

$$G = N[cG_1^{(0)} + (1-c) G_2^{(0)}] + G_M \tag{2.9}$$

† For the $\mathbf{q} \to 0$ limit of $S_{NN}(\mathbf{q})$ etc in a solid alloy, see Bhatia and Thornton (1971).

where $G_1^{(0)}$ and $G_2^{(0)}$ are the Gibbs free energies, per atom, of the two pure species at the same pressure and temperature as the mixture. In the next section we examine the behaviour of $S_{CC}(0)$ on the basis of some simple theoretical expressions for G_M for different types of mixtures. Note that $\partial^2 G/\partial c^2$ (the stability function of Darken 1967) and hence $S_{CC}(0)$ may be evaluated also from the experimental data on G_M, or more conveniently, from the thermodynamic activities a_i (Darken 1967, McAlister and Turner 1972). The formula for $S_{CC}(0)$ becomes

$$S_{CC}(0) = (1 - c_i)\ [(\partial \ln a_i/\partial c_i)_{T,P}]^{-1}. \tag{2.10}$$

Once $S_{CC}(0)$ is known, the evaluation of $S_{NC}(0)$ and $S_{NN}(0)$ and hence of $a_{ij}(0)$ is straightforward from equations (2.5–2.8) and (2.3), if the volume and compressibility are specified (§4).

3. $S_{CC}(0)$ for different types of mixtures

3.1. Ideal and conformal solutions

The simplest realistic model for liquid mixtures is perhaps the conformal solution model of Longuet-Higgins (1951) or the regular solution model in the zeroth approximation (e.g. see Guggenheim 1952) for which G_M is given by†

$$G_M = Nk_BT[c \ln c + (1 - c) \ln (1 - c)] + N\omega c(1 - c) \tag{3.1}$$

where the first term is just $(-T)$ times the entropy of random mixing and ω is an interchange energy, assumed to be concentration independent, such that if a nearest-neighbour A–A atoms pair and a similar B–B atoms pair are replaced by two A–B pairs, the increase in the energy of the alloy is $2\omega/z$, where z is the number of nearest neighbours of an atom. A positive ω implies that like-atom nearest-neighbour pairs are energetically preferred over unlike-atom pairs and *vice versa* for ω negative. When $\omega = 0$, we have the 'ideal' solution.

For the expression (3.1) for G_M to be valid for a liquid (or solid) solution two conditions are necessary. First the size difference between the two types of atoms is small, such that, *roughly* (Guggenheim 1952)

$$\tfrac{1}{2} < \beta < 2, \qquad \beta = v_1^{(0)}/v_2^{(0)} \tag{3.2}$$

where $v_1^{(0)}$ and $v_2^{(0)}$ are the volumes, per atom, of the two pure species. Secondly, ω should be small (strictly only for $(\omega/zk_BT) \to 0$, since otherwise the assumption of random distribution of atoms used in obtaining (3.1) is not valid). The regular solution theory has been extended to higher approximations (e.g. the quasichemical approximation; for reviews, see Guggenheim 1952, Prigogine 1957, Rowlinson 1969) and their inspection shows that (3.1) can be expected to hold only for $|\omega/k_BT| < 2$ and may become a poor approximation even before this limit is reached. Using (3.1) this implies that at $c = \frac{1}{2}$, where $|G_M|$ is largest, G_M must lie within the limits

$$-1{\cdot}2 < G_M/Nk_BT < -0{\cdot}2 \qquad \text{at } c = \tfrac{1}{2}. \tag{3.3}$$

We return to these conditions later.

† Note that $G_M \simeq A_M$, where A_M is the Helmholtz free energy of mixing – which is the quantity evaluated in regular solution theories.

Using (3.1) and (2.9) in (2.7), one has (Bhatia and Thornton 1970, Bhatia *et al* 1973)

$$S_{\rm CC}(0) = \frac{c(1-c)}{1-(2\omega/k_{\rm B}T)\,c(1-c)}, \tag{3.4}$$

which, like $G_{\rm M}$, is symmetric about $c = \frac{1}{2}$. For an ideal solution ($\omega = 0$), $S_{\rm CC}(0)$ is just

$$S_{\rm CC}{}^{\rm id}(0) = c(1-c), \tag{3.5}$$

corresponding to solutions in which the two types of atoms are distributed at random. When unlike-atom pairs are preferred as nearest neighbours, $\omega < 0$, and $S_{\rm CC}(0) < S_{\rm CC}{}^{\rm id}(0)$. For $\omega > 0$, $S_{\rm CC}(0) > S_{\rm CC}{}^{\rm id}(0)$. In particular, the solution now has a critical point of mixing, the critical temperature being $T_{\rm c} = \omega/2k_{\rm B}$, near which $S_{\rm CC}(0)$ and hence the fluctuations in concentration become very large. Expressions (3.1) and (3.4) for such (partially miscible) mixtures are, of course, applicable only above the critical temperature.

Figure 1 illustrates the behaviour of $S_{\rm CC}(0)$ for several values of ω and compares the experimental values for Na–K alloy with the theoretical curve for $\omega/k_{\rm B}T = 1{\cdot}1$ – this value being indicated by data on $G_{\rm M}$. We have not made a detailed comparison for other alloys, but a glance at the tabulated data on $G_{\rm M}$ (Hultgren *et al* 1963, 1973) shows that for several alloys, for which the conditions (3.2) and (3.3) are met, the conformal solution model holds at least approximately, for example Ag–Au, Cd–Zn. However, it is to be noted that even when β and $G_{\rm M}/Nk_{\rm B}T$ are well within the limits set by (3.2) and (3.3), $G_{\rm M}$ has considerable asymmetry about $c = \frac{1}{2}$ for some alloys, for example Cd–Pb. Within the framework of the conformal solution model, this asymmetry can be formally accounted for by assuming ω to depend appropriately on concentration. According to current ideas in electron theory (Heine and Weaire 1970), an asymmetry in energy of mixing can arise if the two types of atoms in the alloy differ in their size, valence, and electronegativity. Calculations elucidating the relative role of these factors would be of interest.

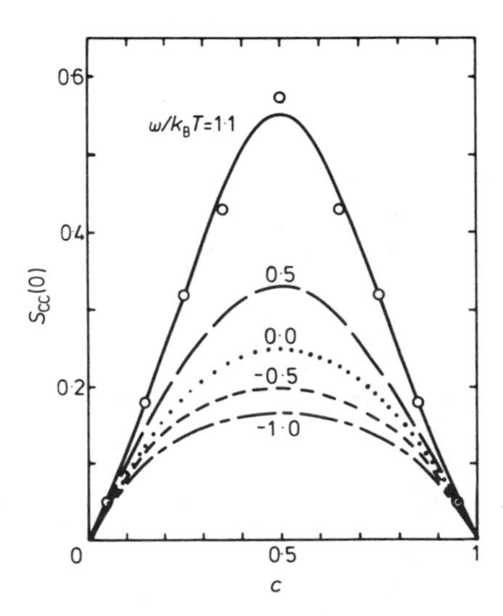

Figure 1. $S_{\rm CC}(0)$ as a function of concentration for conformal solution for different values of $\omega/k_{\rm B}T$. $\omega = 0{\cdot}0$ is the ideal solution curve. o data for Na–K with $c \equiv c_{\rm Na}$.

3.2. Effects of size difference (Flory's approximation)

When the atomic volumes $v_1^{(0)}$ and $v_2^{(0)}$ differ considerably from each other, a more valid approximation, in a similar vein to (3.1), is given by Flory's formula (Flory 1942)

$$G_M = Nk_BT\,[c\ln\phi + (1-c)\ln(1-\phi)] + N\omega\phi(1-\phi)\,[c + (1-c)/\beta] \tag{3.6}$$

where ϕ is the concentration by volume:

$$\phi = \frac{cv_1^{(0)}}{cv_1^{(0)} + (1-c)\,v_2^{(0)}}, \qquad \beta = \frac{v_1^{(0)}}{v_2^{(0)}}. \tag{3.7}$$

Equation (3.6) gives for $S_{CC}(0)$ (Bhatia and March 1975)

$$S_{CC}(0) = \frac{c(1-c)}{1 + c(1-c)\delta^2\,\{1 - [2\beta\delta\,\omega/(\beta-1)^3 k_BT]\}} \tag{3.8}$$

where

$$\delta = \frac{v_1^{(0)} - v_2^{(0)}}{cv_1^{(0)} + (1-c)\,v_2^{(0)}} = \frac{\beta - 1}{(1-c) + c\beta}. \tag{3.9}$$

We observe from (3.8) that unlike the case of regular solutions (δ small), $S_{CC}(0)$ now depends on δ also. In particular, even for an athermal mixture ($\omega = 0$), $S_{CC}(0)$ is not just $c(1-c)$ as for a truly random (ideal) solution and is asymmetric about $c = \frac{1}{2}$ (Bhatia and Thornton 1970, Turner *et al* 1973). Equations (3.6) and (3.8) have been applied by Bhatia and March to explain the observed concentration dependences of activity and $S_{CC}(0)$ (Ichikawa *et al* 1974) and of the liquidus curve (Kim and Letcher 1971) for Na–Cs alloy for which $\beta = \frac{1}{3}$ ($v_1^{(0)} \equiv v_{Na}^{(0)}$). The comparison for $S_{CC}(0)$ is shown in figure 2. The explanation of the discrepancy near $c_{Na} \sim 0{\cdot}6$ is at present lacking.

3.3. Compound-forming alloys

There are many molten alloys (Mg–Bi, Tl–Te, Na–Hg etc) for which the difference in the pairwise interaction energies is not small as is evidenced by the magnitude of

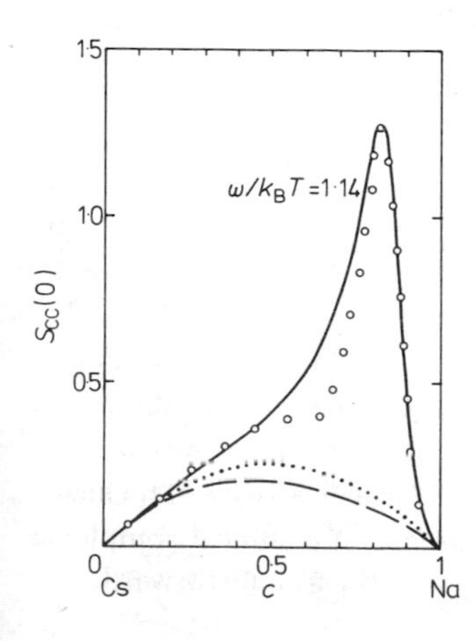

Figure 2. $S_{CC}(0)$ as a function of concentration in Flory's approximation for volume ratio $\beta = \frac{1}{3}$. o data for NaCs from Ichikawa *et al* (1974). For reference, the ideal solution curve (. . . .) is also given.

G_M/Nk_BT. Moreover, G_M and $S_{CC}(0)$ are not symmetric about $c = \frac{1}{2}$ even when the volume ratio β is not appreciably different from unity, so that one cannot explain the results on the higher approximations of the regular solution theory. Such alloys usually have the characteristic that in the solid phase they form compounds at one or more well defined stoichiometric composition. Following an idea originally by Bent and Hilderbrand (1927), recently Bhatia and Hargrove (1973, 1974) and Bhatia *et al* (1974) – jointly referred to here as I – have discussed the thermodynamic properties of some of these alloys on the assumption of formation of appropriate chemical complexes (see also McAlister and Crozier 1974).

Briefly, the idea is that if an A–B alloy forms, in the solid phase, a compound at the composition specified by $A_\mu B_\nu$ (μ, ν small integers), then in the liquid phase there exist, at a given P, T and c, certain numbers of separate A and B atoms and chemical complexes $A_\mu B_\nu$ in chemical equilibrium with one another. We consider for simplicity here the case where it can be assumed that only one type of chemical complexes (one pair of μ, ν) is formed; for generalization see I. Then if n_1, n_2 and n_3 respectively denote the numbers of A, B and $A_\mu B_\nu$ in the mixture, we have from the conservation of atoms (total number of A atoms $= N_A \equiv N_1 = Nc$ etc)

$$n_1 = Nc - \mu n_3, \qquad n_2 = N(1-c) - \nu n_3 \tag{3.10}$$

and

$$n = n_1 + n_2 + n_3 = N - (\mu + \nu - 1)n_3.$$

Now let G_1 and G_2 denote the chemical potentials, per atom, of the species A and B in the mixture and G_3 the chemical potential, per $A_\mu B_\nu$, of the complexes, and let $G_i^{(0)}$, $i = 1, 2, 3$, be the corresponding chemical potentials of the three species in their respective pure states (at the temperature and pressure of the mixture). Then since the Gibbs free energy $G = n_1G_1 + n_2G_2 + n_3G_3$, the free energy of mixing G_M for the binary alloy may be written, using (3.10),

$$G_M = G - NcG_1^{(0)} - N(1-c)\,G_2^{(0)} = -n_3g + G' \tag{3.11}$$

where

$$g = \mu G_1^{(0)} + \nu G_2^{(0)} - G_3^{(0)}$$

$$G' = G - (n_1G_1^{(0)} + n_2G_2^{(0)} + n_3G_3^{(0)}). \tag{3.12}$$

The equilibrium value of n_3 at a given P, T, c is given by

$$(\partial G_M/\partial n_3)_{P,T,N,c} = 0. \tag{3.13}$$

In equation (3.11), the first term $(-n_3g)$ represents the lowering of the (free) energy of mixing due to the formation of the chemical complexes. The second term (G') is the free energy of mixing of a ternary mixture of fixed n_1, n_2, n_3, whose constituents A, B and $A_\mu B_\nu$ are assumed to interact only weakly with one another – the strong bonding interactions between the A and B atoms having been taken care of via the formation of chemical complexes. We can thus take for G' expressions known from statistical mechanics in different approximations, for instance ideal, conformal, Flory's etc. Once the form of G' is specified, the calculation of equilibrium number of chemical complexes and the various thermodynamic properties, activity, $S_{CC}(0)$ etc, is a straightforward, though lengthy, matter. All the three approximations referred to above are described in I and here we discuss only the main results.

The simplest approximation, namely the ideal solution approximation, has been used in different contexts (theory of chemical equilibria etc) for many years. It implies that

$$G' = k_B T \sum_{i=1}^{3} n_i \ln n_i/n,$$

so that one obtains from (3.11) and (3.13), the usual equilibrium equation

$$(n_1/n)^\mu (n_2/n)^\nu = (n_3/n) \exp(-g/k_B T). \tag{3.14}$$

The value of g, which is the only unknown in the equations, may be estimated from the observed G_M. One may distinguish two limiting cases.

First, if there is no tendency to form chemical complexes ($g \to -\infty$), $n_3 = 0$, and the equations just give results appropriate to a binary ideal solution ($S_{CC}(0) = c(1-c)$), as they should. The opposite limiting case is where the tendency to form chemical complexes is very strong ($g \to +\infty$), although physically g can never, of course, be infinitely large as this implies $G_M \to -\infty$ (see below). In this limit

$$n_3 \simeq Nc/\mu, \qquad \text{for} \quad c < c_c, \; c_c = \mu/(\mu+\nu)$$

$$n_3 \simeq N(1-c)/\nu, \qquad \text{for} \quad c > c_c, \tag{3.15}$$

c_c being the concentration at which the compound is formed. One gets for $S_{CC}(0)$, for $c < c_c$,

$$S_{CC}(0) = (c/\mu)\,[\mu - (\mu+\nu)c]\,[\mu - (\mu+\nu-1)c], \quad c < c_c, \tag{3.16}$$

and a similar expression for $c > c_c$ obtained from (3.16) by interchanging $\mu \rightleftharpoons \nu$ and $c \to (1-c)$. In practice (3.15) and (3.16) form a good approximation (except very close to $c = 0$, $c = 1$ and $c = c_c$ where the slopes of n_3 and $S_{CC}(0)$ need careful analysis; see I) once g is such that, at c_c,

$$G_M(c_c)/Nk_B T < -3. \tag{3.17}$$

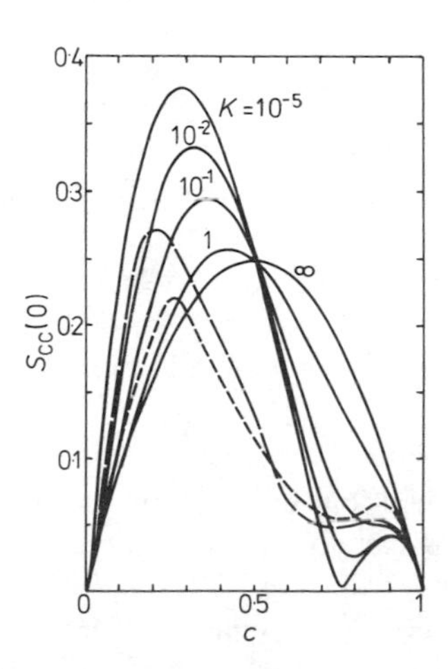

Figure 3. $S_{CC}(0)$ as a function of concentration $c \equiv c_A$ in a mixture forming A_3B_1 complexes for different values of the reaction constant K. Broken curves, data for Ag–Al with $c \equiv c_{Ag}$ (from Bhatia *et al* 1974).

We thus see that when the tendency to form chemical complexes is very strong, $S_{CC}(0)$, apart from being zero at $c = 0$ and $c = 1$, is (nearly) zero also at $c = c_c$ and has two peaks, one between $0 < c < c_c$ and the other between $c_c < c < 1$, their positions and heights depending on the values of μ and ν and the magnitude of g. Figure 3 illustrates the behaviour of $S_{CC}(0)$ for $\mu = 3, \nu = 1$, for several values of the reaction constant $K = \exp(-g/k_B T)$. For comparison the experimental $S_{CC}(0)$ for Ag–Al system which forms the compound Ag_3Al in the solid phase is also given. Note that even though the tendency to form chemical complexes in this system is not very strong at the temperature of observation $(G_M(c_c)/Nk_B T \simeq -1{\cdot}2)$, $S_{CC}(0)$ is highly asymmetric about $c = \frac{1}{2}$ and that this asymmetry cannot be due to volume differences between Ag and Al since $v_{Ag}/v_{Al} \simeq 1{\cdot}0$.

To obtain quantitative agreement even for the simplest of compound-forming systems, we have, however, to go beyond the ideal solution approximation and introduce (weak) pairwise interactions between A, B and $A_\mu B_\nu$. If this is done in the conformal solution approximation for the ternary mixture, one has

$$G' = k_B T \sum_{i=1}^{3} n_i \ln n_i/n + \sum_{i<j} \sum \omega_{ij} (n_i n_j/n) \tag{3.18}$$

where ω_{ij} ($\equiv 0$ for $i = j$) are interaction energies similar to ω of the binary mixture. (The use of Flory's approximation, which allows one to take into account also the possible volume differences between A, B and $A_\mu B_\nu$ is somewhat more complex; see I. Since the results are qualitatively similar we omit its discussion for brevity here.) The main effect of using (3.18) for G' is that the heights and positions of the peaks now depend also on the magnitudes and signs of ω_{ij}. For example, in the limit of no tendency to form chemical complexes one obtains for $S_{CC}(0)$ the expression (3.4), appropriate to a binary conformal solution, as one might expect. In the opposite limit of strong tendency to form chemical complexes, one has, for $c < c_c$,

$$S_{CC}(0) \simeq \frac{A}{1 - A(2\omega_{23}/\mu^2 RT)(N/n)^3} \tag{3.19}$$

where A stands for the expression (3.16) for $S_{CC}(0)$ appropriate for when all $\omega_{ij} = 0$. The corresponding expression for $S_{CC}(0)$ for $c > c_c$ is obtainable from (3.19) and (3.16) by interchanging $\mu \rightleftharpoons \nu$, $c \to (1-c)$ and $\omega_{23} \to \omega_{13}$.

The calculated and experimental variation of $S_{CC}(0)$ with concentration is shown in figure 4 for four systems: Mg–Bi, Tl–Te, Ag–Al and Cu–Sn taking respectively for (μ, ν) (3, 2), (2, 1), (3, 1) and (4, 1). The first two systems show a strong tendency to form chemical complexes while the second two a relatively weaker one. The constants g and ω_{ij} were determined from the data on G_M and their values are given in I.

To conclude this section on $S_{CC}(0)$ in compound-forming systems we should make the following remarks.

First, it should be emphasized that our purpose in presenting the model is not so much to say that the chemical complexes are necessarily formed, but rather to provide a phenomenological model to interpret $S_{CC}(0)$ in one class of systems pending a more microscopic approach which is bound to be complex. Recently a start in this direction has been made by Cartier and Barriol (1974) for systems which show asymmetry cor-

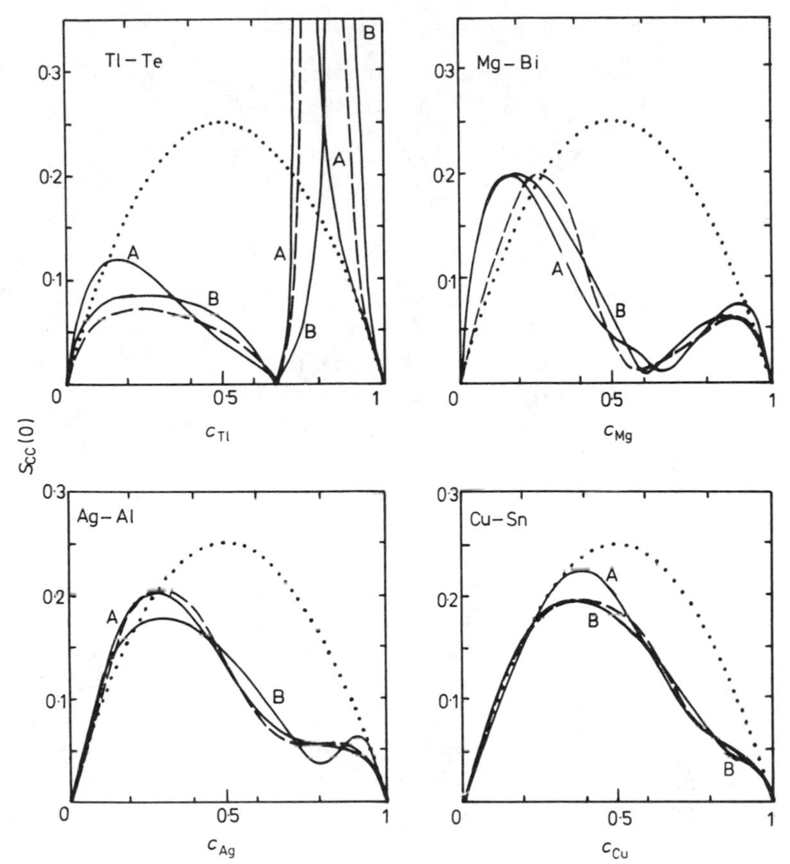

Figure 4. $S_{CC}(0)$ as a function of concentration for four compound-forming systems. Curves A, conformal solution approximation for the ternary mixture; curves B, Flory's approximation. Broken curve, experimental data; dotted curve, ideal solution curve.

responding to formations of A_3B, but it appears to be applicable only to cases where the tendency to form complexes is relatively weak (in our terminology), for example, Ag–Al. The existence, or otherwise, of the chemical complexes can, of course, in principle, be inferred from neutron scattering measurements and we discuss this point later.

Secondly, there exist a number of compound-forming systems which, in the solid phase, form compounds at more than one chemical composition and whose melting points lie close to one another. For such systems the assumption of formation of only one type of chemical complexes (one pair of μ, ν) would clearly be inadequate. Notable examples are the amalgams K–Hg and Na–Hg, where $S_{CC}(0)$, though very small at the concentration specified by Hg_2K, Hg_2Na (compounds having the highest melting point), shows no minimum and has just one peak (see I). An explanation for this is lacking at present.

The behaviour of $S_{CC}(0)$ has been studied for several other compound-forming systems from experimental data, notably by Ichikawa and Thompson (1973), Thompson (1974) and Thompson *et al* (1976), where other references and a discussion of its ($S_{CC}(0)$) relation to various transport properties may also be found.

4. The partial structures a_{ij} in the long wavelength limit

The long wavelength limit of the partial structure factors $a_{ij}(q)$ is obtained by substituting equations (2.5–2.7) in (2.3). The resulting expressions, namely

$$\begin{aligned} a_{11}(0) &= \theta - (c_2/c_1) + (1 - c_1\delta)^2 S_{CC}(0)/c_1^2 \\ a_{22}(0) &= \theta - (c_1/c_2) + (1 + c_2\delta)^2 S_{CC}(0)/c_2^2 \\ a_{12}(0) &= \theta + 1 - (1 - c_1\delta)(1 + c_2\delta) S_{CC}(0)/c_1 c_2 \end{aligned} \tag{4.1}$$

have been used to calculate $a_{ij}(0)$ for a number of cases, some of the references being Bhatia and Thornton (1970), McAlister and Turner (1972), Ichikawa and Thompson (1973), McAlister *et al* (1973), Turner *et al* (1973), Bhatia *et al* (1973, 1974), Crozier *et al* (1974), Ebert *et al* (1974), Ruppersberg and Egger (1975), and Durham and Greenwood (1976). Here we do no more than illustrate the behaviour of $a_{ij}(0)$ for a few cases.

First, if the solution is ideal ($S_{CC}{}^{id}(0) = c(1 - c)$) and if $\delta = 0$, then (4.1) gives

$$a_{11}(0) = a_{22}(0) = a_{12}(0) = \theta. \tag{4.2}$$

For such an alloy one may expect that the three $a_{ij}(q)$ are equal to one another for all q. This is the simplest approximation to make for $a_{ij}(q)$ and is frequently referred to as the substitutional hypothesis (Faber 1972). We note that θ (and therefore $a_{ij}(0)$ for this case) is usually no more than 0·03–0·05 at the melting point of an alloy.

To illustrate how a nonzero δ and (or) deviations from ideality affect $a_{ij}(0)$, we depict first, taking $\theta = 0$, the variation of $a_{ij}(0)$ with concentration for a mixture which is ideal but for which the volume ratio $\beta\ (= v_1^{(0)}/v_2^{(0)}) = \frac{1}{2}$ (figure 5). (For an ideal solution G_M is given by the first term in (3.1), so that V is linear in concentration.) These results are qualitatively similar to those obtained by Faber (1972) using probability considerations.

Figure 6 gives $a_{ij}(0)$ for the Na–K alloy for which $\beta \simeq \frac{1}{2}$. The theoretical curves are from Bhatia *et al* (1973) and are based on the conformal solution model with appropriate ω (see §3.1). The broken curves are on the assumption that the volume varies linearly with concentration and may be contrasted with the results of figure 5. The smooth curves are obtained by assuming a small pressure dependence of ω to account

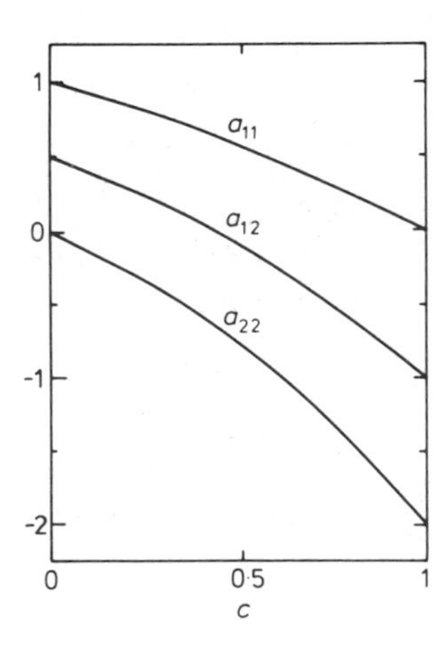

Figure 5. Variation of $a_{ij}(0)$ with concentration in an ideal solution taking $\theta = 0$ and $\beta = \frac{1}{2}$.

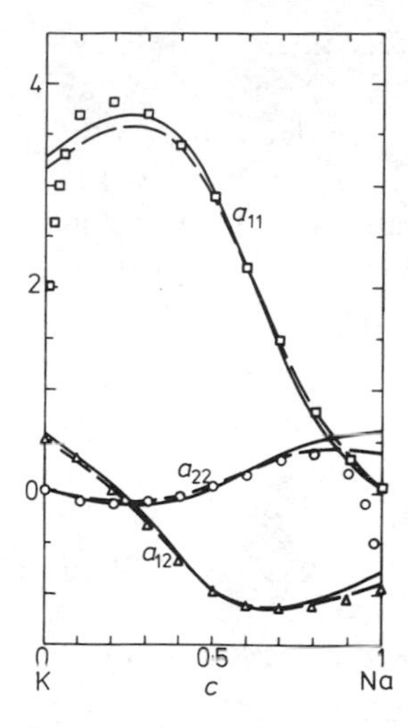

Figure 6. Variation of $a_{ij}(0)$ with concentration for Na–K alloy in conformal solution model. Broken curve, assuming linear concentration dependence of volume ($d\omega/dP = 0$). Full curve, for $d\omega/dP \neq 0$ (see text). □, ○, △ experimental from McAlister and Turner (1972).

for the observed deviation in volume and $V\kappa_T$ from linearity. Experimental points are taken from McAlister and Turner (1972). The discrepancy between theory and experiment in $a_{11}(0)$ near $c = 0$ is due to the fact that the expression for $a_{11}(0)$ in the limit $c = 0$ is

$$a_{11}{}^{0}(0) = \theta^0 - 2\delta^0 + \tfrac{1}{2}\ [2 + d^2 S_{CC}(0)/dc^2]_{c=0}, \tag{4.3}$$

rather than just $\theta^0 - 2\delta^0$ used by McAlister and Turner. Similar remarks apply for $a_{22}(0)$ near $c = 1$; for details see Bhatia *et al* (1973).

For compound-forming systems, Bhatia *et al* (1974) have studied $a_{ij}(0)$ on the assumption of formation of one type of chemical complexes $A_\mu B_\nu$ and treating the resulting ternary mixture as ideal (see §3.3). It is found that when the tendency to form chemical complexes is strong, the concentration dependence of $a_{ij}(0)$ depends characteristically on the values of (μ, ν). Figure 7 shows their results for $\mu = 2$ and $\nu = 1$ where the experimental curves for Hg–K alloy, taken from McAlister and Turner (1972), are also given. Considering the limitations of the model discussed earlier, the agreement between the two is not unsatisfactory.

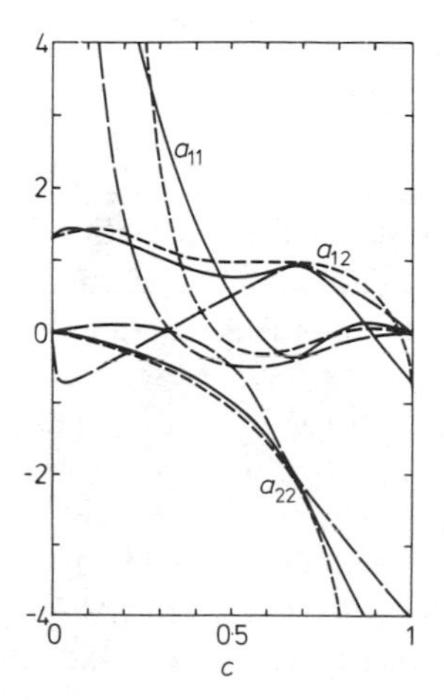

Figure 7. $a_{ij}(0)$ as a function of concentration $c \equiv c_A$ in a mixture with relatively strong tendency to form complexes A_2B_1. ——, theoretical with observed δ; — —, theoretical with $\delta = 0$; – – –, experimental for Hg–K with $c \equiv c_{Hg}$ (from Bhatia *et al* 1974).

We have seen earlier that when the tendency to form chemical complexes $A_\mu B_\nu$ is strong, $S_{CC}(0) \simeq 0$ at the compound-forming concentration $c_1 = \mu/(\mu + \nu)$. This is not surprising since, in the limit when only the complexes are present, a fluctuation in the number of A atoms in any given small (but macroscopic) volume must necessarily be accompanied by the corresponding change in the number of B atoms such that the concentration remains unchanged. By the same token when $S_{CC}(0) = 0$ so also $S_{NC}(0) = -\delta S_{CC}(0) = 0$. Denoting the concentration at which this happens as c_1 ($\neq 0$ or 1), equations (4.1) give for $a_{ij}(0)$ (Bhatia *et al* 1974, Bhatia and Ratti 1976)

$$a_{11}(0) = \theta - c_2/c_1 \tag{4.4}$$

$$a_{11}(0) - a_{12}(0) = -(c_1)^{-1} \tag{4.5}$$

$$a_{22}(0) - a_{12}(0) = -(c_2)^{-1}. \tag{4.6}$$

We remark that for a binary ionic liquid satisfying the condition $c_1 z_1 + c_2 z_2 = 0$ for overall charge neutrality, where z_1 and z_2 are respectively the charges on the two types of ions, equations (4.5) and (4.6) are equivalent to the so-called 'electroneutrality conditions', equations (24) of Stillinger and Lovett (1968). Equations (4.5) and (4.6) also imply

$$a_{11}(0) - a_{22}(0) = (c_2)^{-1} - (c_1)^{-1} \tag{4.7}$$

which has been given previously by Enderby (1974).

The nature of bonding (ionic, covalent, etc) between the atoms in a system cannot, of course, be deduced just from the fact that $S_{CC}(0) \simeq S_{NC}(0) \simeq 0$ or that (4.4)–(4.6) are (approximately) satisfied. Even for metallic bonding, $S_{CC}(0)$ (and hence also $S_{NC}(0)$) can be zero – as in the ordered state of order–disorder type of solid alloys, like CuZn, at the appropriate stoichiometric composition (Krishnan and Bhatia 1944, Bhatia and Thornton 1970). In contrast to compound-forming systems, here the electrical resistivity has a cusp-like minimum (rather than maximum) at the composition at which $S_{CC}(0) \simeq 0$.

5. The structure factors at arbitrary q

5.1. General remarks

We have seen in the preceding sections that the concentration dependence of the long wavelength limit of the structure factors exhibits widely varying features depending on the type of molten system under consideration. In particular we saw that the deviations of $S_{CC}(0)$ from its ideal value $(c_1 c_2)$ reflect rather sensitively the nature of interatomic interactions in the mixture. On the other hand, the only calculations of the structure factors for arbitrary q which seem to be available to date are based on the hard-sphere model in the Percus–Yevick approximation (Ashcroft and Langreth 1967, Enderby and North 1968). In the terminology of §3, such a model corresponds *roughly* to an athermal or non-interacting mixture ($\omega = 0$) and thus cannot be expected to give correctly the $q \to 0$ limits of the structure factors even for the simplest systems. (For example for Na–K alloy at $c = \frac{1}{2}$, the hard-sphere model gives $S_{CC}(0) \simeq 0{\cdot}24$, whereas the experimental value is $S_{CC}(0) \simeq 0{\cdot}55$; see figure 1.) Since the various physical properties, like x-ray and neutron scattering, electron transport in alloys, etc, depend for their quantitative interpretation on the structure factors, it is of interest to enquire:

(i) How $S_{CC}(q)$ and other structure factors would behave at nonzero q if the hard-sphere treatment is modified so that the $q \to 0$ limit of $S_{CC}(q)$ is more in accord with $S_{CC}(0)$ of a weakly interacting (rather than athermal) mixture.
(ii) How these structure factors would behave for compound-forming systems.

We describe briefly here some work which deals with these questions.

5.2. *Weakly interacting systems*

Woodhead-Galloway *et al* (1968) and Woodhead-Galloway and Gaskell (1968) proposed a modification of the hard-sphere treatment based on the random phase approximation (RPA). Using this idea A B Bhatia and W H Hargrove (unpublished) have investigated the behaviour of $S_{CC}(q)$ and other structure factors in weakly interacting mixtures in the following way.

The Ornstein–Zernike direct correlation functions $C_{ij}(r), i,j = 1,2$, are written in the form

$$C_{ij}(r) = C_{ij}^{\text{hs}}(r) - \beta\chi_{ij}(r), \qquad \beta = (k_B T)^{-1} \tag{5.1}$$

where $C_{ij}^{\text{hs}}(r)$ are the direct correlation functions for a mixture of hard spheres of appropriate diameters (σ_1 and σ_2) to represent the strong repulsive cores in the interatomic potentials $\phi_{ij}(r)$. Since it is known from RPA that when the potentials are small, $C_{ij}(r) \simeq -\beta\phi_{ij}(r)$, it can be assumed that for $\chi_{ij}(r)$ that

$$\begin{aligned} \chi_{ij}(r) &= \phi_{ij}(r) && \text{for} \quad r \geqslant r_{ij}^{\text{m}} \geqslant \tfrac{1}{2}(\sigma_i + \sigma_j) \\ &= \phi_{ij}(r_{ij}^{\text{m}}) && \text{for} \quad r < r_{ij}^{\text{m}} \end{aligned} \tag{5.2}$$

where r_{ij}^{m} is the value of r at which $\phi_{ij}(r)$ is minimum. Equations (5.1) and (5.2) imply that $\phi_{ij}(r)$ can be considered to be weak for $r \geqslant r_{ij}^{\text{m}}$. (We note that nonzero χ_{ij} inside the core region are necessary to reduce the effects of the fact that when C_{ij} are modified only outside the core regions, the pair distribution functions $g_{ij}(r)$ are not zero inside the core region as they should be. A fully satisfactory χ_{ij} are hard to calculate and the simple choice made here is necessarily very approximate; for a discussion in pure fluid see Gaskell 1970.) In (5.1) and (5.2), of course, $\sigma_i \simeq r_0^i$, where r_0^i is the position at which $\phi_{ii}(r) = 0$. Their precise values may be chosen so that the structure factors of the pure liquids are reasonably well reproduced.

The structure factors are related to the Fourier transforms $C_{ij}(q)$ of $C_{ij}(r)$ in the standard way. The expression for $S_{CC}(q)$ may be written as

$$S_{CC}(q) = \frac{c_1 c_2}{1 - nc_1c_2(C_{11}(q) + C_{22}(q) - 2C_{12}(q)) - c_1c_2[\Delta^2(q)/\theta(q)]} \tag{5.3}$$

where n is the number density and

$$\begin{aligned} \theta(q) &\equiv S_{NN}(q) - S_{NC}^2(q)/S_{CC}(q) \\ &= \{1 - n[c_1^2 C_{11}(q) + c_2^2 C_{22}(q) + 2c_1c_2C_{12}(q)]\}^{-1} \end{aligned} \tag{5.4}$$

$$\begin{aligned} \Delta(q) &= -S_{NC}(q)/S_{CC}(q) \\ &= n\theta(q)\{c_1[C_{11}(q) - C_{12}(q)] - c_2[C_{22}(q) - C_{12}(q)]\}. \end{aligned} \tag{5.5}$$

Note that $\theta(q)$ and $\Delta(q)$ are defined such that $\theta(0) = \theta$ and $\Delta(0) = \delta$.

It is interesting to consider the simple case where the functional dependence on r of the three interatomic potentials $\phi_{ij}(r)$ is the same (with the same value of r_m and r_0) so that $\phi_{ij}(r) \doteq \epsilon_{ij}\phi(r)$. If the strengths of the potentials ϵ_{11} and ϵ_{22} are not too different from one another, then one can take (approximately) also $\sigma_1 = \sigma_2 = \sigma$, say. Denoting by $\chi_{ij}(q)$ the Fourier transforms of $\chi_{ij}(r)$, noting that χ_{ij}, like ϕ_{ij}, are also proportional to one another and setting

$$2R = (\epsilon_{11} + \epsilon_{22} - 2\epsilon_{12})/\epsilon_{11}, \qquad \gamma = (\epsilon_{11} - \epsilon_{22})/\epsilon_{11}$$

$$\omega(q) = -(n\chi(q))R, \qquad \chi(q) \equiv \chi_{11}(q) \tag{5.6}$$

one gets

$$\theta(q) = \left[1 - nC_{\mathrm{hs}}(q) + \frac{n\chi(q)}{k_{\mathrm{B}}T}(1 - c_2\gamma - 2Rc_1c_2)\right]^{-1} \tag{5.7}$$

$$\Delta(q) = -\frac{n\theta(q)\chi(q)}{k_{\mathrm{B}}T}\left(\tfrac{1}{2}\gamma + R(c_1 - c_2)\right) \tag{5.8}$$

$$S_{\mathrm{CC}}(q) = \frac{c_1c_2}{1 - [2\omega(q)/k_{\mathrm{B}}T]\,c_1c_2 - [\Delta^2(q)/\theta(q)]\,c_1c_2} \tag{5.9}$$

where $C_{\mathrm{hs}}(q)$ is the Fourier transform of the direct correlation function $C_{\mathrm{hs}}(r)$ of a fluid of hard spheres of diameter σ.

We observe that if, in addition, $\gamma = 0$, that is $\epsilon_{11} = \epsilon_{22}$, the volumes $v_1^{(0)}$ and $v_2^{(0)}$ of the two pure species are the same. For this case $\Delta(q)$ has formally similar concentration dependence as the δ for a conformal solution with $v_1^{(0)} = v_2^{(0)}$ (δ here arises from the pressure dependence of ω in the expression (3.1) for G_M). Further, at $q = 0$, $\theta(0) \simeq 0{\cdot}05$ so that, provided $\omega(0)/k_{\mathrm{B}}T$ is not much greater than unity, the term involving Δ in (5.9) can be neglected and (5.9) reduces to the expression (3.4) for $S_{\mathrm{CC}}(0)$ in the conformal solution model.†

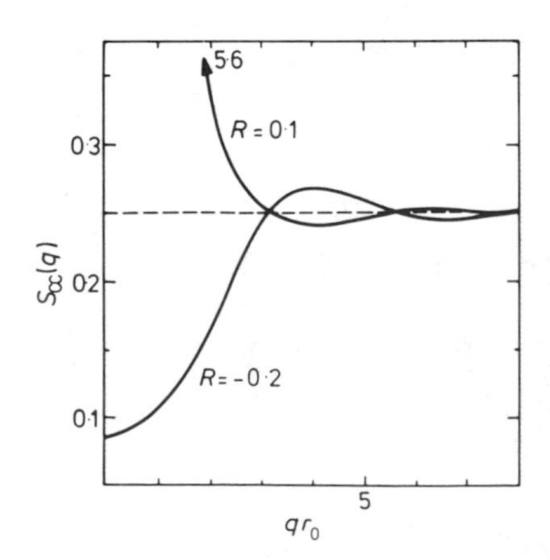

Figure 8. Model $S_{\mathrm{CC}}(q)$ in a weakly interacting mixture for two values of R.

† For another method of calculating the structure factors of conformal solutions see Parrinello *et al* (1974) and Johnson *et al* (1975).

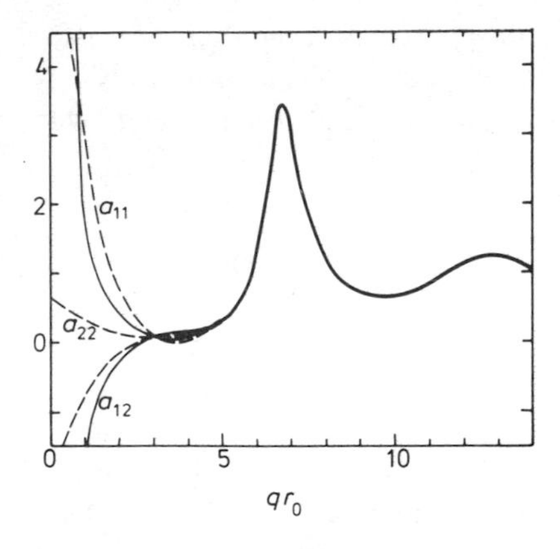

Figure 9. Model $a_{ij}(q)$ for $R = 0{\cdot}1$. Full curve, $c = 0{\cdot}5$ ($a_{11} \equiv a_{22}$); broken curve, $c = 0{\cdot}2$.

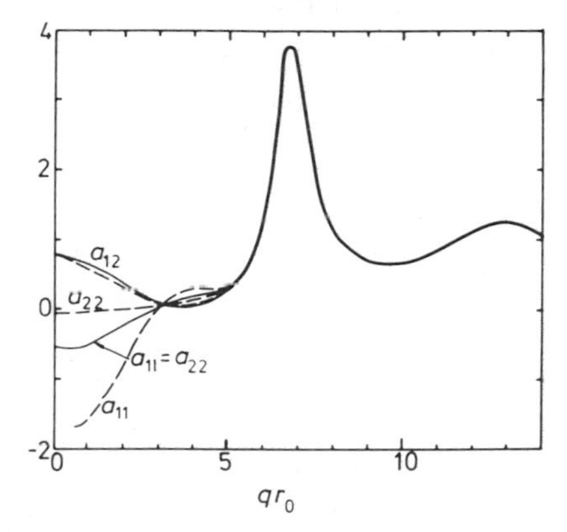

Figure 10. Model $a_{ij}(q)$ for $R = -0{\cdot}2$. Full curve, $c = 0{\cdot}5$ ($a_{11} \equiv a_{22}$); broken curve, $c = 0{\cdot}2$.

Figure 8 illustrates the behaviour of $S_{CC}(q)$ for two values of R or $\omega(0)$ at $c = \frac{1}{2}$. We should note that the case $R = -0{\cdot}2$ in figure 8 implies $\omega(0)/k_B T \simeq -4{\cdot}0$, which is too large an interchange energy for the conformal solution model, and hence the expressions (5.9) etc, to apply. A proper calculation for such a case is likely to give oscillations which are less damped than given in figure 8. (The calculations were made by taking σ and $\phi_{11}(r) = \phi_{22}(r)$ corresponding to liquid argon.) For the same values of R the structure factors $a_{ij}(q)$ are shown in figures 9 and 10 at the concentrations $c = \frac{1}{2}$ and $c = \frac{1}{5}$. Note that, in this example, if $R = 0$ (i.e. the interchange energy $\omega(0) = 0$), the three $a_{ij}(q)$ will be all equal to one another and $a_{ij}(0) \simeq \theta(0) \simeq 0{\cdot}05$. We thus observe that the effects of nonzero R are significant to wavenumbers $q \lesssim \frac{2}{3}q_0$, where q_0 is the position of the main peak in $a_{ij}(q)$. Qualitatively, this conclusion may be expected to hold even when the hard-core diameters are not equal. Scattering and molecular dynamics experiments in weakly interacting mixtures (e.g. Na–K, Na–Li) would be of interest to test the rapid damping of $S_{CC}(q)$ expected on the basis of RPA in such systems.

The calculation of electron-transport properties of metal alloys essentially involves integration of structure factors (with appropriate form and weighting factors) up to $q = 2q_F$, where q_F is the Fermi wavenumber. The structure factors in these calculations are usually taken either on the basis of substitutional hypothesis ($a_{ij}(q)$ all equal) or on hard-sphere model with $\sigma_1 \neq \sigma_2$. The conclusion arrived at above implies that for fully quantitative results one should take proper account of the structure factors at low q for monovalent metals since here $2q_F < q_0$. For polyvalent metals, however, $2q_F > q_0$, and the modifications in the structure factors from the hard-sphere model are not likely to be of much consequence. The effect of these modifications on band structure and Madelung energies of alloys is discussed in the following paper (Ratti and Bhatia 1977).

5.3. Compound-forming systems

If the idea of formation of chemical complexes as outlined in §3.3 is used, the structure factors of a compound-forming system can be approximately calculated as follows (Bhatia and Ratti 1975, 1976).

Let b_A and b_B denote the atom form factors for the A and the B atoms respectively and assume that these are the same whether an A (or B) atom is part of a chemical complex or exists separately – this is the case for neutron scattering but not necessarily so for x-rays. Then, making certain simplifying assumptions regarding orientational correlations etc, the scattering can be expressed in terms of b_A and b_B and either the three structure factors of a binary A–B mixture, or six structure factors of the ternary mixture of A, B and $A_\mu B_\nu$ and the arrangement of the atoms inside a chemical complex. Equating the two gives the desired expressions. Since the constituents of the ternary mixture are assumed to be weakly interacting, its structure factors may be evaluated to a first approximation in the hard-sphere model. If the tendency to form chemical complexes is strong, a further simplification occurs in that the ternary mixture is practically a binary mixture of B atoms and $A_\mu B_\nu$ for $c < c_c$ and of A atoms and $A_\mu B_\nu$ for $c > c_c$, where $c_c = \mu/(\mu + \nu)$ is the compound-forming concentration.

The above investigation was carried out with the aim of interpreting the neutron scattering results of Ruppersberg (1973) and Ruppersberg and Egger (1975) on a ^{7}Li–Pb system. The liquidus curves etc show that this system has a strong tendency to form chemical complexes Li_4Pb although the existence of other complexes cannot be quite ruled out. The calculations were carried out by assuming that the atoms in Li_4Pb are

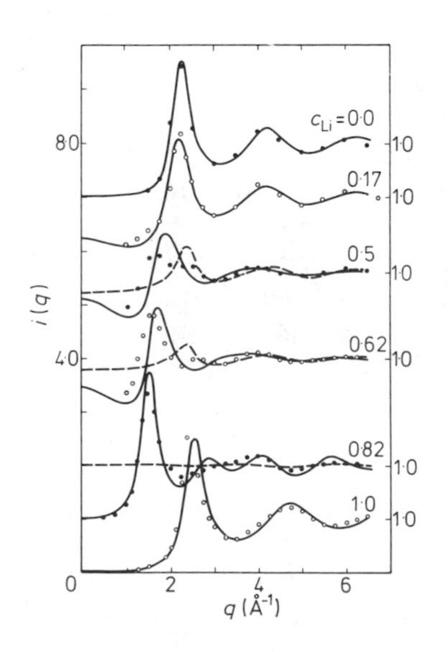

Figure 11. Neutron intensity $i(q)$ for a Li–Pb system at several concentrations: full curve, CFM; broken curve, BRHS; (○, ●) experimental (Ruppersberg 1973, 1975). (Figure from Bhatia and Ratti 1976.)

arranged in the form of a tetrahedron (as in CH_4) with lead at the centre. In terms of $S_{NN}(q)$ etc, the scattering $I(q)$ is given by (Bhatia and Thornton 1970)

$$I(q) = (\bar{b})^2 S_{NN}(q) + 2\bar{b}(\Delta b) S_{NC}(q) + (\Delta b)^2 S_{CC}(q) \tag{5.10}$$

where $\bar{b} = c_A b_A + (1-c) b_B$, $\Delta b = b_A - b_B$. Figure 11 gives the scattering $i(q)$ $[= I(q)/(c b_A^2 + (1-c) b_B^2)]$, normalized to unity for large q, at several concentrations of Li. For comparison $i(q)$ calculated by regarding the system as a binary random hard-sphere (BRHS) mixture of Li and Pb atoms is also given. The BRHS and the complex formation model (CFM) are, of course, identical for pure Li and pure Pb.

We see from figure 11 that the BRHS and CFM give virtually the same result at $c_{Li} = 0{\cdot}17$ also. But at other concentrations, while BRHS shows little agreement with experiment, the CFM gives good overall agreement both in regard to the positions of the major peaks and their heights. A full discussion of these results and of the behaviour of the various individual structure factors is given in Bhatia and Ratti (1976), but it may be noted here that for the ^{7}Li–Pb system, $\bar{b} = 0$ at $c_{Li} \simeq 0{\cdot}8$ so that at this concentration only the $S_{CC}(q)$ contributes to $i(q)$. We observe from the figure that at $c_{Li} = 0{\cdot}82$, the BRHS curve has barely any q dependence whereas both the CFM and experimental curves show large oscillations extending to high values of q. This behaviour of $S_{CC}(q)$ is also in contrast with that in figure 8 for a weakly interacting system where the oscillations in $S_{CC}(q)$ are rapidly damped.

6. Conclusion

In this paper we first examined the concentration dependence of the long wavelength limit of the concentration fluctuation structure factor $S_{CC}(q)$ for different types of mixtures. It is found that $S_{CC}(0)$ depends quite characteristically on the nature of interatomic interactions, for example, whether like atoms are preferred as nearest neighbours or unlike atoms, whether the mixture is weakly or strongly interacting, etc. The concentration dependence of the other two number–concentration structure factors $S_{NN}(0)$ and $S_{NC}(0)$ depends on $S_{CC}(0)$ and the dilatation factor δ in an obvious way. The behaviour of the partial structure factors $a_{ij}(0)$ is more complex and their dependence on $S_{CC}(0)$ and δ is illustrated by considering several cases. Finally, some approximate calculations of the structure factors, at arbitrary q, are described. One finds that $S_{CC}(q)$ for a strongly interacting compound-forming system shows large oscillations (about the ideal value $c_1 c_2$) which extend to high values of q, while for a weakly interacting system the oscillations are expected (in the random phase approximation employed here) to have barely any amplitude for $q \gtrsim \frac{2}{3} q_0$, where q_0 is the position of the main peak in $S_{NN}(q)$ or $a_{ij}(q)$. The significance of these results to related physical properties and areas where more work is required for better understanding are also pointed out.

References

Ashcroft N W and Langreth D C 1967 *Phys. Rev.* **156**, 685–92
Bent H E and Hilderbrand J H 1972 *J. Am. Chem. Soc.* **49** 3011
Bhatia A B and Hargrove W H 1973 *Lett. Nuovo Cim.* **8** 1025–30
—— 1974 *Phys. Rev.* **B10** 3186–96

Bhatia A B, Hargrove W H and March N H 1973 *J. Phys. C: Solid St. Phys.* **6** 621–30
Bhatia A B, Hargrove W H and Thornton D E 1974 *Phys. Rev.* **B9** 435–44
Bhatia A B and March N H 1975 *J. Phys. F: Metal Phys.* **5** 1100–6
Bhatia A B and Ratti V K 1975 *Phys. Lett.* **51A** 386–8
—— 1976 *J. Phys. F: Metal Phys.* **6** 927–41
Bhatia A B and Thornton D E 1970 *Phys. Rev.* **B2** 3004–12
—— 1971 *Phys. Rev.* **B4** 2325–8
Cartier A and Barriol J 1974 *C. R. Acad. Sci., Paris* **279C** 389–91
Crozier E D, McAlister S P and Turner R 1974 *J. Chem. Phys.* **61** 126–8
Darken L S 1967 *Trans. Metall. Soc. AIME (Am. Inst. Min. Metall. Pet. Engng)* **239** 80–9
Durham P J and Greenwood D A 1976 *Phil. Mag.* **33** 427–40
Ebert H, Höhler J and Steeb S 1974 *Z. Naturf.* **29a** 1890–7
Enderby J E 1974 *J. Physique Coll.* **C4** 309–12
Enderby J E and North D M 1968 *Phys. Chem. Liquids* **1** 1–11
Faber T E 1972 *Introduction to the Theory of Liquid Metals* (London: Cambridge UP)
Faber T E and Ziman J M 1965 *Phil. Mag.* **11** 153–73
Flory P J 1942 *J. Chem. Phys.* **10** 51–61
Gaskell T 1970 *J. Phys. C: Solid St. Phys.* **3** 240–7
Guggenheim A E 1952 *Mixtures* (Oxford: Oxford UP)
Heine V and Weaire D 1970 *Adv. Solid St. Phys.* **24** 250–65 (New York: Academic Press)
Hultgren R R, Desai P D, Hawkins D T, Glesier M and Kelley K K 1973 *Selected Values of the Thermodynamic Properties of the Elements* (Metals Park, Ohio: Am. Soc. Metals)
Hultgren R R, Orr R L, Anderson P D and Kelly K K 1963 *Selected Values of Thermodynamic Properties of Metals and Alloys* (New York: Wiley)
Ichikawa K, Granstaff S M and Thompson J C 1974 *J. Chem. Phys.* **61** 4059
Ichikawa K and Thompson J C 1973 *J. Chem. Phys.* **59** 1680–92
Johnson M W, March N H, Page D I, Parrinello M and Tosi M P 1975 *J. Phys. C: Solid St. Phys.* **8** 751–60
Kim M G and Letcher S V 1971 *J. Chem. Phys.* **35** 1164
Krishnan K S and Bhatia A B 1944 *Proc. Natn. Acad. Sci., India* **14** 153–80
Longuet-Higgins H C 1951 *Proc. R. Soc.* **A205** 247
McAlister S P and Crozier E D 1974 *J. Phys. C: Solid St. Phys.* **7** 3509–19
McAlister S P, Crozier E D and Cochran J F 1973 *J. Phys. C: Solid St. Phys.* **6** 2269–78
McAlister S P and Turner R 1972 *J. Phys. F: Metal Phys.* **2** L51–4
Parrinello M, Tosi M P and March N H 1974 *Proc. R. Soc.* **A341** 91–104
Prigogine I 1957 *The Molecular Theory of Solutions* (New York: North-Holland)
Ratti V K and Bhatia A B 1977 this volume
Rowlinson J S 1969 *Liquids and Liquid Mixtures* 2nd edn (London: Butterworth)
Ruppersberg H 1973 *Phys. Lett.* **46A** 75–6
Ruppersberg H and Egger H 1975 *J. Chem. Phys.* **63** 4095–103
Stillinger F H Jr and Lovett R 1968 *J. Chem. Phys.* **49** 1991–4
Thompson J C 1974 *J. Physique* **35** 367–9
Thompson J C, Ichikawa K and Granstaff S M Jr 1976 *J. Chem. Phys. Liquids* **6**
Turner R, Crozier E D and Cochran J F 1973 *J. Phys. C: Solid St. Phys.* **6** 3359–71
Woodhead-Galloway J and Gaskell T 1968 *J. Phys. C: Solid St. Phys.* **1** 1472–5
Woodhead-Galloway J, Gaskell T and March N H 1968 *J. Phys. C: Solid St. Phys.* **1** 271–85

Stability and structure in binary alloys†

N W Ashcroft

Laboratory of Atomic and Solid State Physics and Materials Science Center, Cornell University, Ithaca, New York 14853, USA

Abstract. While there are large volume-dependent contributions to the thermodynamic functions of elemental metals it is also clear that a description of their ionic structure in the liquid state can be achieved in terms of effective pairwise interactions that possess both short-range repulsive and long-range attractive features. When such metals are mixed in arbitrary proportions and at temperatures and pressures compatible with the liquid state, the question of absolute- and meta-stability of the resulting alloy against concentration fluctuations must be resolved by examining, in the free energy, both the volume-dependent terms and the structural terms (directly associated with the pair interactions). For simple metals, whose single-particle electron structure can be adequately described within a local pseudopotential framework, this can be achieved in a model which views the alloy as a two-component classical system. The average potential energy of interaction between ions in this system (including the Madelung energy) is shown to be very dependent on the degree of correlation between ions, the latter being reflected most clearly in the partial structure factors $S_{\alpha\beta}(k)$. In turn, the partial structure factors depend on temperature, relative concentrations (c_α) and of course on the effective interactions $V_{\alpha\beta}(\mathbf{r})$ between ions. Absolute instability of the mixture against concentration fluctuations is marked by a long-wavelength divergence in $S_{\alpha\beta}(k)$ as the temperature is reduced from that prevailing in the homogeneous alloy. Far above such temperatures, the $S_{\alpha\beta}(k)$ are reasonably well described by correlations resulting from an equivalent assembly of hard spheres, and indeed the entropy of a mixture of hard spheres approximates the entropy of an alloy rather well. But phase separation is also closely linked to potential energy effects associated with the long-range part of $V_{\alpha\beta}$ and these dual aspects (the short-range correlations of the hard-sphere system and the long-range interactions between ions) can be approximately accounted for by writing $V_{\alpha\beta} = \bar{V}_{\alpha\beta} + V^1_{\alpha\beta}$ where $\bar{V}_{\alpha\beta}$ is an equivalent hard-sphere potential selected in such a way that $V^1_{\alpha\beta}$ is to be regarded as a perturbation. To first order in the $V^1_{\alpha\beta}$, the divergences in $S_{\alpha\beta}(k)$ can be clearly demonstrated in a perturbation theory that exploits the theory of weakly inhomogeneous fluids and goes somewhat beyond the more conventional mean-field approach.

1. Introduction

The liquid state of any substance occupies a rather small portion of its overall phase diagram. For simple single-component systems, the liquid regime is normally terminated by the onset of two instabilities. At the high-temperature extremity there is a liquid–vapour transition. Viewed from the vapour side we infer from the condensation process that there are long-range attractive interactions between constituent atoms (and from the similarity of densities in all condensed phases of the substance that there are short-range repulsive interactions inhibiting complete collapse). In the limit of low temperatures of the liquid range there is usually freezing. The liquid–vapour transition can be viewed as a long wavelength instability; the transition to a regular solid, for example, is a shorter wavelength phenomenon.

† This work has been supported by the National Science Foundation through the facilities of the Materials Science Center (Grant DMR-72-03029, Technical Report No. 2682).

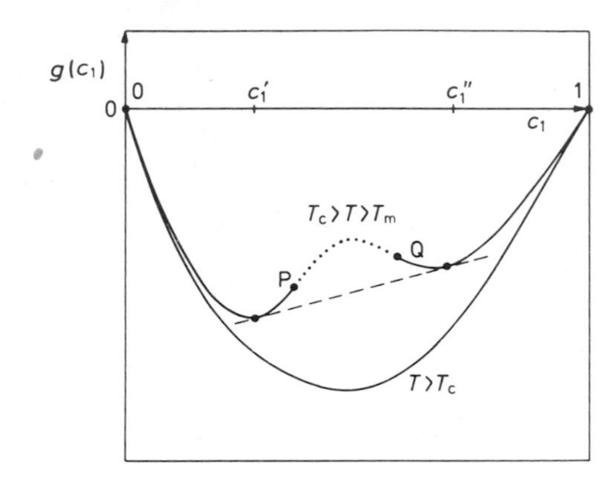

Figure 1. Typical excess Gibbs energy per particle for a partially miscible alloy that is liquid throughout the entire concentration range for the chosen temperature. The curve for which $\partial^2 g/\partial c_1{}^2 < 0$ for all c_1 corresponds to temperatures larger than the critical temperature T_c. The remaining curve shows, by the common tangent construction, the concentrations c_1', c_1'' of alloys, into which the system may separate for temperatures less than T_c (but greater than the melting temperature T_m). In the neighbourhood of c_1', c_1'', where $\partial^2 g/\partial c_1{}^2 > 0$, the system is metastable and, between P and Q, unstable against concentration functions.

When we consider mixtures, whether liquid or solid, there are additional degrees of freedom associated with the relative concentrations (c_α) of the constituents and as a consequence there is a likelihood of additional phases. In binary mixtures we must consider the possibility of phase separation or the onset of a miscibility gap. Such an instability, although possible in principle, need not always be observed in a given mixture since it can be overtaken by the instability of freezing. Thus in the liquid state for example, Na and K appear to mix in all proportions but Na and Li do not. Evidently the excess Gibbs energy $g(c_1c_2)$ per particle for this system must have the general form shown in figure 1, at least in the temperature range of interest. Providing the temperature is less than the critical temperature T_c (but greater than the melting temperature T_m) a system having an excess Gibbs energy of this type will separate into two phases with concentrations $c_1'(T)$, $c_1''(T)$ determined by the standard common-tangent construction. In the vicinity of these particular concentrations there are regions of *metastability* against concentration fluctuations, and a region (the dotted portion of figure 1) of *absolute* instability where $\partial^2 g/\partial c_1{}^2 < 0$. It is the region of absolute instability which we are concerned with here: the locus of points in the temperature-concentration plane determined by the condition $\partial^2 g/\partial c_1{}^2 = 0$. But as is well known, this condition can also be expressed as a long wavelength singular property of the partial structure factors $S_{\alpha\beta}(k)$ of the fluid mixture.

In pure fluids, the structure factor $S(k)$ (which is proportional to the density–density response function) contains considerable information on the nature of the interactions between particles. Provided these interactions can be adequately represented by pair terms, $S(k)$ usually reveals only the range of the repulsive, or hard-sphere part of the potential, at least with any numerical certainty. The weaker, longer-ranged part is not so easily elucidated. On the other hand, given a binary system in which a miscibility gap *is* observed, it is evident that features of the pair-potentials between ions *beyond* the hard-core region can be important. This is because transitions such as phase separation are driven by both potential energy and entropy considerations: the long-range portions of the interactions are important in the former but it is mainly correlations resulting from the short-range interactions that determine the latter.

In developing a structural theory of phase separation in binary alloys both of these aspects must be treated. But alloys, even alloys of simple metals, are really three-component systems comprising assemblies of electrons and two types of ions. (The hamiltonian for the entire system will be written down in the next section.) It is customary to calculate the electron distribution in such systems as an adiabatic response to the ionic distributions. Indeed, if the electron distribution is calculated at the level of linear response, then as shown in §2 the energy of the system can be divided between terms that are wholly dependent on volume, and pair terms that involve the partial structure factors† $S_{\alpha\beta}(k)$ of an effective two-component system (Ashcroft and Langreth 1976a,b). But as noted, phase separation is a long-wavelength phenomenon and the question that naturally arises is to what extent such instabilities reflect the behaviour of the volume-dependent terms or the structural terms (representing ionic correlations in the mixtures). These problems are discussed in §3 for alloys of simple metals in which the effective pair interactions are assumed to be spherically symmetric and other non-essential complications (such as core polarizations and attendant effects) are neglected. The major physical points can be made by considering homovalent alloys.

Correlations between point ions are very important, of course, in the calculation of the average Madelung energy, a term that is shown in §3 to be of some consequence in the possibility of phase separation. When electron screening is taken into account, correlation remains important: the effective ion–ion interactions $V_{\alpha\beta}$ that result and which enter in the determination of the $S_{\alpha\beta}(k)$, are characterized, as noted above, by short-range repulsive and long-range attractive regions. Since the gross structure of simple liquids is so well described by assemblies of hard spheres it is natural to approximate such interactions by $V_{\alpha\beta} = \bar{V}_{\alpha\beta} + V^1$ where $V_{\alpha\beta}$ are taken as a hard-sphere *reference* system underlying the basic correlations in the liquid and $V^1_{\alpha\beta}$ is regarded as a perturbation. The objective is to understand the onset of structural instabilities in terms of the long-range effect of $V^1_{\alpha\beta}$. Instead of the standard approach to this problem via thermodynamic perturbation theory, we describe in §4 a recent method (Henderson and Ashcroft 1976) which approaches the problem from the standpoint of the theory of weakly-inhomogeneous fluids. This is achieved by setting up a perturbation theory in which the one-particle densities are taken as the fundamental variables in the problem of calculating the linear response of classical inhomogeneous fluids. For the phenomenon of phase separation it is possible to extract approximate solutions for the $S_{\alpha\beta}(k)$ at small k, by exploiting the structural properties of a homogeneous fluid at the local mean density. We can show that at first order in $V^1_{\alpha\beta}$, such solutions can differ noticeably from the more conventional mean-field treatment. More important perhaps is the indication that the *details* of the pair interactions can be discerned in the location and asymmetry of the line of instabilities and the approach of §4 generalized to include higher-order

† If viewed as an ionic hamiltonian, this corresponds to a resolution of the potential energy into pair potentials: three- and higher-body forces may well play a role in phase separation but are not considered here.

corrections, may offer a way of extracting from such data details of ion–ion potentials distinctly beyond the hard-core parts.

2. Mixtures of simple metals

In what follows, we shall suppose that the alloy systems to be described retain complete metallic character throughout the entire concentration range. The constituents will be taken from the class of simple metals; the interactions between electrons and the corresponding ions will be taken as consistent with a description in terms of weak pseudopotentials $u_\alpha(\mathbf{k})$, $(\alpha = 1, 2, \ldots)$. No important physics is lost if we assume these to be local, as written. Suppose we consider an open volume Ω of the mixture with a mean total number of ions N where in terms of the mean constituent numbers N_α

$$N = \sum_\alpha N_\alpha, \tag{1}$$

or in terms of concentrations

$$1 = \sum_\alpha c_\alpha = \sum_\alpha (N_\alpha/N).$$

In subsequent manipulations an ultimate thermodynamic limit will be implied ($N \to \infty$, $\Omega \to \infty$, $N/\Omega \to n$).

Let $\mathbf{r}_l$ index electron coordinates, and let $\mathbf{R}_i^\alpha$ $(i = 1, \ldots, N_\alpha)$ denote the coordinate of an ion of type α. Then the hamiltonian for the assembly of electrons and ions is

$$H = \sum_l \frac{p_l^2}{2m} + \frac{1}{2}\sum_{l,m}{}' \frac{e^2}{|\mathbf{r}_l - \mathbf{r}_m|} \tag{2a}$$

$$+ \sum_{i,\alpha} \frac{P^2(\mathbf{R}_i^\alpha)}{2M_\alpha} + \frac{1}{2}\sum_{i,j,\alpha,\beta}{}' \frac{Z_\alpha Z_\beta e^2}{|\mathbf{R}_i^\alpha - \mathbf{R}_j^\beta|} \tag{2b}$$

$$+ \sum_l \sum_{i,\alpha} u_\alpha(\mathbf{r}_l - \mathbf{R}_i^\alpha) \tag{2c}$$

for ions of masses M_α and valences Z_α. In writing down equation (2) it is important to note that only the Coulomb interaction between ions has been retained, as is standard. (For certain systems the contributions resulting from direct and indirect core-electron polarization and exchange effects may well have structural consequences (Mon *et al* 1976).)

For volume Ω the average electron density in the mixture is

$$\langle \hat{\rho}^e \rangle = (N/\Omega)\bar{Z}, \tag{3}$$

where

$$\bar{Z} = \sum_\alpha c_\alpha Z_\alpha$$

may be regarded as a mean valence. Preceding the limit $N \to \infty$ etc, we use equation (3) to convert equation (2) to standard form. From term (2*a*) the self energy of a continuum with a charge density of $-eN\bar{Z}/\Omega$ is subtracted and a compensating quantity added to term (2*b*). To term (2*b*) is also added the interaction energy of the (point) ions with a background of charge density $-eN\bar{Z}/\Omega$ and the same factor subtracted from term (2*c*). In the limit of large systems, these rearrangements have the effect of accumulating together the long-wavelength limits of the divergent interactions. In particular, the first line of equation (2) becomes the familiar hamiltonian for the interacting electron gas H_{eg}. We may then write for H (Pethick 1970, Hammerberg and Ashcroft 1974);

$$H = H_{\mathrm{eg}} \tag{4a}$$

$$+\sum_{\alpha i}\frac{P^2(\mathbf{R}_i^\alpha)}{2M_\alpha}+\frac{N}{2}\sum_{\mathbf{k}\neq 0}\sum_{\alpha\beta}\frac{4\pi Z_\alpha Z_\beta e^2}{\Omega k^2}(c_\alpha c_\beta)^{1/2} \times\left[(N_\alpha N_\beta)^{-1/2}\hat{\rho}_\alpha^{\mathrm{i}}(-\mathbf{k})\hat{\rho}_\beta^{\mathrm{i}}(\mathbf{k})-\delta_{\alpha\beta}\right] \tag{4b}$$

$$+\sum_{\mathbf{k}\neq 0}\sum_{\alpha}\frac{u_\alpha(\mathbf{k})}{\Omega}\hat{\rho}_\alpha^{\mathrm{i}}(\mathbf{k})\,\hat{\rho}^{\mathrm{e}}(-\mathbf{k})+N\bar{Z}\sum_\alpha N_\alpha\lim_{\mathbf{k}\to 0}\left(\frac{u_\alpha(\mathbf{k})}{\Omega}+\frac{4\pi Z_\alpha e^2}{\Omega k^2}\right), \tag{4c}$$

where we have introduced the ionic density operators

$$\hat{\rho}_\alpha^{\mathrm{i}}(\mathbf{r}) = \sum_{i=1}^{N_\alpha}\delta(\mathbf{r}-\mathbf{R}_i^\alpha)$$

and their Fourier transforms

$$\hat{\rho}_\alpha^{\mathrm{i}}(\mathbf{k}) = \sum_{i=1}^{N_\alpha}\exp(\mathrm{i}\mathbf{k}\cdot\mathbf{R}_i^\alpha),$$

with the electronic density operator

$$\hat{\rho}^{\mathrm{e}}(\mathbf{r}) = \sum_l \delta(\mathbf{r}-\mathbf{r}_l)$$

and its Fourier transform

$$\hat{\rho}^{\mathrm{e}}(k) = \sum_i \exp(\mathrm{i}\mathbf{k}\cdot\mathbf{r}_l).$$

Apart from the ionic kinetic energy, the term (4*b*) will be recognized after ionic configuration averaging as the Madelung energy. The last term in (4*c*) expresses, in the long-wavelength limit, the departure of the pseudopotential form pure coulombic behaviour. For any ensemble it can always be written in the form

$$E_0 = N\left[\frac{4\pi N\bar{Z}}{\Omega}\sum_\alpha c_\alpha Z_\alpha (R_{c\alpha})^2\right](e^2/2a_0) \tag{5}$$

which is obviously true for pseudopotentials $u_\alpha(\mathbf{r})$ which vanish for $r < R_{c\alpha}a_0$ and are otherwise coulombic, but for more general potentials is just a defining relation for the

$R_{c\alpha}$ and conveniently introduces an atomic length $R_{c\alpha} a_0$ which distinguishes real ions from point-ion systems. Given H, the properties of the alloy are ultimately determined by the partition function $Z = \mathrm{Tr}\,[\exp(-\beta H)]$. By invoking an adiabatic approximation for the response of the electron system (which for almost all practical purposes can be taken in its ground state), we may carry out the trace over electron coordinates.

This requires us to evaluate $\langle\rho^{\mathrm{e}}(-\mathbf{k})\rangle_{\mathrm{e}}$ for a given configuration of ions. Providing the $u_\alpha(\mathbf{k})$ are sufficiently weak, we may accomplish this as noted above by allowing the electron system to respond linearly to the total (pseudo) potential of the ionic system. If $\epsilon(k)$ is the dielectric function of the interacting electron gas, then for a given arrangement of ions we have

$$\langle\hat{\rho}^{\mathrm{e}}(-\mathbf{k})\rangle_{\mathrm{e}} = \sum_{\beta} \frac{k^2}{4\pi e^2}\left[\frac{1}{\epsilon(k)} - 1\right]\hat{\rho}^{\mathrm{i}}_{\beta}(-\mathbf{k})\, u_\beta(k).$$

Denoting the ground state energy of the electron system by E_{eg}, the effective ionic hamiltonian can now be written

$$\begin{aligned} H^{\mathrm{ion}} = {} & E_{\mathrm{eg}} + E_0 + \sum_{\alpha i} \frac{P_\alpha^2(\mathbf{R}_i^\alpha)}{2M_\alpha} \\ & + \frac{N}{2}\sum_{\mathbf{k}\neq 0}\sum_{\alpha\beta}\left\{\frac{4\pi Z_\alpha Z_\beta e^2}{\Omega k^2} \times \left[1 + \left(\frac{u_\alpha(k)}{-4\pi Z_\alpha e^2/k^2}\right)\left(\frac{u_\beta(k)}{-4\pi Z_\beta e^2/k^2}\right)\left(\frac{1}{\epsilon(k)} - 1\right)\right]\right. \\ & \left. \qquad \times (c_\alpha c_\beta)^{1/2}\left[(N_\alpha N_\beta)^{-1/2}\hat{\rho}^{\mathrm{i}}_\alpha(-\mathbf{k})\,\hat{\rho}^{\mathrm{i}}_\beta(\mathbf{k}) - \delta_{\alpha\beta}\right]\right\} \\ & + \frac{N}{2}\sum_{\mathbf{k}\neq 0}\sum_{\alpha}\frac{4\pi Z_\alpha^2 e^2}{\Omega k^2}\, c_\alpha \left(\frac{u_\alpha(k)}{-4\pi Z_\alpha e^2/k^2}\right)^2\left(\frac{1}{\epsilon(k)} - 1\right) \end{aligned} \qquad (6)$$

where the last term denoted here by E_{s}, is the self energy of the induced electron distribution around the ions and is a one-body term. The term involving the ionic densities may also be written as

$$\frac{N}{2}\sum_{\mathbf{k}\neq 0}\sum_{\alpha\beta}\frac{V_{\alpha\beta}(k)}{\Omega}(c_\alpha c_\beta)^{1/2}\left[(N_\alpha N_\beta)^{-1/2}\hat{\rho}^{\mathrm{i}}_\alpha(-\mathbf{k})\,\hat{\rho}^{\mathrm{i}}_\beta(\mathbf{k}) - \delta_{\alpha\beta}\right]$$

where except for $\mathbf{k} = 0$, the effective interaction $V_{\alpha\beta}(\mathbf{r})$ between ions has the Fourier transform

$$V_{\alpha\beta}(k) = \frac{4\pi Z_\alpha Z_\beta e^2}{k^2}\left[1 + \left(\frac{u_\alpha(k)}{-4\pi Z_\alpha e^2/k^2}\right)\left(\frac{u_\beta(k)}{-4\pi Z_\beta e^2/k^2}\right)\left(\frac{1}{\epsilon(k)} - 1\right)\right] \qquad (7)$$

and explicitly demonstrates the important effects of electron screening.

For point ions equation (7) reduces, as expected, to $V_{\alpha\beta}(k) = 4\pi Z_\alpha Z_\beta e^2/\epsilon(k)k^2$ but for more realistic models the effects of u_α and u_β are quite pronounced, as shown in figure 2 where the pair interactions $V_{\alpha\beta}$ for the Na–K system are reproduced. On the scale of thermal energies the rise in the potential at short range is quite steep: beyond the first minimum the potentials are noticeably attractive. These two significant aspects will feature prominently in the structural perturbation theory (see §4).

In proceeding from equation (6) it is normal to treat the ions classically, at least from

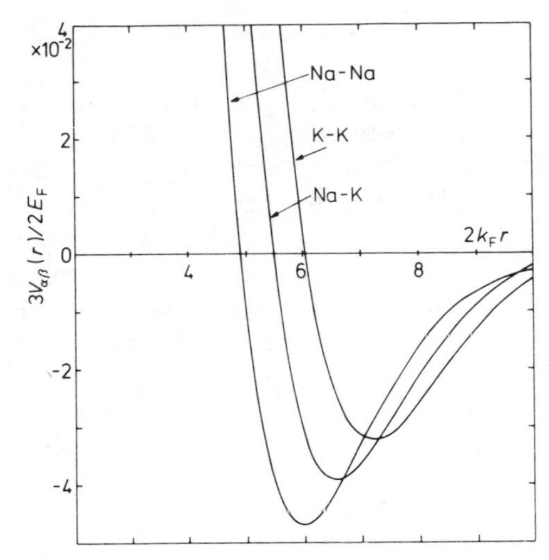

Figure 2. Effective pair interactions $V_{\alpha\beta}(r)$ in the Na–K system at 32·4% Na (from Ashcroft and Langreth 1967b). What is plotted is $V_{\alpha\beta}$ normalized to $2/3\,E_F$ (where E_F is the Fermi energy) against separation scaled by $2k_F$, where $2k_F$ is the Fermi wave-vector. These interactions include electron response contributions but not direct core–core interactions. The electron response contribution depends on the mean density of the electron system, so that the $V_{\alpha\beta}$ are functions of volume as well as of separation. Room temperature corresponds to about one unit on the vertical scale: on this basis the $V_{\alpha\beta}$ exhibit both short-range hard-core and long-range attractive regions common in many pair potentials.

the standpoint of equilibrium properties. Then for a fixed density, the appropriate average over ionic coordinates gives

$$E = E_{eg} + E_0 + E_s + \bar{E}_{kin} + \frac{N}{2} \sum_{k \neq 0} \sum_{\alpha\beta} \frac{V_{\alpha\beta}(k)}{\Omega} (c_\alpha c_\beta)^{1/2} [S_{\alpha\beta}(k) - \delta_{\alpha\beta}] \tag{8}$$

where the $S_{\alpha\beta}(k)$ are partial structure factors for the uniform mixture (these are defined in appendix 1) and $\bar{E}_{kin}$ is the mean kinetic energy of the ions. Given a pair potential possessing a Fourier transform, the last term in equation (8) is simply the mean potential energy of interaction:

$$\bar{E}_{pot} = \frac{N}{2} \sum_{k \neq 0} \sum_{\alpha\beta} \frac{V_{\alpha\beta}(k)}{\Omega} (c_\alpha c_\beta)^{1/2} [S_{\alpha\beta}(k) - \delta_{\alpha\beta}] \tag{9}$$

which may be separated as

$$\begin{aligned}\bar{E}_{pot} &= [E_M] + (\tilde{E}_{BS}^{(2)}) \\ &= \left[\frac{N}{2} \sum_{k \neq 0} \sum_{\alpha\beta} \frac{4\pi Z_\alpha Z_\beta e^2}{\Omega k^2} (c_\alpha c_\beta)^{1/2} (S_{\alpha\beta}(k) - \delta_{\alpha\beta})\right] \\ &\quad + \left[\frac{N}{2} \sum_{k \neq 0} \sum_{\alpha\beta} \frac{4\pi Z_\alpha Z_\beta}{\Omega k^2} \left(\frac{u_\alpha(k)}{-4\pi Z_\alpha e^2/k^2}\right) \left(\frac{u_\beta(k)}{-4\pi Z_\alpha e^2/k^2}\right) \left(\frac{1}{\epsilon(k)} - 1\right) \right. \\ &\qquad \left. \times (S_{\alpha\beta}(k) - \delta_{\alpha\beta})\right]\end{aligned} \tag{10}$$

where $E_{BS}^{(2)} = \tilde{E}_{BS}^{(2)} + E_s$ is frequently referred to as the band-structure energy calculated to second order in the pseudopotential.

The quantity E_M is the Madelung energy and although known to be weakly dependent on structural details is nevertheless strongly influenced by the gross degree of correlation in the ionic system. In principle the partial structure factors $S_{\alpha\beta}(k)$ can be obtained from experiment so that the calculation of both band structure and Madelung energies reduces to a problem of numerical integration. This is somewhat simplified by the observation that the effective interactions $V_{\alpha\beta}$ have, as emphasized, a resemblance to hard-sphere potentials and for these the exact solutions for $S_{\alpha\beta}$ of the Percus–Yevick equations are known. Note that for the one case of Thomas–Fermi screened point-ion interactions (and the same Percus–Yevick approximations for the $S_{\alpha\beta}(k)$) $\bar{E}_{pot}$ can be calculated completely in closed form (Firey and Ashcroft 1976).

Equation (8) forms the basis for pair-force theories of the thermodynamics of liquid metal alloys. It also provides the starting point for the discussion of stability of metallic mixtures.

3. Ionic correlations and phase separation

The simplest model of an alloy includes band-structure energy to the extent that it appears in E_s and calculates the Madelung energy by building up around each ion a sphere of uniform negative charge (at a density of $-eN\bar{Z}/\Omega$) containing sufficient total charge to neutralize the charge of the ion. In this uniform-sphere approximation, the Madelung energy per ion is readily estimated provided we may assume the spheres to be minimally overlapping. In regular solids, the more symmetric the crystal structure then the more reasonable is this approximation. It follows that the calculation of E_M by the contrivance of uniform non-overlapping spheres carries with it an implicit assumption of high correlations in the system. The uniform-sphere approximation may be regarded as the opposite extreme to a free gas of completely random ions ($S_{\alpha\beta} - \delta_{\alpha\beta} = 0$) for which the Madelung energy vanishes and which in any case gives a very unrealistic model of a simple metal. For a homovalent, monovalent alloy (to which without undue loss in generality we restrict attention) the Madelung energy can always be written

$$E_M = -N\frac{\alpha}{r_s}\left(\frac{e^2}{2a_0}\right) \tag{11}$$

where we have defined the standard electron spacing parameter r_s by

$$\frac{4\pi}{3}(a_0 r_s)^3 = \frac{\Omega}{NZ} = \frac{\Omega}{N}.$$

The value of α for the uniform-sphere approximation is 9/5. Other values of α for systems with less correlation can be estimated by starting with

$$E_M = \frac{N}{2}\sum_{\mathbf{k}\neq 0}\sum_{\alpha\beta}\frac{4\pi e^2}{\Omega k^2}(c_\alpha c_\beta)^{1/2}[S_{\alpha\beta}(k) - \delta_{\alpha\beta}],$$

using the substitutional model for the $S_{\alpha\beta}$ and taking the Percus–Yevick hard-sphere form for the resulting $S(k)$ (see appendix 1). The degree of correlation in the system

Table 1.

η (see equation (A7))	α
0	0
0·1	1·0567
0·2	1·4129
0·3	1·5948
0·4	1·6937
0·46	1·7302
0·5	1·7481
Uniform sphere	1·8

can be assigned through the choice of packing fraction η (or $\pi n\sigma^3/6$, where σ is the hard-sphere diameter). Table 1 shows that for $\eta \sim 0{\cdot}46$ (a representative value for liquid metals) the value of α is close to the uniform-sphere result, but that α falls quite rapidly as the system approaches an ideal gas ($\eta = 0$).

In constructing the internal energy we also need E_{eg}, E_0 and E_{s}. For the electron-gas energy we have the well known result

$$E_{\text{eg}} = N\left[\frac{2{\cdot}21}{r_{\text{s}}^2} - \frac{0{\cdot}916}{r_{\text{s}}} + \epsilon_{\text{corr}}(r_{\text{s}})\right]\left(\frac{e^2}{2a_0}\right) \tag{12}$$

where the correlation energy ϵ_{corr} is numerically small and is very weakly dependent on electron density in the standard metallic range. In terms of r_{s}, E_0 is easily seen to be

$$E_0 = N\frac{3}{r_{\text{s}}^3}\sum_{\alpha} c_\alpha R_{c\alpha}^2\left(\frac{e^2}{2a_0}\right). \tag{13}$$

Finally, the self energy may be estimated with sufficient accuracy (for the next point to be made) by taking an empty-core pseudopotential and Thomas–Fermi screening. Then ($Z_\alpha = 1$)

$$E_{\text{s}} = -N\sum_{\alpha} c_\alpha\left(\frac{a_0k_{\text{F}}}{\pi}\right)^{1/2}\left(1 + \exp\left\{-4\left[\left(\frac{a_0k_{\text{F}}}{\pi}\right)^{1/2} R_{c\alpha}\right]\right\}\right)\left(\frac{e^2}{2a_0}\right) \tag{14}$$

where $a_0k_{\text{F}}r_{\text{s}} = (9\pi/4)^{1/3}$. It follows that E_{s} has an explicit *linear* dependence on concentration c_α and an implicit dependence through r_{s}. The latter, however, is very weak, as can be readily verified in any given case. From the point of view of possible phase separation in the liquid state, the important terms in the internal energy

$$\frac{E}{N} = \frac{2{\cdot}21}{r_{\text{s}}^2} - \frac{0{\cdot}916 + \alpha}{r_{\text{s}}} + \frac{3}{r_{\text{s}}^3}\left(\sum_{\alpha} c_\alpha R_\alpha^2\right) + \epsilon_{\text{corr}}(r_{\text{s}}) + \frac{E_{\text{s}}}{N}(r_{\text{s}}, \{c_\alpha\}) \quad \text{(Ryd)} \tag{15}$$

are the first three. As we shall now see, the approximation

$$\frac{E}{N} = \frac{\gamma}{r_{\text{s}}^2} - \frac{\beta}{r_{\text{s}}} + \frac{3}{r_{\text{s}}^3}\left(\sum_{\alpha} c_\alpha R_\alpha^2\right) \quad \text{(Ryd)} \tag{16}$$

has some simple but instructive consequences. First, define quantities

$$\bar{r}_s = -1 + \beta r_s/\gamma$$

$$\bar{r}_{s\alpha} = -1 + \beta r_{s\alpha}/\gamma$$

where the r_{si} are the *measured* equilibrium r_s values for the pure constituents. Then, applying the standard zero-pressure boundary condition, it is easy to show that

$$\bar{r}_s^2 = \sum_\alpha c_\alpha \bar{r}_{s\alpha}^2 \tag{17}$$

which gives the density of the monovalent alloy in terms of the densities of its constituents.

[For a heterovalent system we have, in the uniform-sphere approximation,

$$(\bar{r}_s^2 - 1) = \sum_\alpha d_\alpha c_\alpha (\bar{r}_{s\alpha}^2 - 1)$$

where

$$\bar{r}_s = -1 + \gamma^{-1} r_s \left(0{\cdot}916 + \sum_\alpha 9 c_\alpha Z_\alpha^{5/3} \Big/ \sum_\alpha 5\, c_\alpha Z_\alpha\right)$$

$$\bar{r}_{s\alpha} = -1 + \gamma^{-1} r_{s\alpha} (0{\cdot}916 + 9 Z_\alpha^{2/3}/5)$$

and

$$d_\alpha = \left(0{\cdot}916 + \sum_\alpha 9\, c_\alpha Z_\alpha^{5/3}\right) \Big/ (0{\cdot}916 + 9 Z_\alpha^{2/3}/5).$$

A more general form of this, involving the correct Madelung energy, can easily be written down.]

For a binary system ($c_1 + c_2 = 1$),

$$\begin{aligned}\bar{r}_s^2(c) &= c_1 \bar{r}_{s1}^2 + c_2 \bar{r}_{s2}^2 \\ &= (c_1 \bar{r}_{s1} + c_2 \bar{r}_{s2})^2 + c_1 c_2 (\bar{r}_{s1} - \bar{r}_{s2})^2\end{aligned} \tag{18}$$

and when $\beta = 2{\cdot}716$ (as for the uniform-sphere approximation) equation (18) is a quite satisfactory interpolation for the densities of miscible alkali metals.

Secondly, given equation (17), it follows that

$$\frac{1}{N} E(\{c_\alpha\}) = \left(\frac{-\beta^2}{\gamma}\right) \left[1 - \left(\frac{\bar{r}_s}{1 + \bar{r}_s}\right)^2\right] \tag{19}$$

with $\bar{r}_s$ given by equation (18).

Again, for a binary system we may compare equation (19) with a linear average

$$\frac{1}{N} [\bar{E}(c_1 c_2)] = -\frac{\beta^2}{\gamma} \left\{ c_1 \left[1 - \left(\frac{\bar{r}_{s1}}{1 + \bar{r}_{s1}}\right)^2\right] + c_2 \left[1 - \left(\frac{\bar{r}_{s2}}{1 + \bar{r}_{s2}}\right)^2\right]\right\} \tag{20}$$

to find the excess function

$$\frac{1}{N} e(c_1 c_2) = \frac{\beta^2}{\gamma} \left[\left(\frac{\bar{r}_s}{1 + \bar{r}_s}\right)^2 - c_1 \left(\frac{\bar{r}_{s1}}{1 + \bar{r}_{s1}}\right)^2 - c_2 \left(\frac{\bar{r}_{s2}}{1 + \bar{r}_{s2}}\right)^2\right]. \tag{21}$$

Some properties of e are considered in appendix 2. It is always positive, generally asymmetric, and its magnitude (typically of the order 10^{-2}) tends to increase the more disparate are the values of $\bar{r}_{s1}$ and $\bar{r}_{s2}$ (i.e. R_{c1} and R_{c2}). Equations (19) and (20) are plotted in figure 3; the curves make the point that such mixtures will be unstable against fluctuations in concentration unless the temperature is sufficiently high that the entropy, principally the mixing entropy

$$S = Nk_B (c_1 \ln c_1 + c_2 \ln c_2), \tag{22}$$

stabilizes the system. But the central point is this: it is apparent from equation (21) and the definitions of $\bar{r}_{s1}$ etc that in terms of thermal energies, the separation between the curves in figure 3 can depend quite sensitively on the value of β. In the present model, correlation between the ions has entered β only through the Madelung energy, that is, through direct Coulomb interactions. It has not entered in a self-consistent way, of course, since near phase separation we expect the $S_{\alpha\beta}(k)$ to deviate at small k from those appropriate to a fully mixed system. But the model nevertheless suffices to

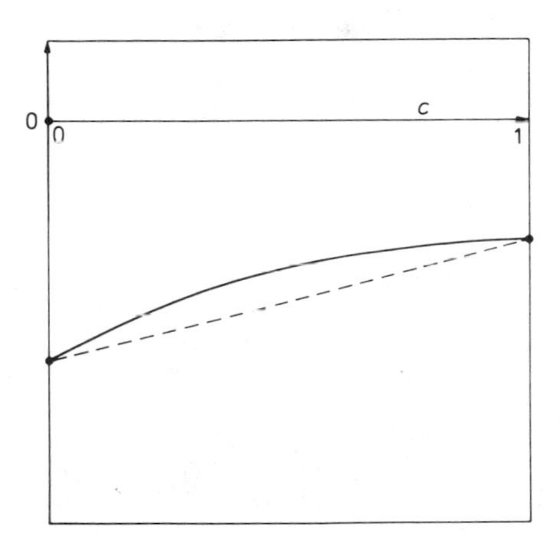

Figure 3. For an alloy of simple metals, $E(\bar{r}_s)$ the energy per ion according to equation (19), is represented by the solid curve. The broken curve is a linear combination (with weights c and $1 - c$) of the energies of the constituents, representing the function $cE(\bar{r}_{s1}) + (1 - c)E(\bar{r}_{s2})$.

demonstrate the importance of correlations in discussing the possibility of phase separation of metallic systems: it ignores the fact that the real interactions are more like those shown in figure 2 and display both short-range hard-core behaviour, and long-range attractive parts. These features are quite consistent with the interpretation of structural data (e.g. the measured $S_{\alpha\beta}(k)$) which show liquid metals to be as highly correlated as simple insulating liquids with comparable n. The computed $S(k)$ for Coulomb systems are surprisingly similar to those for assemblies of hard spheres (Hansen 1973). But mixtures of hard spheres evidently do not phase-separate (Alder 1964, Rotenberg 1965, Lebowitz and Rowlinson 1964) and the long-range parts of the pair interactions are manifestly important (as they indeed are for purely coulombic interactions). We are therefore led to consider the effects of the long-ranged part of the ion–ion interactions as perturbations on the structure resulting from the short-range parts, the latter being well approximated by the structure appropriate to a mixture of hard spheres. The following perturbation theory of this problem is largely taken from the paper by Henderson and Ashcroft (1976).

4. Structural perturbation theory

In classical liquid mixtures the structure factors $S_{\alpha\beta}(k)$ are proportional to the static density–density response functions. Divergent response indicates an instability in the system. Suppose a set of externally applied fields $\phi_\alpha(\mathbf{r})$ (coupling, by fiat, to ions of type α) induce non-uniformities in the mean singlet densities ρ_α. Then in linear response,

$$\delta\rho_\alpha(k) = \sum_\beta \chi_{\alpha\beta}(k)\,\phi_\beta(k), \tag{23}$$

or more compactly,

$$\delta\rho(k) = \chi(k)\,\phi(k), \tag{24}$$

where

$$\chi_{\alpha\beta}(k) = -\beta(\rho_\alpha\,\rho_\beta)^{1/2}\,S_{\alpha\beta}(k) \qquad (\rho_\alpha = N_\alpha/\Omega).$$

In equation (23) it is understood that

$$\delta\rho_\alpha^\phi(r) = \rho_\alpha^\phi(r) - \rho_\alpha \tag{25}$$

is the departure from uniformity accompanying imposition of the corresponding external field ϕ_α.

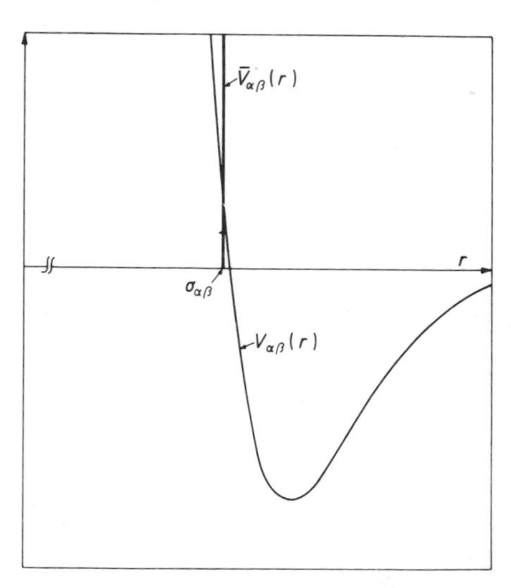

Figure 4. Representation of the real ion–ion interactions $V_{\alpha\beta}(r)$ in terms of hard-sphere interactions $\bar{V}_{\alpha\beta}(r)$ (heavy vertical line) and a perturbation. Although the perturbation may be regarded as that part of $V_{\alpha\beta}$ lying outside the corresponding hard-sphere distance $\sigma_{\alpha\beta}$, in practice the perturbation may formally be taken as the entire $V_{\alpha\beta}(r)$. This follows because the pair distributions $\bar{g}_{\alpha\beta}(r)$ vanish identically for $r < \sigma_{\alpha\beta}$ and in perturbation theory there can be no contribution to any thermodynamic function from $V_{\alpha\beta}(r)$ extended into this domain.

As has been noted, the $S_{\alpha\beta}(k)$ (and hence the $\chi_{\alpha\beta}$) are a measure of correlations in the system and in turn mirror the form of the pair interactions $V_{\alpha\beta}(k)$. In approaching phase separation from the point of view of perturbation theory, we first establish the structure of the reference system as that appropriate to a mixture described by hard-sphere interactions $\bar{V}_{\alpha\beta}$. The entire interaction between particles is then approximated (see figure 4) by setting

$$V_{\alpha\beta} = \bar{V}_{\alpha\beta} + V^1_{\alpha\beta}$$

with $V^1_{\alpha\beta}$ regarded as the perturbation.†

† The perturbation may be taken as the entire effective interaction between ions since any finite portion of it for distances within hard-sphere dimensions can have no effect.

Let the structure factors corresponding to real ($V_{\alpha\beta}$) and reference ($\bar{V}_{\alpha\beta}$) liquids be $S_{\alpha\beta}$ and $\bar{S}_{\alpha\beta}$ respectively. The corresponding density–density response functions will be $\chi_{\alpha\beta}$ and $\bar{\chi}_{\alpha\beta}$. For real and reference systems define

$$\mathbf{f} = -\chi^{-1} \tag{26}$$

and

$$\bar{\mathbf{f}} = -\bar{\chi}^{-1}. \tag{27}$$

(The $f_{\alpha\beta}$ and $\bar{f}_{\alpha\beta}$ are closely related to the Ornstein–Zernike direct correlation functions.) Evidently one way of representing the effect of the perturbation $V^1_{\alpha\beta}$ is to construct the difference

$$\mathbf{f}^1 = \mathbf{f} - \bar{\mathbf{f}}. \tag{28}$$

We shall shortly see that these **f** enter into certain expansions for the free energy of a liquid, which therefore indicates a possible approach to the actual *calculation* of the $f_{\alpha\beta}$.

Using equations (26), (27) and (28) we find

$$\chi = (\mathbf{I} - \bar{\chi}\,\mathbf{f}^1)^{-1}\,\bar{\chi} \tag{29}$$

or from equation (24)

$$\delta\boldsymbol{\rho} = \chi\boldsymbol{\phi}^{\text{eff}}.$$

In other words, the response $\delta\boldsymbol{\rho}$ of the real fluid to external fields $\boldsymbol{\phi}$ can formally be written in terms of the response of the reference fluid to an *effective* field $\boldsymbol{\phi}^{\text{eff}}$, where

$$\boldsymbol{\phi}^{\text{eff}} = \boldsymbol{\phi} + \mathbf{f}^1\,\delta\boldsymbol{\rho}.$$

Since this contains the required $\delta\boldsymbol{\rho}$, progress has only been made to the extent that the application of a physical model permits an independent identification of $\boldsymbol{\phi}^{\text{eff}}$. For instance, consider a one-component liquid in which the reference system is taken as an ideal gas ($\bar{V} = 0$). The simplest approximation for the effective field ϕ^{eff} to which the ideal gas responds when there are interactions V between particles, is the Hartree field; that is, the external field plus the average interaction associated with the induced density:

$$\phi^{\text{eff}}(\mathbf{r}) = \phi(\mathbf{r}) + \int \mathrm{d}\mathbf{r}'\, V(\mathbf{r} - \mathbf{r}')\,\delta\rho(\mathbf{r}')$$

from which it follows that

$$f^1 = V$$

and subsequently that

$$S(k) = \frac{1}{[1 + \beta\rho_0 V(k)]}$$

which is the standard random-phase approximation result. If instead the reference fluid is taken as a one-component hard-sphere system ($\bar{V}$) and the potential written as $V = \bar{V} + V^1$, then a corresponding Hartree approximation suggests

$$\phi^{\text{eff}}(\mathbf{r}) = \phi(\mathbf{r}) + \int \mathrm{d}\mathbf{r}'\, V^1(\mathbf{r} - \mathbf{r}')\,\delta\rho(\mathbf{r}'), \tag{30}$$

leading to the identification

$$f^1 = V^1$$

and hence to

$$S(k) = \frac{\bar{S}(k)}{1 + \beta\rho_0 \bar{S}(k)\, V^1(k)}. \tag{31}$$

This result, also well known (Brout 1965), does not meet the test that $S(k)$ given by equation (31) should be independent of arbitrary additions to V^1 within the hard-sphere dimension. The difficulty is that equation (30) does not fully take into account, even at first order, the effects of correlations in the reference system. We therefore need a more complete solution for f^1 (or $\mathbf{f}^1$ for the two-component system). For given such a solution, we find from equation (29) that the $S_{\alpha\beta}$ have a denominator D in common, where

$$D = 1 - \bar{\chi}_{11} f_{11}^1 - \bar{\chi}_{22} f_{22}^1 - 2\bar{\chi}_{12} f_{12}^1 + [\bar{\chi}_{11}\bar{\chi}_{22} - \bar{\chi}_{12}^2]\,[f_{11}^1 f_{22}^1 - (f_{12}^1)^2] \tag{32}$$

and the divergent response referred to above is therefore associated with the roots of D.

Apart from their relationship to the direct-correlation functions, the quantities $\mathbf{f}$ and $\bar{\mathbf{f}}$ have been introduced here through simple considerations of structure. As noted in §1, they are also closely related to the expansion of the free energy of inhomogeneous fluids in terms of the $\delta\boldsymbol{\rho}$ defined earlier. In particular, the quantity $\mathscr{F}$, which is the free energy F diminished by the one-body terms (arising from the external fields that cause the inhomogeneities in density), has an expansion

$$\bar{\mathscr{F}}(\rho) = \bar{\mathscr{F}}(\rho_0) + \tfrac{1}{2}\sum_{\alpha\beta} \int d\mathbf{r} \int d\mathbf{r}'\, \delta\rho_\alpha(\mathbf{r})\, \delta\rho_\beta(\mathbf{r}')\, \bar{f}_{\alpha\beta}(\mathbf{r}-\mathbf{r}') + \ldots \tag{33}$$

here written for the reference system.†

For the real system we may also write

$$\mathscr{F}(\rho) = \mathscr{F}(\rho_0) + \tfrac{1}{2}\sum_{\alpha\beta} \int d\mathbf{r} \int d\mathbf{r}'\, \delta\rho_\alpha(\mathbf{r})\, \delta\rho_\beta(\mathbf{r}')\, f_{\alpha\beta}(\mathbf{r}-\mathbf{r}') + \ldots \tag{34}$$

so that

$$\mathscr{F}^1(\rho) = \mathscr{F}(\rho) - \bar{\mathscr{F}}(\rho) = F(\rho) - \bar{F}(\rho) - \sum_{\alpha} \int d\mathbf{r}\, \rho_\alpha(\mathbf{r})\,[\phi_\alpha^\rho(\mathbf{r}) - \bar{\phi}_\alpha \rho(\mathbf{r})] \tag{35}$$

$$= \mathscr{F}^1(\rho_0) + \tfrac{1}{2}\sum_{\alpha\beta} \int d\mathbf{r} \int d\mathbf{r}'\, \delta\rho_\alpha(\mathbf{r})\, \delta\rho_\beta(\mathbf{r}')\, f_{\alpha\beta}^1(\mathbf{r}-\mathbf{r}') + \ldots. \tag{36}$$

† The details leading to equation (33) are given in Henderson and Ashcroft (1976): that equation (33) contains no linear terms in $\delta\rho$ is guaranteed by the Gibbs–Bogolyubov inequality, from which it also follows that given a predetermined set of densities ρ_α in the system, the external fields ϕ_α^ρ which might induce these densities are, to within constants, determined uniquely. Such constants cancel from

$$\mathscr{F} = F - \sum_{\alpha} \int d\mathbf{r}\, \rho_\alpha(r)\, \phi_\alpha^\rho(r)$$

which is therefore also uniquely determined.

The procedure is clear: we are required to calculate $\mathscr{F}^1$ from equation (35) and then by expansion using equation (36) to identify the $f^1_{\alpha\beta}$ from the quadratic terms. To illustrate for a one-component inhomogeneous fluid, if we could legitimately ignore the entropy terms in the free energy, then from the definition of the pair distribution function $\rho_2(\mathbf{r}, \mathbf{r}')$ we would write for F

$$F(\rho) \simeq \tfrac{1}{2} \int d\mathbf{r} \int d\mathbf{r}' \, V(\mathbf{r}-\mathbf{r}') \, \rho_2(\mathbf{r}-\mathbf{r}') + \int d\mathbf{r} \, \phi^\rho(\mathbf{r}) \, \rho(\mathbf{r})$$

and for $\bar{F}$,

$$\bar{F}(\rho) \simeq \tfrac{1}{2} \int d\mathbf{r} \int d\mathbf{r}' \, \bar{V}(\mathbf{r}-\mathbf{r}') \, \bar{\rho}_2(\mathbf{r}-\mathbf{r}') + \int d\mathbf{r} \, \bar{\phi}^\rho(\mathbf{r}) \, \rho(\mathbf{r}).$$

It follows that

$$\mathscr{F}(\rho) \simeq \mathscr{F}(\rho_0) + \tfrac{1}{2} \int d\mathbf{r} \int d\mathbf{r}' \, V(\mathbf{r}-\mathbf{r}') \, [\rho_2(\mathbf{r}-\mathbf{r}') - \rho_2^0(\mathbf{r}-\mathbf{r}')] \tag{37}$$

and

$$\bar{\mathscr{F}}(\rho) \simeq \bar{\mathscr{F}}(\rho_0) + \tfrac{1}{2} \int d\mathbf{r} \int d\mathbf{r}' \, \bar{V}(\mathbf{r}-\mathbf{r}') \, [\bar{\rho}_2(\mathbf{r}, \mathbf{r}') - \bar{\rho}_2^0(\mathbf{r}-\mathbf{r}')] \tag{38}$$

where ρ_2^0 and $\bar{\rho}_2^0$ are pair distribution functions in the homogeneous systems. Combining equations (37) and (38) we find

$$\begin{aligned} \mathscr{F}^1(\rho) - \mathscr{F}^1(\rho_0) &\simeq \tfrac{1}{2} \int d\mathbf{r} \int d\mathbf{r}' \, V(\mathbf{r}-\mathbf{r}') \, [\rho_2(\mathbf{r}-\mathbf{r}') - \rho_2^0(\mathbf{r}-\mathbf{r}')] \\ &\quad - \tfrac{1}{2} \int d\mathbf{r} \int d\mathbf{r}' \, \bar{V}(\mathbf{r}-\mathbf{r}') \, [\bar{\rho}_2(\mathbf{r}, \mathbf{r}') - \bar{\rho}_2^0(\mathbf{r}-\mathbf{r}')] \\ &\simeq \tfrac{1}{2} \int d\mathbf{r} \int d\mathbf{r} \, \delta\rho(\mathbf{r}) \, \delta\rho(\mathbf{r}') \, f^1(\mathbf{r}-\mathbf{r}'). \end{aligned} \tag{39}$$

This indicates the type of problem to be solved but it also exposes an impediment, for from equation (39) we see that in order to extract the $f^1(\mathbf{r}-\mathbf{r}')$, we will certainly require the pair distribution $\rho_2(\mathbf{r}, \mathbf{r}')$ whose structure, in fact, we are attempting to determine. The problem can be partially overcome, however, provided we are content to work at first order and with weakly inhomogeneous systems. To first order in V^1 it can be shown (Henderson and Ashcroft 1976) that

$$\begin{aligned} &\int d\mathbf{r} \int d\mathbf{r}' \, V^1_{\alpha\beta}(\mathbf{r}-\mathbf{r}') \, [\bar{\rho}_{\alpha\beta}(\mathbf{r}, \mathbf{r}') - \bar{\rho}^0_{\alpha\beta}(\mathbf{r}-\mathbf{r}')] \\ &\quad = \sum_{\alpha\beta} \int d\mathbf{r} \int d\mathbf{r}' \, \delta\rho_\alpha(\mathbf{r}) \, \delta\rho_\beta(\mathbf{r}') \, a^1_{\alpha\beta}(\mathbf{r}-\mathbf{r}') \end{aligned} \tag{40}$$

which approximates equation (39) and is now written for a binary system. In equation (40), $a^1_{\alpha\beta}(\mathbf{r}-\mathbf{r}')$ is that part of $f^1_{\alpha\beta}$ that is entirely first order in $V^1_{\alpha\beta}$. Otherwise, the significant aspect of this relation is that only distribution functions of the reference fluid are involved; a modest advance. To be sure, we need to know the distribution function of an *inhomogeneous reference* fluid, but provided the inhomogeneities are weak (i.e. concentrated around $k \to 0$) we may approximate $\bar{\rho}_2(\mathbf{r}, \mathbf{r}')$ by the distribution function of a locally *homogeneous* fluid. (We note that the long-wavelength region is precisely the region of interest for the problem of instability against concentration fluctuations.) This last approximation is again most easily illustrated for a one-component fluid: in the mean density approximation (MDA) we set

$$\bar{\rho}_2(\mathbf{r}, \mathbf{r}') - \bar{\rho}_2^0(\mathbf{r} - \mathbf{r}')$$

$$= \tfrac{1}{2}\,[\delta\rho(\mathbf{r}) + \delta\rho(\mathbf{r}')]\,\frac{\partial}{\partial\rho_0}\,\bar{\rho}_2^0(\mathbf{r} - \mathbf{r}') + \tfrac{1}{2}\,\delta\rho(\mathbf{r})\,\delta\rho(\mathbf{r}')\,\frac{\partial^2}{\partial\rho_0^2}\,\bar{\rho}_2(\mathbf{r} - \mathbf{r}'). \tag{41}$$

It is to be compared with the consequences of the mean-field approximation (MFA) which sets

$$\bar{\rho}_2(\mathbf{r}, \mathbf{r}') = \rho(\mathbf{r})\,\rho(\mathbf{r}') \tag{42}$$

as for an ideal gas.

In the binary system the solution for $a^1_{\alpha\beta}(k)$ can be extracted by writing, for small $\mathbf{k}$,

$$\delta\rho_\alpha(\mathbf{r}) = \delta\rho_\alpha \exp(\mathrm{i}\mathbf{k}\cdot\mathbf{r})$$

which yields†

$$a^1_{\alpha\beta}(\mathbf{k}) = \tfrac{1}{2}\sum_{\gamma\delta}\int\frac{\mathrm{d}\mathbf{k}'}{(2\pi)^3}\,V^1_{\gamma\delta}(\mathbf{k}')\,\frac{\partial^2}{\partial\rho_\alpha^0\,\partial\rho_\beta^0}\,\bar{\rho}^0_{\gamma\delta}(\mathbf{k} - \mathbf{k}') \tag{43}$$

and is the first-order solution for $f^1_{\alpha\beta}$ to be used in the search for zeros of equation (32).

From equation (42) it is straightforward to show that the mean-field solution of equation (40) can be written

$$\begin{aligned} a^1_{\alpha\beta}(r) &= 0 \qquad & r < \sigma_{\alpha\beta} \\ &= V^1_{\alpha\beta} \qquad & r > \sigma_{\alpha\beta} \end{aligned} \tag{44}$$

for a hard-sphere reference system with diameters $\sigma_{\alpha\beta}$. Given analytic forms for the $V^1_{\alpha\beta}$ (e.g. equation (7) with the pseudopotentials u_α chosen for the particular alloy) the $a^1_{\alpha\beta}(k)$ can then be directly computed. For the same $V^1_{\alpha\beta}$ we may also compute equation (43) with the Percus–Yevick hard-sphere structure factors taken for the reference system Given the same basic $V^1_{\alpha\beta}$ we are now able to compare mean-field and mean-density approximations at least to the extent that they lead to different sequences of roots of equation (32). The results for the Na–Li system, treated in this manner by Henderson and Ashcroft (1976), are shown in figure 5.

The main point that emerges from a comparison of the curves of figure 5 is that even at first order it is important to properly treat the correlations that are present in the underlying reference system. It is evident that phase separation can be described by a model of hard-sphere interactions augmented by weak attractive interactions. It is equally evident that effects beyond those just of first order in $V^1_{\alpha\beta}$ will ultimately

† In terms of structure factors the one-component version of equation (43) reads

$$a^1(\mathbf{k})^1(k) = V^1(k) + \rho_0^{-1}\int\frac{\mathrm{d}\mathbf{k}'}{(2\pi)^3}\,V^1(\mathbf{k})\left[\rho_0\,\frac{\partial}{\partial\rho_0}\,\bar{S}(\mathbf{k} - \mathbf{k}') + \tfrac{1}{2}\rho_0^2\,\frac{\partial^2}{\partial\rho_0^2}\,\bar{S}(\mathbf{k} - \mathbf{k}')\right].$$

whose two-component equivalent is given in detail by Henderson and Ashcroft (1976).

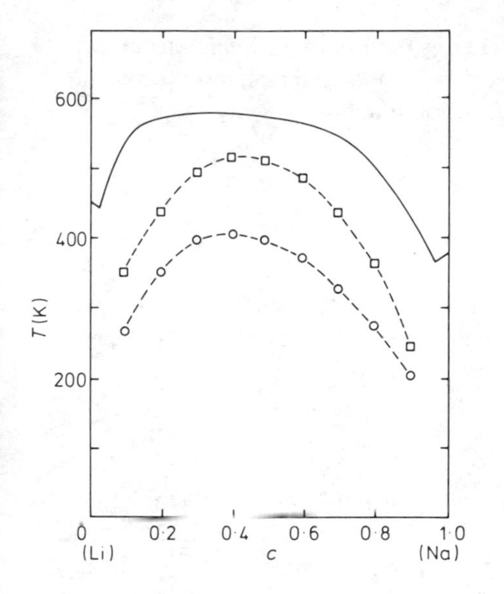

Figure 5. Phase-separation curve in the sodium–lithium system (from Hansen 1965). Also plotted is the line of singularities in $S_{\alpha\beta}(k \to 0)$ according to the mean-density approximation □ (equation (43)), and the mean field approximation ○ (equation (44)). Note that at the critical point, the line of singularities, if determined correctly, should meet the phase separation curve in a point of osculation.

require attention. (For metals there may be additional corrections associated with the intrinsic density dependence of effective ion–ion interactions.) However, the central conclusion remains valid; this is that the lines of instability of fluids and ultimately the phase-separation curves themselves, contain much information on the nature of effective interactions between ions. Some of this information, of course, is found in a different guise in the melting curves of alloys, possibly in the details of miscibility gaps in solids, or even in order–disorder transitions, should they occur. In this context we note that the discussion of §3 is largely applicable to solid alloys to the extent that the band-structure energy is dominated by E_s. However, the structure actually assumed by a solid alloy of simple metals will mainly be determined by the remaining band-structure term $\tilde{E}_{BS}$ (Stroud and Ashcroft 1971) and to a lesser degree by the ionic dynamics. The entropy associated with the latter augments any contributions from mixing entropy and the resulting sum enters into the total Gibbs energy for the solid. For the melting transition this must be compared with the corresponding quantity for the liquid phase. Both phases have their contributions from band-structure energy and for both these can be cast, as shown, in terms of effective interactions between pairs.

As the temperature is lowered from that of the fully mixed liquid, freezing and phase separation are two possible instabilities for the liquid phase. In either case the location of the transition reflects information on the nature of the pair interactions which enter in the phonon dynamics (and entropy) of the liquid structure. The Li–Na system shown in figure 5 is a case in point: near $c = 0$ and $c = 1$ we notice regions where the effect of dilute quantities of one metal in the other is simply to depress the melting point of the alloy, an effect that is quickly overtaken by phase separation at points that may well have special importance.

Finally, within the two-component approximation to a three-component system, the instability of the liquid has been traced to a divergence in the long-wavelength limit of the density response functions. (In principle the procedure can be extended by treating electrons on an equal footing as the ions and examining the response functions of the three-component system.) The approach which calculates $S_{\alpha\beta}(k)$ from $\bar{S}_{\alpha\beta}$ and $V^1_{\alpha\beta}$ by means of the properties of a weakly-inhomogeneous fluid simply exploits the known properties of the pair distribution in hard-sphere assemblies and circumvents, for the $k \to 0$ properties, the need for the higher distribution functions which emerge in standard thermodynamic perturbation theory.

Acknowledgments

I am grateful to Dr R L Henderson and Mr D Straus for many helpful discussions connected with the subject matter in this paper. Professor N D Mermin is profusely thanked for sharing with me a bout of flu which did much to impede progress in putting pen to paper.

Appendix 1

We summarize here the definitions of the low-order distribution functions for liquid mixtures. Let the species be labelled by α, and their positions by $\mathbf{R}_i^\alpha$, ($i = 1, 2, \ldots, \hat{N}_\alpha$, where $\hat{N}_\alpha$ is one possible ensemble number with corresponding chemical potential μ_α). The singlet density operator is

$$\hat{\rho}_\alpha(\mathbf{r}) = \sum_{i=1}^{\hat{N}_\alpha} \delta(\mathbf{r} - \mathbf{R}_i^\alpha) \qquad \text{(A1)}$$

where for open volume Ω,

$$\int_\Omega \mathrm{d}\mathbf{r}\, \delta(\mathbf{r} - \mathbf{R}_i^\alpha) = \hat{N}_\alpha.$$

When averaged over all possible ensembles $\hat{N}_\alpha$ (with probability of occurrence proportional to

$$\exp\left(-\beta \sum_\alpha \hat{N}_\alpha \mu_\alpha\right)),$$

we have

$$\rho_\alpha(\mathbf{r}) = \langle \hat{\rho}_\alpha(\mathbf{r}) \rangle$$

so that

$$\int_\Omega \mathrm{d}\mathbf{r}\, \rho_\alpha(r) = \langle \hat{N}_\alpha \rangle = N_\alpha$$

and for isotropic systems $\rho_\alpha(r) = \rho_\alpha = N_\alpha/\Omega$.

The pair density is defined by

$$\hat{\rho}_{\alpha\beta}(\mathbf{r}, \mathbf{r}') = \sum_{i=1}^{\hat{N}_\alpha} \sum_{j=1}^{\hat{N}_\beta} \delta(\mathbf{r} - \mathbf{R}_i^\alpha)\, \delta(\mathbf{r}' - \mathbf{R}_j^\beta) \quad \text{where} \quad i \neq j, \qquad \text{(A2)}$$

which satisfies

$$\int_\Omega \mathrm{d}\mathbf{r} \int \mathrm{d}\mathbf{r}'\, \hat{\rho}_{\alpha\beta}(\mathbf{r}, \mathbf{r}') = \hat{N}_\alpha \hat{N}_\beta - \delta_{\alpha\beta} \hat{N}_\alpha.$$

Its average is

$$\rho_{\alpha\beta}(\mathbf{r}, \mathbf{r}') = \langle \hat{\rho}_{\alpha\beta}(\mathbf{r}, \mathbf{r}') \rangle = \rho_{\beta\alpha}(\mathbf{r}, \mathbf{r}')$$

from which it follows that

$$\int_{\Omega} d\mathbf{r} \int_{\Omega} d\mathbf{r}' \, \rho_{\alpha\beta}(\mathbf{r}, \mathbf{r}') = \langle \hat{N}_\alpha \hat{N}_\beta \rangle - \delta_{\alpha\beta} \langle \hat{N}_\alpha \rangle.$$

In an isotropic system

$$\rho_{\alpha\beta}(\mathbf{r}, \mathbf{r}') = \rho_{\alpha\beta}(\mathbf{r} - \mathbf{r}')$$

so that

$$\int_{\Omega} d\mathbf{r} \, \rho_{\alpha\beta}(\mathbf{r}) = \frac{1}{\Omega} \langle \hat{N}_\alpha \hat{N}_\beta \rangle - \delta_{\alpha\beta} \, \rho_\alpha.$$

The radial distribution functions $g_{\alpha\beta}(r)$ are then defined by

$$\rho_{\alpha\beta}(r) = \rho_\alpha \rho_\beta g_{\alpha\beta}(r).$$

Fourier transforms

For the singlet density we find

$$\hat{\rho}_\alpha(\mathbf{q}) = \sum_{i=1}^{\hat{N}_\alpha} \exp(i\mathbf{q} \cdot \mathbf{R}_i^\alpha)$$

so that for isotropic systems

$$\langle \hat{\rho}_\alpha(\mathbf{q}) \rangle = N_\alpha \, \delta_{\mathbf{q},0}. \tag{A3}$$

Forming the product $\hat{\rho}_\alpha(\mathbf{q}) \, \hat{\rho}_\beta(-\mathbf{q})$ we find

$$\langle \hat{\rho}_\alpha(\mathbf{q}) \, \hat{\rho}_\beta(-\mathbf{q}) \rangle = \left\langle \sum_{i=1}^{\hat{N}_\alpha} \sum_{j=1}^{\hat{N}_\beta} \exp[i\mathbf{q} \cdot (\mathbf{R}_i^\alpha - \mathbf{R}_j^\beta)] \right\rangle.$$

But for an isotropic system

$$\Omega \int_{\Omega} d\mathbf{r} \exp(i\mathbf{q} \cdot \mathbf{r}) \, \rho_{\alpha\beta}(\mathbf{r}) = \left\langle \sum_{i=1}^{\hat{N}_\alpha} \sum_{j=1}^{\hat{N}_\beta} \exp[i\mathbf{q} \cdot (\mathbf{R}_i^\alpha - \mathbf{R}_j^\beta)] \right\rangle - \delta_{\alpha\beta} \langle \hat{N}_\alpha \rangle \tag{A4}$$

or using the definition of $g_{\alpha\beta}$

$$(N_\alpha N_\beta)^{1/2} \int \frac{d\mathbf{r}}{\Omega} \exp(i\mathbf{q} \cdot \mathbf{r}) \, (g_{\alpha\beta}(\mathbf{r}) - 1) + \delta_{\alpha\beta} = \frac{1}{(N_\alpha N_\beta)^{1/2}} \langle \rho_\alpha(\mathbf{q}) \, \rho_\beta(-\mathbf{q}) \rangle - (N_\alpha N_\beta)^{1/2} \, \delta_{\mathbf{q},0}.$$

The right hand side of this last equation defines the partial structure factor for the system:

$$S_{\alpha\beta}(\mathbf{q}) = (N_\alpha N_\beta)^{-1/2} \langle \hat{\rho}_\alpha(\mathbf{q}) \, \hat{\rho}_\beta(-\mathbf{q}) \rangle - (N_\alpha N_\beta)^{1/2} \, \delta_{\mathbf{q},0}. \tag{A5}$$

For large $\mathbf{q}$, $S_{\alpha\alpha} \to 1$, $S_{\alpha\beta} \to 0$, the latter merely being a statement that for very small separations unlike particles are not correlated. In the substitutional model for a liquid

$$g_{\alpha\beta}(\mathbf{r}) = g(r)$$

from which it immediately follows that

$$(S_{\alpha\beta}(k) - \delta_{\alpha\beta}) = (c_\alpha c_\beta)^{1/2}\,[S(k) - 1]. \tag{A6}$$

Provided all ions have the same valence (Z) it is straightforward to show that the Madelung reduces to

$$E_{\mathrm{M}} = \frac{N}{2} \sum_{\mathbf{k} \neq 0} \frac{4\pi Z^2 e^2}{\Omega k^2}\,[S(k) - 1]$$

which Jones (1971) has shown to be equal to

$$-(NZ)\,6\eta^{2/3}\,(1 - \eta/5 + \eta^2/10)\,Z^{2/3}\,(e^2/2a_0)/r_{\mathrm{s}} \tag{A7}$$

if $S(k)$ is the Percus–Yevick structure factor for hard spheres of diameter σ at a packing fraction η, where $\eta = (\pi/6)n\sigma^3$.

Appendix 2

As a notational simplification we set $x = \bar{r}_{\mathrm{s}1}$, $y = \bar{r}_{\mathrm{s}2}$, and $z = \bar{r}_{\mathrm{s}}$, where for a two-component system

$$z^2 = c_1 x^2 + c_2 y^2 = cx^2 + (1 - c)\,y^2. \tag{A8}$$

From equation (21) we are led to consider the function

$$f(c) = \left(\frac{z}{1+z}\right)^2 - c\left(\frac{x}{1+x}\right)^2 - (1-c)\left(\frac{y}{1+y}\right)^2 \tag{A9}$$

whose derivative with respect to concentration is

$$f'(c) = \frac{(x^2 - y^2)}{(1+z)^3} - \left(\frac{x}{1+x}\right)^2 - \left(\frac{y}{1+y}\right)^2, \tag{A10}$$

and whose second derivative is

$$f''(c) = -\frac{3(x^2 - y^2)^2}{2z(1+z)^4} \tag{A11}$$

which is always negative.

From (A10) we can show that

$$f'(0) = \left(\frac{(x-y)^2}{(1+x)^2\,(1+y)^2}\right)\left(\frac{x(x+2y) + 2x + y}{1+y}\right) > 0, \tag{A12}$$

and

$$f'(1) = -\frac{(x-y)^2}{(1+x)^2\,(1+y)^2}\left[\frac{y(y+2x) + 2y + x}{1+x}\right] < 0, \tag{A13}$$

while from (A10) it is also clear that for real z, $f'(c)$ vanishes *once* in the interval $0 < c < 1$ at a concentration $\bar{c}$ and a corresponding $\bar{z}$ $(= [\bar{c}x^2 + (1 - \bar{c})y^2]^{1/2})$ given by

$$(1 + \bar{z}) = \frac{(1 + x)^{2/3}(1 + y)^{2/3}}{(x + y)^{1/3}(x + 2xy + y)^{2/3}}. \tag{A14}$$

From these last three results we conclude that the excess function $f(c)$ is always positive as indicated in figure 3. It is also asymmetric to a degree determined by the magnitudes of x and y which in turn depend on R_{c1} and R_{c2}. For typical values of x and y (say $x = 3, y = 4$), the maximum value of $f(c)$ is $0{\cdot}65 \times 10^{-2}$: with the factor $\beta^2/\gamma \sim 3$ Ryd, this difference in e then approaches the range of thermal energies.

References

Alder B J 1964 *J. Chem. Phys.* **40** 2724
Ashcroft N W and Langreth D C 1967a *Phys. Rev.* **155** 682
—— 1967b *Phys. Rev.* **159** 500
Brout R 1965 *Phase Transitions* (Menlo Pk., Calif: W A Benjamin)
Firey B and Ashcroft N W 1976 *Phys. Rev.* submitted
Hammerberg J and Ashcroft N W 1974 *Phys. Rev.* **B9** 409
Hansen J P 1973 *Phys. Rev.* **A8** 3096
Hansen M 1965 *Constitution of Binary Alloys Suppl. No. 1* (New York: McGraw Hill)
Henderson R L and Ashcroft N W 1976 *Phys. Rev.* **A13** 859
Jones H 1971 *J. Chem. Phys.* **55** 2640
Lebowitz J L and Rowlinson J S 1964 *J. Chem. Phys.* **41** 133
Mon K K, Ashcroft N W and Chester G V 1976 *Phys. Rev.* submitted
Pethick C 1970 *Phys. Rev.* **B2** 1789
Rotenberg A 1965 *J. Chem. Phys.* **43** 4377
Stroud D and Ashcroft N W 1971 *J. Phys. F: Metal Phys.* **1** 113

The entropies and structure factors of liquid simple metals†

R Kumaravadivel‡ and R Evans

H H Wills Physics Laboratory, Royal Fort, University of Bristol, Bristol BS8 1TL

Abstract. Some results of a systematic study of the entropies and structure factors of 14 liquid metals are presented. The *ab initio* model potentials of Shaw and the dielectric function of Vashista and Singwi have been used to construct effective pairwise interatomic potentials for these liquid metals. These pairwise potentials have then been employed in both the Weeks–Chandler–Anderson (WCA) and the variational thermodynamic perturbation theories to determine appropriate effective hard-sphere diameters. In the variational scheme the excess entropy of the liquid metal consists solely of a hard-sphere contribution while in the WCA approach there are important additional contributions. For 10 of the metals, the excess entropies calculated in the WCA theory are in reasonably good agreement with experiment. However, for Li, In, Tl and Pb the calculated hard-sphere diameters are too large and lead to packing fractions and entropies which are non-realistic.

We examine in detail the short-range repulsive part of the pairwise potential in several liquid metals and show, within the WCA theory, how the softness of this influences the form of the liquid structure factor. For Mg and Al, in particular, the combination of our calculated pairwise potentials and the WCA theory leads to structure factors which agree well with those measured experimentally. In the alkali metals, however, the repulsive potentials are softer and the hard-sphere potential is a poor starting approximation for the WCA procedure. We discuss briefly the calculation of the compressibility from the long wavelength limit of the structure factor.

† To be published in *J. Phys. C: Solid St. Phys.*

‡ Present address: Department of Physics, University of Sri Lanka, Peradeniya, Sri Lanka.

The surface tension and ion density profile of a liquid metal†

R Evans and R Kumaravadivel‡
H H Wills Physics Laboratory, Royal Fort, University of Bristol, Bristol BS8 1TL

Abstract. A simple scheme for determining the ion density profile and the surface tension of a liquid metal is described. Assuming that the interaction between metallic pseudo-ions is of the form introduced in our earlier papers an approximate expression for the excess free energy of the system is derived using the thermodynamic perturbation theory of Weeks, Chandler and Anderson. This excess free energy is then minimized with respect to a parameter which specifies the ion density profile and the surface tension is given directly.

From our consideration of the dependence of the interionic forces on the electron density distribution we predict that the ions should take up a very steep density profile at the liquid metal surface. We contrast this behaviour with that to be expected for rare-gas fluids in which the interatomic forces are density independent.

The values of the surface tension calculated for liquid Na, K and Al from a simplified version of the theory are in reasonable agreement with experiment.

In the final section we attempt to relate our work to that of other workers who have considered the surface energies of solid metals.

† Published in *J. Phys. C: Solid St. Phys.* 1976 9 1891–905.
‡ Present address: Department of Physics, University of Sri Lanka, Peradeniya, Sri Lanka.

Three-ion interaction and structures of liquid metals†

M Hasegawa‡
Department of Electronic Engineering, Shibaura Institute of Technology, Minato-ku, Tokyo 108, Japan

Abstract. Linear screening theory for the interionic interaction in simple liquid metals is extended by including nonlinear contributions due to the third-order perturbation term in the electron–ion pseudopotential. The theory of Anderson, Weeks and Chandler and the variational method based on the Gibbs–Bogoliubov inequality are used to estimate the influences of additional pair interaction $\phi_2^{(3)}(r)$ on the liquid structures of Na and K. The effect of three-ion interaction $\phi_3^{(3)}(\mathbf{r}_1, \mathbf{r}_2, \mathbf{r}_3)$ is estimated semi-quantitatively through the effective pair potential defined in the presence of the three-body interaction. It is found that $\phi_2^{(3)}(r)$ is a strong and attractive potential for small r and affects strongly the position of the hard core of the pair potential. On the other hand, $\phi_3^{(3)}$ is found to have only a slight effect on the liquid structure. A full report of this work can be found in M Hasegawa 1976 *J. Phys. F: Metal Phys.* 6 649.

† Work supported by the Grant for Scientific Research, Ministry of Education, Japan, No 974070.
‡ Present address: School of Mathematics and Physics, University of East Anglia, Norwich NR4 7TJ, UK.

A study of corresponding states for the liquid alkali metals

Raymond D Mountain

Heat Division, National Bureau of Standards, Washington, DC 20234, USA

Abstract. The pseudopotential pair potentials developed by Price *et al* and by Dagens, Rasolt and Taylor are used to investigate the microscopic basis for a law of corresponding states for the liquid alkali metals. Both sets of potential functions show small departures from corresponding states. Monte Carlo simulation is used to show that the temperature-dependent part of the equations of state for Na and K scale with an error of the order of 10%. The pair distribution functions for Na are in good agreement with the results of x-ray diffraction measurements. These studies suggest that corresponding states is a reasonable, but not completely accurate, way of describing the thermodynamic properties of the liquid alkali metals.

1. Introduction

The pseudopotential functions, which have been developed in recent years for the alkali metals, have been used satisfactorily to predict a number of fluid state properties (Price 1971, Murphy and Klein 1973, Rahman 1974). In this paper we shall examine the possibility of providing a microscopic basis for a law of corresponding states for the configuration-dependent part of the equation of state of the liquid alkali metals.

Two classes of effective pair potentials will be examined. The first is the one developed by Price (1971) and Price *et al* (1970) and the second is the one developed by Dagens *et al* (1975) and by Rasolt and Taylor (1975). We shall refer to them as the Price potential and the DRT potential respectively. These potentials will be examined to see how well they can be scaled for the different alkali metals. Two potentials $\phi(r)$ are said to scale if φ, the reduced form of the potential, scales. Here

$$\varphi(r/\sigma, n\sigma^3) = \frac{1}{\epsilon}\,\phi(r/\sigma) \qquad (1)$$

where ϵ is the magnitude of ϕ at its absolute minimum, σ is the smallest value of r for which $\phi = 0$ and n is the number density of the ions.

As we shall see in §2, the scaling property for these potentials is only approximate. However, the departures from scaling, over the region of the potential which is significant for the structural properties of liquids, is not so large as to preclude the possibility of corresponding states being reasonably accurate. Monte Carlo simulation is then employed to determine the pair distribution function, $g_2(r)$, and the configuration-dependent parts of the pressure and the internal energy. The scaling properties of these quantities are examined in §3 and, where possible, these quantities are compared with experiment.

The electron-gas contribution to the pressure and internal energy may preclude the possibility of a corresponding states rule for both the complete and the temperature (configuration)-dependent parts of the equation of state and the internal energy. We examine this point briefly in §4.

Section 5 contains a discussion of our results and an assessment of the utility of corresponding states arguments for comparing the thermodynamic properties of the liquid alkali metals.

Before we begin the examination of the potential functions let us recall briefly the relation between the potential functions and the equation of state and the internal energy of the fluid.

The energy and equation of state of a liquid alkali metal described by a pairwise additive pseudopotential $\phi(r, n)$ can be expressed as the sum of several contributions (Price 1971). The energy per ion is given by

$$E = E_K + E_1 + E_2 + E_3, \tag{2a}$$

where E_K is the kinetic energy of the ion,

$$E_1 = \tfrac{1}{2} \int \mathrm{d}\mathbf{r} g_2(r)\, \phi(r) \tag{2b}$$

is the usual pair potential contribution to the energy,

$$E_2 = \tfrac{1}{2} \phi_{\mathrm{bs}}(r = 0) \tag{2c}$$

is the self-energy of the band-structure term in the pseudopotential and E_3 is the electron-gas energy not included in E_1 and E_2. The radial distribution function for the ions is $g_2(r)$; E_2 and E_3 depend on the density and are independent of the temperature of the fluid.

The corresponding terms in the equation of state are

$$p = nkT - \frac{1}{6} n \int \mathrm{d}\mathbf{r} g_2(r) \left(r \frac{\partial \phi}{\partial r} + a \frac{\partial \phi}{\partial r} \right) - \frac{1}{6} na \frac{\partial}{\partial a} [\phi_{\mathrm{bs}}(r = 0)] - \frac{1}{3} na \frac{\partial E_3}{\partial a}, \tag{3}$$

where a is the lattice constant for the BCC lattice of the same density as the liquid, namely

$$n^{-1} = a^3/2. \tag{4}$$

2. Effective pair potentials

The pseudopotential formalism leads to expressions for the effective potential, ϕ, of the form

$$\phi(r) = \frac{(Ze)^2}{r} - \frac{(Ze)^2}{2\pi^2} \int \frac{F(q)}{q^2} \exp(-\mathrm{i}\mathbf{q} \cdot \mathbf{r})\, \mathrm{d}\mathbf{q} \tag{5}$$

where Ze is the ionic charge and $F(q)$ is the energy–wavenumber characteristic which contains both the effects of the screening of the bare ions by the electrons and the model potential for the bare ion–ion interaction. In addition to being density dependent, $\phi(r)$ contains several adjustable constants which are specific to the substance involved. Phonon dispersion measurements were used by Price *et al* (1970) for adjusting these constants in the effective potentials for the alkali metals. This potential contains one 'empty-core' term and uses the electron screening treatment developed by Singwi *et al* (1970). Rasolt and Raylor (1975) took a different approach using the Hohenberg–Kohn (1963) theorem and Dagens' (1972) charge density calculations to adjust the constants appearing in their version of the effective potential. The resulting potential contains three 'empty-core' terms, uses the Geldart–Taylor (1970) treatment of electron screening and includes non-local effects. The reader interested in the details should consult the papers referenced.

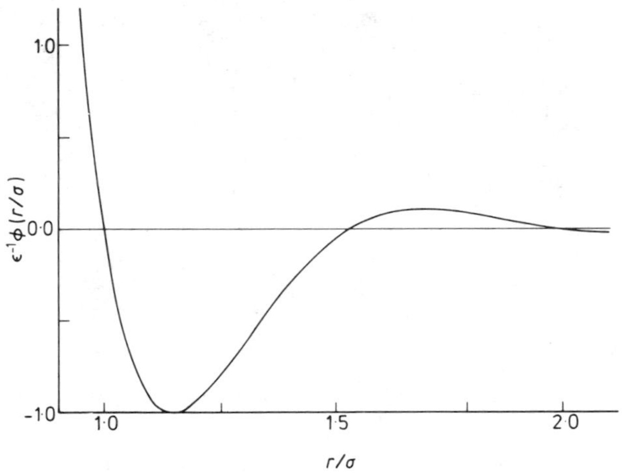

Figure 1. The reduced DRT potential for liquid Na at a density of 0·0243 $Å^{-3}$. The Price reduced potential differs mainly in that the bowl is wider and the oscillations are less pronounced.

Our interest here is in the scaling properties of the effective potentials $\phi(r)$. Potential functions for Li, Na, K and Rb† have been constructed numerically for a range of liquid state densities. Figure 1 exhibits the DRT form for Na near the melting point (a = 4·35 Å). The qualitative features of this curve are typical of all states considered. Figure 2 shows the density dependence of σ and ϵ for Na for the two model potentials.

The scaling features of the reduced potentials, $\varphi(r/\sigma)$, are best examined using a difference plot since the deviations are small. The differences between the corresponding states for K and Na are shown in figure 3 for the DRT potential when $n\sigma^3 = 0{\cdot}927$. This corresponds to liquid Na at a density of 0·0243 atoms/$Å^3$(a = 4·35 Å) and a temperature of 373 K (zero pressure). One should keep in mind that the pressure and the pair distribution function are mainly determined by the strongly repulsive portion of the potential. Thus the relatively large departures from scaling found for $r/\sigma > 1{\cdot}5$ are not significant for our purposes (Schiff 1969).

The departures from scaling found for $r/\sigma < 1{\cdot}5$ are indicative that the equation of state and the pair distribution functions for liquid K and Na predicted by these model potentials will not satisfy a law of corresponding states exactly. However, the departures are not so large as to rule out an approximate correspondence with, say, 10%

† The Rb parameters for the DRT potential are not available.

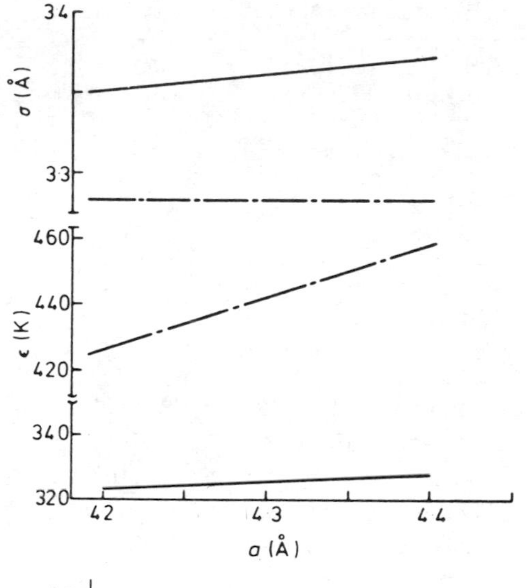

Figure 2. The density dependence of σ and ϵ for the DRT (full curve) and Price (chain curve) potentials for Na. The equivalent BCC lattice parameter a is related to the number density through equation (4).

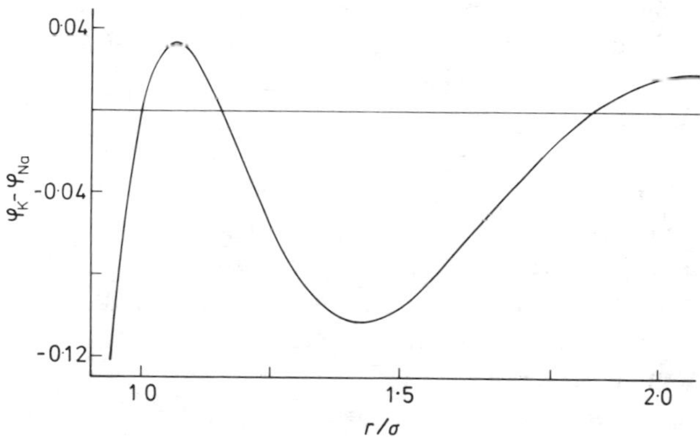

Figure 3. The difference between the DRT reduced potentials at a reduced density $n\sigma^3 = 0{\cdot}927$. The Price potentials for the equivalent reduced density show smaller differences.

deviations. It is also evident that for a given density, the departures from corresponding states will increase with increasing temperature. Only the K–Na comparison is presented here. For the Price potential, these results are typical of the alkali metals over a substantial range of liquid densities. The situation is more complicated for the DRT potential. As noted above, the constants for Rb are not available. Also, the potentials constructed for Li have unusually large values for ϵ. It is unlikely that such large values for ϵ are appropriate for a liquid modelled by a pair potential. Dagens *et al* (1975) have suggested that three- and four-body forces may be much more significant for Li than for Na or K. If that is so, Li would be excluded from a corresponding states rule.

3. Configuration-dependent terms

A direct test of corresponding states is made here by using the effective potentials discussed above to determine the configuration-dependent terms in equations (2) and (3) and to determine the pair distribution function $g_2(r)$. The six model fluid states

Table 1. Densities and temperatures of the liquid states studied using Monte Carlo simulation.

	Physical value		Reduced value	Price potential	Reduced value	DRT potential
Substance	n(Å^{-3})	T(K)	$n\sigma^3$	kT/ϵ	$n\sigma^3$	kT/ϵ
Na	0·0243	373	0·868	0·831	0·927	1·142
K	0·0125	352	0·872	0·831	–	–
K	0·0128	401	–	–	0·921	1·142
Rb	0·0104	333	0·931	0·842	–	–
Rb	0·0104	473	0·931	1·120	–	–

listed in table 1 were studied using Monte Carlo simulation to construct the fluid properties. The K states were selected to have the same reduced temperature and density as the Na state. The parameters for Na were selected to model one of the zero pressure states studied by Greenfield *et al* (1971) in their x-ray diffraction investigation of liquid structure in Na and K. In this way we may not only see how well corresponding states applies, but also see how well the effective potentials are able to reproduce an experimentally determined pair distribution function. The Rb states were selected so that the isochoric changes in the pressure and in the internal energy could be compared with the equation of state data of Rosenbaum (1972).

The Monte Carlo simulation employed the procedure of Metropolis *et al* (1953) and was applied to 250 particles with periodic boundary conditions. Considerable care was taken to ensure that the results were independent of the initial configuration of the particles, an important consideration at the relatively high reduced densities and low reduced temperatures involved in this study. Once this independence of the results was achieved, between 5×10^5 and 1×10^6 configurations were generated. Stable results with good statistics were obtained from runs of this length.

First, we consider the pair distribution functions of Na and K within the framework of corresponding states. Figure 4 shows $g_2(r)$ for K obtained using the Price potential. The upper curve shows the difference between the K and Na results. Figure 5 contains

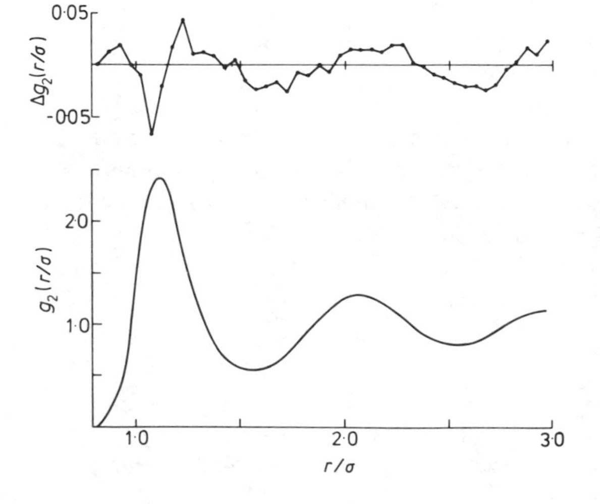

Figure 4. The pair distribution function for K as determined by the Price potential (lower curve) and the difference between the K and Na pair functions (upper curve). The fluid states are listed in table 1.

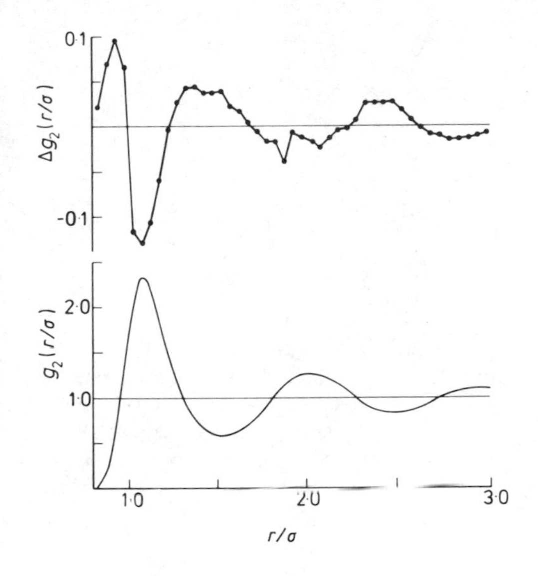

Figure 5. The pair distribution function for K as determined by the DRT potential (lower curve) and the difference between the K and Na pair functions (upper curve). The fluid states are listed in table 1.

the same quantities obtained using the DRT potential. The pair distribution functions are seen to be consistent with a law of corresponding states to within about 10% for both types of potential functions. The smaller departures from corresponding states for $g_2(r)$ determined by the Price potential are consistent with the trends indicated in the caption to figure 3.

The adequacy of the effective potentials for modelling real fluids can be estimated by considering figures 6 and 7. Figure 6 is for the Price potential model and figure 7 for the DRT model. In each case, the pair distribution function for Na, obtained by Monte Carlo simulation, is displayed along with the difference of this quantity from the pair distribution function derived from the liquid structure data of Greenfield *et al* (1971). The errors are less than 10% for both model systems and small adjustments in the

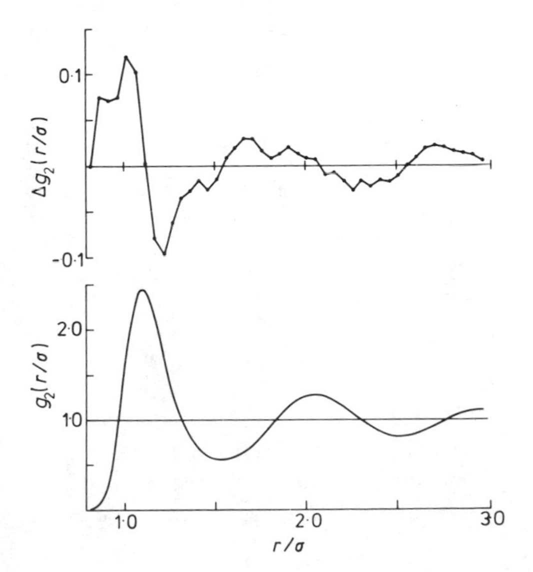

Figure 6. The pair distribution function for Na as determined by the Price potential (lower curve) and the difference between this pair function and the experimentally determined one (upper curve). The fluid states are listed in table 1.

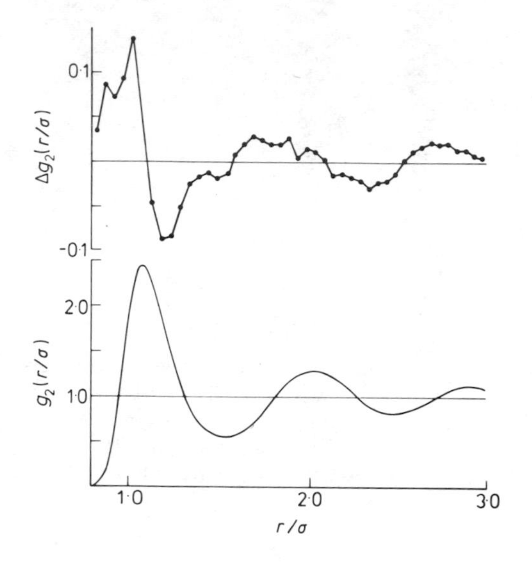

Figure 7. The pair distribution function for Na as determined by the DRT potential (lower curve) and the difference between this pair function and the experimentally determined one (upper curve). The fluid states are listed in table 1.

'crossing points' (values of r where $g_2 = 1$) would eliminate most of the differences between the calculated and measured curves.

With this test, no preference can be established for one potential function over the other. If one were to overlay the Price and DRT reduced potentials one would see that the strongly repulsive parts are nearly identical. Since it is just this portion of the potential which dominates the form of $g_2(r)$, the inability to establish a preference for one potential by this means is not surprising. The two reduced potentials are not similar beyond the first minimum so the temperature-dependent portions of the pressure and the internal energy should also be considered.

The temperature-dependent portions of the equation of state and the energy are

$$P_T = nkT - \frac{1}{6}n\int \mathrm{d}\mathbf{r}g_2(r)\left(r\frac{\partial\phi}{\partial r} + a\frac{\partial\phi}{\partial a}\right) \tag{6}$$

and

$$E_T = \tfrac{3}{2}kT + \tfrac{1}{2}\int \mathrm{d}\mathbf{r}g_2(r)\,\phi(r). \tag{7}$$

The corresponding state quantities to be compared are the dimensionless quantities, the reduced pressure

$$P_T^* = P_T\sigma^3/\epsilon, \tag{8}$$

and the reduced energy,

$$E_T^* = E_T/\epsilon. \tag{9}$$

This comparison is made by examining table 2 where $n^* = n\sigma^3$, $T^* = kT/\epsilon$, and P_T^* and E_T^* are displayed for the Na and K states listed in table 1. The Price potential yields reduced pressures which differ by 10% while the DRT potential yields reduced pressures which differ by 20%. The energy differences amount to 4% and 7% respectively.

Table 2. Reduced thermodynamic properties for liquid Na and liquid K as determined by Monte Carlo simulation using the Price and DRT potentials.

	Price potential				DRT potential			
	n^*	T^*	P_T^*	E_T^*	n^*	T^*	P_T^*	E_T^*
Na	0·868	0·831	6·36	−2·74	0·927	1·142	7·06	−1·42
K	0·872	0·831	5·76	−2·63	0·921	1·142	5·69	−1·49

In order to compare quantities of this sort with experiment, it is necessary to have thermodynamic data where the pressure is a variable. Data of this sort for the liquid alkali metals are scarce. To illustrate the type of comparisons which are possible we compare the Price potential model calculation for liquid Rb with the unpublished experimental results of Rosenbaum (1972). The Rb states listed in table 1 correspond to states with zero pressure at $T = 333$ K and with approximately 75 MPa at 473 K. The change in the Monte Carlo pressure between these states is 75 MPa with an uncertainty of about ±5 MPa. This uncertainty is estimated by determining the variation in the pressure during the course of the Monte Carlo run. The experimentally estimated change in the internal energy (ΔE_1) of liquid Rb between these two states is approximate 220 J mol^{-1}. The Monte Carlo estimate of the change is of the order of 120 J mol^{-1}.

From the comparison of experimental and calculated pressure differences, it is possible to suggest that the repulsive part of the Price potential is probably reliable for liquid Rb. One must be more cautious in drawing inferences from the energy differences. The experimental difference represents a 1% change in the values of internal energy tabulated by Rosenbaum. Since no uncertainties have been assigned to the experimental numbers, it is not possible to say if the factor of 2 between the experimental and theoretical differences is significant. If it is significant, it would suggest that the attractive portion of the effective potential is somehow in error.

The temperature (configuration)-dependent properties of the alkali metals as modelled by the Price effective potential satisfy a law of corresponding states with an uncertainty of about 10%. The DRT model for Na and K indicates that departures from corresponding states of up to 20% are possible. The experimental situation is not sufficiently well developed to determine which of these two model potentials is to be preferred for the estimation of liquid state properties.

4. Configuration-independent terms

The configuration-independent terms in the energy, E_2 and E_3, depend on the fluid density but not on the temperature. The band-structure self-energy E_2 and the corresponding term in the pressure can be determined when the effective potentials are generated. For the states listed in table 2, the dimensionless pressures derived from E_2 are found to be −3·14 for Na and −3·26 for K using the Price potential, −3·91 for Na and −2·46 for K using the DRT potential. If these terms are added to the reduced pressures listed in table 2, our view of the agreement with corresponding states changes drastically. By including the band-structure self-energy term, the DRT potential yields pressures which do correspond closely while the Price model now departs from correspondence by 20%. However, the electron-gas contributions embodied in E_3 have not yet been included in the estimate of the pressure so no overall conclusions can be drawn at this point.

The electron-gas contribution is the least accurately known term in equation (3). Price (1971) has noted the need to make adjustments in the calculated values of the electron-gas energy in order to obtain observed lattice constants. Cohen *et al* (1976) employed a semi-empirical procedure to estimate E_3 and the corresponding term in the pressure. These procedures are probably reliable, but the absence of a theoretical justification for them makes it difficult to determine how E_3 scales with density given the limited number of states studied.

5. Summary

The Price and DRT potentials for the alkali metals have been shown to depart from a corresponding states form. The departures from such a form are not so large that an approximate scaling of configuration-dependent properties is unlikely. Monte Carlo calculations have shown that the pair distribution function for liquid Na can be accurately reproduced using these potentials and that the correspondence between Na and K holds at the 10% level for the ionic terms in the equation of state. Also, we have seen that the Price potential leads to pressure changes, but not to internal energy changes which are consistent with thermodynamic data for liquid Rb.

From the results of this study, we suggest that the ionic contribution to the thermal properties of the alkali liquid metals satisfies an approximate (10%) law of corresponding states when determined using the Price or DRT potentials. We also suggest that these model potentials lead to thermodynamic properties which are not grossly different from those of the real liquids. The latter statement needs verification using a considerably larger body of experimental data than is currently available. Finally, we note the need for improved estimates for E_3 to complete the evaluation of corresponding states on the microscopic level.

Acknowledgments

The DRT potentials were constructed using a program kindly supplied by R Taylor of NRC. This work was supported by a grant of computer time from the Energy Research and Development Administration AT(49–16)3003.

References

Cohen S S, Klein M L, Duesbery M S and Taylor R 1976 *J. Phys. F: Metal Phys.* **6** 337
Dagens L 1972 *J. Phys. C: Solid St. Phys.* **5** 2333
Dagens L, Rasolt M and Taylor R 1975 *Phys. Rev.* **B11** 2726
Geldart D J W and Taylor R 1970 *Can. J. Phys.* **48** 155
Greenfield A J, Wellendorf J and Wiser N 1971 *Phys. Rev.* **A4** 1607
Hohenberg P and Kohn W 1963 *Phys. Rev.* **B136** 864
Metropolis N A, Rosenbluth M N, Rosenbluth A W, Teller A H and Teller E 1953 *J. Chem. Phys.* **21** 1087

Murphy R D and Klein M L 1973 *Phys. Rev.* **A8** 2640
Price D L 1971 *Phys. Rev.* **A4** 358
Price D L, Singwi K S and Tosi M P 1970 *Phys. Rev.* **B2** 2983
Rahman A 1974 *Phys. Rev.* **A9** 1667
Rasolt M and Taylor R 1975 *Phys. Rev.* **B11** 2717
Rosenbaum I J 1972 *Naval Ordnance Laboratory Report* No. NOLTR 72-107 (unpublished)
Schiff D 1969 *Phys. Rev.* **186** 151
Singwi K S, Sjölander A, Tosi M P and Land R H 1970 *Phys. Rev.* **B1** 1044

Pressure and entropy calculations for liquid Na, K and Rb along the melting line

M Silbert
School of Mathematics and Physics, University of East Anglia, Norwich NR4 7TJ, UK

Abstract. The Gibbs–Bogoliubov variational principle is used to derive an equation of state for liquid metals, following a procedure indicated by Watabe and Young. Results for liquid Na, K and Rb along the melting curve are presented. It is found that for liquid Na there is overall agreement with experimental results up to 80 kbar, but for liquid K and Rb only up to 25–30 kbar.

1. Introduction

This work was prompted by a remark made by Watabe and Young (1974) that a liquid metal, described by volume and pairwise forces only, using a hard-sphere reference system and the Gibbs–Bogoliubov variational principle, will satisfy the following equation of state:

$$P = \rho kT + \rho^2 \frac{du_0(\rho)}{d\rho} - \frac{1}{2}\rho^2 \int dr \left(\frac{r}{3} \frac{\partial u(r,\rho)}{\partial r} - \rho \frac{\partial u(r,\rho)}{\partial \rho} \right) g_{hs}(r), \tag{1}$$

if the packing fraction, η, is chosen as the variational parameter.

In this equation, k is the Boltzmann constant, T the temperature and $u_0(\rho)$ and $u(r,\rho)$ are, respectively, the configuration-independent and the pair interaction parts of the effective potential energy of the ion system. Both of the latter depend on the number density of conduction electrons and, through the charge neutrality requirement, on the number density of ions ρ. The radial distribution function $g_{hs}(r)$ is that for the hard-sphere reference system. If $g_{hs}(r)$ is replaced by the radial distribution function of the actual system, equation (1) is the exact virial equation of state obtained by Hasegawa and Watabe (1974).

I use equation (1) to write down explicitly an equation of state for a liquid metal – for a given choice of pseudopotential and dielectric function – which offers the distinct advantage of evaluating the pressure directly. Consequently, many thermodynamic coefficients may be evaluated as first- rather than second-order derivatives of an appropriately chosen thermodynamic potential.

A very brief discussion of the formalism used here is given in the next section, where the equation of state is also presented. In §3 I report results for the pressure and the entropy of liquid Na, K and Rb, along the melting curve, as well as $(\partial P/\partial T)_\rho$ along the melting point isochore. Results for other thermodynamic properties such as internal energy, isothermal compressibility, specific heat, etc, will be reported elsewhere. Finally, a discussion and interpretation of results is offered in §4.

2. Formalism

According to the Gibbs–Bogoliubov inequality (see, for example, Isihara 1968), if the Hamiltonian for a system is regarded as that for a reference system plus a perturbation, then the Helmholtz free energy for the reference system plus the expectation value of the perturbation averaged over the reference system is an upper bound of the Helmholtz free energy of the actual system. That is,

$$F \leqslant F_{\mathrm{ref}} + \Delta E \tag{2}$$

where

$$\Delta E = u_0(\rho) + \frac{\rho}{2}\int \mathrm{d}\mathbf{r}\, u(r)\, g_{\mathrm{ref}}(r). \tag{3}$$

If the reference system is now chosen as a hard-sphere system and η as the variational parameter, then

$$P = \rho^2 \left.\frac{\partial F}{\partial \rho}\right|_{T,\eta}$$

Alternatively, the free energy may be characterized by its various components

$$\Delta E = E_{\mathrm{eg}} + E_0 + E_{\mathrm{m}} + E_{\mathrm{bs}}. \tag{4}$$

Here, I am using a formalism described by, amongst others, Stroud and Ashcroft (1972), Edwards and Jarzynsky (1972), Jones (1973) and Silbert *et al* (1975), and I refer the reader to the above papers for a detailed account of the theory.

I only need to add that I have characterized the metallic system in the following way:

(i) The electron gas is described by the Nozières–Pines (1958) interpolation formula for the total energy, E_{eg}, and a modified Hubbard (Geldart and Vosko 1966) dielectric screening function $\epsilon(q)$.
(ii) The electron–ion interaction is specified via the empty core pseudopotential (Ashcroft 1966).
(iii) The hard-sphere dynamics are described in the Percus–Yevick approximation (Wertheim 1964).
(iv)

$$E_0 = \alpha z^2 \rho \tag{5}$$

where z is the valency and α is to be treated here as an adjustable parameter to give the experimental value of the pressure at the melting point.

The derivatives under the integral sign in equation (1) are now carried out explicitly to obtain

$$P = \rho kT + p_{\mathrm{eg}} + \rho E_0 + \tfrac{1}{3}\rho E_{\mathrm{m}} + \tfrac{1}{3}\rho E_{\mathrm{bs}} - \tfrac{1}{3}\Delta p_{\mathrm{I}} - \tfrac{1}{3}\Delta p_{\mathrm{II}} \tag{6}$$

where

$$p_{\mathrm{eg}} = \rho^2 \left(\frac{\partial E_{\mathrm{eg}}}{\partial \rho}\right)_{T,\eta},$$

$$\Delta p_{\mathrm{I}} = \frac{2z^2}{\pi}\rho \int_0^\infty q r_{\mathrm{c}} \sin q r_{\mathrm{c}} \left(\frac{1}{\epsilon(q)} - 1\right) S_{\mathrm{hs}}(q)\, \mathrm{d}q,$$

and

$$\Delta p_{\rm II} = \frac{2z^2}{\pi}\rho\int_0^\infty \frac{1}{\epsilon(q)}\left[1+\frac{h(y)}{1-h(y)}\left(1+\frac{(8y^2-0{\cdot}153\,g^2\lambda^2)}{(2y^2+g)}\right)\right]\cos^2 qr_{\rm c}\left(\frac{1}{\epsilon(q)}-1\right) S_{\rm hs}(q)\,{\rm d}q,$$

where $r_{\rm c}$ is the empty-core radius of the model potential and $S_{\rm hs}(q)$ the hard-sphere structure factor.

By writing down explicitly the expression for the dielectric screening function:

$$\epsilon(q) = 1 + \frac{\lambda^2 f(y)}{y^2(1-h(y))} \qquad h(y) = \frac{\lambda^2 f(y)}{(2y^2+g)}$$

with

$$f(y) = \frac{1}{2} + \frac{1}{4y}(1-y^2)\ln\left|\frac{1+y}{1-y}\right|,$$

$$g = \frac{1}{1+0{\cdot}153\,\lambda^2}, \qquad \lambda^2 = \frac{1}{\pi k_{\rm F}}, \qquad y = \frac{q}{2k_{\rm F}}$$

and $k_{\rm F}$ the Fermi wavenumber, it is now easy to identify the terms in the integrand of $\Delta p_{\rm II}$.

Equation (6) constitutes the basis for the calculations performed in this work.

3. Calculations

In order to proceed with the calculations of the pressure for liquid Na, K and Rb, I need to state how the two adjustable parameters are chosen. (i) The empty-core radius of the model potential is evaluated such that it reproduces the observed electrical resistivity at the melting point. These values are shown in table 1 where, for comparison, those values obtained by fitting the shape of the Fermi surface are also included (Cohen and Heine 1970). (ii) As stated previously, the parameter α in equation (5) is so adjusted as to give the experimental value of the pressure at the melting point. Values for α are also given in table 1.

Table 1. Values of the empty-core radius $r_{\rm c}$ of the empty-core model potential which reproduces experimental values of the electrical resistivity at the melting point, and values of the α coefficient introduced in equation (5) obtained from reproducing the experimental value of the pressure at the melting point.

Metal	$r_{\rm c}$(a.u.)		α(a.u.)
	This work	Fermi surface	
Na	1·72	1·66	42·629
K	2·21	2·13	77·533
Rb	2·26	2·61	88·281

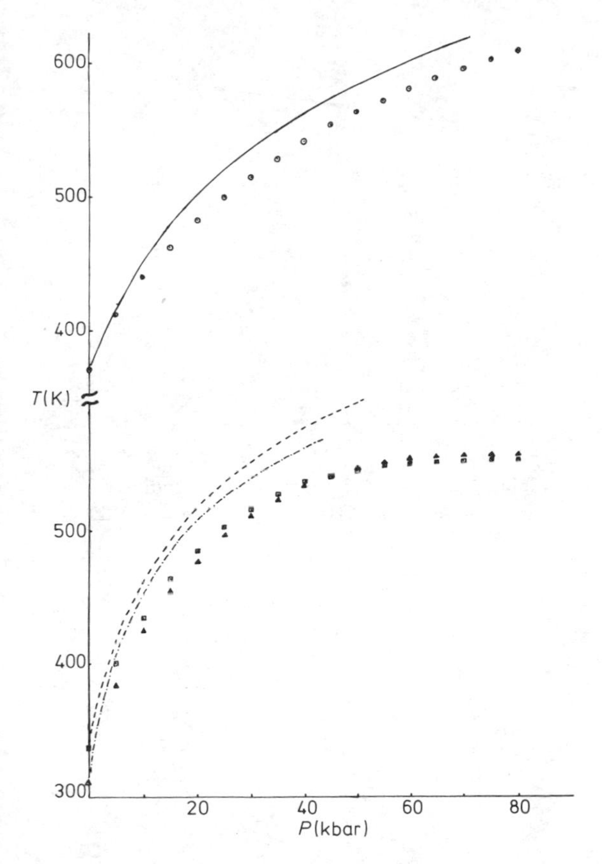

Figure 1. Calculated and experimental melting lines. Theory: —— Na; ---- K; -·-·-·- Rb. Experimental (Luedemann and Kennedy 1968): ⊙ Na; ⊡ K; △ Rb.

The reason for choosing α as an adjustable parameter, instead of writing $\alpha = 4\pi r_c^2$, is that no single r_c, consistent with the electron states at the Fermi level and therefore giving a satisfactory resistivity, is also consistent with electron energies throughout the band.

Once the adjustable parameters have been determined, the calculations are straightforward, provided the temperature and density are specified. I have used the empirical relations between temperature and volume along the melting curve given by Kraut and Kennedy (1966) and, with the appropriate changes, assumed their validity in the liquid state. Once the thermodynamic state is specified and the value of the variational parameter obtained for the state, the values for the Helmholtz free energy, pressure and Gibbs free energy are evaluated with relative ease. The P–T diagrams for liquids Na, K and Rb thus obtained are shown in figure 1 and compared with the experimental values obtained by Luedemann and Kennedy (1968).

It can be shown (Silbert *et al* 1975) that the entropy, S, of a liquid metal, within the formalism used in this work, is the entropy of the reference system of hard spheres. Thus

$$S - S_{\text{id}} = S(\eta) \tag{7}$$

where S_{id} is the entropy of the ideal gas, while $S(\eta)$ is only a function of the packing fraction. Figure 2 shows how $S(\eta)$ (in units of R, the gas constant) for liquid Na, K and

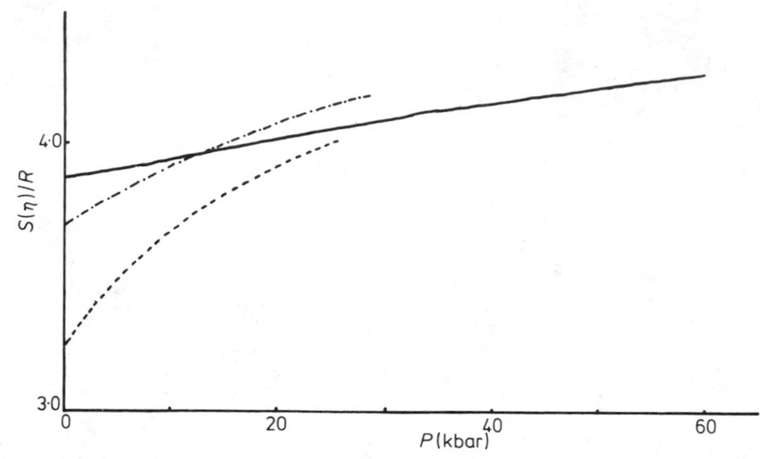

Figure 2. Calculated values for the nonideal part of the entropies of liquid sodium, potassium and rubidium along the melting curve: ——— Na; —·—·—·— K; ------ Rb.

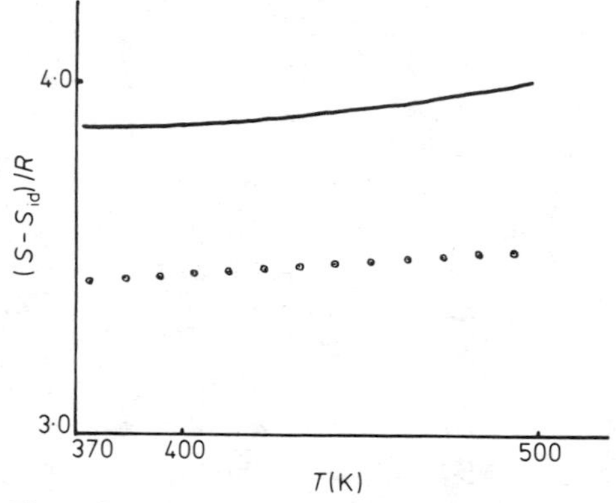

Figure 3. Calculated and experimental values of the nonideal part of the entropy of liquid sodium along the melting curve: ——— theory; ⊙ experiment (Stishov 1975).

Rb varies with pressure along the melting curve. Figure 3 gives the variation of $S(\eta)$ against T, again along the melting curve, for liquid Na and the results are compared with the experimental values for $S-S_{\mathrm{id}}$ reported by Stishov (1975). I have also calculated the melting point isochore which gives $\partial P/\partial T\,|_{\rho}$. The values are given in table 2 and appear to agree well with the experimental values of Endo (1963).

4. Discussion

More as a matter of convenience than principle, I divide the discussion of results into two parts. First I shall consider the $P-T$ diagrams and finally the entropy calculations.

It is clear from figure 1 that only the calculations for liquid Na are in good agreement with experimental results. Calculations for liquid K and Rb, while giving satisfactory agreement with experimental results up to 25–30 kbar are completely at variance with them at higher pressures. In order to understand these results it is important to realize what may be expected from assuming a local pseudopotential in evaluating the effec-

Table 2. Calculated and experimental values of the pressure coefficient at the melting point. The experimental values are those reported by Endo (1963).

Metal	$(\partial P/\partial T)_\rho$(bar K^{-1})	
	This work	Experimental
Na	15·6	13·7
K	8·3	7·8
Rb	7·1	6·9

tive interionic forces. This problem is discussed in a recent review by Stishov (1975) and, restricting ourselves to the $P-T$ diagram, it states that the thermodynamics of melting (in the limit of high pressures) is determined by

$$P \propto T^2. \tag{8}$$

This is, in fact, the type of relation I have obtained not only for Na but for K and Rb as well. However, the experimental curve for K flattens markedly above 50 kbar and essentially no increase in melting temperature is seen between 60 and 80 kbar. Presumably it indicates the onset of a maximum in the melting temperature analogous to that known to exist in Cs (Kennedy *et al* 1962, Stishov 1975).

Similar features are observed in Rb, for which experimental results published by Brundy (1959) show a sharp maximum in melting temperature is the region of 35 to 40 kbar. Irrespective of where the onset of a maximum in temperature takes place, it does in fact indicate that there have been some important changes in the electronic structure of the system under consideration which are not being taken into account by such *ab initio* calculations as are presented in this work. The problem undoubtedly lies with the pseudopotential used. In fact, the heavier alkalis are known to have non-local pseudopotentials as the pressure is increased (Dickey *et al* 1967).

We now turn to the entropy calculations shown in figures 2 and 3, which I would like to discuss within the context of Lindemann's melting criterion. It can be shown (Stishov 1975, and references therein) that for a system of particles with repulsive interaction $u(r) \sim 1/r^n$, the Lindemann melting criterion in the liquid state entails the constancy of the nonideal part, $S-S_{id}$, of the entropy and that the structure factor along the melting curve remains unchanged.

Within the formalism used in this paper, the above two results are equivalent to the requirement that the packing fraction should be essentially constant along the melting curve. In fact, this point was already made by Stroud and Ashcroft (1972) for Na. Here, I merely confirm their conclusions for this metal and figure 3 compares $S-S_{id}$ with the experimental work reported by Stishov (1975). If the quantitative difference between the experimental and theoretical results is ignored, which is presumably due to the softness of the repulsive part of the effective potential (Silbert *et al* 1975), then it is clear that the present calculations qualitatively follow the experimental results. In terms of the packing fraction it means that η varies from 0·44 to 0·46 over a range of 60 kbar. Figure 2 shows that the situation is very different for liquid K and Rb, where $S(\eta)$ for these liquids is seen to rise more steeply with pressure than Na, largely confirming the remarks made in connection with the $P-T$ diagram.

In conclusion, if we are to define a simple liquid metal as a metallic system whose thermodynamic properties along the melting curve follow a relation such as in equation (8) (i.e. such that it may be characterized in the $P-T$ diagram as a system whose temperature increases indefinitely on compression), then amongst the liquid alkali metals studied in this work, only Na satisfies the above criterion over the whole range that is experimentally known to date. Moreover, the present calculations suggest that 25–30 kbar is an upper limit below which thermodynamic properties for liquid K and Rb may be evaluated within the formalism used here.

Acknowledgments

I am very grateful to Professor W H Young for valuable discussions while this work was being carried out, to Professor M Watabe for useful comments and to Dr R G Ross for his help in connection with experimental information.

References

Ashcroft N W 1966 *Phys. Lett.* **23** 48–50
Brundy F P 1959 *Phys. Rev.* **115** 274–7
Cohen M L and Heine V 1970 *Solid St. Phys.* **24** 37–248 (New York: Academic Press)
Dickey J M, Meyer A and Young W H 1967 *Proc. Phys. Soc.* **92** 460–75
Edwards D J and Jarzynsky J 1972 *J. Phys. C: Solid St. Phys.* **5** 1745–56
Endo H 1963 *Phil. Mag.* **8** 1403–15
Geldart D J W and Vosko S H 1966 *Can. J. Phys.* **44** 2137–71
Hasegawa M and Watabe M 1974 *J. Phys. Soc. Japan* **36** 1510–5
Isihara A 1968 *J. Phys. A: Gen. Phys.* **1** 539–43
Jones H 1973 *Phys. Rev.* **A8** 3215–26
Kennedy G C, Jayaraman A and Newton R C 1962 *Phys. Rev.* **126** 1363–6
Kraut E A and Kennedy G C 1966 *Phys. Rev.* **151** 668–75
Luedemann H D and Kennedy G C 1968 *J. Geophys. Res.* **73** 2795–805
Nozières P and Pines D 1958 *Phys. Rev.* **111** 442–54
Silbert M, Umar J H, Watabe M and Young W H 1975 *J. Phys. F: Metal Phys.* **5** 1262–76
Stishov S M 1975 *Sov. Phys.– Usp.* **17** 625–43
Stroud D and Ashcroft N W 1972 *Phys. Rev.* **5** 371–83
Watabe M and Young W H 1974 *J. Phys. F: Metal Phys.* **4** L29–31
Wertheim M S 1964 *J. Math. Phys.* **8** 927–51

Equation of state of liquid alkali metals: sodium, potassium and caesium

I N Makarenko, A M Nikolaenko and S M Stishov
Institute of Crystallography, Academy of Sciences of USSR, Moscow, USSR

Abstract. The volume measurements of liquid and solid Na, K and Cs have been performed with an accuracy of 0·1–0·2% at pressures up to 22 kbar and temperatures from the melting point of the metal at atmospheric pressure to 220 °C. A piston piezometer method was used. In the stated *P–T* region, volume data for alkali metals are obtained for the first time. On the basis of the experimental data the compressibilities of the metals in the liquid and solid states are calculated and the behaviour is analysed in detail.

Since the equation of state is one of the most fundamental characteristics of condensed matter, it is determined by atomic properties as well as by statistical material properties. It is not surprising that a theoretical calculation of the equation of state is a very difficult task which has not yet been solved completely, even for the simplest systems such as the noble gases. In the case of a metal this problem is complicated by its essentially quantum character and the necessity to take account of many various interactions. The recently developed pseudopotential method has provided for the calculation of various properties of solid non-transition metals, including *P–V* and *P–V–T* relations (Ashcroft and Langreth 1967, Brovman *et al* 1970, Stroud and Ashcroft 1972). It has turned out also that the pseudopotential approach is flexible enough to be applied to the liquid state (Stroud and Ashcroft 1972, Price 1971, Hasegawa and Watabe 1972), but the theoretical determination of the liquid metal equation of state requires a knowledge of the ionic correlation functions and these can not be reliably calculated at present. Therefore precise measurements of the *P–V–T* properties of metals seem to be necessary both for checking the reliability of the calculations of cohesion energy and of the equation of state and for testing various approximations used for determination of the liquid metal structure.

This paper reports the volume measurements of liquid and solid sodium and caesium at high pressure up to 22 kbar and potassium up to 14 kbar. The measurements were carried out using the piston piezometer technique (Makarenko *et al* 1974, 1975), which is illustrated by figures 1 and 2. As one can see from the figures, a piston piezometer consists of a cylindrical vessel supplied with a sliding piston and a displacement gauge for the determination of piston position. During the taking of measurements, the piezometer is situated inside a thick-walled high-pressure bomb, in which pressure is created by an outer device. The piezometer piston plays the role of a divider, separating the metal under study from the liquid pressure medium, and serves as an indicator of the level of metal. Details of the technique are described in the cited literature. The accuracy of volume data obtained was better than ±0·2%. As an example, the typical isotherms of compression of sodium in the region of melting are shown in figure 3.

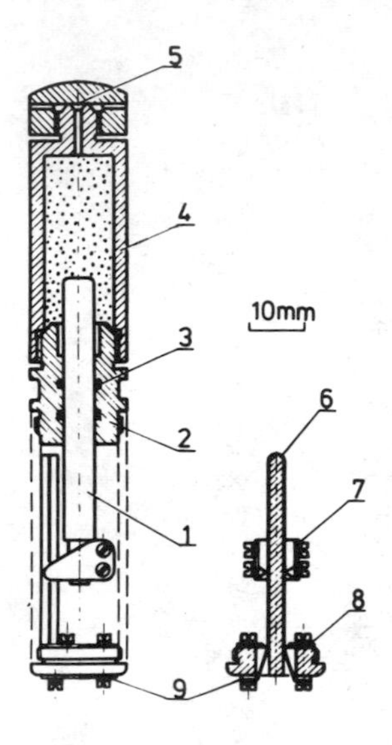

Figure 1. Piezometer for measuring the volume of liquid and solid sodium: 1, 2, piston pair; 3, viscous grease; 4, piezometer body; 5, orifice for filling; 6, constantan wire; 7, mobile potential contacts; 8, fixed potential contacts; 9, current contacts.

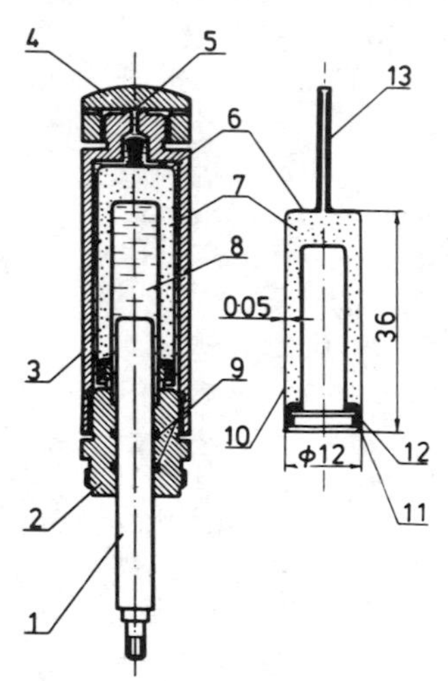

Figure 2. Piezometer for measuring the volume of potassium and caesium in liquid and solid states (Makarenko *et al* 1974). This piezometer is analogous in many respects to the piezometer for measuring the volume of sodium (see figure 1), but differs from the latter by the presence of an inner hermetic ampoule, containing metal. The space between the walls of the piezometer cavity and the ampoule is filled with mercury. To calculate the volume of the substance being investigated at high pressures it is necessary to know the equation of state of mercury: 1, 2, piston pair; 3, casing; 4, sealing nut; 5, orifice for filling the piezometer with mercury; 6, ampoule containing metal; 8, mercury; 9, viscous grease; 10, deformable part of the ampoule; 11, welded joint; 12, base of the ampoule; 13, orifice for filling the ampoule.

To determine the absolute magnitudes of volume, we used atmospheric pressure data given by Gol'tsova (1966) for Na, Stokes (1966) for K and Basin (1970) for Cs. The experimental data both for liquid and solid metal were approximated by the polynomials $P = \Sigma_{ij} V^{-i} T^{j}$, which have been used to calculate pressure, compressibility

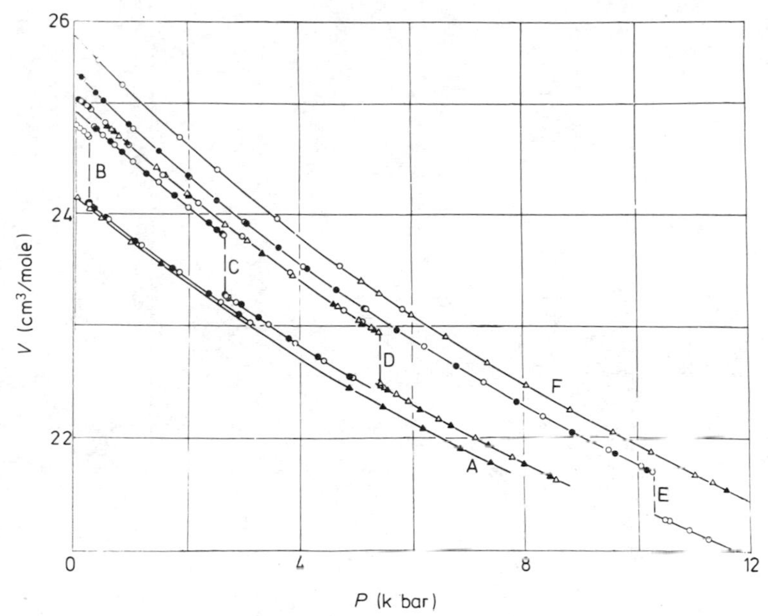

Figure 3. Compression isotherms for sodium: A, 367·4 K; B, 372·5 K; C, 392·5 K; D, 412·3 K; E, 442·5 K; F, 492·6 K.

and other thermodynamic functions (see tables 1–6). Here we shall restrict our discussion to the isothermal compressibility $\beta_T = -(1/V)(\partial V/\partial P)_T$. As one can see from figures 4, 5 and 6, β_T for liquid and solid phases, firstly, differs by not more than 2% for Na and less than 1% for K and Cs, and secondly, varies very weakly with temperature at constant volume. The latter, in general, is not unexpected for solids (Swenson 1970), but in the case of liquids the temperature dependence of compressibility, as a rule, is more manifest.

It is very interesting that these results can be described satisfactorily within the limits of the uniform electron distribution model. This could be seen using the results of numerical studies on thermodynamic properties of a classical one-component plasma (Hansen 1973, Pollock and Hansen 1973). At first sight such a fact does not seem to agree with well known results of pseudopotential calculations. For example, the results of Brovman *et al* (1970) show that the second-order term of the electron energy, which describes the inhomogeneity of the electron gas, gives a contribution to the compressibility of sodium of the order of 25%. However, since the second-order term includes the ionic structure factor, one could expect the difference in compressibilities of liquid and solid phases to be greater than the value that is determined by experiment.

This suggests the idea that the higher-order contributions to the compressibility of sodium are given only by the long-range component of the effective ion–ion potential. In this connection we should remember that in second-order perturbation theory, commonly used in pseudopotential calculations, the indirect ion–ion interaction can be described by introducing the effective pair potential, the parameters of which, however, depend on density and, in part, on temperature.

Separating the long-range, constant components of the effective potential, we can write its contribution to the energy of the system as α/V, where α is a constant and

Table 1. Pressure P (kbar) of liquid (l) and solid (s) sodium.

V_l	T (K)							V_s
(cm^3/mole)	493·15	473·15	453·15	433·15	413·15	393·15	373·15	(cm^3/mole)
19·5	22·642	—	—	14·181	13·914	13·646	13·378	20·5
20·0	19·459	19·166	—	11·659	11·392	11·126	10·859	21·0
20·5	16·620	16·332	—	9·409	9·143	8·877	8·610	21·5
21·0	14·082	13·798	13·510	—	7·132	6·867	6·602	22·0
21·5	11·809	11·530	11·247	—	5·334	5·069	4·805	22·5
22·0	9·772	9·497	9·218	8·934	—	3·459	3·195	23·0
22·5	7·942	7·671	7·396	7·117	—	—	1·753	23·5
23·0	6·296	6·029	5·758	5·483	5·203	—	0·458	24·0
23·5	4·815	4·552	4·284	4·013	3·736	3·456	—	—
24·0	3·480	3·220	2·957	2·688	2·416	2·139	—	—
24·5	2·276	2·020	1·759	1·494	1·225	0·952	0·674	—
25·0	1·188	0·936	0·679	0·417	0·151	—	—	—
25·5	0·206	—	—	—	—	—	—	—

Table 2. Isothermal compressibility, $-V^{-1}(\partial V/\partial P)_T \times 10^6$ (bar^{-1}), of liquid and solid sodium.

V_l	T (K)							V_s
(cm^3/mole)	493·15	473·15	453·15	433·15	413·15	393·15	373·15	(cm^3/mole)
19·5	7·61	—	—	9·14	9·14	9·14	9·14	20·5
20·0	8·32	8·34	—	10·00	10·00	10·01	10·01	21·0
20·5	9·09	9·11	—	10·94	10·94	10·95	10·95	21·5
21·0	9·92	9·94	9·96	—	11·96	11·97	11·97	22·0
21·5	10·82	10·84	10·86	—	13·07	13·07	13·08	22·5
22·0	11·78	11·80	11·83	11·85	—	14·27	14·28	23·0
22·5	12·81	12·84	12·87	12·90	—	—	15·58	23·5
23·0	13·93	13·97	14·00	14·04	14·07	—	17·00	24·0
23·5	15·14	15·18	15·22	15·26	15·30	15·34	—	—
24·0	16·44	16·49	16·53	16·58	16·62	16·67	—	—
24·5	17·85	17·90	17·95	18·00	18·06	18·11	18·17	—
25·0	19·36	19·42	19·49	19·55	19·61	—	—	—
25·5	21·01	—	—	—	—	—	—	—

Table 3. Pressure P (kbar) of liquid (l) and solid (s) potassium.

V_l	T (K)						V_s
(cm³/mole)	473·15	453·15	413·15	373·15	333·15	293·15	(cm³/mole)
—	—	16·722	16·423	16·125	15·828	—	34·0
—	—	12·317	12·020	11·724	11·427	11·130	36·0
36·0	13·187	—	8·561	8·265	7·969	7·673	38·0
38·0	9·742	9·591	—	5·532	5·237	4·942	40·0
40·0	6·994	6·848	6·549	3·364	3·070	2·774	42·0
42·0	4·799	4·657	4·365	—	1·345	1·050	44·0
44·0	3·045	2·906	2·621	2·327	—	—	—
46·0	1·643	1·506	1·227	0·939	—	—	—
48·0	0·522	0·388	0·115	—	—	—	—

Table 4. Isothermal compressibility, $-V^{-1}(\partial V/\partial P)_T \times 10^6$ (bar^{-1}), of liquid and solid potassium.

V_l	T (K)						V_s
(cm³/mole)	473·15	453·15	413·15	373·15	333·15	293·15	(cm³/mole)
—	—	11·85	11·85	11·85	11·85	—	34·0
—	—	14·29	14·29	14·29	14·30	14·30	36·0
36·0	14·43	—	17·17	17·17	17·18	17·18	38·0
38·0	17·15	17·17	—	20·60	20·60	20·61	40·0
40·0	20·41	20·44	20·51	24·68	24·69	24·70	42·0
42·0	24·32	23·37	24·45	—	29·58	29·59	44·0
44·0	29·03	29·09	29·21	29·33	—	—	—
46·0	34·73	34·81	34·98	35·14	—	—	—
48·0	41·68	41·79	42·01	—	—	—	—

Table 5. Pressure P(kbar) of liquid (l) and solid (s) caesium.

V_l	T (K)						V_s
(cm³/mole)	493·15	473·15	433·15	393·15	353·15	313·15	(cm³/mole)
—	—	—	18·401	18·239	18·076	17·914	44·0
44·0	—	18·661	13·386	13·220	13·053	12·886	48·0
48·0	—	13·864	—	9·322	9·152	8·981	52·0
52·0	10·095	10·032	9·899	6·344	6·170	5·997	56·0
56·0	7·135	7·066	6·920	—	3·902	3·725	60·0
60·0	4·871	4·797	4·644	4·479	—	1·994	64·0
64·0	3·146	3·070	2·912	2·743	2·564	0·671	68·0
68·0	1·831	1·754	1·595	1·424	1·244	—	—
72·0	0·829	0·752	0·592	0·423	0·243	0·054	—
76·0	0·065	—	—	—	—	—	—

Table 6. Isothermal compressibility, $-V^{-1}(\partial V/\partial P)_T \times 10^6$ (bar^{-1}), of liquid and solid caesium.

V_l	T (K)						V_s
(cm³/mole)	493·15	473·15	433·15	393·15	353·15	313·15	(cm³/mole)
—	—	—	16·2	16·2	16·1	16·1	44·0
44·0	17·3	17·3	18·9	18·9	18·8	18·8	48·0
48·0	19·4	19·4	—	22·6	22·6	22·5	52·0
52·0	22·8	22·8	22·7	27·5	27·5	27·5	56·0
56·0	27·6	27·6	27·5	—	33·8	33·8	60·0
60·0	33·9	33·8	33·7	33·5	—	40·5	64·0
64·0	41·6	41·6	41·5	41·4	41·4	51·0	68·0
68·0	51·4	51·4	51·4	51·4	51·3	—	—
72·0	63·6	63·7	63·6	63·8	63·9	64·0	—
76·0	79·1	—	—	—	—	—	—

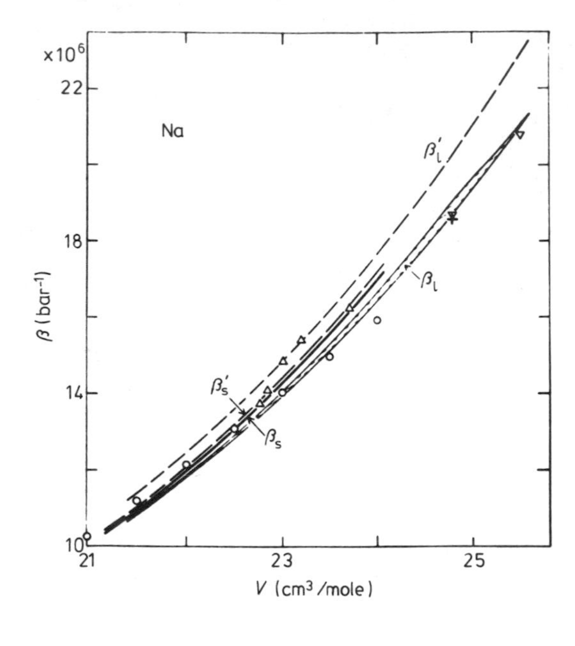

Figure 4. Compressibilities of liquid (β_l) and solid (β_s) sodium as functions of molar volume. Liquid: ▽ (Pasternak 1968/69), $T = 373-473$ K; + (Webber and Stephens 1968), $T = T_m$. Solid: ○ (Beecroft and Swenson 1961, Swenson 1966), 20 K$<T<T_m$ <350 K; △ (Diederich and Trivisonno 1966), 78 K$<T<$300 K. For notation β', see text.

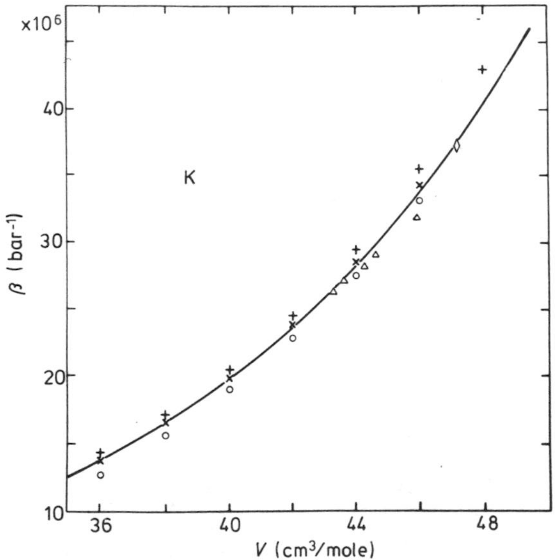

Figure 5. Compressibilities of liquid (β_l) and solid (β_s) potassium as functions of molar volume. Our data on β_l and β_s are shown by a unique solid line as they very weakly depend on temperature and practically coincide. Liquid: ◇ (Webber and Stephens 1968), $T = T_m$. Solid: ○ (Monfort and Swenson 1965), $T = 20-320$ K; △ (Marquardt and Trivisonno 1965), $T = 4{\cdot}2-295$ K. Values of β'_l (+) and β'_s (X) are also given.

V is the volume. (Such a result in the case of long-range potentials dates back to the time of Van der Waals.). It should be noted here that the functional form of the part of the energy defined by the long-range average component of the ion–ion potential is essentially the same as in the case of the first-order term of electron energy. As for the residual, non-constant component of the effective potential, we shall consider the contribution to the compressibility of the metal due to this component to be equal to zero. To verify these ideas we shall use the present experimental results on the equation of state for sodium, potassium and caesium.

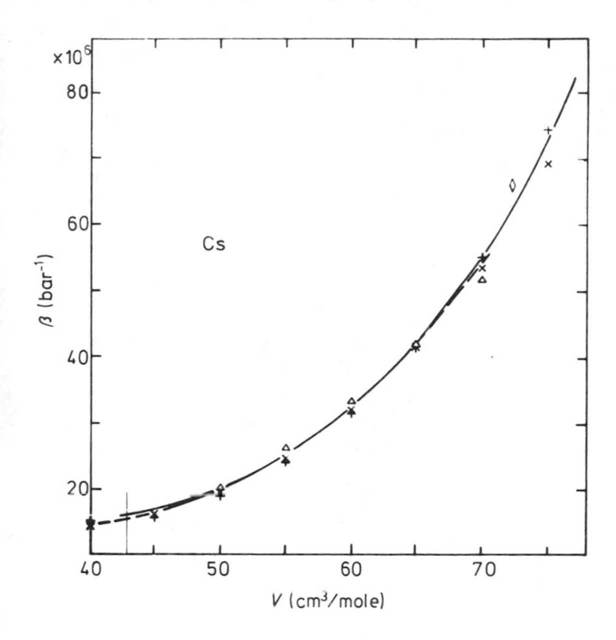

Figure 6. Compressibilities of liquid (β_l) and solid (β_s) caesium as functions of molar volume. Liquid: ——— our data; ◇ (Webber and Stephens 1968), $T - T_m$. Solid: ------- our data; △ (Anderson *et al* 1969), $T = 20-280$ K. Values of β_l' (+) and β_s' (×) are also given.

To the second order of perturbation theory for the pressure of a non-transition metal, we have

$$P = P_i + P_e^0 + P_e^1 + P_e^2, \tag{1}$$

where P_i is the pressure of a system of point ions in a neutralizing background, P_e^0 is the pressure of an interacting electron gas and P_e^1 and P_e^2 are the pressures due to the first- and second-order terms in an expansion of electron energy in the pseudopotential.

Equation (1) is rewritten in the form:

$$P - P_i - P_e^0 = C/V^2 + \delta P_e^2. \tag{2}$$

Here C/V^2 is the sum of the contributions to pressure from the first-order term and the long-range component of the ion–ion interaction, which is a part of the second-order term. C is a constant. The remainder of the pressure due to the second-order term is designated as δP_e^2.

In accordance with our ideas, the remainder δP_e^2 could depend only on temperature, and not on volume. Then, $\Delta P = P - P_i - P_e^0$, where P stands for the experimental pressure, in a general case should yield a family of straight parallel lines, when plotted against V^{-2}, corresponding to different isotherms of compression.

Such a representation of experimental data for Na, K and Cs is illustrated in figure 7. The values of P_i were calculated using the expressions of Hansen (1973) and Pollock and Hansen (1973) for a classical system of point charges in a neutralizing background. To calculate P_e^0, we used the well known expression for the energy of an interacting electron gas with the correlation energy obtained by Nozieres and Pines (1958). In practice, the treatment of experimental data was carried out by choosing a value of C in equation (2), to yield the least mean square root deviation of the experimental values of compressibilities from the calculated values.

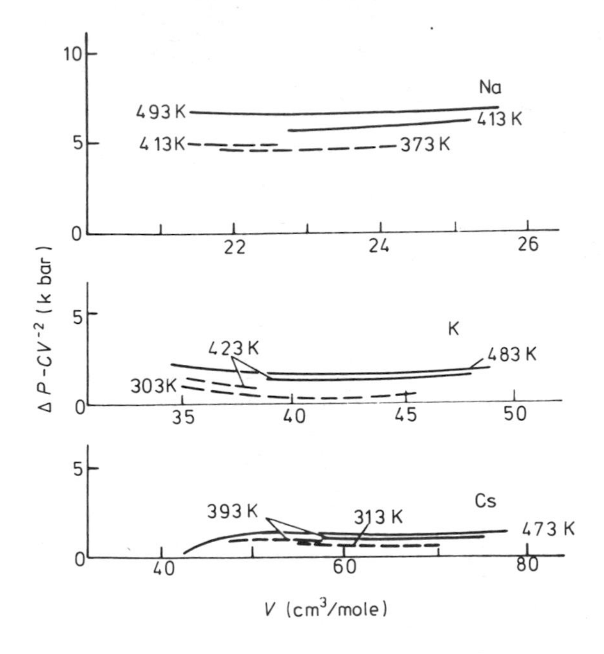

Figure 7. A plot of $\Delta P - C/V^1$ against molar volume V for the liquid (——) and the solid (------) phases of sodium, potassium and caesium. $\Delta P = P - P_i - P_e$, where the variables are explained in the text. For Na, K and Cs, $C = 4{\cdot}263 \times 10^7$, $7{\cdot}736 \times 10^7$ and $11{\cdot}088 \times 10^7$ bar cm^6 mole^{-2}, respectively.

As one can see from figure 7, the above ideas indeed seem to be true. The dependences $\Delta P(V^{-2})$ at constant temperature are linear within the limit of errors of compressibility determination.† Moreover, the coefficient C turns out to be the same for both the liquid and solid phases of the metals under investigation, which also supports our ideas. It should be added here that the deviation of $\Delta P(V^{-2})$ (see figure 7) from a straight line is of the same character for all the metals. But the accuracy of our measurements and some theoretical results is not high enough to discuss this point in detail.

We should like to note that the specific functional form of the second-order term, or more precisely, of the sum of the higher-order terms in the electron energy of the system, does not allow the pseudopotential parameters to be uniquely fitted to the macroscopic characteristics of the metal. At the same time, this feature of higher-order terms evidently does provide for an excellent agreement between the experimental and calculated values of the compressibility, which is demonstrated in papers where the fitting of pseudopotential parameters is used.

Another interesting fact that follows from the experimental equation of state is that the temperature dependence of compressibility for liquid and solid metals is defined only by the entropy part of the free energy, and the energy part of compressibility $(\beta')^{-1} = [\beta_T^{-1} + TV(\partial^2 S/\partial V^2)]^{-1}$ turns out to be temperature independent (see figures 4, 5 and 6). The differences in β' between liquid and solid metals is also small and makes up in average about 3%. It should be noted that β' for crystalline metals is constantly lower than the values for liquids. We would expect such a behaviour of β' by taking into account the possible differences in values of ionic energies for solid and liquid metals (Stroud and Ashcroft 1972). Obviously this result supports our idea that

† The bend of $\Delta P(V^{-2})$ for liquid Cs at $T = 473$ K and volumes up to 50 cm^3/mole probably is connected with s–d electron transition (Sternheimer 1950).

the contribution of the higher-order terms to the compressibility of alkali metals does not depend on ion positions.

In conclusion we would like to emphasize that many static properties of liquid alkali metals are very close to the corresponding ones of solids and this fact needs a careful theoretical analysis.

References

Anderson M S, Gutman E J, Packard J R and Swenson C A 1969 *J. Phys. Chem. Solids* **30** 1587
Ashcroft N W and Langreth D C 1967 *Phys. Rev.* **155** 682
Basin A S 1970 *Thermophysical Properties of Substances* (Novosibirsk: Nauka) pp60–80
Beecroft R I and Swenson C A 1961 *J. Phys. Chem. Solids* **18** 329
Brovman E G, Kagan Yu and Kholas A 1970 *Fiz. Tverd. Tela* **12** 1001
Diederich M E and Trivisonno J 1966 *J. Phys. Chem. Solids* **27** 637
Gol'tsova E I 1966 *Teplofiz. Vys. Temp.* **4** 364
Hansen J P 1973 *Phys. Rev.* **A8** 3096
Hasegawa M and Watabe M 1972 *J. Phys. Soc. Japan* **32** 14
Makarenko I N, Ivanov V A and Stishov S M 1974 *Prib. Tekh. Eksp.* **2** 223
Makarenko I N, Nikolaenko A M, Ivanov V A and Stishov S M 1975 *Sov. Phys.–JETP* **69** 1724
Marquardt W R and Trivisonno J 1965 *J. Phys. Chem. Solids* **26** 273
Monfort C E and Swenson C A 1965 *J. Phys. Chem. Solids* **26** 291
Nozieres P and Pines D 1958 *Phys. Rev.* **111** 442
Pasternak A D 1968/69 *Mater. Sci. Eng.* **3** 65
Pollock E L and Hansen J P 1973 *Phys. Rev.* **A8** 3110
Price D L 1971 *Phys. Rev.* **A4** 358
Sternheimer R 1950 *Phys. Rev.* **78** 235
Stokes R H 1966 *J. Phys. Chem. Solids* **27** 51
Stroud D and Ashcroft N W 1972 *Phys. Rev.* **B5** 371
Swenson C A 1966 *J. Phys. Chem. Solids* **27** 33
—— 1970 *Les proprietes physiques des solides sous pression* (Paris: CNRS) p27
Webber G M B and Stephens R W B 1968 *Physical Acoustics* vol 4B ed W P Mason (New York: Academic Press) ch 2

On anomalous entropies of fusion

Göran Grimvall
Institute of Theoretical Physics, Chalmers University of Technology, Fack, S-402 20 Göteborg 5, Sweden

Abstract. Most elements follow Richard's rule and have an entropy of fusion ΔS_f close to k_B per atom. Silicon and germanium are exceptions with $\Delta S_f \sim 3{\cdot}6\ k_B$/atom. This anomaly is mainly due to the higher vibrational entropy in the metallic liquid state. Some elements with a high ΔS_f also have a solid lattice structure of low symmetry, but this is not by itself a cause of deviations from Richard's rule. Iron, cobalt and nickel all seem to have a considerable entropy in their solid paramagnetic phases which is due to local spin disorder. Since their entropy of fusion is not anomalously low, there is thermodynamic evidence for the existence of spin disorder also in the liquid phases.

1. Introduction

Richard's rule states that the entropy of fusion ΔS_f for the elements is approximately equal to k_B (Boltzmann's constant) per atom. Table 1 shows that there are some notable exceptions. The origin of these irregularities has been a matter of constant interest (see for instance Cusack and Enderby (1960), Chakraverty (1969), Glazov *et al* (1969), Van Vechten (1973), Regel *et al* (1974) and Lasocka (1975)). In this paper, we consider three different aspects of ΔS_f. The anomalous values of ΔS_f observed for Si and Ge are shown to originate mainly from the increased vibrational entropy in the metallic liquid phase. We next discuss whether a low-symmetry crystal structure gives rise to a tendency for a large vibrational component in ΔS_f. Finally, the fact that Fe, Co and Ni obey Richard's rule will provide information regarding persistent spin fluctuations in the liquid phases. As in all other treatments of ΔS_f referred to above, we will have to rely mainly on qualitative arguments.

2. The entropy of a normal metallic liquid

At the melting temperature T_m one has $\Delta S_f = \Delta E/T_m$ where ΔE is the energy difference between the solid and the liquid at T_m. A large value of ΔS_f is therefore qualitatively equivalent to a large energy difference ΔE, for instance due to radical changes in the electronic structure on melting. However, studies of ΔE are less informative since then T_m has also to be considered. A direct study of the entropy of the solid and liquid phases will give a better understanding of ΔS_f. We first consider elements that obey Richard's rule. They will be said to have a 'normal' metallic liquid state, with an entropy given by the solid phase entropy $S_s(T = T_m)$ plus approximately k_B per atom. For a (non-magnetic) solid, S_s can be written

$$S_s = S_{vib} + S_{anh} + S_{el} \simeq 3k_B\,[1 - \ln(h\nu/k_B T)] + S_{anh} + S_{el}. \tag{1}$$

Table 1. The entropy of fusion ΔS_f for some elements (Hultgren *et al* 1973). The averages do not include rare earths, actinides and those data that are considered uncertain by Hultgren *et al.* The crystal structures are those just prior to melting.

Element	ΔS_f (k_B/atom)
FCC + HCP, average (10 elements)	1·18 ± 0·11
BCC, average (11 elements)	0·90 ± 0·10
Si	3·61
Ge	3·67
Ga	2·22
Sn(w)	1·67
Sb	2·65
Bi	2·50
Pu (BCC)	0·37
Fe (BCC)	0·92
Co (FCC)	1·10
Ni (FCC)	1·22

This approximate equality results from the high-temperature ($T \gg \theta$) form of the harmonic vibrational entropy, ν being a characteristic phonon frequency and θ the associated Debye temperature. The last two terms in equation (1), giving the anharmonic and the electronic contributions to the entropy, are often approximately linear in T. In that case, they are also numerically equal to the corresponding contributions to the heat capacity C_p at temperature T. The discontinuity in C_p at T_m is usually less than $k_B/2$ per atom. Hence the last two terms in equation (1) are not expected to give rise to any appreciable deviations from Richard's rule. Exceptions may occur for transition metals if the electron density of states $N(E)$ has a sharp peak at the Fermi level or if $N(E_F)$ depends strongly on the crystal structure. Plutonium and several rare earths probably fall in this category (Kmetko and Hill 1976). Iron, cobalt and nickel will be discussed separately.

We now consider ΔS_f for the ideal case of a solid semiconductor which transforms on fusion to a 'normal' metallic liquid. We can estimate the entropy of the liquid from equation (1) for a hypothetical solid metallic phase plus k_B per atom. The Debye temperature of the solid metallic and semiconducting phases are θ_m and θ_s respectively. If we neglect the difference in $S_{anh} + S_{el}$ between the various phases, we find that ΔS_f for the semiconductor should be anomalously high by an amount ΔS_a per atom, with

$$\Delta S_a = 3k_B \ln(\theta_s/\theta_m). \qquad (2)$$

3. Application to the entropy of fusion of some elements

3.1. Silicon and germanium

Silicon and germanium are metallic in their liquid states (Glazov *et al* 1969). The volume decreases on fusion by 10% and 5% respectively, to be compared with the increase by

2–4% observed for most metals (Lasocka 1975). The discontinuity in C_p on fusion is small, although the data are probably rather uncertain (Hultgren *et al* 1973). It is therefore reasonable to apply equation (2) in an attempt to account for the anomalous ΔS_f-values for Si and Ge. We consider hypothetical solid metallic phases, with atomic volumes equal to those observed for Si and Ge at room temperature, but reduced by 13% and 8% respectively. Rather than relying on some elaborate, but yet uncertain, first-principles calculation of the phonon spectrum, we will start from a well known estimate for free-electron-like metals, $c_s^2 \sim (m/M)V_F^2$, (Kittel 1963), which relates the longitudinal sound velocity to the electron and ion masses and to the Fermi velocity. This formula is not used directly, but it will be assumed that the elastic-limit Debye temperature θ_0 scales as $c_s q_d$ where q_d is the radius of the Debye sphere. We thus consider the ratio $\alpha = \theta_0/(c_s q_d)$ with experimental values θ_0 taken from Gschneidner (1973). For the three polyvalent FCC or HCP metals (Al,Tl,Pb) which are nearest to Si and Ge in the periodic table, we get $\alpha = 1{\cdot}00 \pm 0{\cdot}10$ where the average α has been arbitrarily normalized to unity. With the same normalization, the typical free-electron-like metal sodium has $\alpha = 1{\cdot}07$. Let us therefore take $\alpha = 1$ as a reasonable approximation also for hypothetical solid metallic Si and Ge. The Debye temperatures are then predicted as $\theta_m(\text{Si}) = 412$ K and $\theta_m(\text{Ge}) = 221$ K, to be compared with the experimental values $\theta_s(\text{Si}) = 647$ K and $\theta_s(\text{Ge}) = 378$ K for the semiconducting phases. When inserted into equation (2), this gives ΔS_a of the order of $1{\cdot}5\ k_B$ per atom, which is about 60% of the amount by which ΔS_f is anomalously high for Si and Ge. Thus the main reason for the anomalous entropy of fusion is that the metallic liquid has weaker interatomic forces and hence a higher vibrational entropy. If this effect is to account for the full observed $\Delta S_a \simeq 2{\cdot}5\ k_B$/atom, the Debye temperature θ_m has to be lower still by some 40%. We refrain from speculating about this possibility or any other sources of additional anomalous entropy.

3.2. Gallium, white tin, antimony and bismuth

The crystal structures of Ga and Sn(w) are distorted relative to the common high-symmetry lattices, and we first ask the question whether a low lattice symmetry will give rise to a high ΔS_f. Heine and Weaire (1970) and others have discussed in detail how a lowering of the lattice symmetry leads to a lower total energy, if the pseudopotential of the metal happens to have a node in the vicinity of the nearest reciprocal lattice vectors of the undistorted structure. The formulae for the total crystal energy and for the average phonon frequency $\langle\omega^2\rangle$ both involve summations over the reciprocal lattice vectors (cf for instance Heine and Weaire 1970, equation (5.20) and Cohen and Heine 1970, equation (16.11)). Thus a distortion of the lattice which lowers the total energy will also tend to increase the Debye temperature. This is indeed observed experimentally. Cd, the HCP metal with the largest c/a ratio, has $\alpha = 1{\cdot}83$. Further, $\alpha(\text{Ga}) = 1{\cdot}48$ and $\alpha(\text{Sn(w)}) = 1{\cdot}35$. If the liquid structure factor resembles that of fused BCC or close-packed metals, we would expect an anomalously high ΔS_f. However, a reminiscence of the low solid-phase lattice symmetry has been seen in the liquid structure factors of Hg, Ga, Cd, Bi, Sb, Ge and Sn (Heine and Weaire 1970, p456), so a distorted solid lattice is not by itself a reason for an anomalous ΔS_f.

The elements Ga,Sn(w), Sb and Bi show metallic conduction. Still their electronic structure is not entirely free-electron-like but shows some degree of covalency. For instance, covalency has to be included in a calculation of phonon frequencies for Sn(w) (Heine and Weaire 1970, p413). The atomic volume decreases on melting for Ga, Sb and Bi (Lasocka 1975) which is a sign of unusual changes in the electronic structure. This is expected to change the vibrational entropy and thus account for the high ΔS_f values, although the simple approach used for Si and Ge will fail in this case.

3.3. Iron, cobalt and nickel

The thermodynamics of solid iron have been considered in detail elsewhere (Grimvall 1976) and we first briefly review some important results. The heat capacity has a characteristic lambda-shaped peak around the Curie temperature. If we subtract a smooth background heat capacity, the entropy of the magnetic transformation is about 0·6 k_B/atom. In the paramagnetic state, this entropy can be related either to a large and narrow peak in the electron density of states or to persistent but directionally disordered magnetic moments. In the latter case, $N(E)$ is expected to retain much of its shape from the ferromagnetic state with spin-split bands and with no strong peak at the Fermi level. Heat capacity data alone cannot distinguish between these two models for the paramagnetic state, but there is overwhelming recent evidence that the idea of persistent spin fluctuations is correct (Grimvall 1976 and references therein). Iron follows Richard's rule and there is no large discontinuity in C_p on fusion. These facts provide strong thermodynamic evidence for the existence of persistent spin fluctuations also in the liquid phase, and with roughly the same associated entropy. Without any such magnetic entropy in the liquid, ΔS_f would come out as low as approximately 0·3 k_B/atom. (It is very unlikely that the structurally disordered liquid phase has a strong and narrow peak in $N(E)$ at the Fermi level.)

The same argument will now be applied to cobalt. Heat capacity data (Hultgren *et al* 1973) suggest a magnetic entropy of about 0·6 k_B/atom for the FCC phase. No band-structure calculation for FCC Co shows a sharp peak in $N(E)$ at the Fermi level (Ballinger and Marshall 1973, Snow and Waber 1969, Ishida 1972). Thus there is a spin-disorder entropy of the order of 0·6 k_B/atom. The fact that ΔS_f is not anomalously low shows that a magnetic entropy is present also in liquid cobalt.

For nickel the situation is not so clear, since the unsplit paramagnetic $N(E)$ has enough structure to give rise to the observed magnetic entropy, which is of the order of 0·4 k_B/atom (Jones *et al* 1974). Still, it seems most natural to assume that there are persistent spin fluctuations also in solid paramagnetic nickel. The quite normal value of ΔS_f then indicates that such fluctuations are present in liquid nickel too.

4. Conclusions

It has been argued that most of the anomalously high entropy of fusion for Si and Ge can be explained from the weaker restoring forces and hence higher vibrational entropies of the liquid states, which are metallic. A low crystal symmetry in a solid phase tends to be associated with a low vibrational entropy, but the low symmetry is in general reflected also in the liquid structure factor. A low crystal symmetry is therefore not by itself a reason for a high entropy of fusion. Finally, the fact that ΔS_f for paramagnetic Fe, Co and Ni is not unusually small provides thermodynamic evidence for the existence of persistent spin fluctuations also in the liquid phases of these metals. This is an important complement to the sometimes ambiguous interpretations of resistivity (Güntherodt *et al* 1975) or magnetic susceptibility data (Dovgopol *et al* 1974).

References

Ballinger R A and Marshall C A W 1973 *J. Phys. F: Metal Phys.* **3** 735
Chakraverty B K 1969 *J. Phys. Chem. Solids* **30** 454
Cohen M L and Heine V 1970 *Solid State Physics* **24** 37 (New York: Academic Press)
Cusack N and Enderby J E 1960 *Proc. Phys. Soc.* **75** 395
Dovgopol S P, Radovskii I Z and Gel'd P V 1974 *Sov. Phys.-Dokl.* **18** 614
Glazov V M, Chizhevskaya S N and Glagoleva N I 1969 *Liquid semiconductors* (New York: Plenum)
Grimvall G 1976 *Phys. Scr.* **13** 59
Gschneidner K A 1964 *Solid State Physics* **16** 275 (New York: Academic Press)
Güntherodt H-J, Hauser E, Künzi H U and Müller R 1975 *Phys. Lett.* **A 54** 291
Heine V and Weaire D 1970 *Solid State Physics* **24** 249 (New York: Academic Press)
Hultgren R, Desai P D, Hawkins D T, Gleiser M, Kelley K K and Wagman D D 1973 *Selected values of the thermodynamic properties of the elements* (Metals Park Ohio: American Society for Metals)
Ishida S 1972 *J. Phys. Soc. Japan* **33** 369
Jones R W, Knapp G S and Maglic R 1974 *Int. J. Magn.* **6** 63
Kittel C 1963 *Quantum theory of solids* (New York: Wiley) p144
Kmetko E A and Hill H H 1976 *J. Phys. F: Metal Phys.* **6** 1025
Lasocka M 1975 *Phys. Lett.* **A 51** 137
Regel' A R, Glazov V M and Aivazov A A 1974 *Sov. Phys.-Semicond.* **8** 335
Snow E C and Waber J T 1969 *Acta Metall.* **17** 623
Van Vechten J A 1973 *Phys. Rev.* **B7** 1479

Ab initio thermodynamic calculations for liquid alloys

I Yokoyama†, M J Stott‡, I H Umar§ and W H Young†

† School of Mathematics and Physics, University of East Anglia, Norwich, UK
‡ Department of Physics, Queen's University, Kingston, Ontario, Canada
§ State College of Advanced Studies, Kano, Nigeria

Abstract. The Gibbs–Bogoliubov method based on a hard-sphere reference system can be used to calculate the entropies of simple liquid metals and their binary alloys. The results, however, are sensitive to the pseudopotential and density input information. Moreover the efficiency of the perturbation theory is diminished for highly heterovalent systems.

With the above remarks in mind, we have studied every simple binary alloy of which we are aware for which the necessary (density and entropy) data have been measured and where the valence difference between the constituent ions is zero or unity. A search showed that it was possible to assign a single empty core pseudopotential to each ion type so as to build up a rather satisfactory overall description of the entropies for the pure metals and excess entropies for the alloys.

The core radii thus deduced correlate quite satisfactorily with earlier results found by fitting Fermi surface data (though direct use of the latter can produce poor results). They also lead, for the pure metals, to a good description of the internal energies and some account of the compressibilities. The heat of mixing remains, however, an outstanding problem.

1. Introduction

In the spirit of the Born–Oppenheimer approximation, one can use pseudopotential theory to define the Hamiltonian for the ionic motion in a liquid metal or alloy. Then one can apply the Gibbs–Bogoliubov method to the ions so as to determine the thermodynamics. This method was first applied to alloys by Stroud (1973) and the formalism and procedure has been given in full by Umar *et al* (1974a).

In view of the latter paper, we refrain from giving the detailed formulae. Instead, we remark that the aim of this variational method is to approximate the real system by an effective hard-sphere one. The 'best' such reference system is obtained by selecting those diameters which minimize the expression

$$F = F_{eg} + F_1 + F_2 + F_M + \frac{3}{2}kT - TS_{hs} \tag{1}$$

for the Helmholtz free energy per ion‖. Here F_{eg} is a uniform electron-gas contribution in a compensating positive background, F_1 is the first-order electronic term arising from pseudopotentials from purely coulombic forms, F_2 and F_M are respectively the second-order (band structure) and Madelung terms both containing the hard-sphere partial structure factors, $\frac{3}{2}kT$ is the mean kinetic energy per ion and S_{hs} is the entropy expression for a binary hard-sphere mixture.

§ Present address: Department of Physics, Bayero University College, Kano, Nigeria
‖ All thermodynamic quantities are expressed in the appropriate units per ion

Denoting the sphere diameters by σ_1 and σ_2 and the mean volume per ion by Ω, equation (1) is minimized when

$$\left(\frac{\partial F}{\partial \sigma_i}\right)_{\Omega,T} = 0 \qquad (i=1,2). \tag{2}$$

We now assume that this condition applies to equation (1) which thereby becomes the free-energy estimate for the real system.

In view of equation (2), it is readily proved that the entropy of the real system is

$$S = S_{\text{hs}} + S_{\text{elec}}\,. \tag{3}$$

S_{elec} is a small first-order term arising from F_{eg} and F_2 as a result of the departure of the electron gas from complete degeneracy. For simple systems it is sufficiently well described by the free-electron result. S_{hs}, chosen in a form consistent with the accurate results of Mansoori *et al* (1971), is given explicitly in the paper by Young (1977).

Other results may also be obtained from equation (1). For example, the internal energy is given by $U = F + TS$, the pressure by $P = -(\partial F/\partial \Omega)_T$ and the bulk modulus by $B = -\Omega(\partial P/\partial \Omega)_T$. In contrast with the entropy, however, these quantities depend on the rather large and not very well defined F_1 term.

In the detailed calculations on which we report below, the theory of Nozières and Pines (1958) was invoked to describe F_{eg} at $T = 0$ and the Geldart and Vosko (1966) approach to describe the screening used in the definition of F_2.

2. Procedure

If one attempts to use Ashcroft (1966) pseudopotentials with core radii tabulated by Cohen and Heine (1970) some poor results are obtained for the excess entropies of mixing defined by

$$\Delta S_{\text{E}} = S_{\text{alloy}} - c_1 S_1 - c_2 S_2 - S_c. \tag{4}$$

Here the c_i are the atomic concentrations, the S_i the pure metal entropies at the same temperatures and (one atmosphere) pressure and $S_c = -k(c_1 \ln c_1 + c_2 \ln c_2)$ is the ideal mixing term. Furthermore, even if core radii are chosen to fit the S_i exactly, the predictions for ΔS_{E} are, on the whole, unsatisfactory. Consequently a search procedure was conducted by Yokoyama *et al* (1976) to see if any single set of core radii could be found to satisfy (more or less) the observed entropy data.

Now it is known (i) that excess entropies are very sensitive to the volumes of mixing (Umar *et al* 1974b, Umar *et al* 1976) and (ii) that pseudopotential perturbation theory for alloys decreases in efficiency as the valence difference between the ions increases.

Consequently, in Yokoyama *et al* (1976), all 50–50 alloys were studied for which the entropies and volumes have been measured and where the valence differences are not more than unity.

Details of the volume data used as well as of the very long search for suitable pseudopotentials will be given in Yokoyama *et al*; in the present work we simply state the outcome. A single set of core radii were found which yield the results shown later

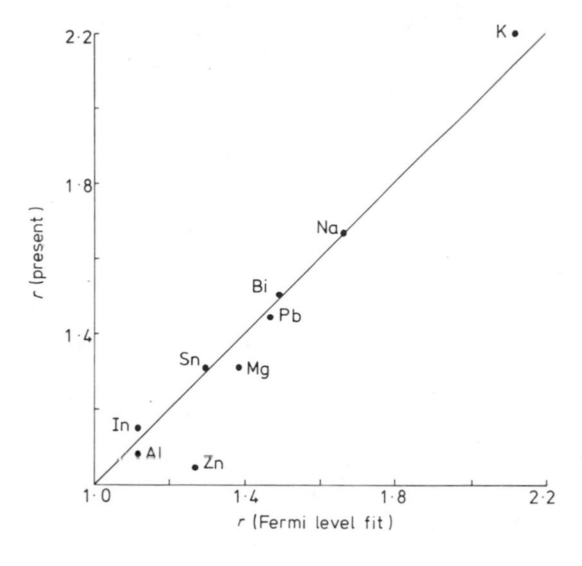

Figure 1. Empty core radii fitted by the present (band average) method versus those fitted to Fermi level data (and tabulated by Cohen and Heine).

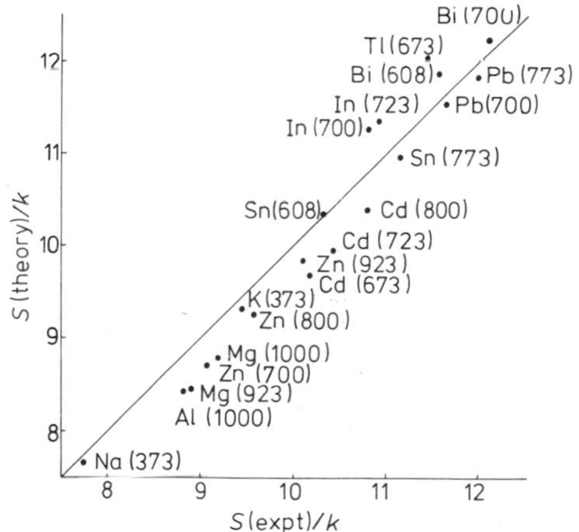

Figure 2. Measured entropies versus those calculated using the data of figure 1.

in figures 2 and 3. Furthermore, only the most minor variations from the chosen radii are possible without producing serious disagreements between theory and experiment.

The comparison between the radii obtained by Yokoyama *et al* and the Fermi level fits listed by Cohen and Heine is shown in figure 1. Only the result for Zn is seriously out of line. If a true pseudopotential is energy dependent, the Ashcroft core radius fitted to Fermi level data will be somewhat different from the present one which is more representative of the whole occupied band. So, perhaps, for Zn, there is enough d-character low in the band to be responsible for the discrepancy indicated.

3. Entropies

3.1. Pure metals

The calculated pure metal entropies shown in figure 2 are in every case within 5% of experiment (a criterion arbitrarily imposed by Yokoyama *et al*). They also tend to lie

Table 1. Heat capacities.

Metal	C_p/k (theory)	C_p/k (expt)
Mg	4·3	3·9
Zn	4·1	3·8
Cd	4·1	3·6
In	3·1	3·5
Sn	2·6	3·6
Pb	2·9	3·7
Bi	2·3	3·7

below the corresponding observed values, in agreement with the conclusions of Young (1977) and of Kumaravadivel and Evans (1977), though the present fitting was performed without regard to such work. The latter authors attribute this discrepancy to an explicit contribution due to non-hard-sphere behaviour.

For a given metal, the locus of points in figure 2 as T varies, will be parallel to the 45° line if the heat capacity is perfectly described and the slope will be greater or less than this depending upon whether theory overestimates or underestimates experiment. A glance at the diagram shows that only for Sn and Bi are the alignments seriously wrong. These, however, are the two cases included in the study where the structure-factor evidence points to significant departures from hard-sphere character. For this reason a simple hard-sphere reference system can be expected to be less satisfactory quantitatively (Young 1977). The numerical results are summarized in table 1.

3.2. *Alloys*

Suppose the (container) volumes and the sphere volumes, before alloying, are Ω_1, Ω_2 and ω_1, ω_2. The mean volumes, again before alloying, are thus $\Omega_0 = c_1\Omega_1 + c_2\Omega_2$ and $\omega_0 = c_1\omega_1 + c_2\omega_2$. Then, if $\Delta\Omega$ and $\Delta\omega$ are the respective changes in Ω_0 and ω_0 on alloying, to an excellent approximation (essentially a first-order expansion of equation (4)),

$$\Delta S_{\mathrm{E}} \simeq \Delta S_{\mathrm{E}_0} + (k - \eta_0 S'_{\eta_0})\frac{\Delta\Omega}{\Omega_0} + \eta_0 S'_{\eta_0}\frac{\Delta\omega}{\omega_0}\,. \tag{5}$$

Here $\eta_0 = \omega_0/\Omega_0$ and likewise ΔS_{E_0} is calculated from equation (4) by using η_0, Ω_0 etc in S_{alloy} instead of η, Ω etc. S_η contributes to the definition of the latter and is of the Carnahan and Starling (1969) form. Equation (5) represents all the points shown in figure 3 to graphical accuracy.

When selecting the pseudopotential parameter (figure 1), that for Tl is subject to fewer constraints than most, since there is only information available for the one alloy, Cd Tl. For this reason, ΔS_{E} could have been improved somewhat by a modified choice of Tl core radius. However, this would have meant a disagreement between the calculated and observed pure metal entropies of more than 5%, so this was not done. The only poor result is for Cd Zn and we have no explanation for this though it will be noted that, of the alloys studied, this seems the most likely case where d-banding might play some role in both constituent metals.

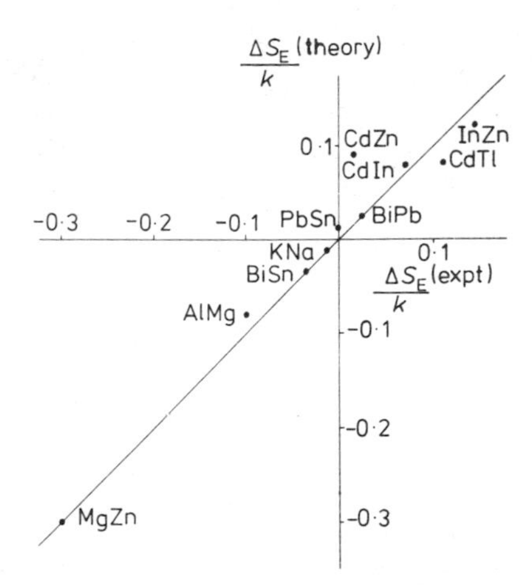

Figure 3. Measured excess entropies versus those calculated using the data of figure 1.

The full details of the parameters calculated for each system will be given by Yokoyama *et al* and we limit ourselves to a couple of general remarks. First, $-\eta_0 S'_{\eta_0}/k$ varies from about 8 to about 6 as η_0 decreases from 0·45 to 0·40 and, in view of equation (5), this describes the sensitivity already noted of ΔS_E to $\Delta\Omega$. Second, the calculations suggest that $\Delta\omega/\omega_0$ is, in general, as important as $\Delta\Omega/\Omega_0$. However, it is probable that the starting point for pseudopotential perturbation theory (a uniform electron gas) leads to an exaggeration in the final result. Perhaps in truth $\Delta\omega/\omega_0$ plays, usually, a non-negligible but nevertheless subsidiary role. This would be more in line with the empirical study of Umar *et al* (1976) which suggested that in most (though not all) cases $\Delta\omega/\omega_0$ can be neglected.

4. Internal energies

Using the core radii indicated in figure 1, we have calculated the pure metal internal energies indicated in figure 4. The pressures corresponding to this parametrization will not be zero, however. Therefore, we also investigated the procedure adopted by Ashcroft and Langreth (1967) for solids and adjusted the zeroth Fourier component

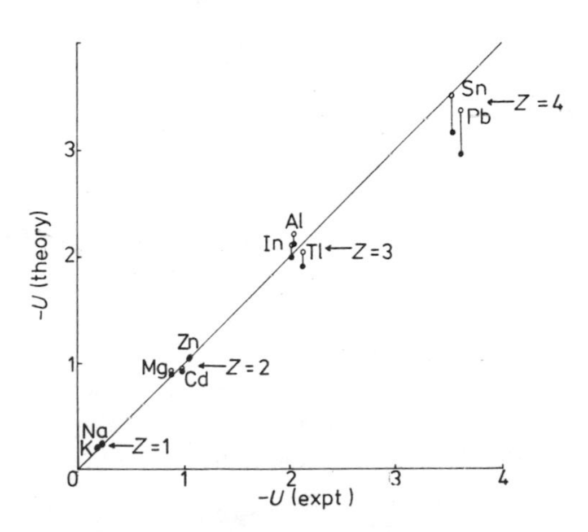

Figure 4. Measured internal energies versus those calculated using the data of figure 1. The open circles denote those calculated without modification of the zeroth Fourier component of the pseudopotentials; the closed circles denote the revised points found by making the adjustments to yield zero pressures.

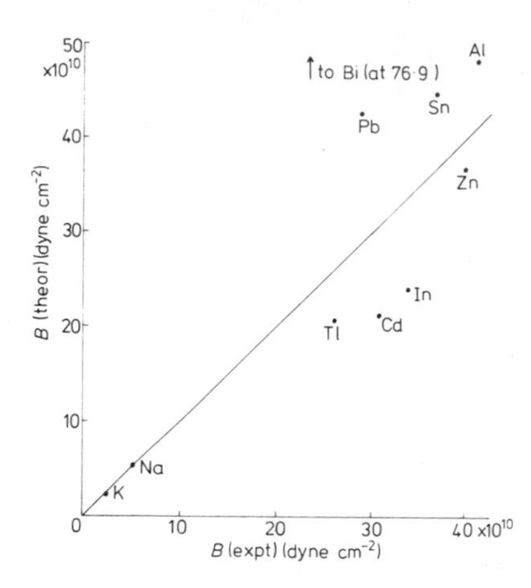

Figure 5. Bulk moduli, B; measured versus calculated. The result for Bi (measured = 24, calculated = 77) is so poor that it is off the graph. The calculated values for Na, K, Mg, Zn, Cd, Al, In, Tl, Sn, Pb and Bi are for 373, 373, 923, 700, 673, 1000, 700, 673, 773, 773 and 700 K respectively whereas the measurements took place near the melting temperatures (not always close to the above). An effective core radius characterizing the zeroth Fourier component enters into the calculation of B. It is $(F_1\,\Omega/2\pi\bar{Z})^{1/2}$ where $\bar{Z}$ is the mean number of free electrons per atom. These values are respectively 1·818, 2·486, 1·420, 1·086, 1·358, 1·182, 1·330, 1·424, 1·548, 1·722 and 1·841 and may be compared with the values shown in figure 1 which are used for defining the nonzero Fourier components.

of the pseudopotential (contained in the definition of F_1) so as to obtain the observed (zero) pressure. The results thereby obtained for the internal energies move to the new points indicated in figure 4. The revised values for Pb and Sn are compatible with the notion that covalency has some role to play in these cases.

With the latter fit, it is possible to examine the bulk moduli and these are indicated in figure 5. As with the heat capacities, the poorer results are for Sn, Pb and Bi, perhaps for much the same reasons. For the other cases, there is some suggestion of a trend.

5. Conclusions

It seems possible to obtain a fairly satisfactory description of entropies for pure metals and for their alloys provided pseudopotential parameters are chosen with sufficient care.

The parameter choice (in F_1) is also critical for describing the pressures but not for the internal energies. Thus a fit to the former produces agreement with the latter for pure metals. The same parameters also provide some description of the bulk moduli. At the present time we have not succeeded in obtaining a convincing picture of the heats of mixing but we are working on the problem.

Acknowledgments

We wish to thank the Science Research Council and the Canada Council both of whom, at various times, have supported this research.

References

Ashcroft N W 1966 *Phys. Lett.* **23** 48
Ashcroft N W and Langreth D C 1967 *Phys. Rev.* **155** 682
Carnahan N F and Starling K E 1969 *J. Chem. Phys.* **51** 635
Cohen M L and Heine V 1970 *Solid State Physics* **23** 186 (New York: Academic Press)
Geldart D J W and Vosko S H 1966 *Can. J. Phys.* **44** 2137

Kumaravadivel R and Evans R 1977 this volume
Mansoori G A, Carnahan N F, Starling K E and Leland T W 1971 *J. Chem. Phys.* **54** 1523
Nozières P and Pines D 1958 *Phys. Rev.* **111** 442
Stroud D 1973 *Phys. Rev.* **B7** 4405
Umar I H, Meyer A, Watabe M and Young W H 1974a *J. Phys. F: Metal Phys.* **4** 1691
Umar I H, Watabe M and Young W H 1974b *Phil. Mag.* **30** 957
Umar I H, Yokoyama I and Young W H 1976 *Phil. Mag.* in the press
Yokoyama I, Meyer A, Stott M J and Young W H 1976 to be published
Young W H 1977 this volume

Interatomic forces in binary alloys

J Hafner

Max-Planck-Institut für Festkörperforschung, D7 Stuttgart 80, Fed. Rep. Germany

Abstract. The description of the interatomic forces in an AB alloy system requires the construction of potentials for A–A, B–B and A–B interactions of each concentration of A in B. This is achieved by using a first-principles pseudopotential theory which is optimized specifically for the case of a binary alloy. In the pure metal limit the resulting pseudopotentials yield an accurate description of a wide range of different properties for the solid as well as for the liquid state. The theory allows one to calculate the interatomic potentials in the binary alkali–metal systems and in the Li–Mg, Li–Al, and Al–Mg systems. These potentials are used to calculate the heat of formation for liquid and solid alloys and for intermetallic compounds.

1. Introduction

After the successful application of the pseudopotential method to the theory of non-transition metals, there is now an increasing interest in the application of this scheme to study the alloying behaviour of these metals. This problem, however, involves additional complications due to contributions from small wavevectors (corresponding to large pseudo potential matrix-elements) which appear because of the violation of translational symmetry in disordered alloys (or because of the complex crystal structure of intermetallic phases).

2. Optimized pseudopotential for binary systems

As is well known the pseudopotential (PP) may be optimized in the sense that the pseudo-wavefunction is smooth and that the perturbation series has optimal convergence (Harrison 1966). Very recently, this optimized first-principles PP scheme has been generalized to binary systems (Hafner 1976b).

Subdividing the electronic eigenstates into a set of tight-binding core states (which are assumed to be identical with the corresponding free-ion states) and a set of quasi-free valence states, we can formulate a generalized orthogonalized plane wave (OPW) for a binary system, given by

$$|\mathbf{k}\rangle - \sum_{j(A)} \sum_{t} |\mathbf{r}_j, At\rangle\langle \mathbf{r}_j, At|\mathbf{k}\rangle - \sum_{j(B)} \sum_{s} |\mathbf{r}_j, Bs\rangle\langle \mathbf{r}_j, Bs|\mathbf{k}\rangle \equiv (1 - P_\mathrm{A} - P_\mathrm{B})|\mathbf{k}\rangle. \qquad (1)$$

Here $|\mathbf{k}\rangle$ stands for a plane wave and $\langle \mathbf{r}|\mathbf{r}_j, At\rangle = \psi_t^\mathrm{A}(\mathbf{r} - \mathbf{r}_j)$ for core states of type A(B), centred at $\mathbf{r}_j$ and labelled by the set of quantum numbers $t(s)$. The last identity serves to define the projection operators. Expanding the valence states in terms of these OPWs, inserting in the Schrödinger equation and re-arranging terms, one obtains a generalized Phillips and Kleinmann (1959) equation for the PP. As in the case of a pure metal, we have to reformulate the Phillips–Kleinmann equation before proceeding to actual calculations. This may be done by applying the optimization criterion of the smoothest possible wavefunction (Cohen and Heine 1961). The procedure widely parallels that for a pure metal (for details see Hafner 1976b). Finally we arrive at the following factorized form of the pseudopotential matrix element

$$\langle \mathbf{k}+\mathbf{q}|W|\mathbf{k}\rangle = S_{\mathrm{A}}(\mathbf{q})\langle \mathbf{k}+\mathbf{q}|w_{\mathrm{A}}|\mathbf{k}\rangle + S_{\mathrm{B}}(\mathbf{q})\langle \mathbf{k}+\mathbf{q}|w_{\mathrm{B}}|\mathbf{k}\rangle \tag{2}$$

with the partial structure factors $S_{\mathrm{A}}, S_{\mathrm{B}}$ describing the spatial arrangement of A and B ions, and with the form factors

$$\langle \mathbf{k}+\mathbf{q}|w_{\mathrm{A}}|\mathbf{k}\rangle = u_{\mathrm{A}}(q) + \sum_t (k^2 + \langle \mathbf{k}|W|\mathbf{k}\rangle - E_t^{\mathrm{A}})\langle \mathbf{k}+\mathbf{q}|0,At\rangle\langle 0,At|\mathbf{k}\rangle \tag{3a}$$

$$\langle \mathbf{k}+\mathbf{q}|w_{\mathrm{B}}|\mathbf{k}\rangle = u_{\mathrm{B}}(q) + \sum_s (k^2 + \langle \mathbf{k}|W|\mathbf{k}\rangle - E_s^{\mathrm{B}})\langle \mathbf{k}+\mathbf{q}|0,Bs\rangle\langle 0,Bs|\mathbf{k}\rangle. \tag{3b}$$

For the diagonal matrix element we obtain

$$\langle \mathbf{k}|W|\mathbf{k}\rangle = (1-c)\langle \mathbf{k}|w_{\mathrm{A}}|\mathbf{k}\rangle + c\langle \mathbf{k}|w_{\mathrm{B}}|\mathbf{k}\rangle \tag{4a}$$

$$\langle \mathbf{k}|w_{\mathrm{A}}|\mathbf{k}\rangle = u_{\mathrm{A}}(0) + (1-\langle \mathbf{k}|P|\mathbf{k}\rangle)^{-1}\sum_t (k^2 + \bar{U} - E_t^{\mathrm{A}})|\langle 0,At|\mathbf{k}\rangle|^2 \tag{4b}$$

$$\langle \mathbf{k}|w_{\mathrm{B}}|\mathbf{k}\rangle = u_{\mathrm{B}}(0) + (1-\langle \mathbf{k}|P|\mathbf{k}\rangle)^{-1}\sum_t (k^2 + \bar{U} - E_s^{\mathrm{B}})|\langle 0,Bs|\mathbf{k}\rangle|^2 \tag{4c}$$

with

$$\bar{U} = (1-c)\,u_{\mathrm{A}}(0) + c\,u_{\mathrm{B}}(0) \tag{5a}$$

and

$$\langle \mathbf{k}|P|\mathbf{k}\rangle = (1-c)\sum_t |\langle 0,At|\mathbf{k}\rangle|^2 + c\sum_s |\langle \mathbf{k}|0,Bs\rangle|^2 \tag{5b}$$

for the concentration averages of the $q = 0$ component of the crystal potentials $U_{\mathrm{A}}, U_{\mathrm{B}}$ and of the projection operators (c is the concentration of the B-component).

In practice, the self-consistent crystal potential is not known. Instead, we have to start with the bare electron–ion potentials U_{A}^0, U_{B}^0, construct a bare PP w^0 by using U_{A}^0 and U_{B}^0 in equations (2) to (5). This PP is then made self-consistent by linearly screening it by a homogeneous electron gas, exchange and correlation corrections being included in the Vashishta and Singwi (1972) form.

The construction of the electron–ion potential and the calculation of the core shifts appropriate to an alloy have been described elsewhere (Hafner 1975, 1976a, b), but the following points should be emphasized: (i) the only disposable parameter in the potential describes the valence–core many-body interactions. It has been demonstrated that its introduction at this stage is natural and necessary. It allows to include the fundamental nonlinearity of the PP in the valence-electron density in a simple way and to give an accurate description of a wide range of different properties (Hafner 1975, 1976a, b, Hafner and Eschrig 1975). (ii) The calculation of the core shift is straightforward in principle, but care must be exercised when very different structures or ordering transitions are considered: the interchange of neighbouring ions may have a non-negligible influence on the core shift, and this can significantly affect the final results. It is important to note that the form factor $\langle \mathbf{k}+\mathbf{q}|w_{\mathrm{A(B)}}|\mathbf{k}\rangle$ of the component A(B) in the alloy is different from that in the pure metal due to (a) the change in the

Fermi level, (*b*) the change in the core shift and (*c*) the change in the average crystal potential $\bar{U}$ and finally (*d*) due to the presence of a second kind of core state which appears in the projection operator *P*. The last point is also important for the definition of the effective valencies Z_A^*, Z_B^*.

3. Interatomic forces and total energy

The calculation of the total energy of the binary system is somewhat lengthy, but straightforward (Hafner 1976b). The structure-dependent parts of the total energy are the Madelung energy of the ions (for its calculation in liquid alloys see Umar *et al* 1974) and the band-structure energy E_{bs} given by

$$E_{bs} = \sum_{i,j} \sum_{\mathbf{q}}{}' S_i^*(\mathbf{q})\, S_j(\mathbf{q})\, F_{ij}(q) \qquad i, j = A, B \tag{6a}$$

with the energy wavenumber characteristics

$$F_{ij}(q) = \frac{2\Omega}{(2\pi)^3} \int_{|\mathbf{k}| \leqslant k_F} \frac{\langle \mathbf{k}+\mathbf{q}|w_i|\mathbf{k}\rangle\langle \mathbf{k}|w_j|\mathbf{k}+\mathbf{q}\rangle}{k^2 - |\mathbf{k}+\mathbf{q}|^2}\, \mathrm{d}^3k - \frac{q^2}{16\pi} \frac{w_i^{\mathrm{scr}}(q)\, w_j^{\mathrm{scr}}(q)}{(1 - G(q))} \tag{6b}$$

which describe an indirect ion–electron–ion interaction. Fourier transforming the F_{ij} and adding the direct Coulomb potentials, we arrive at the effective interionic pair potentials. Again it is very important to emphasize that the interionic potential depends strongly on both components. As an example we present in figure 1 the effective pair potential between two Li ions in pure Li metal and in equiatomic alloys of Li with Na, Mg and Al. We see that the pair potential is drastically changed both in its attractive part and in the oscillatory region.

4. Applications

4.1. *Liquid metals*

Within the past few years a method has been developed for calculating thermodynamic properties of liquid metals and alloys. The technique is based on the hard-sphere

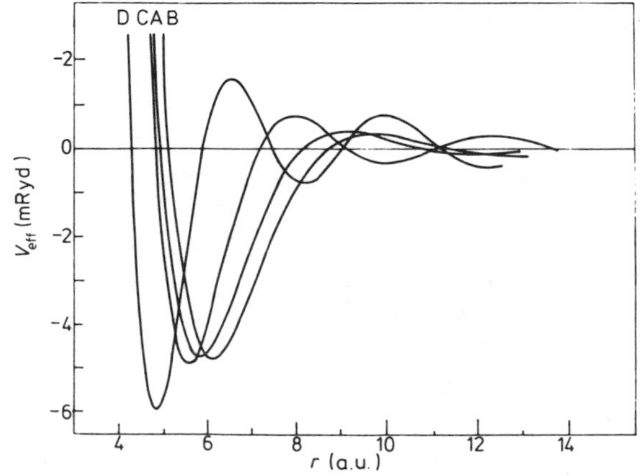

Figure 1. Interionic potential between two lithium ions in pure Li metal (curve A) and in equiatomic alloys of Li with Na (curve B), Mg (curve C) and Al (curve D).

Table 1. Thermodynamic properties near their melting point: r_0 is zero-pressure atomic radius, η is packing density, B_T is isothermal bulk modulus, α is thermal expansion coefficient, and S_E is excess entropy. Experimental values are given in parentheses.†

	T (°C)	r_0 (a.u.)	η‡	B_T (kbar)	α (10^{-4} K^{-1})	$-S_E/k$
Li	180	3·341 (3·318)	0·430 (0·43)	88·7	3·7 (1·73)	3·60 (3·61)
Na	100	4·30 (4·048)	0·365 (0·37)	30·7 (52·3)	3·7 (2·44)	2·56 (3·45)
K	65	5·340 (5·017)	0·358 (0·37)	14·8 (26·2)	4·0 (2·8)	2·40 (3·45)
Mg	665	3·408 (3·460)	0·460 (0·46)	235 (204)§	1·5 (1·75)	4·03 (3·30)
Al	665	2·944 (3·119)	0·490 (0·50)	510 (430)	1·5 (1·16)	4·74 (3·49)

† If not otherwise indicated, experimental values have been taken from the compilations of Hultgren *et al* (1963), Allen (1972) and Webber and Stephens (1968).
‡ Numbers in parentheses are calculated from the pair potential using equation (7).
§ Maier and Steeb (1973).

description of liquids and on the Gibbs–Bogoliubov inequality which provides a variational principle for selecting a best hard-sphere reference system (see for example Umar *et al* 1974, Silbert *et al* 1975 and further references cited therein).

Some representative thermodynamic quantities calculated using this technique and the optimized PP are collected in table 1. The agreement between theory and experiment is reasonably good throughout. An important point to note is that the packing fractions determined by the variational technique and by using the relation

$$V(\sigma) - V_{\rm min} = \tfrac{3}{2}kT \qquad (7)$$

are in very good agreement (σ is the hard-core diameter and $V_{\rm min}$ is the minimum value of the pair potential, Ashcroft and Langreth 1967).

4.2. *Alloys*

4.2.1. Lithium–magnesium. Very successful calculations on solid Li–Mg alloys have been reported in a previous publication (Hafner 1976b). The theory predicts correctly the HCP → BCC → HCP sequence of alloy phases and the observed deviations from Vegard's rule. For the calculation of the ordering energy it turns out to be important to take into account the influence of the local environment on the core shift. This reduces the ordering energy of the CsCl phase to a very reasonable value of $\Delta H = -0{\cdot}92$ mRyd/ion. The change in the coreshift affects both the volume-dependent contribution to the total energy and the pair potentials. Hence it is not possible to describe the ordering process in terms of a simple ordering pair potential.

4.2.2. Lithium–aluminium. A similar calculation has been performed for the Li–Al system. The pair potentials shown in figure 2 demonstrate that an important factor in the alloy formation is the contraction of the more compressible Li ion. However, even with this type of potential, a CsCl type of ordering is always energetically more favourable than the NaTl structure. The Zintl phase is stabilized if the leading third-order terms are taken into account. This corroborates Inglesfield's (1971) ideas on the physical principles governing the Zintl-phase formation.

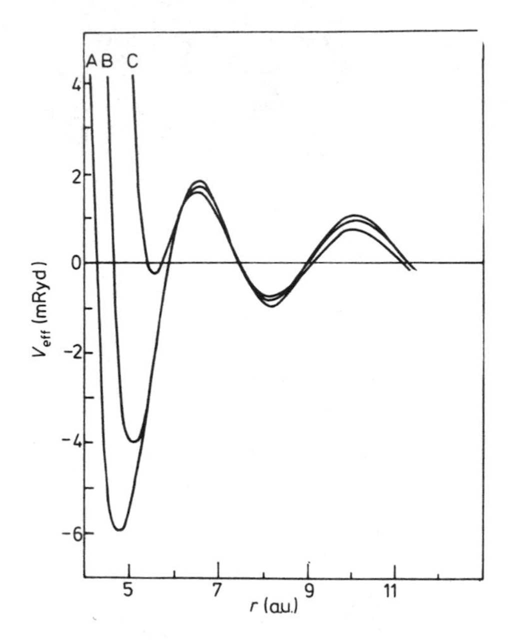

Figure 2. Interionic potentials in an equiatomic Li–Al alloy. Curve A, Li–Li; curve B, Li–Al; curve C, Al–Al.

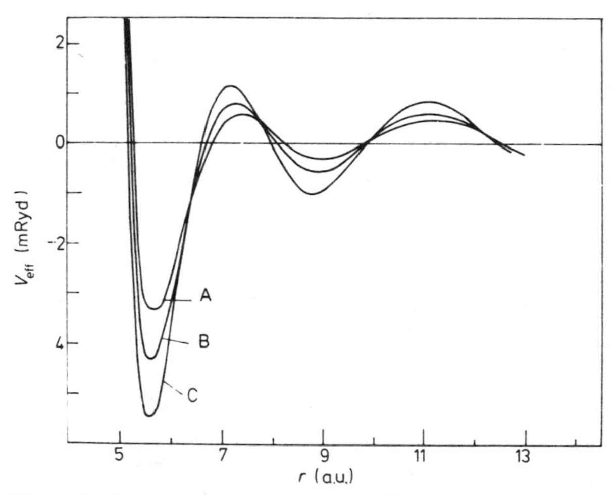

Figure 3. Interionic potentials in an equiatomic Al–Mg alloy. Curve A, Al–Al; curve B, Al–Mg; curve C, Mg–Mg.

Since it is difficult to obtain a reliable estimate of the free energy of Li at temperatures far above the melting point, no determination of the enthalpies of formation of liquid Li–Mg and Li–Al alloys was possible.

4.2.3. Aluminium–magnesium. The pair potentials for an equiatomic Al–Mg alloy are shown in figure 3. Both the Al–Al and the Mg–Mg potentials are quite drastically changed against their shape in the pure metal. From the pair potentials one would

Table 2. Thermodynamic properties of alloys: $\Delta\Omega/\Omega$ is volume change on alloying, ΔH is enthalpy of formation, ΔS is entropy of formation. Experimental values are given in parentheses.

A–B	T (°C)	η	σ_A/σ_B	$\Delta\Omega/\Omega$ (%)	ΔH (cal/g atom)	ΔS (cal/g atom deg.)
Li–Na	200	0·39	0·86	+ 2	860 (>0)	1·80
Na–K	100	0·37	0·86	−1 (−1·2)†	500 (172)¶	1·26 (1·28)§
Al–Mg	665	0·48	0·98	−1·3 (−3)‡	−2660 (−900)§	1·26 (0·90)§

† Faber (1972).
‡ Steeb and Woerner (1965).
§ Hultgren *et al* (1963).
¶ Yokokawa and Kleppa (1964).

expect the ratio of the hard-core diameters σ_{Mg}/σ_{Al} to be nearly equal to unity. Application of the rule $V(\sigma) - V_{min} = \frac{3}{2}kT$ yields $\sigma_{Mg}/\sigma_{Al} = 0{\cdot}98$, in good agreement with the variational calculation. Generally, this rule yields $\sigma_{AB} \simeq 0{\cdot}5(\sigma_A + \sigma_B)$, which is a prerequisite for the hard-sphere description of the alloy systems. The calculated value for the change of enthalpy, entropy and volume on alloying (relative to the theoretical results for the pure metals) are listed in table 2. Full quantitative agreement with experiment was certainly not to be expected, but the semiquantitative agreement is very encouraging.

4.2.4. Alkali–metals. The calculated enthalpies of formation of alkali–metal solid solutions demonstrate the typical behaviour expected from the Hume–Rothery 15% rule. For the K–Rb, Rb–Cs and K–Cs systems, the positive sign and magnitude of ΔH is in reasonably good agreement with the ΔH estimated by Yokokawa and Kleppa (1964) from liquid metal data: K–Rb: ΔH = 156(46) cal/g atom; RbCs: ΔH = 120 (13) cal/g atom; K–Cs: ΔH = 700 (166) cal/g atom. ΔH is somewhat overestimated because the difference in the theoretical densities are somewhat greater than the corresponding experimental differences. The intermetallic phases between alkali–metals (Na_2K, Na_2Cs and K_2Cs with $MgZn_2$-type Laves structure and K_7Cs_6 with a crystal structure closely related to the Fe_7W_6-type μ-phase) have recently been studied by Simon *et al* (1976). Our theory explains the observed structures successfully in terms of a delicate balance between electrostatic and electronic forces. The nearest-neighbour distances in the Laves phase correspond exactly to the minima of the interionic pair potentials of Na_2K (figure 4) and K_2Cs. For the Na_2Cs phase, however, s–d hybridization is necessary to explain the occurrence of the Laves phase. For the same reason, the calculated K_7Cs_6 structure is slightly different from the observed structure (the Cs–Cs distances are enhanced by about 10%). The agreement between the calculated and the experimental enthalpy of formation of Na_2K is surprisingly good: $\Delta H_{theor} = -125$ cal/g atom, $\Delta H_{exp} = -$ 145 cal/g atom. In the liquid state, the alloys Li–Na and Na–K have been investigated. In agreement with experiment, Li–Na is found to be immiscible at temperatures near the melting point of the elements. For the Na–K alloy, the magni-

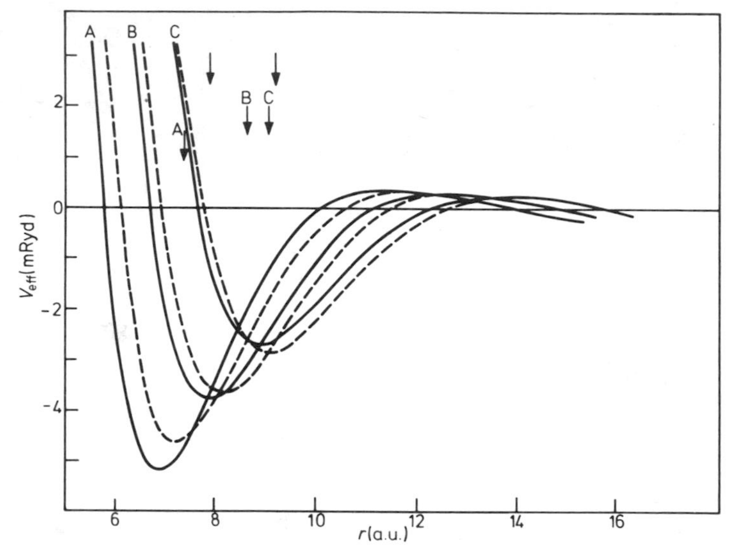

Figure 4. Interionic potentials in Na–K alloys with composition Na_2K (solid lines) and NaK (broken lines). The vertical arrows indicate the nearest-neighbour positions in a BCC solid solution (upper) and in the Laves phase (lower). Curve A, Na–Na; curve B, Na–K; curve C, K–K.

tude of the positive enthalpy of formation is overestimated. This is again connected with the different densities.

5. Conclusions

We have presented an optimized pseudopotential theory for binary alloys of non-transition metals. This method has been employed to construct the interatomic potentials and to investigate the alloying behaviour of these metals, both in the solid and in the liquid state. The interatomic potentials display a behaviour compatible with a hard-sphere description of the alloys. The theory predicts the thermodynamic properties of the alloy in at least semiquantitiative agreement with experiment.

References

Allen B C 1972 *Liquid Metals, Chemistry and Physics* ed S Z Beer (New York: Dekker) p 186
Ashcroft N W and Langreth D C 1967 *Phys. Rev.* **156** 685–91
Cohen M H and Heine V 1961 *Phys. Rev.* **122** 1821–6
Faber T E 1972 *Introduction to the Theory of Liquid Metals* (London: Cambridge UP) p 470
Hafner J 1975 *Z. Phys.* **B22** 351–7
—— 1976a *Z. Phys.* **B24** 41–52
—— 1976b *J. Phys. F: Metal Phys.* **6** 1243–57
Hafner J and Eschrig H 1975 *Phys. St. Solidi* **B72** 179–88
Harrison W A 1966 *Pseudopotentials in the Theory of Metals* (New York: Benjamin)
Hultgren R, Orr R L, Andersen P D and Kelley K K 1963 *Selected Values of Thermodynamic Properties of Metals and Alloys* (New York: Wiley)
Inglesfield I E 1971 *J. Phys. C: Solid St. Phys.* **4** 1003–12

Maier U and Steeb S 1973 *Phys. Condens. Matter* **17** 1–10
Phillips J C and Kleinman L 1959 *Phys. Rev.* **116** 972–80
Silbert M, Umar I H, Watabe M and Young W H 1975 *J. Phys. F: Metal Phys.* **5** 1262–76
Simon A, Brämer W, Hillenköter B and Kullmann H J 1976 *Z. Anorg. Allg. Chem.* **419** 253–73
Steeb S and Woerner S 1965 *Z. Metallk.* **56** 771–5
Umar I H, Meyer A, Watabe M and Young W H 1974 *J. Phys. F: Metal Phys.* **4** 1691–706
Vashishta P and Singwi K S 1972 *Phys. Rev.* **B6** 872–82
Webber G M B and Stephens R W B 1968 *Physical Acoustic* ed W P Mason (New York: Academic Press) vol 4B p 53–97
Yokokawa T and Kleppa O J 1964 *J. Chem. Phys.* **40** 46–54

Temperature dependence of the interference function (structure factor) of liquid metals

C N J Wagner

Materials Department, School of Engineering and Applied Science, University of California, Los Angeles, California 90024, USA

Abstract. The interference functions (structure factors) $I(K)$ have been evaluated from the x-ray scattering patterns of Zn, Cd, Ga and Sn, measured in transmission with Mo or Ag Kα radiation, and corrected for sample and cell absorption, and incoherent and multiple scattering. The temperature of measurement ranged from 450 to 550 °C for Zn, from 350 to 650 °C for Cd, from 25 to 800 °C for Ga, and from 250 to 1100 °C for Sn. The interference functions $I(K)$ of liquid Ga and Sn show the characteristic shoulder on the high K side of the first peak. This shoulder gradually diminishes with increasing temperature, and disappears above 500 °C in liquid Ga, and 1100 °C in liquid Sn. However, the shape of the first peak still remained rather asymmetric even at these temperatures, i.e. the high K side is less steep than the low K side. The opposite asymmetry was found for the first peak of liquid Zn and Cd. The heights of all peaks of $I(K)$ decrease monotonically with increasing temperature. In particular, the ratio of the height of the first peak observed at temperature T (K) to that observed at the melting point T_m follows a universal curve when plotted as a function of T/T_m. Using the Ziman theory, the electrical resistivities ρ_r were calculated from the experimental interference functions $I(K)$ and the Animalu–Heine pseudopotential form factors $U(K)$. Good agreement was found between the calculated and experimental temperature coefficients α_ρ only for liquid Sn. In order to test the accuracy of $I(K)$ at low values of K, the pair-potential $\phi(r)$ was calculated for liquid Cd, and it was found that within the experimental error, $\phi(r)$ was temperature-independent, yielding a minimum value at $r = 3\cdot1$ Å in close agreement with the interatomic distance $r_1 = 3\cdot1$ Å.

1. Introduction

In order to calculate the electrical resistivity and the pair-potential of a liquid metal, the precise knowledge of the atomic distribution is required which can best be described by the atomic distribution function $g(r) = \rho(r)/\rho_0$ where $\rho(r)$ specifies the number of atoms per unit volume at the distance r from a given atom, and ρ_0 is the average atomic density. This distribution function $g(r)$ permits us to evaluate the total correlation function $h(r) = g(r) - 1$ and the direct correlation function

$$c(r) = h(r) - \rho_0 \int c(r - r')\, h(r')\, dr'.$$

The functions $g(r)$, $h(r)$ and $c(r)$ can be deduced from the x-ray or neutron scattering of the liquid. The intensity $I_N(K)$, scattered by N atoms in the liquid of volume V, permits us to calculate the interference function $I(K)$, also called structure factor or structure function, which represents the Fourier transform of the atomic distribution function $g(r)$ (Wagner 1972, Enderby 1972),

$$I_N(K)/(Nf^2) = I(K) = 1 + [(4\pi\rho_0)/K] \int_0^\infty r(g(r) - 1) \sin Kr\, dr \qquad (1)$$

where $K = 4\pi \sin\theta/\lambda$ is the length of the diffraction vector, 2θ being the scattering angle, $\rho_0 = N/V$, and f is the atomic scattering factor.

Thus, the functions $g(r)$, $h(r)$ and $c(r)$ can be calculated from $I(K)$ (Enderby 1972):

$$h(r) = g(r) - 1 = [1/(2\pi^2 r\rho_0)] \int_0^\infty K(I(K) - 1) \sin Kr \, dK \tag{2}$$

and

$$c(r) = [1/(2\pi^2 r\rho_0)] \int_0^\infty [K(I(K) - 1)/I(K)] \sin Kr \, dK. \tag{3}$$

From $g(r)$, we can determine the interatomic distances and the number of neighbours about a given atom in the liquid. As shown by Barker and Gaskell (1975), the direct correlation function $c(r)$ might be related to the pair-potential through the empirical equation

$$\phi(r) = -k_B T c(r) \tag{4}$$

where k_B is the Boltzmann constant and T is the absolute temperature.

The theory of the electronic transport properties as formulated by Ziman (1961), utilizes directly the experimentally accessible interference function $I(K)$. The electrical resistivity ρ_r can be expressed as follows (Halder *et al* 1969);

$$\rho_r = [(\pi^3 \hbar z)/(e^2 k_F)] \langle F(K) \rangle \tag{5}$$

where z is the valence of the metal and k_F is the Fermi wave number. $F(K)$ represents the resistivity integral.

$$\langle F(K) \rangle = 4 \int_0^1 |U'(K)|^2 I(K)(K')^3 \, dK' \tag{6}$$

where $K' = K/(2k_F)$ and $U'(K) = U(K)/(2E_F/3)$, $U(K)$ being the pseudopotential form factors, and E_F the Fermi energy.

In principle, it should be possible to calculate the pair-potential using equation (4), and the electrical resistivity of liquid metals with equations (5) and (6). In practice, however, one finds that exact values of $I(K)$, particularly for small values of K, and $U(K)$ as a function of temperature are only scarcely available.

In this paper we will discuss the experimental results of the temperature dependence of the interference functions of liquid Zn, Cd, Ga and Sn, which have been determined recently in our laboratory. These data have been used to evaluate the electrical resistivities in these metals, and to calculate the pair-potential in liquid Cd.

2. Experimental techniques

In order to determine reliable values of the interference function $I(K)$, particularly at small values of K, the transmission method (North and Wagner 1969a) has been employed. The data on liquid Zn and Cd were obtained with Mo Kα radiation using a monochromator in the primary beam, whereas the data on liquid Ga were obtained with Ag Kα radiation using a silicon solid state detector in conjunction with a narrow energy window. Liquid Sn was measured with both experimental arrangements to ensure that they yield identical results.

Samples were chosen to correspond to a thickness $t = 1/\mu$ (~ 50 to 100 μm) where μ is the linear absorption coefficient, and were held between two thin sheets of Be, pyrolitic graphite, or mica (each 0·12 mm thick). These sandwiches were placed in a stainless steel or graphite cell (25 mm × 20 mm × 6 mm, with a vertical window 5 mm wide and 16 mm high), and then heated in a graphite tube furnace under a He–20%H atmosphere up to 1100 °C.

The measured intensity $I_m(2\theta)$ were corrected for sample holder scattering and absorption, for polarization and absorption in the liquid sample, and then normalized to the scattering $I_a(K)$ from a single atom. Thus, the scattered intensity $I_a(K)$ expressed in electron units is given by

$$I_a(K) = f^2 I(K) + Q(K) I_{inc}(K) + I_{ms}(K).$$

The values of the atomic scattering factors f were taken from Cromer and Waber (1965) and those of the Compton scattering $I_{inc}(K)$ from Cromer and Mann (1967). The multiple scattering $I_{ms}(K)$ was calculated for each sample with a computer program compiled by D M North (unpublished). The resolution function $Q(K)$ was determined experimentally for the solid state detector arrangement using the procedure given by Ruland (1964), and $Q(K) = 1$ when using a monochromator in the primary beam. All calculations, as well as the Fourier transforms $g(r)$ and $c(r)$, were made on the IBM 360/91 computer using a program written in FORTRAN IV.

3. Experimental results

The interference functions $I(K)$ of liquid Sn and Ga are shown in figure 1 for different temperatures. The data for liquid Sn measured at 250 °C agree very well with the neutron data of North *et al* (1968), and the x-ray data of Kaplow *et al* (1966), Waseda and Suzuki (1972), and North and Wagner (1969a). The scatter among the values of $I(K)$ beyond $K = 3$ Å^{-1} is about ±0·02. However, there exists a slight discrepancy in

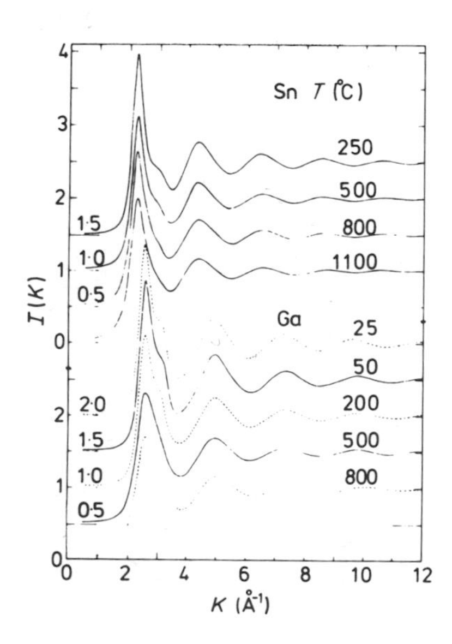

Figure 1. Interference function (structure factor) $I(K)$ of liquid Sn and Ga.

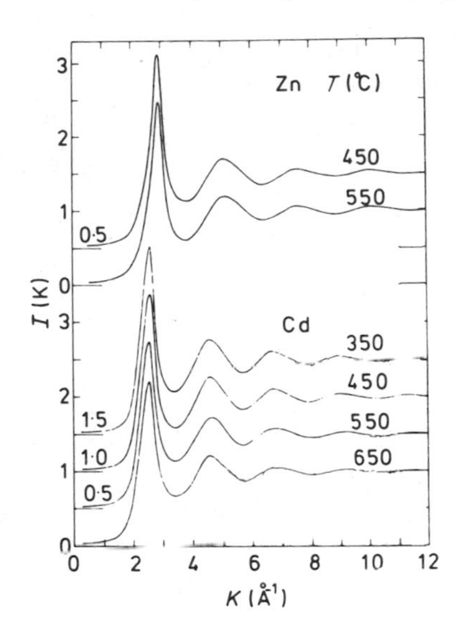

Figure 2. Interference function (structure factor) $I(K)$ of liquid Zn and Cd.

the peak height of the first peak, which varies from 2·35 to 2·60 in the various investigations. Very good agreement was found with respect to the position $K_1 = 2{\cdot}26\ \text{Å}^{-1}$ of the first peak, which does not change with temperature.

The same observation can be made for liquid Ga measured close to room temperature (20 to 50 °C). Our data are in good agreement with those of Narten (1972), Bizid *et al* (1974) and Suzuki *et al* (1975) with respect to the positions of the first peak ($K_1 = 2{\cdot}53\ \text{Å}^{-1}$), and the scatter of the data points beyond $K = 3{\cdot}5\ \text{Å}^{-1}$. Again, the height of the first peak varies between 2·36 and 2·54 in the different investigations. It is interesting to note that the x-ray reflection data of liquid Sn and Ga show a greater height of the first peak than the corresponding transmission data (either obtained by neutrons or x-rays).

Both Sn and Ga exhibit the well known shoulder on the high-angle side of the first peak. This shoulder gradually diminishes and is not visible at 1100 °C in Sn, and 500 °C in Ga, although the shape of the first peak remains rather asymmetric, i.e. the high K side is less steep than the low K side of the first peak.

The interference functions of liquid Zn and Cd are plotted in figure 2 for different temperatures as indicated in the figure. The data on Zn are in reasonable agreement with those of North *et al* (1968) and Knoll (1977) at values of K beyond the first peak. Again, there is a discrepancy in the value of the height $I(K_1)$ of the first peak.

The position K_1 of the first peak for liquid Zn is $K_1 = 2{\cdot}90\ \text{Å}^{-1}$, which is in good agreement with the previous investigations. The positions K_1 for liquid Cd decrease with increasing temperature from $K_1 = 2{\cdot}57\ \text{Å}^{-1}$ at 350 °C to $K_1 = 2{\cdot}54\ \text{Å}^{-1}$ at 650 °C (North and Wagner 1969b). It should be noted that the profiles of the first peaks of Zn and Cd are also asymmetric. However, in these cases the low K side is less steep than the high K side.

The values of the interference function $I(K)$ in the range of K between 0·5 and 3·5 Å^{-1} are of particular interest for the evaluation of the electrical resistivity (equation (5)), and the pair-potential (equation (4)). When we normalize the height $I(K_1)$ of the

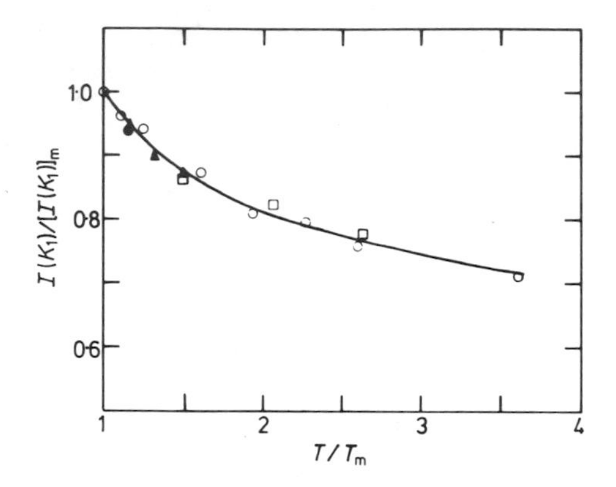

Figure 3. Ratios of the height $I(K_1)$ of the first peak of $I(K)$ to that observed close to the melting point, T_m, plotted as a function of the reduced temperature T/T_m. ○ represent the data for liquid Ga, □ for liquid Sn, ▲ for liquid Cd, and ● for liquid Zn.

first peak to that observed close to the melting point, i.e. $[I(K_1)]_m$ of the liquid metal and plot the ratio $I(K_1)/[I(K_1)]_m$ as a function of the reduced temperature T/T_m where T_m is the melting temperature of the metal, we obtain a single curve through the data point of liquid Sn and Ga as shown in figure 3.

It has been noted by Wingfield and Enderby (1968), and by North and Wagner (1969b) that the decrease of $I(K_1)$ with increasing temperature in liquid Zn and Cd, respectively, is relatively small. However, if one plots $I(K_1)/[I(K_1)]_m$ of these metals also in figure 3, one notices that the data points fall on the common curve.

The electrical resistivities of liquid Sn, Ga, Cd and Zn have been calculated using the pseudopotential form factors of Animalu and Heine (1965), as given by Harrison (1966). The calculated values of ρ_r using equations (5) and (6), are shown in figure 4 together with the experimental values of Roll and Motz (1957), and Cusack and Kendall (1960).

4. Discussion

The most noticeable effect of the temperature of measurement on the interference function $I(K)$ is the reduction of the height of the first peak, and the corresponding increase of the values of $I(K)$ on either side of the first peak. Careful measurements reveal that the position K_1 of the first peak does not change with increasing tempera-

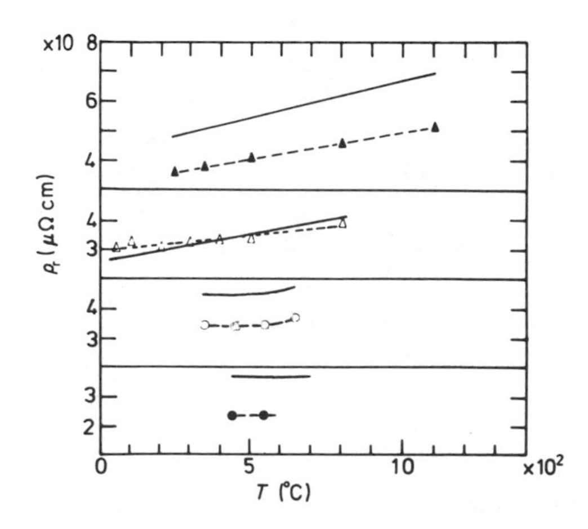

Figure 4. Electrical resistivities ρ_r of liquid Sn (▲), Ga (△), Cd (○) and Zn (●) plotted as a function of temperature T. The solid curves are the experimental data, and the broken curves are drawn through the calculated values using equation (5).

ture in liquid Sn and Ga. However, the positions of the third and subsequent maxima of $I(K)$ of Sn and Ga move slowly towards larger K values with increasing temperature, which results in a slight decrease in the interatomic distances r_1 as deduced from the Fourier transform of $I(K)$, i.e. from the pair-probability function $g(r)$ (equation (2)). In liquid Cd, the position K_1 moves to smaller values of K, whereas no change in position was observed for the higher peaks.

The asymmetry of the first peak of $I(K)$ of liquid Zn, Cd, Sn and Ga must have structural origin Both Zn and Cd crystallize in the hexagonal close packed structure with a relatively large c/a ratio (~ 1·86), whereas Sn and Ga possess more complicated structures at temperatures close to the melting point. In all of these materials, the main peak at K_1, and the asymmetry or the subsidiary peak occur in the K-space where there are strong reflections in the crystalline stable or metastable phases (Bizid *et al* 1974, Suzuki *et al* 1974, Peemüller 1970). The distorted structures in the crystalline and the liquid phases have been thought by Heine and Weaire (1966) to arise from an interplay between the electronic and Ewald term contributions to the total energy of the system.

The electrical resistivities were calculated with equation (5) using the pseudo-potential form factors $U(K)$ of Animalu and Heine (1965), which were assumed to be energy independent. Changes in the volume of the melt were taken into account when normalizing $U(K)$ to $(2/3)E_F$ at $K = 0$.

As shown in figure 4, the calculated values of ρ_r are usually smaller than the experimental values (except for Ga) because of small errors in $I(K)$, either in position or in height, and in $U(K)$, which may lead to large errors in the magnitude of ρ_r. Therefore, we evaluated the temperature coefficient $\alpha_\rho = [\rho(T_2) - \rho(T_1)] / [(T_2 - T_1)\,\rho(T_1)]$ of the resistivity. The following values were obtained for Zn, Cd, Sn and Ga, respectively: 0·0, 2·8. 5·0 and $3{\cdot}7 \times 10^{-4}\,°C^{-1}$ which compare as follows to the experimental values of α_ρ: −1·3, 1·3, 5·2 and $7{\cdot}3 \times 10^{-4}\,°C^{-1}$. The calculated and experimental values of α_ρ for liquid Sn are in close

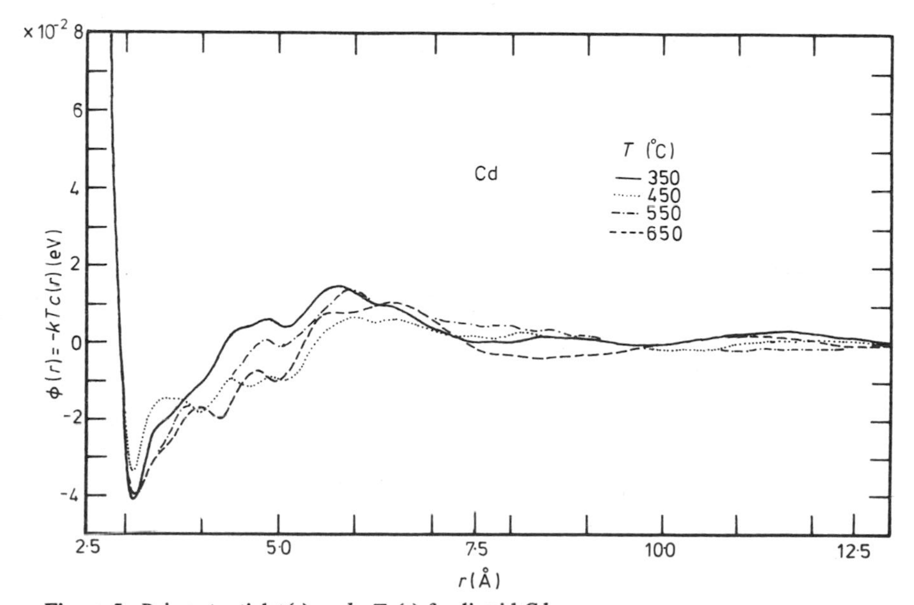

Figure 5. Pair-potential $\phi(r) = -k_BTc(r)$ for liquid Cd.

agreement with each other. However, the discrepancy between the theoretical and experimental values in the other metals is not very large, and might still be a consequence of errors in $I(K)$ and $U(K)$.

In order to test the accuracy of the interference function $I(K)$ at small values of K, the direct correlation functions $c(r)$ were calculated for liquid Cd using equation (4), from the data obtained at different temperatures. Following the suggestion made by Barker and Gaskell (1975) the values of $\phi(r) = -k_B Tc(r)$ are plotted in figure 5. It should be pointed out that $c(r)$ is very sensitive to the values below the first peak (North *et al* 1968). The scatter of the data for the four curves is certainly caused by the slight experimental errors in $I(K)$. However, the minimum in $\phi(r)$ at $r = 3{\cdot}10$ Å agrees reasonably well with the position of the first peak in $g(r)$ at $r = 3{\cdot}00$ Å.

Acknowledgment

The author wishes to thank the National Science Foundation for the support of this investigation, D Lee and N Mardesich for the measurements of liquid Ga, and H Heinisch for the calculation of the pair-potentials.

References

Animalu A O E and Heine V 1965 *Phil. Mag.* **12** 1249
Barker M I and Gaskell T 1975 *Phys. Lett.* **A53** 285
Bizid A, Bosio L, Curien H, Defrain A and Dupoint M 1974 *Phys. Stat. Solidi* **23** 135
Cromer D T and Mann J B 1967 *J. Chem. Phys.* **47** 1892
Cromer D T and Waber J T 1965 *Acta Cryst.* **18** 104
Cusack N and Kendall P 1960 *Proc. Phys. Soc.* **75** 309
Enderby J E 1972 *Liquid Metals, Chemistry and Physics* ed S Z Beer (New York: Marcel Dekker) p 585
Halder N C, North D M and Wagner C N J 1969 *Phys. Rev.* **177** 47
Harrison W 1966 *Pseudopotentials in the Theory of Metals* (New York: Benjamin)
Heine V and Weaire D 1966 *Phys. Rev.* **145** 82
Kaplow R, Strong S L and Averbach B L 1966 *Local Atomic Arrangement Studied by X-Ray Diffraction* ed J B Cohen and J E Hilliard (New York: Gordon and Breach) p 159
Knoll W 1977 this volume
Narten A H 1972 *J. Chem. Phys.* **56** 1185
North D M, Enderby J E and Egelstaff P A 1968 *J. Phys. C: Solid St. Phys.* **1** 1075
North D M and Wagner C N J 1969a *J. Appl. Cryst.* **2** 149
—— 1969b *Phys. Lett.* **A30** 440
Peemüller H 1970 *PhD Thesis* Freie Universität, Berlin
Ruland W 1964 *Br. J. Appl. Phys.* **15** 1301
Suzuki K, Misawa M and Fukushima Y 1975 *Trans. JIM* **16** 297
Wagner C N J 1972 *Liquid Metals, Chemistry and Physics* ed S Z Beer (New York: Marcel Dekker) p 257
Waseda Y and Suzuki K 1972 *Phys. Stat. Solidi* **B49** 339
Wingfield B F and Enderby J E 1968 *Phys. Lett.* **A27** 704
Ziman J M 1961 *Phil. Mag.* **6** 1013

The structure factor for liquid zinc at different temperatures

W Knoll

Institut Laue–Langevin, 156X Centre de Tri, 38042 Grenoble Cedex, France

Abstract. The structure factor of liquid Zn has been determined by means of neutron diffraction at three temperatures in a Q-range of $0{\cdot}2 \leqslant Q \leqslant 15\ \text{Å}^{-1}$. The structure factor shows a clear decrease in the height and a broadening of the first maximum when the temperature rises from 460 °C to 650 °C. The temperature coefficient of the electrical resistivity has been calculated using the Ziman theory of electron transport in liquid metals. The calculated value, $\alpha_c = -1{\cdot}35 \times 10^{-4}/°\text{C}$, corresponds quite well with the experimental result, $\alpha_e = -1{\cdot}75 \times 10^{-4}/°\text{C}$. The pair correlation functions $g(r)$ give a nearest-neighbour distance of 2·73 Å for all three temperatures.

The structure factor $S(Q)$ of liquid zinc is of particular interest in the calculation of transport properties using the Ziman theory of electron transport in liquid metals. Due to the two conduction electrons per atom, the diameter of the Fermi sphere $2k_F$ corresponds approximately to the position of the main peak in $S(Q)$. The temperature dependence of the structure factor should lead (e.g. Faber 1972) to a negative temperature coefficient of the electrical resistivity, as found experimentally by Roll and Motz (1957).

Wingfield and Enderby (1968) tried to calculate the temperature coefficient from structure data, stating that $S(Q)$ of liquid Zn in the range of 450–620 °C is almost independent of temperature between k_F and $2k_F$. This would be a very unconventional behaviour of $S(Q)$ for a liquid metal. There are no other measurements of the temperature dependence of $S(Q)$ available; only one x-ray (Gamertsfelder 1941) and some neutron experiments (North *et al* 1968, Dasannacharya *et al* 1968, Caglioti *et al* 1967) have been performed at about 450 °C and these show significant differences, especially in the region of the first peak.

An attempt was therefore made to obtain more accurate data on liquid zinc by extending the measured Q-range to $0{\cdot}2 \leqslant Q \leqslant 15\ \text{Å}^{-1}$, using several samples and much better counting rates (statistical error less than 1%)†. The samples have been measured in quartz tubes of 6 mm diameter with a wall thickness of 0·3 mm. After the full corrections described in an earlier paper (Knoll and Steeb 1973), the structure factor has been obtained for three temperatures; 460, 550 and 650 °C (figure 1). It can be stated that, as already found by other authors, the low-angle side of the first peak of $S(Q)$ is less steep than the high-angle side at 460 and 550 °C. But at 650 °C the first peak becomes fairly symmetrical and further investigations are continuing at higher temperatures in order to verify whether this is related to the change in the temperature dependence of isothermal compressibility at 650 °C reported by Schubert (1968). In addition, quite a considerable temperature dependence of the height of the first peak has been observed, the value of $S(Q)$ at the maximum, $S_p(Q)$, being $2{\cdot}85 \pm 0{\cdot}05$, 2·67 and

† Work performed on the D4 apparatus at the Institut Laue–Langevin, Grenoble, France.

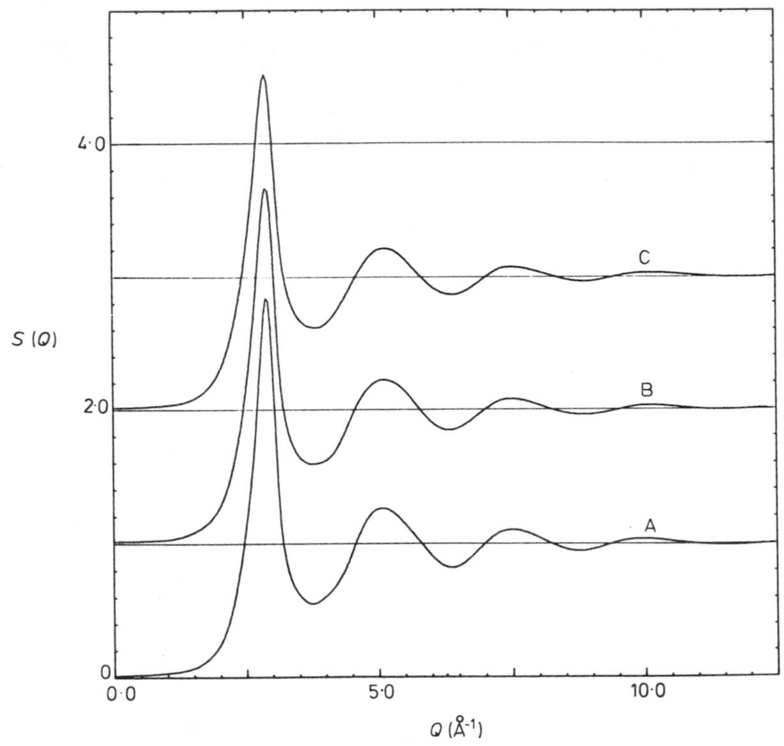

Figure 1. Structure factor, $S(Q)$, of zinc at temperatures: A, 460 °C; B, 550 °C; C, 650 °C.

2·53 for 460, 550 and 650 °C respectively. Also a broadening can be observed, although the position of the peak does not vary within the experimental error (i.e. $Q_p = 2{\cdot}89 \pm 0{\cdot}02$ Å^{-1}). Comparison with hard-sphere calculations shows good agreement for the whole structure factor except for the height of the first peak.

Using the pseudopotentials of Animalu and Heine (1965) and correcting for volume changes only, the calculated temperature coefficient of the electrical resistitivity

$$\alpha = \frac{\rho(T) - \rho(460)}{(T - 460)\,\rho(460)}$$

is $\alpha_c = -1{\cdot}35 \times 10^{-4}/°C$. This agrees well with the experimental value, $\alpha_e = -1{\cdot}75 \times 10^{-4}/°C$ (Roll and Motz 1957), and lends support to the Ziman theory for liquid zinc.

Finally, the pair correlation function $g(r)$ was calculated by Fourier transformation of $S(Q)$ without any smoothing of the experimental points (figure 2 for 460 °C). Only small ripples occur at $r \leqslant 2$ Å, not exceeding 0·1 in height, and it should be pointed out that this is one of the best results ever obtained. The $g(r)$ therefore can be considered as very precise functions giving a nearest-neighbour distance of $r_I = 2{\cdot}73 \pm 0{\cdot}02$ Å for the temperatures of 450, 550 and 650 °C, and the number of nearest neighbours $N_I = 9{\cdot}8 \pm 0{\cdot}4$ atoms. The r_I value agrees with the result of Dasannacharya *et al* (1968) but not with the value of 2·6 Å at 450 °C given by Caglioti (1967).

As a conclusion it can be stated that measurements of $S(Q)$ in liquid zinc show that the height of the first peak decreases and broadens with increasing temperature. This

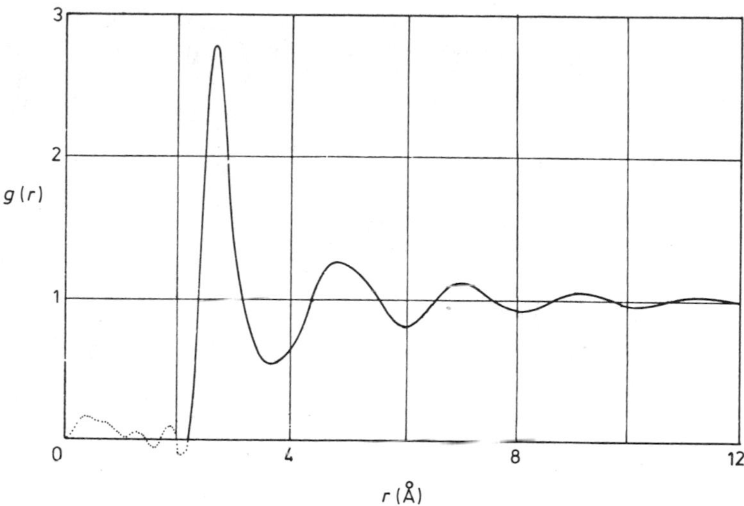

Figure 2. Pair correlation function, $g(r)$, of zinc at 460 °C.

had been questioned previously. As an anomaly, the low-angle side of the first peak is less steep than the high-angle side for 460 and 550 °C, but at 650 °C the main peak becomes symmetrical. The temperature coefficient of the electrical resistivity was calculated from the Ziman theory of electron transport in liquid metals and was found to be negative and in good agreement with the experimental value..

References

Animalu A O E and Heine V 1965 *Phil. Mag.* **12** 1249
Caglioti G, Corchia M and Rizzi G 1967 *Nuovo Cim.* **49B** 222
Dasannacharya B A *et al* 1968 *Phys. Rev.* **173** 241
Faber T E 1972 *An Introduction to the Theory of Liquid Metals* (London: Cambridge Univ. Press)
Gamertsfelder C 1941 *J. Chem. Phys.* **9** 450
Knoll W and Steeb S 1973 *J. Phys. Chem. Liquids* **4** 39
North D M, Enderby J E and Egelstaff P A 1968 *J. Phys. C: Solid St. Phys.* **1** 1075
Roll A and Motz H 1957 *Z Metallk.* **48** 272
Schubert F 1968 *Diplomarbeit* Univ. des Saarlandes, Saarbruecken
Wingfield B F and Enderby J E 1968 *Phys. Lett.* **27A** 704

Structure factor and radial distribution function in liquid aluminium

D Jović†, I Padureanu† ‡ and S Rapeanu§

† Institute of Nuclear Sciences 'Boris Kidrič', Beograd, Yugoslavia

§ Institute of Atomic Physics, Bucharest, Romania

Abstract. The structure factor of liquid aluminium has been determined by neutron diffraction measurements, at three temperatures for a momentum transfer range up to 13·5 Å^{-1}. The structure factor data were used to calculate the radial distribution function.

1. Introduction

In the last few years a lot of work has been devoted towards understanding the nature of the properties and interatomic forces in liquid metals (March 1968, Egelstaff 1967). To calculate thermodynamic properties of liquid metals, the structure factor $S(Q)$ and the correlation functions have to be known. The knowledge of $S(Q)$ within a large range of Q provides information on both macroscopic properties, obtained in the limits of very small Q, and the microscopic structure at higher momenta.

The aim of the present work was to obtain, with a good statistical accuracy, the structure factor of liquid aluminium for several temperatures. The radial distribution function and the density distribution function have been calculated from the experimental data. Coordination numbers have been obtained from the latter function.

2. Experimental procedure

The measurements were carried out on a diffractometer at the Vinča reactor, RA-1, operating at 6·5 MW. The diffraction patterns were obtained from aluminium of 99·99% purity at 666 °C, 707 °C and 800 °C. The sample was inside a container (40 × 18 × 4 mm) made of magnesium oxide (0·5 mm thick) with an active surface of 35 × 17 mm. Heating was accomplished by using molybdenum wire with three concentric cylinders as radiation shields. A vacuum higher than 10^{-5} Torr was maintained during the experiment. The temperature was measured to an accuracy of ±0·5 °C using chromel–alumel thermocouples.

The experimental runs were taken for scattering angles, 2θ, from 3° to 100° in steps of 15′. The neutron beam was monitored by a low efficiency BF_3 counter mounted between the monochromator and the sample. To determine the structure factor $S(Q)$ over a wide momentum transfer range of Q, three incident neutron wavelengths were necessary. For this purpose an (002) pyrolytic graphite crystal was used. The incident wavelengths and full widths at the half maximum are given in table 1.

‡ Visiting scientist from the Institute of Atomic Physics, Bucharest, Romania, on an IAEA fellowship.

Table 1.

Wavelength (Å)	2·513	1·168	0·715
ΔQ(Å^{-1})	0·03	0·10	0·14

3. Data analysis and structure factor

The total scattered intensities from the liquid aluminium sample, $I_{c+s}(Q)$, and the empty sample holder, $I_c(Q)$, were measured using the neutron diffraction technique. To obtain the effective structure factor from the diffraction patterns, corrections for background, transmission, effective sample volume, counter efficiency and multiple and inelastic scattering were applied to the measurements. A detailed study of these corrections was made by North *et al* (1968). According to their procedure we obtained $S(Q)$ from the relation:

$$S(Q) = \frac{1}{\alpha(\theta)} \frac{I_{c+s}(Q) - I_c(Q)\, T_s^{\sec\theta}}{\gamma(\theta)} - \frac{M(Q)}{\gamma(\theta)} - F(Q), \tag{1}$$

where T_s is the sample transmission, $F(Q)$ the inelastic scattering, $M(Q)$ the multiple scattering, $\gamma(\theta)$ the absorption within the sample and $\alpha(\theta)$ the normalization factor.

In order to calculate the contribution of the inelastic scattering, $F(Q)$, different methods for estimating the departure from the static approximation have been suggested by Placzek (1952) and Ascarelli and Caglioti (1966). Yarnell *et al* (1973) extended Placzek's corrections by assuming a detector with an arbitrary energy-dependent efficiency. Following the same approach, together with our experimental conditions, we calculated $F(Q)$ from the equation:

$$F(Q) = AkT - \frac{Q^2}{K_0^{\,2}}B + CkT + D\frac{Q^4}{K_0^{\,4}} \tag{2}$$

where

$$A = \frac{m}{2M}\,\frac{1}{E_0},$$

$$B = 2 - 0{\cdot}00514,$$

$$C = 3{\cdot}5054\,A,$$

$$D = -0{\cdot}00116.$$

The efficiency of our detector is given by the relation:

$$\epsilon(K) = 1 - \exp\,(-1{\cdot}93\,K_0/K). \tag{3}$$

The corrections for multiple scattering were made according to the methods of Vineyard (1954) and Cocking and Heard (1965). It was found that the average contribution from multiple scattering is 12%. The absorption correction, $\gamma(\theta)$, was calculated from the measured transmission and the effective volume of the sample. In order to obtain the

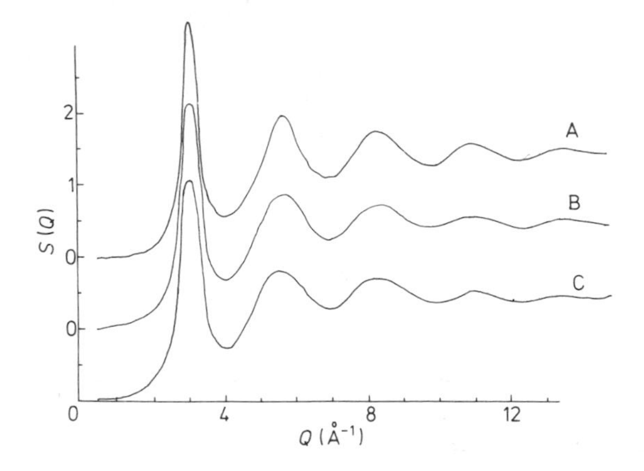

Figure 1. Structure factors of liquid aluminium at: A, 666 °C; B, 707 °C; C, 800 °C.

normalization factor, $\alpha(\theta)$, we used the calculated values of $F(Q)$, $M(Q)$, $\gamma(\theta)$ and the condition

$$\lim S(Q) \to 1 \quad \text{as} \quad Q \to \infty. \tag{4}$$

The corrected structure factors of liquid aluminium at the three temperatures are shown in figure 1. The numerical values of $S(Q)$ are tabulated in table 2.

It is interesting to compare liquid aluminium data from this experiment with the previous ones measured by different authors using neutron and x-ray diffraction. For example, Larsson *et al* (1965) and Egelstaff *et al* (1966) observed that the structure factor for aluminium is symmetrical in the region of the main peak. In a recent experiment by Stallard and Davis (1973) the symmetry of the main peak was confirmed. On the other hand, Okazaki *et al* (1970) have found a shoulder on the right side of the main peak which decreases with increasing temperature. They also observed some structure on the left side of the first main peak which is in disagreement with our results. Fessler *et al* (1966) reported a detailed analysis of x-ray diffraction at 665 °C and 750 °C and they concluded that the effect of temperature on the structure is small over the studied temperature range and this is in agreement with our conclusion.

The radial distribution function, $g(r)$, and the density distribution function, $n(r) = 4\pi r^2 n g(r)$, shown in figure 2, were calculated from the Fourier transform of the experimental structure factor. For mono-atomic, $g(r)$ is related to the structure factor $S(Q)$ by the relation:

$$g(r) - 1 = (2\pi^2 nr)^{-1} \int_0^\infty [S(Q) - 1]\ Q \sin Qr \,\mathrm{d}Q, \tag{5}$$

where n is the mean atomic density.

The positions and intensities of the $g(r)$ peaks, at determined temperatures, are given in table 3. As can be seen, the temperature does not affect the positions of the maxima of the coordination sphere. Qualitatively, our results on $g(r)$ are in agreement with the earlier calculations by Fessler *et al* (1966), Sommer and Werber (1968) and Brauneck *et al* (1970). In order to determine the coordination number, we have integrated the areas under the first peaks of the density distribution function. The results are given in table 4. An important change in the coordination number of atoms from the first shell

Table 2.

Q(Å^{-1})	666 °C	707 °C	800 °C	Q(Å^{-1})	666 °C	707 °C	800 °C	Q(Å^{-1})	666 °C	707 °C	800 °C
0·5	0·021	0·024	0·029	4·7	1·201	1·230	1·240	8·8	0·910	0·970	0·965
0·6	0·022	0·026	0·031	4·8	1·310	1·271	1·260	8·9	0·949	0·979	0·970
0·7	0·022	0·028	0·033	4·9	1·380	1·301	1·270	9·0	0·980	1·000	0·980
0·8	0·023	0·030	0·037	5·0	1·410	1·310	1·271	9·1	1·010	1·018	1·000
0·9	0·025	0·033	0·040	5·1	1·390	1·311	1·260	9·2	1·041	1·038	1·018
1·0	0·030	0·041	0·045	5·2	1·312	1·280	1·240	9·3	1·079	1·060	1·038
1·1	0·035	0·045	0·052	5·3	1·190	1·251	1·190	9·4	1·098	1·065	1·040
1·2	0·040	0·050	0·060	5·4	1·099	1·170	1·140	9·5	1·111	1·071	1·050
1·3	0·055	0·065	0·071	5·5	1·010	1·110	1·110	9·6	1·120	1·080	1·050
1·4	0·060	0·091	0·100	5·6	0·961	1·030	1·040	9·7	1·110	1·079	1·058
1·5	0·076	0·110	0·110	5·7	0·890	0·971	1·010	9·8	1·100	1·074	1·055
1·6	0·088	0·131	0·140	5·8	0·871	0·930	0·965	9·9	1·099	1·070	1·050
1·7	0·110	0·150	0·179	5·9	0·830	0·888	0·930	10·0	1·077	1·060	1·041
1·8	0·138	0·189	0·210	6·0	0·805	0·860	0·910	10·1	1·056	1·055	1·035
1·9	0·170	0·241	0·269	6·1	0·800	0·859	0·899	10·2	1·036	1·031	1·015
2·0	0·200	0·299	0·340	6·2	0·810	0·860	0·899	10·3	1·001	1·011	1·000
2·1	0·285	0·389	0·430	6·3	0·815	0·879	0·910	10·4	0·998	1·000	0·999
2·2	0·420	0·470	0·531	6·4	0·863	0·910	0·940	10·5	0·975	0·999	0·980
2·3	0·600	0·680	0·790	6·5	0·910	0·939	0·979	10·6	0·960	0·985	0·978
2·4	0·789	1·110	1·120	6·6	0·941	0·989	1·010	10·7	0·939	0·975	0·976
2·5	1·390	1·620	1·800	6·7	0·110	1·030	1·060	10·8	0·940	0·965	0·970
2·6	2·000	2·180	2·130	6·8	1·050	1·070	1·100	10·9	0·945	0·960	0·970
2·7	2·340	2·200	2·150	6·9	1·110	1·110	1·130	11·0	0·950	0·964	0·972
2·8	2·020	2·160	2·030	7·0	1·160	1·151	1·150	11·1	0·961	0·970	0·975
2·9	1·200	1·600	1·750	7·1	1·210	1·179	1·179	11·2	0·979	0·981	0·978
3·0	0·800	1·001	1·300	7·2	1·252	1·189	1·185	11·3	0·998	1·010	0·981
3·1	0·621	0·750	0·860	7·3	1·260	1·210	1·185	11·4	1·010	1·017	1·000
3·2	0·540	0·620	0·690	7·4	1·255	1·200	1·180	11·5	1·031	1·021	1·011
3·3	0·489	0·550	0·601	7·5	1·230	1·200	1·179	11·6	1·042	1·032	1·020
3·4	0·430	0·501	0·540	7·6	1·200	1·181	1·178	11·8	1·060	1·040	1·030
3·5	0·421	0·481	0·510	7·7	1·161	1·160	1·160	11·9	1·059	1·039	1·030
3·6	0·420	0·482	0·500	7·8	1·110	1·131	1·130	12·0	1·058	1·038	1·025
3·7	0·431	0·501	0·531	7·9	1·075	1·099	1·110	12·1	1·056	1·037	1·020
3·8	0·458	0·530	0·572	8·0	1·030	1·059	1·080	12·2	1·044	1·036	1·010
3·9	0·510	0·578	0·640	8·1	0·990	1·038	1·050	12·3	1·033	1·025	1·000
4·0	0·558	0·640	0·690	8·2	0·961	1·010	1·030	12·4	1·020	1·020	0·999
4·1	0·630	0·710	0·770	8·3	0·929	0·999	1·000	12·5	1·011	1·015	0·998
4·2	0·710	0·780	0·840	8·4	0·911	0·982	0·999	12·6	1·011	1·010	1·000
4·3	0·789	0·891	0·960	8·5	0·900	0·970	0·971	12·7	0·999	1·000	0·999
4·4	0·880	0·980	1·040	8·6	0·899	0·966	0·965	12·8	1·002	0·998	1·001
4·5	0·980	1·065	1·130	8·7	0·898	0·965	0·960	13·0	1·000	0·999	1·000
4·6	1·081	1·150	1·200								

is observed at the melting point. Above the melting point this number remains practically constant for increasing temperature within our measured temperature interval.

4. Conclusions

The structure factor for liquid aluminium has been obtained at three temperatures. A temperature dependence is observed in the intensity of the peaks, while their positions remain nearly unchanged. The radial distribution function has been calculated from the structure factor data. A noticeable change in the coordination number of atoms in the first shell has been observed at the melting point. An attempt has been

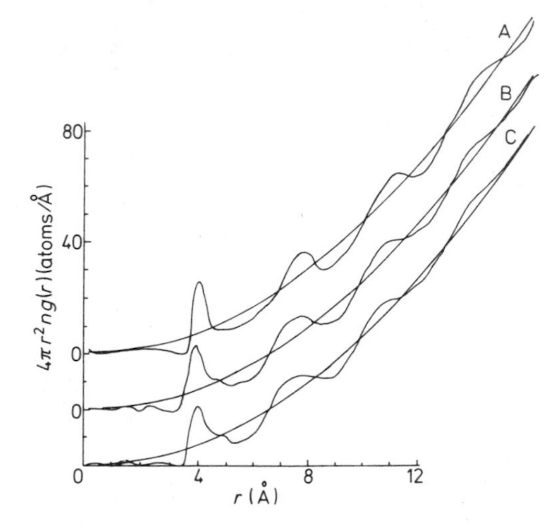

Figure 2. The density distribution functions of liquid aluminium at: A, 666 °C; B, 707 °C; C, 800 °C.

Table 3.

Position (Å)	Intensities of the maxima		
	666 °C	707 °C	800 °C
2·8	3·38	3·01	2·82
5·3	1·30	1·25	1·23
7·7	1·10	1·09	1·08
9·9	1·05	1·04	1·02
12·1	1·03	1·01	–

Table 4.

Temperature (°C)	Area under first peak
666	7·2
707	7·0
800	7·0

made to obtain some information on the pair interaction potential from the Percus–Yevick (1958) and hypernetted chain theories, but the result is not satisfactory.

Acknowledgments

One of us (IP) would like to express his gratitude to the IAEA in Vienna for sponsoring his fellowship in Yugoslavia and to the Institute of Nuclear Sciences 'Boris Kidrič'-Vinča. It is a pleasure to thank Mr P Balabanović for his technical assistance. This work was supported by the Scientific Association of SR Serbia.

References

Ascarelli P and Caglioti G 1966 *Nuovo Cim.* **43B** 376
Brauneck W, Gahn U, Sommer F, Werber K and Willee C 1970 *Phys. Lett.* **31A** 237

Cocking S J and Heard C 1965 *AERE Report No.* R5016
Egelstaff P A 1967 *An Introduction to the Liquid State* (London: Academic Press)
Egelstaff P A, Duffill C, Rainey V, Enderby J E and North D M 1966 *Phys. Lett.* **21** 286
Fessler R R, Kaplow R and Averbach B L 1966 *Phys. Rev.* **150** 34
Larsson K E, Dahlborg H and Jović D 1965 *Proc. Symposium on Inelastic Scattering of Neutrons* (Vienna: IAEA) vol 1 p117
March N H 1968 *Liquid Metals* (London: Pergamon)
North D M, Enderby J E and Egelstaff P A 1968 *Proc. Phys. Soc.* **1** 784
Okazaki H, Iida K and Tamaki S 1970 *J. Phys. Soc. Japan* **29** 1396
Percus J K and Yevick G Y 1958 *Phys. Rev.* **110** 1
Placzek G 1952 *Phys. Rev.* **86** 377
Sommer F and Werber K 1968 *Phys. Lett.* **27A** 425
Stallard J M and Davis C M 1973 *Phys. Rev.* 8 368
Vineyard G H 1954 *Phys. Rev.* **95** 93
Yarnell J L, Katz M J, Wenzel R G and Koening S H 1973 *Phys. Rev.* **A7** 2130

Structure factor of expanded liquid rubidium up to 1400 K and 200 bar

R Block, J-B Suck†, W Freyland‡, F Hensel‡ and W Gläser§
Institut für Angewandte Kernphysik, Kernforschungszentrum, Karlsruhe
‡ Institut für Physikalische Chemie der Universität, Marburg
§ Physik Department der Technischen Universität, München

Abstract. The structure factor of liquid rubidium has been measured for densities between 1·42 and 0·98 g cm^{-3} and temperatures up to 1400 K in a region of momentum transfer between 0·2 and 2·5 Å^{-1}. The first maximum of the structure factor is shifted to lower momentum transfers and broadened with decreasing densities. The comparison of the results with a hard-core model shows that a temperature-dependent hard-core radius is needed to fit the data. Using the measured structure factors the conductivity of the liquid was calculated within the NFE model. The results agree satisfactorily with those of a recent measurement.

1. Introduction

The structural properties of the rare gas liquids have been known in an extended region of densities and temperatures for more than 30 years (Eisenstein and Gingrich 1942). Ten years ago the density and temperature dependence of the structure factor of argon were studied in detail near the critical point (Mikolaj and Pings 1966). There is as yet no equivalent information for a liquid metal, probably due to the high temperatures and pressures needed in such experiments.

From the difference in the interatomic potential one would expect a different behaviour for the structure factor of a liquid metal compared to that of rare gas liquids. In addition the temperature and density dependence of the electrical transport properties are correlated with the corresponding variations of the structure in a liquid metal (see for example Faber 1972). Knowledge of the latter is therefore essential for understanding the changes in electrical transport properties of a liquid metal in the range between its triple and its critical point. We have therefore investigated the density dependence of the structure factor of liquid rubidium by neutron diffraction techniques. Measurements were made near the saturation line of the liquid–gas system down to a density of about 65% of that at the triple point. For experimental reasons it was convenient in this first experiment to limit the range of Q values to the region of the first maximum.

Rubidium was chosen because of its favourable properties for thermal neutron scattering, namely low absorption and incoherent scattering cross section. In addition, Rb is suitable for investigations over a wide range of densities and temperatures of the liquid, because of its low critical temperature of about 1850 °C (Bhise and Bonilla 1973). Nevertheless in this first experiment it was necessary to restrict the temperature to 1400 K and the pressures to 190 bar for experimental reasons. Much higher pressures

† Present address: Institut Laue–Langevin, Grenoble

would have been needed to separate the influence of temperature and of density on the structure factor. We have considered mainly the variation in density in our analysis as its influence on the structure factor is expected to be greater than that of the temperature (Eisenstein and Gingrich 1942).

2. Experiment and data evaluation

Rubidium of 99·9% purity was kept under argon pressure in a molybdenum cylinder 25 mm in height, with 11 mm outer and 9·6 mm inner diameter. The temperature was controlled by three Pt/PtRh thermocouples. The multidetector instrument D7 of the Institut Laue–Langevin in Grenoble was used with an incoming wavelength of 4·906 Å. The scattering angles were between 9·5° and 149·5° corresponding to momentum transfers between 0·2 and 2·5 $Å^{-1}$.

Varying the temperature and pressure from 150 K to 1400 K, and from 6 to 190 bar respectively, we measured 17 structure factors at 16 different densities between 1·42 and 0·98 g cm^{-3}. A detailed description of the experiment is given elsewhere (Block *et al* 1976). Data were taken during several repeated runs on the sample, on the empty molybdenum cylinder at different sample temperatures, and on a vanadium calibration sample. Corrections were made for scattering from the empty sample holder, for multiple scattering in the rubidium (Blech and Averbach 1965), and for the absorption of the neutrons in the rubidium and the molybdenum cell (Paalman and Pings 1962), and the usual Placzek corrections were applied to the data (see for example Enderby 1968). The relative efficiency of the 32 detector tubes was accounted for by the vanadium runs and owing to the cooled beryllium filter in the incoming neutron beam no corrections were necessary for higher-order beam contamination. No corrections were made for the resolution of the instrument or for the multiple scattering between cell and sample. This latter effect was calculated and found to double the multiple-scattering corrections. As these corrections were small compared to all others we neglected them. Owing to the low incoming energy of 3·4 meV the integration of the scattering law on the energy loss side of the spectrum is not complete. According to scattering law measurements near the triple point (e.g., Suck 1975) this may cause a lack of intensity of less than 10% in the worst cases (lowest and highest momentum transfers). The Placzek corrections were larger than any others due to the large incoming wavelength and the high temperatures used.

3. Experimental results and discussion

The structure factors measured at nine different densities are shown in figure 1. They show a continuous change with decreasing density and increasing temperature. We remark first on the increase in intensity of the structure factor to the right of the main peak, observed as the density decreases. This effect seems to be stronger than in the rare gas liquids, where three well defined maxima and minima are found even near the critical point. The decrease and shift of the first maximum corresponds to a broader and shifted first maximum of the pair correlation function. This means that the 'shell of nearest neighbours' of a given atom is less well defined at higher temperatures and the radius of this 'shell' increases with decreasing density. Starting with the highest

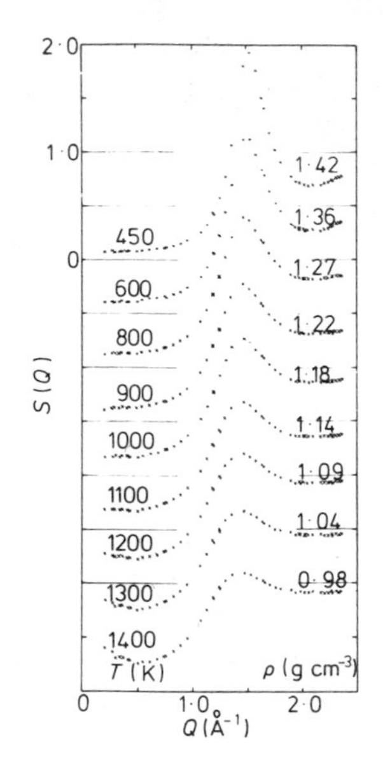

Figure 1. Measured structure factors of liquid rubidium at nine different densities and temperatures for momentum transfers between 0·2 and 2·5 $Å^{-1}$.

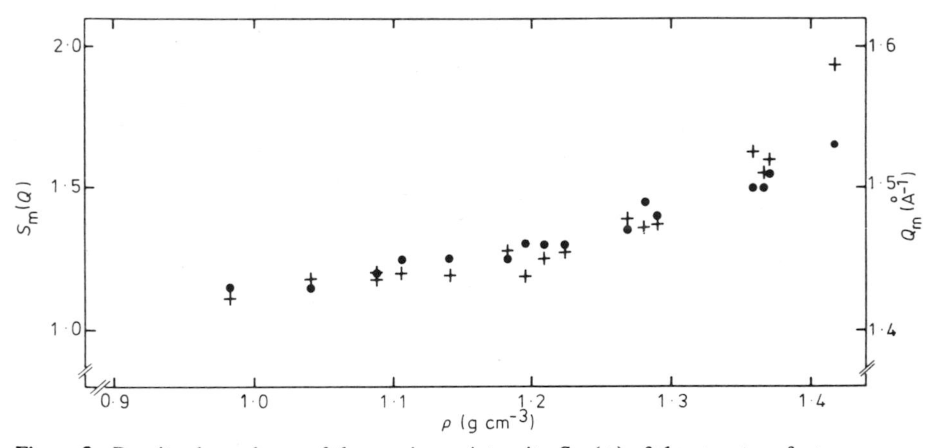

Figure 2. Density dependence of the maximum intensity S_m (+) of the structure factor and of the position Q_m (•) of these maxima at corresponding values of temperatures.

momentum transfers we find the minimum between the first and the second maximum greatly reduced with decreasing densities. In the region of the main peak, which has its maximum value S_m at the momentum transfers Q_m, the structure factor is considerably broadened and Q_m is shifted to lower Q values. This and the peak heights are shown in detail in figure 2. In the region of low momentum transfers we find a Q-independent intensity for Q values below 0·5 $Å^{-1}$ down to densities of approximately 1·1 g cm^{-3}. The intensity at low Q values increases by more than a factor of three as the density is decreased to 1·1 g cm^{-3}. For lower densities a slight increase in the scattered intensity with decreasing Q values was observed. The largest change in the structure factor takes place in the highest density region, as can be seen from figure 2, whereas in the next region of densities and temperatures an almost linear variation with density seems to occur. Thus the most rapid change in the structure of liquid rubidium is observed at temperatures not too far from its triple point.

Ashcroft and Lekner (1966) used the solution of the Percus–Yevick equation with a hard-core potential to describe the structure of liquid metals. They found remarkable agreement between their results and experiment, particularly in the region of the first maximum where they adjusted the hard-core radius r_{hc} to reproduce the height to the measured structure factor. We therefore tried to determine whether this simple model could describe the data in the whole range of densities covered in our experiment. In this model the static structure factor is described by the Fourier transform $C(Q)$ of the Ornstein–Zernike direct correlation function

$$S(Q) = \frac{1}{1 - C(Q)} \tag{1}$$

$$\begin{aligned} C(Q) = & -\frac{24\eta}{x^6(1-\eta)^4}\{[\alpha x^3(\sin x - x\cos x) \\ & +\beta x^2[2x\sin x - (x^2-2)\cos x - 2] \\ & +\gamma[(4x^3-24x)\sin x - (x^4-12x^2+24)\cos x + 24]\} \end{aligned} \tag{2}$$

where $\eta = n\pi r_{hc}{}^3/6$, $x = Qr_{hc}$, $\alpha = (1+2\eta)^2$, $\beta = -6\eta(1+\eta/2)^2$, and $\gamma = \eta(1+2\eta)^2/2$.

We made a least squares fit of equation (1) to the data by varying r_{hc}, the only parameter of this model. The qualities of fit were acceptable only for momentum transfers between 0·5 and 1·25 Å^{-1}. A typical result is shown in figure 3. The increase of the scattered intensity at low Q values is not included in this simple model. The resulting hard-core radii r_{hc} are shown in figure 4. We find a nearly linear decrease for densities of 1·35 and 1·05 g cm^{-3}, in other words for temperatures between 700 K and 1300 K. This is not surprising as the hard-sphere radius is expected to be temperature-dependent because it is mainly an average of the repulsive part of the pair interaction.

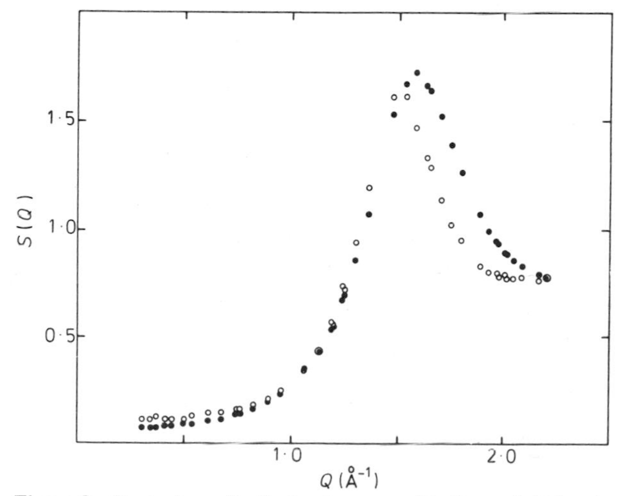

Figure 3. Typical result of a least squares fit of a model structure factor (•) (Ashcroft and Lekner 1966) to experimental results (○) taken at 600 K and 1·36 g cm^{-3} for liquid rubidium.

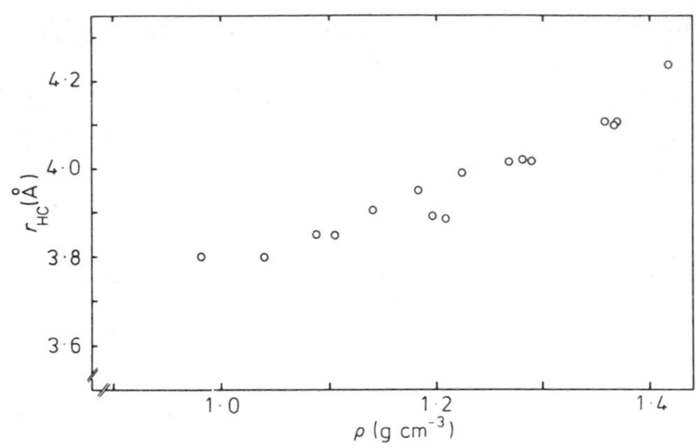

Figure 4. Hard-core radii r_{hc} resulting from the least squares fit of the model structure factor (Ashcroft and Lekner 1966) to the measured data.

The heights of the peak in the model function are slightly greater than the measured peak heights. As the peak position and its height are determined by the same parameter r_{hc} they are not independent of one another. It is difficult to fit the peak position and its height with the same hard-core radius simultaneously, as has been already noticed by other authors (see for example Yarnell *et al* 1973).

As a second step we evaluated the electrical conductivity of liquid rubidium using Ziman's formula (Ziman 1961):

$$\frac{1}{\sigma} = \frac{3\pi m^2 \Omega}{4\hbar^3 e^2 k_F^6} \int_0^{2k_F} S(Q)|U(Q)|^2 Q^3 \,dQ \tag{3}$$

where $U(Q)$ is the screened pseudopotential of a single ion (Faber 1972), $S(Q)$ is the structure factor of the liquid, m is the electron mass, k_F is the wavenumber of the electrons at the Fermi surface, and the other symbols have their usual meaning. For the bare-ion form factor in this potential we took the Ashcroft potential (Ashcroft 1966):

$$u(Q) = -\frac{4\pi Z e^2}{\Omega Q^2} \cos(Q r_c) \tag{4}$$

where Z is the valency of the ion and the core radius r_c is the only parameter of this model. For r_c we used $2{\cdot}2a_0$, a value which is in accordance with others given in the literature (a_0 is the Bohr radius). Following Shyu and Gaspari (1968) we combined this bare-ion potential with a dielectric function used by Heine and Abarenkov (1964) and Sham (1965):

$$\epsilon(Q) = 1 + \frac{1 - f(Q)}{Q^2} g_H \tag{5}$$

and

$$g_H = \frac{2k_F m e^2}{\pi \hbar^2}\left(1 + \frac{4k_F^2 - Q^2}{4k_F Q} \ln\left|\frac{2k_F + Q}{2k_F - Q}\right|\right) \tag{6}$$

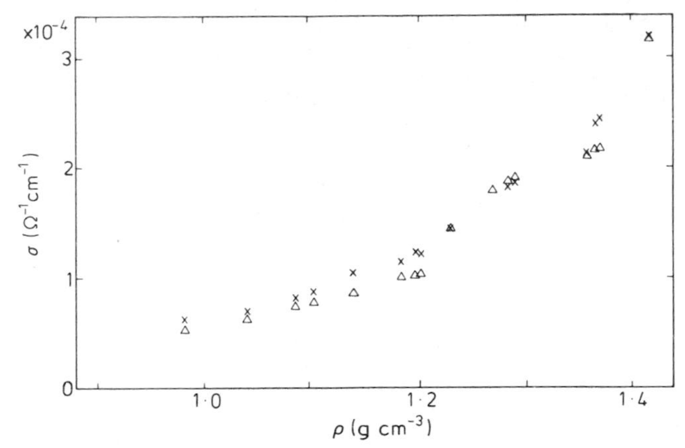

Figure 5. The electrical conductivity of liquid rubidium: (×) calculated with the Ziman formula using a screened Ashcroft pseudopotential and experimental structure factors; (Δ) recently measured values (Pfeifer *et al* 1976).

where g_H is the Hartree function derived for a non-interacting Fermi gas. To improve this approximation, $f(Q)$ takes account of exchange and correlations in the electron gas:

$$f(Q) = \frac{Q^2}{2(Q^2 + k_F{}^2 + Q_s{}^2)}$$

$$Q_s{}^2 = \frac{2k_F}{\pi a_0}. \tag{7}$$

For the structure factor of the liquid we used the experimental results extrapolated to $Q = 0$. The results of these calculations are compared with values of the electrical conductivity measured recently (Pfeifer *et al* 1976) in figure 5. The agreement between the calculated and the experimental results is satisfactory.

Acknowledgment

We are grateful to Dr W S Howells who made available to us two of his evaluation programs. We are indebted to the ILL at Grenoble for the use of the D7 diffractometer and for support during this experiment. Expert technical assistance in the construction of the high-pressure apparatus by W Baltz and H Schüssler is gratefully acknowledged.

References

Ashcroft N W 1966 *Phys. Lett.* **23** 48
Ashcroft N W and Lekner J 1966 *Phys. Rev.* **145** 83
Bhise V S and Bonilla C F 1973 *Proc. 6th Symp. on Thermophysical Properties* (New York: ASME)
Blech I A and Averbach B L 1965 *Phys. Rev.* **137** A1113
Block R, Suck J-B, Gläser W, Freyland W F and Hensel F 1976 *Ber. Bunsenges. Phys.* 8 718
Eisenstein A and Gingrich N S 1942 *Phys. Rev.* **62** 261
Enderby J E 1968 *Physics of Simple Liquids* (Amsterdam: North-Holland)
Faber T E 1972 *Theory of Liquid Metals* (London: Cambridge UP)

Heine V and Abarenkov I 1964 *Phil. Mag.* **9** 451
Mikolaj P G and Pings C J 1967 *J. Chem. Phys.* **46** 1401
Paalman H H and Pings C J 1962 *J. Appl. Phys.* **33** 2635
Pfeifer H P, Freyland W F and Hensel F 1976 *Ber. Bunsenges. Phys.* 8 716
Sham J C 1965 *Proc. R. Soc.* **A283** 33
Shyu Wei-Mei and Gaspari G D 1968 *Phys. Rev.* **170** 687
Suck J-B 1975 *Commun. Kernforschungszentrum Karlsruhe* 2231 (and references therein)
Yarnell J L, Katz M J, Wenzel R G and Koenig S H 1973 *Phys. Rev.* **A7** 2130
Ziman J M 1961 *Phil. Mag.* **6** 1013

Density, compressibility and diffraction data of LiAg alloys

H Reiter, H Ruppersberg and W Speicher
Fachbereich Angewandte Physik der Universität des Saarlandes,
D-6600 Saarbrücken 11, West Germany

Abstract. For liquid LiAg alloys the density, the compressibility and the mutual distribution of the two atomic species deviate from an ideal behaviour. The observed deviations are similar but less pronounced than in the case of liquid LiPb. The short-range order, which persists almost unchanged through the melting region of the FCC α-solid solutions, is still present in dilute liquid solutions of silver in lithium. There is some evidence that the LiAg distance is smaller than the mean distance between like atoms.

1. Introduction

The phase diagram of LiAg is known from the work of Freeth and Raynor (1953). Lithium is easily soluble in FCC silver. The range of the α-solid solution is reported to extend from 0 to 46 at.% Li at room temperature, and to 61 at.% Li in the region between about 330 °C and the solidus line. At the lithium side of the diagram several incongruently melting intermetallic phases exist with large solubility ranges. The structure of these phases is related to the structure of γ-brass and CsCl. The volume per atom of the compounds and of the α-solid solutions is smaller than the mean volume of the pure components. Also, for solid solutions of Ag in BCC Li, a strong lattice contraction has been reported by Firth *et al* (1974). Such volume contractions are typical for alloys of group I and group II metals with a more noble metal and are said (Biltz and Weibke 1935) to be due to a transfer of charge from the less noble towards the more noble ion. This transfer reduces the size of the former to a value which is nearly independent of the noble metal (rule of additive constant increments; see Ruppersberg and Speicher 1976). In the α-range Ruppersberg (1975) observed superlattice lines by neutron diffraction, which transformed at about 340 °C into short-range order (SRO) scattering. According to these qualitative diffraction patterns, the SRO scattering persisted through the melting region. SRO in the liquid phase has also been observed by x-ray diffraction (Goebbels and Ruppersberg 1973). This finding agrees with the observation of G Schwitzgebel and W Becker (private communication 1976) who showed that the excess stability function E^{xs} is positive.

In the present paper the density, compressibility and additional diffraction data obtained for LiAg alloys will be compared with those observed for PbLi alloys. All the known properties of the latter deviate strongly from an ideal behaviour (Ruppersberg and Egger 1975). Extreme deviations are observed in the range of about 20 at.% Pb, suggesting an interaction of four electronic states per lead atom with one lithium electron. As can be concluded from the E^{xs} values, the interaction between the components is much stronger in LiPb than in LiAg.

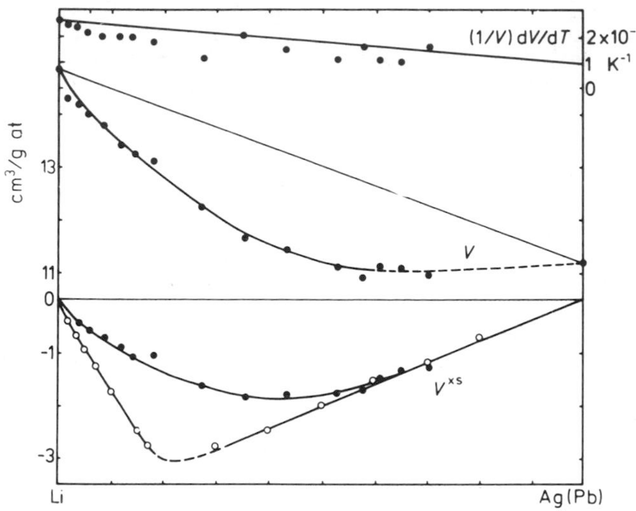

Figure 1. Volume per atom V of liquid LiAg alloys at 600 °C and its relative temperature derivative $(1/V)\,\mathrm{d}V/T$. Excess volume V^{xs} of liquid LiAg (full circles) and LiPb (open circles) alloys at 600 °C.

2. Experiments and results

The samples were prepared from 99·99% Ag and from 99·9% Li. The samples for neutron diffraction were prepared from ^{7}Li. The density was measured using the maximum bubble pressure technique. The adiabatic compressibility χ_{a} was calculated from the velocity of sound, which has been measured interferometrically using the pulse-echo technique. The details of the experiments have been described by Ruppersberg and Speicher (1976). The highest attainable temperature was about 800 °C. The variation with composition of the molar volume V, of the excess volume

$$V^{\mathrm{xs}} = V - x_{\mathrm{Li}}V_{\mathrm{Li}} - x_{\mathrm{Ag}}V_{\mathrm{Ag}},$$

both at 600 °C, and of the thermal expansion coefficient are depicted in figure 1. In the dotted region, the temperature of the liquidus is higher than 600°C. The molar volume of pure silver has been calculated by extrapolating the data of Veazey and Roe (1972). The diagrams of χ_{a}, $\chi_{\mathrm{a}}^{\mathrm{xs}}$ and $1/\chi_{\mathrm{a}}(\mathrm{d}\chi_{\mathrm{a}}/\mathrm{d}T)$ against x are shown in figure 2. The values for pure silver were taken from the work of Pronin and Filippov (1963). The neutron diffraction experiments were carried out on the spectrometer D4 located at the Institute Max von Laue–Paul Langevin, Grenoble, France. Details of the experiments and of the data processing are described by Ruppersberg and Egger (1975). Contrary to the case of ^{7}Li–Pb a Placzek fall-off has been observed for all the liquid ^{7}Li–Ag alloys studied. This fall-off could be corrected by inserting, for the Li atoms, an atomic mass of 7·0 into the corresponding equation of Yarnell *et al* (1973). The diffraction patterns were normalized using the known coherent and incoherent scattering cross sections of ^{7}Li and of Ag. Multiple scattering has been calculated according to Blech and Averbach (1965). $S^{\mathrm{n}}(0)$ obtained by extrapolating the scattered intensity from $k = 0{\cdot}2$ to $k = 0$ is in quite good agreement with the values calculated from the preliminary thermodynamic data of G Schwitzgebel and W Becker (private communication 1976). $S^{\mathrm{n}}(k)$

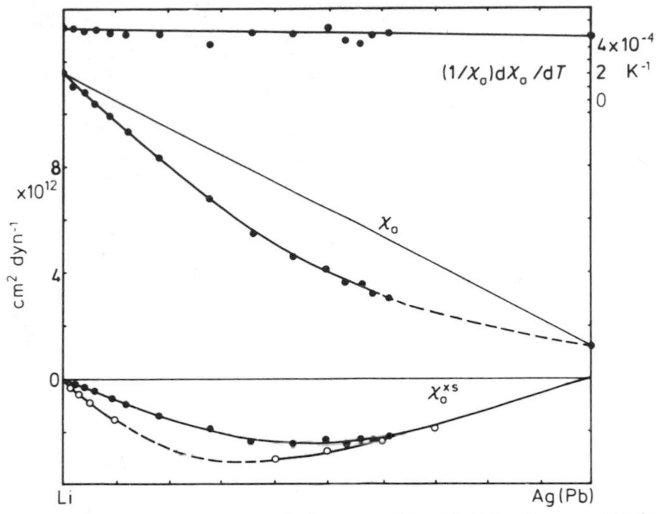

Figure 2. Adiabatic compressibility χ_a of liquid LiAg alloys at 600 °C and its relative temperature derivative $(1/\chi_a)\,d\chi_a/dT$. Deviation χ_a^{xs} of the adiabatic compressibilities of liquid LiAg (full circles) and LiPb (open circles) alloys at 600 °C, from the mean value of the pure components.

curves for liquid and solid alloys are shown in figure 3. The x-ray diffraction patterns of the liquid alloys were obtained with a non-focusing reflection arrangement using Mo K_α-radiation and balanced filters for monochromatization. With the same technique Ruppersberg and Reiter (1972) have obtained satisfactory results for liquid mercury. Coherent x-ray scattering intensities $S^x(k)$ normalized according to equations (1) and (3) are shown in figure 3. The results between $k = 0$ and $0{\cdot}8$ are obtained by interpolation.

3. Discussion

As in the case of LiPb alloys, V^{xs} of LiAg (figure 1) is negative and its absolute value does not diminish with rising temperature. The V against x diagram of LiAg is smoother than the diagram of LiPb which could be separated into two almost linear portions. Nevertheless, at low Li content the two V^{xs} diagrams are nearly identical which shows that the rule of additive constant increments holds also for LiAg. In the Li rich solutions, the variation of V^{xs} per added lead atom is twice stronger than that for an added silver atom. Correspondingly the two tangents of V^{xs} from the lithium and the silver side of the diagram do not cut at $x_{Ag} = 0{\cdot}5$ but at about $0{\cdot}35$. The compressibility diagrams of LiAg (figure 2) are quite similar to the volume diagrams (figure 1). A comparison of the χ_a^{xs} curves for LiAg and LiPb with the V^{xs} curves reveals analogous deviations of volume and compressibility from the ideal behaviour.

$S(k)$ of binary alloys A, B with composition x_A, x_B is related to the partial structure factors $a_{ij}(k)$ $(i, j = \text{A, B})$ by the following equation:

$$S(k) = [x_A x_B(\Delta b)^2 + x_A{}^2 b_A{}^2 a_{AA} + x_B{}^2 b_B{}^2 a_{BB} + 2x_A x_B b_A b_B a_{AB}]/\langle b^2\rangle \qquad (1)$$

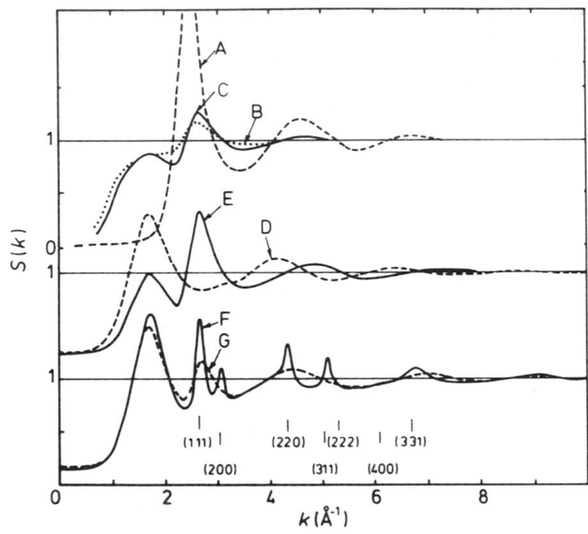

Figure 3. Normalized coherent scattering intensities $S(k)$: A, Li at 320 °C (neutrons); B, $Li_{0\cdot95}Ag_{0\cdot05}$ at 230 °C (x-rays); C, $Li_{0\cdot90}Ag_{0\cdot10}$ at 230 °C (x-rays); D, $Li_{0\cdot715}Ag_{0\cdot285}$ at 320 °C (neutrons); E, $Li_{0\cdot715}Ag_{0\cdot285}$ at 320 °C (x-rays); F, solid $Li_{0\cdot55}Ag_{0\cdot45}$ at 360 °C (neutrons); G, liquid $Li_{0\cdot55}Ag_{0\cdot45}$ at 500 °C (neutrons).

where b_A and b_B are the coherent scattering amplitudes, $\Delta b = b_A - b_B$, and $\langle b^2 \rangle = x_A b_A{}^2 + x_B b_B{}^2$. The next equation defines the symbol T_F (Fourier transform):

$$T_F\{a_{ij} - 1\} = \frac{1}{2\pi^2} \int_0^\infty k(a_{ij} - 1) \sin(kr)\, dk = \left(\frac{\rho_{ij}}{x_j} - \rho_0\right) r \qquad (2)$$

where $\rho_{ij}(r)$ is the number of j-type atoms per unit volume at the distance r from an i-type atom and ρ_0 is the mean number density. In terms of the number-concentration structure factors $S_{mn}(k)$ $(m, n = \mathrm{N}, \mathrm{C})$ which have been introduced by Bhatia and Thornton (1970), $S(k)$ is given by

$$S(k) = [\langle b \rangle^2 S_{NN} + (\Delta b)^2 S_{CC} + 2\Delta b \langle b \rangle S_{NC}]/\langle b^2 \rangle. \qquad (3)$$

$S_{NN}(k)$ oscillates around 1 and approaches 1 at high k-values. $T_F\{S_{NN} - 1\}$ is a measure of the distance-correlation of local A and B atom number-density fluctuations. It describes the overall structure and is directly measured by a scattering experiment which does not distinguish A from B atoms (i.e., one for which $\Delta b = 0$). $S_{CC}(k)$ oscillates around $x_A x_B$ and approaches this value at high k-values. The term containing $S_{CC}(k)$ in equation (3) is called Laue diffuse scattering. $S_{CC}(k) = x_A x_B$ if the mutual distribution of A and B atoms is random. $4\pi r^2 \rho_{CC}(r) = 4\pi r T_F\{S_{CC}/x_A x_B - 1\}$ is negative for distances with preference for unlike pairs and positive for like pairs. From this term the Warren SRO parameters are calculated in the case of solid disordered solutions. The scattering from a so-called zero alloy, for which $\langle b \rangle = 0$, measures $S_{CC}(k)$ directly. S_{NC} oscillates around zero and describes the distance-correlation between density and concentration fluctuations. $S_{NC}(0) = 0$ if $(1/V)\, dV/dx = 0$. The total $S_{NC}(k)$ vanishes if, in addition, the local number density of A and B atoms around an A atom is the same as around a B

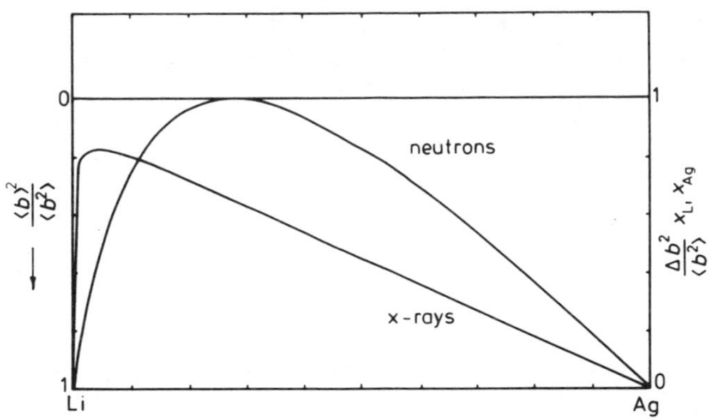

Figure 4. Mean contributions of the S_{NN} (left) and the S_{CC} scattering (right) to the total normalized intensity $S(k)$ of LiAg alloys. The x-ray curve has been calculated at $k = 2$.

atom. In the limit $x_j = 1$, $S_{NN}(k)$ becomes equal to the structure factor of pure j while $S_{CC}(k)$ and $S_{NC}(k)$ vanish. The $a_{ij}(0)$, $S_{mn}(0)$ and $S(0)$ are related to E^{xs}, V, dV/dx and χ_T.

The x-dependence of the mean contributions to $S(k)$, of the S_{NN} and of the S_{CC} scattering respectively, is shown in figure 4. For neutrons this curve is independent of k and for x-rays it has been calculated for $k = 2$. At the zero-alloy composition, with $x_{Li} = 0{\cdot}72$, the neutron diffraction pattern reveals directly the diffuse Laue scattering. The x-ray diffraction patterns are most sensitive to concentration fluctuations at $x_{Li} = 0{\cdot}95$, where the overall structure contributes only about 15% to the total

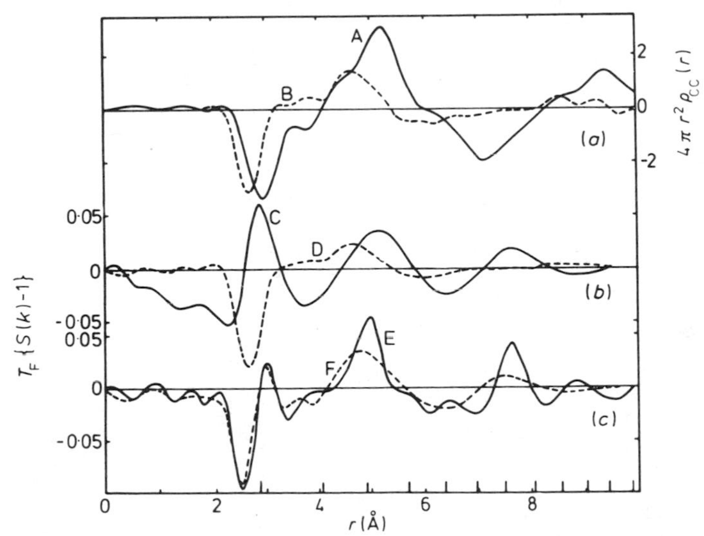

Figure 5. (*a*) Radial concentration correlation curves of the liquid zero alloys: A, $Li_{0{\cdot}80}Pb_{0{\cdot}20}$ at 800 °C; B, $Li_{0{\cdot}715}Ag_{0{\cdot}285}$ at 320 °C. (*b*) Fourier transforms (T_F) of liquid $Li_{0{\cdot}715}Ag_{0{\cdot}285}$ at 320 °C: C, x-ray diffraction pattern; D, neutron diffraction pattern. (*c*) Fourier transforms of the neutron diffraction patterns of $Li_{0{\cdot}55}Ag_{0{\cdot}45}$: E, solid at 360 °C; F, liquid at 500 °C. The bars at 2·85 Å, 4·05 Å, etc, indicate the distance from the origin of the first, second, etc, coordination shell.

Table 1. Coefficients of equation (4) at $k = 0$ and $k = 9$.

	a		*b*		*c*	
k	0	9	0	9	0	9
$S_{CC}/x_{Li}x_{Ag}$	0·03	0·03	1·0	1·0	0	0
S_{NN}	5·7	6·3	−1·7	−2·0	2·7	3·0
a_{LiLi}	8·5	9·0	−1·3	−1·6	2·7	3·0
a_{AgAg}	−1·4	−0·8	0·9	0·6	2·7	3·0
a_{LiAg}	3·5	4·0	−2·7	−3·0	2·7	3·0

scattering Correspondingly the peaks at $k = 1·7$ in figure 3 are S_{CC} peaks, which had also been observed at the same k value with LiPb alloys and which are due to SRO with preference for unlike nearest neighbours. This SRO is still present in dilute solutions of Ag in Li, as can be seen from $S^x(k)$ for the 90 and 95% Li alloys in figure 3. It should be noted that a hump at small k values is still visible in the 97·5 and 99% Li curves, but not in the curve of pure Li. The corresponding $S^x(k)$ curves could not be calculated because of the low x-ray absorption of the samples.

A comparison of the $S^n(k)$ curves of solid and liquid $Li_{0·55}Ag_{0·45}$ (figure 3) and of the corresponding curves in r-space (figure 5) shows that the SRO persists almost unchanged through the melting region. A quantitative description of the SRO is possible in the case of the zero alloy, for which the curve $4\pi r^2\rho_{CC}(r)$ is depicted in figure 5. The negative peak at about 2·7 Å indicates preference for unlike nearest neighbours. If it is supposed that each atom in the melt is surrounded by ten nearest neighbours inside a

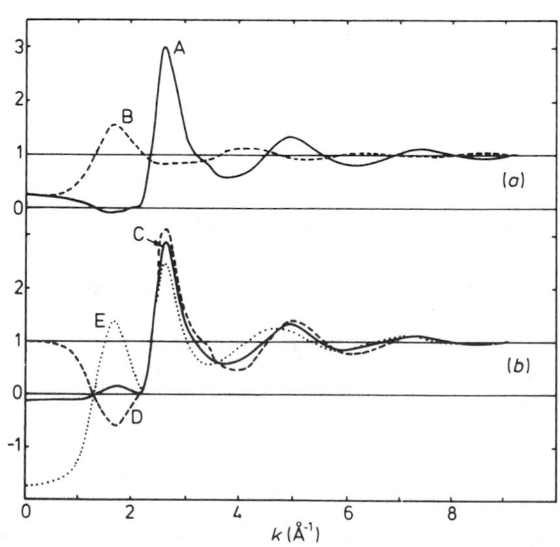

Figure 6. (*a*) Curves $S_{mn}(k) - aS_{NC}(k)$ and (*b*) curves $a_{ij}(k) - aS_{NC}(k)$ of liquid $Li_{0·715}Ag_{0·285}$ at 320 °C, calculated from x-ray and neutron diffraction data according to equation (4): A, $S_{NN} - 6·0S_{NC}$; B, $S_{CC}/x_{Li}x_{Ag} - 0·03S_{NC}$; C, $a_{LiLi} - 8·8S_{NC}$; D, $a_{LiAg} - 3·8S_{NC}$; E, $a_{AgAg} + 1·1S_{NC}$. The coefficients, a, are given for $k = 2$.

sphere of about 3·5 Å, the Warren SRO parameter is about −0·15, compared with −0·25 for liquid $Li_{0\cdot80}Pb_{0\cdot20}$ at 800 °C. For LiPb the SRO is much more pronounced and, as can be observed from figure 5, extends over a larger distance.

Because $S^n(k)$ and $S^x(k)$ of liquid $Li_{0\cdot715}Ag_{0\cdot285}$ (figure 3) are very different, the two sets of data were combined in order to obtain further information about partial structure factors F_{PS}. From equations (1) and (3), expressions of the following type were calculated:

$$F_{PS} - aS_{NC} = bS^n(k) + cS^x(k) + 1 - b - c \tag{4}$$

where a, b and c are given in table 1 for $k = 0$ and 9. The calculated curves and their Fourier transforms are depicted in figures 6 and 7 respectively. In figure 7, $f_{ij}(S_{NC})$ is approximately equal to $T_F\{S_{NC}\}$ multiplied with the corresponding value of a from table 1. Because of the similar size of Li and Ag in the alloy, $S_{NC}(k)$ is probably small and therefore these curves are quite similar to the true F_{PS} and their Fourier transforms. Of course, in reality S_{NC} does not vanish. From $S^n(0)$ and from thermodynamic data it follows that $S_{NC}(0) \simeq -0\cdot02$, $S_{CC}(0) \simeq 0\cdot05$, $S_{NN}(0) \simeq 0\cdot04$, $a_{LiLi}(0) \simeq -0\cdot33$, $a_{AgAg}(0) \simeq -1\cdot75$ and $a_{LiAg}(0) \simeq 0\cdot86$. And, because $S_{NN}(k)$ is always positive, one should conclude from figure 6 that $S_{NC}(k)$ stays negative till $k \simeq 1\cdot2$ and then changes its sign. Thus, for small k values the F_{PS} except $a_{Ag\,Ag}$ will be somewhat below the curves of figure 6 and somewhat higher beyond $k \simeq 1\cdot2$. But their general shape will not deviate very much from these curves. The same is true if the negative values of $S_{NN}(k)$ are not due to $S_{NC}(k)$ but to $S^x(k)$ which is very sensitive to experimental errors in this region of k.

From the curves in figure 7 it can be observed that the Li–Li distance is smaller than in pure Li, for which it is about 3·1 Å. The Li–Ag distance seems to be smaller than the

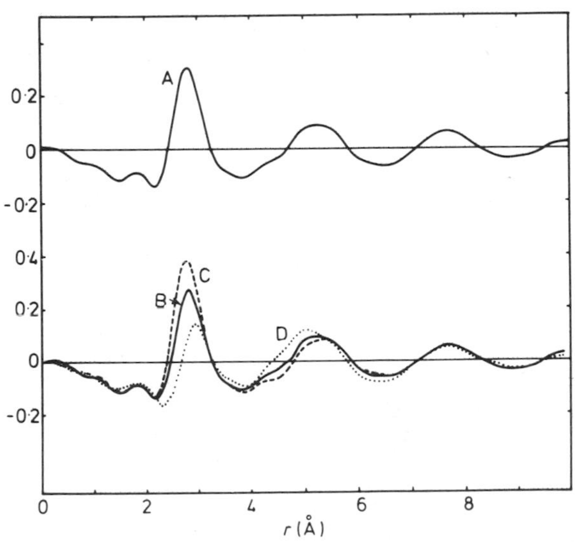

Figure 7. Fourier transforms of the curves shown in figure 6. $f_{ij}(S_{NC})$ and $f_{mn}(S_{NC})$ are approximately equal to $T_F\{S_{NC}\}$ multiplied with the corresponding value of a from table 1: A, $T_F\{S_{NN}-1\} - f_{NN}(S_{NC})$; B, $T_F\{a_{LiLi}-1\} - f_{LiLi}(S_{NC})$; C, $T_F\{a_{LiAg}-1\} - f_{LiAg}(S_{NC})$; D, $T_F\{a_{AgAg}-1\} - f_{AgAg}(S_{NC})$.

mean distance between like atoms. Because of the unknown S_{NC}, this conclusion is somewhat doubtful but it is confirmed by the $T_F\{S^n(k) - 1\}$ curve of the solid alloy (figure 5) where the negative peak at about 2·5 Å does not coincide with the nearest-neighbour distance of 2·85 Å, calculated from the lattice parameter.

Acknowledgments

The financial support by the Deutsche Forschungsgemeinschaft, by the Institut Max von Laue–Paul Langevin (ILL), by the Institut de Recherches de la Siderurgie (IRSID), and by the Verein Deutscher Eisenhüttenleute (VdeH) are gratefully acknowledged. We would like to thank Dr G Schwitzgebel and Mr W Becker for supplying us with unpublished thermodynamic data and Mr P Rathjens for his help in the data processing.

References

Bhatia A B and Thornton D E 1970 *Phys. Rev.* **132** 3004
Biltz W and Weibke F 1935 *Z. Anorg. (Allg.) Chem.* **223** 321
Blech I A and Averbach B L 1965 *Phys. Rev.* **A137** 1113
Firth L D, Nowaira N H A and Scott W 1974 *J. Phys. F: Metal Phys.* **4** L200
Freeth W E and Raynor G V 1953 *J. Inst. Metals* **82** 569
Goebbels K and Ruppersberg H 1973 *The Properties of Liquid Metals* ed S Takeuchi (London: Taylor and Francis) p 63
Pronin L A and Filippov S I 1963 *Izv. Vyssh. Ucheb. Zaved.* **5** *10*
Ruppersberg, H 1975 *Phys. Lett.* **54A** 151
Ruppersberg H and Egger H 1975 *J. Chem. Phys.* **63** 4095
Ruppersberg H and Reiter H 1972 *Acta Crystallogr.* **A28** 233
Ruppersberg H and Speicher W 1976 *Z. Naturf.* **31a** 47
Veazey S D and Roe W C 1972 *J. Mater. Sci.* **7** 445
Yarnell J L, Katz M J and Wenzel R G 1973 *Phys. Rev.* **A7** 2130

Structure factors of liquid caesium and sodium–caesium alloys

M J Huijben and W van der Lugt
Solid State Physics Laboratory, Materials Science Center, University of Groningen, Melkweg 1, Groningen, The Netherlands

Abstract. X-ray transmission diffraction experiments have been carried out for liquid sodium, potassium, caesium and some sodium–caesium alloys. The structure factor of liquid pure caesium at the melting point can be scaled down to those of sodium and potassium with reasonable accuracy.

Adopting the methods developed by Brady and Greenfield (1967) and by Greenfield *et al* (1971) an x-ray goniometer for transmission experiments in liquid alkali metals has been constructed. Special attention was paid to solving one of the experimental problems mentioned by these authors: that of obtaining a thin liquid metal layer with a well defined, homogeneous thickness. For this purpose, the sample holder is provided with two parallel beryllium windows with continuously adjustable distance. A set of rings prevents air from entering the sample holder during adjustment as well as during the actual diffraction experiment. The step scanning method was used with intervals of $1/4°$ for 2θ.

Structure factors of liquid sodium, potassium and caesium have been determined in addition to the x-ray diffraction pattern of some sodium–caesium alloys. For sodium and potassium, the overall agreement with the work by Greenfield *et al* (1971) is satisfactory, although, particularly for low q values, some discrepancies, as yet not explained, exist.

The x-ray results for liquid caesium are, as far as the authors know, the first of this kind. They are presented for different temperatures between the melting point (approximately 30°C) and 150°C in figure 1. With increasing temperature the peaks become lower and broader, the oscillations are progressively damped. This effect, of course, also shows up in the pair-correlation functions, one of which was published elsewhere (Huijben and van der Lugt 1976). On the other hand, the position of the first peak of the structure factor is independent of temperature within the limits of accuracy.

Using model potentials as calculated by Hallers *et al* (1974), the electrical resistivity and its temperature derivative have been computed from the measured structure factors. They are listed in table 1. The overall agreement is satisfactory.

Figure 2 demonstrates the result of an attempt to make the structure factors of sodium, potassium and caesium coincide as precisely as possible by scaling the q values. Evidently this attempt is rather successful, except for the low q values. All data in figure 2 pertain to the melting points.

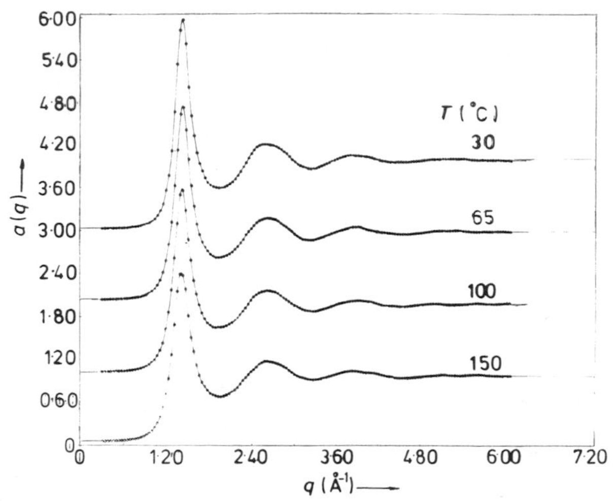

Figure 1. The experimental structure factors $a(q)$ of liquid caesium at four different temperatures.

The measurements have been extended to liquid alloys of caesium with sodium. This system was chosen because the atomic volumes of the components differ widely (by a factor of 3), so that, of all binary alkali systems, these alloys are the most likely to exhibit structural effects. Several investigations of different physical properties provide indications in favour of such effects occurring on alloying, whereas others do not. A review was given by Feitsma *et al* (1976) at this conference. X-ray investigations on a number of alloys throughout the entire composition range have been carried out. An example is given in figure 3. In the limit of $q \to 0$, the intensity of the diffracted beam increases rapidly if sodium is added to caesium and falls suddenly for sodium concentrations higher than at least 90%. Simultaneously the oscillations in the diffraction pattern are progressively damped. A clear-cut interpretation of the results has not yet been obtained. As is well known, for obtaining the three partial structure

Table 1. The electrical resistivity, ρ, and its first derivative with respect to the temperature ($d\rho/dT$), at four different temperatures calculated for liquid caesium using the experimental structure factors (*a*). For comparison, the values of ρ calculated with the hard-sphere Percus–Yevick structure factor (*b*) as well as the experimental results of Feitsma *et al* (1974) and Hennephof *et al* (1972) are added (*c*).

	(*a*)		(*b*)		(*c*)	
T (°C)	ρ ($\mu\Omega$ cm)	$(d\rho/dT)$ ($\mu\Omega$ cm/°C)	ρ ($\mu\Omega$ cm)	$(d\rho/dT)$ ($\mu\Omega$ cm/°C)	ρ ($\mu\Omega$ cm)	$(d\rho/dT)$ ($\mu\Omega$ cm/°C)
30	40·374				38·346	
		0·0842				0·1083
65	43·322				42·136	
		0·1111				0·1096
100	47·210		52·8		45·973	
		0·1222				0·1112
150	53·323				51·535	

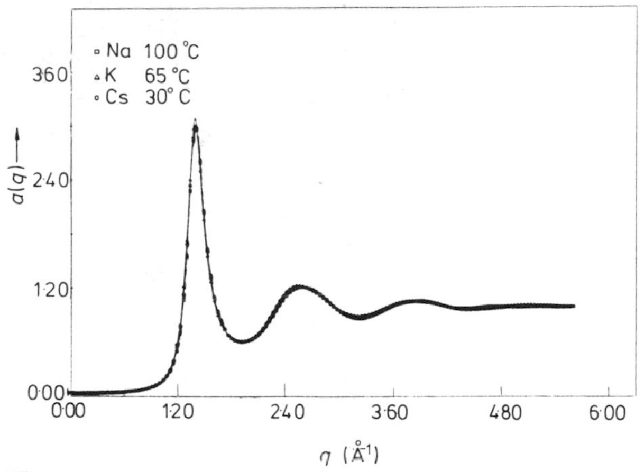

Figure 2. The structure factors $a(q)$ of sodium, potassium and caesium after scaling the q values.

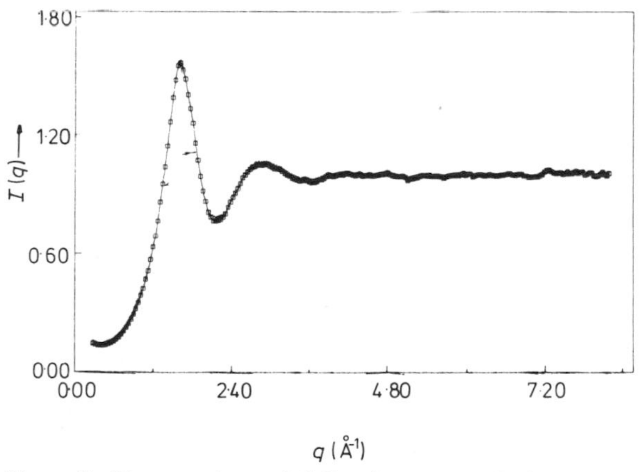

Figure 3. The experimental diffraction pattern $I(q)$ at T = 100 °C of a liquid sodium–caesium alloy containing 75·18 at.% sodium. $I(q)$ is obtained from the measurements by dividing the normalized intensity curves, after subtraction of the incoherent scattering, by $\langle f^2(q)\rangle$, f being the atomic scattering factor.

factors, three independent measurements are required. We are attempting to get at least two of them by carrying out neutron diffraction experiments in parallel with the x-ray experiments, using the facilities of the RCN (Reactor Centrum Nederland). As this neutron experiment is still in an initial stage, more definite results cannot be presented here.

Acknowledgments

The authors want to thank Mr M R Leenstra for making the first design of the sample holder. They acknowledge the technical assistance of Mr J Roode in constructing the x-ray diffractometer and Mr F van der Horst for many useful discussions.

This work is part of the research programme of the 'Stichting voor Fundamenteel Onderzoek der Materie' (FOM) and has been made possible by financial support from the 'Nederlandse Organisatie voor Zuiver Wetenschappelijk Onderzoek' (ZWO).

References

Brady G W and Greenfield A J 1967 *Rev. Sci. Instrum.* **38** 736–9
Feitsma P D, Hennephof J and van der Lugt W 1974 *Phys. Rev. Lett.* **32** 295–7
Feitsma P D, Lee T and van der Lugt W 1976 this volume
Greenfield A J, Wellendorf J and Wiser N 1971 *Phys. Rev.* **4** 1607–16
Hallers J J, Mariën T and van der Lugt W 1974 *Physica* **78** 259–72
Hennephof J, van der Lugt W, Wright G W and Mariën T 1972 *Physica* **61** 146–51
Huijben M J and van der Lugt W 1976 *J. Phys. F: Metal Phys.* **6** 225–9

Structure of liquid alloys from Cu–Ge and Cd–Sb alloy systems

B R Orton

Physics Department, Brunel University, Uxbridge, Middlesex

Abstract. A double hard-sphere model, containing features similar to those proposed by Ashcroft (1977), has been successfully applied to reproduce the structure factors of liquid Ge (Orton 1975) and Sb (Orton 1976). The essential feature of this model is that the liquid is composed of two atomic species having different distances of atomic separation and the interaction between unlike species is similar to that within one set of species. If it is assumed that the interactions between the like components of a liquid alloy are similar to those measured in the pure state, it is found that the observed structure factors for Cu–Ge alloys (Isherwood and Orton 1972) can be computed using a set of hard-sphere structure factors. It must, however, be assumed that the Cu–Ge interaction is similar to that of the pure component which has the largest concentration. In the case of Cd–Sb alloys (Animashaun 1975), using the same set of assumptions, it is only possible to compute certain features of the structure factors of this alloy system.

References

Animashaun R O 1975 *PhD Thesis* Brunel University
Ashcroft N W 1977 this volume
Isherwood S P and Orton B R 1972 *J. Phys. C: Solid St. Phys.* **5** 2977
Orton B R 1975 *Z. Naturf.* **30A** 1500
—— 1976 *Z. Naturf.* **31A** 397

Structural study of liquid eutectic Ag–Ge alloy by means of neutron diffraction

M C Bellissent-Funel and P J Desre

ENSEEG, 38401 Saint Martin d'Heres, France

Abstract. In this paper, we present our results concerning neutron scattering studies of the structure of the liquid eutectic Ag–Ge. This study has been justified by the abnormal behaviour of the thermodynamic properties of alloys of noble metals and semiconductors. First, by means of classical diffraction at 0·7 Å and 1·13 Å neutron wavelengths (0·1 to 15 $Å^{-1}$ wavevector values) we have determined the partial interference functions of the eutectic system. The former experiments were performed using the concentration method. The latter ones were achieved using the isotopic method which allows us to test the invariance of partial interference functions with concentration. The partial pair correlation functions were deduced from these partial interference functions and the scattered intensity for small angles in the wavevector range 0·01 to 0·2 $Å^{-1}$ was measured for alloys of different concentrations.

A specific model for the eutectic Ag–Ge alloy is proposed. It fits both small-angle and large-angle scattering measurements. The microscopic physical properties which have been experimentally determined in this study are interpreted on the basis of this model.

Determination of effective interionic potentials from experimental data

S K Mitra†

Theoretical Physics Division, AERE Harwell, Oxfordshire OX11 0RA, UK

Abstract. In the present work, new theoretical methods are suggested to extract an effective interionic potential from the structure and thermodynamic data of liquid metals. These methods combine the perturbation theory and thermodynamic consistency approaches with the calculations of molecular dynamics (MD) simulation. Several metals including rubidium, copper and nickel have been studied in this way. A comparison of the calculated potentials reveals many interesting facts; for example, the strength of the repulsion increases from rubidium to copper and nickel and an oscillatory tail is present for all three potentials. For copper and nickel these potentials were used to calculate the solid state properties, that is, the phonon dispersion relation, stacking fault energy and Debye–Waller factor. The calculations show excellent agreement with experimental values.

1. Introduction

In the calculation of various properties of liquid metals, a knowledge of interionic forces is essential. At present, there are two principal ways of determining this effective two-body potential for liquid metals: (i) from the electron theory of metals in the framework of the generalized pseudopotential approximation (Faber 1972) and (ii) from molecular theories of liquids (Croxton 1974).

So far, the electron theory approach has proved to be successful mainly for the alkali metals. In noble and transition metals, the pseudopotential framework seems to be less promising because of the complications introduced due to the d-level electrons. The molecular theory approach, which evolved from various statistical approximations (e.g., Percus–Yevick (PY), Born–Green (BG) and hypernetted chain (HNC)) to its present, fully thermodynamically consistent (FTC) integral equation form (Brennan *et al* 1974), has been successfully used in rare-gas liquids and to some extent in simple liquid metals (Mitra *et al* 1976). In principle one can use this method for any system as long as all the relevant data of the system are available (i.e., the structure factor, $S(k)$ and the thermodynamic parameters). But the pure molecular theory in its present form deals only with particles interacting among themselves: for liquid metals it does not account for the cohesive effect of the conduction electrons and their contribution to thermodynamic properties. Also, it is difficult to determine experimentally the contribution from the ions alone.

In the present work we have tried yet another method, that is, molecular dynamics simulation in conjunction with the FTC equation. Essentially, the calculation of the pair distribution function, $g(r)$ and other thermodynamic properties by MD simulation,

† Present address: Institute of Theoretical Physics, S402-20 Göteborg, Sweden.

tests the calculated effective potential obtained from the FTC equation. Then $g(r)$ from MD can be compared with $g(r)$ calculated by Fourier transforming the experimental structure factor $S(k)$ (henceforth we shall call this the experimental $g(r)$) (Schiff 1969).

2. Calculation procedure

2.1. Fully thermodynamically consistent integral equation

It is well known that the structure factor and the pair distribution are themselves insensitive to details of the potential. It is only when these are combined with the potential to calculate pressure or internal energy that sensitivity of the potential is manifest. Thus any statistical theory to be used in the calculation of the potential must include pressure, energy and compressibility directly or indirectly as basic constraints. One of the recent theories proposed on these lines is the FTC theory. The details of its derivation are given elsewhere (Brennan 1974) and here we describe only the FTC equation.

We define the following quantities in addition to the pair distribution function $g(r)$ and the effective pair potential $\phi(r)$:

$$h(r) = g(r) - 1, \tag{1}$$

$$y(r) = g(r) \exp(\beta\phi(r)), \tag{2}$$

$$\epsilon(r) = \beta\phi(r) g(r), \tag{3}$$

and

$$\Delta(r) = (y^s(r) - 1)/s, \tag{4}$$

where β is the inverse of temperature in energy units.

Denoting the Fourier transform of these functions by a tilde, the FTC equation is:

$$\tilde{\Delta}(k) = \tilde{h}(k) \frac{n\tilde{h}(k) + nt\tilde{\epsilon}(k)}{1 - t + n\tilde{h}(k)} \tag{5}$$

where n is the number density and s and t are parameters which, in theory, are to be determined by the conditions of thermodynamic consistency. For $t = 0$, equation (5) reduces to the pressure consistency theory of Hutchinson and Conkie (1972); $s = 0$ and 1 then correspond to the HNC and PY theories respectively. It is important to note that these parameters are volume and temperature dependent and thus it is impossible to calculate any volume or temperature dependence of the thermodynamic properties by direct use of equation (5) for a given set of s and t.

It is not very straightforward to use the FTC equation directly to extract the potential for liquid metals because thermodynamic data on ions are not available. However, in some special cases, where the scattering experiments are performed at different temperatures but at the same density, one can use these $S(k)$ and the specific heat at constant volume, C_V, as the thermodynamic constraint (regarded as the difference in internal energy at two different temperatures) in the FTC equation, because the electronic contribution to the specific heat is only a few per cent in the neighbourhood of the melting point. The details of this type of calculation are given by Mitra *et al* (1976) and in the next section we shall discuss their results briefly.

2.2. FTC equation in conjunction with the mean spherical model (MSM) and molecular dynamics simulation (MD)

The above-mentioned difficulties, together with the lack of elaborate scattering experiments and thermodynamic data for many liquid metals, seriously limits the use of the FTC equation in extracting the effective interionic potential. One way to overcome this is first to approximate the effective potential by the MSM approximation and then refine it by the FTC equation and MD simulation. Here we briefly describe the MSM and justify its use in the present calculation.

It has been observed that a rigid sphere potential explains the main features of the structure factor of noble and transition liquid metals (Waseda and Tamaki 1975). The attractive tail to this potential, which is needed to account for the detailed thermodynamics, can be added from the direct correlation function $C(r)$. That is:

$$\phi(r) = \phi_0(r) \qquad 0 \leqslant r \leqslant \sigma \tag{6}$$

$$= -\frac{1}{\beta}\, C(r) \qquad r > \sigma. \tag{7}$$

To a first approximation, this potential ensures that the thermodynamic properties of the model fluid will not be too unreasonable, but it has to be improved with a softer core and a better attractive tail in order that the desired pair distribution function and thermodynamics are reproduced. This is achieved by refining the model potential using the FTC equation and MD simulation. The softness of the core and depth of the potential (at the first minima) reveal themselves in MD simulation through a smooth rise in $g(r)$ near the core radius and through the height of the peaks in $g(r)$ respectively. Similarly in the FTC equation, the change in the parameters s and t changes the core radius, softness and depth of potential. Thus under the general guideline of the MSM calculation, one can calculate a desired effective interionic potential for the given set of $S(k)$ data from the FTC equation, which will reproduce the experimental $g(r)$ and the thermodynamic properties of the system.

3. Results

Using the procedure described in §2 we report here calculations of the effective interionic potential in three metals; rubidium, copper and nickel. These calculated potentials, when used in the calculation of the MD simulation, reproduced the main features of the experimental $g(r)$ and the specific heat. The details for each liquid now follow.

3.1. Rubidium

As described earlier, and in the paper of Mitra *et al* (1976), the calculation procedure needs structure factors at two different temperatures and at fixed density (assuming that the temperature dependence of the effective potential is negligible). We have used the neutron diffraction data of Wingfield and Enderby (quoted by Howells 1973) for liquid rubidium and these satisfy the above requirement in that measurements were made at 45 °C and 200 °C at a fixed density of 0·01058 atm $Å^{-3}$. The value of specific heat at constant volume was taken from Vilcu and Misdolea (1968). For this set of data,

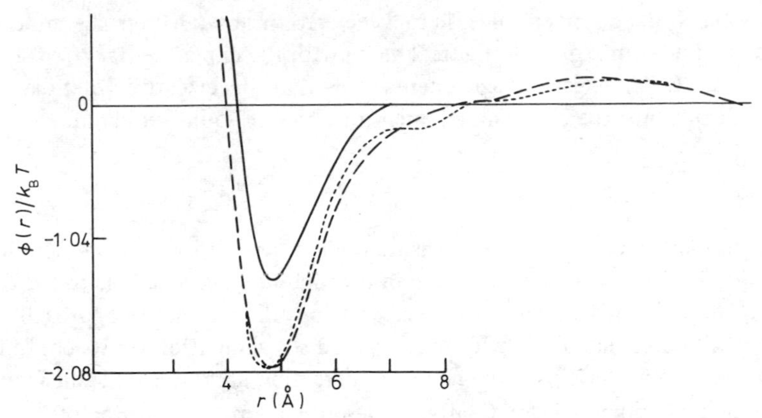

Figure 1. Effective interionic potential for liquid rubidium: solid line, from Price *et al* (1971); broken line, present calculation at 200 °C; dotted line, present calculation at 45 °C.

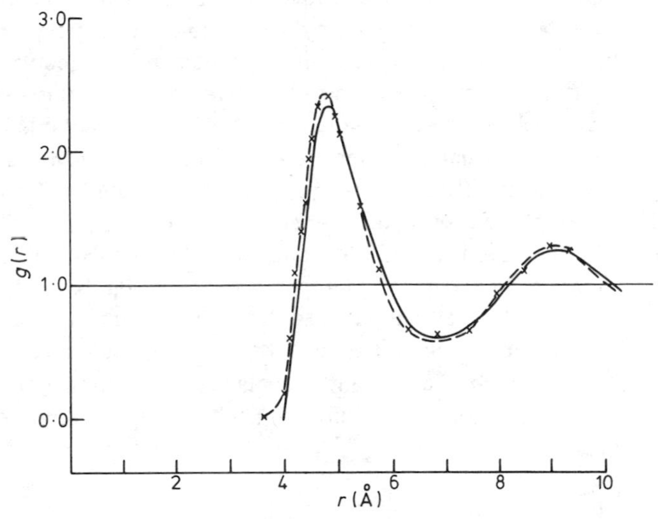

Figure 2. Pair distribution function for liquid rubidium at 45 °C: solid line, from Fourier transformation of $S(k)$; broken line, from MD simulation.

FTC equations have been solved for different sets of s and t parameters so as to get the correct specific heat and similar looking potentials at these temperatures. As a further test of the calculated potential, we have used it in MD simulation to calculate the pair distribution function, $g(r)$.

In figure 1 we have plotted the calculated potential for liquid rubidium along with a potential derived from electron theory by Price *et al* (1971). Although at first sight they look very different, the first derivative of the potential (i.e., the interionic force) for these two look remarkably similar.

In MD simulation of the liquid rubidium phase, the melting point was observed to be near the experimental melting point and the specific heat also agreed well with the observed value (because of the small size, the melting point from MD calculation does not always match the observed value; in fact the error is about 10%). The molecular dynamics $g(r)$ has been plotted in figure 2 along with the experimental $g(r)$ for liquid rubidium at 45 °C. These general agreements show that this effective potential is adequate in explaining the constant volume properties of liquid rubidium.

3.2. Copper

So far, no attempt has been made to measure the neutron or x-ray diffraction data for liquid copper at different pressures, which in turn inhibits our attempts to use directly the FTC equation for this system. Thus here we adopt the alternative approach, using the FTC equation in conjunction with the MSM and MD simulation approach. In the present calculation we have used the neutron diffraction data of Breuil and Tourand (1970) for liquid copper at 1150 °C, after some refinement in the region of $k < 2{\cdot}5$ Å^{-1} to fit the compressibility at $k = 0$. The value of specific heat is taken from Tamaki and Waseda (1976).

We first approximate the effective interionic potential by a hard-core repulsive potential and approximate the soft attractive tail by the scaled direct correlation function $C(r)$ outside the core radius. This core radius was calculated by requiring the MSM structure factor to fit the observed $S(k)$ at small wavevectors. Then this MSM potential was treated as a guideline for FTC calculation to find a set of suitable s and t parameters. Such an effective potential generated from the FTC equation was further refined by MD simulation as mentioned earlier. In figure 3 the calculated effective interionic potential has been plotted for liquid copper. The simulated pair distribution function for this potential has reproduced all the main features of the experimental $g(r)$ which can be seen in figure 4. The melting temperature of the simulation system has been observed to be near 1000 °C as compared to the melting point of copper around 1100 °C. The specific heat at constant volume obtained from this calculation is $3{\cdot}37\,NK_\beta$ as compared to $3{\cdot}03\,NK_\beta$ reported by Tamaki and Waseda (1976). Assuming this effective potential to be valid in the solid phase also, we have used it to calculate some well

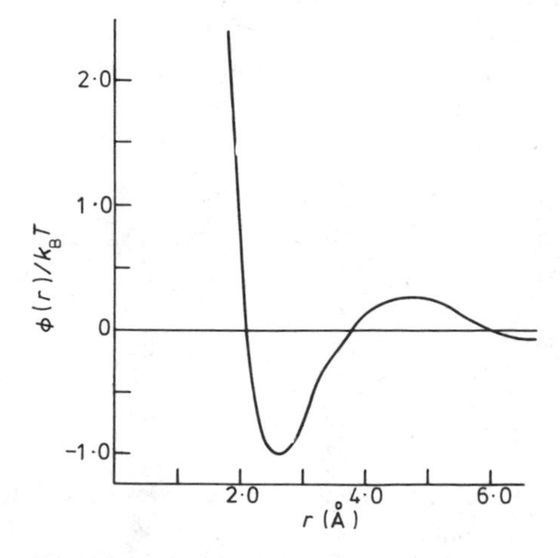

Figure 3. Effective interionic potential for liquid copper at 1150 °C.

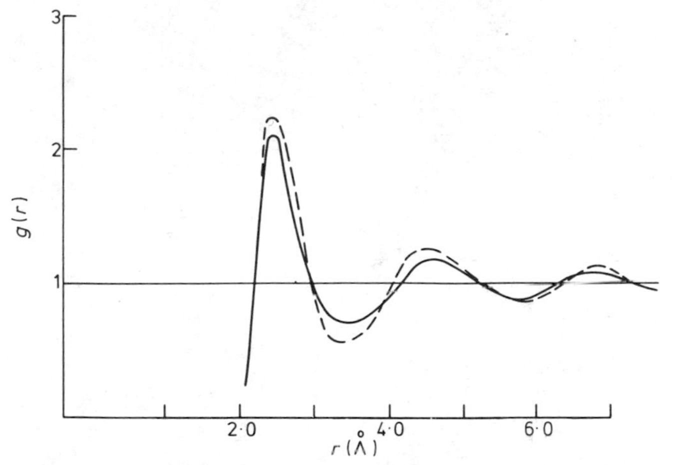

Figure 4. Pair distribution function for liquid copper at 1150 °C: solid line, from Fourier transformation of $S(k)$; broken line, from MD simulation.

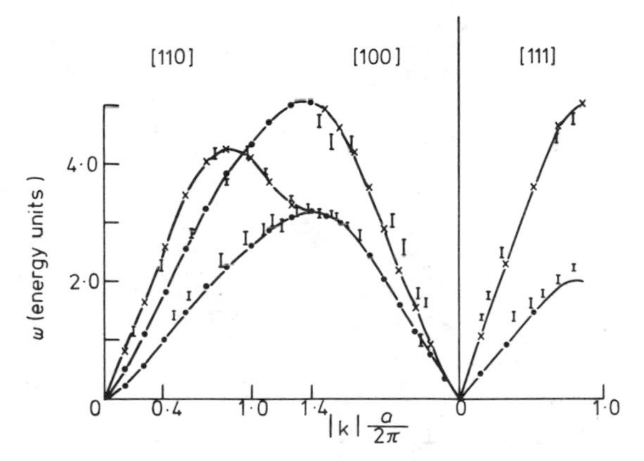

Figure 5. Phonon dispersion relation for copper at 300 K: solid line, present calculation; points with error bar, from neutron scattering results of Sinha (1966).

known solid state properties of copper, namely, the phonon dispersion relation and the stacking fault energy. Phonon dispersion relations for this potential have been compared with the neutron scattering results of Sinha (1966) in figure 5, and are in good agreement. While this is not a sensitive test of the potential it is nevertheless a necessary check that the potential is reasonable. The stacking fault energy (which is a sensitive test of the long-range part of potential) calculated for the present potential is about 20% higher than the observed one (Englert *et al* 1970). This discrepancy may be due to the uncertainty involved in numerical Fourier transformation, which introduces an error in the potential beyond fifth-neighbour distance.

3.3. Liquid nickel

In the present work we have used the neutron diffraction data of Johnson *et al* (1976) for liquid nickel at 1600 °C. Here also we have adopted the same approach as that

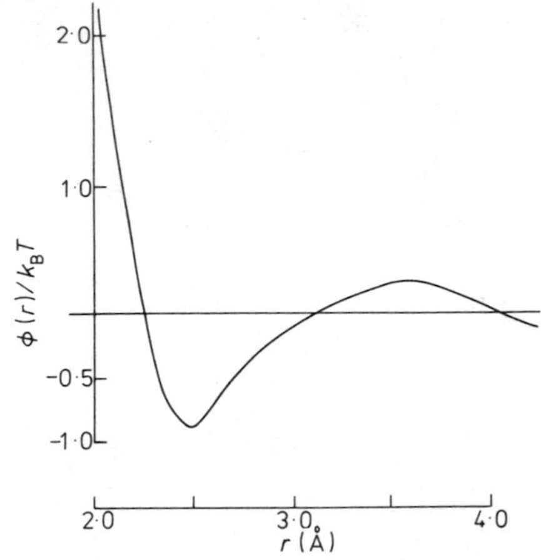

Figure 6. Effective interionic potential for liquid nickel at 1600 °C.

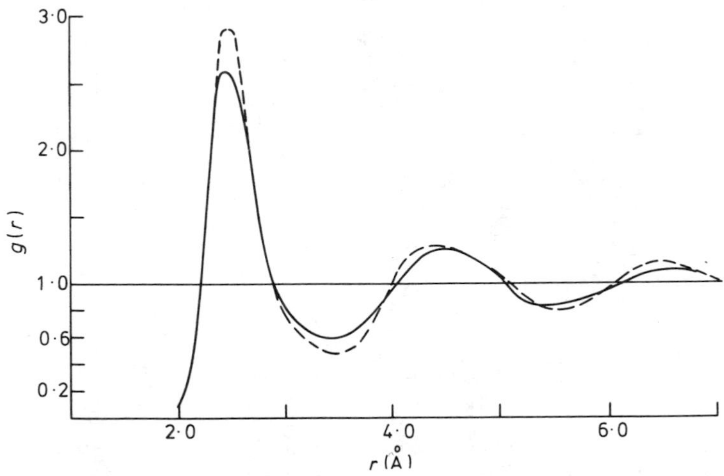

Figure 7. Pair distribution function of liquid nickel at 1600 °C: solid line, from the Fourier transformation of $S(k)$; broken line, from MD simulation.

discussed in the last section for copper (and for the same reasons). After going through the whole process of approximating and refining the effective interionic potential for this system we have achieved a reasonable potential (plotted in figure 6) which reproduced the main features of the experimental $g(r)$ when used in the simulation program. In figure 7 we have compared the $g(r)$ from MD with the experimental $g(r)$. The simulation calculation also satisfies the melting criteria and the specific heat at constant volume calculated from this potential agrees with the reported value of C_V for this system (Tamaki and Waseda 1976).

As the change in volume due to melting is very small, we have assumed that this effective potential will be valid for solids also and we simulated the solid phase of nickel to calculate the Debye–Waller factor at room temperature. In figure 8 we have compared these with the recent observed values of Windsor and Sinclair (1976), and they are in good agreement.

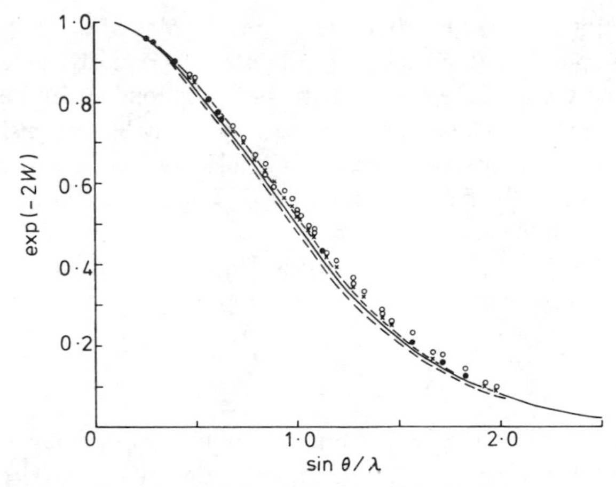

Figure 8. Debye–Waller factor of solid nickel at room temperature: solid line, the experimental results (and broken lines, the experimental uncertainty) of Windsor and Sinclair (1976); full circles, from the MD simulation.

3.4. Comparison of effective interionic potentials

The present selection of liquid metals are each from a different periodic group (namely rubidium from the alkali metal group, copper from the noble metals and nickel from the transition metal group) and this is a fairly good selection for understanding the basic nature of the effective potential from the point of view of their electronic configuration. Unfortunately there is no detailed knowledge of electron theory yet available to us to carry out such a study. Thus here we shall attempt to find some possible reasons to explain the present results of the effective interionic potentials.

From figures 1, 3 and 6, the following are the common features we observe in the potentials: (i) all these potentials have a rather softer repulsive core in comparison to rare gas liquids and (ii) each has a long-range damped oscillatory tail.

The other important features of these potentials are given in table 1, where σ is the hard-core radius of the reference hard-core system (as determined by the MSM procedure described earlier), η is the packing fraction of this reference system and p, the repulsivity, is defined as $\phi(r) = C(1/r)^p$ near the core. From table 1 we note that the packing fractions η of all these three liquids are approximately equal, around 0·5, which suggests that characteristics of the reference system structure factors remain the same for these liquids. The repulsivity, p, increases from rubidium to copper and nickel. One of the

Table 1. Important features of interionic potentials.

Elements	Hard-core radius σ (Å)	Packing fraction η	Repulsivity p	Depth of potential (eV)
Rubidium	4·52	0·505	7·315	0·064
Copper	2·393	0·49	10·67	0·13
Nickel	2·290	0·498	12·87	0·118

reasons for this change in repulsivity may be attributed to overlapping of ion charge clouds. In rubidium the overlap is very small and thus produces less repulsivity, whereas in copper and nickel, the weakly bound d-electrons have wavefunctions spread out to large distances, and direct interaction between d-electrons may partly be responsible for the stronger repulsive potential. The observed increase in repulsivity from Cu to Ni, which can be explained by recalling the fact that in Cu the d-electrons are more tightly bound than Ni, is compatible with the above argument.

The oscillatory tail of the potential can be explained in terms of Friedel oscillations due to a sharp cut-off of the electron distribution at the Fermi surface.

4. Conclusions

The calculations reported here demonstrate recent progress that has been made in determining interionic potentials for metals using entirely experimental data. The important ingredients of the calculation procedure are (i) the necessity to incorporate thermodynamic data (specifically compressibility, specific heat etc) because of the insensitivity of structure factor alone to details of the potential and (ii) the use of simulation methods to test and refine the results.

The results we have obtained for the three different metals are in accord with what one might expect from their different electronic structures. We find that copper and nickel have rather strong repulsive cores (see table 1), compared to rubidium. This is reflected in the thermal expansion coefficient which is much smaller for the former. Although the effective potential is expected to be volume dependent, the difference in volume between the solid and that for which the potential is derived, is less than 3% in copper and nickel, and we have therefore used it to calculate solid state properties with some success. For rubidium the difference in volume between the solid and 310 K liquid is of the order of 6% and one would not expect much success in this case. We conclude therefore that with more accurate liquid structure factor measurements over a range of temperature and pressure, it should now be possible to obtain reliable interionic potentials for metals for use in defect and other calculations in the solid state (except possibly for the alkali metals where they may not be needed since for these we have got a well established electron theory).

Acknowledgments

I am grateful to Dr P Schofield for his interest in this work and for his useful suggestions. I would also like to thank Drs P Hutchinson, M W Finnis and R C Perrin for their help.

References

Brennan M 1974 *PhD Thesis* Reading University
Brennan M, Hutchinson P, Sangster M L J and Schofield P 1974 *J. Phys. C: Solid St. Phys.* 7 L411
Breuil M and Tourand G 1970 *J. Phys. Chem. Solids* **31** 549
Croxton C A 1974 *Liquid State Physics: a Statistical Mechanical Introduction* (London: Cambridge UP) ch 2

Englert A, Tompa H and Bullough R 1970 *National Bureau of Standards, Special Publications No.* 317 vol 1 p 273
Faber T E 1972 *Introduction to the Theory of Liquid Metals* (London: Cambridge UP)
Howells W S 1973 *The Properties of Liquid Metals* ed S Takeuchi (London: Taylor and Francis) p 43
Hutchinson P and Conkie W R 1972 *Molec. Phys.* **27** 567
Johnson M W, Page D I, March N H, McCoy B, Mitra S K and Perrin R C 1976 *Phil. Mag.* **33** 203
Mitra S K, Hutchinson P and Schofield P 1976 *Phil. Mag.* to be published
Price D L, Singwi K S and Tosi M P 1971 *Phys. Rev.* **B2** 2983
Schiff D 1969 *Phys. Rev.* **186** 151
Sinha S K 1966 *Phys. Rev.* **143** 422
Tamaki S and Waseda Y 1976 *J. Phys. C: Solid St. Phys.* **6** L89
Vilcu R and Misdolea C 1968 *J. Chem. Phys.* **49** 3179
Waseda Y and Tamaki S 1975 *Phil. Mag.* **32** 273
Windsor C and Sinclair R N 1976 *Acta Crystallogr.* **A32** 395

Study of the structure and dynamics of simple classical liquids by the self-consistent field method: application to liquid rubidium †

Narinder K Ailawadi
Institut für Theoretische Physik, Freie Universität Berlin, Arnimallee 3, 1 Berlin 33, Germany

Abstract. The self-consistent field method for calculating the collective motions in classical liquids is modified to account for the fact that the radial distribution function vanishes in the highly repulsive hard-core region of the potential. With this modified theory, both the static structure and dynamics of classical liquids can be calculated within one single framework. The zeroth moment of the dynamic structure factor $S(q, \omega)$ is satisfied; however, the fourth moment is only approximately satisfied. The problem reduces to the study of the dynamic self-correlation function $S_s(q, \omega)$. The theory is expected to be valid for q and ω encountered in neutron-scattering experiments and molecular dynamics calculations.

1. Introduction

In liquid theory a description of both the structure and dynamics of liquids within one single framework is not yet available. The collective motions in liquids are studied typically by using the formalism of Kadanoff and Martin (1963) or by the Zwanzig–Mori projection operator technique (Zwanzig 1961, Mori 1965). Both of these techniques involve memory functions which can either be modelled or else calculated in a very approximate way (Götze and Lücke 1975). Alternatively, they can also be studied by the self-consistent field method developed by Singwi *et al* 1968 (hereafter referred to as I) for an interacting electron gas and later applied to classical liquids by Singwi *et al* 1970 (hereafter referred to as II) and by Pathak and Singwi (1970). In both these schemes, *a priori* knowledge of the static structure of the liquids is necessary. This is commonly obtained from molecular dynamics, Monte Carlo calculations or x-ray and neutron-diffraction experiments. Theoretically, the static structure is typically obtained by the well known Percus–Yevick hard-sphere equation (Wertheim 1963, Thiele 1963) or by the more successful WCA (Weeks *et al* 1971, Anderson *et al* 1971) perturbation theory.

In the last few months, the author and co-workers(Ailawadi *et al* 1976a, b) have been able to modify the self-consistent field method in I such that $g(r)$ vanishes sufficiently rapidly as $r \rightarrow 0$. This modification enables us to calculate the pair correlation function and the static structure factor in a self-consistent manner.

In this paper we propose to incorporate this modification of the self-consistent field method in II such that both the static structure and dynamics of simple classical liquids can be reasonably well described in one single theoretical framework. In §2, the theory

† Supported in part by the DFG.

of collective motions in classical liquids discussed by Pathak and Singwi (1970) and in II is briefly reviewed. Our proposed modification of this procedure is discussed in §3. The shortcomings and possible improvements presented in §4 conclude this paper.

2. Theory of collective motions

In I, the dynamical density response function $\chi(q, \omega)$ is expressed as

$$\chi(q,\omega) = \frac{\chi_0(q,\omega)}{1 - \psi(q)\,\chi_0(q,\omega)} \tag{1}$$

where $\chi_0(q, \omega)$ is the dynamical density response function of a free-particle (non-interacting) system and $\psi(q)$ is the Fourier transform of the polarization potential $\psi(r)$ defined by

$$\frac{\mathrm{d}}{\mathrm{d}r}\,\psi(r) = g(r)\,\frac{\mathrm{d}\phi(r)}{\mathrm{d}r}, \tag{2}$$

$\phi(r)$ being the pair potential of the system. The fluctuation–dissipation theorem relates the dynamic structure factor $S(q, \omega)$ to the imaginary part of $\chi(q, \omega)$. In particular, in the classical limit

$$S(q,\omega) = -\frac{k_{\mathrm{B}}T}{n\pi\omega}\,\mathrm{Im}\,\chi(q,\omega) \tag{3}$$

where k_{B} is Boltzmann's constant, T is the temperature and n is the number density. Using the Kramers–Kronig relation,

$$\chi(q,\omega) = \frac{1}{\pi}\int_{-\infty}^{+\infty} \mathrm{d}\omega'\,\frac{\mathrm{Im}\,\chi(q,\omega')}{\omega' - \omega - \mathrm{i}\eta}, \qquad \eta \to 0^{+}. \tag{4}$$

The zeroth moment of $S(q, \omega)$, commonly known as the static structure factor $S(q)$, is given by

$$S(q) = -\frac{k_{\mathrm{B}}T}{n}\chi(q,0). \tag{5}$$

Substituting equation (1) into (5) we obtain the relationship between $\psi(q)$ and $S(q)$:

$$[S(q)]^{-1} = \frac{n}{k_{\mathrm{B}}T}\{\psi(q) - [\chi_0(q,0)]^{-1}\}. \tag{6}$$

In particular,

$$\chi_0(q,0) = -\,n/k_{\mathrm{B}}T \tag{7}$$

and therefore equation (6) becomes

$$S(q) = \frac{1}{1 + (n/k_{\mathrm{B}}T)\,\psi(q)} \tag{8}$$

where $\psi(q)$ is equation (2). Formally, the polarization potential $\psi(r)$ is just $-k_BTC(r)$ where $C(r)$ is the direct correlation function. For a given pair potential equations (2) and (8) can in principle be solved self-consistently to calculate the polarization potential $\psi(q)$ and the static structure factor $S(q)$. However, such an iterative procedure for a Lennard-Jones liquid yields, in fact, a non-convergent and oscillatory $g(r)$ (A Sjölander 1973 private communication).

In the application of this approach to the study of the dynamics of simple classical liquids (II, Pathak and Singwi 1970, Bansal and Pathak 1975), the free-particle dynamic response function $\chi_0(q,\omega)$ is replaced by screened single-particle response function $\chi_{sc}(q,\omega)$ and both $\psi(q)$ and $\chi_{sc}(q,\omega)$ are considered to be *a priori* unknown. Pathak and Singwi (1970) and Bansal and Pathak (1975) assume $\mathrm{Im}\,\chi_{sc}(q,\omega)$ to be of the form

$$\mathrm{Im}\,\chi_{sc}(q,\omega) = -\frac{2n\omega q^2}{2q^2k_BT+m\Gamma(q)}\left(\frac{\pi m}{2q^2k_BT+m\Gamma(q)}\right)^{1/2}\exp\left(-\frac{m\omega^2}{2q^2k_BT+m\Gamma(q)}\right) \tag{9}$$

so that the free-particle response function is broadened by the width $\Gamma(q)$ of the gaussian. From equation (4)

$$\chi_{sc}(q,0) = -\frac{2nq^2}{2q^2k_BT+m\Gamma(q)}. \tag{10}$$

Equation (6) can now be written as

$$S(q) = \left(1+\frac{n}{k_BT}\psi(q)+\frac{m}{2k_BTq^2}\Gamma(q)\right)^{-1}. \tag{11}$$

If the width $\Gamma(q)$ of the gaussian in (9) reduces to zero, equation (8) and hence the theory described in I is obtained. Furthermore, the fourth moment of the dynamic structure factor $S(q,\omega)$ is given by

$$\begin{aligned}\langle\omega^4\rangle &= \int_{-\infty}^{+\infty}\omega^4S(q,\omega)\,\mathrm{d}\omega\\ &= q^2\frac{k_BT}{m}\left[6\beta(q)+\frac{nq^2}{m}\psi(q)\right]\end{aligned} \tag{12}$$

where

$$\beta(q) = -\frac{m}{6\pi nq^2}\int_{-\infty}^{+\infty}\mathrm{d}\omega\,\omega^3\,\mathrm{Im}\,\chi_{sc}(q,\omega). \tag{13}$$

For $\chi_{sc}(q,\omega)$ given by (9), equation (12) is expressed as

$$3q^2k_BT+\tfrac{3}{2}m\Gamma(q)+nq^2\psi(q) = \frac{m^2}{q^2k_BT}\langle\omega^4\rangle. \tag{14}$$

Using the molecular dynamics data on $S(q)$ and $\langle\omega^4\rangle$, the parameters $\psi(q)$ and $\Gamma(q)$ are determined uniquely from equations (11) and (14).

On the other hand, in II, equation (8) is preserved by assuming $\chi_{sc}(q,\omega)$ to be either the free-particle response function $\chi_0(q,\omega)$ as in I, or the single-particle response function

$$\operatorname{Im}\chi_{sc}(q,\omega) = -\frac{\pi n\omega}{k_B T} S_s(q,\omega) \tag{15}$$

where $S_s(q,\omega)$ is the dynamic self-correlation function representing the self-motion of the atoms. From equations (13) and (15),

$$\beta(q) = \frac{q^2}{6m}\left(3k_B T + \frac{n}{q^2}\int d^3r g(r)\frac{\partial^2\phi(r)}{\partial x^2}\right). \tag{16}$$

Thus equation (12) becomes

$$\langle\omega^4\rangle = \frac{k_B T}{m^2} q^4\left(3k_B T + \frac{n}{q^2}\int d^3r g(r)\frac{\partial^2\phi(r)}{\partial x^2} + n\psi(q)\right). \tag{17}$$

Comparing equation (17) with the exact fourth moment (de Gennes 1959)

$$\langle\omega^4\rangle = \frac{k_B T}{m^2} q^4\left(3k_B T + \frac{n}{q^2}\int d^3r\, g(r)\frac{\partial^2\phi(r)}{\partial x^2} - \frac{1}{q^2}\int d^3r g(r)\cos(qx)\frac{\partial^2\phi(r)}{\partial x^2}\right), \tag{18}$$

the 'non-self' contribution to the fourth moment is approximated by the term $n\psi(q)$. The polarization field $\psi(q)$ is still considered to be *a priori* unknown and different approximations are tried for $\psi(q)$.

3. Proposed modification

We propose to modify the theory presented in the previous section such that the static structure as well as the dynamics of simple classical liquids can be described reasonably well by the self-consistent field method. Since the Pathak and Singwi (1970) and Bansal and Pathak (1975) approaches are rather *ad hoc* and involve a parameter $\Gamma(q)$ which cannot be determined theoretically, we concentrate on the formulation as set out in II.

We restrict ourselves to that class of $\chi_{sc}(q,\omega)$ which have

$$\chi_{sc}(q,0) = -n/k_B T. \tag{19}$$

Two particular cases where this equation is satisfied are the free-particle response function $\chi_0(q,\omega)$ (as in the original formulation of I) and equation (15) relating $\chi_{sc}(q,\omega)$ to the self-motion of the atom. The static structure factor $S(q)$ is in the latter case also given by equation (8). The polarization field $\psi(q)$ is no longer considered to be unknown. However, $\psi(r)$ cannot be assumed to be written simply in the form of equation (2) since a direct iteration for the Lennard-Jones liquid has not proved fruitful. Equation (2) is derived from a physically incorrect ansatz in I: as the particle moves in a liquid, it feels the presence of distant particles in the same manner as the next-nearest neighbours. From a purely computational viewpoint, whenever the potential is harshly repulsive, the derivative $d\phi/dr$ of the potential is very large for small r and difficulties are encountered in the self-consistent procedure.

We assume that for the polarization field $\psi(r)$, equation (2) is valid only for distances $r > r_0$ and r_0 is such that the large derivatives $d\phi/dr$ of the potential in equation (2) are omitted. For $r < r_0$, the polarization field $\psi(r)$ is assumed to be of the form

$$\psi(r) = A + B[1 - (r/r_0)] + [1 - (r/r_0)]^2 \sum_{n=0}^{\infty} C_n P_n[(2r/r_0) - 1]; \quad r < r_0. \tag{20}$$

Here, the unknown coefficients A and B are fixed from the continuity condition for $\psi(r)$ in the two regions $r < r_0$ and $r > r_0$. The coefficients C_n are calculated by minimizing a functional $J(\psi)$ given by

$$J(\psi) = \int d^3r[g(r) - 1]\,\psi(r) + \frac{k_B T}{n^2(2\pi)^3}\int d^3q\left[\frac{n}{k_B T}\psi(q) - \ln\left(1 + \frac{n}{k_B T}\psi(q)\right)\right]. \tag{21}$$

The details of the numerical procedure and the physical significance of the function $J(\psi)$ are discussed elsewhere (Ailawadi *et al* 1976a, b). With an iteration of equations (2), (20) and equation (8), coupled with the minimization of (21), the self-consistent structure factor $S(q)$, the pair correlation function $g(r)$ and the polarization field $\psi(q)$ are obtained.

The dynamics of liquids is now calculated by assuming that the screened dynamical response function $\chi_{sc}(q, \omega)$ is given by equation (15) in terms of the self-motion of the particle.

4. Results and discussion

Based on this modified self-consistent field method, the static pair correlation function $g(r)$ and the polarization field $\psi(q)$ for liquid Rb at a temperature of 319 K and a density of 1·502 g cm^{-3} are presented in figures 1 and 2 respectively. The calculated pair correlation function is compared with the molecular dynamics data of Rahman (1974). The experimental data of Suzuki *et al* (1975) at a slightly different temperature (55 °C) obtained by pulsed neutron-diffraction techniques are also shown in figure 1. These results, as well as those presented elsewhere on liquid sodium and argon (Ailawadi *et al* 1976a, b), show that the theory is quite adequate in describing the static structure of classical liquids. As shown elsewhere (Ailawadi and Naghizadeh 1976a, b) equations (2) and (8) can also be inverted to obtain the effective pair potentials when the static structure factor is known from the experimental data.

If one now calculates the collective motions in liquids by assuming $\chi_{sc}(q, \omega)$ to be the free-particle dynamic response function $\chi_0(q, \omega)$, the results shown in figure 3 are obtained. In figure 3, the symmetrized scattering function defined by

$$\tilde{S}(q, \omega) = \exp\left(-\frac{\hbar\omega}{2k_B T}\right) S(q, \omega) \tag{22}$$

is plotted as a function of ω for three values of q. The calculations of Bansal and Pathak (1975), in which $\chi_0(q, \omega)$ is modified by a gaussian of unknown width $\Gamma(q)$ discussed in §2, and the neutron-scattering data of Copley and Rowe (1974a, b) are also shown for comparison. This is also the case for liquid Ar discussed in II. The results are generally not very good except at large values of q and ω, where the single-particle motion can be approximated by the free-gas $\chi_0(q,\omega)$ limit. On the other hand, assumption (15) for $\chi_{sc}(q,\omega)$ seems reasonable because the complicated diffusive and vibratory self-motion of the atom is being taken exactly into account.

The zeroth moment of the dynamic structure factor $S(q,\omega)$ is now satisfied. However, the fourth moment $\langle\omega^4\rangle$ is expected to be only approximately satisfied. This modified self-consistent field method reduces the problem to a study of the dynamic

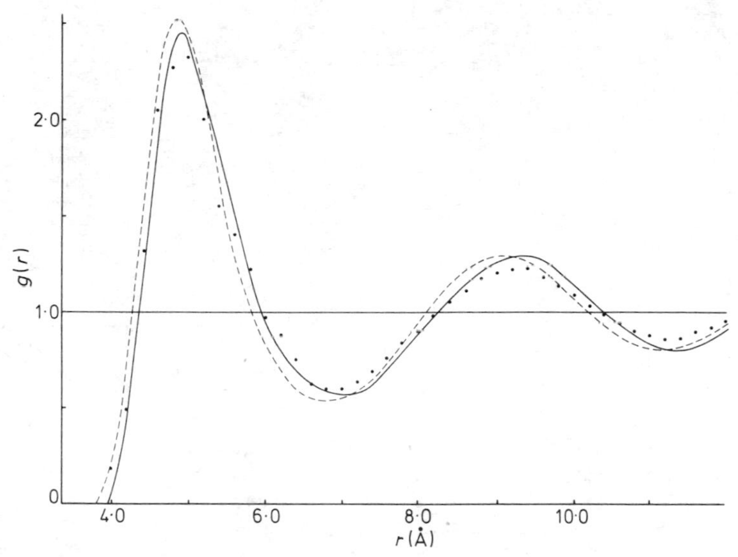

Figure 1. Pair correlation function $g(r)$ for liquid rubidium at a temperature of 319 K and a density of 1·502 g cm^{-3}. Full curve, present work; broken curve, molecular dynamics data (Rahman 1974); the dots are experimental data (Suzuki *et al* 1975). Note that the experimental data are at a slightly different temperature (55 °C) than the other two curves.

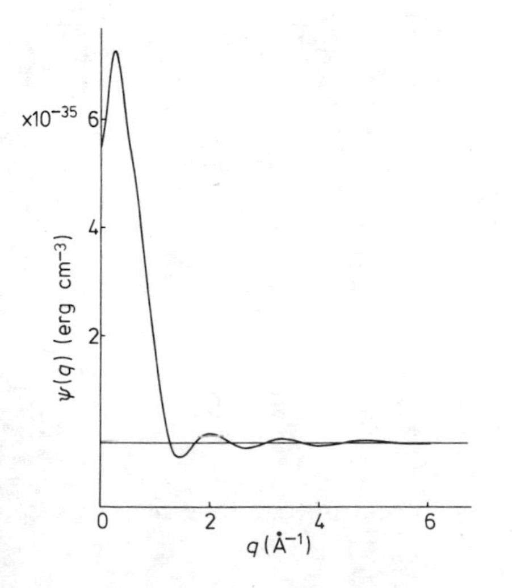

Figure 2. The polarization field $\psi(q)$ as a function of q obtained from the self-consistent solution of equations (2), (8), (20) and (21) with $r_0 = 3{\cdot}202$ Å.

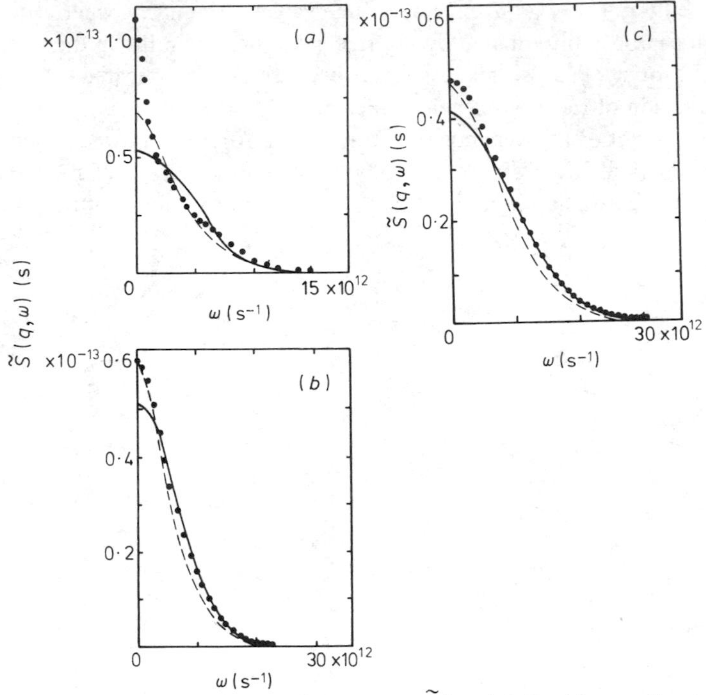

Figure 3. Symmetrized scattering function $\tilde{S}(q, \omega)$ defined in equation (22) as a function of frequency ω for: (*a*) $q = 2{\cdot}0$ Å^{-1}; (*b*) $q = 3{\cdot}5$ Å^{-1}; (*c*) $q = 5{\cdot}0$ Å^{-1}. Full curve, present work; broken curve, Bansal and Pathak (1975); the dots are experimental data of Copley and Rowe (1974).

self-correlation function. As is well known, this theory does not yield the hydrodynamic limit as $q \to 0$ and $\omega \to 0$. Moreover, the static structure factor in the limit $q \to 0$ does not go to the compressibility limit. This becomes quite obvious when the ion contribution to the pressure

$$P_\mathrm{i} = nk_\mathrm{B}T - \frac{n^2}{6}\int g(r) r \frac{\mathrm{d}\phi(r)}{\mathrm{d}r} \mathrm{d}^3 r \tag{23}$$

is examined. From equations (2) and (8), equation (23) can be expressed as

$$P_\mathrm{i} = nk_\mathrm{B}T - \frac{n}{2}\int C(r) \mathrm{d}^3 r \tag{24}$$

which reduces to

$$P_\mathrm{i} = \frac{1}{2} n \left[k_\mathrm{B}T + \left(\frac{\partial P}{\partial n}\right)_{T, V}\right]. \tag{25}$$

However, for wavevectors and frequencies of interest in neutron scattering and molecular dynamics, this is not a serious limitation.

Recently, Aldrich *et al* (1976) have proposed the following equation for the dynamical density response function (instead of equation (1)):

$$\chi(q,\omega) = \frac{\chi_{sc}(q,\omega)}{1 - [\psi(q) + (\omega^2/q^2)f(q)]\,\chi_{sc}(q,\omega)} \tag{26}$$

to explain zero sound in liquid ^{3}He. However, both $\psi(q)$ and $f(q)$ are parameters in the theory. Also, in a recent preprint, 'Kinetic Theory of Classical Liquids: Basic Theory', L Sjögren and A Sjölander (1976 unpublished) have developed a microscopic theory in which the polarization field $\psi(q)$ is also a function of ω. These theories need to be studied further.

References

Ailawadi N K, Miller D E and Naghizadeh J 1976a *Ber. Bunsenges. Phys. Chem.* **80** 704
—— 1976b *Phys. Rev. Lett.* **36** 1494
Ailawadi N K and Naghizadeh J 1976a *Ber. Bunsenges. Phys. Chem.* **80** 708
—— 1976b *Solid St. Commun.* **20** 45
Aldrich C H, Pethick C J and Pines D 1976 *Phys. Rev. Lett.* **37** 845
Anderson H C, Weeks J D and Chandler D 1971 *Phys. Rev.* **A4** 1597
Bansal R and Pathak K N 1975 *Phys. Rev.* **A11** 1450
Copley J R D and Rowe J 1974a *Phys. Rev. Lett.* **32** 49
—— 1974b *Phys. Rev.* **A9** 1656
de Gennes P G 1959 *Physica* **25** 825
Götze W and Lücke M 1975 *Phys. Rev.* **A11** 2173
Kadanoff L P and Martin P C 1963 *Ann. Phys., NY* **24** 419
Mori H 1965 *Prog. Theor. Phys., Kyoto* **33** 423
Pathak K N and Singwi K S 1970 *Phys. Rev.* **A2** 2427
Rahman A 1974 *Phys. Rev.* **A9** 1667
Singwi K S, Sköld K and Tosi M P 1970 *Phys. Rev.* **A1** 454 (II in text)
Singwi K S, Tosi M P, Land R H and Sjölander A 1968 *Phys. Rev.* **176** A589 (I in text)
Suzuki K, Misawa M and Fukushima Y 1975 *Trans. Jap. Inst. Met.* **16** 297
Thiele E 1963 *J. Chem. Phys.* **39** 474
Weeks J D, Chandler D and Andersen H C 1971 *J. Chem. Phys.* **54** 5237
Wertheim M S 1963 *Phys. Rev. Lett.* **10** 321
Zwanzig R 1961 *Lectures in Theoretical Physics* ed W E Brittin (New York: Wiley Interscience) vol 3

Stability of the Born–Green equation for effective pair potential in dense liquids

M Tanaka

Department of Applied Science, Faculty of Engineering, Tohoku University, Sendai 980, Japan

Abstract. The Johnson–March scheme of calculating an effective pair potential from the Born–Green (BG) equation in real space is examined to elucidate (i) how to truncate the BG equation to form simultaneous linear equations, allowing for the cut-off of the integral, when accurate data of $g(r)$ are given over a reasonably wide range, and (ii) how its solution approximates the true pair potential known to generate the input $g(r)$ at a given density and temperature. The truncated linear equations are naturally ill-conditioned with the condition number of the order of 100, and a smooth solution can be obtained to approximate the true potential only at smaller r, when $g(r)$ is accurate around its initial ascent as well as up to the fifth or sixth ripple. The BG scheme should be applied to liquid metals with allowance made for the ill-conditioning at different temperatures.

1. Introduction

Owing to the fact that the accuracy in measurement of the structure factor $S(k)$ in liquid metals has improved substantially, a number of investigations have appeared which compare several methods of calculating the effective pair potentials from $S(k)$, or its transform $g(r)$, through well known approximate equations in the statistical mechanical theories of fluids.

The Percus–Yevick (PY) and the hypernetted chain (HNC) approximations involve no more than the Fourier transform of $S(k)$ and $[S(k)-1]/S(k)$, and some qualitative conclusions have been drawn about the application of the PY and HNC pair-potentials to liquid metals (North *et al* 1968, Howells 1973, Ballentine and Jones 1973). The third scheme, based on the Born–Green (BG) equation, has also been applied to many liquid metals using several methods of computing the integro-differential equation. However, it has yet to be determined whether or not the use of this scheme is justified.

Kumaravadivel *et al* (1974) have carefully re-examined several methods of solving the BG equation for the case of liquid sodium. They conclude that the linearized simultaneous equation (LSE) method (Waseda and Suzuki 1971) cannot give a unique solution for the effective potential in Na at a known temperature, and that the Fourier space form of the BG equation (Gaskell 1966, Gehlen and Enderby 1969) is favourable at this stage of the investigation, but gives solutions which are substantially dependent on temperature. They also criticize the iteration scheme (Ailawadi *et al* 1974) because of the ill-conditioned character of the BG equation. They support the conclusion of Howells and Enderby (1972) that the BG theory is not suitable for liquid metals since the derived effective pair potentials have an unacceptable temperature variation. Unlike the PY and HNC schemes, we consider that the ill-conditioned character is inevitable in the BG scheme even in the Fourier space method. This can be clearly demonstrated by varying the input data for $S(k)$ (Gehlen and Enderby 1969). One can suspect that the unacceptable temperature variation of the solutions may be partly due to a substantial change with temperature in the ill-conditioning of the truncated equations. With regard to the real-space BG scheme, on the other hand, it is still not clear how an optimum solution is obtained, even at a single temperature with the input $g(r)$ given as accurately as possible.

In the next section, we examine (i) how to truncate the BG equation to a single system of linear equations when accurate data of $g(r)$ are given over a reasonably wide range $r_l \leqslant r \leqslant r_{max}$, and (ii) how the solution approximates the true pair potential that is known to generate the input $g(r)$ at a given density and temperature. In order to avoid the problem concerning the applicability of the pair-potential theory to liquid metals, we take the model of a liquid using molecular dynamics simulation with the same LRO-II potential (Tanaka and Fukui 1975) as for the input $g(r)$. In §3 we examine the case of argon at 85 K with the input $g(r)$ of Yarnell *et al* (1973), and discuss the ill-conditioned character of the solution with the lack of reliable data at the edge of the first peak of $g(r)$.

2. Stability of the BG scheme in real space

We start with the following form of the BG equation; the notation is standard (Kumaravadivel *et al* 1974, Ailawadi *et al* 1974).

$$\frac{\phi(r)}{kT} + \ln g(r) = \int_0^\infty \mathrm{d}s\, K(r,s)\, \frac{\phi(s)}{kT}, \tag{1}$$

where

$$K(r,s) = -\left(\frac{n\pi}{r}\right)\left\{g'(s) \int_{-s}^{s} \mathrm{d}t(s^2-t^2)(t+r)\,[g(|t+r|) - 1] + 2sg(s)\int_{-s}^{s} \mathrm{d}t(t+r)\,[g(|t+r|)-1]\right\}. \tag{2}$$

Equation (1) can be written as an integral equation for $v(r) = [\phi(r)/kT] + \ln g(r)$ to correspond with the Fourier space method:

$$v(r) - \int_0^\infty \mathrm{d}s\, K(r,s) = v(s)\, u(r), \tag{3}$$

where

$$u(r) = -\int_0^\infty \mathrm{d}s\, K(r,s)\, \ln g(s). \tag{4}$$

First, we assume that the input $g(r)$ of the simulated LRO-II liquid at 377·01 K, with number density $n = 0{\cdot}0241\ \text{Å}^{-3}$, is given exactly for $r_l\ (=2{\cdot}4\ \text{Å}) \leqslant r \leqslant r_{max}\ (=16{\cdot}8\ \text{Å})$

with steps of 0·1 Å (Tanaka and Fukui 1975)†, and that

$$g(r) = g'(r) = 0 \qquad r < r_1 ,$$
$$g(r) = 1, \qquad g'(r) = 0 \qquad r_{max} < r . \tag{5}$$

The kernel function $K(r, s)$ is smooth but oscillates rapidly (roughly with the period of $g'(s)$ along s for a given r) and obviously, from equation (5), $K(r, s) = 0$ for $s < r_1$ and $|r-s| > r_{max}$. $K(r, s)$ goes to zero smoothly at the boundaries and its maximum amplitude, against s, for a given r occurs, roughly speaking, around the line $r = s$ and decays slowly along it: $\max|K(r, s)| = 3{\cdot}271$ at $s = 3{\cdot}4$ Å for $r = 2{\cdot}4$ Å, and 0·362 at $s = 28{\cdot}0$ Å for $r = 28{\cdot}0$ Å. Therefore the convergence of the integral in the RHS of equation (1) with the upper limit s_{max} is expected to be non-uniform with respect to r. In other words, a solution of the Fredholm-type equation truncated from equation (1) or (3) to the finite domain $r_1 \leqslant r$ and $s \leqslant s_{max}$ can approximate the solution of equation (1) or (3) only in the range $r_1 \leqslant r < r_c \ll s_{max}$. We can estimate a reasonable value of s_{max} by examining the convergence of the integral with the true potential $\phi(r)$. The value of the integral is constant to within one part in 10^6 in the interval $2{\cdot}4 \leqslant r \lesssim 10{\cdot}0$ Å when $s_{max} \gtrsim 30{\cdot}0$ Å. With a suitable choice of s_{max}, the equation (3) is truncated to a set of N simultaneous linear equations over the discrete points $r_1 \leqslant r_i$ and $s_j \leqslant s_{max}$ as

$$V_i - \sum_{j=1}^{N} Q_{ij} V_j = U_i \qquad 1 \leqslant i \leqslant N \tag{6}$$

where $\Delta r = \Delta s = 0{\cdot}1$ Å and

$$V_i = v(r_i), \qquad U_i = u(r_i), \qquad Q_{ij} = \Delta_j K(r_i, s_j). \tag{7}$$

We use the Newton–Cotes five-point formula for the weight Δ_j and also to evaluate $g'(s_j)$.

It is instructive to examine the cut-off effects in equation (6). We examined the change with N in the distribution of the eigenvalues of the matrix Q_{ij}, especially around 1·0 on the real axis. If we choose N to be smaller, say, $N \leqslant 145$ ($s_{max} \leqslant 16{\cdot}8\,^{\circ}$Å), only a few λ_n are found to be real, but a large number of λ_n become real for $N \geqslant 249$ ($s_{max} \geqslant 27{\cdot}2$ Å). With the change of N from 277 to 301, the number of λ_n with which $|1 - \lambda_n| < 0{\cdot}1$ remains constant. Because the matrix Q_{ij} is diagonalized with no degenerate eigenvalues in our cases, the solution V_i can be written formally with the normalized right and left eigenvectors, $x_i^{(n)}, y_i^{(n)}$, in the following form:

$$V_i = \sum_{n=1}^{N} \frac{1}{(1-\lambda_n)} \sum_j (y_j^{(n)} U_j) x_i^{(n)}, \tag{8}$$

and we define the measure of the ill-conditioning of equation (6), that is, the condition number, as

$$N_c = \frac{|1-\lambda_m|_{max}}{|1-\lambda_n|_{min}} \tag{9}$$

(see, for example, Hartree 1952).

† We used the smoothed values of $g(r)$ given in table 2 of Tanaka and Fukui (1975).

Thus the cut-off s_{max} or, equivalently, the dimension N, must be as large as possible to ensure that the number of the most important eigenvectors in the expansion of equation (8) becomes stationary in practice. Because $|Q_{ij}|_{max}$ is roughly $3{\cdot}27\,\Delta_j$, then $|1-\lambda_m|_{max}$ is always of the order of unity and N_c can be of the order of 100. Equation (6) is really ill-conditioned. It should be noted also that the determinant of $(\delta_{ij}-Q_{ij})$ happens to change sign at some values of N due to the oscillatory nature of the elements Q_{ij} introducing the critical ill-conditioning to equation (6). With such a choice of N, only one term in equation (8) governs the i-dependence of V_i to result in an unexpected solution $\phi(r)$.

In figure 1 we show the results of $[\phi(r)/kT]$ obtained by the gaussian elimination method from equation (6) with two choices of N:

(i) $N=301$ ($s_{max}=32{\cdot}4$ Å), $N_c=144$; curve C,
(ii) $N=249$ ($s_{max}=27{\cdot}2$ Å), $N_c=91$; curve D.

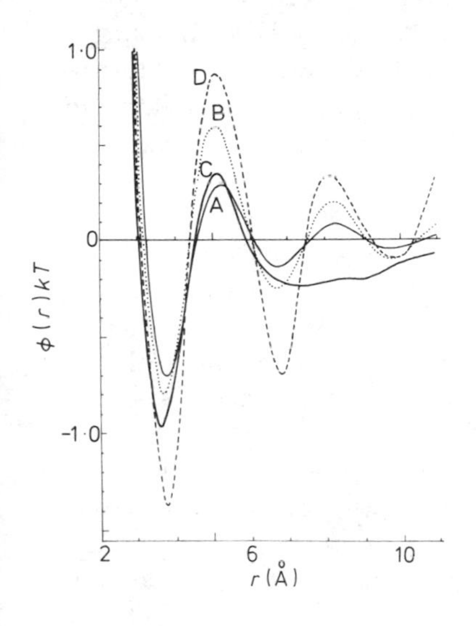

Figure 1. Solution of equation (6) for the LRO-II liquid (377·01 K and 0·0241 Å^{-3}). Curve A: LRO-II potential, $\phi(r)/kT$. Curve B: the potential of mean force divided by kT, $-\ln g(r)$. Curve C: solution with $N=301$ ($s_{max}=32{\cdot}4$ Å). Curve D: solution with $N=249$ ($s_{max}=27{\cdot}2$ Å). Solutions have unacceptable r-dependence at larger r due to the truncation effect.

In both cases the output V_i in equation (6) is quite smooth with respect to i.

As is clearly seen, the solution with $N=249$ (curve D) seems to be an exaggerated potential of mean force, but when $N=301$ (curve C) it approximates the true LRO-II potential qualitatively up to $r<r_c\simeq 7$ Å. Due to the truncation effects as stated above, the solutions generally have unacceptable r-dependence at larger values of r. We doubt the proposition that one can analyse the long-range part of the effective pair potential in a dense liquid through the BG scheme.

We solved equation (6) in the alternative way for the case $N=301$ with $U_i=-\ln g(r_i)$ to obtain $[\phi(r_i)/kT]$ directly, and the solution is almost identical to curve C in figure 1 within an absolute error of $\pm 2{\cdot}0\times 10^{-4}$ over all the points $2{\cdot}4\leqslant r_i\leqslant 32{\cdot}4$ Å. We consider that the solution with $s_{max}=32{\cdot}4$ Å is optimum with respect to the ill-conditioning of equation (6) for the input $g(r)$ given up to $r_{max}=16{\cdot}8$ Å.

If we truncate the input data at $r'_{max}=15{\cdot}2$ Å, that is, make $g(r)$ flat beyond the end of the fourth minimum, the matrix elements for $|r_i-s_j|\leqslant r'_{max}$ change only within

±0·2% but there occurs a crossover in the order of eigenvalues of $|1-\lambda_n| < 0{\cdot}1$. The solution can be altered substantially with the truncation of the input data.

3. Uncertainties due to the lack of data at the steep ascent of $g(r)$

As an example of real, dense liquids, we have examined argon at 85 K (number density 0·02125 $Å^{-3}$) and compared the solution to the ordinary Lennard-Jones (6, 12) potential ($\epsilon = 119{\cdot}8$K, $\sigma = 3{\cdot}405$Å). Yarnell *et al* (1973) transformed the smoothed data of $S(k)$ into $g(r)$, and showed that, for $r \geqslant 3{\cdot}2$Å, the result is identical with that obtained from molecular dynamics using the Lennard-Jones potential. The data of $g(r)$ given in their table 2 for $0{\cdot}0681 \leqslant r \leqslant 27{\cdot}24$Å have anomalous ripples at $r < 3{\cdot}2$ Å, and the oscillations in $[g(r)-1]$ decay to within ±0·005 beyond $r = 20{\cdot}5$Å.

We assumed $g(r)$ to be zero at $r \leqslant 3{\cdot}1$ Å and interpolated their data smoothly to every 0·1 Å in r in the interval $r_1 = 3{\cdot}2 \leqslant r \leqslant r_{max} = 20{\cdot}5$ Å. The result is taken as the input $g(r)$ for the linear equation (6). We chose $N = 301$ ($s_{max} = 33{\cdot}2$ Å), and solved equation (6) as before. The condition number is about 200 and the ill-conditioning is worse than the previous case. We could, however, obtain the solution $[\phi(r)/kT]$ as being quite smooth from point to point, as shown in figure 2 (curve C). Although the solution is not an exaggerated potential of mean force, it fails to approximate the magnitude of the true potential at smaller r.

It is interesting to note that curve C in figure 2 is quite similar at $r \lesssim 6$Å to the solution of equation (1) using the substitution method (Johnson and March 1963, Johnson *et al* 1964), except for its scale (the minimum of curve C at $r = 3{\cdot}6$ Å is about $-4{\cdot}01 \times 10^{-2}$eV). We consider that the solution in figure 2 fails partly because the straight interpolation of $g(r)$ at the edge of the first peak is not strictly valid, although this was carried out in conventional manner ($g(r) = 0{\cdot}0720$ at $r = 3{\cdot}2$Å and 0·4925 at 3·3 Å, being compared to the LRO-II case: $g(r) = 0{\cdot}0020$ at $r = 2{\cdot}4$Å, 0·0111 at 2·5Å and 0·0324 at 2·6 Å). This has a substantial effect on the values of $K(r, s)$ in the truncated domain where $r_1 \leqslant r$ and $s \leqslant s_{max}$. In this case, in fact, the kernel $K(r, s)$ has discontinuities at the boundary $s = r_1$ (for example, $K(r, s)$ jumps from zero to 1·002 at $r = 3{\cdot}2$ Å and to 0·697 at $r = 3{\cdot}7$ Å along the boundary). We need reliable data of $g(r)$

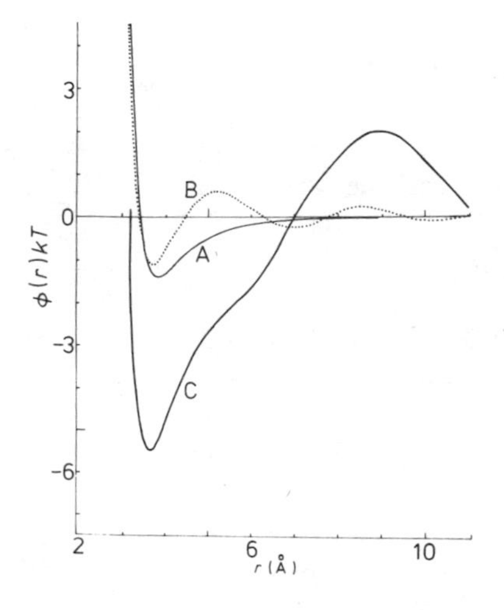

Figure 2. Solution for liquid Ar (85 K and 0·02125 $Å^{-3}$). Curve A: Lennard-Jones potential, $\phi(r)/kT$, ($\epsilon = 119{\cdot}8$K, $\sigma = 3{\cdot}405$Å). Curve B: the potential of mean force, $-\ln g(r)$, from the experimental data of Yarnell *et al* (1973). Curve C: solution with $N = 301$ ($s_{max} = 33{\cdot}2$Å). The input $g(r)$ are linearly interpolated to zero at $r = 3{\cdot}2$Å inside the first peak.

at the edge of the main peak in order that the short-range part of the pair potential can be examined.

4. Concluding remarks

The BG scheme in real space is shown to be ill-conditioned, and the condition number of the simultaneous linear equations, truncated from the integral equation over the infinite interval, is of the order of 100. The ill-conditioned character depends strongly upon the choice of cut-off parameter for the integration. When the input $g(r)$ is given accurately over a reasonably wide interval, there could exist a stable solution which only qualitatively approximates the true pair potential at smaller r. We must have reliable data of $g(r)$ around its initial ascent, as well as at distant r, so that the solution can be improved.

We have not examined the change with temperature in the ill-conditioning of the simultaneous linear equations, and have not analysed the unacceptable temperature variation of the BG solutions that is reported by Kumaravadivel *et al* (1974) and others.

Acknowledgments

The author is grateful to colleagues for discussions. The numerical computation was performed on the NEAC 2200 computer of the Computer Center, Tohoku University. A part of this work was supported by the Grant in Aid of Scientific Research from the Ministry of Education.

References

Ailawadi N K, Benerjee P K and Choudry A 1974 *J. Chem. Phys.* **60** 2571
Ballentine L E and Jones J C 1973 *The Properties of Liquid Metals* ed S Takeuchi (London: Taylor and Francis) p51
Gaskell T 1966 *Proc. Phys. Soc.* **89** 231
Gehlen P C and Enderby J E 1969 *J. Chem. Phys.* **51** 547
Hartree D R 1952 *Numerical Analysis* (Oxford: Clarendon Press) ch 8.12
Howells W S 1973 *The Properties of Liquid Metals* ed S Takeuchi (London: Taylor and Francis) p43
Howells W S and Enderby J E 1972 *J. Phys. C: Solid St. Phys.* **5** 1277
Johnson M D, Hutchinson P and March N H 1964 *Proc. R. Soc.* **A282** 283
Johnson M D and March N H 1963 *Phys. Lett.* **3** 313
Kumaravadivel R, Evans R and Greenwood D A 1974 *J. Phys. F: Metal Phys.* **4** 1839
North D M, Enderby J E and Egelstaff P A 1968 *J. Phys. C: Solid St. Phys.* **1** 1075
Tanaka M and Fukui Y 1975 *Prog. Theor. Phys.* **53** 1547
Waseda Y and Suzuki K 1971 *Phys. Stat. Solidi* **B49** 643
Yarnell J L, Katz M J, Wentzel R G and Koenig S H 1973 *Phys. Rev.* **A7** 2130

Higher-order partial structure factors for multicomponent systems

M Silbert, P Gray and S M Johnson
School of Mathematics and Physics, University of East Anglia, Norwich NR4 7TJ, UK

Abstract. Gray and Silbert (1976) have proposed a definition of higher-order structure factors for multicomponent systems which generalizes that of Ballentine and Lakshmi (1975) for pure systems, but differs from that of Parrinello and Tosi (1975). The structure factors are the Fourier transforms of a class of correlation functions which are closely connected with cumulant averages of number fluctuations in the Grand Ensemble. The functional differentiation approach used to introduce the correlation functions allows the definition of their inverses, which are shown to be related to higher-order generalizations of the direct correlation functions.

Moreover, it is possible to construct a hierarchy of equations relating partial structure factors of different orders, which may be expressed in terms of either the number of particles of different species or the number-concentration formalism (Bhatia and Thornton 1970). A corresponding hierarchy also exists for the inverse functions.

We have calculated the long wavelength limits of the third-order partial structure factors in both species–species and number-concentration modes, and also of the direct correlation functions, for Na–K at 100°C. We have assumed the validity of conformal solution theory to compensate for the lack of sufficient experimental data. To compare our calculations with those of Parrinello and Tosi we first calculated the structure factors according to their definition. We obtained peak heights about 20% smaller, and did not find the discrepancy, shown in their figure 1, between the two different ways of approaching the long wavelength limit for the inter-species structure factor. We find, however, that omission of the second term in their equation (29) leads to results apparently the same as theirs. The species–species structure factors are generally much larger than the number-concentration ones, though for most concentrations both types of function are less than unity. The direct correlation functions, being inverses of the former type, are generally very large. All these functions possess significant features at a potassium concentration of approximately 60%, which is close to the eutectic concentration, although these data are for a temperature (100°C) well above the eutectic temperature.

Full details of these calculations are being submitted to *J. Phys. F: Metal Physics.*

References

Ballentine L E and Lakshmi A 1975 *Can. J. Phys.* **53** 372–6
Bhatia A B and Thornton D E 1970 *Phys. Rev.* **B2** 3007–12
Gray P and Silbert M 1976 *Phys. Lett.* **56A** 192–4
Parrinello M and Tosi M P 1975 *Nuovo Cim.* **25B** 242–50

Photoelectron spectroscopy of liquid metals and alloys

C Norris

Department of Physics, University of Leicester, Leicester LE1 7RH, UK

Abstract. Photoelectron spectroscopy is now established as an important technique in the study of condensed matter. With appropriate experimental arrangements information may be obtained about the electronic structure, surface effects and local ionic order. Practical difficulties limit the range of liquid metallic systems which may be conveniently studied with photoelectron spectroscopy. Nevertheless, data on a series of liquid pure metals and alloys are now available and some general trends may be discerned.

With the exception of Al, low-energy photoelectron spectra of liquid simple metals show consistent deviation from the picture predicted by weak-scattering formalisms. This is particularly noticeable for the heavy metals Hg, Pb and Bi where features associated with the 6s and 6p levels can be distinguished. Calculations of the emission spectra which account for matrix elements in a nearly-free-electron approximation fail to resolve the discrepancy. Photoemission results for liquid noble metals are dominated by bands associated with the strong d-resonances. For liquid copper comparison is made with the results of multiple scattering calculations.

Liquid binary alloys are often characterized by non-random short-range ordering and preferential surface enrichment. Measurements of the intensity and kinetic energy of electrons photoejected from localized levels enable these effects to be monitored. Results for several alloy systems including Cu–Ag and Au–Sn are discussed.

1. Introduction

The application of photoelectron spectroscopy (PES) to the study of the liquid phase in general has received increasing attention in recent years (Siegbahn 1974). Measurements have now been reported on a wide range of liquid metallic systems and it is the aim of this paper to summarize the results obtained so far, to outline the problems as well as the possibilities.

The photoelectron process is a conceptually simple and direct probe of the electronic structure of a conductor. An incoming photon lifts up an electron from either a valence or a core level to a final state in which it may escape and be recorded in the surrounding vacuum. The spectral and angular distributions of the photocurrent are determined by several factors including the electronic structure, the local ionic microstructure, matrix elements, surface effects and many-body effects. By suitable experimental arrangement and measurement of parameters such as the energy distribution of photoemitted electrons $N(E, \omega)$, the total photocurrent

$$Y(\omega) = \int N(E, \omega)\,\mathrm{d}E,$$

and the workfunction ϕ, information concerning these phenomena may be derived. For liquid metals two areas are of primary interest: the local atomic arrangement in alloys which may be inferred indirectly from the binding energies of core electrons and the valence electronic structure.

The theoretical description of electron states is of central importance in the understanding of liquid metallic systems. The subject was reviewed at the Tokyo conference by Cusack (1973) and Ehrenreich (1973) and recently by Ballentine (1975). The problem is two-fold: choice of an effective potential and solution of the Schrödinger equation for a disordered array which is not entirely random but exhibits a large degree of short-range order. A commonly used approach to the latter is through multiple-scattering theory. Thus the scattering matrix of the entire system is used to define the configurationally averaged Green's function

$$G(E) = \langle (E - H)^{-1} \rangle \tag{1}$$

from which properties such as the one-electron density of states function $\rho(E)$ follow as a summation over a continuum of states

$$\rho(E) = -\frac{1}{\pi} \sum_{k} I_m \langle k | G | k \rangle. \tag{2}$$

Experimental determinations, using for example PES, of parameters such as $\rho(E)$ over a wide energy range allow an important check to be made on the validity of the approximations necessary in the practical solution of equations (1) and (2). Other techniques used in this field are often sensitive only to electrons at the Fermi level or, like optical absorption, give integrated effects.

The range of possible systems which may be studied by PES is limited by the constraints imposed by the temperature, containment and in particular vapour pressures. High evaporation rates lead to loss of specimen, contamination of electron and photon optics and scattering of emergent electrons with subsequent loss of information. As a rough guide, electrons in the 1 keV range will traverse l mm in a vapour of pressure p Torr provided $lp < 1$. This physically restricts the range of systems to those with vapour less than 0·1 Torr.

The interpretation of photoelectron spectra in terms of the density of states function relies on a number of simplifications. Following Schaich and Ashcroft (1970) the time-averaged photocurrent may be expressed in the 'golden rule' form as

$$\langle J(\mathbf{r}) \rangle = \frac{2\pi e}{\hbar} \sum_{mu} n(E_m) \left| \left\langle m \left| \frac{ie\hbar}{mc} \mathbf{A} \cdot \nabla \right| u \right\rangle \right|^2 \delta(E_m + \hbar\omega - E_u) \tag{3}$$

where $\hbar\omega$ is the photon energy, $|m\rangle$ is the initial one-electron state of the unperturbed distribution with occupation $n(E_m)$ and $|u\rangle$ is the final outgoing state which includes the effects of scattering and transmission through the barrier. We follow the traditional (phenomenological) path and factorize out the scattering and transmission effects whereupon the energy distribution of primary photoelectrons is given by

$$N(E,\omega) = c \left[\sum_{mf} n(E_m) |\langle m | \mathbf{A} \cdot \nabla | f \rangle|^2 \, \delta(E_m + \hbar\omega - E_f) \, \delta(E - E_f) \, T(E) \right] \tag{4}$$

where f is the final state of the unperturbed system and $T(E)$ is the probability of escape without scattering. For ordered systems the matrix element enforces conservation of the one-electron momentum vector $\mathbf{k}$ during the optical transition with the consequence that $N(E, \omega)$ is (at low energies $\hbar\omega \leqslant 40$ eV) strongly dependent on both E and ω. For highly disordered systems this restriction is relaxed. Assuming the matrix element is a constant, equation (4) reduces to a product of initial and final densities of states

$$N(E, \omega) \propto \rho(E - \hbar\omega)\, \rho(E)\, T(E). \tag{5}$$

Final-state scattering (for liquids we must consider electron–ion as well as electron–electron interactions) reduces the effective escape depth to between 20 and 5 Å. For the study of specifically bulk phenomena it is important to work with final-state energies E either less than 30 eV or greater than 1000 eV, which is readily feasible with laboratory sources. The complications of slowly varying matrix elements, surface effects and many-body phenomena imply that equation (5) should be regarded as a first step to the interpretation of PES spectra of liquid metals, most applicable in cases such as the noble metals where $\rho(E - \hbar\omega)$ is a dominant function. It would be wrong to attempt to deduce precise results such as the density of states at the Fermi level $\rho(E_F)$ for example.

2. Noble metals

The valence states of the noble metals are dominated by semi-localized d-bands which should be well described by tight binding methods taking account only of the nearest-neighbour shell. Short-range order as indicated by the number and spacing of nearest neighbours changes little in the crystalline to liquid transition. Only a small increase occurs in the specific volume, both phases being characterized by close-packed structures. Intuitively we might therefore expect relatively small modifications to the electronic structure. Effects observed in the photoelectron spectrum should be attributable to the removal of long-range order and possibly the occurrence of non-crystalline local arrangements. Low-energy ($\hbar\omega \leqslant 40$ eV) photoemission measurements by Eastman (1971), Williams and Norris (1974, 1976a) largely confirm this view.

Figure 1 shows spectra obtained at two photon energies with liquid and solid Cu. Unlike the solid case which possesses translational symmetry, conservation of **k** is not an important optical selection rule for the liquid. The liquid spectrum shows little variation with ω as expected from equation (5) and the band between −2 and −6 eV thus reflects the density of occupied 3d states.

Because of this change in the nature of the optical transitions comparison of liquid and solid spectra at a single photon energy is not very meaningful. An experimental density of states for comparison with the liquid spectra may, however, be obtained from EDCs of the solid using the expression

$$\rho(E - \hbar\omega) = \text{const.} \int_0^\infty \frac{N(E, \omega)}{\omega}\, d\omega \tag{6}$$

derived by Eastman (1972) from consideration of the f-sum rule. Figure 2 shows liquid spectra obtained at 21·2 eV for the noble metals together with curves deduced from equation (6) using spectra obtained at close photon-energy intervals over the range 12 to 45 eV with synchrotron radiation (Eastman 1972, Norris and Williams 1976a,b). For the liquid there is a loss of fine structure as would be expected with an infinity of local

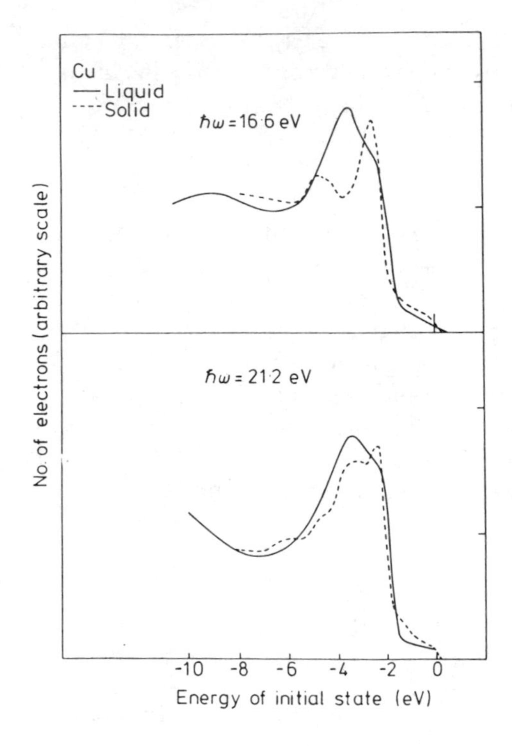

Figure 1. Photoelectron spectra of liquid copper (Williams and Norris 1974) and of a polycrystalline copper film.

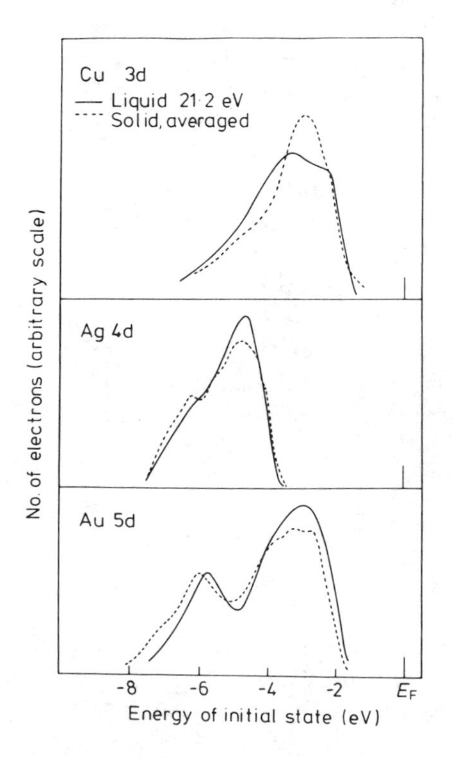

Figure 2. Comparison of the density of states for the d-bands in solid noble metals derived from the low-energy average (sum rule) with the 21·2 eV spectra for the liquid (background substracted). For references see text.

ionic configurations but overall, particularly for Ag and Au, there is little difference between the liquid and solid curves, consistent with the similar *average* local arrangements in the two phases. It suggests that the main features of the density of states curve for liquid noble metals may be obtained with a calculation which takes account of the two-body ionic correlation function $g_2(r)$ but not of higher-order terms.

Theoretical calculations of the density of states for the noble metals using realistic rather than model potentials have been reported only in the case of Cu. Attempts to extend the highly successful coherent-potential approximation to the case of liquid metals have been hampered by the problem of creating an effective medium without losing information about the ionic structure (Roth 1974). Figure 3 compares the photoelectron energy distribution curve for liquid Cu with the theoretical density of states derived by Keller *et al* (1974) using a multiple-scattering cluster calculation and also by Chang *et al* (1975) using the prescription of Anderson and McMillan (1967).

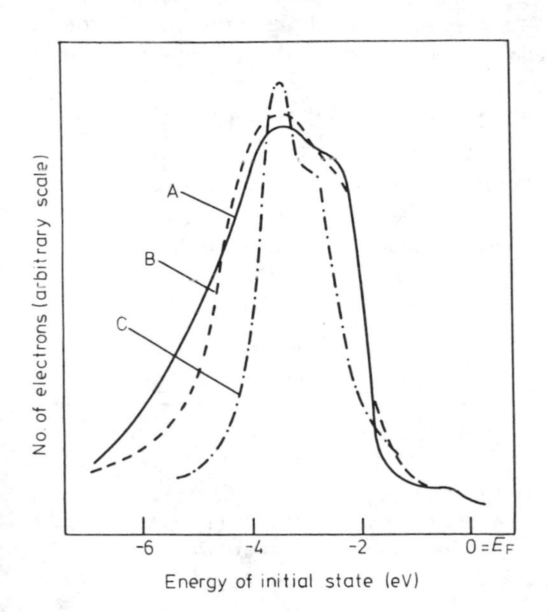

Figure 3. Comparison of the 21·2 eV spectra for liquid Cu (A) with theoretical calculations of Keller *et al* (1974) (B) and Chang *et al* (1975) (C). The theoretical curves have been broadened by a gaussian of 0·4 eV halfwidth.

The latter technique attracted attention after the original study on liquid Fe as it appeared to encompass the resonant nature of the d-bands without reference to the geometric structure. Chang *et al* (1975) and Olsson (1975) have questioned this approach since it predicts an anomalously small band width as confirmed in figure 3; the cluster calculation, on the other hand, gives a noticeably good fit to experiment.

3. Simple metals

The theoretical description of liquid simple metals (those metals without tightly bound d or f orbitals close to the Fermi level) has changed little since the nearly-free-electron (NFE) theory was used to explain the transport properties (Ziman 1961). Calculations of the electronic structure have largely followed the work of Edwards (1958) who used a perturbation expansion to obtain the Green's function as

$$G(E) = \left(E - \frac{\hbar^2 k^2}{2m} - \Sigma(k)\right)^{-1} \tag{7}$$

where $\Sigma(k)$ is the complex self-energy. This procedure is strictly valid only for weak pseudopotentials and, not surprisingly, theoretical densities of states have tended to follow the free-electron $E^{1/2}$ rule (Cusack 1973).

The low melting points and vapour pressures of several of these metals makes them a particularly attractive group for study using PES. Measurements have been reported for $\hbar\omega < 12$ eV, using LiF windows, of the yield and the workfunction as well as energy-resolved spectra. The results may be summarized as follows.

(i) There is relatively little change (often only a few per cent) in the total yield per absorbed photon on taking a metal from room temperature through the melting point. Where changes are found they may be related to differences in the scattered background and uncertainty in the values of reflectivity.
(ii) The workfunction as determined from a Fowler plot of the yield $Y(\omega)^{1/2}$ against $\hbar\omega$ is also little affected by melting (table 1). Lang and Kohn (1971) have stressed the dependence of ϕ on the surface electron density. Between (111) and (110) surfaces of the same single crystal, variations of up to 10% may be expected. The observed invariance suggests metal surfaces are sharp discontinuities similar to polycrystalline surfaces.

Table 1. Workfunctions of simple metals as determined from photoemission yield.

Metal	ϕ liquid (eV)	ϕ solid (eV)	Reference
Hg	$4{\cdot}49 \pm 0{\cdot}01$	No change	Cotti *et al* (1973)
In	$3{\cdot}94 \pm 0{\cdot}07$	$4{\cdot}06 \pm 0{\cdot}10$	Rodway (1976, unpublished)
Al	$4{\cdot}02 \pm 0{\cdot}04$	No change	Rodway (1976, unpublished)
Sn	$4{\cdot}27 \pm 0{\cdot}02$	No change	Rodway (1976, unpublished)
Pb	$3{\cdot}94 \pm 0{\cdot}03$	No change	Rodway (1976, unpublished)

(iii) Contrary to the view supported by transport measurements the photoelectron spectra of liquid metals are not readily explained on a free-electron-like density of states. Liquid indium has been measured by several groups (Koyama and Spicer 1971, Stevenson and Enderby 1972, Norris *et al* 1972). Although the results differ in detail due to the photon energies used, resolution and sample preparation technique, they indicate a dip at $-2{\cdot}6$ eV which corresponds in position with the minimum in the density of states for the solid due to the overlap of the first and second conduction bands. By contrast little agreement is found with predictions based on the liquid density of states curve computed by Shaw and Smith (1969). A similar correspondence between structure in the liquid spectra and details of the band structure of the solid have also been argued for Al and Hg (Norris *et al* 1974).

The clearest disagreement with the simple picture is found with the heavy polyvalent metals Pb and Bi. Two separate bands superposed on a background of secondary electrons are observed (figure 4): a spin–orbit split 6p band extending from the Fermi edge to −4 eV (Bi) and −3 eV (Pb) and a 6s band centred at −12 eV (Bi) and −7 eV (Pb).

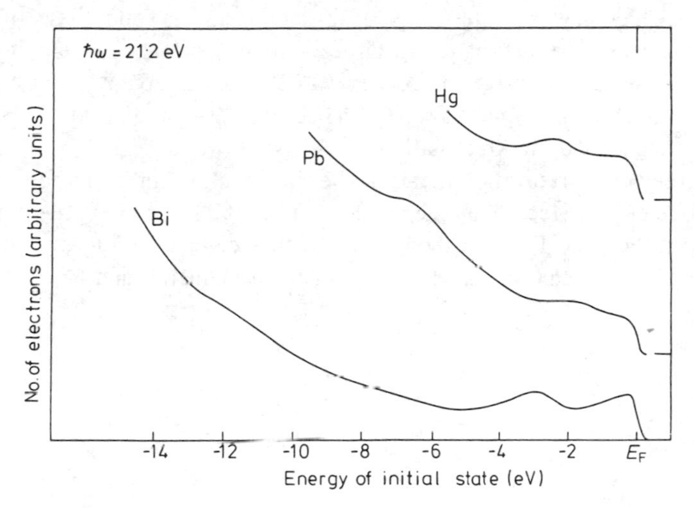

Figure 4. Photoelectron spectra obtained at 21·2 eV photon energy for the heavy 6s6p metals Hg, Pb and Bi in the liquid phase (Norris *et al* 1974, Norris *et al* 1976).

Similar features are found with solidified specimens in agreement with calculations which reveal a large energy gap in the band structure. From Bi to Hg there is a shift of structure towards E_F consistent with the decrease in valence from 5 to 2. For Hg the photoelectron spectra (see also Cotti *et al* 1973) indicate a minimum in the density of states at the Fermi level as argued by Mott (1966).

The significance of matrix elements may be judged by comparing spectra obtained at greatly differing photon energies. Thus, for example, the structure observed for Pb and Bi is correctly assigned to density of states since the curves broadly agree with solid spectra excited by Al Kα radiation (1486 eV, Ley *et al* 1974). An attempt to incorporate matrix elements within a NFE description of photoemission and thereby ascertain the influence of short-range order was made by Williams and Norris (1975). The results compared favourably with low-energy spectra of liquid Al but could not explain the more pronounced structure observed with the heavier metals. The possible existence of non-homogeneous diffuse or semi-ordered surface layers has been discussed by several authors (Bloch and Rice 1969, Croxton 1973). The evidence of the work-function measurements (table 1) and the fact that photoelectron spectra are not normally measured under conditions of liquid–vapour equilibrium, necessary for the creation of stable ionic density oscillations would, however, argue against the significance of these effects.

From photoemission measurements we would conclude therefore that the electronic structures of liquid simple metals beyond groups I and II are not well described by a weak-scattering model. The strength of the ionic potential and the local atomic order are such that the principal features in the density of states for the solid are retained in the liquid phase.

4. Alloys

Mixtures of metals which have a limited range of miscibility in the solid phase, often are completely miscible above the melting point. Photoelectron spectra for several liquid alloy systems have been reported: Hg–In (Norris *et al* 1974), Cu–Ag (Williams and Norris 1976a), Cu–Ni (Williams and Norris 1976b) using 21·2 eV radiation and Au–Sn (Ichikawa 1975) using Al and Mg Kα radiation.

One complication with mixtures of metals in thermal equilibrium is that in the surface region probed by the technique the concentration of a given constituent may be quite different from the bulk. Figure 5 shows the effective concentration as determined from the strength of the 3d and 4d bands of Cu–Ag alloys (Williams and Norris 1976a).

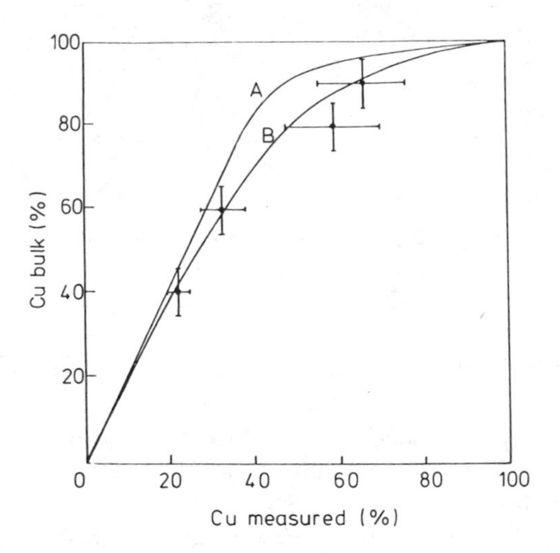

Figure 5. The variation of the surface concentration as a function of bulk concentration for liquid AgCu alloys: A, without surface relaxation of bond enthalpy; B, with surface relaxation of bond enthalpy. The experimental points were determined from low-energy ($\hbar\omega$ = 21·2 eV) photoelectron spectra (Williams and Norris 1976a).

The results indicate preferential surface enrichment of Ag at the expense of Cu. The continuous curves are values computed on the basis of the pair-wise bonding theory of Williams and Nason (1974). Better agreement with experiment is found by allowing the surface bonds between Ag atoms to assume an increased enthalpy.

$$H^R_{AA} = 1{\cdot}2\,H_{AA}$$

where H_{AA} is the enthalpy of the Ag–Ag bond in the bulk.

When dissimilar atoms are mixed together the modification which occurs in the valence electron distribution may be seen in two ways. First, the valence density of states function $\rho(E)$ assumes a different profile. Thus as In is added to Hg the dip observed in the photoelectron distribution at E_F is replaced by a small peak (Norris *et al* 1974), a result not inconsistent with the arguments of Mott that $\rho(E_F)$ should rise with increasing In concentration. Second, changes in the coulombic interaction between valence and core electrons will perturb the binding energy (BE) of core electrons (a chemical shift). For example, the 5d levels of Hg in HgIn have a greater BE than in pure Hg which may be understood in terms of charge transfer from the In to Hg sites.

The sensitivity of core binding energies to the presence of 'foreign' atoms suggests that this parameter can be used to investigate the occurrence of non-random local environments in alloys. Terms such as 'bonds' and 'molecular clusters' have frequently been invoked to explain, for example, the dramatic changes that take place in the electrical properties of the so-called liquid semiconductors such as TeTl. If well defined molecular complexes do exist in the melt we might expect characteristic valence band structure for the molecular species and characteristic core electron BEs for atoms within the cluster. To date no data is available for liquid semiconducting systems, Au–Sn, however, shows comparable behaviour. Figure 6 shows the BE of the Au $4f_{7/2}$ and Sn $3d_{5/2}$ levels as a function of concentration for both the liquid and solid phase by Ichikawa (1975). Friedman *et al* (1973) have analysed the results for solid Au–Sn

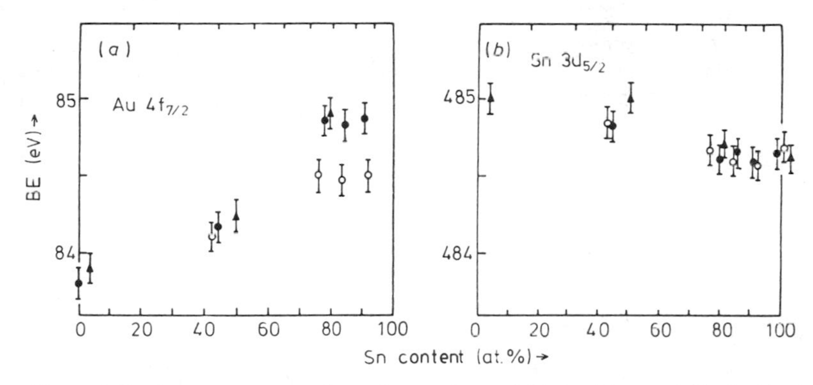

Figure 6. Surface composition dependence of BEs of (*a*) Au $4f_{7/2}$ and (*b*) Sn $3d_{5/2}$ peaks in liquid Au–Sn (Ichikawa 1975, Friedman *et al* 1973). The BEs of the peaks of liquid Au–Sn alloys are represented by open circles and those of solid by closed circles and triangles.

in terms of a charge compensation model which involves little or no charge transfer between Au and Sn sites. The movement of the BE to higher values with increasing impurity concentration is a consequence of covalent bonding between Au and Sn atoms and subsequent reduction in coulombic interaction between core and valence electrons. The fact that the BE of the Au $4f_{7/2}$ level increases up to 50% Au and is then constant indicates that for Sn concentration >50% the local environment of the Au atom is unchanged. This agrees with diffraction studies (Kaplow *et al* 1966) which show that the atomic arrangements are not random, but a portion with short-range order as found in the AuSn solid intermetallic exists in the melt.

5. Summary

In spite of the problems associated with matrix elements which are not readily quantified in the disordered phase and of surface phenomena, photoelectron spectroscopy has added to our understanding of the electronic structure of liquid metals and alloys. Within the limits of available technical knowledge we may envisage future measurements on systems such as liquid Se, the structure of which is not well described by a hard-sphere model. It is, however, in the field of alloys, particularly those with semi-conducting properties, that a very real contribution may be expected:

Acknowledgments

I am indebted to my colleagues at the University of Leicester, in particular to Professor J E Enderby, Dr G P Williams, Mr D C Rodway and Mr J T M Wotherspoon for many useful discussions and for the provision of hitherto unpublished experimental results. Financial support of the Science Research Council is also gratefully acknowledged.

References

Anderson P N and McMillan W L 1967 *Proc. Int. School of Physics, Enrico Fermi* ed W Marshall (New York: Academic Press)
Ballentine L E 1975 *Non Simple Liquids* ed I Prigogine and S A Rice (New York: Wiley) p 263
Bloch A N and Rice S A 1969 *Phys. Rev.* **185** 933
Chang, K S, Sher A, Petzinger K G and Weisz G 1975 *Phys. Rev.* **B12** 5506
Cotti P, Guntherodt H J, Munz P, Oelhafen P and Willschleger J 1973 *Solid St. Commun.* **12** 635–8
Croxton C A 1973 *The Properties of Liquid Metals* ed S Takeuchi (London: Taylor and Francis) p451
Cusack N E 1973 *The Properties of Liquid Metals* ed S Takeuchi (London: Taylor and Francis) p157
Eastman D E 1971 *Phys. Rev. Lett.* **26** 1108
—— 1972 *Proc. Conf. on Electron Spectroscopy, Asiloma, California* ed D A Shirley (Amsterdam: North-Holland) p487
Edwards S F 1958 *Proc. R. Soc.* **A267** 518
Ehrenreich H 1973 *The Properties of Liquid Metals* ed S Takeuchi (London: Taylor and Francis) p173
Friedman R M, Hudis J, Perlman M L and Watson R E 1973 *Phys. Rev.* **B8** 2433
Ichikawa T 1975 *Phys. St. Solidi* **A32** 369
Kaplow R, Strong S L and Averbach B L 1966 *Local Atomic Arrangements studied by X-ray Diffraction* ed J B Cohen and J E Hilliard (New York: Gordon and Breach)
Keller J, Fritz J and Garritz A 1974 *J. Physique* **C4** 379
Koyama R Y and Spicer W E 1971 *Phys. Rev.* **B4** 4318
Lang N D and Kohn W 1971 *Phys. Rev.* **B3** 1215
Mott N F 1966 *Phil. Mag.* **13** 989
Norris C, Rodway D C and Williams G P 1973 *Properties of Liquid Metals* ed S Takeuchi (London: Taylor and Francis) p181
—— 1974 *J. Physique* **C4** 61
Norris C, Rodway D C and Wotherspoon J T M 1976 to be published
Norris C and Williams G P 1976a *J. Phys. F: Metal Phys.* **6** L167
—— 1976b to be published
Olsson J J 1975 *Phys. Rev.* **B12** 2908
Roth L M 1974 *J. Physique* **C4** 317
Schaich W L and Ashcroft N W 1970 *Solid St. Commun.* 8 1959
Shaw R N and Smith N V 1969 *Phys. Rev.* **178** 985
Siegbahn K 1974 *J. Electron Spectrosc. and Related Phemomena.* **5** 3
Stevenson A and Enderby J E 1972 quoted by Enderby in 'Liquid Metals' edited by S J Beer (New York: Decker Inc.)
Williams F L and Nason D 1974 *Surface Sci.* **45** 377
Williams G P and Norris C 1974 *J. Phys. F: Metal Phys.* **4** L175
—— 1975 *Proc. Int. Conf. on the Electronic and Magnetic Properties of Liquid Metals, Mexico City* ed J Keller
—— 1976a *Phil. Mag.* in the press
—— 1976b *Commun. Phys.* in the press
Ziman J M 1961 *Phil. Mag.* **6** 1013

γ-ray Compton profiles of liquid alkali metals

K Suzuki, F Itoh, M Kuroha and T Honda
The Research Institute for Iron, Steel and Other Metals, Tohoku University, Sendai 980, Japan

Abstract. Compton profiles of lithium and sodium metals in both the liquid and solid states were measured by using a Ge(Li) detector and 59·54 keV γ-rays emitted from ^{241}Am. The theoretical Compton profiles were calculated for conduction electrons in the liquid metals by means of a free-electron model. Careful comparisons were then made between the experimental profile deconvoluted with an instrument resolution function and the theoretical profile convoluted with a residual instrument function. It was found that the experimental Compton profile for liquid sodium is in good agreement with the theoretical free-electron model but for liquid lithium, the experimental deviates a little from the theoretical. A small difference of Compton profiles between the liquid and solid states may be interpreted in terms of the free-electron model based on a change in the electron density upon melting.

1. Introduction

Compton scattering is the inelastic scattering of photons by electrons. The energy and momentum conservation during scattering lead to a relation connecting the energy of scattered photons to the initial momentum of the electrons:

$$\omega_1 - \omega_2 = \frac{|\mathbf{k}|^2}{2m} + \frac{\mathbf{k}\cdot\mathbf{p}_0}{m}, \tag{1}$$

where ω_1 and ω_2 are the incident and scattered photon energies ($\hbar = 1$), $\mathbf{k} = \mathbf{k}_1 - \mathbf{k}_2$ is the scattering vector, and $\mathbf{p}_0$ the initial momentum of the electrons. Since the second term on the right-hand side of equation (1) indicates the projection of the initial electron momentum on to the scattering vector, the Compton scattering experiment can directly provide the initial electron momentum distribution; namely the Compton profile.

The differential Compton cross section $\mathrm{d}^2\sigma/\mathrm{d}\Omega\,\mathrm{d}\omega_2$ is proportional to the Compton profile $J(q)$ as follows:

$$\frac{\mathrm{d}^2\sigma}{\mathrm{d}\Omega\,\mathrm{d}\omega_2} = C(\omega_1, \omega_2, \theta, q)\,J(q), \tag{2}$$

where $C(\omega_1, \omega_2, \theta, q)$ is the energy dependence of the relativistic Compton cross section (Manninen *et al* 1974). The Compton profile $J(q)$ is defined by

$$J(q) = \int\int_{-\infty}^{\infty} n(p_0)\,\mathrm{d}p_x\,\mathrm{d}p_y, \tag{3}$$

where

$$q = \mathbf{k}\cdot\mathbf{p}_0/|\mathbf{k}| \qquad \text{and} \qquad \int_{-\infty}^{\infty} J(q)\,\mathrm{d}q = 1 \text{ per electron} \tag{4}$$

and $n(p_0)$ is the momentum probability distribution for an electron to have an initial momentum p_0.

The free-electron model is the simplest model of the conduction electrons but there are electron–ion and electron–electron interactions in real metals and these interactions must give rise to deviations from the free-electron behaviour of the momentum probability distribution function of the conduction electrons. The measurement of the Compton profile may therefore be one of the most powerful experiments that can be used to investigate the electronic state in metals.

From both the experimental and theoretical points of view, lithium and sodium have been regarded as typical metal samples for the Compton scattering experiment (Eisenberger *et al* 1972, Wachtel *et al* 1975). Hence it is an interesting problem to study a modification of the electronic state due to the disordered ionic arrangement in liquid lithium and sodium metals by measuring the Compton profiles over a wide momentum range using high-energy γ-rays.

2. Experiment

A schematic diagram of the spectrometer is shown in figure 1. All measurements of the Compton scattering intensity were carried out by using 59·54 keV γ-rays from a 45 mCi ^{241}Am point source (RCC, Amersham; active diameter, 3 mm) and a Ge(Li) detector (Ortec model 8113-06200) having a total resolution width of 384 eV (FWHM) at 59·54 keV. The incident and scattered γ-rays were collimated to ±3° and ±4°, respectively, and the scattering angle θ was fixed at 165°. Counts of scattered photons were stored in the memory of a multi-channel pulse height analyser (Tracor Northern model NS-720) after passing through ordinary electronic amplifiers (Ortec model 452). A channel interval was set at 29·7 eV and about 20 000 counts per channel were accumulated at the Compton peak in every run. The energy scale in the detector system was calibrated by using x-ray fluorescence emitted from Ba in $BaCO_3$ and Eu in $EuCO_3$.

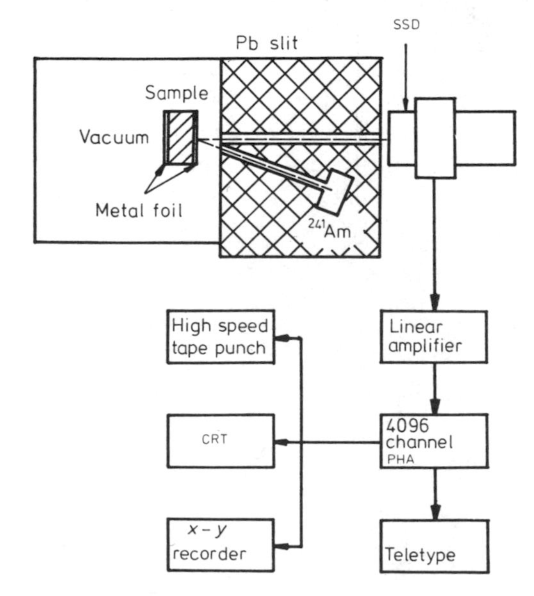

Figure 1. Schematic diagram of the Compton scattering spectrometer.

Lithium and sodium metal samples were contained in an iron cylindrical vessel of 20 mm inner diameter and 5·2 mm depth, located in a vacuum box (5×10^{-6} Torr) with an Al foil (0·1 mm thick) window. Both of the opening sides of the vessel were sealed with an Fe foil (0·005 mm thick) through which the incident γ-ray entered into and the scattered γ-ray came out from the sample. Based on the experimental observation of Compton profiles for aluminium single crystals, as a function of the thickness of the crystal, the authors (Kuroha *et al* 1976) have reached the conclusion that the contribution of the multiple scattering may be regarded as negligible if the value of $\mu_s t_s$ is less than about 0·1 (where μ_s is the linear absorption coefficient and t_s is the thickness of the sample). Therefore, a sample depth of 5·2 mm was chosen in this work, since the value of $\mu_s t_s$ was 0·04 for lithium and 0·11 for sodium.

Scattered γ-ray intensities were first measured for liquid metals (lithium at 240 °C and sodium at 150 °C) and then for polycrystalline metals (solidified in the same vessel at 25 °C) without changing the initial geometry of the arrangement of the detector and sample. Background measurements were carried out separately using the same vessel with the Fe foil on the front or the rear window, before filling with the sample.

3. Data processing

An example of the raw observed data $I_{\text{total}}(\omega_2)$ for liquid lithium metal is shown in figure 2, together with an instrument resolution function at 59·54 keV. After correcting the counting efficiency of the Ge(Li) detector, background subtractions were carried out according to the following procedures:

$$I_{\text{total}}(\omega_2) = B_n(\omega_2) + B_f(\omega_2) + f_w\,[I_s(\omega_2) + f_s\,B_r(\omega_2)]\,, \tag{5}$$

where $B_n(\omega_2)$ is the natural background, $B_f(\omega_2)$ and $B_r(\omega_2)$ the γ-ray intensities scattered from the Fe foil on the front and on the rear window of the vessel, and $I_s(\omega_2)$ the γ-ray intensity scattered from the sample. The correction factors, f_w and f_s, are for

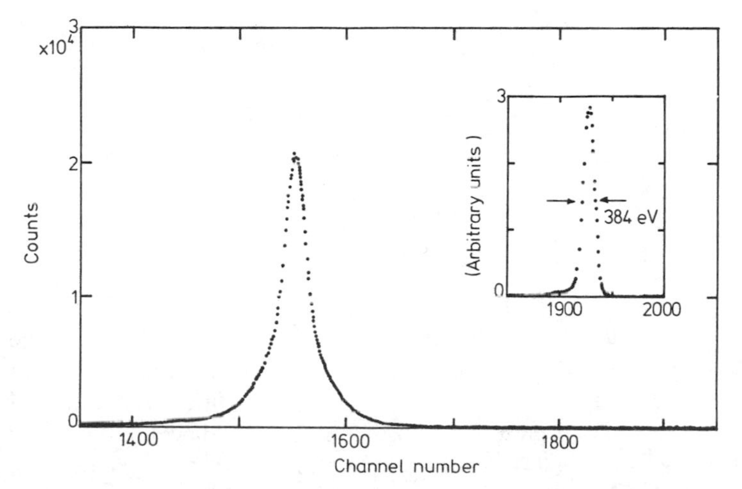

Figure 2. Raw data of the Compton scattering intensity for a liquid lithium metal run at 250 °C. The inset shows the instrument resolution function at 59·54 keV.

the energy dependence of the absorption of the γ-rays and are given by

$$f_w = \exp\{-t_w[\text{cosec}\,\alpha\mu_w(\omega_1) + \text{cosec}\,\beta\mu_w(\omega_2)]\} \tag{6}$$

and

$$f_s = \exp\{-t_s[\text{cosec}\,\alpha\mu_s(\omega_1) + \text{cosec}\,\beta\mu_s(\omega_2)]\}, \tag{7}$$

where the inferior letters w and s indicate the Fe foil windows and the sample, t means the thickness and α and β are the angles between the sample surface and the incident and scattered γ-rays, respectively. Ratios of $I_{total}(\omega_2)/B_n(\omega_2)$ and $I_{total}(\omega_2)/B_f(\omega_2)$ ($\sim I_{total}(\omega_2)/B_r(\omega_2)$) at the Compton peak were about 300 and 200 for both lithium and sodium.

The $I_s(\omega_2)$, obtained from the $I_{total}(\omega_2)$ by correcting the absorption effects described above, was deconvoluted by means of the generalized least-squares method of Fourier analysis (Paatero *et al* 1974) using the instrument resolution function as shown in figure 2. The deconvoluted $I_s(\omega_2)$ was then further corrected for the absorption effects of the γ-rays in the sample itself and the relativistic effect in the Compton cross section (Manninen *et al* 1974). The scale conversion was made from an energy unit (ω_2; eV) of scattered photons to a momentum unit (q; a.u.) of electrons in the sample (Eisenberger and Reed 1972). Finally the Compton profile was normalized by equating the area under the observed profiles over the range of q from $-6{\cdot}0$ to $+6{\cdot}0$ a.u. to the same area under the theoretical profiles (2·989 electrons for Li and 10·648 for Na) calculated from Clementi's (1965) free-atom wavefunctions.

4. Results and discussion

Paatero *et al* (1974) have pointed out that even for an ideal deconvolution procedure, a residual instrument function (RIF) is unavoidably retained in a resultant Compton profile. Therefore, for a proper comparison between theory and experiment, it is necessary first of all to convolute the theoretical Compton profiles with the RIF and then to compare them with experimental Compton profiles. The theoretical convoluted Compton profile $J_{th}(q)$ is related to the original theoretical profile $J^0_{th}(q)$ through the RIF $\tilde{g}(q)$ as

$$J_{th}(q) = \int \tilde{g}(q - q')\, J^0_{th}(q')\, dq', \tag{8}$$

where $\tilde{g}(q)$ is defined by

$$\tilde{g}(q) = \frac{2}{a}\int_0^{t_0} \frac{\cos(2\pi qt/a)}{1 + \lambda(2\pi t/a)^{2k}/(a^2|G(t)|^2)}\, dt, \tag{9}$$

where $G(t)$ is the tth Fourier coefficient for the instrument resolution function, a the range of the integration used in obtaining the Fourier coefficient and λ and k the parameters in the generalized least-squares method. A combination of values, $\lambda = 300$ and $k = 3$, has been adopted as a satisfactory choice in this work. These values have been empirically determined from a criterion that in the experimental Compton profiles deconvoluted with the instrument resolution function, the oscillations in the high momentum region are effectively damped, while the heights of the profiles in the low momentum region are hardly affected. The value of λ is not too critical, provided that relatively large values are given. Paatero *et al* (1974) have suggested that the value $k = 3$ is satisfactory. The RIF used in this work is shown in figure 3, in comparison with a RIF which has another choice of parameter values.

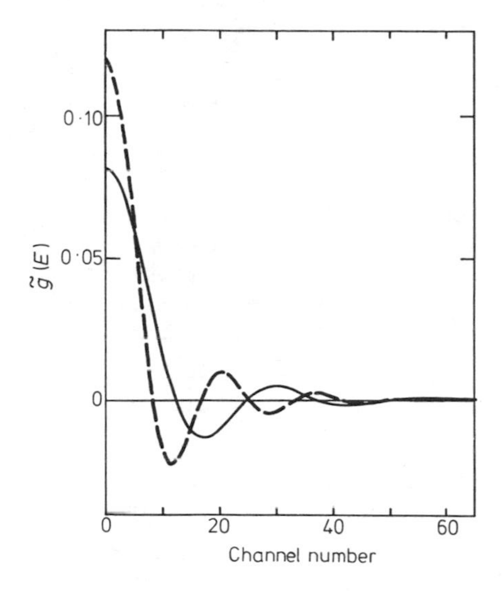

Figure 3. Residual instrument function (RIF) used in this work (solid line) for $k = 3$, $\lambda = 300$, $a = 1000$. The broken line shows a variation in RIF for a different choice of the parameter $\lambda = 3$. Practical calculations of the convolution were carried out in photon energy space.

In the case of both lithium and sodium in either the liquid or solid states, the experimental Compton profiles $J_{\text{exp}}(q)$, for values of q greater than 2 a.u., are in very good agreement with the theoretical Compton profiles $J_{\text{th}}^{\text{core}}(q)$ obtained by convoluting the $J_{\text{th}}^{\text{core, 0}}(q)$ which are calculated for corresponding atomic core configurations ($1s^2$ for Li and $1s^2 2s^2 2p^6$ for Na) using the free-atom wavefunctions of Clementi (1965).

Figures 4 and 5 show the Compton profiles for liquid lithium and sodium when only the contribution from the conduction electrons is considered. Here, the experimental

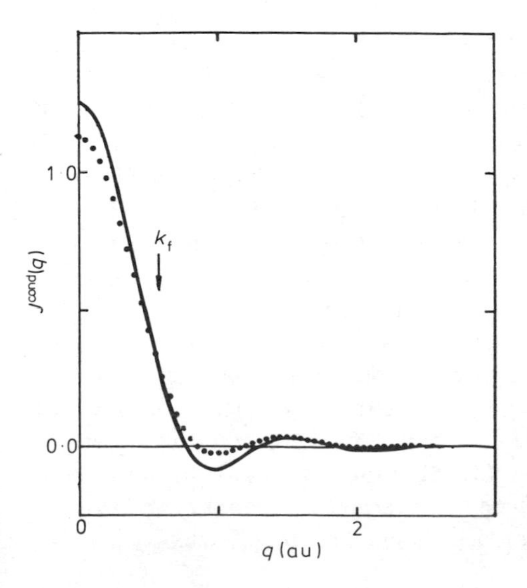

Figure 4. Compton profiles for conduction electrons in liquid lithium metal. The solid line is the free-electron model and the dotted line gives the experimental results. The Fermi momentum in the free-electron model, $k_{\text{F}} = 0{\cdot}576$ a.u., is indicated.

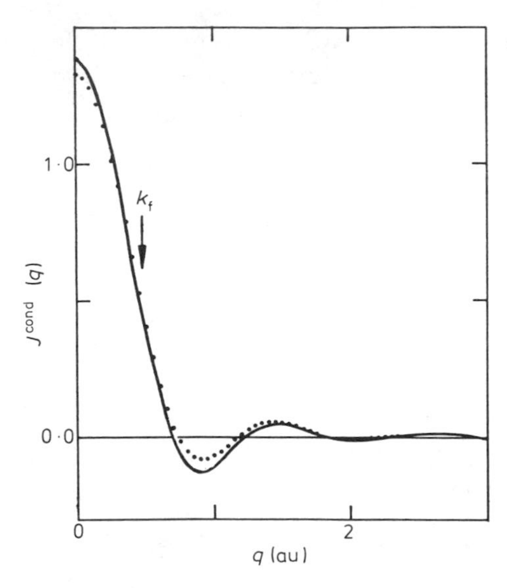

Figure 5. Compton profiles for conduction electrons in liquid sodium metal. The solid line is the free-electron model and the dotted line gives the experimental results. The Fermi momentum in the free-electron model, k_F = 0·472 a.u., is indicated.

Compton profiles for the conduction electrons, $J_{exp}^{cond}(q)$, are obtained by subtracting the $J_{th}^{core}(q)$ from the $J_{exp}(q)$. For a comparison, the theoretical Compton profiles, $J_{th}^{cond}(q)$, convoluted with the RIF, are shown for a free-electron model. Oscillatory tails appearing in the Compton profiles originate from the convolution effect due to the RIF.

From figures 4 and 5, we can conclude that $J_{exp}^{cond}(q)$ for liquid sodium is well reproduced by the free-electron model. Liquid lithium, on the other hand, has $J_{exp}^{cond}(q)$ a little lower than that calculated by the free-electron model for q below k_F (Fermi momentum) and higher for q beyond k_F.

It is quite interesting to compare values of $J_{exp}(q)$ in the liquid and solid states, for both lithium and sodium metals, because a change in the electronic state may be expected due to the disordered ionic arrangement, particularly in lithium which has a significant electron–ion interaction. In order to make this comparison, highly precise measurements are needed with statistical errors less than 1%. Systematic errors contained in the $J_{exp}(q)$ are likely to be the same in both the liquid and solid states, and would cancel each other if the measurements were carried out under the same experimental conditions.

Figure 6 shows the differential Compton profiles, defined as $\Delta J(q) = J^{liq}(q) - J^{sol}(q)$, between the liquid and solid states for the conduction electrons in lithium and sodium. Statistical errors in $J^{liq}(q)$ and $J^{sol}(q)$ are about ±0·5% at the Compton peak because two experimental runs have been bunched-up to result in around 40 000 counts per channel at the peak. In the low q region centred around $q = 0$, the Compton profiles in the liquid state seem to have larger values than those in the solid state. Conversely, the solid profiles may become larger than the liquid ones in the range of q centred around $q = k_F$. However, it is impossible to determine whether the magnitudes of the $\Delta J(q)$ are significant or not, since the error bars in figure 6, indicating uncertainties of ±1%, are almost as large as the amplitudes of the oscillations. As shown in figure 6, the overall behaviour of the $\Delta J(q)$ may be interpreted in terms of the free-electron model based on a change in electron density upon melting (Suzuki *et al* 1976).

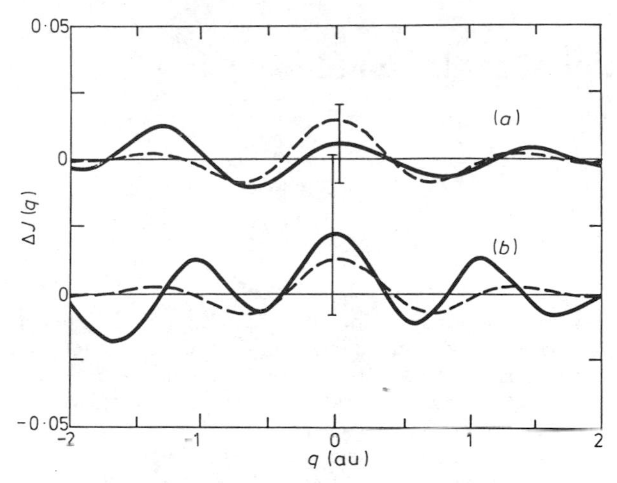

Figure 6. Comparison of Compton profiles of (*a*) lithium and (*b*) sodium metals in the liquid and solid states. Solid lines indicate the experimental observations and dashed lines the theoretical calculations based on a free-electron model. Vertical bars indicate the errors corresponding to ±1%.

References

Clementi E 1965 *IBM J. Res. Dev. Suppl.* **9** 2

Eisenberger P, Lam L, Platzman O M and Schmidt P 1972 *Phys. Rev.* **B6** 3671

Eisenberger P and Reed W A 1972 *Phys. Rev.* **A5** 2085

Kuroha M, Itoh F, Honda T and Suzuki K 1976 *Proc. 31st Ann. Meeting of Phys. Soc. of Japan (No. 3)* p284 (in Japanese)

Manninen S, Paakari T and Kajantie K 1974 *Phil. Mag.* **29** 167

Paatero P, Manninen S and Paakari T 1974 *Phil. Mag.* **30** 1281

Suzuki K, Itoh F, Kuroha M and Honda T 1976 *Phys. Lett.* **57A** 95

Wachtel S, Felsteiner J, Kahane S and Opher R 1975 *Phys. Rev.* **B12** 1285

Theory of the Hall effect in liquid metals

L E Ballentine

Department of Physics, Simon Fraser University, Burnaby, British Columbia, Canada V5A 1S6

Abstract. No progress was made for a very long time in understanding the deviations of the Hall coefficient in liquid metals from the free-electron value because the Boltzmann equation predicted zero deviation. All density of states or effective mass corrections cancel out if the system is isotropic and the scattering can be represented by an energy-dependent relaxation time. The Kubo formula is difficult to evaluate, and has not led to useful results for this problem. Random force correlation theory has shown that the corrections to the free-electron value are of at least third order in the scattering. Recently the possibility of skew scattering (lack of symmetry of the scattering probability between initial and final states) due to spin–orbit interaction has been treated by means of the Boltzmann equation. Preliminary results for liquid Bi are encouraging.

1. Introduction

The Hall effect in liquid metals is a remarkable problem in that it is very easy to construct the zero-order theory but extremely difficult to make further progress. In this paper we shall try to indicate why the problem has resisted theoretical effort for so long, and shall summarize the progress that has recently been achieved. In such a difficult subject as this there have been some false starts, and in the Appendix we point out the flaws in some unsuccessful attempts.

If a steady electric current flows perpendicularly to a magnetic field **B**, the Lorentz force on the charge carriers tends to deflect them sideways, and must be balanced by a transverse electric field. The Hall coefficient is defined to be

$$R_{\mathrm{H}} = E_y/(j_x B_z) \tag{1}$$

with boundary conditions $j_y = j_z = 0, B_x = B_y = 0$. Here **j** is the current density and **E** is the electric field. If one assumes that all conduction electrons move with a common drift velocity, then the balance of electric and magnetic forces leads to the zero-order value,

$$R_{\mathrm{H}}^0 = 1/nec \tag{2}$$

where n is the number of conduction electrons per unit volume and $e = -|e|$ is the electron charge.

The experimental situation has been summarized by Busch and Güntherodt (1974). For most normal liquid metals the free-electron value (2) is observed, but substantial (~20–30%; agreement among different experimenters is only fair) deviations occur for Tl, Pb and Bi. For transition metals equation (2) does not even give the correct sign.

2. Boltzmann equation

If the collision term of the Boltzmann equation can be represented by a relaxation time then the electrical conductivity tensor can be explicitly evaluated as a Fermi surface integral involving the constant energy surfaces $E = E(\mathbf{k})$, the group velocity $\mathbf{v} = \nabla_{\mathbf{k}} E/\hbar$, and the relaxation time $\tau = \tau(\mathbf{k})$. In the absence of any applied magnetic field the conductivity tensor is

$$\sigma_{\mu\nu} = \frac{e^2}{4\pi^3} \int \frac{\tau v_\mu v_\nu}{|\nabla_{\mathbf{k}} E(\mathbf{k})|} \, \mathrm{d}S_{\mathrm{F}}. \tag{3}$$

The term linear in the magnetic field which is responsible for the Hall effect is

$$\sigma_{xy} = B_z \frac{e^3}{4\pi^3 \hbar c} \int \frac{v_x \tau \omega_z v_y \tau}{|\nabla_{\mathbf{k}} E(\mathbf{k})|} \, \mathrm{d}S_{\mathrm{F}} \tag{4}$$

where $\omega = \mathbf{v} \times \nabla_{\mathbf{k}}$ is an operator in k space. The Hall coefficient is given by

$$R_{\mathrm{H}} = \sigma_{xy}/(\sigma_{xx}\sigma_{yy}B_z). \tag{5}$$

Different portions of the Fermi surface may contribute positively or negatively to (5), depending upon the local curvature, and so there is no difficulty in principle in explaining the variation of R_{H} in magnitude and sign for crystalline substances.

Strictly speaking there is no $E(\mathbf{k})$ function for an electron in a liquid metal, and the assumption that one exists can even lead to absurd results in some cases (see section VI of Ballentine 1975). But if, nevertheless, we apply this theory to a liquid then $E(k)$ and τ must be independent of the direction of $\mathbf{k}$. However, no assumption need be made about the form of $E(k)$ as a function of $k = |\mathbf{k}|$.

The dynamics of band electrons can be characterized by an inverse effective mass tensor,

$$\left(\frac{1}{m^*}\right)_{\mu\nu} = \frac{1}{\hbar^2} \frac{\partial^2 E}{\partial k_\mu \partial k_\nu} = \frac{1}{\hbar} \frac{\partial v_\mu}{\partial k_\nu} \tag{6}$$

which relates the rate of change of velocity to the external force,

$$\frac{\mathrm{d}v_\mu}{\mathrm{d}t} = \sum_\nu \left(\frac{1}{m^*}\right)_{\mu\nu} F_\nu. \tag{7}$$

For a *spherical band*, $E = E(|\mathbf{k}|)$, this tensor has two independent components (Jan 1962):

$$\left(\frac{1}{m^*}\right)_{\mu\nu} = \frac{1}{\hbar^2} [u_\mu u_\nu (E'' - E'/k) + \delta_{\mu\nu} E'/k] \tag{8}$$

where $u_\mu = k_\mu/k$ is a direction cosine of $\mathbf{k}$, $E' = \mathrm{d}E/\mathrm{d}k$, and $E'' = \mathrm{d}^2E/\mathrm{d}k^2$. This tensor is diagonal if and only if the function $E(k)$ is a parabola.

If the external force is resolved into its components, $\mathbf{F}_{\mathrm{n}}$ normal and $\mathbf{F}_{\mathrm{t}}$ tangential, to the sphere of constant energy, then (8) becomes

$$\frac{\mathrm{d}\mathbf{v}}{\mathrm{d}t} = \frac{1}{m_{\mathrm{n}}^*} \mathbf{F}_{\mathrm{n}} + \frac{1}{m_{\mathrm{t}}^*} \mathbf{F}_{\mathrm{t}} \tag{9}$$

where, following Jan (1962), we have defined the *normal* and *tangential* (k dependent) effective masses

$$m_n^* = \hbar^2/E'', \qquad m_t^* = \hbar^2 k/E'. \tag{10}$$

It is clear by inspection of (3) and (4) that only m_t^*, involving the first derivative of $E(k)$, is relevant to steady state transport properties.

Thus for a spherical energy band the zero magnetic field conductivity (3) becomes $\sigma_{\mu\nu} = \sigma\delta_{\mu\nu}$,

$$\sigma = \frac{e^2\tau k_F{}^3}{3\pi^2 m_t^*} = \frac{ne^2}{m_t^*}\tau \tag{11}$$

where k_F is the Fermi wavevector and n is the conduction electron density. The Hall conductivity (4) becomes

$$\sigma_{xy} = \frac{e^3}{3\pi^2}\frac{\tau^2}{(m_t^*)^2 c}k_F{}^3\alpha B_z \tag{12}$$

where

$$\alpha = +1 \quad \text{if} \quad dE/dk > 0,$$
$$\alpha = -1 \quad \text{if} \quad dE/dk < 0.$$

Thus the Hall coefficient is

$$R_H = \alpha/nec = 1/nec \tag{13}$$

in the normal situation ($dE/dk > 0$). (Recall that $e = -|e|$ for electrons.)

We see that the standard transport theory which is so successful in treating crystals offers no possibility for explaining the deviations of the Hall coefficient of liquid metals from the classical value (2). It is therefore natural to turn to theories that are more fundamental and of more general validity than the Boltzmann equation.

3. Kubo formula

A general expression for the conductivity of any system, derived by Kubo (1956, 1957), may be written as

$$\sigma_{\mu\nu}(\omega) = \lim_{\eta\to 0^+}\int_0^\infty \exp\left[-i(\omega + i\eta)t\right]\ (J_\mu(t); J_\nu(0))\ dt/\Omega. \tag{14}$$

Here $J_\mu(t)$ is the total current operator (for all electrons) in the Heisenberg picture, and Ω is the volume of the system. We use the notation

$$(X; Y) = \mathrm{Tr}\left(\exp(-\beta H)\int_0^\infty \exp(\lambda H)\, X^\dagger \exp(-\lambda H)\, d\lambda Y\right)\Big/\mathrm{Tr}\left[\exp(-\beta H)\right] \tag{15}$$

which may be interpreted as a quantum-mechanical correlation function. As was shown by Ballentine and Heaney (1974), it has the mathematical properties of an inner product between X and Y.

If we specialize to *non-interacting electrons*, then the current–current correlation function becomes

$$(J_\mu(t); J_\nu(0)) = \sum_{r,s} \langle r|j_\mu(t)|s\rangle\langle s|j_\nu(0)|r\rangle \frac{f^0(E_r) - f^0(E_s)}{E_s - E_r} \tag{16}$$

where j_μ is a single-particle current operator, $|s\rangle$ (E_s) is an eigenvector (eigenvalue) of the one-electron Hamiltonian, and $f^0(E)$ is the Fermi–Dirac distribution function. The zero-frequency conductivity tensor for independent electrons then becomes

$$\sigma_{\mu\nu} = \frac{i}{\Omega} \sum_{r,s} \frac{\langle r|j_\mu|s\rangle\langle s|j_\nu|r\rangle}{E_r - E_s + i\eta} \frac{f^0(E_r) - f^0(E_s)}{E_s - E_r} \tag{17}$$

where the sums over states must include the spin variable. This can be separated into two terms by means of the standard substitution

$$\lim_{\eta \to 0^+} \frac{1}{E_r - E_s + i\eta} = \frac{P}{E_r - E_s} - i\pi\delta(E_r - E_s). \tag{18}$$

Now it is clear that the symmetric part of $\sigma_{\mu\nu}$ comes entirely from the delta function term and the antisymmetric part, which includes the Hall effect, comes entirely from the principal value term. (For nonzero frequencies this statement should refer to the Hermitean, $(\sigma_{\mu\nu} + \sigma_{\nu\mu}^*)/2$, and anti-Hermitean, $(\sigma_{\mu\nu} - \sigma_{\nu\mu}^*)/2$, parts of the conductivity tensor.) If there is no magnetic field present then time-reversal invariance implies that the wavefunctions may be chosen to be real, and the antisymmetric term will vanish.

The zero magnetic field conductivity has been calculated from (17) in the weak scattering limit by several authors (Langer 1960, Neal 1970, Bringer and Wagner 1971). Their work is easily extended (appendix D of Schaich 1970) to allow for deviations of the density of states from the free-electron value. By the 'weak scattering limit' we mean lowest order in the quantity Γ/E_F, where the ensemble average of the Green function is

$$G(k, E) = [E - k^2 - \Sigma(k, E)]^{-1},$$

and $\Gamma = -\mathrm{Im}\Sigma$. This is somewhat less restrictive than a lowest-order perturbation expansion in powers of the scattering potential because no assumption need be made about $\mathrm{Re}\Sigma(k,E)$, which primarily determines the density of states.

When a magnetic field $\mathbf{B}$ is present, the current operator in (17) becomes

$$\mathbf{j} = \frac{e}{m}\mathbf{p} - \frac{e^2}{mc}\mathbf{A},$$

with $\mathbf{B} = \mathrm{curl}\,\mathbf{A}$. In expanding (17) to first order in B one obtains contributions from the current operators and from the wavefunctions, and it is important to take account of gauge invariance. This has been done in two ways. Fukuyama *et al* (1969) took the vector potential to be $\mathbf{A}(\mathbf{r}) = \mathbf{A}_0 \exp(\mathrm{i}\mathbf{q}\cdot\mathbf{r})$, finally letting q tend to zero, and they worked only to lowest (second) order in the scattering potential. Their result was

$$R_\mathrm{H} = \frac{1}{nec}\left(\frac{n_\mathrm{fe}(E_\mathrm{F})}{n(E_\mathrm{F})}\right)^2 \tag{19}$$

where $n_\mathrm{fe}(E)$ and $n(E)$ are the free-electron and true density of states functions respectively (the latter of course evaluated by second-order perturbation theory). In a lengthy calculation, the details of which are not published, I have used $\mathbf{A}(\mathbf{r}) = \frac{1}{2}\mathbf{B}\times\mathbf{r} + \nabla\chi(\mathbf{r})$ where $\chi(\mathbf{r})$ is an arbitrary scalar. I made no expansion in the scattering potential, but worked to lowest order in Γ/E_F. The result is

$$R_\mathrm{H} = \frac{\alpha}{nec}\left(\frac{n_\mathrm{fe}(E_\mathrm{F})}{n(E_\mathrm{F})}\right)^2 + \mathrm{O}(\Gamma/E_\mathrm{F}). \tag{20}$$

The factor $\alpha = \pm 1$ has the same meaning as in (12), so that this result would apply not only to nearly free electrons, but also to 'nearly free holes' if such were to exist in a liquid metal (which seems unlikely). In those metals for which the relevant calculations have been done (see Table III of Ballentine 1975) the quantity Γ/E_F has the same order of magnitude as does the deviation of the density of states factor from unity, so the fact that (19) failed to account for the systematic deviations of $R_\mathrm{H}/R_\mathrm{H}^0$ from unity can now be understood. But this is a very small reward for the very large amount of work that went into deriving (20).

4. Random force correlation theory

Four years ago in Tokyo, there was considerable interest in a newly proposed force–force correlation formula for the electrical resistivity, although it would be fair to say that the subject was controversial. Today the controversy has, hopefully, been resolved with the following conclusions (Ballentine and Heaney 1974):

(i) The original arguments were incomplete and in some cases incorrect.
(ii) The formula expressing proportionality between the total force–force correlation function and the resistivity is not correct. Indeed it is easy to show that the zero frequency limit of the (Fourier transform of the) total force autocorrelation function must vanish, and so can not yield the DC resistivity.
(iii) There is a similar correct formula for the resistivity, involving the autocorrelation function of the 'random' part of the force. Generalized to arbitrary frequency and magnetic field, it is of the form

$$r_{\mu\nu}(\omega) = \frac{m}{ne^2}\left(\mathrm{i}\omega\delta_{\mu\nu} - \frac{e}{mc}\epsilon_{\mu\nu\gamma}B_\gamma + \frac{1}{mn\Omega}\int_0^\infty \exp(-\mathrm{i}\omega t)(R_\mu(t); R_\nu(0))\,\mathrm{d}t\right) \tag{21}$$

where the notation (15) is used once more. The precise definition of the random force $R(t)$ is given by Ballentine and Heaney (1974), but it can be illustrated by the one-dimensional Langevin equation

$$\begin{aligned} m\,\mathrm{d}v(t)/\mathrm{d}t &= F(t) \\ &= -m\int_{t_0}^{t} \gamma(t-t')\,v(t')\,\mathrm{d}t' + R(t). \end{aligned} \tag{22}$$

The random force $R(t)$ remains after the frictional force has been removed from the total force $F(t)$.

(iv) The weak scattering limit is non-uniform with respect to time. That is to say lowest-order perturbation theory is always valid for short enough times, but no matter how weak the scattering may be there is a time beyond which it must fail. This almost self-evident fact has a surprising consequence. Since the total force and the random force are equal for short times their autocorrelation functions are equal in lowest-order perturbation theory, even though the two functions are very different for long times. Thus both the correct random force and the incorrect total force correlation formulae lead to the same lowest-order expression for the resistivity (with $\omega = 0, B = 0$),

$$\frac{2\pi}{e^2n^2\Omega}\sum_{\mathbf{k}}\sum_{\mathbf{k}'}\left(-\frac{\partial f^0}{\partial E}\right)\langle\mathbf{k}|\nabla_\mu V|\mathbf{k}'\rangle\langle\mathbf{k}'|\nabla_\nu V|\mathbf{k}\rangle\delta(E_k - E_{k'}) \tag{23}$$

which becomes the familiar Ziman formula in the degenerate limit ($kT \ll E_\mathrm{F}$).

While they argued the relative merits of various formulations, most of the participants in the debate overlooked the fact that (23) is *not* the correct lowest-order expression for the resistivity. It is equivalent to averaging the scattering rate, $1/\tau(E)$, over the energy distribution, whereas the lowest-order solution to the Boltzmann equation averages $\tau(E)$. It is clear that the latter is correct because the conductivity of electrons with energy E is proportional to $\tau(E)$, and non-interacting electrons are parallel conductors. In fact (23) is the correct weak scattering limit only at $T = 0$, when $\partial f^0/\partial E$ becomes $-\delta(E - E_\mathrm{F})$.

The cause of the trouble was exposed by Argyres and Sigel (1973, 1974) and by Huberman and Chester (1975). Although the correlation function in (21) vanishes as $t \to \infty$, a perturbation expansion of it contains terms that diverge as $t \to \infty$. Thus, although (21) is the only term of order λ^2 (λ being a measure of the strength of the scatterers), there are higher-order terms of order $\lambda^2(\lambda^2/\omega)^n$ that may not be neglected in the limit $\omega \to 0$. The infinite series can be summed, and its limit as $\lambda^2/\omega \to \infty$ yields the correct weak scattering formula in agreement with the Boltzmann equation.

Returning to the Hall effect, the attractiveness of (21) lies in the fact that the second term yields the free-electron value (2) and the last term yields a correction of second order in the scattering force,

$$R_\mathrm{H} = \frac{1}{nec}\left(1 + \frac{1}{en\Omega}\int_0^\infty \frac{\partial}{\partial B_z}(R_y(t); R_x(0))\bigg|_{B_z = 0} \mathrm{d}t\right). \tag{24}$$

Since it required considerable labour to extract the free-electron value from the Kubo formula in the weak scattering limit, this equation seems much more attractive.

To evaluate the correction term in (24) to second order in the scattering strength we simply drop the scattering potential from the Hamiltonian H that appears in (15) and that is implicit in the time development of $R_y(t)$. The magnetic field dependence likewise comes from two sources: the statistical factors in (15) and the dynamical evolution of $R_y(t)$. Neither the statistical nor the dynamical contribution to (24) are zero, but their sum vanishes. Thus there is no second-order correction to R_H. (We have not summed the singular λ^2/ω terms, but presumably they only restore agreement with the

Boltzmann equation for non-degenerate energy distributions, as was the case for the resistivity.) The reason for this unforeseen cancellation can be understood as follows: by dropping the scattering potential from H we are calculating the force on an electron as it moves with constant velocity along a free-particle trajectory, but neglecting the effect that the force would have on the trajectory. In zero magnetic field that trajectory is a straight line; in a weak field it becomes the arc of a large circle (see figure 1).

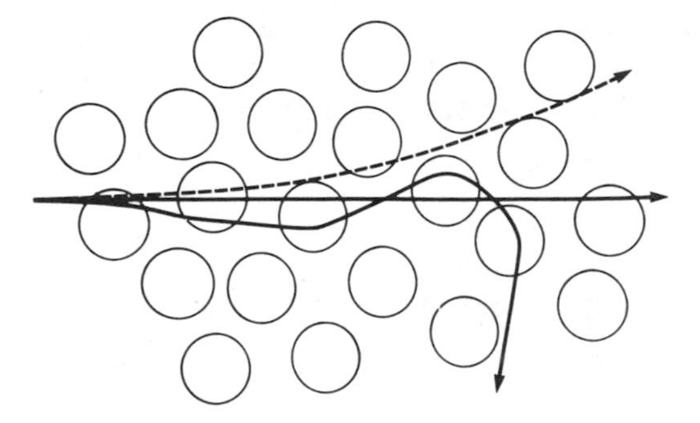

Figure 1. Schematic path of electron for calculating the force autocorrelation function: crooked full line, actual path; straight full line, free electron approximation; broken line, free electron in magnetic field.

The integrand in (24) is an averaged measure of the difference between the force experienced by an electron on these two trajectories, and since the separation between the trajectories increases as t^2 it follows that the integrand is proportional to t^2. But by time-reversal symmetry a correlation function like $(R_x(0); R_y(t))$ must be invariant under the simultaneous transformation: $B \to -B, t \to -t$. Hence the integrand of (24) must be an odd function of t. This is possible only if the second-order contribution (proportional to t^2) vanishes.

We conclude that the lowest-order correction to the free-electron Hall coefficient is at least of third order in the scattering. This result provides a useful check of any proposed theory. In particular, since both terms of (20) contain contributions of second order in the scattering, this result proves that there must be cancellation between them, as was suspected from the disagreement of (19) with experiment.

5. Skew scattering

In §2 we argued that the Boltzmann equation could yield only the classical value for the Hall coefficient of a liquid metal. We did not stress the implicit assumption that the scattering term could be represented by a relaxation time τ, but it is well known that this is true for elastic scattering that depends only on the relative angle θ between initial and final momenta, or equivalently on $\cos\theta = \hat{\mathbf{k}} \cdot \hat{\mathbf{k}}'$. However, in the presence of an external magnetic field the microreversibility of the transition probability $W(\mathbf{k}', \mathbf{k})$ need not hold and there may be antisymmetric contributions that depend on $\hat{\mathbf{k}}' \times \hat{\mathbf{k}}$. The term 'skew scattering' refers to the inequality $W(\mathbf{k}, \mathbf{k}') - W(\mathbf{k}', \mathbf{k}) \neq 0$.

There may be several possible sources of skew scattering†, but the one that seems

† Chambers (1973) has proposed a mechanism that deserves further study. It is a modification of the scattering by the magnetic field. However, it raises some conceptual problems, such as whether the quantal treatment of the magnetic field within the scatterer plus the classical Lorentz force in the Boltzmann equation might double count some magnetic effects.

most important is the spin–orbit (SO) interaction. The SO interaction has the form

$$V_{SO} = \frac{-e}{2mc^2}\mathbf{E}\times\mathbf{v}.\mathbf{s} \tag{25}$$

where $e = -|e|$, $\mathbf{E}$ is the electric field, $\mathbf{v}$ is the velocity operator, and $\mathbf{s} = \hbar\sigma/2$ is the spin operator. According to ten Bosch (1973) the SO interaction contributes to the Hall effect in two ways.

(i) An electron moving in the x direction through the inhomogeneous electric field of an ion will experience a transverse force, the sign of which will depend on the spin direction. Since the spin-up and spin-down populations are unequal in an external magnetic field, there will be a net transverse electric polarization, proportional to the current and to the magnetic field and perpendicular to both. Thus the transverse electric field set up by this polarization will be measured as part of the Hall effect. This effect is very important in transition metals, according to the results of ten Bosch (1973), but it contributes only about 1% of R_H^0 for Pb.
(ii) Skew scattering: although ten Bosch made only a rough estimate of this effect, her result suggests that it can be many times larger than the polarization effect in liquid metals with short mean free paths.

In Born approximation the SO scattering amplitude is proportional to $i\mathbf{k}'\times\mathbf{k}$, and being antisymmetric it must be pure imaginary in order to be Hermitean. The ordinary potential scattering amplitude is real and symmetric. Therefore, in Born approximation, we have

$$\langle\mathbf{k}|(V+V_{SO})\mathbf{k}'\rangle = \langle\mathbf{k}'|(V+V_{SO})|\mathbf{k}\rangle^*$$

hence the scattering probability is symmetric. To obtain skew scattering we must go beyond the Born approximation.

Let us define three Hamiltonians relevant to a single atom†: $H_0 = p^2/2m$, $H_1 = H_0 + V_{MT}(r)$, and $H = H_1 + V_{SO}$. Here V_{MT} is the muffin-tin (MT) potential of the atom (see Loucks 1967). Instead of the Born approximation which treats $V_{MT} + V_{SO}$ as a perturbation on H_0, we use the *distorted wave Born approximation* (DWBA) (chapter 19 of Messiah 1964) which treats V_{SO} as a perturbation on H_1. The scattering amplitude from a state of momentum and z component of spin $(\mathbf{k}, \sigma)$ to a state $(\mathbf{k}', \sigma')$ with angle θ between $\mathbf{k}$ and $\mathbf{k}'$ is

$$f^{sc}_{\sigma',\sigma}(\theta) = f^{MT}(\theta)\,\delta_{\sigma',\sigma} + f^{SO}_{\sigma',\sigma}(\theta). \tag{26}$$

The first term is the usual scattering amplitude for H_1,

$$f^{MT}(\theta) = \frac{1}{\kappa}\sum_l (2l+1)\exp(\mathrm{i}\delta_l)\sin\delta_l P_l(\cos\theta) \tag{27}$$

where the energy measured from the MT zero is $E = \hbar^2\kappa^2/2m$. The DWBA asserts

$$f^{SO}_{\sigma',\sigma}(\theta) \cong -\frac{m}{2\pi\hbar^2}\langle\psi^{(-)}_{\mathbf{k}'\sigma'}|V_{SO}|\psi^{(+)}_{\mathbf{k}\sigma}\rangle \tag{28}$$

† The remainder of this paper is a preliminary report of work being done in collaboration with Dr M Huberman.

where $\psi^{(+)}$ and $\psi^{(-)}$ are scattering states of H_1 satisfying outgoing and incoming boundary conditions, respectively. Thus to first order in V_{SO} we have

$$f^{SO}(\theta) = \frac{2m}{\hbar^2} \mathrm{i}\sigma.\hat{\mathbf{k}}' \times \hat{\mathbf{k}} \sum_l (2l+1) \exp(2\mathrm{i}\delta_l) \frac{\mathrm{d}P_l(\cos\theta)}{\mathrm{d}(\cos\theta)} \Lambda_l \tag{29}$$

where $\sigma = (\sigma_x, \sigma_y, \sigma_z)$ are Pauli spin matrices. The spin–orbit coupling parameter is

$$\Lambda_l = \frac{\hbar^2}{4m^2c^2} \int_0^a \frac{1}{r} \frac{\mathrm{d}V_{MT}}{\mathrm{d}r} |R_l(r)|^2 r^2 \,\mathrm{d}r \tag{30}$$

where a is the MT radius beyond which $V_{MT}(r)$ vanishes. The wavefunctions $R_l(r)$ are radial eigenfunctions of H_1 which satisfy the boundary condition

$$R_l(r) = \cos\delta_l j_l(\kappa r) - \sin\delta_l n_l(\kappa r), \qquad r \geqslant a. \tag{31}$$

The scattering probability is given by the square of the modulus of the amplitude (26). For $\sigma = \sigma'$ (no spin flip) the cross terms between f^{MT} and f^{SO} will yield skew scattering. If we neglect terms of second order in the SO interaction then we may neglect spin-flip scattering.

The Boltzmann equation for a homogeneous steady state electron distribution is

$$\frac{\mathbf{F}}{\hbar} \frac{\partial f(\mathbf{k})}{\partial \mathbf{k}} = \int [W(\mathbf{k},\mathbf{k}') f(\mathbf{k}') - W(\mathbf{k}',\mathbf{k}) f(\mathbf{k})] \,\mathrm{d}\Omega_{\hat{k}'}, \tag{32}$$

where $\mathbf{F} = e(\mathbf{E} + \mathbf{v} \times \mathbf{B}/c)$ is the external force, $W(\mathbf{k}',\mathbf{k})$ is the scattering rate from $\mathbf{k}$ to $\mathbf{k}'$ (per unit solid angle of $\mathbf{k}'$), and the integration is over the angles of $\mathbf{k}'$. The spin dependence of f and W is not explicitly indicated. It should be emphasized that there are no factors of $(1-f)$ for the final states in the scattering terms. This fact, which runs contrary to many heuristic textbook presentations, was demonstrated by Kohn and Luttinger (1957). The inclusion of the extraneous $(1-f)$ factors makes no difference if $W(\mathbf{k}',\mathbf{k})$ is symmetric, but this is not true if there is skew scattering.

The total scattering amplitude is obtained by adding up the scattering amplitude from each atom with the appropriate relative phase, and the scattering rate W is equal to the flux associated with the initial state multiplied by the square of the modulus of the total scattering amplitude,

$$W(\mathbf{k}',\mathbf{k}) = \frac{\hbar k}{m} n_{\mathrm{i}} S(|\mathbf{k} - \mathbf{k}'| f^{sc}(\theta)|^2 \tag{33}$$

where $S(q)$ is the liquid structure factor and n_{i} is the number density of ions. If SO scattering were neglected, this scattering rate would become identical with that used by Drierach *et al* (1972). We can separate the normal (symmetric) scattering rate from the skew-scattering rate by writing (33) in the form

$$W(\mathbf{k}',\mathbf{k}) = W_0(\theta) + \hat{\sigma}_z.\hat{\mathbf{k}}' \times \hat{\mathbf{k}} W_s(\theta). \tag{34}$$

Here $\hat{\sigma}_z$ is the unit vector in the $\pm z$ direction (+ for spin up, – for spin down), where the z axis is the direction of the magnetic field $\mathbf{B}$. W_0 and W_s may depend on the energy as well as the angle θ between $\hat{\mathbf{k}}'$ and $\hat{\mathbf{k}}$.

The electron distribution function $f(\mathbf{k})$ can be expanded in spherical harmonics, and only the $l = 1$ components need be considered since the others do not contribute to the electric current. Therefore we try the familiar ansatz $f(\mathbf{k}) = f^0(\mathbf{k}) + \hat{\mathbf{k}} \cdot \mathbf{A}$, where the vector $\mathbf{A}$ is independent of the direction of $\hat{\mathbf{k}}$. This still yields an exact solution (to lowest order in $\mathbf{E}$) even in the presence of skew scattering. With this ansatz the Boltzmann equation reduces to

$$\frac{e\hbar}{m} k \frac{\partial f^0}{\partial E_k} \mathbf{E} = \mathbf{A} \times \left(\frac{e}{mc} \mathbf{B} + \frac{\hat{\sigma}_z}{\tau_s} \right) - \frac{\mathbf{A}}{\tau_0}. \tag{35}$$

The normal relaxation time τ_0 and the skew-scattering time are given by

$$\frac{1}{\tau_0} = 2\pi \int_{-1}^{1} W_0(\theta)(1 - \cos\theta)\, d(\cos\theta) \tag{36}$$

$$\frac{1}{\tau_s} = \frac{2\pi}{3} \int_{-1}^{1} W_s(\theta)\,[1 - P_2(\cos\theta)]\, d(\cos\theta) \tag{37}$$

where $P_2(\cos\theta)$ is a Legendre polynomial. It can be seen that the effect of skew scattering is the same (for a given spin orientation) as that of the magnetic field, and hence it will modify the Hall coefficient.

The conductivity tensor for electrons of one spin orientation is found to be

$$\sigma_{xx}^{(s)} = \frac{e^2 \tau_0 k_F^3}{m 6\pi^2} \frac{1 + (\tau_0/\tau_s)^2 - 2\sigma_z \omega_c \tau_0^2/\tau_s}{[1 + (\tau_0/\tau_s)^2]^2} \tag{38}$$

$$\sigma_{xy}^{(s)} = \sigma_{xx}^{(s)} (\omega_c \tau_0 + \sigma_z \tau_0/\tau_s) \tag{39}$$

where $\omega_c = eB/mc < 0$. We have made the fully justifiable assumption that $|\omega_c \tau_0| \ll 1$. In these expressions k_F depends on σ_z, and hence so do τ_0 and τ_s. The total conductivity tensor is the sum of the spin-up and spin-down conductivities, and the Hall coefficient is found to be, neglecting $O(B^2)$,

$$R_H = \frac{1}{nec} \left(1 - r^2 + \frac{\hbar}{\epsilon_F \tau_s} \frac{\chi_p}{\chi_p^0} [\tfrac{3}{4}(1 + r^2) + \gamma_0 - \tfrac{1}{2}(1 - r^2)\gamma_s] \right) \tag{40}$$

where $r = \tau_0/\tau_s$, $\epsilon_F = \hbar^2 k_F^2/2m$, $\gamma_0 = (\epsilon_F/\tau_0)\, \partial\tau_0/\partial E$, and $\gamma_s = (\epsilon_F/\tau_s)\, \partial\tau_s/\partial E$.

The corrections to the classical value are of two kinds. Substitution of (38) into (39) yields a term involving $(\sigma_z)^2$, which leads to the term $-r^2$ in (40). The remaining correction terms are linear in σ_z, and their contribution to (40) is proportional to the net spin polarization, that is, to the paramagnetic susceptibility χ_p. The exchange and correlation modification of the energy required to flip the spin of an electron is taken into account by the factor χ_p/χ_p^0, the ratio of the paramagnetic susceptibilities of interacting and non-interacting electrons. According to the calculations of Hedin and Lundqvist (1969) this factor is about 1·3 to 1·4 for simple metals.

Since this work is very recent, only preliminary numerical results can be given. The largest deviation of R_H/R_H^0 from unity among the normal metals is that of Bi, so we have examined it first. We used the phase shifts of Ratti (1972), which are $\delta_0 = 1{\cdot}14$,

$\delta_1 = 1{\cdot}26$, and $\delta_2 = 0{\cdot}03$ at an energy $E_F = 0{\cdot}433$ Ryd above MT zero. The SO coupling parameter Λ_l (30) can be estimated from the atomic parameters (Herman and Skillman 1963) if we neglect the difference between V_{MT} and the atomic potential, and renormalize the atomic wavefunctions to satisfy (31) at $r = a(= 3{\cdot}2$ Bohr units). This yields $\Lambda_1 = 0{\cdot}29$ Ryd, and Λ_l negligible for $l > 1$. We have used the Percus–Yevick hard-sphere structure factor for convenience.

The dimensionless parameters that determine the corrections to R_H are $r = -0{\cdot}24$ and $\hbar/E_F\tau_s = -0{\cdot}082$. We have not yet evaluated the derivative terms γ_0 and γ_s, but these should be small since there is no scattering resonance in Bi. Taking the susceptibility ratio to be about 4/3, this yields $R_H/R_H^0 = 0{\cdot}86$. The experimental values are: 0·69 (Greenfield 1964); 0·60 (Busch and Tièche 1963); 0·95 (Takeuchi and Endo 1961). The resistivity is calculated to be 137 $\mu\Omega$ cm, compared with the experimental value of 128 $\mu\Omega$ cm.

Although this result for Bi is only preliminary, it encourages us to believe that our theory is on the right track. The atomic spin–orbit interaction strength increases approximately in proportion to the square of the atomic number within a column of the periodic table, but much more rapidly across a row of the table. This provides a natural explanation of why Bi, Pb, and Tl are the only pure normal metals that exhibit substantial deviations of R_H from R_H^0. It remains to be investigated whether this skew-scattering effect will seriously modify the results of ten Bosch (1973) for transition metals.

In conclusion, we hope that experimentalists will now be encouraged to perform accurate measurements of the Hall coefficients of liquid metals so as to reduce the confusion among the conflicting published values.

Appendix. Some false starts and where they went wrong

Since a theory of the Hall effect in liquid metals is just beginning to emerge, it may be useful to explain why some results appearing in the literature are incorrect.

Güntherodt and Künzi (1973) gave a formula for R_H involving the energy derivative of the density of states, and it is quoted in the review by Busch and Güntherodt (1974). It was obtained by incorrectly using the effective mass m_n^* (equation 10) instead of m_t^*. The source of the confusion seems to lie in Ziman's book (1960), in which he set out to treat an arbitrary spherical $E(k)$ relation but unfortunately assumed that the effective mass tensor was diagonal, not noticing that this is true only for a parabolic $E(k)$.

Szabo (1972) derived an expression for the Hall coefficient from the total force–force correlation formula that is known to give the DC resistivity to be identically zero when evaluated exactly. However, that error is not fatal, since the weak scattering expansion derived from that formula is correct to lowest order for a degenerate Fermi gas. The fatal error may be his assumption that the angular momentum operator commuted with the Hamiltonian of the disordered system. We note that his result would yield a correction of second order in the scattering potential, which should not exist according to §4.

Jones (1975) has derived another expression which would yield second-order corrections to R_H. His derivation seems to contain a technical error in the evaluation of the force correlation function. Just as the antisymmetric part of the conductivity tensor comes entirely from the principal value term of the current correlation function (equations 17, 18), so the antisymmetric part of the resistivity tensor comes entirely from the principal value term of the force correlation function. The delta function term in the correlation functions can only yield a symmetric tensor. Now Jones' expression for the yx component of resistivity, R^0_{yx}, is expressed in terms of energy delta functions, and indeed one can see that his expression for R^0_{yx} is symmetric in x and y, instead of antisymmetric.

We note that Chambers (1973) and ten Bosch (1973) both pass the test of our §4, their corrections to R_H being of third order in the scattering (second order in the phase shift times first order in SO coupling in the latter case).

References

Argyres P N and Sigel J L 1973 *Phys. Rev. Lett.* **31** 1397–400
—— 1974 *Phys. Rev.* **B9** 3197–206
Ballentine L E 1975 *Adv. Chem. Phys.* **31** 263–327
Ballentine L E and Heaney W J 1974 *J. Phys. C: Solid St. Phys.* **7** 1985–98
ten Bosch A 1973 *Phys. Kondens. Mater.* **16** 289–318
Bringer A and Wagner D 1971 *Z. Phys.* **241** 295–307
Busch G and Güntherodt H J 1974 *Solid St. Phys.* **29** 235–313 (New York: Academic Press)
Busch G and Tièche Y 1963 *Phys. Kondens. Mater.* **1** 78–104
Chambers W G 1973 *J. Phys. C: Solid St. Phys.* **6** 2441–5.
Dreirach O, Evans R, Güntherodt H J and Künzi H U 1972 *J. Phys. F: Metal Phys.* **2** 709–25
Fukuyama H, Ebisawa H and Wada Y 1969 *Prog. Theor. Phys.* **42** 494–511
Greenfield A J 1964 *Phys. Rev.* **135A** 1589–95
Güntherodt H-J and Künzi H U 1973 *Phys. Kondens. Mater.* **16** 117–46
Hedin L and Lundqvist S 1969 *Solid St. Phys.* **23** 1–181 (New York: Academic Press)
Herman F and Skillman S 1963 *Atomic Structure Calculations* (Englewood Cliffs, NJ: Prentice-Hall)
Huberman M and Chester G V 1975 *Adv. Phys.* **24** 489–514
Jan J P 1962 *Am. J. Phys.* **30** 497–9
Jones W 1975 *J. Phys. F: Metal Phys.* **5** 1365–7
Kubo R 1956 *Can. J. Phys.* **34** 1274–7
—— 1957 *J. Phys. Soc. Japan* **12** 570–86
Kohn W and Luttinger J M 1957 *Phys. Rev.* **108** 590–611
Langer J S 1960 *Phys. Rev.* **120** 714–25
Loucks T L 1967 *Augmented Plane Wave Method* (New York: Benjamin)
Messiah A 1964 *Quantum Mechanics* (Amsterdam: North-Holland)
Neal T 1970 *Phys. Fluids* **13** 249–62
Ratti V K 1972 *PhD Thesis* University of Bristol
Schaich W L 1970 *PhD Thesis* Cornell University
Szabo N 1972 *J. Phys. C: Solid St. Phys.* **5** L241–6
Takeuchi S and Endo H 1961 *Trans. Japan Inst. Metals* **2** 243
Ziman J M 1960 *Electrons and Phonons* (Oxford: Clarendon)

On the sign of the Hall effect in liquid metals

L M Roth
State University of New York at Albany, Albany, NY 12222, USA

Abstract. We calculate the conductivity, Hall coefficient and thermoelectric power for simple tight-binding models of liquid metals and liquid alloys, using an extension of the Ishida–Yonezawa theory of the electronic structure for which vertex corrections can be ignored. The thermoelectric power changes sign on going from hole to electron conduction but the Hall coefficient remains negative, in qualitative agreement with experimental observations.

1. Introduction

There are a number of instances in liquid metals in which the Hall constant R is negative or electron-like while other indications imply conduction in a nearly filled band. Thus, on varying the composition of liquid alloys such as Tl–Te, near the stoichiometric composition the thermoelectric power α changes sign suggesting a change from electron to hole conduction, but while R has a peak, its sign remains negative throughout (Donnally and Cutler 1971, Enderby and Collings 1970). Other examples are given by Allgair (1969) who has suggested that the absence of hole pockets near a zone edge accounts for the negativity of R in most liquid metals.

The first calculation bearing on this problem is that of Friedman (1971) who used a one-orbital tight-binding model and a random-phase approximation and obtained a negative R. In the present calculation we shall also use a tight-binding model, and apply the recently developed self-consistent single-site approximations to study the transport coefficients. Recently, we have shown (Roth 1976) that our effective medium approximation (EMA) (Roth 1974), which is equivalent to the coherent potential approximation (CPA) for solid substitutional alloys, can be well approximated for a wavefunction overlap that is not too great, by a simpler theory (IY) developed by Ishida and Yonezawa (1973) which has a wavevector-independent self energy. We shall use the IY theory here and also neglect non-orthogonality of the wavefunctions. Use of the IY theory leads to the simplification that 'vertex corrections' can be neglected. The calculation is closely related to the CPA calculation of Levin *et al* (1970) for transport coefficients in an alloy model. It is also a generalization of the conductivity calculation for the random case by Matsubara and Toyazawa (1961).

2. Derivation of results

We begin with the exact Kubo–Greenwood equations for the conductivity tensor which consist of both the familiar symmetric term

$$\sigma_{xx} = \frac{e^2\hbar}{\pi}\int d\omega\left(-\frac{\partial f}{\partial\omega}\right)\langle \mathrm{Tr}[v_x\mathscr{G}''(\omega)v_x\mathscr{G}''(\omega)]\rangle = \int d\omega\left(-\frac{\partial f}{\partial\omega}\right)\sigma(\omega) \quad (1)$$

and the less familiar antisymmetric term

$$\sigma_{xy} = \frac{-\mathrm{i}e^2\hbar}{\pi} \int \mathrm{d}\omega\, f(\omega) \left\langle \mathrm{Tr} \left[v_x \frac{\partial \mathcal{G}'(\omega)}{\partial \omega} v_y \mathcal{G}''(\omega) - v_x \mathcal{G}''(\omega)\, v_y \frac{\partial \mathcal{G}'(\omega)}{\partial \omega} \right] \right\rangle. \quad (2)$$

Here $\mathcal{G} = \mathcal{G}' + \mathrm{i}\mathcal{G}''$ is the one-electron Green's function $(\omega - \mathcal{H})^{-1}$, $\mathbf{v}$ is the velocity operator, and f is the Fermi function for energy ω. We have included a configurational average. These expressions are given by Friedman (1971) and by Levin *et al* (1970). Equation (1) is evaluated for zero magnetic field B, and equation (2) to first order in B which is assumed to be in the z direction. The conductivity is $\sigma = \sigma_{xx}$ and the Hall constant is $R = \sigma_{xy}/\sigma_{xx}{}^2 B$. Levin *et al* also obtain the exact result for the thermoelectric power

$$\alpha = \frac{1}{e\sigma T} \int \mathrm{d}\omega (\omega - \omega_\mathrm{F})\, \sigma(\omega) \left(- \frac{\partial f}{\partial \omega} \right) \simeq \frac{\pi^2 k^2 T}{3e\sigma} \frac{\mathrm{d}\sigma(\omega_\mathrm{F})}{\mathrm{d}\omega_\mathrm{F}} \quad (3)$$

where ω_F is the Fermi energy and the latter form is valid for low temperatures.

For the one-orbital tight-binding model, we assume for the Green's function

$$\mathcal{G}(\mathbf{r}, \mathbf{r}', \omega) = \sum_{ij} \phi_i(\mathbf{r}) \mathcal{G}_{ij} \phi_j^*(\mathbf{r}') \quad (4)$$

where ϕ_i is an atomic orbital on site i. Then either equation (1) or (2) is obtained from a quantity of the form

$$P_{\alpha\beta} = \langle \mathrm{Tr}[v_\alpha \mathcal{G}^{(1)} v_\beta \mathcal{G}^{(2)}] \rangle = \left\langle \sum_{ij} v_{ij}^\alpha \mathcal{G}_{jl}^{(1)} v_{lm}^\beta \mathcal{G}_{mi}^{(2)} \right\rangle \quad (5)$$

where $\mathcal{G}^{(1)}$ and $\mathcal{G}^{(2)}$ correspond to appropriate Green's functions. We make the usual approximation for the velocity operator

$$v_{ij}^\alpha = (\mathrm{i}/\hbar)(\mathbf{R}_i - \mathbf{R}_j)_\alpha \mathcal{H}_{ij} \quad (6)$$

where $\mathcal{H}_{ij} = H(\mathbf{R}_i - \mathbf{R}_j)$ is a transfer integral between sites i and j, with $\mathcal{H}_{ii} = 0$. The Green's function in this representation has a perturbation expansion (neglecting non-orthogonality):

$$\mathcal{G}_{ij} = \omega^{-1}\delta_{ij} + \omega^{-2}\mathcal{H}_{ij} + \omega^{-3} \sum_l \mathcal{H}_{il}\mathcal{H}_{lj} + \dots \quad (7)$$

We can now obtain a perturbation expansion for equation (5), and the first few terms are represented by the diagrams in figure 1, similar to those used in Roth (1975). Here the vertices represent distinct sites, transfer integrals in the expansion of $\mathcal{G}$ are given by directed lines, and lines marked α and β are velocity matrix elements. Ionic correlation is included in a Kirkwood sense, with in-chain correlations included via a pair distribution function g (dashed lines) while out-of-chain correlations are represented by $h = g - 1$ (dotted lines).

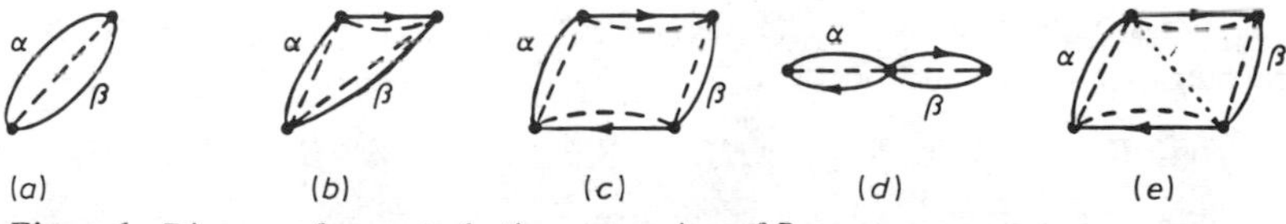

Figure 1. Diagrams for perturbation expansion of $P_{\alpha\beta}$.

In figures 1(*a*), (*b*) and (*c*), which occur in the IY theory, the Green's function portions of each term (i.e. the parts between α and β) are uncorrelated, so that we can write $\langle \mathcal{G}^{(1)} \mathcal{G}^{(2)} \rangle = \langle \mathcal{G}^{(1)} \rangle \langle \mathcal{G}^{(2)} \rangle$ for these terms. Figures 1(*d*) and (*e*) give departures from this situation, that is, vertex corrections. In a perturbation treatment of transport, Edwards (1958) showed that vertex corrections correspond to the '$\cos\theta$' in the '$(1-\cos\theta)$' factor in the usual resistivity formula. Levin *et al* (1970) showed that for the tight-binding alloy CPA, these vertex corrections vanish for a lattice with inversion symmetry because of the oddness of the velocity operator. Actually the term in (*e*) includes out-of-chain correlation of the type found in the EMA or CPA, but not in the IY theory. For the lattice case, h is site diagonal, so (*e*) vanishes in the tight-binding model as well as (*d*). For the liquid metal, however, the argument goes through for (*d*) but not for (*e*) so that using the EMA would require a consideration of vertex corrections. For the present, we shall restrict ourselves to the IY theory for which we can neglect vertex corrections and so we obtain

$$P_{\alpha\beta} = \int \mathrm{d}\mathbf{R}\,\mathrm{d}\mathbf{R}'\,\mathrm{d}\mathbf{R}''\,\mathrm{d}\mathbf{R}'''\; v_\alpha(\mathbf{R},\mathbf{R}')\, g(\mathbf{R}-\mathbf{R}')\,G^{(1)}(\mathbf{R},\mathbf{R}'')\, v_\beta(\mathbf{R}'',\mathbf{R}''') \times g(\mathbf{R}''-\mathbf{R}''')\,G^{(2)}(\mathbf{R}''',\mathbf{R}') + P'_{\alpha\beta}. \tag{8}$$

Here, the G correspond to the IY continuum Green's function, and, for example, the first velocity matrix element is between orbitals at $\mathbf{R}$ and $\mathbf{R}'$. $P'_{\alpha\beta}$ is a correction term for diagrams of the form of figure 1(*a*), which should have only one factor of g rather than two.

For zero magnetic field, the Fourier transform of G is given for the IY theory (Roth 1975) by

$$G_{\mathbf{k}} = n(\omega - \Sigma - \epsilon_{\mathbf{k}})^{-1} \tag{9}$$

$$\epsilon_{\mathbf{k}} = n\tilde{H}_{\mathbf{k}} = \int \mathrm{d}\mathbf{R}\, ng(\mathbf{R})\,H(\mathbf{R}) \exp(\mathrm{i}\mathbf{k}.\mathbf{R}) \tag{10}$$

$$\Sigma = (\omega - \Sigma)^{-1} \int \frac{\mathrm{d}\mathbf{k}}{8\pi^3}\, n\tilde{H}_{\mathbf{k}}(H_{\mathbf{k}} + \tilde{H}_{\mathbf{k}}^2 G_{\mathbf{k}}). \tag{11}$$

The result for the zero-field conductivity is

$$\sigma_{xx} = \frac{2e^2\hbar}{\pi} \int \mathrm{d}\omega \left(-\frac{\partial f}{\partial\omega}\right) \int \frac{\mathrm{d}\mathbf{k}}{8\pi^3} \left\{ v_x^2 \left[\frac{G_{\mathbf{k}}''}{n}\right]^2 + v_x(v_{0x} - v_x) \left[\mathrm{Im}\,\frac{1}{\omega - \Sigma}\right]^2 \right\}. \tag{12}$$

The velocity matrix elements are given from equations (6), (8) and (10) to be: $\mathbf{v} = \hbar^{-1}\nabla_{\mathbf{k}}\epsilon_{\mathbf{k}}$ and $\mathbf{v}_0 = \hbar^{-1}\nabla_{\mathbf{k}}(nH_{\mathbf{k}})$ where $H_{\mathbf{k}}$ is the Fourier transform of $H(R)$ (*without* a pair distribution function). The solid alloy CPA Green's function has the same form as equation (9) with $\epsilon_{\mathbf{k}}$ the band energy function, but with Σ differently defined. Equation (12) applies to the alloy CPA with $\mathbf{v}_0 = \mathbf{v}$.

For σ_{xy} we must go to first order in the magnetic field. It is well known (Friedman 1971, Roth 1962) that it is appropriate to modify the wavefunction ϕ_i by including the Peierls phase factor which is $\exp(\mathrm{i}\mathbf{b}.\mathbf{r}\times\mathbf{R}_i)$ in the symmetric gauge where $b = eB/2\hbar c$. The transfer matrix element becomes

$$\bar{H}(\mathbf{R},\mathbf{R}') = \exp(\mathrm{i}\mathbf{b}.\mathbf{R}\times\mathbf{R}')\,H(\mathbf{R}-\mathbf{R}'), \tag{13}$$

where to first order in the magnetic field $H(\mathbf{R})$ is the zero-field transfer matrix element.

We can treat G and $\mathbf{v}$ in equation (8) in a similar manner. In his calculation Friedman expanded the exponents in equation (13) to first order in b. We find it simpler, however, to Fourier transform and use an algorithm for multiplication of such magnetic matrix elements. Thus if $\bar{A}\bar{B} = \bar{C}$, the Fourier transform of C can be shown (Roth 1962) to be given by

$$C_{\mathbf{k}} = \exp(-\,\mathrm{i}\mathbf{b}\cdot\nabla_{\mathbf{k}} \times \nabla_{\mathbf{k}'})\; A_{\mathbf{k}}B_{\mathbf{k}'}|_{\mathbf{k}=\mathbf{k}'}, \tag{14}$$

where A, B, C are related to $\bar{A}, \bar{B}, \bar{C}$ as in equation (13). We apply this to equation (8) and use the fact that $\mathrm{Tr}\,\bar{A} = \mathrm{Tr}\,A$, to obtain from equation (2) the result

$$\sigma_{xy} = 4\,\frac{e^2 b\hbar^3}{3\pi}\int \mathrm{d}\omega\left(-\frac{\partial f}{\partial\omega}\right)\int\frac{\mathrm{d}\mathbf{k}}{8\pi^3}\,\frac{(v_x{}^2+v_y{}^2)}{m_\mathrm{t}}\left(\frac{G''_{\mathbf{k}}}{n}\right)^3, \tag{15}$$

where m_t is a transverse effective mass;

$$\frac{1}{m_\mathrm{t}} = \frac{1}{\hbar(v_x{}^2+v_y{}^2)}\left(\frac{\partial v_x}{\partial k_x}v_y{}^2 + \frac{\partial v_y}{\partial k_y}v_x{}^2 - 2\frac{\partial v_x}{\partial k_y}v_xv_y\right). \tag{16}$$

As in equation (12) this result is valid for both the alloy CPA and the liquid IY theories.

3. Comparison between solid and liquid

We are now in a position to compare the alloy CPA result, which also applies to the nearly perfect solid, with the liquid or amorphous case. This is of particular interest with respect to the sign of R which depends according to equation (15) on the sign of the transverse effective mass m_t. For the liquid metal case, we use the fact that $\epsilon_{\mathbf{k}}$ depends only on the magnitude of $\mathbf{k}$ to obtain

$$\frac{1}{m_\mathrm{t}} = (\hbar^2 k)^{-1}\frac{\partial\epsilon}{\partial k} = \frac{v_{\mathbf{k}}}{\hbar k}. \tag{17}$$

For a typical case, ϵ_k has the form $\sin(ka)/ka$, so that m_t is positive for the region in which $v_{\mathbf{k}}{}^3$ is largest. Thus it appears unlikely that R could ever have a hole-like sign for this model. For the solid alloy case on the other hand, we have, for example, in a simple cubic model with $\epsilon_k = -2H_1(\cos k_xa + \cos h_ya + \cos k_za)$,

$$\frac{1}{m_\mathrm{t}} = \frac{2H_1a^2}{\hbar^2}\left(\frac{\cos k_aa\sin^2 k_ya + \cos k_ya\sin^2 k_xa}{\sin^2 k_xa + \sin^2 k_ya}\right). \tag{18}$$

This is positive for $k \sim 0$, but becomes negative near the top of the band. Thus, if the Fermi surface (as represented by $G''(\omega_\mathrm{F})$) is not too smeared out by the scattering, we can expect the Hall constant to change sign on going from a nearly empty to a nearly filled band.

Comparing our result with Friedman's calculation, the random-phase approximation corresponds in our model to neglecting ϵ_k compared with $\omega - \Sigma$ in the expression for the Green's function, that is, neglecting the off-diagonal matrix elements of G. Since the density of states is given by

$$n\rho(\omega) = -\frac{1}{\pi}\,\mathrm{Im}\,\frac{n}{\omega-\Sigma},$$

this corresponds to approximating G'' by $-\pi n\rho(\omega)$. We then find an exact reproduction of Friedman's result. If we make the same approximation for an alloy, we must average $v_{\mathbf{k}}^2/m_t$ over the Brillouin zone. For the above simple cubic case this vanishes, but for a FCC model it is positive.

4. Numerical results

We apply our theory first to a one-orbital tight-binding model of a liquid metal, using for $H(\mathbf{R})$ a constant H_1 times an overlap of exponential wavefunctions exp $(-\lambda r)$, and a simple site exclusion for ionic correlation. The model is described in more detail elsewhere (Roth 1976). In figure 2 we display the results for the conductivity and Hall constant at 0 K (i.e. for degenerate statistics), and the thermoelectric power, all as a function of the Fermi level. We also include the density of states at the Fermi level $\rho(\omega)$. We see that in going from a nearly empty to a nearly filled band, α changes sign while R remains negative throughout the range of Fermi levels.

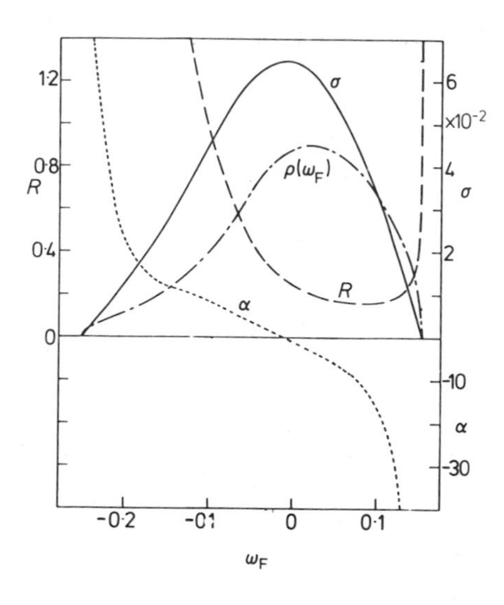

Figure 2. Conductivity σ, Hall coefficient R, thermoelectric power α, and density of states per ion at the Fermi level $\rho(\omega_F)$, as a function of ω_F for tight-binding model of text for hard-sphere packing fraction $\eta = 0{\cdot}125$, $H_1 = -1$, $\lambda = 4{\cdot}8/r_0$, with r_0 an equivalent sphere radius. Units of σ, R and α are $e^2hr_0^{-1}$, r_0^3/ec, and $\pi^2k^2T/3e|H_1|$, with $e < 0$.

We have also done a calculation for a simple alloy, in which a splitting is introduced between the atomic levels of the two components, chosen so that there is a gap in the density of states. Figure 3(*a*) shows the density of states for several compositions. The transport coefficients are shown as a function of composition in figure 3(*b*). We see that σ goes to zero at the stoichiometric composition $x = 0{\cdot}5$. In this region α changes sign but R remains negative. For a finite temperature the conductivity at $x = 0{\cdot}5$ will be activated, and R and α are finite.

5. Discussion

We see that the present calculation is in qualitative agreement with the observation of negative Hall coefficients in liquid metals, and figure 3(*b*) gives the general behaviour observed in liquid alloys. On looking at the curves we notice an asymmetry in R which

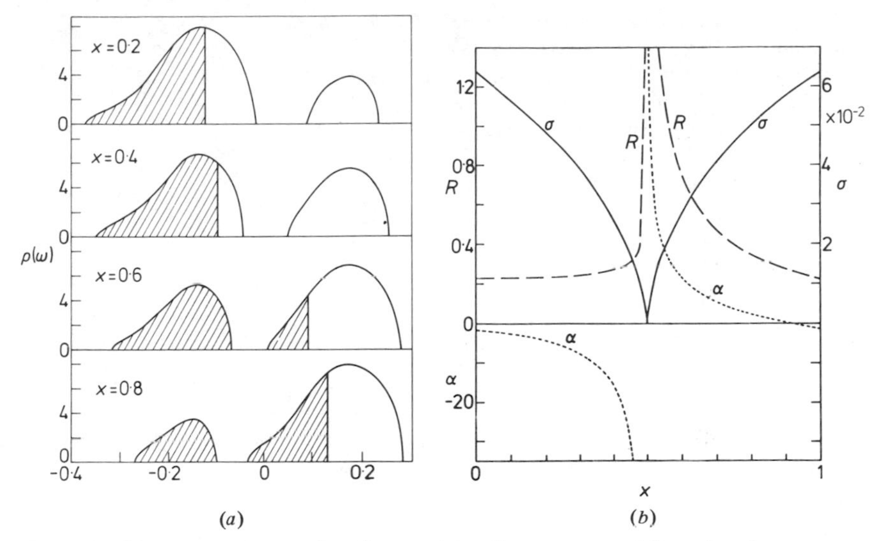

Figure 3. (*a*) Density of states for alloy model with parameters of figure 2 and splitting $\delta = 0{\cdot}3$, for various values of x, the concentration of the higher component. Filled part of band for one electron per atom is cross-hatched. (*b*) Transport coefficients as a function of x.

is larger for regions of electron conduction (i.e. nearly empty band in figure 2, and $x \gtrsim 0{\cdot}5$ in figure 3(*b*)) than for regions of hole conduction (nearly filled band in figure 2, $x \lesssim 0{\cdot}5$ in figure 3(*b*)). In fact we find for the hole regions that as the hole concentration approaches zero, the Hall mobility $R\sigma$ approaches a $1/R$ dependence corresponding to Friedman's strong scattering regime, while in the electron regions, $R\sigma$ is approximately constant, corresponding to weak scattering. Hence the model encompasses both weak scattering and strong scattering regimes.

The best data on the Hall coefficients in liquid semiconductors are those for Tl–Te, and we find (Donnally and Cutler 1971) that while there is an asymmetry in the R versus concentration data, it is in the opposite direction, R being larger for the 'hole' regions. Thus our simple model does not apply in detail here. Possibly, multiband effects are important here, or molecules are being formed as discussed by Cohen and Jortner (1974). It would be interesting if Hall measurements would be made on Mg–Bi alloys for which an ionic model seems to be appropriate (Enderby 1974).

Finally we remark that our equations (12) and (15) apply with only slight changes ($G_{\mathbf{k}}/n \to G_{\mathbf{k}}$, $v_{\mathbf{k}} = v_{0\mathbf{k}} = \hbar k/m$) to the more general multiple scattering model of a liquid metal, provided we neglect vertex corrections as well as the dependence of the self-energy on wave vector.

References

Allgair R S 1969 *Phys. Rev.* **185** 227
Cohen M H and Jortner J 1974 *J. Physique* **35** C4–345
Donnally J M and Cutler M 1971 *J. Phys. Chem. Solids* **33** 1017
Edwards S F 1958 *Phil. Mag.* **3** 1020
Enderby J E 1974 *J. Physique* **35** C4–309

Enderby J E and Collings E M 1970 *J. Non. Cryst. Solids* **4** 161
Friedman L 1971 *J. Non. Cryst. Solids* **6** 329
Ishida Y and Yonezawa F 1973 *Prog. Theor. Phys.* **49** 731
Levin K, Velicky B and Ehrenreich H 1970 *Phys. Rev.* **B2** 1771
Matsubara T and Toyazawa Y 1961 *Prog. Theor. Phys.* **26** 739
Roth L M 1962 *J. Phys. Chem. Solids* **23** 433
—— 1974 *Phys. Rev.* **B9** 2476
—— 1975 *Phys. Rev.* **B11** 3769
—— 1976 *J. Phys. F: Metal Phys.* **6** 2267–88

Lorenz number and thermal conductivity of liquid metals

W Haller†, H-J Güntherodt† and G Busch‡

† Institut für Physik, Universität, Basel, Switzerland
‡ Laboratorium für Festkörperphysik ETH, Zürich, Switzerland

Abstract. A new apparatus has been developed to measure directly the Lorenz ratio of liquid metals up to 500 °C. The well known Kohlrausch method has been applied to a sample in the shape of a sphere with a constriction. Our first measurements on liquid tin are reported and the results are in a good agreement with the Wiedemann–Franz law.

1. Introduction

Considerable experimental and theoretical work has been done in the field of the Lorenz ratio of liquid normal metals, but published results show substantial inconsistencies in the measurements reported by different authors. Moreover, the results of the experiments deviate drastically from the Wiedemann–Franz law. Typical deviations are 30% at the melting point and up to 100% at higher temperatures and no theoretical explanation of such large deviations is known. In fact, theories predict that the Lorenz ratio should be essentially a temperature-independent constant with the Sommerfeld value, $L_0 = 2{\cdot}443 \times 10^{-8}\,\mathrm{V^2\,K^{-2}}$.

In view of the present situation there is clearly a need for new, independent and consistent measurements. Therefore we have designed a new apparatus to measure the Lorenz ratio directly for temperatures up to 500 °C.

2. Experimental method

Most of the recent determinations of the Lorenz ratio $L = \lambda/\sigma T$ have been made by separate measurements of the electrical conductivity σ and the thermal conductivity λ. It is well known that the thermal conductivity is extremely difficult to measure in the liquid state: consequently we tried to determine the ratio λ/σ directly. This can be done by applying the Kohlrausch method (electrical method) which has been used very successfully for measurements in the solid state (Kohlrausch 1900, Flynn and O'Hagan 1967).

The liquid metal is self-heated by a current passing through the electrodes S_1 and S_2 (see figure 1). In our experiment, the sample has the shape of a sphere with a constriction, formed by the two electrodes S_1 and S_2 and by a thin disc with surface S. If S is thermally and electrically insulated, that is,

$$\nabla_n T = 0, \qquad \nabla_n \phi = 0 \tag{1}$$

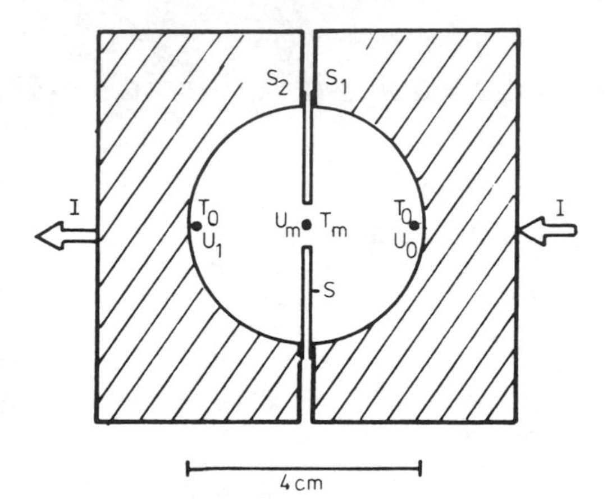

Figure 1. Kohlrausch method applied to a spherically shaped sample with a constriction.

and λ and σ are homogeneous functions of temperature, that is,

$$\lambda = \lambda(T), \qquad \sigma = \sigma(T), \tag{2}$$

then simple solutions for the temperature and potential distribution ϕ (Greenwood and Williamson 1958) can be obtained from:

$$\nabla(\lambda\nabla T) - \mathrm{i}\nabla\phi = 0. \tag{3}$$

It turns out that every equipotential is also an isothermal and that there exists a maximum temperature T_m. Another result is given by:

$$(U_m - U_0)^2 = 2\int_{T_0}^{T_m} \frac{\lambda}{\sigma}\, \mathrm{d}T = 2\int_0^{\theta_m} \frac{\lambda}{\sigma}\, \mathrm{d}\theta, \tag{4}$$

where T is the absolute temperature, T_0 the temperature of an arbitrary isothermal within the sample at the potential U_0 and T_m is the maximum temperature at potential U_m. It is convenient to introduce relative temperatures $\theta = T - T_0$, $\theta_m = T_m - T_0$ and it is assumed that λ/σ may be described for small θ by

$$\frac{\lambda}{\sigma/T} = L + L_1\theta + L_2\theta^2 + \ldots = \mathrm{LF}. \tag{5}$$

There is experimental evidence (figure 2) that the coefficients $L_1(T_0)$, $L_2(T_0)$, . . . mainly describe the influence of convection in the liquid state. Since convection is assumed to vanish in the limit $\theta \to 0$, the constant L in the polynomial LF is a good measure of the electronic contribution to the Lorenz ratio.

The integral in equation (4) can be solved easily for small θ:

$$\frac{(U_m - U_0)^2}{2T_0\theta + \theta^2} = L + \frac{L_1}{2}\theta + \frac{L_2}{3}\theta^2 + \ldots. \tag{6}$$

The quantity on the left-hand side of equation (6) can be measured: we need only to know the potential difference $U_m - U_0$ and the temperature difference θ_m between any point on the isothermal T_m and any point on the isothermal T_0. In practice it is easier to measure the voltage $U_1 - U_0$ between two symmetrical points both on T_0 rather than measure $U_m - U_0$.

We note that there is no geometrical factor in equation (4). The determination of the the geometry, which is necessary for most of the other conductivity measurements, is replaced by the measurement of a temperature difference between two points.

In our experiment the temperature difference θ_m has been measured by a differential thermocouple, calibrated *in situ.* A dynamic technique which measures θ_m as a function of time by switching the current I on and off, prevents errors caused by drifts in the θ_m measurement. For the determination of T_0, a calibrated 100 Ω platinum resistance was used. The experimental realization of the thermal insulation of the disc, which is one of the boundary conditions of the Kohlrausch method (equation (1)), is difficult to fulfil. In our experiment this problem has been reasonably solved by the use of a thin disc of a low thermal conductivity ceramic. For symmetry reasons the temperature distributions on the left- and on the right-hand side of the disc are the same. Therefore there is no heat flow perpendicular through the disc. Heat losses in other directions seem to be negligible, since the use of different disc materials, whose thermal conductivities differ by a factor of two, has no influence on the result of the Lorenz ratio within the experimental error.

3. Experimental results

Using the above method, we were able to measure the Lorenz ratio of liquid Ga, Hg and Hg–In alloys up to 80 °C. We have found good agreement with the Wiedemann–Franz law (Busch *et al* 1972, 1973). In the present work we have extended the temperature range up to 500 °C.

Figure 2 shows some typical measurements on liquid and solid tin. In the liquid state LF depends drastically on θ, but in the solid state LF is a constant, independent of θ. Only convection can cause such a drastic influence on LF in the liquid state and in our experiment we certainly will have convection because the isothermal of maximum

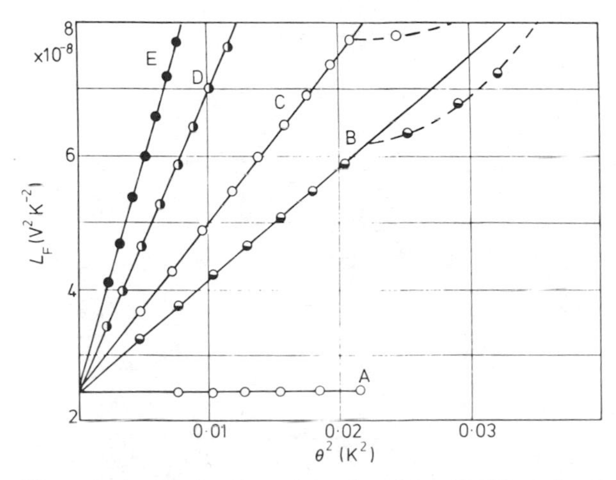

Figure 2. Lorenz function of tin: A, solid at 215 °C; B, liquid at 234 °C; C, 301 °C; D, 350 °C; E, 400 °C.

temperature lies within the sample. The important experimental fact is that the influence of the convection can be described by a simple function up to a critical θ_m. Therefore it is possible to make an extrapolation for $\theta_m \to 0$ and, as mentioned above, the extrapolated value is believed to be a reliable measure of the Lorenz ratio. It must be pointed out that in the presence of convection, the interpretation of the values of LF for $\theta_m > 0$ is not simple, because the condition of the Kohlrausch theory given in equation (2) is no longer fulfilled. However, we think that these values give the correct order of magnitude for the contribution of the convection to LF. Our data were analysed by a least-squares fit with a polynomial of second order. For every fit about 50 points were used.

In figure 3 our results for the Lorenz ratio are compared with recently published measurements. The open circles are the extrapolated values of LF for $\theta_m \to 0$. Assuming a statistical distribution, they are in good agreement with the Sommerfeld value L_0.

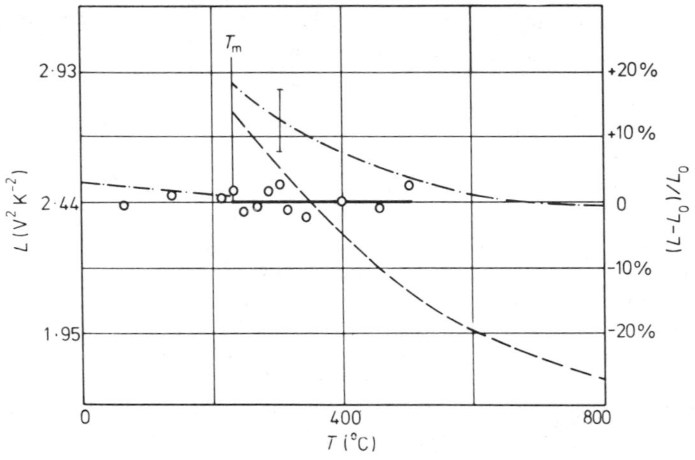

Figure 3. Lorenz ratio of tin: chain line, TPRC recommended values (Ho *et al* 1972); broken line, Filippov (1968); full line and open circles, present work.

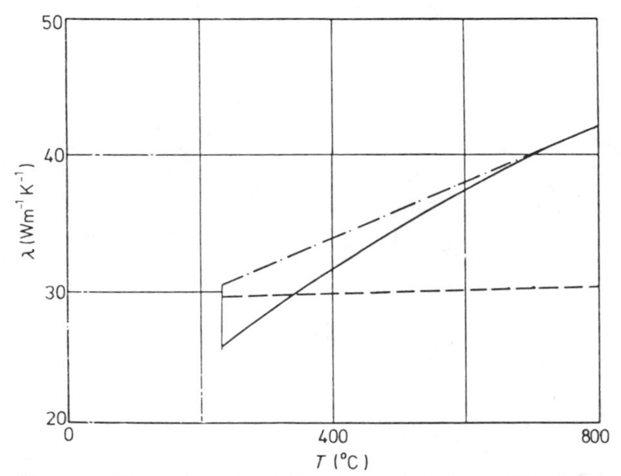

Figure 4. Thermal conductivity of tin: chain line, TPRC (Ho *et al* 1972); broken line, Filippov (1968); full line, $L_0 \sigma T$.

The TPRC recommended values are deduced from all available thermal conductivity experiments (Ho *et al* 1972, Filippov 1968).

In the solid state the interpretation of the data is not simple because tin is reported to have anisotropic conductivities and even two possible phases at low temperatures. The data in figure 3 are presented for polycrystalline white tin.

Having confirmed experimentally the validity of the Wiedemann–Franz law we are able to deduce the thermal conductivity from the Lorenz number L_0 and the measured electrical conductivity σ according to:

$$\lambda = L_0 \sigma T. \qquad (7)$$

In figure 4 the calculated thermal conductivity of liquid Sn is shown.

4. Discussion

We have found that the Kohlrausch method is well suited to measuring the Lorenz ratio of liquid metals up to 500 °C. A much better agreement with the Wiedemann–Franz law has been achieved than in previous reported measurements. There is still a statistical error of a few per cent due mainly to the enormous contribution of convection to the thermal conductivity in the liquid state, and at present, therefore, it is not worthwhile to draw any final conclusion as to whether or not the Wiedemann–Franz law is obeyed to within, say, 1%. Such small deviations have been predicted by theories as arising from a direct ionic contribution or from electron–electron scattering (Rice 1970).

More experimental effort has to be undertaken in order to reduce the experimental error. We believe that our experiment can be developed to higher accuracy by using a different geometry which leads to a smaller influence of the convection: it should then be feasible to investigate even very small deviations from the Wiedemann–Franz law.

Acknowledgments

We would like to thank the 'Eidgenössische Stiftung zur Förderung Schweizerischer Volkswirtschaft durch wissenschaftliche Forschung' and the 'Schweizerische National-fonds zur Förderung der wissenschaftlichen Forschung' for financial support.

References

Filippov L P 1968 *Int. J. Heat Mass Transfer* **11** 331–45
Flynn D R and O'Hagan M E 1967 *J. Res. Natl. Bur. Stand.* **71C** 255–84
Greenwood J A and Williamson J B P 1958 *Proc. R. Soc.* **A246** 13–31
Ho C Y, Powell R W and Liley P E 1972 *J. Phys. Chem. Ref. Data* **1** 406
Kohlrausch F 1900 *Ann. Phys., Leipzig* **1**
Rice M J 1970 *Phys. Rev.* **B2** 4800

The electrical resistivity and thermopower of high-purity barium in the solid and liquid phases

J B Van Zytveld
Physics Department, Calvin College, Grand Rapids, Michigan, USA

Abstract. We have measured the electrical resistivity, ρ, and thermopower, S, of high-purity barium to temperatures above the melting point. We find that in the solid, increased purity results in decreased ρ and increased S up to about 550 °C. At this temperature a fairly sharp break occurs in S, and the value drops in magnitude by about 35 μV/°C within a range of about 100 °C. In the liquid, increased purity leaves S unchanged, but raises ρ at the melting point to 338 $\mu\Omega$ cm and increases $-\mathrm{d}\rho/\mathrm{d}T$ to 0·08 $\mu\Omega$ cm/°C.

1. Introduction

There has been considerable interest recently in the electronic properties of the liquid divalent metals. Much of this interest results from the fact that, for the alkaline earth metals and also for europium and ytterbium, an unfilled d-band intersects the Fermi energy and dominates the conduction properties. Attempts have been made to calculate the electrical resistivities and thermopowers of some of these liquid metals. For example, Moriarty (1972) used a modification of the Harrison (1969) transition metal pseudopotential, localizing the d-states by adding an appropriate attractive potential of variable strength to the free-ion Hamiltonian, and Ratti and Evans (1973) have developed a muffin tin potential to study these metals. This latter method is basically a t-matrix formulation cast in terms of the various phase shifts (η_l for the lth partial wave). Both methods use the basic Ziman formulae to calculate the physical parameters. In the approach of Ratti and Evans, if one assumes that d-scattering dominates, then the resistivity can be approximated as

$$\rho \simeq \frac{30\pi^3\hbar^3 \sin^2 \eta_2(E_\mathrm{F})\, a(2K_\mathrm{F})}{me^2\Omega_0 K_\mathrm{F}{}^2 E_\mathrm{F}}, \qquad (1)$$

where $a(K)$ is the structure factor, K is the wavevector corresponding to momentum transfer, and the other symbols have their usual meanings. The thermopower can in general be cast as

$$S = -\frac{\pi^2 k_\mathrm{B}{}^2 T}{3|e|E_\mathrm{F}}\, x, \qquad (2)$$

where k_B is Boltzmann's constant, T is absolute temperature, and

$$x \equiv -\left.\frac{\partial \ln \rho}{\partial \ln E}\right|_{E_\mathrm{F}}. \qquad (3)$$

We therefore expect the thermopower to be more sensitive to the energy dependence of the scattering matrix, since this energy dependence enters S directly as the energy derivative. This is reflected in the fact that attempts to calculate ρ for the liquid alkaline earth metals have generally been more successful than attempts to calculate S.

On the experimental side, progress in studying these metals has been hampered by their extreme reactivity. Such difficulties have been pointed out by Rashid and Kayser (1971), Katerberg *et al* (1975), Güntherodt *et al* (1976), and Cook and Laubitz (1976). Metallic impurities in general do affect the electronic properties of these metals at high temperature; however, gaseous impurities (H_2 especially, but also N_2 and O_2) appear to have an even greater effect. For this reason, great care must be taken in the preparation and handling of these metals to ensure the highest possible purity.

Cook and Laubitz (1976) have recently succeeded in preparing samples of barium which are apparently of very high purity. These workers report substantial decreases in resistivity and increases in thermopower in the temperature range from about 300 °C to about 500 °C, the upper limit of their investigations. In the present paper we report an extension of the measurements of Cook and Laubitz to above the melting point, using samples of barium provided by these authors. These new data are then discussed in terms of recent theoretical work.

2. Experimental procedure

The barium samples used in these experiments were loaned to us by Dr J Cook of NRC, Ottawa, Canada. They were prepared by sublimation from commercial barium stock; subsequently, we further purified them by sealing them in welded tantalum containers and baking these for three days at above 900 °C with an outer pressure of 10^{-4} to 10^{-5} Torr. This latter step removed most of the H_2 impurities in the samples, since H is readily absorbed interstitially by Ta at high temperatures. Furthermore, at the temperatures used, H that is absorbed in Ta is virtually completely purged in a vacuum of better than 10^{-4} Torr (cf. Borgucci and Verdini 1965), leaving the Ta container at its initial purity. Thus resistivities could be measured without removing the barium from the cylindrical Ta containers, and the resistivity of pure Ta could be used in extracting the sample resistivity from the total resistivity of this parallel-conductor arrangement. For these resistivity measurements a four-probe DC technique was used; contacts were made by means of stainless steel knife-edges applied to the Ta tubes.

For measurements of thermopower, a modification of the method of small gradients was used (see Van Zytveld *et al* 1973). (In this method each data point is independent of all others, and the data are not smoothed.) In the present modification, the sample metal was contained in a horizontal alumina (Al_2O_3) boat of very high density, which in turn was held in nesting furnace tubes of high purity tantalum and stainless steel (see figure 1). The volume between the Ta and stainless steel furnace tubes was evacuated to a pressure of less than 10^{-4} Torr. In this way the purity of the helium atmosphere over the sample was maintained during the course of the measurements. Prior to mounting each sample the entire high-temperature portion of the apparatus (including the inner volume of the Ta furnace tube and the alumina crucible) was baked out under high vacuum at temperatures above 900 °C.

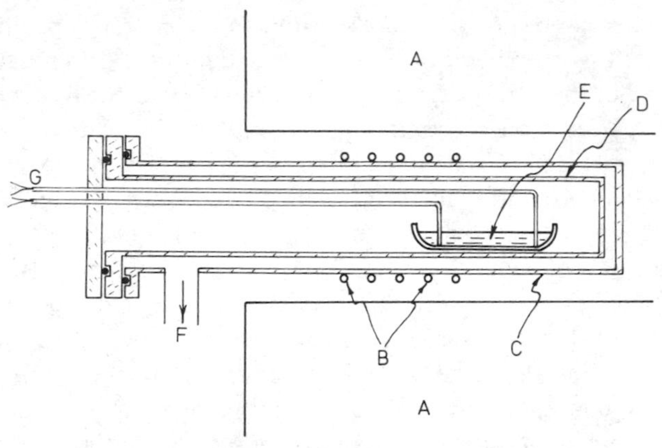

Figure 1. Schematic of the thermopower measurement apparatus: A, main furnace; B, thermal gradient control winding; C, stainless steel furnace tube; D, high-purity tantalum furnace tube; E, sample; F, port to the diffusion pump; G, measurement thermocouples sheathed in 347 stainless steel.

The thermopower samples were prepared in the same way as those for the resistivity measurements. The Ta sample containers were then opened in an evacuable glove box under a He atmosphere. (This atmosphere was continuously recirculated through a trap cooled to liquid N_2 temperature to maintain its purity.) The barium samples were mounted in the experimental container while in this glove box; this container was then sealed before removal to the experimental measuring apparatus. In this way the introduction of gaseous impurities into the samples could be minimized.

3. Results

We have measured the electrical resistivity and thermopower of purified barium from room temperature to above the melting point. The results of these measurements are shown in figures 2 and 3 respectively. Within experimental error, the room temperature value measured for ρ (see figure 2) is as low as any yet found, agreeing with both Rashid and Kayser (1971) and Cook and Laubitz (1976). At higher temperatures in the solid, our data does not show any of the breaks which Cook and Laubitz have associated with hydrogen contamination, but rather has a very gentle curvature falling well below the recent data of Güntherodt *et al* (1976), and slightly below that of Cook and Laubitz as well. We conclude that this barium retained higher purity throughout the measurements than any barium examined to date in this temperature range.

We can fit a parabola to our resistivity data below 400 °C to well within experimental error, obtaining the curve

$$\rho(\mu\Omega\ \mathrm{cm}) = 33{\cdot}09 + 0{\cdot}17903\, T(^\circ\mathrm{C}) + 1{\cdot}586 \times 10^{-4}\, T^2(^\circ\mathrm{C}^2).$$

We find the values for ρ at 0 °C and at the melting temperature of barium (714 °C) to be:

$$\rho(0\ ^\circ\mathrm{C}) = 33{\cdot}1\ \mu\Omega\ \mathrm{cm} \qquad \text{and} \qquad \rho(714\ ^\circ\mathrm{C}) = 213\ \mu\Omega\ \mathrm{cm}.$$

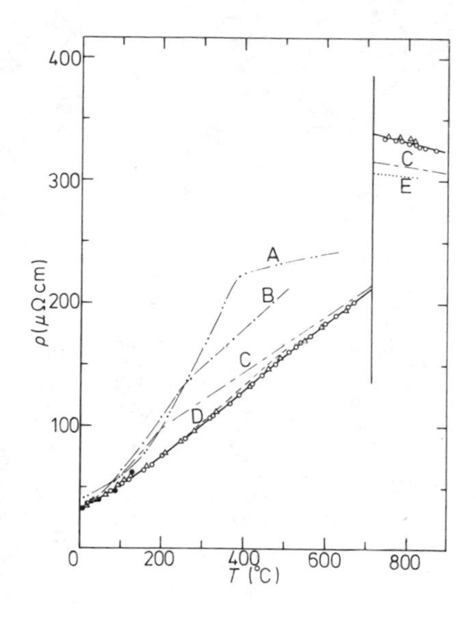

Figure 2. Electrical resistivity of barium: A, Rinck (1931); B, Katerberg *et al* (1975); C, Güntherodt *et al* (1976); D, Cook and Laubitz (1976); E, Van Zytveld *et al* (1972); •, Rashid and Kayser (1971); ○, △, present data, separate runs.

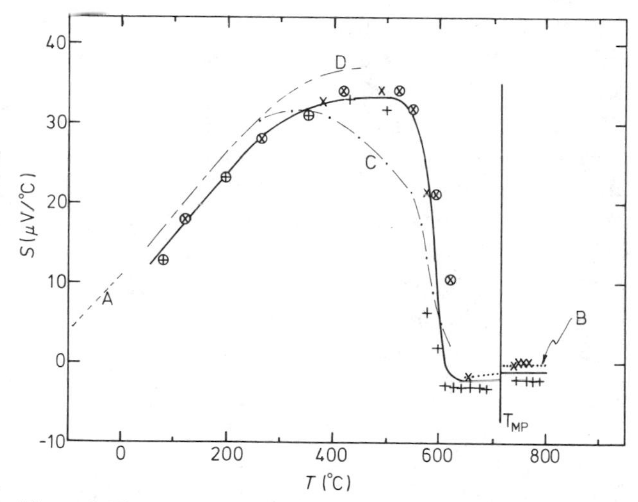

Figure 3. Thermopower of barium: A, Cook and Van der Meer (1973); B, Van Zytveld *et al* (1973); C, Katerberg *et al* (1975); D, Cook and Laubitz (1976); ×, ⊗, +, ⊕, present data (two runs), circled data increasing in temperature, uncircled data decreasing in temperature.

For the liquid, we see that increasing purity apparently results in *increasing* resistivity. We also note that the negative $\mathrm{d}\rho/\mathrm{d}T$ in the liquid also appears to increase in magnitude as the purity increases. (The data of Van Zytveld *et al* (1972) were taken on barium of nominally 99·5% metallic purity, and would have reproduced largely the same curve as that of Rinck (1931) in the solid, indicating a fairly high hydrogen concentration.) For the liquid, we obtain values:

$$\rho(\text{melting point}) = 338\,\mu\Omega\,\text{cm} \qquad \text{and} \qquad \frac{\mathrm{d}\rho}{\mathrm{d}T} = -0{\cdot}08\,\mu\Omega\,\text{cm}/^\circ\text{C},$$

yielding

$$\frac{1}{\rho}\frac{d\rho}{dT}\text{(melting point)} = -2{\cdot}4 \times 10^{-4}/^{\circ}\text{C}.$$

We also measure a larger change in ρ at the melting point than previously reported:

$$\frac{\rho_L - \rho_S}{\rho_S} = 0{\cdot}59.$$

The present results for thermopower (see figure 3) follow the general features of the recent data of Cook and Laubitz (1976), but fall about 3 μV/°C lower over the entire common temperature range. In an attempt to identify the origin of this difference (which is just outside experimental error) we re-calibrated our chromel counter-electrodes after the completion of these experiments, but found the calibration unchanged. Our calibration is also in agreement with the calibration of Cook and Laubitz to within a maximum difference of 0·3 μV/°C. In their work, Cook and Laubitz have shown that charging their barium samples with hydrogen lowered the thermopower but only in the range from about 300 °C to 500 °C (the upper limit of their measurements). We conclude that, while the origin of this difference in data is uncertain, the general features of $S(T)$ are well preserved, and both sets of data are characteristic of material of higher purity than that of Katerberg *et al* (1975) and also than that of Van Zytveld *et al* (1973) (data near the melting point). We observe, then, that increasing the purity has not altered the thermopower either in the liquid or in the solid just below the melting point, nor has it removed the generally parabolic curvature noted by Katerberg *et al*, but that it has sharpened considerably the 'transition' at about 600 °C. These data were highly reproducible, showing no progressive contamination: the uncircled data points were taken with decreasing temperature, and the circled points were measured the next day with increasing temperature. We observe a change in thermopower of about 35 μV/°C over the temperature range of only about 100 °C centred at 575 °C.

4. Discussion

While the measured dependence on temperature of the resistivity for solid barium now looks rather similar to that for many simple metals, the thermopower of the solid looks far from normal. In fact, it appears very much as if an electronic transition occurs at about 575 °C. Corresponding to this, we do see a small change in slope of ρ in the range from about 500 °C to 550 °C, above which ρ is linear in temperature. It is unlikely that this small change in slope is related to disassociation of any residual BaH_2, since we see (figure 2) that with increasing purity the break in ρ originating from this effect moves to *lower* temperatures. We are reminded by the $S(T)$ for solid Ba of the behaviour of the thermopower of liquid cesium as a function of pressure. The thermopower of liquid caesium has a maximum (of nearly the same magnitude as that of barium) at about 16 kbar; at this same pressure, the resistivity of liquid caesium begins to increase rather rapidly (Oshima *et al* 1974). This behaviour in liquid caesium has been explained by Ratti and Jain (1973) as being due to changing d-character with increasing pressure.

In fact, at very high pressures, solid caesium may actually become a transition metal (see Louie and Cohen 1974).

Because of their proximity in the periodic table and their similar crystal structures, we would expect many of the electronic properties of caesium to be shared by barium. If, however, the maximum in $S(T)$ for barium has the same origin as the $S(p)$ for caesium, then, by comparison, the change in the scattering properties of barium must be very abrupt. That the effect is more pronounced in the thermopower of Ba than in resistivity is not surprising, since ρ enters S through the energy derivative of the resistivity (see equation (3)); that the effect is barely visible in the resistivity is a bit more surprising.

We see that the increase in purity of the sample has not greatly altered the measured characteristics of either ρ or S for liquid barium. We see no change in S to within our estimated error, and note also that the change in S upon melting is small. We do note, however, that increasing purity appears to have two observable effects on the resistivity: both ρ and $|d\rho/dT|$ (and $(1/\rho)|d\rho/dT|$) increase with increasing purity. The increase in ρ is consistent with the assumption that gaseous impurities in liquid barium contribute little to the scattering of conduction electrons; the removal of these impurities will then do little more than raise the Fermi energy and increase the magnitude of $\eta_2(E_F)$, resulting in an increase in ρ (see equation (1)). The total effect from this mechanism of adding 5 at.% of H to liquid barium would be to lower its resistivity by about 5–10 $\mu\Omega$ cm. This is certainly of the right order of magnitude.

Both Moriarty (1972) and Ratti and Evans (1973) have calculated ρ for liquid barium. The values Moriarty obtains range from 84 to 444 $\mu\Omega$ cm for his most realistic models; Ratti and Evans find $\rho = 415$ $\mu\Omega$ cm. These calculations do indicate that taking proper account of the strong d-scattering does provide a reasonable estimate of the resistivity of liquid barium. On the other hand the value of S for liquid Ba calculated by Ratti and Evans (1973) (11 μV/°C) differs significantly from the experimental values of S (0 to -1 μV/°C). Ratti and Evans do, however, predict that the thermopowers of liquid calcium, strontium, and barium should be positive and that their magnitudes should decrease as one moves through the sequence from Ca to Ba. Liquid Ca is observed to have a fairly large positive thermopower (Van Zytveld *et al* 1973) which is rather sensitive to impurity content. Now that techniques for preparing and handling these reactive materials in high-purity form are available, a re-examination of the resistivity and thermopower of liquid (and solid) Ca and Sr would be of interest. These investigations are under way in this laboratory.

Acknowledgments

It is our pleasure to acknowledge financial support provided by Research Corporation through a Cottrell College Science Grant. We also appreciate very much the willingness of the High Temperature Physics Group, NRC, Ottawa, Canada, to loan us the barium samples used in the present investigation. And we thank Dr John Cook, NRC, for several helpful suggestions and much stimulating correspondence.

References

Borgucci M V and Verdini L 1965 *Phys. Stat. Solidi* **9** 243–50
Cook J G and Laubitz M J 1976 *Can. J. Phys.* **54** 928–37

Güntherodt H J, Hauser E, Künzi H U, Evans R, Evers J and Kaldis E 1976 *J. Phys. F: Metal Phys.* **6** 1513–22
Harrison W A 1969 *Phys. Rev.* **181** 1036–53
Katerberg J, Niemeyer S, Penning D and Van Zytveld J B 1975 *J. Phys. F: Metal Phys.* **5** L74–9
Louie S G and Cohen M L 1974 *Phys. Rev.* **B10** 3237–45
Moriarty J A 1972 *Phys. Rev.* **B6** 4445–58
Oshima R, Endo H, Shimomura O and Minomura S 1974 *J. Phys. Soc. Japan* **36** 730–8
Rashid M S and Kayser F X 1971 *J. Less-Common Metals* **24** 253–7
Ratti V K and Evans R 1973 *J. Phys. F: Metal Phys.* **3** L238–43
Ratti V K and Jain A 1973 *J. Phys. F: Metal Phys.* **3** L69–74
Rinck E 1931 *C.R. Acad. Sci., Paris* **193** 1328–30
Van Zytveld J B, Enderby J E and Collings E W 1972 *J. Phys. F: Metal Phys.* **2** 73–8
—— 1973 *J. Phys. F: Metal Phys.* **3** 1819–27

Thermoelectric power and electrical resistivity of liquid Na–Cs, Li–Mg and Li–Na alloys

P D Feitsma, T Lee and W van der Lugt
Solid State Physics Laboratory, Materials Science Center, University of Groningen, Melkweg 1, Groningen, The Netherlands

Abstract. Experimental and theoretical values of the electrical resistivity and the thermopower of liquid Na–Cs, Li–Mg and Li–Na alloys are discussed. The influence of the term derived from non-locality on the thermopower appears to be puzzling.

1. Sodium–caesium alloys

In an earlier paper (Feitsma *et al* 1974), we reported on the experimental determination of the resistivities ρ of liquid sodium–caesium alloys. It was shown that the temperature dependence of the resistivity, $(\partial\rho/\partial T)_p$, plotted as a function of caesium concentration c, exhibits a distinct maximum at $c \cong 0{\cdot}25$ and a minimum at $c \cong 0{\cdot}6$. This minimum is fairly deep at 50 °C but fills up at higher temperatures and disappears at 300 °C. These data for $(\partial\rho/\partial T)_p$ could be correlated with the pressure dependence $(\partial\rho/\partial p)_T$ measured as a function of concentration by Tamaki *et al* (1973). The resistivity results, and those obtained for other properties of the same alloy system, have given rise to a good deal of discussion (Kim and Letcher 1971, Ichikawa and Thompson 1974, Ichikawa *et al* 1974, Bhatia and March 1975). These results formed a stimulus for also determining the thermoelectric power Q and its temperature dependence $(\partial Q/\partial T)_p$ of the same alloys in order to establish whether or not an effect connected with the resistivity anomaly would show up in the results for this related transport property.

Experimentally, we have used the method with small temperature differences ΔT between the 'hot' and 'cold' junctions. For each temperature T of the 'cold' junction, thermoelectric voltages were measured (using pure copper leads) at five regularly chosen temperatures $T + \Delta T$ of the 'hot' junction (ΔT never exceeded 5 °C). The thermoelectric power is determined by the slope of the straight line relating the thermoelectric voltages and the corresponding temperature differences ΔT. The advantage of this method has been discussed by Valiant and Faber (1974).

For each concentration, the thermoelectric power Q was measured as a function of temperature T at temperature intervals of 20 °C. $Q(T)$ was fitted by a polynomial of degree two. In all cases the standard deviation was less than $0{\cdot}05\ \mu V\ °C^{-1}$ and the total error in the absolute thermoelectric power was estimated to be $\pm 0{\cdot}2\ \mu V\ °C^{-1}$. Our data for the pure metals are in agreement with those obtained by Kendall (1968).

The experimental results for the thermopower are given as a function of the caesium concentration c in figure 1 for a few selected temperatures between 50 and 200 °C. Our data at 100 °C agree on the whole with the less detailed measurements of Tamaki *et al*

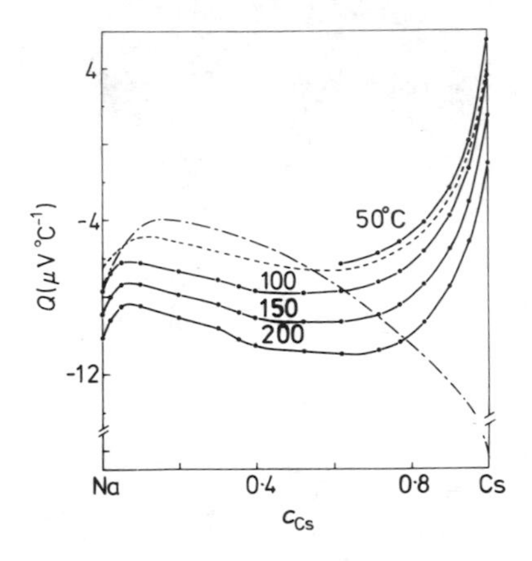

Figure 1. Thermopower isotherms for liquid Na–Cs alloys. Full curves, experiment; broken curve, theory (r excluded for 100 °C); chain curve, theory (r included for 100 °C). Here, and in all subsequent figures, c denotes the atomic fraction.

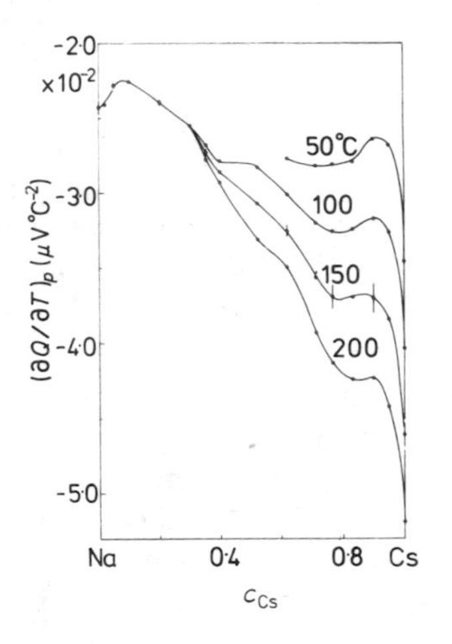

Figure 2. Experimental isotherms for the temperature derivative of the thermopower for liquid Na–Cs alloys. Error bars indicate the standard deviations.

(1973) but exhibit a maximum at $c \approx 0{\cdot}075$ not observed by these authors. The temperature dependence $(\partial Q/\partial T)_p$ is given in figure 2 for the same temperatures. Perhaps apart from a rather sudden change in $\mathrm{d}^2Q/\mathrm{d}T^2$ near $c = 0{\cdot}25$, the thermopower results provide no clear evidence of special effects occurring at $c = 0{\cdot}25$ and $c = 0{\cdot}60$. In the search for related effects, we have also carefully determined the densities (Huijben *et al* 1975) and the Knight shifts (Van Hemmen *et al* 1974) of these alloys, but the effects did not occur in either of them. Hallers *et al* (1974) carried out theoretical calculations of the resistivities ρ of these alloys applying model potential theory with Toigo–Woodruff screening and appropriate corrections for effective masses and for core shifts. For the structure factors, the hard-sphere solution of the Percus–Yevick equation was used. For ρ they obtained fair agreement (within 25%) with experiment over the whole range of concentration, but they were unable to reproduce the minimum of $(\partial\rho/\partial T)_p$ at $c \approx 0{\cdot}60$.

Recently, Huijben *et al* (1976, 1977) have determined the structure factor of liquid caesium experimentally by applying the x-ray transmission method. Substitution of this experimental structure factor into the resistivity formula reduces the discrepancy between experiment and theory for pure caesium to 3%, and, additionally, $(\partial\rho/\partial T)_p$

could be calculated with an error of only 10%. This has considerably strengthened our belief in the accuracy of the form factors used by Hallers *et al* and gives an indication that the oscillations in $(\partial\rho/\partial T)_p$ may be due to structural effects.

The standard formula for the thermopower Q for a binary alloy in terms of partial structure factors $a_{ij}(q)$ and non-local model potential form factors $w_q^{\,i}(k_f)$ is given by

$$Q = \frac{\pi^2 k_B^{\,2} T}{3e\,E_f}\,(3 - 2s - \tfrac{1}{2}r) \tag{1}$$

where

$$s = \frac{F(2k_f, k_f)}{\langle F(q, k_f)\rangle} \qquad \text{and} \qquad r = \frac{\langle k_f(\partial F(q,k)/\partial k)_{k=k_f}\rangle}{\langle F(q, k_f)\rangle}$$

with

$$F(q, k_f) = c_1\,[w_q^{\,1}(k_f)]^2 a_{11}(q) + c_2\,[w_q^{\,2}(k_f)]^2 a_{22}(q) + 2\,(c_1c_2)^{1/2}\, w_q^{\,1}(k_f)\, w_q^{\,2}(k_f)\, a_{12}(q)$$

and

$$\langle F(q, k_f)\rangle = \int_0^1 F(q, k_f)\, 4(q/2k_f)^3\, \mathrm{d}(q/2k_f).$$

The partial structure factors a_{ij} are defined by Ashcroft and Langreth (1967). The term r has its origin in the non-locality of the model potential operator. When applying the model potential used by Hallers *et al* to the calculation of the thermoelectric power of of Na–Cs alloys, a remarkable result is obtained. When the term r is excluded, excellent agreement with experiment is obtained for all concentrations while inclusion of this term seriously deteriorates the result and even leads to the wrong sign (see figure 1).

2. Lithium–magnesium alloys

The resistivities of lithium–magnesium alloys have been determined as functions of temperature and composition. The experimental equipment and procedure have been

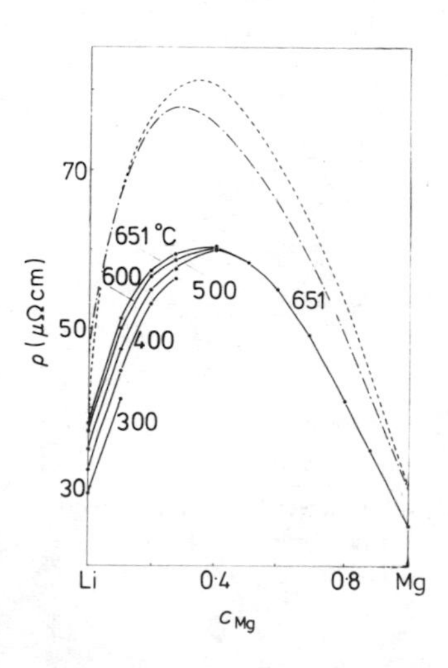

Figure 3. Resistivity isotherms for liquid Li–Mg alloys. Full curves, experiment; broken curve, theory (parameter set II for 651 °C); chain curve, theory (parameter set I for 651 °C; see text).

described earlier by Feitsma *et al* (1975). Resistivity isotherms are shown in figure 3; they exhibit the usual parabolic behaviour as a function of concentration. For 15 and 25 at.% Mg, our resistivities in liquid alloys at the liquidus are substantially lower than those measured by Van Zytveld (1975). However, as will be discussed below, the position of the liquidus is not yet firmly established.

The temperature dependence $(\partial\rho/\partial T)_p$ is plotted in figure 4 as a function of concentration. It is well known that for pure magnesium, $\partial\rho/\partial T$ is small in accordance with the result of the diffraction model for bivalent metals. Earlier (Feitsma *et al* 1975), we observed that for lithium $\partial\rho/\partial T$ is also small, that is, smaller than for any of the other alkali metals. In the lithium–magnesium alloys, $\partial\rho/\partial T$ is even slightly negative for concentrations between 50 and 85% Mg.

In the course of this investigation the liquidus of the lithium–magnesium phase diagram was determined (figure 5). Our points deviate by no more than a few degrees Celsius from those obtained by Grube *et al* (1934), but in the lithium-rich part the more recent results of Freeth and Raynor (1954) are vastly different from ours.

Hallers *et al* (see Feitsma *et al* 1975) have calculated the resistivities of lithium–sodium alloys along the same lines as for the sodium–caesium alloys. They showed clearly that, for pure lithium (and *a fortiori* for its alloys with sodium), the lithium potential should be modelled for $l = 0$ and 1. Moreover, the calculated lithium resistivities are very sensitive to the choice of the hard-sphere parameters – as the lithium form factor at $q = 2k_f$ is very large – and to the kind of screening of the model potential.

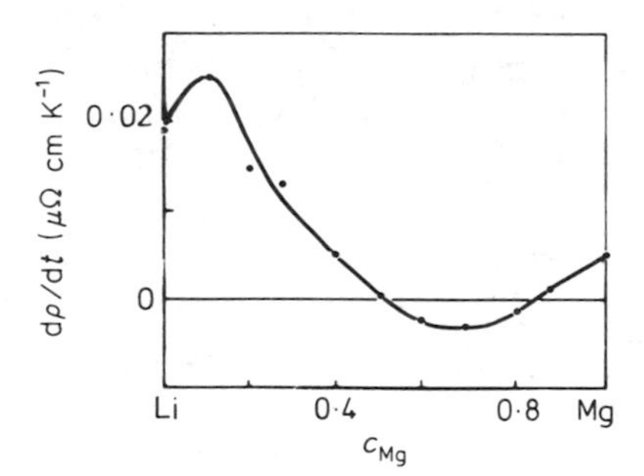

Figure 4. Experimental isotherm (at 650 °C) for the temperature derivative of the resistivity of liquid Li–Mg alloys.

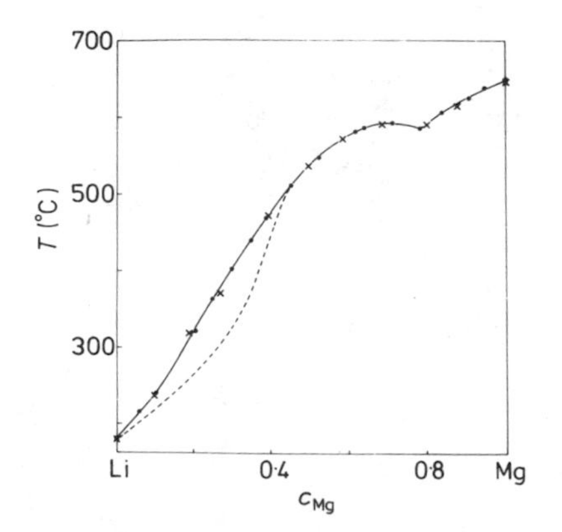

Figure 5. Li–Mg liquidus curve. × this experiment; • Grube *et al* (1934); broken curve, Freeth and Raynor (1954).

For the calculation of the resistivities of the Li–Mg system, we followed Hallers *et al* in modelling the lithium potential for $l = 0$ and $l = 1$. For magnesium, modelling is carried out for $l = 0$, 1 and 2. The model radii were taken as 2·8 and 2·6 a.u. for lithium and magnesium respectively. Throughout, screening according to Toigo and Woodruff (1970) was applied.

Since the present calculations relate to much higher temperatures (651 °C) than those of Hallers *et al*, the structure factors have to be adjusted accordingly. Again, the hard-sphere solution of the Percus–Yevick equation is used. We have chosen two sets of (η, σ)

Table 1. Parameters for the hard-sphere structure factors for Li and Mg. Set I and set II are defined in the text.

	Li(180 °C)		Li(651 °C)		Mg(651 °C)	
	η	σ(Å)	η	σ(Å)	η	σ(Å)
Parameter set I			0·329	2·490	0·450	2·794
Parameter set II	0·492	2·757	0·391	2·636	0·440	2·764

parameter couples (see table 1). The first set is obtained by determining η from the isothermal compressibility χ_T (McAlister *et al* 1974, Novikov *et al* 1969) and, subsequently, σ from η and the density (Crawly 1974) with the help of the relation

$$\eta = \frac{1}{6}\frac{\pi\sigma^3}{\Omega} \tag{2}$$

where Ω is the atomic volume. The second set of parameters was obtained by adjusting the Percus–Yevick hard-sphere structure factors to the available experimental structure factors at their first peak positions. For magnesium, we used the experimental structure factor, obtained by Woerner *et al* (1965) close to the appropriate temperature (651 °C), which was listed by Stoll (1969). The values of η and σ for magnesium determined in this way were found to satisfy the relation (2) only approximately. For lithium such experimental data are not available for the temperature considered, but at 320 °C neutron diffraction experiments have been performed by H Ruppersberg (1976, private communication; see also Ruppersberg and Egger 1975).

From the results above, we estimated the value of η. The change of η from 320 °C to the appropriate temperature was derived from the corresponding change of χ_T. Subsequently, using the relation (2), σ was obtained. As the densities of the liquid alloys are not known, additivity of atomic volumes was assumed throughout.

The calculated isothermal resistivities are presented in figure 3. We have also calculated resistivities at the liquidus; η and σ were derived according to the procedure described for parameter set II. For magnesium, the density and the compressibility were found by extrapolation of the values in the liquid. As for the isotherms, the resulting resistivities are definitely higher than the experimental ones, in contrast to the values calculated by Van Zytveld.

In figure 6 the calculated values of the thermopower Q at the liquidus are compared with the experimental data of Van Zytveld. The same set of parameters as used for the resistivities at the liquidus were employed. Again we observe that exclusion of the term

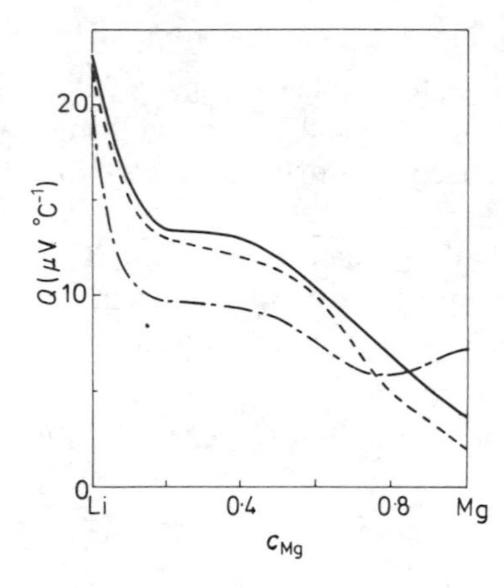

Figure 6. Thermopowers at the liquidus of liquid Li–Mg alloys. Broken curve, experiment (Van Zytveld 1975); full curve, theory (r excluded); chain curve, theory (r included).

r in equation (1) leads to significantly better results than those obtained with the inclusion of r. The same observation has been made for the sodium–caesium system.

3. Lithium–sodium alloys

In addition to the resistivity measurements in this alloy system (Feitsma *et al* 1975) we have determined the thermopower for a few concentrations at both ends of the concentration range. At intermediate concentrations we encountered some experimental difficulties as most isolating refractories are attacked by lithium at high temperatures.

Theoretically, the influence of r is much smaller than for the other two systems mentioned, and exclusion or inclusion of r does not have the same remarkable consequences. The agreement between the two theoretical results (with and without r) and the few experimental data is satisfactory.

In conclusion, we may say that the role of the term r derived from the non-locality of the model potential in the calculation of the thermopower is still not well understood and needs more extensive consideration.

Acknowledgments

This work is part of the research programme of the Stichting voor Fundamenteel Onderzoek der Materie (FOM) and has been made possible by financial support from the Nederlandse Organisatie voor Zuiver Wetenschappelijk Onderzoek (ZWO). The authors would like to thank Professor N E Cusack for supplying a table of experimental data, and Dr H Ruppersberg for data from the neutron diffraction experiments at 320 °C.

References

Ashcroft N W and Langreth D C 1967 *Phys. Rev.* **156** 685–92
Bhatia A B and March N H 1975 *J. Phys. F: Metal Phys.* **5** 1100–6
Crawly A F 1974 *Int. Metall. Rev.* **19** 32–48
Feitsma P D, Hallers J J, Van der Werff F and Van der Lugt W 1975 *Physica* **79B** 35–52
Feitsma P D, Hennephof J and Van der Lugt W 1974 *Phys. Rev. Lett.* **32** 295–7
Freeth W E and Raynor G V 1954 *J. Inst. Metals* **82** 575–80

Grube G, Von Zeppelin H and Bumm H 1934 *Z. Elektrochem.* **40** 160–4
Hallers J J, Mariën T and Van der Lugt W 1974 *Physica* **78** 259–72
Huijben M J, Klaucke H, Hennephof J and Van der Lugt W 1975 *Scr. Metall.* 9 653–6
Huijben M J and Van der Lugt W 1976 *J. Phys. F: Metal Phys.* **6** L225–9
—— 1977 this volume
Ichikawa K, Granstaff S M Jr and Thompson J C 1974 *J. Chem. Phys.* **61** 4059–62
Ichikawa K and Thompson J C 1974 *J. Phys. F: Metal Phys.* **4** L9–12
Kendall P W 1968 *Phys. Chem. Liquids* **1** 33–48
Kim M G and Letcher S V 1971 *J. Chem. Phys.* **55** 1164–70
McAlister S P, Crozier E D and Cochran J F 1974 *Can. J. Phys.* **52** 1847–51
Novikov I I, Trelin Yu S and Tsyganova T A 1969 *High Temp.* 7 1140–1
Ruppersberg H and Egger H 1975 *J. Chem. Phys.* **63** 4095–103
Stoll E 1969 *Eidg. Institut für Reaktorforschung, Würenlingen, Report* AF-SSP-31
Tamaki S, Ross R G, Cusack N E and Endo H 1973 *The Properties of Liquid Metals* ed S Takeuchi (London: Taylor and Francis) pp 289–83
Toigo F and Woodruff T O 1970 *Phys. Rev.* **B2** 3959–66
Valiant J C and Faber T E 1974 *Phil. Mag.* **29** 571–83
Van Hemmen J L, Van der Molen S B and Van der Lugt W 1974 *Phil. Mag.* **29** 493–511
Van Zytveld J B 1975 *J. Phys. F: Metal Phys.* **5** 506–14
Woerner S, Steeb S and Hezel R 1965 *Z. Metallk.* **56** 684

The electrical resistivity of liquid alkali–thallium and alkali–mercury alloys

Masahiro Kitajima† and Mitsuo Shimoji
Department of Chemistry, Faculty of Science, Hokkaido University, Sapporo, Japan

Abstract. Measurements of the electrical resistivity have been made for liquid Na–Tl, K–Tl, Cs–Tl and Cs–Hg alloys. The observed resistivity of these systems has a large maximum around the equiatomic composition; its magnitude considerably increases going from Na to Cs. The temperature coefficient of the resistivity in the same region of concentration is strongly negative for K–Tl and Cs–Tl and weakly positive or negative for Na–Tl, though it is always positive for Na–Hg, K–Hg and Cs–Hg. These results would appear to be explained in terms of the Faber–Ziman theory, except for the Cs–Tl system which should be regarded as a strong-scattering metal.

1. Introduction

We describe electrical resistivity measurements on some alloys of liquid alkali and polyvalent metals: Cs–Hg, Na–Tl, K–Tl and Cs–Tl. No detailed experiments have been reported on any samples which cover the whole concentration range of the alloy systems mentioned, except for K–Tl due to Aronson and Rimler (1973) who used an open-type cell which might have allowed vaporization of the metals under measurement. In view of the considerable difference between the electronegativities of the constituent atoms, these alloys can be regarded as being the so-called 'compound-containing systems' (Wilson 1965), which are usually associated with large negative excess enthalpy and liquidus maxima. The resistivity–composition results are characterized by curves of an upward concave shape; in particular a strongly peaked anomalous structure is observed for K–Tl and Cs–Tl. These are discussed in terms of current theories of electronic transport in liquids.

2. Experimental results

The experimental procedures of the resistivity measurements were essentially the same as in previous work (Itami and Shimoji 1970). Examples of the resistivity–composition relations of three Tl alloy systems are plotted in figure 1; positive deviations from a linear relation connecting the values for the pure metals are clearly observed. In particular, the curves for both K–Tl and Cs–Tl have a strongly peaked structure near the 1 : 1 composition, although results for the latter system may still be questionable because of the relative scarcity of experimental points. The magnitude of the maximum value is approximately 500 $\mu\Omega$ cm at 340 °C for K–Tl and 900 $\mu\Omega$ cm at 420 °C for Cs–Tl. The present results for K–Tl are qualitatively similar to those measured by Aronson

† Present address: National Research Institute for Metals, Nakamegro, Megro, Tokyo, Japan.

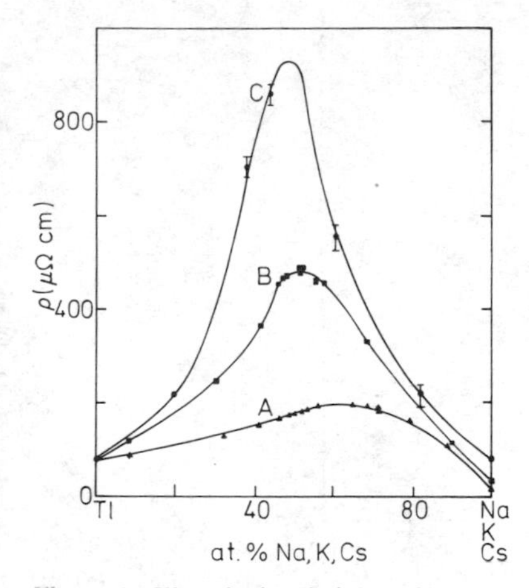

Figure 1. Electrical resistivity of liquid Na–Tl, K–Tl and Cs–Tl alloys: A, Na–Tl at 310 °C; B, K–Tl at 340 °C; C, Cs–Tl at 420 °C.

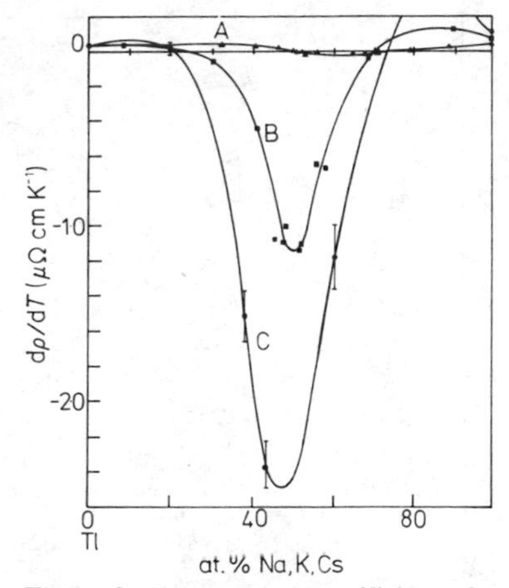

Figure 2. Temperature coefficient of resistivity of liquid Na–Tl, K–Tl and Cs–Tl alloys: A, Na–Tl at 310 °C; B, K–Tl at 340 °C; C, Cs–Tl at 420 °C.

and Rimler (1973). The corresponding maximum value for Na–Tl is around 200 μΩ cm at 310 °C, which is relatively small. Figure 2 shows the experimental temperature coefficients of the resistivity of the Tl alloy systems: it can be seen that for K–Tl and Cs–Tl the temperature coefficients are extremely negative around the 1 : 1 composition.

The results for Cs–Hg are summarized in figures 3 and 4, together with earlier results for Na–Hg and K–Hg (Müller 1910). No anomalous peaked structure can be observed in the resistivity–composition curve, though the degree of positive deviations from the linear plot is much larger in Cs–Hg than in Na–Hg and K–Hg. For all three Hg systems, the temperature coefficient of the resistivity is positive and the magnitude is by no means anomalous.

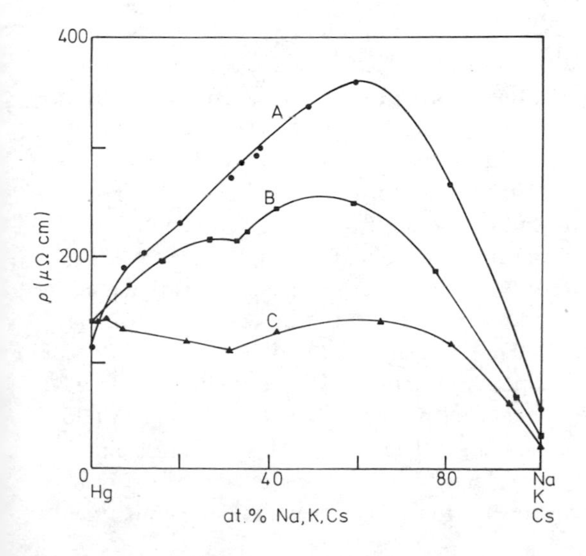

Figure 3. Electrical resistivity of liquid Cs–Hg, K–Hg and Na–Hg alloys: A, Cs–Hg at 220 °C; B, K–Hg at 350 °C; C, Na–Hg at 350 °C. The results for Na–Hg and K–Hg are from Müller (1910).

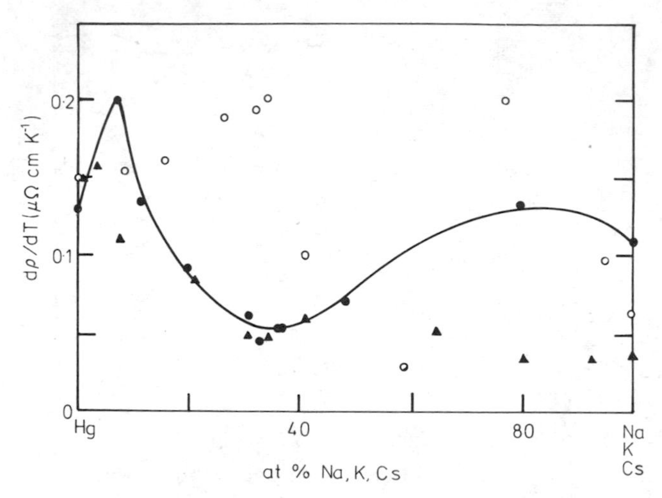

Figure 4. Temperature coefficient of resistivity of liquid Cs–Hg, K–Hg and Na–Hg alloys: ▲, Na–Hg (Müller 1910); ○, K–Hg (Müller 1910); ●, Cs–Hg (this work).

3. Discussion

The weak-scattering theory of the resistivity of liquid alloys due to Faber and Ziman (1965) does not always appear to be suitable for explaining the considerably large values of the resistivity near to the maximum reported here. For Na–Tl and K–Tl some preliminary resistivity calculations were made using the Faber–Ziman theory. Although the details of the calculations are omitted here, optimized non-local model potentials (Shaw 1968, Evans 1970) and analytical partial structure factors of hard-sphere mixtures (Ashcroft and Langreth 1967, Enderby and North 1968) were used. Thus, for example, the use of Toigo and Woodruff's (1970) many-electron factor was found to yield a promisingly good order of magnitude for the resistivity maximum of these systems. Of course, large uncertainties are involved depending on the structure factors employed. For Cs–Tl alloys such a weak-scattering approach is clearly invalid in view of the extremely large values of the resistivity maxima; if Mott's (1966) strong-scattering model is applied to this system, the density of states of the equicomposition alloy at the Fermi level becomes about 0·6 of the free-electron value.

We note that the electron–atom ratio of the alkali–Tl alloy group varies from 1 to 3 while that of the alkali–Hg alloy group goes from 1 to 2. Large differences in the temperature coefficient of the resistivity of both groups (figures 2 and 3) could be accounted for if they are regarded as quasi-single component systems where the negative temperature coefficient can occur when the effective valence is nearly equal to 2 (or $2k_F$ approaches the position of the main peak of the structure factor (Bradley *et al* 1962, Busch and Güntherodt 1967)). Thus the negative temperature coefficient of the resistivity of alkali–Tl alloys would appear to be attributable to the temperature variation of the structure factor in the Faber–Ziman theory. But since the magnitude of $(1/\rho)(d\rho/dT)$ of K–Tl and Cs–Tl is much larger than that of alloys of noble and polyvalent metals, the weak-scattering theory may be inappropriate for K–Tl as well as Cs–Tl. In view of the large difference in the electronegativity values of the constituent atoms, the liquid structure of these systems may be very different from that of simple liquid alloys. On the other hand, for alkali–Hg alloys the electron–atom ratio is always less than 2 so that the temperature coefficient may be positive and may be insensitive to the temperature change of the structure factor. To justify such a conjecture quantitatively it would be necessary to obtain more accurate knowledge of the structure factors.

References

Aronson S and Rimler B 1973 *J. Less-Common Metals* **31** 317–20
Ashcroft N W and Langreth D C 1967 *Phys. Rev.* **156** 685–91
Bradley C C, Faber T E, Wilson E G and Ziman J M 1962 *Phil. Mag.* 7 865–87
Busch G and Güntherodt H-J 1967 *Adv. Phys.* **16** 651–6
Enderby J E and North D M 1968 *Phys. Chem. Liquids* **1** 1–11
Evans R 1970 *J. Phys. C: Solid St. Phys. (Suppl)* **3** S137–52
Faber T E and Ziman J M 1965 *Phil. Mag.* **11** 153–73
Itami T and Shimoji M 1970 *Phil. Mag.* **21** 1193–9
Mott N F 1966 *Phil. Mag.* **13** 989–1014
Müller P 1910 *Metallurgie* **7** 755–71
Shaw R W 1968 *Phys. Rev.* **174** 769–81
Toigo F and Woodruff T O 1970 *Phys. Rev.* **B2** 3958–66
Wilson J R 1965 *Metall. Rev.* **10** 381–590

The structure of liquid transition metals and their alloys

Y Waseda
X-ray Diffraction Laboratory, The Research Institute of Mineral Dressing and Metallurgy, Tohoku University, Sendai 980, Japan

Abstract. An attempt is made in this review to cover the characteristics of the structure of liquid transition metals, including the lanthanide elements, and alloys with a transition metal as a constituent. One difference in the structure of liquid transition metals from that of normal metals is that, as the atomic number increases from Ti to Ni or from Ce to Yb, the oscillations in the structure factor increase in amplitude. This gives qualitative support for the suggestion that the partial overlap between atoms which have a nearly empty 3d-shell or 4f-shell, such as Ti or Ce, is larger than in those having a nearly filled 3d-shell or 4f-shell, such as Ni or Yb. Similar behaviour is found in the structure of liquid Al–Ti and Al–Fe alloys. Results are also given for liquid transition metal–metalloid systems, such as Ni–P alloys.

1. Introduction

A number of experimental studies have been reported on various properties of metals and alloys in the liquid state, and much theoretical effort has been devoted to the atomic properties of fluids and to the electronic properties of liquid metals and alloys. For quantitative discussion of these problems, a knowledge of the structure is essential.

The main purpose of this paper is to review the structural information obtained experimentally for liquid transition and rare earth metals and their alloys. With respect to the presence of the 3d and 4f states, simple explanations are given for the observed behaviour.

2. Experimental techniques

X-rays and neutrons have been used to determine the structure of liquid metals and alloys by the techniques of both transmission and reflection. There are many processes used in the analysis of measured intensity and experimentalists have made significant technical progress. Most of these have been reviewed already (Wagner 1972, Enderby 1967, 1973) and need no description here although, of course, a few new techniques have been reported since the 1972 Tokyo conference. One is the energy-scanning x-ray diffraction method in which white radiation and a solid state detector are used (Prober and Schultz 1975). Another is the application of x-ray anomalous scattering for the separation of the three partial structure factors in binary disordered alloys. The fundamental relations were given by Ramesh and Ramaseshan (1971) and the first application of this method was carried out by Waseda and Tamaki (1975b) in the study of liquid Ni–Si alloys. This technique can give supplementary information to the results of the common x-ray and neutron diffraction measurements as well as the isotope enrichment technique first reported by Enderby *et al* (1966) and the polarized-neutron technique suggested by Bletry and Sadoc (1975). This technique should be a useful method for the study of binary disordered alloys, particularly binary transition metal alloys, because the wavelengths of commercial x-ray targets such as Cu, Co, Ni and Fe are located near the absorption edge of 3d-transition metals. There are, however, two disadvantages: the wavevector range is restricted up to a value of about 7 $Å^{-1}$ and the result is open to experimental uncertainty because of the small difference between the anomalous dispersion terms (Cromer 1965, 1976) and the original atomic scattering factor. Other aspects of this technique have already been described in the work of Waseda and Tamaki (1975b) and a more detailed account will be given at this conference by Waseda *et al* (1977).

3. Experimental structural data for liquid metals and alloys

3.1. Pure metals

Although the pair potential in liquid metals, the so-called effective interionic potential, is of a long-range oscillatory type, it is well known that the hard-sphere solution of the Percus–Yevick equation moderately reproduces the experimental structure factor of most liquid metals, as long as the hard-sphere diameter, σ, or the packing density, $\eta = \frac{1}{6}\pi\rho_0\sigma^3$, is suitably selected (where ρ_0 is the average number density of atoms). This interesting result, first suggested by Ashcroft and Lekner (1966), corresponds to the existence of neutral pseudo-atoms in metals (Ziman 1964). The structure factor of liquid metals is dominated by the repulsive core part of the effective interionic potential $\phi(r)$ and it is known that the hard-core size in $\phi(r)$ is nearly equal to that chosen to fit the experimental data using the Ashcroft–Lekner method (Waseda and Tamaki 1976c). The hard-sphere structure factor shows the following characteristics: (i) the first peak of the structure factor is symmetrical; (ii) at temperatures just above the melting point, the best agreement with experimental data is found with a packing density $\eta = 0{\cdot}45$; and (iii) the ratio (Q_2/Q_1) of the position of the second peak (Q_2) to that of the first peak (Q_1) is about 1·86. Thus the measurement of deviation from the hard-sphere structure factor gives useful information in the structural study of liquid metals. From this point of view, the results of our serial works for 39 metallic elements are listed in table 1 together with data for Li, Rb and Cs (Gingrich and Heaton 1961).

As an example, a comparison between the experimental data and the hard-sphere structure factor is shown in figure 1 for liquid Al, Zn and Sn near their melting points. The results on liquid Al satisfy the above characteristics of the hard-sphere structure factor (classification *a*), but the structure factor of liquid Zn has an asymmetry of the first peak (classification *b*) while that of liquid Sn has a small hump on the high-angle side of the first peak (classification *c*). With respect to the ratio (Q_2/Q_1), both Zn and Sn have large deviations from the value of 1·86. As shown in table 1, all of the metallic elements can be put into one of these three types, that is, having structures similar to Al, Zn or Sn.

However, the present author maintains the view that a random distribution of hard spheres of appropriate diameter is better for a fundamental understanding of atomic distributions in liquid metals. In this respect, most metallic elements seem to be consistent with the fact that the characteristics of their respective crystal structures

Table 1. Structural information of liquid metals near the melting point†.

	Temp. (°C)	Density (g cm^{-3})	Q_1	Hump (Å^{-1})	Q_2	Q_2/Q_1	r_1	Hump (Å)	r_2	r_2/r_1	n_1 (atoms)	η_m	Classification
Li‡	180	0·504	2·49		4·55	1·83	3·15		6·0	1·90	9·5	0·46	*a*
Na	105	0·928	2·03		3·75	1·85	3·81		7·1	1·86	10·4	0·46	*a*
K	70	0·826	1·61		2·98	1·85	4·63		8·5	1·84	10·5	0·46	*a*
Rb‡	40	1·47	1·53		2·80	1·83	4·97		9·6	1·93	9·5	0·43	*a*
Cs‡	30	1·84	1·47		2·68	1·82	5·31		9·8	1·85	9·0	0·43	*a*
Mg	680	1·545	2·42		4·40	1·82	3·21		6·0	1·87	10·9	0·46	*a*
Ca	850	1·37	1·95		3·63	1·85	3·83		7·3	1·91	11·1	0·46	*a*
Sr	780	2·38	1·78		3·30	1·85	4·23		8·0	1·89	11·1	0·46	*a*
Ba	730	3·32	1·73		3·21	1·86	4·31		8·1	1·88	10·8	0·46	*a*
Zn	450	6·91	2·93		5·14	1·75	2·68		4·9	1·83	10·5	0·46	*b*
Cd	350	7·954	2·62		4·63	1·77	3·11		5·8	1·86	10·3	0·45	*b*
Hg	20	13·55	2·32		4·55	1·96	3·07	§	5·9	1·92	10·0	0·45	*b*
Al	670	2·37	2·68		4·96	1·85	2·82		5·3	1·88	11·5	0·45	*a*
Ga	50	6·082	2·52	3·12	4·90	1·94	2·82	3·4	5·6	1·99	10·4	0·43	*c*
In	160	7·03	2·30		4·32	1·88	3·23	§	6·1	1·89	11·6	0·45	*a*
Tl	315	11·27	2·26		4·25	1·88	3·28	§	6·2	1·89	11·6	0·45	*a*
Si	1460	2·59	2·72	3·40	5·62	2·07	2·50	3·8	5·7	2·28	6·4	0·38	*c*
Ge	980	5·56	2·56	3·24	5·11	2·00	2·82	4·2	5·8	2·06	6·8	0·38	*c*
Sn	250	6·93	2·21	2·98	4·33	1·96	3·23	3·8	6·3	1·95	10·9	0·43	*c*
Pb	340	10·66	2·28		4·23	1·86	3·33		6·4	1·92	10·9	0·46	*a*
Sb	660	6·48	2·15	3·00	4·21	1·96	3·33	4·5	6·5	1·95	8·7	0·40	*c*
Bi	300	10·03	2·11	2·85	4·12	1·95	3·38	4·1	6·6	1·95	8·8	0·40	*c*
Cu	1150	7·97	3·00		5·46	1·82	2·57		4·8	1·87	11·3	0·46	*a*
Ag	1000	9·27	2·61		4·86	1·86	2·87		5·5	1·92	11·3	0·45	*a*
Au	1150	17·2	2·66		4·90	1·84	2·86		5·4	1·89	10·9	0·46	*a*
Ti	1700	4·15	2·45		4·41	1·80	3·17		6·0	1·89	10·9	0·44	*a*
V	1900	5·36	2·71		4·98	1·84	2·82		5·3	1·88	11·0	0·44	*a*
Cr	1900	6·27	3·01		5·48	1·83	2·58		4·8	1·84	11·2	0·45	*a*
Mn	1260	5·97	2·83		5·19	1·83	2·67		5·0	1·87	10·9	0·45	*a*
Fe	1550	7·01	2·98		5·46	1·83	2·58		4·8	1·86	10·6	0·44	*a*
Co	1550	7·70	3·02		5·60	1·85	2·56		4·7	1·84	11·4	0·45	*a*
Ni	1500	7·72	3·10		5·70	1·84	2·53		4·7	1·86	11·6	0·45	*a*
Pd	1580	10·5	2·81		5·23	1·86	2·71		5·1	1·90	10·9	0·47	*a*
Pt	1780	18·7	2·78		5·17	1·86	2·73		5·2	1·89	11·1	0·47	*a*
Zr	1900	5·93	2·32		4·36	1·88	3·19		6·0	1·89	10·6	0·44	*a*
Sc	1560	2·92	2·49		4·83	1·94	2·92	§	5·6	1·92	10·3	0·43	*a*
La	970	5·95	2·10		3·86	1·84	3·87		7·5	1·94	11·1	0·43	*a*
Ce	870	5·92	2·14		4·06	1·90	3·75		7·4	1·97	10·9	0·42	*a*
Eu	830	4·61	1·87		3·71	1·98	4·05		7·9	1·95	10·7	0·42	*a*
Gd	1330	6·91	2·07		3·93	1·90	3·65	§	7·0	1·92	10·7	0·43	*a*
Tb	1380	7·24	2·08		3·97	1·91	3·62	§	6·9	1·92	11·1	0·43	*a*
Yb	850	6·20	2·02		3·75	1·86	3·85		7·3	1·90	10·3	0·43	*a*

† η_m; packing density at melting point: Q_i; position of the ith peak in the structure factor: r_1 and n_1 are the nearest-neighbour distance and its coordination number, respectively.

‡ From Gingrich and Heaton (1961). § There seems to be a slight asymmetry in the first peak.

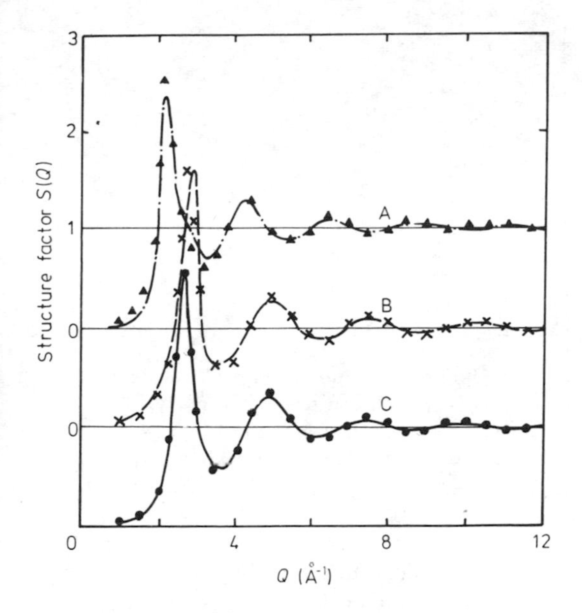

Figure 1. Structure factors of liquid Al, Zn and Sn near the melting point: A, Sn at 250 °C; B, Zn at 450 °C; C, Al at 670 °C. Points indicate results obtained using the hard-sphere model.

become obscure on melting; then the increase in the freedom of atomic configuration contributes to the construction of universal short-range order, which mainly depends upon the size factor similar to that of the hard-sphere model. This is supported by the isotropic metallic bonding. In other words, those elements (e.g., the covalent elements) which have anisotropic bonding show a deviation from the above hierarchy, and this is related, in a crude approximation, to a crystal structure with a lower degree of symmetry than the close-packed structures such as FCC, HCP and BCC. Deviation of this type decreases with an increase of fluid volume arising from the elevation of temperature and then disappears at temperatures of 300–500 °C above the melting point. The problem of deviations from the hard-sphere structure factor has been discussed in detail (Weaire 1968, Franchetti 1973, March *et al* 1976).

The following feature for various liquid metals, deduced from the data listed in table 1, is worthy of note. Although transition metals and rare earth metals include the incomplete 3d or 4f states, their structure factors can be put in the same class as simple metals such as Na and Al. The present author believes that this experimental fact gives a surprisingly simple explanation for the structure of liquid transition metals and rare earth metals. The oscillations in the structure factor increase in amplitude in a series from Ti to Ni as shown in figure 2 (Waseda and Tamaki 1975a). It is also well known that the value of the effective hard-sphere diameter is approximately 87% of the nearest-neighbour distance r_1 for most liquid metals. According to this rule, one obtains the values listed in table 2. The difference between the estimated value and the normal value of 0·45 for Ti and V implies that the hard-sphere diameters of liquid Ti and V are considerably smaller than the value of the fourth column in table 2. The best agreement for peak height is in fact obtained for a packing density of 0·44 for both liquid metals, as shown in the ninth column of table 1. This shows that the hard-sphere diameter is close to 2·53 Å for liquid Ti and 2·37 Å for liquid V. From these results, it appears that the gradient of the repulsive core part of $\phi(r)$ for liquid Ti and V is relatively

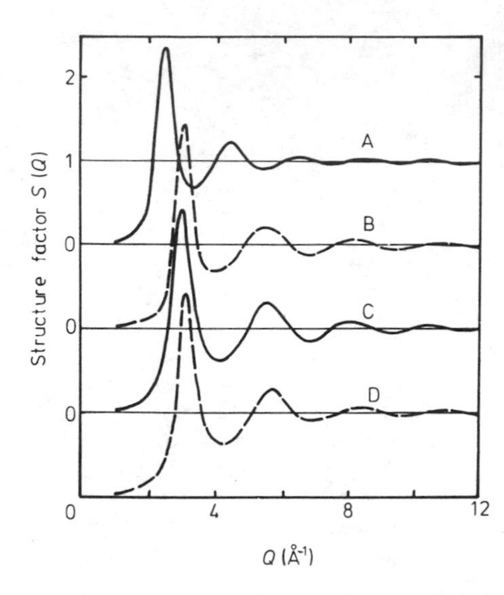

Figure 2. Structure factors of liquid Ti, Cr, Fe and Ni near the melting point: A, Ti at 1700 °C; B, Cr at 1900 °C; C, Fe at 1550 °C; D, Ni at 1500 °C.

small compared with that of other 3d-transition metals. Consequently, the pair distribution function of Ti and V should have a rather high value in the region of small r below the first peak (corresponding to the repulsive core part). This is confirmed in figure 3. This experimental fact gives qualitative support for the suggestion that a partial overlap between atoms having a nearly empty d-shell (such as Ti and V) is larger than that for the elements having a nearly filled or filled d-shell (such as Ni and Cu). This characteristic must be related to the difference in the electronic structure with an incomplete 3d-band of 3d-transition metals. This characteristic structure plays a significant role in a discussion of the excess entropy (Meyer *et al* 1976, Waseda and Tamaki 1976a). A similar behaviour in the liquid structure factor, related to the electronic structure, is also found in liquid rare earth metals. Figure 4 shows the structure factor of liquid lanthanide elements at 870 °C (Waseda 1976). In this case, the samples were put into a tantalum cell and the x-rays emerged through a beryllium window (0·10 mm) protected by a thin molybdenum (0·01 mm) sheet from attack by

Table 2. Effective hard-sphere diameter σ and packing density η.

Elements	Temperature (°C)	Electronic structure of outer shells	σ (Å)	Estimated packing density, η
Ti	1700	$3d^2 4s^2$	2·77	0·58
V	1900	$3d^3 4s^2$	2·46	0·50
Cr	1900	$3d^5 4s^1$	2·25	0·44
Mn	1260	$3d^5 4s^2$	2·33	0·44
Fe	1550	$3d^6 4s^2$	2·25	0·45
Co	1550	$3d^7 4s^2$	2·24	0·46
Ni	1500	$3d^8 4s^2$	2·21	0·45
Cu	1150	$3d^{10} 4s^1$	2·25	0·45
Ce	870	$4f^2 6s^2$	3·28	0·53
Eu	830	$4f^7 6s^2$	3·54	0·43
Yb	850	$4f^{14} 6s^2$	3·36	0·43

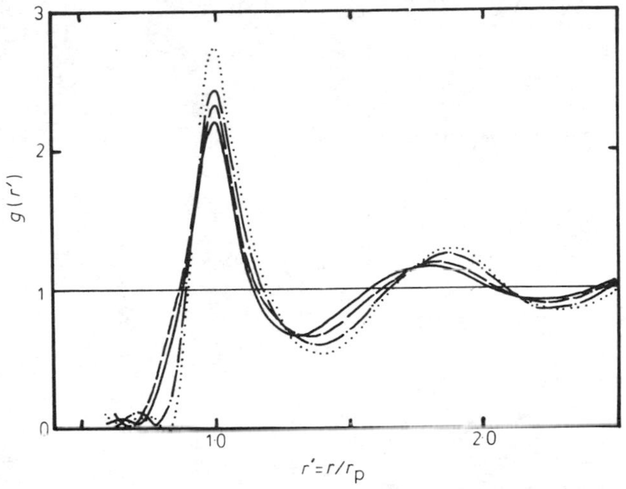

Figure 3. Reduced pair distribution function $g(r')$ of liquid Ti (solid line) at 1700 °C, V (broken line) at 1900 °C, Ni (chain line) at 1500 °C and Cu (dotted line) at 1150 °C. $r' = r/r_p$, where r_p is the position of the first peak in $g(r)$.

the metal sample. On progressing in atomic number from Ce to Yb, the oscillations systematically increase in amplitude. This characteristic, on Fourier transformation, gives the physical property that the repulsive core part of $g(r)$ for Yb is relatively steep compared with that of Ce. This is supported by the packing density listed in table 2 as well as by $g(r)$ in figure 4. As the effective hard-sphere diameters of lanthanide elements are larger than those of 3d-transition metals, the overlap effect seems to be important in the case of rare earth metals. The above inference, related to the electronic structure, agrees with experimental information on the structure of liquid rare earth metals.

3.2. Alloys

There are various methods used in the classification of alloys; for example, on the basis of thermodynamic or electronic properties. The classification based on structural information obtained from diffraction experiments has already been discussed (Waseda 1976) and, for convenience, the essential features are given below. The metallic alloy systems are divided into the following three groups.

(i) The total structure factor shows the behaviour of random mixing without a subpeak or asymmetry of the first peak. The partial structure factor of unlike-atom pairs (i–j) has maxima which lie in between those of two like-atom pairs (i–i and j–j). (For example, Na–K and Pb–Bi.)
(ii) The total structure factor shows the behaviour of compound-forming with a subpeak below the first peak. The partial structure factor of unlike-atom pairs has a very sharp first peak with a subpeak below the main peak, in comparison to those of two like-atom pairs. (For example, Ag–Mg and Cu–Mg.)
(iii) The structural information shows intermediate behaviour between those observed in both type (i) and type (ii). Although the subpeak below the first peak is not found,

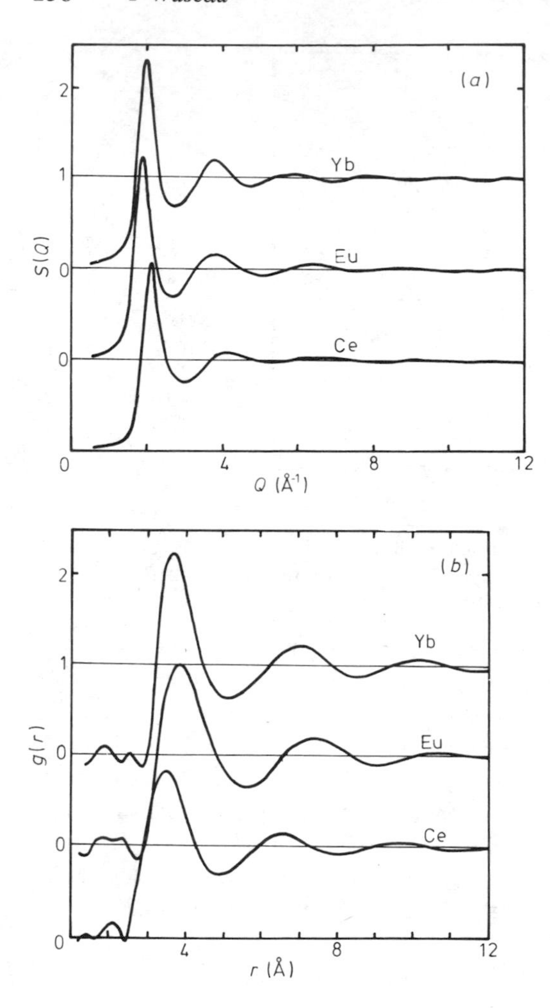

Figure 4. (*a*) Structure factors and (*b*) pair distribution function of liquid Ce, Eu and Yb at 870 °C.

the partial structure factor of unlike-atom pairs is very close to either of two like-atom pairs. (For example, Cu–Sn and Ag–Sb.)

According to this classification, the structural information obtained in our serial works on liquid transition metal alloys can be classified in the following manner:

Alloys of group (i): Al–Ti, Fe–Mn, Fe–Cr, Cu–Ni.
Alloys of group (iii): Al–Fe, Fe–Ni, Fe–Co, Ge–M (M: Mn, Fe, Co, Ni, Cu).
Alloys of other types: Ni–P, Fe–P, Fe–C, Fe–Si, Ni–Si.

There is no alloy of group (ii) within the results obtained in our work.

As an example, the partial structure factors of liquid Al–Ti and Al–Fe alloys are given in figure 5; these are obtained under a moderate assumption (Halder and Wagner 1967) with the anomalous scattering technique given in the previous section. The error in these results, including the system of Ni–P (figure 6), due both to alloy concentration and to the residual experimental uncertainty, is almost the same as in the case of liquid Ni–Si or Fe–Si alloys (Waseda and Tamaki 1975b, 1976b). The partial structure factors

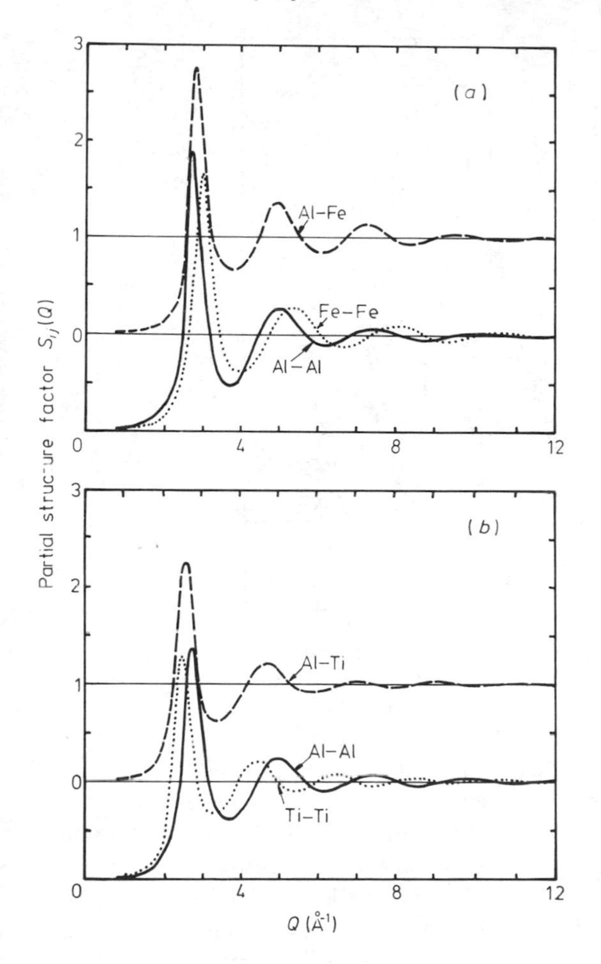

Figure 5. Partial structure factors of (*a*) liquid Al–Ti and (*b*) Al–Fe alloys at temperatures about 50 °C above the liquidus.

in liquid Al–Ti alloys have no sharp peak in comparison to those in liquid Al–Fe alloys. This may be related to a nearly empty 3d-band in Ti atoms. Sharp peaks in liquid Al–Fe alloys suggest that the rigid repulsive forces between Fe–Fe atoms play a main role in the atomic distribution of this alloy system. The partial structure factor of Al–Ti pairs is approximately expressed by the average of those of Al–Al and Ti–Ti pairs. On the other hand, the partial structure factor of Al–Fe pairs is close to that of Al–Al pairs. Although this structural information on liquid transition metal alloys must be affected more or less by the characteristic electronic structure, quantitative discussion is not available at the present time.

From the practical metallurgical aspect, it is desirable to determine the structure of liquid transition metal–metalloid alloys. Figure 6 shows the three partial structure factors for liquid Ni–P alloys (metalloid contents of 15, 20, 25 and 30 at.%) obtained using the anomalous scattering technique. In the case of liquid Ni–Si (Waseda and Tamaki 1975b) and Fe–Si alloys (Waseda and Tamaki 1976b), similar behaviour to that of group (iii) is found. In contrast to this, one of the most interesting results is the experimental fact that the significant distance of P–P pairs, deduced from the partial pair distribution function, is about 3·2 Å. This is larger than the value (2·2 Å) predicted

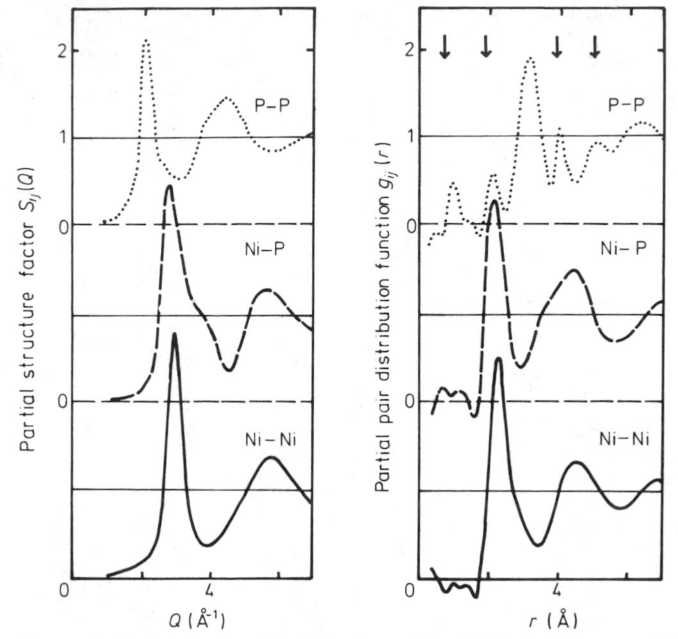

Figure 6. Partial structural information of liquid Ni–P alloys at 1200°C.

from the size factor of P atoms and the P atoms are never at the first nearest-neighbour position in liquid Ni–P alloys for the present experimental compositions (up to 30 at.%). Here it was confirmed that the peaks at the position of about 1, 2, 4 and 5 Å are spurious ripples due to the termination effect, using the results of Finback (1949) and Sugawara (1951). This experimental result implies that the fundamental configuration of atoms in the short-range order of liquid Ni–P alloys is similar to that of Ni_3P (see table 3) and that the atomic configuration in liquid Ni–P alloys consists mainly of the disordered distribution of Ni and P atoms, like that in the dense random packing model based on tetrahedral units. This is consistent with the fact that the short-range order of close-packed structures in liquids is also composed of tetrahedrons as well as that in the amorphous state. As is easily predicted from the size factor of Ni and P atoms (Ni: 2·5 Å; P: 2·2 Å), the expansion of the Ni–Ni distance is unavoidable in configurations of the above type. This inference is supported by the results given in figure 7. A

Table 3. Comparison between the correlation of nearest-neighbour atoms in liquid Ni–P alloys and crystalline Ni_3P.

	Origin atom	Ni		P	
		r (Å)	n (atoms)	r (Å)	n (atoms)
Liquid Ni–P	Ni	2·60	10·6	2·33	2·6
alloys at 1200 °C	P	2·33	8·1	3·20	3·4
Crystalline Ni_3P†	Ni	2·68	10	2·28	3
	P	2·28	9	3·44	4

† Rundqvist *et al* 1962.

similar behaviour was also found in liquid Fe–C alloys (Maier and Steeb 1973, Waseda *et al* 1975) and Fe–P alloys (Y Waseda, unpublished). The increase of fluid volume on melting seems to contribute to the construction of these compulsory configurations of atoms for liquid alloy systems, in which the difference in the size factor of constituent elements is larger than about 15%. Configurations of this type can not exist in the normal solid state (except for metastable states of amorphous metals). In addition, this characteristic of liquid transition metal–metalloid alloys must be related to the interesting phenomena of activity in metallurgical processes (Elliot *et al* 1963).

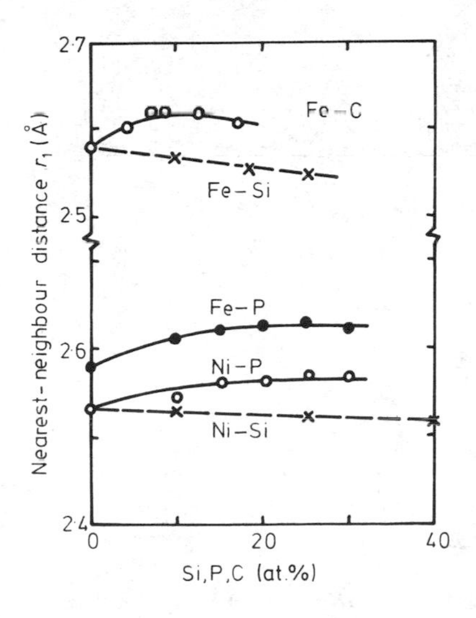

Figure 7. Concentration dependence of the nearest-neighbour distance in liquid transition metal–metalloid alloys at temperatures about 50 °C above the liquidus.

4. Concluding remarks

To summarize the results in our serial works, the static structure of liquid transition metals and rare earth metals can be interpreted as a simple fluid in which the arrangement of the atoms is similar to the hard-sphere model, although these metallic elements have the complicated electronic structure related to the 3d or 4f states. But the fundamental units of the structure in these liquid metals are positively charged ions with screening clouds of electron gas, and one ought to use the hard-sphere diameter which is chosen to fit the experimental data. We may further add that this concept is applicable to about 80% of the metallic elements (see table 1).

Liquid binary alloys which involve transition metals as constituents can, in the first approximation, be classified in the same way as suggested for normal metal alloys. A partial overlap effect, observed in liquid Al–Ti alloys, which is due to the electronic structure in the outer shell, may be one of the most interesting factors in the detailed atomic distribution of disordered alloys, especially for the liquid transition metal alloys.

In the transition metal–metalloid systems, such as the Ni–P alloys, the size factor difference of more than 15% plays an important role in the atomic distribution, and this seems to be related to the interesting phenomena in the metallurgical processes.

Acknowledgments

The author is grateful for valuable discussions with Professor S Tamaki. I would like to thank Professors M Ohtani, Y Shiraishi, T Masumoto and K Suzuki for their support and encouragement in these serial works. The financial assistance of the Sakkokai Foundation is very gratefully acknowledged.

References

Ashcroft N W and Lekner J 1966 *Phys. Rev.* **145** 83
Bletry J and Sadoc J F 1975 *J. Phys. F: Metal Phys.* **5** L110
Cromer D T 1965 *Acta Crystallogr.* **18** 17
—— 1976 *Acta Crystallogr.* **A32** 339
Elliot J F, Gleisser M and Ramakrishna V 1963 *Thermochemistry for Steelmaking* (New York: Addison-Wesley)
Enderby J E 1967 *Physics of Simple Liquids* ed H N Tempeley, J S Rowlinson and G B Rushbrooke (Amsterdam: North-Holland) p611
—— 1973 *The Properties of Liquid Metals* ed S Takeuchi (London: Taylor and Francis) p3
Enderby J E, North D M and Egelstaff P A 1966 *Phil. Mag.* **14** 961
Finback C 1949 *Acta Chem. Scand.* **3** 1279
Franchetti S 1973 *Nuovo Cim.* **18B** 247
Gingrich N S and Heaton L 1961 *J. Chem. Phys.* **34 873**
Halder N C and Wagner C N J 1967 *J. Chem. Phys.* **47** 4385
Maier U and Steeb S 1973 *Phys. Condens. Matter* **17** 11
March N H, Parrinello M and Tosi M P 1976 *Phys. Chem. Liquids* **5** 39
Meyer A, Stott M J and Young W H 1976 *Phil. Mag.* **33** 381
Prober J M and Schultz J M 1975 *J. Appl. Crystallogr.* 8 405
Ramesh J G and Ramasehan S 1971 *J. Phys. C: Solid St. Phys.* **4** 3029
Rundqvist S, Hassler E and Lundvik L 1962 *Acta. Chem. Scand.* **16** 242
Sugawara T 1951 *Sci. Rep. Res. Inst., Tohoku Univ.* **3A** 39
Wagner C N J 1972 *Liquid Metals Physics and Chemistry* ed S Z Beer (New York: Marcel Dekker) p257
Waseda Y 1976 *Metal Phys. Seminar (Japan)* **1** 302
Waseda Y, Masumoto T and Tamaki S 1977 this volume
Waseda Y and Tamaki S 1975a *Phil. Mag.* **32** 273
—— 1975b *Phil. Mag.* **32** 951
—— 1976a *J. Phys. F: Metal Phys.* **6** L89
—— 1976b *Commun. Phys.* **1** 3
—— 1976c *Metal Phys. Seminar* (Japan) **1** 133
Waseda Y, Tokuda M and Ohtani M 1975 *Tetsu to Hagane (Japan)* **61** 54
Weaire D 1968 *J. Phys. C: Solid St. Phys.* **1** 210
Ziman J M 1964 *Adv. Phys.* **13** 89

Investigation of the structure of liquid La, Ce and Pr by neutron diffraction

H Rudin†, A H Millhouse‡, P Fischer‡ and G Meier‡
† Institut für Physik, Universität, Basel, Switzerland
‡ Eidgenössisches Institut für Reaktortechnik, Würenlingen, Switzerland

Abstract. The structure factor $S(Q)$ for liquid La, Ce and Pr has been determined by neutron scattering. The position of the first peak is $Q_p = 2{\cdot}05\ \text{Å}^{-1}$, $2{\cdot}13\ \text{Å}^{-1}$ and $2{\cdot}10\ \text{Å}^{-1}$ for La, Ce and Pr respectively. The paramagnetic contribution to the scattering of Pr is consistent with the assumption of a $^3H_4$ spectroscopic state of Pr^{3+}. The magnetic scattering of Ce for $Q \rightarrow 0$ is only 70% of what is to be expected for the $^2F_{5/2}$ state of the Ce^{3+} ion.

The liquid structure factors $S(Q)$ of the light rare earth elements have been investigated by neutron diffraction at Saclay (Bellissent and Tourand 1975, Breuil and Tourand 1969) and also at Leicester (Enderby and Nguyen 1975). However, the results obtained by these two groups were inconsistent in the sense that the Saclay group did not observe a second maximum in the structure factor as a function of momentum transfer Q. We therefore decided to repeat these experiments and also to extend the study to neodymium and ytterbium with the aim of determining $S(Q)$ as well as the amount of magnetic scattering due to localized magnetic moments.

To ensure the validity of the static approximation neutrons of 1·05 Å wavelength were employed. We used cylindrical samples of 5 cm height contained in tantalum cells of 2 cm in diameter for La, Ce and Pr and of 0·96 cm in diameter for Nd and Yb. The scattered neutron intensity, after correction for background and self-absorption in the sample can be written as

$$I_c = \alpha[S(Q) + CF^2(Q) + D],$$

the three terms on the right hand side representing coherent nuclear scattering, paramagnetic scattering and incoherent nuclear scattering plus multiple scattering respectively. The calibration constant α was determined with a vanadium sample as a standard scatterer. An independent determination of α was made by using as reference a sample of nickel powder, which yielded the same calibration constant within 5% accuracy. In order to be able to evaluate the Q-dependent part of the scattering cross section we made the assumption that D, which includes the multiple scattering contribution, is approximately independent of scattering angle (Blech and Averbach 1965).

In the case of lanthanum there is no magnetic contribution to the scattered intensity and $S(Q)$ can be determined directly by using the condition $S(Q) = 1$ for large values of Q. The full line in figure 1 shows $S(Q)$ for La at a temperature of 1012°C. The first peak at $Q_p = 2{\cdot}05$ Å^{-1} is clearly followed by an oscillating part which was not observed by the Saclay group (Bellissent and Tourand 1975).

For cerium there is in addition to nuclear scattering also magnetic scattering. In order to separate these two contributions we assumed the magnetic form factor $F(Q)$ to be the same as for the $^2F_{5/2}$ spectroscopic state of Ce^{3+}. However, if we extrapolate the measured intensity toward the long wavelength limit ($Q \to 0$) we observe only 70% of the scattering to be expected from the magnetic moment of 2·54 Bohr magnetons of the assumed state. Within the experimental uncertainty of 10% this result confirms the observations of Enderby and Nguyen (1975). We used the dipole approximation for the magnetic form factor of Ce^{3+} and by subtraction of the magnetic scattering part we found the structure factor $S(Q)$ for liquid cerium at 1012°C as shown by the broken line of figure 1. The position of the first peak lies at $Q_p = 2{\cdot}13$ Å^{-1} and the overall behaviour of $S(Q)$ is quite similar to that of lanthanum.

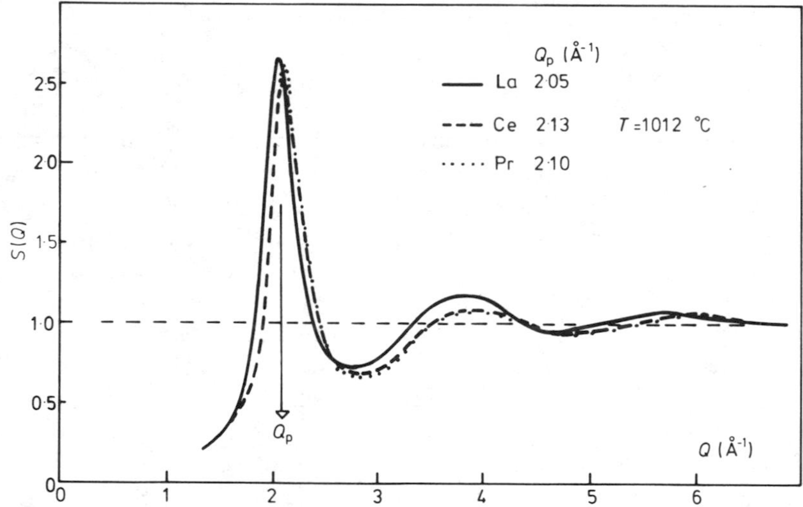

Figure 1. Liquid structure factor $S(Q)$ of La (full line), Ce (broken line) and Pr (dotted line) at 1012°C.

In the case of praseodymium we performed the same kind of analysis. The strong magnetic scattering was attributed to the $^3H_4$ spectroscopic state of Pr^{3+} and again the corresponding dipole approximation to $F(Q)$ was taken in order to evaluate $S(Q)$. But now for $Q \to 0$ the magnetic scattering is consistent with the full magnetic moment of 3·58 Bohr magnetons of the Pr^{3+} ground state. The structure factor $S(Q)$ is given by the dotted line in figure 1. The position of the first peak is at $Q_p = 2{\cdot}10$ Å^{-1} and for larger values of Q the oscillating part of $S(Q)$ is the same for Pr as for Ce within the experimental accuracy.

The results for neodymium at a temperature of 1100°C again showed a magnetic contribution consistent with the $^2I_{9/2}$ state of Nd^{3+} and the structure factor $S(Q)$ could not be distinguished from that of Pr.

With ytterbium no magnetic scattering was observed. $S(Q)$ is very similar to the structure factor of La, except that the first peak position of Yb lies at $Q_p = 1{\cdot}99$ Å^{-1}.

The structure factors for liquid rare earth metals obtained by x-ray diffraction by Waseda *et al* (1977) and reported at this conference also show oscillating behaviour at large values of Q and are in agreement with our data.

References

Bellissent R and Tourand G 1975 *J. Physique* **36** 97
Blech I A and Averbach B L 1965 *Phys. Rev.* **A137** 1113
Breuil M and Tourand G 1969 *Phys. Lett.* **A29** 506
Enderby J E and Nguyen V T 1975 *J. Phys. C: Solid St. Phys.* 8 L112
Waseda Y, Masumoto T and Tamaki S 1977 this volume

Structure and compressibility of Au–Co melts †

Siegfried Steeb and Richard Bek

Max-Planck-Institut für Metallforschung, Institut für Werkstoffwissenschaften, Seestrasse 92, 7000 Stuttgart 1, West Germany

Abstract. Alloys of the Au–Co system containing 0, 20, 27, 30, 32, 36, 41, 55, 70 and 100 at.% Co were investigated at temperatures between 10 and 20 °C above the liquidus by means of thermal neutron (wavelength 1·20 Å) and x-ray (Mo–Kα) diffraction. The interference functions show no additional maxima or deformations. Radial distribution functions were obtained and from these the coordination numbers N^{I} and nearest-neighbour distances r^{I} were calculated. The N^{I} against concentration curve shows a deviation to higher values, thus indicating a tendency to segregation within these melts.

The velocity of ultrasound was measured in molten Au–Co alloys containing 0, 10, 20, 27, 30, 40, 50, 60, 70, 80, 90 and 100 at.% Co within the temperature range from 995 °C to 1700 °C. The adiabatic compressibility was calculated from the velocity, and from this the partial structure factors $a_{ij}(0)$ and $S_{kl}(0)$ for zero momentum transfer were obtained. According to the results, the melts of the Au–Co system show a weak tendency to segregation.

† To be published in *Z. Naturf.* **31A** (1976)

Neutron and x-ray diffraction measurements of liquid Fe–C alloys

Y Kita†, M Ueda†, Z Morita†, K Tsuji‡ and H Endo‡

† Department of Metallurgical Engineering, Faculty of Engineering, Osaka University, Suita, Osaka 565, Japan

‡ Department of Physics, Faculty of Science, Kyoto University, Kyoto 606, Japan

Abstract. Neutron and x-ray diffraction measurements have been made on liquid Fe–11 at.% C and Fe–17 at.% C alloys at 1450 and 1350 °C respectively. No shift in the main peak was observed for either alloy. The position of the main peaks of the partial structure factor, $a_{Fe-C}(K)$, were found to be at 3·3 Å^{-1} for both alloys. These experimental results are discussed on the basis of a random mixture model.

1. Introduction

Attention has been focused on the liquid Fe–C alloy not only because of its interest as a transition metal–nonmetal system but also because of its metallurgical importance in relation to the iron and steelmaking industries. Although various properties have been measured by a number of workers on this system, there exist significant discrepancies in the data, caused by experimental errors due to technical difficulties at high temperatures. Published reports have suggested anomalous changes of the properties with carbon concentration, from which some investigators proposed the existence of strong interaction between iron and carbon in the liquid state (Filippov and Samarin 1968, Maier and Steeb 1973, Waseda *et al* 1975).

In the present work, in order to obtain some information on the liquid structure of this system, neutron and x-ray diffraction measurements were carried out on liquid Fe–11 at.% C and Fe–17 at.% C alloys and the partial structure factor $a_{Fe-C}(K)$ was deduced from the results.

2. Experimental

Neutron diffraction was carried out by the use of a diffractometer, KUR-ND, at the Research Reactor Institute, Kyoto University. The neutron beam, of wavelength 1·006 Å and cross section 28 × 49 mm, was irradiated into the liquid specimen after being monochromatized by a copper single crystal. Scattered neutrons were collimated with a Soller slit (0·5°) and counted by a BF_3 detector. About 250 g of Fe–C was used as a specimen and this was melted in an alumina cell (ID, 16 mm; wall thickness, 1·3 mm) which was inside a wound tungsten heater (0·5 mm diameter wire) and set in an aluminium vessel. A constant vacuum of 2×10^{-5} Torr was maintained and the temperature controlled with an accuracy of ±5 °C. Neutron diffraction measurements were performed for liquid Fe–11 at.% C at 1450 °C and Fe–17 at.% C at 1350 °C and the background intensity diffracted from the empty cell and heater was measured at the same temperatures. The total counts per point were about 40 000 for the cell with the specimen and about 10 000 for the empty cell, in the region near the main peak. Corrections were made to the scattering intensities for the absorption by the cell and specimen and for the relatively weak paramagnetic scattering (Freeman and Watson 1961), both having an angular dependence, and also for the incoherent and multiple scattering (Vineyard 1954) which had little angular dependence. Thus, the coherent scattering intensity from the specimen $I_{coh}(K)$ was obtained, and the total structure factor $a(K)$ determined by normalizing $I_{coh}(K)$ according to the usual procedure (Enderby 1968).The statistical error in $a(K)$ was less than 1·5% at the main peak position and 3% at large values of K.

A θ–θ diffractometer with a Mo target x-ray tube was used for the x-ray diffraction. The x-rays passed through a divergence slit ($\frac{1}{6}^\circ$), were reflected on the free surface of the liquid specimen, then monochromatized by a graphite single crystal after passing through a scattering slit (1°) and a receiving slit (0·6 mm width), and finally counted by a scintillation counter. In this experiment, the specimens were heated by a tungsten heater and melted in an alumina cell (meniscus area, 25 × 35 mm) in an Ar–10% H_2 atmosphere, and held at the given temperatures with an accuracy of ±5 °C. Scattering intensities for liquid Fe–11 at.% C at 1450 °C and Fe–17 at.% C at 1350 °C were accumulated to give 12 000 counts at the main peak position. After the scattering intensities were normalized by the high-angle region method (Gingrich 1943) and corrected for the incoherent scattering (Lonsdale *et al* 1962), the total structure factor $a(K)$ was determined by taking the values for the atomic scattering factor $f(K)$ from the literature (Cromer and Waber 1965) and correcting for anomalous dispersion (Cromer 1965). The statistical error in $a(K)$ was less than 0·9% at the main peak position and 2·2% at large K.

Throughout both the neutron and x-ray diffraction measurements, very little change was observed in the chemical composition of the specimens before and after the experiments.

3. Results and discussion

The total structure factor can be expressed as follows:

$$a(K) = (c_{Fe}f_{Fe} + c_C f_C)^{-2}\,[c_{Fe}^2 f_{Fe}^2 a_{Fe-Fe}(K) + c_C^2 f_C^2 a_{C-C}(K) + 2c_{Fe}c_C f_{Fe} f_C a_{Fe-C}(K)] \tag{1}$$

where c_{Fe}, c_C, f_{Fe} and f_C represent the atomic concentrations and scattering lengths of iron and carbon, and $a_{Fe-Fe}(K)$, $a_{C-C}(K)$ and $a_{Fe-C}(K)$ are the three partial structure factors required to characterize a binary system.

Figure 1 shows the total structure factor $a(K)$ obtained from the results of neutron diffraction. The position of the main peak appears at 2·99 ± 0·02 Å^{-1} for Fe–11 at.% C and at 3·00 ± 0·02 Å^{-1} for Fe–17 at.% C alloys and these values are hardly different from the value for liquid pure iron, 3·00 ± 0·02 Å^{-1} (Morita and Kita 1974). It should be noted, however, that the main peak for these alloys shows remarkable asymmetry, spreading to the higher angle side, and it becomes more broad with increasing carbon concentration.

Figure 2 represents the total structure factor $a(K)$ obtained from the x-ray diffraction measurement. The position of the main peak can be found at 2·97 ± 0·02 Å^{-1} for

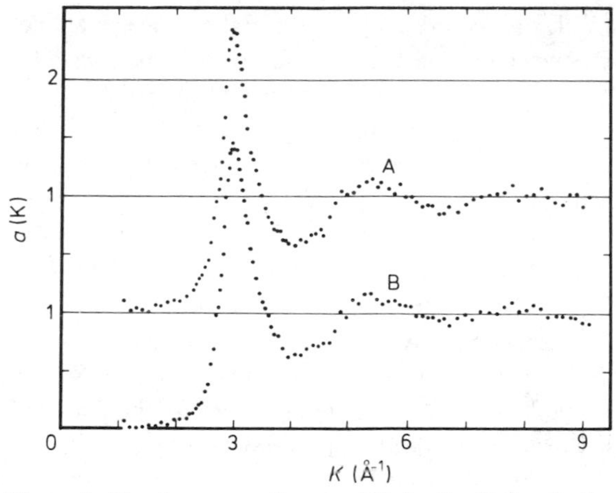

Figure 1. Total structure factors $a(K)$ for liquid Fe–C alloys obtained by neutron diffraction: A, Fe–11 at.%C at 1450 °C; B, Fe–17 at.%C at 1350 °C.

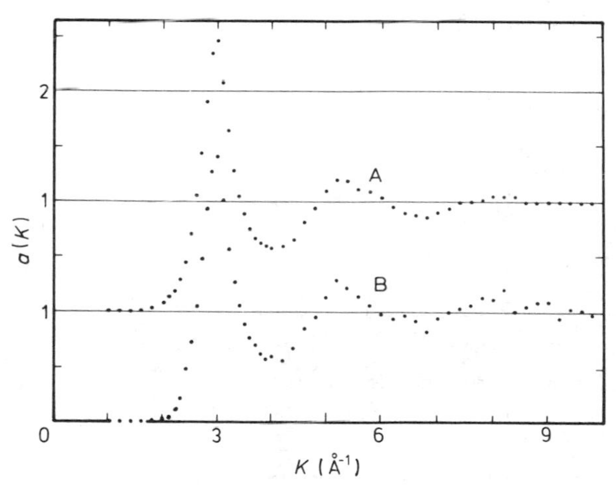

Figure 2. Total structure factors $a(K)$ for liquid Fe–C alloys obtained by x-ray diffraction: A, Fe–11 at.%C at 1450 °C; B, Fe–17 at.%C at 1350 °C.

Fe–11 at.% C and at 2·96 ± 0·02 Å^{-1} for Fe–17 at.% C, and these values again are almost the same as that for liquid pure iron. As shown in the figure, it is obvious that the main peak is practically symmetrical, in contrast with that obtained by neutron diffraction. From the well known fact that the scattering length of carbon for x-rays (i.e., f_C) is quite small, then in such a low carbon concentration region the total structure factor $a(K)$ may be regarded as a partial structure factor $a_{Fe-Fe}(K)$, as is obvious from equation (1).

The partial structure factors $a_{Fe-C}(K)$, deduced from equation (1) by using the results of the neutron and x-ray diffractions, are shown in figure 3. The vertical lines show the uncertainty which is an inevitable consequence of the experimental error in $a(K)$. The peak position is found at 3·3 Å^{-1} for both Fe–11 at.% C and Fe–17 at.% C alloys, and seems not to vary with carbon concentration. It can be seen that the amplitude of oscillation of $a_{Fe-C}(K)$ is damped rapidly on the higher angle side, and this suggests that the interaction between Fe and C appears not to be so strong. If accurate results were available for $a_{Fe-C}(K)$ over the whole significant range of K, it would be possible to invert them to the interatomic potential $\phi_{Fe-C}(r)$. As this is not yet possible, then as a first step, the results may be compared with the hard-sphere model (Ashcroft and Langreth 1967).

Adopting 6·92 g cm^{-3} and 6·97 g cm^{-3} (Lucas 1964) as the density values of Fe–11 at.% C and Fe–17 at.% C at their experimental temperatures, we use 2·25 Å for the hard-sphere diameter, σ, of iron and assume the total packing fraction $\eta = 0{\cdot}45$ to obtain σ of carbon for both alloys to be 1·4 Å. The solid lines in figure 3 show the $a_{Fe-C}(K)$ calculated using the above values for the hard-sphere model, and these follow the obtained results with reasonable accuracy. It is interesting that the pair interaction between the carbon and iron atoms can be described in such a simple model, in spite of

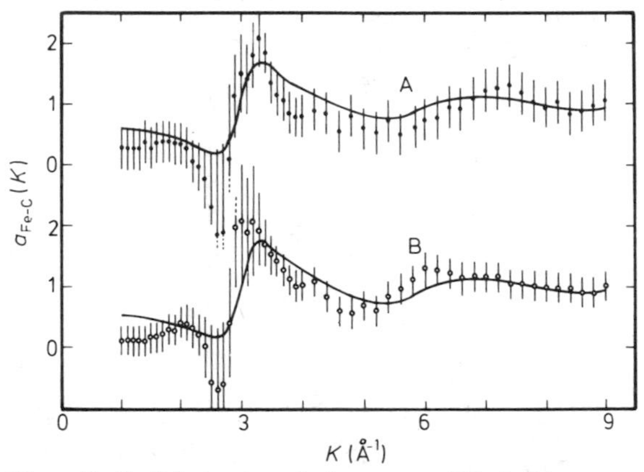

Figure 3. Partial structure factors $a_{Fe-C}(K)$. Solid curves are calculated from the PY hard-sphere model. Curve A, Fe–11 at.% C at 1450 °C; curve B, Fe–17 at.% C at 1350 °C.

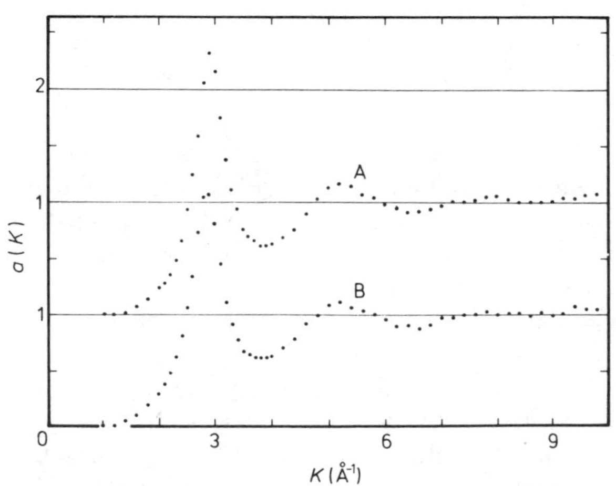

Figure 4. Total structure factors $a(K)$ for liquid Fe–Sn alloys obtained by x-ray diffraction: A, Fe–11 at.% Sn at 1450 °C; B, Fe–17 at.% Sn at 1350 °C.

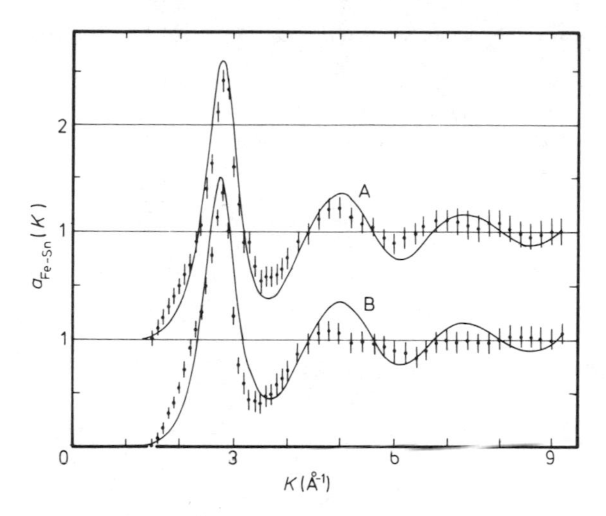

Figure 5. Partial structure factors $a_{Fe-Sn}(K)$. Solid curves are calculated from the PY hard-sphere model. Curve A, Fe–11 at.% Sn at 1450 °C; curve B, Fe–17 at.% Sn at 1350 °C.

the fact that the s- and p-orbitals of carbon mix with the d-orbitals surrounding the iron atoms.

The hard-sphere diameter of carbon is much smaller than that of iron, while the diameter of tin, which belongs to the same IVb group as carbon, is much larger than that of iron. In order to investigate the effect of a solute element with a different diameter on the structure of transition metal alloys, the x-ray diffraction measurements were made for liquid Fe–Sn alloys at the same concentrations of the solute and the same temperatures as those of Fe–C: the results are shown in figure 4. Figure 5 gives the partial structure factor $a_{Fe-Sn}(K)$ deduced from the experimental results, assuming that $a_{Fe-Fe}(K)$ is regarded in the same manner as that evaluated from the Fe–C alloys of equivalent concentrations. It should be recognized that for the Fe–Sn alloys, the obtained results are also consistent with the same hard-sphere model that was applied to the Fe–C alloys, as shown in figure 5 by the solid lines. It is concluded that the simple hard-sphere model may be applicable to these liquid transition metal alloys.

References

Ashcroft N W and Langreth D C 1967 *Phys. Rev.* **156** 685–92
Cromer D T 1965 *Acta Crystallogr.* **18** 17–23
Cromer D T and Waber J T 1965 *Acta Crystallogr.* **18** 104–9
Enderby J E 1968 *Physics of Simple Liquids* ed H N V Temperley *et al* (Amsterdam: North-Holland) pp611–44
Filippov E S and Samarin A M 1968 *Fiziko-Khimicheskie Osnovy Proizvodstva Stali* (Moskva: Nauka) p3 (in Russian)
Freeman A J and Watson R E 1961 *Acta Crystallogr.* **14** 231–4
Gingrich N S 1943 *Rev. Mod. Phys.* **15** 90–110

Lonsdale K *et al* (ed) *International Tables for X-ray Crystallography* 1962 Vol 3 (Birmingham: Kynoch Press) pp247–53
Lucas L D 1964 *Mem. Sci. Rev. Metall.* **61** 97–116
Maier U and Steeb S 1973 *Phys. Condens. Matter* **17** 11–6
Morita Z and Kita Y 1974 *Report of the Japan Society for the Promotion of Science (19th Committee) No.* 9767 (in Japanese)
Vineyard G H 1954 *Phys. Rev.* **96** 93–8
Waseda Y, Tokuda M and Ohtani M 1975 *Tetsu to Hagane (J. Iron and Steel Inst. Japan)* **61** 54–70 (in Japanese)

Transition metals in the amorphous state

J G Wright
School of Mathematics and Physics, University of East Anglia, Norwich NR4 7TJ, England

Abstract. The preparation and resulting structure of amorphous samples, both single elements and alloys, is considered. Structural models are discussed in the light of experimental results and comparisons made with the structure of liquid phases. The measurement of physical properties of amorphous transition metals is reviewed and where possible contrasted with similar work and theoretical expectations for liquid samples. The feasibility of using amorphous samples as quasi-low-temperature liquids is discussed.

1. Introduction

During the past decade the study of amorphous solids has been gathering momentum and has become an important area of condensed media physics. Systems exhibiting an amorphous structure may be prepared in several ways with a large number of materials. The purpose of this paper is to review the present situation in the amorphous state in transition elements and their alloys and, where possible, to compare or contrast the structure and properties of the amorphous and liquid states of the materials. Initially we shall consider the preparation and resulting structure together with interpretations of the structural data. In the second part of the paper, the physical properties, particularly those which may be compared with the liquid state, will be discussed. In general the study of amorphous transition metals has received considerable attention on account of the fundamental interest in ferromagnetism in disordered structures and more recently, on account of the application to the storage of information, in amorphous transition metal/rare-earth, ferromagnetic-bubble devices. It is not the intention of this paper to deal at any length with the magnetic properties.

2. Preparation of the amorphous phase

Essentially the preparation of amorphous systems is effected by preventing the process of crystallization. Although the number of techniques seems at first sight to be large, they may be grouped under:

(i) rapid freezing of the liquid state;
(ii) growth of thin films by the serial deposition of atoms under conditions which inhibit surface mobility.

Alternative (ii) is achieved by (*a*) growth onto very low-temperature substrates, (*b*) the co-deposition of at least two atomic species which are either of significantly different atomic radii or are normally immiscible, or (*c*) by a combination of both.

The process of cooling from the liquid state by firing liquid metal onto low-temperature surfaces gives cooling rates of up to 10^9 K s^{-1} (Davies and Hull 1976). The samples produced are thin foils with thicknesses in the range 10 μm to 2 mm and are therefore quasi-bulk. However, cooling is, generally speaking, non-uniform and consequently some of the slower cooling material at the centre of the sample may well crystallize.

All other techniques produce thin-film samples which should be microscopically homogeneous. As far as the pure transition elements are concerned only very thin amorphous films may be obtained and then only by deposition onto substrates at 4 K (Bennett and Wright 1972a, Leung and Wright 1974a, b). However, once significant impurity levels of the order of 1% are occluded in the form of deliberately introduced impurities such as Si, Ge (Felsch 1969), or in the form of accidentally trapped gases (Davies and Grundy 1971, Grundy *et al* 1975), the amorphous phase may be stabilized to both greater thicknesses and higher temperatures. Nominally pure films which appear to be stable to high temperatures (Tamura and Endo 1969, Ichikawa 1973), probably have low levels of the order of 0·2% (Nagakura *et al* 1963) of filament material such as Mo or W occluded. Certainly 0·2% of occluded gas is sufficient to affect the stability of the amorphous phase (Bennett 1974). However, since such impurities are in very low concentrations the samples may well be suitable for the measurement of structural, electronic and magnetic properties. The fact that impurities may stabilize the amorphous phase is deliberately used to good effect to produce a wide range of amorphous samples of transition elements with such non-metallic additives as P, C and Si (e.g. Ni–P, Fe–C–P etc) and transition metals with noble or rare-earth elements (e.g. Co–Au, Fe–Gd etc). The range of such alloys is large and those systems which have been studied structurally have recently been reviewed by Cargill (1975). The alloy systems may be made by a variety of deposition techniques such as vacuum deposition (Mader *et al* 1967, Taylor 1976), sputtering (Hasegawa 1974), electroless deposition (Simpson and Brambley 1972) and RF decomposition (Stirling *et al* 1966), using substrate temperatures between 2 K and 400 K depending upon the alloy composition and mode of deposition.

3. Structure

3.1. Experimental measurements

3.1.1. Transition elements. Early work on the deposition of metals at low temperatures showed that as far as simple and noble metals were concerned, only crystalline deposits could be formed (e.g. see Yoshida *et al* 1972). At the same time the semi-metals were found to be capable of exhibiting a non-crystalline structure with consequently interesting physical properties such as the superconductivity seen only in the amorphous structure of bismuth (Buckel and Hilsch 1954). More recently work began in the area of amorphous transition elements, motivated by the problem of collective magnetism in disordered systems. To date amorphous, but not necessarily pure, samples of Co, Cr, Fe, Ni, Mn, Pd, V and Ti have been produced and examined by electron diffraction (Fujime 1966, 1967). More accurate structural investigations on purer samples have been carried out on Co, Cr, Fe, Ni and Mn (Leung and Wright 1974a, b, Ichikawa 1973) using electron diffraction, either with energy filtering or rotating sectors for high resolution. These studies have shown that amorphous nickel may not be formed by low-temperature condensation unless some impurity, probably in the form of under 1% of refractory metal is occluded. On the other hand splat-cooled nickel has been shown to be amorphous even when cooled to only 77 K (Davies and Hull 1976). This could be explained either in terms of the impurities occluded in the samples or, if the samples were pure, it would suggest that bulk self-diffusion to crystalline sites is much less easily activated than surface diffusion to such sites during the growth process of a thin film. This latter suggestion may partly be supported by the fact that once amorphous nickel films have been stabilized (Tamura and Endo 1969) by what must be assumed to be a very low level of impurity (Nagakura *et al* 1963), the recrystallization temperature is much higher than that observed in other transition elements such as Fe or Co. Studies of Fe and Co have shown (Suits 1963, Leung and Wright 1974a, b, Lazarev *et al* 1961) that the stabilization of an amorphous structure is a thickness-dependent property. This may be because of poor thermal conductivity through the film as the sample grows or more likely because of surface energy effects.

Electron diffraction studies (Ichikawa 1973, Leung and Wright 1974a, b) yield characteristic structure factors for the amorphous elements, the main features of which are a sharp well defined first peak with a height of about 3·5 – considerably higher than in liquid metals – and a second peak which is sharply split, unlike the single second peak seen in liquid metals. Figure 1 shows the structure factors for amorphous iron (the figure is also typical of Co, Cr and Mn (see Leung and Wright 1974b)) together with that for a microcrystalline nickel film. The general level of experimental accuracy is also demonstrated in figure 1 which shows the structure factor for Fe obtained in two entirely different experiments. At the same time the difference between the amorphous and liquid states is demonstrated. It is of considerable importance to note that the very characteristic split second peak observed in amorphous transition elements is seen not only in thin films but also in samples obtained by splat cooling from the liquid state (Davies and Hull 1976).

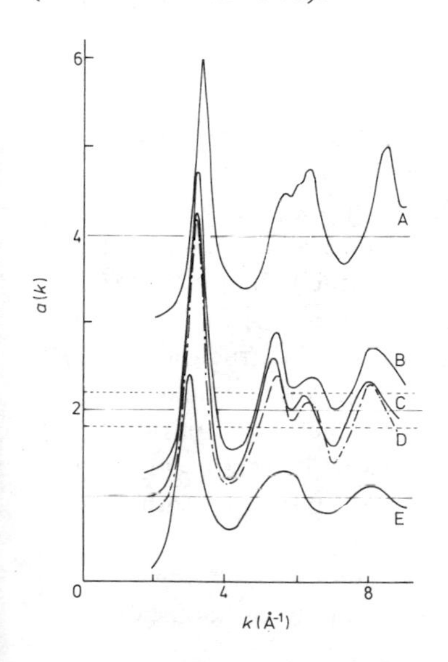

Figure 1. Experimental structure factors for: A, microcrystalline nickel deposited at 4 K; B, amorphous iron at 4 K (Leung and Wright 1974b); C, amorphous iron at 77 K (Ichikawa 1973); D, amorphous iron calculated from a hard-sphere model with surface effects excluded (Ichikawa 1975); and E, liquid iron (Waseda and Ohtani 1974).

The first impression is that amorphous structures have somewhat more order than liquid metals, a fact which is confirmed by inverting the structure factor in the usual way (see Cargill 1975) to obtain the radial distribution function shown in figure 2.

A summary of the basic structural features of amorphous transition metals is given in table 1 where comparison may be made with similar data for liquid transition metals and for noble-metal amorphous films which may be impurity stabilized (Grundy and Davies 1974). The very similar nature of the results for these amorphous metals indicates that a single model might be expected to describe the structure. This is further demonstrated by the close similarities in the first peak of the radial distribution functions for these systems (see figure 2*b*).

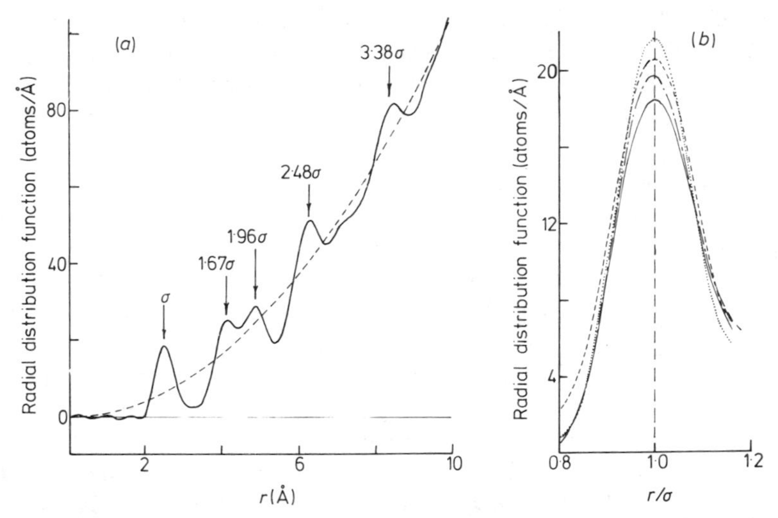

Figure 2. (*a*) Radial distribution function for amorphous manganese. The peak positions in terms of atomic diameter have been obtained from the pair correlation function; (*b*) the first peak of the radial distribution functions for Co (···········), Cr (– – – – – – –), Fe (–·–·–·) and Mn (———).

3.1.2. Alloys. There has been considerably more activity in the field of structural investigations of amorphous alloys than in the case of elements. In particular the structural work has largely been undertaken by x-ray diffraction (Cargill 1975). The large class of alloys consisting of a transition element with non-metallic additives (e.g. Fe–C–P, Co–P or Ni–P) yield a diffraction intensity almost entirely due to the metal (in general the diffracted intensity from the transition element is in excess of 75% of the total). Thus the derived structure factor is essentially a single-element structure factor. It would appear from figure 3 that the similarity between the alloy structure factors for Ni–P and those for pure elements may be attributed to the non-metallic impurities taking up quasi-interstitial positions. The transition element structure factor only begins to suffer significant deviations from that expected for the pure element once substantial levels of impurities have been reached.

The data (Logan 1975) for amorphous Fe–P in table 1 demonstrate the slow expansion of the Fe–Fe distance together with the fact that the ratio of the major

Table 1. Structural data for both amorphous and liquid metals with comparisons between a random tetrahedral model for the amorphous case and a very random model for the liquid case. Amorphous data from †Leung and Wright (1974b), ‡Davies and Grundy (1972), and Grundy and Davies (1974), amorphous alloy data from Logan (1975) and liquid metal data from Waseda and Tamaki (1975) and Y Waseda (1976 private communication). For the purposes of comparison r_2 and r_3 for the liquid system are assumed to be the single peak referred to as r_{2+3}.

	Crystalline interatomic separation	Amorphous						Liquid				
		r_1(Å)	r_2/r_1	r_3/r_1	r_4/r_1	r_5/r_1	N	r_1(Å)	r_{2+3}/r_1	r_4/r_1	r_5/r_1	N
†Co	2·50	2·48	1·69	1·93	2·49	3·35	11·4	2·56	1·87	2·73	3·61	11·3
†Cr	2·49	2·55	1·66	1·91	2·49	3·36	11·9	2·58	1·83	2·68	3·51	11·2
†Fe	2·53	2·46	1·67	1·96	2·50	3·41	11·3	2·58	1·86	2·73	3·55	10·8
†Mn	2·61	2·47	1·67	1·96	2·48	3·38	10·7	2·67	1·85	2·72	3·60	10·9
‡Ni	2·49	2·55	1·66	1·90	2·56	3·46		2·53	1·81	2·71	3·57	11·7
‡Au	2·87	2·89	1·65	1·90	2·55	3·46		2·80	1·92			
‡Ag	2·88	2·90	1·65	1·90	2·55	3·46		2·83	1·92			
Random tetrahedral model			1·67	2·00	2·52	3·34						
Very random model									1·87			
$Fe_{85.5}P_{14.5}$		2·599	1·68	1·95								
$Fe_{84}P_{16}$		2·606	1·68	1·95								
$Fe_{81.6}P_{18.4}$		2·613	1·68	1·95								
$Fe_{78.6}P_{21.6}$		2·618	1·68	1·95								

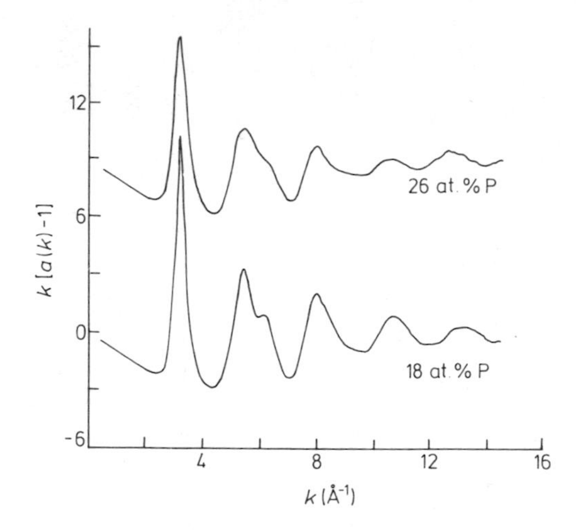

Figure 3. The function $F(k) = k[a(k) - 1]$ for two compositions of Ni–P showing the disappearance of the splitting of the second peak as the phosphorous concentration is increased (Chi and Cargill 1976).

interatomic separations to the first Fe–Fe distance remains virtually constant as though the iron structure is simply being slowly expanded. The quasi-interstitial nature of the phosphorus position is in agreement with Mössbauer studies (Logan and Sun 1976).

The class of alloy having two or more components contributing to the measured structure factor is essentially dealt with in the same manner as liquid metal alloys. In the case of liquid metals the partial structure factors may in some cases be obtained by using neutron diffraction and isotopic substitution. The use of neutron diffraction with amorphous samples has so far been limited mainly on account of the lack of sufficient sample volume and certainly no experiments using isotopic substitution have so far been performed except for very thin cobalt films prepared for Mössbauer studies (Kwan and Hoffmann 1974). The neutron experiments performed to date (Rhyne *et al* 1972, Bletry and Sadoc 1974) have been aimed at obtaining magnetic rather than structural data. At present the best approach to the problem is to obtain a radial distribution function and attempt to correlate the observed peaks with known interatomic separations (Cargill and Kirkpatrick 1976). In some cases an inhomogeneity is observed between the in-plane and normally observed results as shown in figure 4 – a fact which may be explained by the

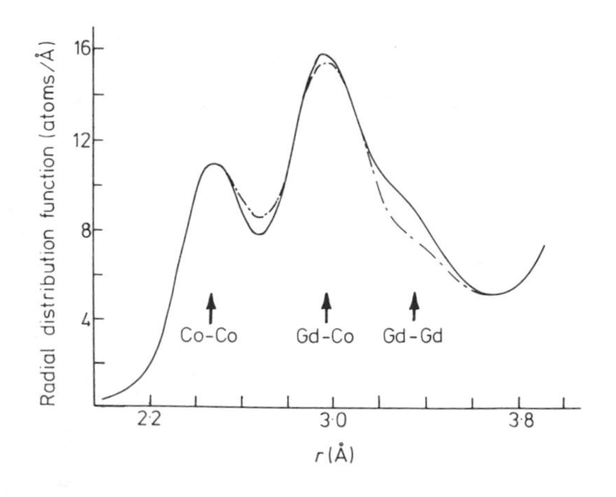

Figure 4. First peak in the radial distribution function for Gd–Co obtained from x-ray diffraction experiments (Cargill and Kirkpatrick 1976) showing the structural anisotropy between the in-plane (_._._) and perpendicular (——) directions in a thin-film sample.

deposition mechanisms (P Chaudhari 1975 private communication) leading to a fibrous deposit and pair ordering (Cargill 1975).

3.2 Interpretation of the structure

In dealing with the structure of liquid metals, experimental results are normally compared with an $a(k)$ curve derived from the Percus–Yevick or Born–Green theories. Although these give tolerably good results in certain cases for liquid metals they never exhibit a splitting of the second peak in $a(k)$ such as is observed in the case of the amorphous elements described in the previous section. An alternative approach developed over the last six years, which in fact sprang from the original Bernal (1964) work on liquid models, is to build hard-sphere models using computer simulation. The early work concentrated on comparisons between observed and calculated radial distribution functions but more recently comparisons have been between $a(k)$ curves. The diameter of the hard spheres is normally the interatomic separation (whereas in Percus–Yevick the hard sphere is typically 10% smaller than the interatomic separation and the packing fraction is of the order of 0·68 compared with the 0·45 common to many liquid metals – which in fact means that the atomic density is much the same in both systems).

The type of model has varied from the well packed system developed by Bernal and extended by Finney (1970) which could be considered as representative of a quenched liquid metal, to systems built by serial deposition (Bennett 1972) and more recently including some interatomic potential which is used to obtain atomic positions consistent with an energy minimum for the system as a whole (Heimendahl 1975).

The serial deposition mechanism which is appropriate to the usual growth mode of amorphous samples may be of two basic types: (*a*) global models where each atom is located at the vacant site closest to the original centre of the cluster; and (*b*) local growth where the next site to be filled is chosen at random. In both systems a site is such that it forms a regular tetrahedron with three already full sites. In (*b*) the growth is uneven and leads to voids but this may be considered physically more reasonable than (*a*) which would require such high surface mobility of atoms that crystallization rather than the retention of the amorphous structure would intuitively be expected. Figure 5 shows a local model constructed from 10 tetrahedral seed clusters on a flat plane (to

Figure 5. Model constructed from ten randomly orientated tetrahedral seeds sited on a flat plane.

simulate a substrate) whilst figure 6, where $l = 0$, shows a structure factor typical of such models calculated from the Debye equation:

$$a(k) = (2/N) \sum_{ij} \sin(sr_{ij})/sr_{ij}.$$

The split second peak is well defined and is a feature of all models in which construction is based upon the principle of building with tetrahedra having an atom at each corner. The peak positions in the pair correlation function obtained from such models may be compared with experiment and the values obtained (Wright 1976) are given in table 1. The characteristic separation of 1·67 atomic diameters seen for the first part of the split peak (Leung and Wright 1974b) may be reconciled with the distance DE in figure 6(*b*).

If the tetrahedra are allowed to become irregular by allowing normal bond lengths to take values of up to $l\sigma$ (where σ is the atomic diameter and l a constant), the peak splitting in $a(k)$ diminishes until by the time $l = 1.4$, the splitting has virtually disappeared, as shown in figure 6. At the same time the density increases and the two peaks in $g(r)$ at $1{\cdot}67\sigma$ and 2σ diminish whilst a new maximum begins to appear at $1{\cdot}9\sigma$ which is in good agreement with the liquid metal case (see table 1).

A further refinement to the model building is to use only the central portion of a large cluster, thereby removing the effects of diminishing density at the cluster edge (Ichikawa 1975, Cargill and Kirkpatrick 1976). The effect is to increase peak heights to a point where agreement between experiment and calculation is extremely good, as

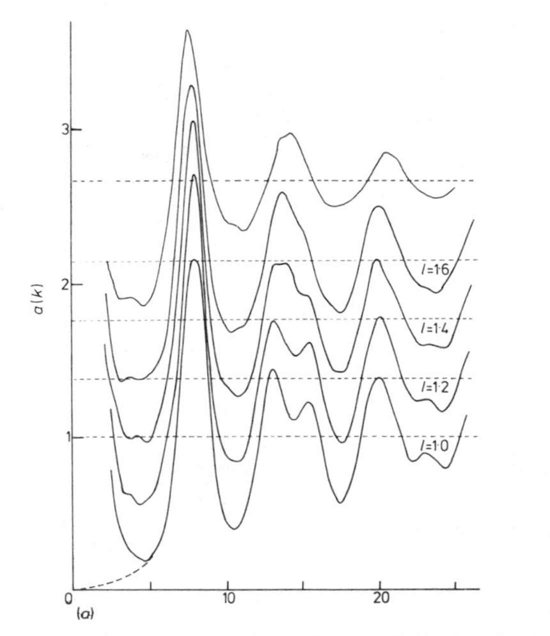

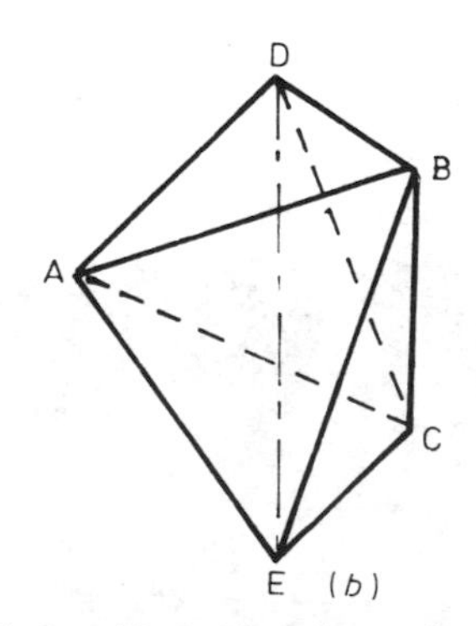

Figure 6. (*a*) Structure factor for a model built using tetrahedral packing with various values of maximum allowed interatomic separation $l\sigma$. Bottom curves are for various values of l, and top curve shows the structure factor for a very random model of 500 atoms. (*b*) Atomic positions ABCD in a regular tetrahedral cluster showing the characteristic separation $1{\cdot}67\sigma$ (DE) observed in experimentally obtained pair correlation functions of amorphous elements when a fifth atom is added at E so that ABCE is also a regular tetrahedron.

shown in figure 1. It is often argued that samples might be crystalline but this agreement shows this to be unlikely, especially as the experimental curve for nickel in figure 1, which is different from the amorphous phases, is in good agreement with a calculation for strained microcrystalline FCC samples of 125 atoms per crystallite (Cargill 1970). In particular, diminution of crystallite size leads to a loss of splitting of the second peak but never leads to the first part being larger than the second.

The introduction of a Morse potential (Heimendahl 1975) with a consequent relaxation of a hard-sphere assembly leads to small improvements between the measured and calculated pair correlation functions (figure 7). However, the potential used was much softer than that deduced for liquid transition metals (figure 8). A harder potential used by Weire *et al* (1971) also gives a broadening of the first peak by about 10% which is in tolerable agreement with experiment (see peak half widths in figure 2*b*).

More random models built by allowing random deposition onto a substrate with only sufficient mobility to move $0{\cdot}2\sigma$ produced constructions with $\eta - 0{\cdot}54$ and a second peak in the pair correlation function at 1·87 (Leung *et al* 1976). The structure factor

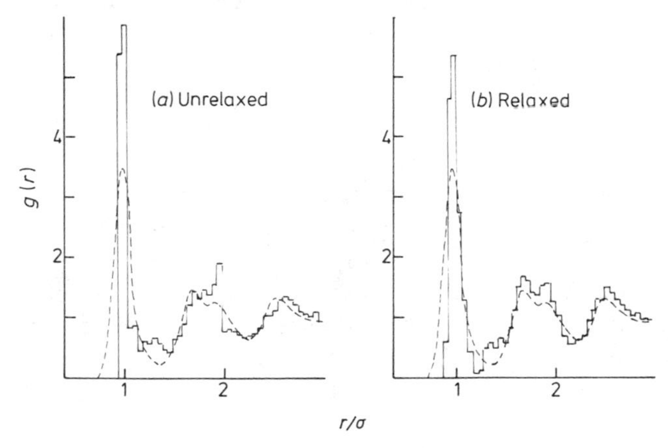

Figure 7. The effect on $g(r)$ of using a Morse pair potential to modify, by energy minimization, a computer-simulated, hard-sphere model (Heimendahl 1975).

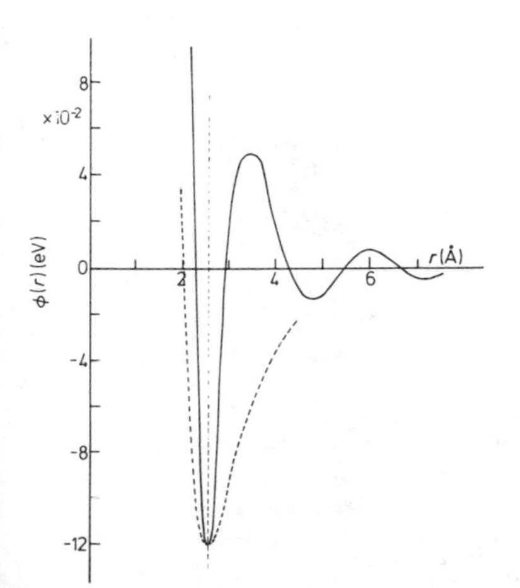

Figure 8. Pair potential curves for liquid (Waseda and Ohtani 1975). The Morse potential, normalized to give the same binding energy as the experimental pair potentials used by Heimendahl (1975), is shown by the broken curve.

for such a 500 atom model is shown in figure 6(*a*). Of particular importance is the absence of splitting in the second peak. Apart from demonstrating differences between liquid metals and amorphous samples (essentially, greater disorder is allowed by way of the smaller hard-sphere diameter for the liquid), the model also shows that some clustering is required to explain the observed structure factor for amorphous samples. Moreover, detailed work (Leung *et al* 1976) shows that the largest degree of order required is the basic tetrahedron with no long-range order being present. The required clustering is probably provided by the heat of condensation, which for transition metals would cause a local temperature of several hundred degrees to exist for several atomic vibrations. During this time the local configuration would become similar to that of a liquid, which now gives a physical reason for the structure of amorphous films and splat-cooled liquids to be so similar and identifies the structure of amorphous films with that of the liquid state.

In using the process of distortions of the regular tetrahedra it is pertinent to consider the interatomic potentials derived by Waseda and Ohtani (1975) from structure data of liquid transition elements using the Born–Green approximation. These potentials may not be exactly right for the amorphous system but should give a fair indication of the potential to be expected (see figure 8). It would seem unlikely from this curve, which is typical for transition elements, that interatomic separations between nearest neighbours are likely to exceed the mean interatomic separation by more than 20%, because of the considerable elastic strain which would then be placed on the system. Indeed the sharpness of the curve would suggest that the hard-sphere model with nearly regular tetrahedra is to be expected at low temperatures, especially since a nearest-neighbour equilibrium can be expected, due to the short period when the heat of condensation is liberated from each arriving atom. The potential curve also shows why, in the case of liquid metals, the hard-sphere diameter would diminish by about 10% with a consequent loss of some order in the structure, that is, the effective tetrahedral lengths for the liquid structure may vary by +20% and −10% with a loss of splitting in the second peak.

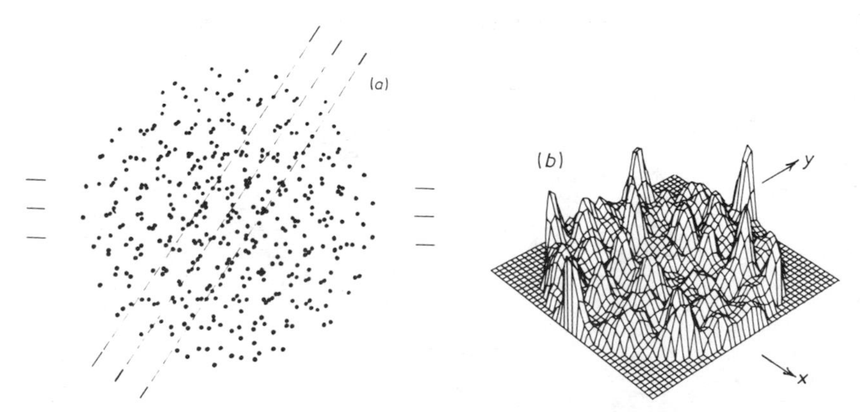

Figure 9. (*a*) Projection of 400 central units of a dense random-packed model onto a plane showing two strong scattering directions for $|k| = 7{\cdot}7\sigma^{-1}$. (*b*) Dependence of the structure factor $a(\mathbf{k})$ on the direction of $\mathbf{k}$ for 890 central units of the Bennett (1972) model evaluated for $|k| = 7{\cdot}7\sigma^{-1}$; spherical angles are related to x and y by $\theta = (\pi/2)(x^2 + y^2)^{1/2}$ and $\phi = \tan^{-1} x/y$ (from Alben *et al* 1976).

It is of considerable importance to note that once models exist of proven reliability, they may be used to deduce not only pair-correlation functions but also higher-order correlation functions which are of considerable importance in such areas as ferromagnetism.

One further point requires discussion. Electron microscopy studies of amorphous films using high-resolution, dark-field illumination show inhomogeneities in intensity as if there were singular scattering regions of very small size. This has been interpreted by some workers to indicate a microcrystalline array. However, it has been very clearly shown (Alben *et al* 1976) that these observations are quite consistent with a random system of the type developed in model building. Figure 9(*a*) shows a projection of atomic positions for a random model onto a plane. Careful examination shows regularly spaced pseudo-planes in the system when viewed along certain directions, with a consequence that inhomogeneous scattering will take place as the direction of the incident beam is rotated round the sample. By evaluating the scattering intensity

$$a(\mathrm{k}) = (1/N) \sum_{ij} |f(k)|^2 \exp[\mathrm{i}(\mathrm{s}r_{ij})]$$

at a constant value of $|k|$ but for different directions of the incident beam, this inhomogeneity may quickly be detected as shown in figure 9(*b*). The non-uniform nature of the scattered radiation from different areas of an amorphous sample is consistent with the random fluctuations in the amorphous structure and explains the nature of the high-resolution electron micrographs.

The present state of understanding is that amorphous transition elements are not microcrystalline, are consistent with frozen liquids and are quite well represented by hard-sphere models.

The structural problems associated with alloy systems are more severe. In the case of the predominant single species scatterer (e.g. Fe–C–P or Ni–P etc) the partial structure factors have not been resolved. However, as a first approximation it appears that such alloys are closely related to the pure elements with some distortion to accommodate the impurities, a view supported by the model-building work of Sadoc *et al* (1973) and Cargill (1975). For the multi-component metal alloys the structural problems are similar to those with liquid metals but are complicated by the anisotropy in the growth process leading to an anisotropic structure and anisotropic physical properties, in particular the special magnetic properties of alloys such as Gd–Fe and Gd–Co used as magnetic-bubble materials, with similar states of the art in both fields.

4. Physical properties

Some of the physical properties of amorphous systems are very different from those for the crystalline phase even where the crystallite size is less than 20 Å, a fact which may be used as further evidence that the amorphous phase is distinct and non-crystalline. Figure 10 indicates some of the main areas of interest in both amorphous and liquid systems and indicates those areas where an overlap of interests exists. Apart from the structure, the most easily compared area is that of electron transport and allied properties and it is unfortunate that:

(i) the investigation of such properties in liquid transition elements is experimentally

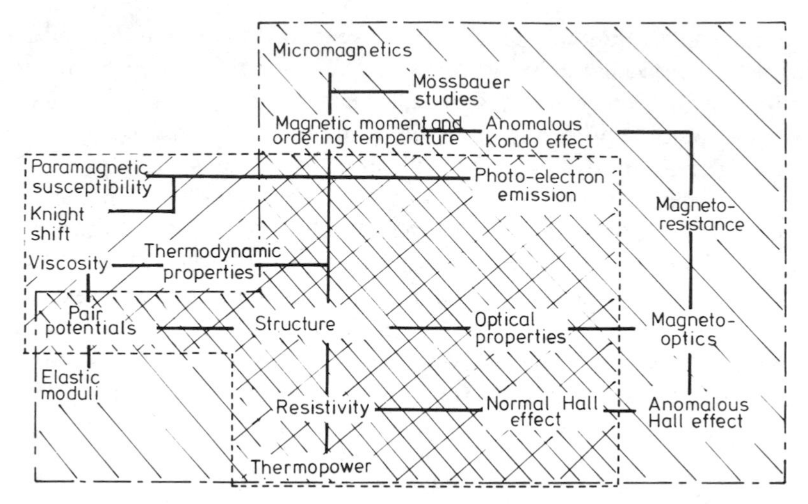

Figure 10. The major properties of interest in liquid transition metal work (hatched area inside broken line) and amorphous transition metal work (hatched area inside chain line), indicating areas where overlap of present and future interests lie.

difficult and in many instances only results obtained by extrapolation from alloy data are available;

(ii) since magnetism has been the driving force behind much of the research in amorphous systems only scant information about the electron-transport properties is so far available.

4.1. *Resistivity*

Figure 11 is typical of an annealing curve for a pure amorphous transition element indicating a resistivity change on crystallization of about 75% (see Wright 1976). Resistivity measurements made to date have been hampered by both an inexact

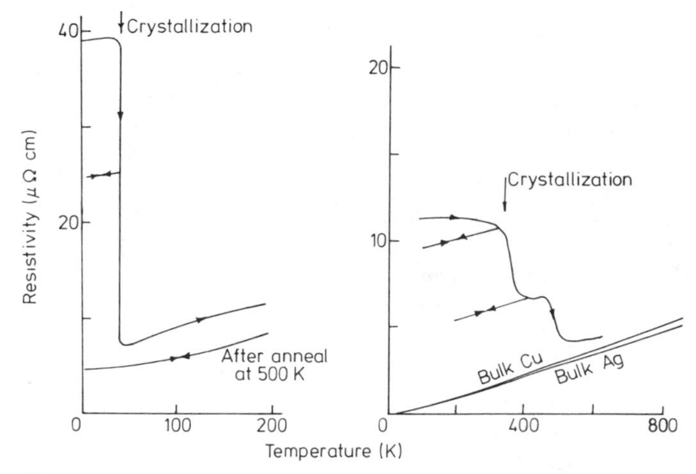

Figure 11. Resistivity curves during annealing for (*a*) Co (Bennett and Wright 1972a), (*b*) amorphous Au–Cu (Mader *et al* 1967).

knowledge of thin-film thicknesses and by possible thin-film effects where either the sample has to be restricted to thicknesses less than 100 Å (as in Fe) or has a possible structure gradient (as in Co). It could be argued that the technique of Evans *et al* (1971) for calculating resistivities of liquid transition metals could well be applied to amorphous elements, provided that suitable interatomic potentials and structure factors may be obtained. By calculating appropriate phase shifts for the amorphous systems and using measured structure factors (Leung and Wright 1974a, b), Y Waseda (1976 private communication) has performed such calculations, the results of which are shown in table 2 together with calculated and measured values for the liquid case. The level of agreement is tolerably good for Fe, Ni and Co but in poor agreement for Cr and Mn where the experimental values are anomalously high. At the present time it seems more likely that a revision of the theory is required rather than that the experimental values are incorrect by a factor of 3, since there are also substantial differences between calculated and measured resistivities for liquid metals.

Typical measurements on alloys are those due to Mader *et al* (1967) which are reproduced in figure 11(*b*).

Table 2. Resistivities in $\mu\Omega$ cm of liquid and amorphous transition elements. The measured resistivities for amorphous chromium and manganese are anomalously high but are in agreement with other experiments (Bennett 1974, Leung 1975). The calculated values for the amorphous case have been obtained using phase shifts obtained from the measured pair correlation functions (Y Waseda 1976 private communication) whilst the values in parentheses have been obtained by assuming liquid phase shifts.

Element	Liquid		Amorphous	
	Calculated	Experiment	Calculated	Experiment
Co	83	100	37 (26)	50
Cr	120	100	67 (84)	324
Fe	182	140	47 (115)	150
Mn	280	180	108 (115)	418
Ni	54	85	25 (26)	100

4.2. *Hall effect*

The measured value of the Hall coefficient in non-transition metals has been found to be similar in the liquid and amorphous phases (see Bergmann 1976). In the case of transition metals, measurements in the amorphous phase are hampered by ferromagnetism which leads to a two-part Hall effect:

$$p = R_0 B + R_1 M.$$

It is often assumed that R_0, the normal Hall coefficient and the part of the expression which persists in the ferromagnets above the Curie temperature, is the quantity which may be compared with the Hall coefficients of the non-magnetic materials. So far, the Hall effect has been measured only in amorphous Co, Fe and NiAu. The results of the normal part, R_0, of these measurements are shown in table 3. In the case of liquid transition elements the only values for cobalt and iron are from extrapolation of alloys

Table 3. Normal Hall coefficients in m³/As × 10¹¹.

Element	Bulk crystalline	Thin films		Liquid metal
		Crystalline	Amorphous	
Co	+13	6 ± 6	6 ± 6	+12†
Fe	+25	+2	+60	+40†
‡ Ni–Au	−14		−11·9 §	−11·5

† The liquid metal values are obtained by extrapolation from alloy measurements (Busch *et al* 1972) whilst the bulk crystalline values are taken from Herd (1972).
‡ From Bergmann (1976). The thin-film values for Co are from Whyman and Aldridge (1974) and those for iron from Aldridge and Raeburn (1976).
§ (−12·6 extrapolation to liquid temperature 1400 K).

with germanium (Busch *et al* 1972) but it is of importance to note that amorphous samples of FeGe may be made (Endoh *et al* 1976) and such samples could well be used for Hall measurements to give comparisons with the liquid alloys. In general the correlation between the liquid and amorphous states is high and it is important and encouraging to note that after crystallization the films exhibit Hall coefficients expected for bulk crystalline material.

4.3. Other measurements

To date very few measurements have been made on properties other than those directly associated with ferromagnetism. Briefly they include:

(i) Positron angular correlation and lifetime studies in amorphous and crystalline Ni–P which yield no change at the phase change (R N West 1976 private communication).
(ii) Measurement of evolution of excess energy on crystallization (~2·5% of the total heat of condensation) using differential thermal analysis (see Sharma *et al* 1975) shows that the amorphous phase is (*a*) energetically distinct from and (*b*) in a metastable state of higher energy than, the crystalline state.
(iii) Optical measurements on amorphous Co (Bennett 1974) at 546 nm which give a change of less than 1% in both n and k on crystallization.
(iv) Photoemission studies on amorphous Ni (Banninger *et al* 1970) indicating a significant shift in the distribution of electrons in the spin-up and spin-down configurations compared with the crystalline case.
(v) The existence of an anomalous Kondo effect in Mn and Cr (Leung *et al* 1974) which has been interpreted (Slechta 1975) to show that 'free' spins may exist in a structurally disordered single-element material. This, however, raises interesting questions in the case of chromium where the magnetic moment is generally thought to exist as charge density waves composed of almost entirely itinerant electrons. The problem may possibly be resolved by an alternative explanation (Cochrane *et al* 1975), which is based upon structural disorder and not Kondo scattering.
(vi) Magnetoresistance measurements in amorphous cobalt (Bennett and Wright 1972b) indicating that the normal magnetocrystalline anisotropy is no longer present. Since this anisotropy is due to the first and second coordination shells, these observations add weight to the arguments against the existence of even short-range crystalline order in amorphous systems.

(vii) Magnetic moment studies both direct and indirect, which show substantial reductions in the amorphous moment compared with the crystalline value in Fe and Ni but not for cobalt (see Wright 1976). In particular the reduction in iron is only observed if the level of impurities is less than about 2% (Felsch 1969). This probably indicates the onset of mixed magnetism at a critical number of nearest neighbours, a view which is supported by recent experiments on strained amorphous iron (Bostanjoglo and Giese 1976). On the other hand the reduction in nickel may be due to a change in density of states as described for liquid transition metals by Bennemann (1976).

4.4. Theoretical work

To date almost all theoretical contributions have been directed at understanding the structure of the systems or at understanding the magnetic properties (e.g. see Ferchmin and Kobe 1974). Small bridges are slowly being forged between areas of interest in both the liquid and amorphous states such as the application of theoretical resistivity expressions as described in §4·1 and the recent work on paramagnetism in the liquid transition elements and magnetic moments in the amorphous systems (Bennemann 1976). Apart from the work on structure it would seem intuitively reasonable to expect that a great deal of the work with density of states and electron-transport properties should be equally applicable to both the amorphous and liquid states.

5. Conclusions

The work of amorphous transition elements and their alloys is now well established. In the case of the elements and some alloys the structure is understood in terms of hard-sphere models based on tetrahedral packing, models which are also consistent with frozen liquids. In these areas, where properties may be compared between the liquid and amorphous phases, there are close relationships and consequently there are very strong arguments for examining a wide range of physical properties in amorphous elements and alloys with a view to expanding our understanding of the structurally disordered state. Stabilization of such samples may be needed by impurity occlusion of the order of 1% or less. This should preferably be a refractory metal rather than a gas in order to give the maximum temperature stability of the structure. It is felt that the theoretical approaches to electron-transport-like properties in liquids should in most cases be equally valid for amorphous systems.

Acknowledgments

The author would like to thank the Science Research Council for funding much of the research in his laboratory which has been referred to in this review. He would also like to thank Dr T Ichikawa for providing numerical data from his structure studies, Dr G S Cargill for supplying figure 9(*b*), Dr Y Waseda for undertaking resistivity calculations, Mr J J Quinn for computational work and Dr R V Aldridge and Professor N E Cusack for useful discussions.

References

Alben R, Cargill G S III and Wentzel J 1976 *Phys. Rev.* **B13** 835–42
Aldridge R V and Raeburn S J 1976 *Phys. Lett.* **A56** 211–2

Banninger U, Busch M, Campagna M and Siegmann H C 1970 *Phys. Rev. Lett.* **25** 585–7
Bennemann K H 1976 *J. Phys. F: Metal Phys.* **6** 43–58
Bennett C H 1972 *J. Appl. Phys.* **43** 2727–34
Bennett M R 1974 *PhD Thesis* University of East Anglia
Bennett M R and Wright J G 1972a *Phys. Stat. Solidi* **A13** 135–44
—— 1972b *Phys. Lett.* **A38** 419–20
Bergmann G 1976 *Solid St. Commun.* **18** 897–9
Bernal J D 1964 *Proc. R. Soc.* **A25** 1752–3
Bletry J and Sadoc J F 1974 *Phys. Rev. Lett.* **33** 172–5
Bostanjoglo O and Giese W 1976 *Phys. Stat. Solidi* **A34** K1–4
Buckel W and Hilsch R 1954 *Z. Phys.* **138** 109–20
Busch G, Güntherodt H J, Künzi H U and Meier H A 1972 *Proc. 2nd Int. Conf. on Liquid Metals, Tokyo* (London: Taylor and Francis) 263–76
Cargill G S III 1970 *J. Appl. Phys.* **41** 12–29
—— 1975 *Solid St. Phys.* **30** 227–320 (New York: Academic Press)
Cargill G S III and Kirkpatrick S 1976 *Proc. Int. Conf. on Structure and Excitations of Amorphous Solids, Williamsburg, Va* (New York: Am. Inst. Phys.) to be published
Chi G C and Cargill G S III 1976 *Proc. Int. Conf. on Structure and Excitations of Amorphous Solids, Williamsburg, Va* (New York: Am. Inst. Phys.) to be published
Cochrane R W, Harris R, Ström-Olson J O and Zuckermann M J 1975 *Phys. Rev. Lett.* **35** 676–9
Davies H A and Hull J B 1976 *J. Mater. Sci.* **11** 215–23
Davies L B and Grundy P J 1971 *Phys. Stat. Solidi* **A8** 189–97
Endoh Y, Yamada K, Beille J, Bloch D, Endo H, Tamura K and Fakushima J 1976 *Solid St. Commun.* **18** 735–7
Evans R, Greenwood D A and Lloyd P 1971 *Phys. Lett.* **A35** 57–8
Felsch W 1969 *Z. Phys.* **219** 280–99
Ferchmin A R and Kobe S 1974 *Bibliothek der Technischen Universität Dresden, Bibliographische Arbeiten* No. 9
Finney J L 1970 *Proc. R. Soc.* **A319** 479–93
Fujime S 1966 *Jap. J. Appl. Phys.* **5** 1029–35
—— 1967 *Jap. J. Appl. Phys.* **6** 305–10
Grundy P J and Davies L B 1974 *J. Phys. F: Metal Phys.* **4** L111–4
Grundy P J, Nandra S S and Ali A 1975 *IEEE Trans. Magn.* **11** 1329–31
Hasegawa R 1974 *J. Appl. Phys.* **45** 3109–12
Heimendahl L V 1975 *J. Phys. F: Metal Phys.* **5** L141–5
Herd C M 1972 *The Hall Effect in Metals and Alloys* (New York: Plenum)
Ichikawa T 1973 *Phys. Stat. Solidi* **A19** 707–16
—— 1975 *Phys. Stat. Solidi* **A29** 293–302
Kwan M M L and Hoffmann R W 1974 *Jap. J. Appl. Phys.* suppl. 2, part 1, 729–32
Lazarev B G, Semenko E E and Sudovtsov A I 1961 *Sov. Phys.–JETP* **13** 75–7
Leung P K 1975 *PhD Thesis* University of East Anglia
Leung P K, Quinn J J and Wright J G 1976 to be published
Leung P K, Slechta J and Wright J G 1974 *J. Phys. F: Metal Phys.* **4** L21–3
Leung P K and Wright J G 1974a *Phil. Mag.* **30** 185–94
—— 1974b *Phil. Mag.* **30** 995–1008
Logan J 1975 *Phys. Stat. Solidi* **A32** 361–8
Logan J and Sun E 1976 *J. Non. Cryst. Solids* **20** 285–98
Mader S, Nowick A S and Widmer H 1967 *Acta Metall.* **15** 203–22
Nagakura S, Kikuchi M, Kaneko Y and Oketani S 1963 *J. Appl. Phys.* **2** 201–5
Rhyne J J, Pickart S J and Alperin H A 1972 *Phys. Rev. Lett.* **29** 1562–4
Sadoc J F, Dixmier J and Guinier A 1973 *J. Non. Cryst. Solids* **12** 46–60
Sharma S K, Geserich H P and Theiner W A 1975 *Phys. Stat. Solidi* **A32** 467–74
Simpson A W and Brambley D R 1972 *Phys. Stat. Solidi* **B49** 685–91
Slechta J 1975 *J. Non. Cryst. Solids* **18** 137–47
Stirling H F, Alexander J H and Joyce R J 1966 *Le Vide No. Special A VI Semin. October* pp80–94

Suits J C 1963 *Phys. Rev.* **131** 588–91
Tamura K and Endo H 1969 *Phys. Lett.* **A20** 52–3
Taylor R 1976 *J. Appl. Phys.* **47** 1164–7
Waseda Y and Ohtani M 1974 *Phys. Stat. Solidi* **A62** 535–46
—— 1975 *Z. Naturf.* **A30** 485–91
Waseda Y and Tamaki S 1975 *Phil. Mag.* **32** 273–81
Weire D, Ashby M F, Logan J and Weins M J 1971 *Acta Metall.* **19** 779–88
Whyman P and Aldridge R V 1974 *J. Phys. F: Metal Phys.* **4** L6–8
Wright J G 1976 *IEEE Trans. Magn.* **12** 95–102
Yoshida N, Natakeyama J, Kiritani M and Fujita F E 1972 *J. Phys. Soc. Japan* **33** 858

A comparison between the structure of amorphous Ni–P and Cu–Zr alloys

Y Waseda, T Masumoto† and S Tamaki‡

X-ray Diffraction Laboratory, The Research Institute of Mineral Dressing and Metallurgy, Tohoku University, Sendai 980, Japan

Abstract. A detailed analysis of the x-ray scattering intensity of two-type metallic amorphous alloys, that is a metal–metalloid system (Ni–P) and a metal–metal system (Cu–Zr), has been carried out and three partial structure factors which are required to characterize the binary alloys have been estimated. The partial-pair distribution functions were also obtained by means of common Fourier analysis. The difference between the structure of both systems is given.

1. Introduction

Structural studies have been carried out on a number of amorphous alloy systems prepared by several methods such as vapour quenching, liquid quenching, electrodeposition and so on (Giessen and Wagner 1972, Cargill 1975, Waseda 1975). Although it is known that almost invariably these alloys involve a metalloid as one of their constituents, exceptions have recently been found in several alloy systems consisting of two metallic elements.

The purpose of this paper is to report on the difference between the structure of an amorphous metal–metalloid system (Ni–P) and that of an amorphous metal–metal system (Cu–Zr).

2. Experimental procedures

The sample preparation, the measurements of the scattered x-ray intensity, operating procedures and correction of the observed intensity were almost identical to those described in our previous work (Waseda and Masumoto 1975, Waseda and Tamaki 1975).

3. Analysis of intensity patterns

The analysis of the measured x-ray intensity was made in the same manner as that described previously (Ramesh and Ramaseshan 1971, Waseda and Tamaki 1975). For convenience, the essential features are given below. The total structure factor $S(Q)$ can be written in the following form using the coherent x-ray scattering intensity

† Permanent address: The Research Institute for Iron, Steel and Other Metals, Tohoku University, Sendai 980, Japan.

‡ Permanent address: Department of Physics, Faculty of Science, Niigata University, Niigata 950-21, Japan.

$I_{eu}^{coh}(Q)$ which is directly obtained by experiment,

$$S(Q) = [I_{eu}^{coh}(Q) - (\langle f^2\rangle - \langle f\rangle^2)] / \langle f\rangle^2$$
$$= w_{ii}S_{ii}(Q) + w_{jj}S_{jj}(Q) + 2w_{ij}S_{ij}(Q), \quad (1)$$

where

$$\langle f^2\rangle = \Sigma c_i f_i^2, \qquad \langle f\rangle = \Sigma c_i f_i, \qquad w_{ij} = c_i c_j f_i f_j / \langle f\rangle^2, \qquad Q = 4\pi \sin\theta/\lambda.$$

The partial structure factor $S_{ij}(Q)$ is defined by the generalized equation,

$$S_{ij}(Q) = 1 + \frac{4\pi\rho_0}{Q}\int_0^\infty r[g_{ij}(r) - 1] \sin(Qr)\,\mathrm{d}r, \quad (2)$$

where $g_{ij}(r)$ is the average distribution of type j atoms found at a radial distance r from the atom i at the origin and ρ_0 is the average number density of atoms. It is well known that when the anomalous scattering of x-rays occurs, the total scattering factor of equation (1) becomes complex in the following form:

$$f_i = f_i^0 + \Delta f_i' + i\Delta f_i'', \qquad f_j = f_j^0 + \Delta f_j' + i\Delta f_j'', \quad (3)$$

where f_i^0 and f_j^0 correspond to the atomic scattering factor for radiation with frequency much higher than any absorption edge, whereas $\Delta f'$ and $\Delta f''$ are the real and imaginary components of the anomalous dispersion term. Using this relation, the Laue monotonic scattering term, $(\langle f^2\rangle - \langle f\rangle^2)$, and the weighting factor w_{ij} in equation (1) may be written as follows,

$$(\langle f^2\rangle - \langle f\rangle^2) = c_i(1 - c_i) f_i^* f_i + c_j(1 - c_j) f_j^* f_j - 2c_i c_j[(f_i^0 + \Delta f_i')(f_j^0 + \Delta f_j') + \Delta f_i''\Delta f_j''], \quad (4)$$

$$w_{ii} = c_i^2 f_i^* f_i / \langle f\rangle^2 \qquad w_{jj} = c_j^2 f_j^* f_j / \langle f\rangle^2, \quad (5)$$

$$w_{ij} = c_i c_j[(f_i^0 + \Delta f_i')(f_j^0 + \Delta f_j') + \Delta f_i''\Delta f_j''] / \langle f\rangle^2. \quad (6)$$

The anomalous dispersion terms $\Delta f'$ and $\Delta f''$ are dependent on the wavelength of the incident radiation. Consequently measurement with several kinds of radiation gives additional items of information concerning the structure of a binary alloy.

4. Results and discussion

All data were obtained with the combination of Mo, Cu, Co and Fe radiation. The anomalous dispersion terms are listed in table 1 (from Cromer 1965, 1976). In this analysis, the significant problems are the normalization and numerical solution of equation (1) but since they have already been discussed in our previous work, they need no description here.

Figure 1 shows the partial structure factors $S_{ij}(Q)$ for amorphous Ni–P and Cu–Zr alloys evaluated from the measured intensity data (Ni–17, 20, 23 and 27 at.% P alloys; Cu–35, 40, 43 and 45 at.% Zr alloys). The vertical lines in this figure denote the difference due both to alloy concentration and to the residual experimental un-

Table 1. The anomalous dispersion terms given by Cromer (1965 and 1976) for the four radiations which were used.

Radiation	Mo		Cu		Co		Fe	
	$\Delta f'$	$\Delta f''$	$\Delta f'$	$\Delta f''$	$\Delta f'$	$\Delta f''$	$\Delta f'$	$\Delta f''$
Ni	0·37	1·20	−3·20	0·67	−1·62	0·67	−1·48	0·93
P	0·11	0·12	0·27	0·46	0·33	0·57	0·32	0·68
Cu	0·36	1·36	−2·15	0·75	−1·34	0·77	−1·31	1·04
Zr	−3·14	0·78	−0·62	2·42	−0·13	2·90	0·54	3·37

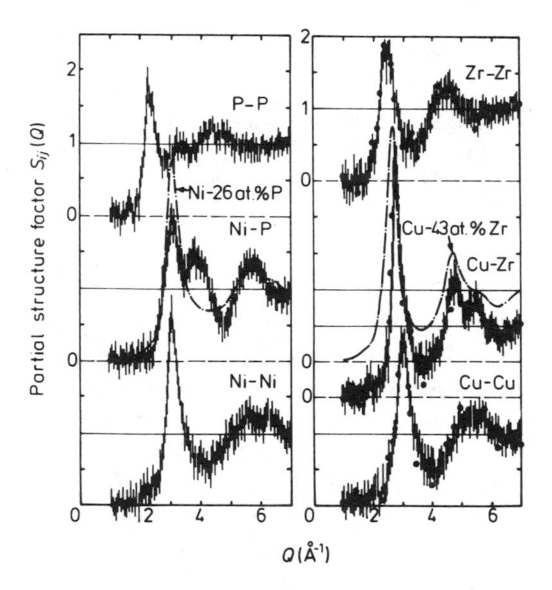

Figure 1. Three partial structure factors derived from measured intensity data with four different alloy compositions. Full circles denote the values derived from the combination of the data of neutron diffraction (Kudo *et al* 1975) with present x-ray diffraction. Chain curves are the observed total structure factors of amorphous Ni–26 at.% P and Cu–43 at.% Zr alloys.

certainty. Because the lines are of the order of ±0·3, the partial structure factors are not sensitively dependent on alloy concentration in these amorphous alloys. The full circles in the Cu–Zr system indicate the partial structure factors derived from the combination of the data from neutron diffraction measurements for amorphous Cu–43 at.% Zr alloy (Kudo *et al* 1975) with our x-ray diffraction measurements. These results agree with those obtained using x-ray anomalous scattering techniques. As shown in figure 1, the obtained values of $S_{ij}(Q)$ are dispersed in places but the general form of the structure factors is a smooth function without sharp changes of slope near the first peak and this seems to be realistic on the basis of the results for various disordered metals and alloys. Consequently, the average values of $S_{ij}(Q)$ which are estimated so as to fit the smooth curve, are used in the calculation of the pair distribution function $g_{ij}(r)$ in equation (2). The obtained partial-pair distribution functions are shown in figure 2. It is well known that analysis by means of x-ray anomalous scattering techniques is possible only up to about $Q = 7{\cdot}0$ Å^{-1} and the spurious ripples in $g_{ij}(r)$ due to the finite termination of $S_{ij}(Q)$ cannot be removed even though quite accurate data of $S_{ij}(Q)$ are used. However, it is relatively easy to trace the positions where the spurious ripples are significant. According to Finback (1949) and Sugawara (1951), they appear

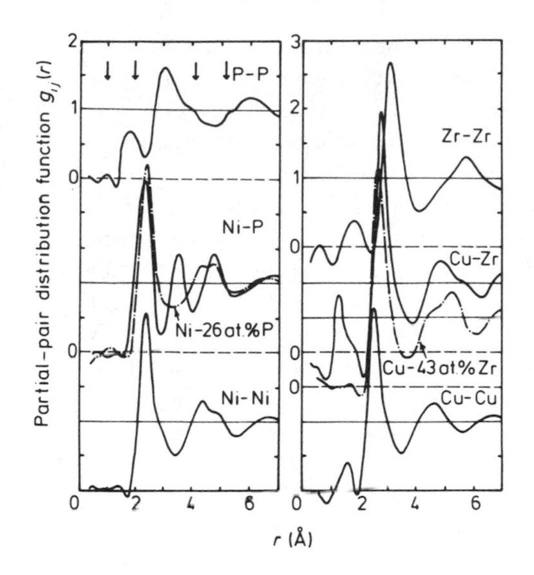

Figure 2. Three partial-pair distribution functions. Chain curves are the total pair distribution functions of amorphous Ni–26 at.% P and Cu–43 at.% Zr alloys.

at $\Delta r \simeq \pm 5\pi/2Q_{\max}$ or $\pm 9\pi/2Q_{\max}$ from the principal peak position, where $Q_{\max}$ is the upper limit of Q which is experimentally observed. This check is important in the case of P–P pairs. The arrows in figure 2 indicate the positions of these spurious ripples. The peak at about 1·9 Å in the pair distribution function of P–P pairs seems to be a spurious one as well as those at about 1·0 and 5·2 Å, although the corresponding spurious peak at 4·2 Å is not distinctly observed in figure 2. Therefore it is probably reasonable to say that the peak of the pair distribution function for P–P pairs at about $r = 3{\cdot}1$ Å is the significant one and P atoms are never at the first, nearest-neighbour distance. As shown in our previous work (Waseda *et al* 1976), this result supports the discussion that the short-range order at the nearest-neighbour distance in amorphous Ni–P alloys seems to be of a Ni_3P-type structure.

The general features of the structure of amorphous Ni–P and Cu–Zr alloys are given in the following:

(i) The form of the partial structure factors of two like-atom pairs is similar to that observed in pure liquid Ni, Cu, Zr and P (Thomas and Gingrich 1938, Waseda and Tamaki 1975, Waseda and Ohtani 1975).
(ii) The premaximum below the first peak corresponding to compound formation or long-range ordering is not found in any of the partial structure factors. Such ordering behaviour has been observed, for example, in an amorphous Cu–65 at.% Mg alloy (Lukens and Wagner 1976).
(iii) The first peak in the partial-pair distribution function of Ni–Ni and Cu–Cu pairs is asymmetric where the left-hand side of the first peak is steeper than that of the right-hand side.
(iv) The splitting of the second peak is observed in the partial structure factors of Ni–Ni and Cu–Zr pairs. Moreover, the third peak in the partial structure factor of Ni–P pairs lies in the region of the split second peak in the total structure factor and similarly in the case of Cu–Zr alloys there is a second peak of Zr–Zr pairs. Such behaviour is also found in the pair distribution functions.

The first feature, as well as the result in figure 1, implies that the partial structure factors in these amorphous alloys are approximately independent of concentration within the measured composition range. The second feature indicates that the structure of these amorphous alloys is consistent with a more or less random mixture of two constituent elements. The third feature may correspond to the situation that the repulsive core part of the pair potential for Ni–Ni and Cu–Cu pairs is more vertical than the attractive part. The fourth feature could be an indication that the distribution of unlike-atom pairs contributes mainly to the splitting of the second peak which is one of the differences encountered in the structure of both disordered atomic distributions (amorphous and liquid states).

The above features apply to both alloy systems, that is the metal–metalloid and metal–metal systems. But the following difference is found in the structural information: the curious peak in the partial structure factor of Ni–P pairs is clearly indicated, whereas this is not found in the partial structure factor of Cu–Zr pairs. In addition, the subsidiary peak is not reproduced by the dense random-packing model with single-size spheres (Cargill 1975). This behaviour may suggest that the detailed distribution in amorphous metal–metalloid systems differs from that in amorphous metal–metal systems. As discussed in our previous work (Waseda *et al* 1976), the atomic distribution in amorphous Ni–P alloys consists mainly of randomly distributed Ni atoms similar to those in the dense random-packing model based on tetragonal units with P atoms. On the other hand, the Cu and Zr atoms act, to a first approximation, as hard spheres packed closely in amorphous, liquid-like Cu–Zr alloys. But from these experimental results no conclusions can be made concerning the difference between the atomic distributions of the metal–metalloid and metal–metal systems because the fundamental unit in the short-range order for close-packed, liquid-like structures is tetragonal as well as that in the disordered structure of the dense random-packing model. Consequently, no definite answer is available at the present time to explain the subsidiary peak of the Ni–P pairs.

Acknowledgments

The authors are indebted to Professor M Ohtani for his support and encouragement in this study. One of us (YW) thanks the Sakkokai Foundation for a grant.

References

Cargill G S III 1975 *Solid St. Phys.* **30** 227 (New York: Academic Press)
Cromer D T 1965 *Acta Crystallogr.* **18** 17
—— 1976 *Acta Crystallogr.* **A32** 339
Finback C 1949 *Acta Chem. Scand.* **3** 1279
Giessen B C and Wagner C N J 1972 *Liquid Metals Physics and Chemistry* ed S Z Beer (New York: Dekker) p 663
Kudo T, Niimura N and Mizoguchi T 1975 *Res. Rep. Lab. Nucl. Sci., Tohoku Univ.* **8** 296
Lukens W E and Wagner C N J 1976 *J. Appl. Crystallogr.* **9** 159
Ramesh J G and Ramaseshan S 1971 *J. Phys. C: Solid St. Phys.* **4** 3029
Sugawara T 1951 *Sci. Rep. Res. Inst., Tohoku Univ.* **3A** 39
Thomas C D and Gingrich N S 1938 *J. Chem. Phys.* **6** 659

Waseda Y 1975 *Jap. J. Solid St. Phys.* **10** 459
Waseda Y and Masumoto T 1975 *Z. Phys.* **B21** 235
Waseda Y and Ohtani M 1975 *Z. Phys.* **B21** 229
Waseda Y, Okazaki H and Masumoto T 1976 *Sci. Rep. Res. Inst., Tohoku Univ.* **26A** 12
Waseda Y and Tamaki S 1975 *Phil. Mag.* **32** 951

The formation of amorphous metallic phases by continuous cooling from the liquid state

B G Lewis and H A Davies
Department of Metallurgy, Sheffield University, UK

Abstract. A generalized kinetic treatment for the formation of metallic glasses is developed and discussed. This involves the construction of time–temperature transformation curves from which the critical cooling rates for the avoidance of a detectable fraction of crystal can be estimated. It is shown that the reduced glass temperature and, to a lesser extent, the viscosity at the melting temperature have a dominating effect in determining the critical cooling rate and hence the glass-forming tendency of an alloy.

The theoretical dependence of the critical cooling rate on the reduced glass temperature is determined for an 'average' model system for different viscosity and crystallite/liquid interface conditions. This gives satisfactory agreement with experimentally estimated cooling-rate data over the whole range pertaining to metallic glasses. These calculations facilitate a reasonable order of magnitude prediction of critical cooling rate for any alloy from a knowledge of its reduced glass temperature.

Uncertainties in the calculations arising from a general lack of experimental data, and the assumptions involved in the theory are discussed, including considerations of heterogeneous nucleation and a temperature dependence of the solid/liquid interfacial energy.

1. Introduction

Amorphous phases have now been observed in a wide range of liquid quenched alloys, semi-metals and a nominally pure metal. Thus the early discussions of the ease of glass formation and stability of the phases in terms of size effects (Bennett *et al* 1971) or chemical bonding factors (Chen and Park 1973) have largely been superseded by kinetic treatments (Uhlmann 1972, Davies *et al* 1974, Davies 1975, Davies and Lewis 1975), where the ease of formation and stability of metallic glasses (although not completely independent) are considered separately. Implicit in this analysis is that any liquid, with the exception of highly pure metals which would lack the necessary barrier to crystallization (Davies and Hull 1976), is a glass former, provided that it is cooled sufficiently quickly to a temperature below which crystallization does not occur.

In this kinetic treatment of glass formation, theories of homogeneous nucleation, crystal growth and transformation kinetics are employed to construct a time–temperature transformation (TTT) curve, expressing the time for a barely detectable fraction of crystallization to occur as a function of temperature. A critical cooling rate (R_c) to avoid detectable crystallization can then be estimated from the TTT curve.

In this paper the general applicability of the kinetic approach to metallic systems is demonstrated. The cooling rate for an 'average' model alloy system is considered as a function of the parameters critical in determining ease of glass formation, and the analysis is extended to also examine the effect on R_c of (i) heterogeneous nucleation and (ii) a temperature-dependent crystal/melt interfacial energy.

2. Kinetics of the liquid–solid transformation

In this section we will briefly outline the kinetic approach to glass formation (Uhlmann 1972) that we adopt. Considering a transformation from liquid to crystal and using the Johnson and Mehl (1939) and Avrami (1939, 1940, 1941) treatment of transformation kinetics, a small volume fraction (X) crystallized in time t is given by

$$X \simeq \frac{\pi}{3} I_v^{\text{hom}} u^3 t^4 \tag{2.1}$$

where I_v^{hom}, the homogeneous nucleation frequency and u, the growth velocity of the crystal/liquid interfaces are assumed to be time independent.

For thermally activated atomic motion at the nucleus/liquid interface,

$$I_v^{\text{hom}} = \frac{N_v^0 D_n}{a_0{}^2} \exp\left(\frac{-\Delta G^*}{kT}\right). \tag{2.2}$$

Here, N_v^0 is the average number density of atoms per unit volume, a_0 is the mean atomic diameter, D_n is the diffusion coefficient across the nucleus/liquid interface at a temperature T and ΔG^* is the free energy of formation of a critical nucleus.

Taking $\Delta G^*/kT$ as about 50 at an undercooling $\Delta T_r = (T_m - T)/T_m$ of 0·2 (Uhlmann 1972), it can be shown (Uhlmann 1969) that

$$I_v^{\text{hom}} = \frac{N_v^0 D_n}{a_0{}^2} \exp\left(\frac{-1{\cdot}024}{T_r^3 \Delta T_r^2}\right) \tag{2.3}$$

where $T_r = T/T_m$, and T_m is the equilibrium melting temperature. When crystal growth involves thermally activated atomic jumps of the order of an interatomic distance, u can be expressed (Turnbull 1962) as

$$u = \frac{f D_g}{a_0}\left[1 - \exp\left(\frac{-\Delta T_r \Delta H_f^m}{RT}\right)\right] \tag{2.4}$$

with D_g the diffusion coefficient at the crystal/liquid interface, ΔH_f^m the molar heat of fusion and f the fraction of sites at the crystal surface where growth can occur ($f = 1$ for a rough interface and $f = 0{\cdot}2\,\Delta T_r$ for a smooth interface). Assuming $D_n = D_g$, and that they can both be related to the bulk liquid viscosity (η) by the Stokes–Einstein relation

$$D = kT/3\pi a_0 \eta \tag{2.5}$$

the time in seconds to a small X (taken to be an arbitrarily small value of 10^{-6}) is

$$t \sim \frac{9{\cdot}3}{kT}\eta\left(\frac{X a_0^9}{f^3 N_v^0} \frac{\exp(1{\cdot}024/T_r^3 \Delta T_r^2)}{[1 - \exp(-\Delta H_f^m \Delta T_r/RT)]^3}\right)^{1/4}. \tag{2.6}$$

3. Estimation of the critical cooling rate

In order to predict the critical cooling rate for the avoidance of a barely detectable fraction of crystal for specific alloys, it is necessary to estimate the temperature dependence of the viscosity in the region $T_m \geqslant T \geqslant T_g$ where T_g is the thermally manifest glass transition temperature. The viscosity in this temperature regime is, however, very largely inaccessible to experimental measurement and must be obtained by use of an empirical Vogel–Fulcher relationship (Vogel 1921, Fulcher 1925) or by smooth interpolation between the liquid viscosity at T_m and a value of 10^{13} poise at T_g. We have found the smooth interpolation more satisfactory in view of the breakdown of the V–F relation for temperatures near T_g and the non-physical nature of the three adjustable parameters in the equation. In our calculations the viscosity at T_m has been taken from or estimated from the literature (Wilson 1965, Ofte and Wittenberg 1963), and the glass transition temperatures (or crystallization temperatures where no T_g was observed) were those obtained from thermal analysis. The critical cooling rate for glass formation R_c is then the cooling rate required to just bypass the nose of the TTT diagram constructed and is given by

$$R_c \sim (T_m - T_N)/t_N \qquad (\mathrm{K\,s^{-1}}) \tag{3.1}$$

where T_N and t_N are the temperature and time, respectively, at the nose of the TTT curve.

Wherever possible, experimental data for the mean atomic diameter, heat of fusion and density were used; however, there is a general lack of relevant experimental data so that estimates are necessary which can lead to uncertainties in the predicted R_c. The uncertainties in R_c arising from this source are of the order of ± 1 orders of magnitude (Davies and Lewis 1975). A further source of uncertainty is likely to be the value assumed for the free energy to form a critical nucleus (ΔG^*). In this work ΔG^* has been taken as $50\,kT$ at an undercooling of 0·2 (Uhlmann 1972). In our experience, and in previous calculations (Onorato and Uhlmann 1976), it has been found that variations of about 20% in ΔG^* at a fixed undercooling adjusts R_c by about one order of magnitude. However, the various experimental data (Buckle 1960) indicate that variations in ΔG^* are about 5%, which would modify R_c by the relatively small factor of two.

The cooling rate, as calculated by equation (3.1), is in fact an overestimate of the required R_c since no account has been taken of the effect of continuous cooling. If continuous cooling curves (CCT) are constructed from the isothermal TTT diagrams, a value of R_c is obtained which is lowered by less than an order of magnitude, and typically by a factor of about two.

4. The reduced glass temperature and viscosity dependence of critical cooling rates

In order to investigate, theoretically, the influence on R_c of continuously varying the reduced glass temperature, T_g/T_m, an 'average' model alloy was chosen having the following properties, considered to be typical values for a metallic glass-forming system: $a_0 = 0{\cdot}26$ nm, $\Delta H_f^m = 12500\ \mathrm{J\,mole^{-1}}$, $\rho = 9{\cdot}0\ \mathrm{g\,cm^{-3}}$, atomic weight = 70.

Two different models of the temperature dependence of viscosity were considered:

(i) $\eta(T = T_m)$ was assumed equal to that of Ni at its melting point, with the activation energy for viscous flow also equal to that of Ni. Variations in T_g/T_m were effected by either keeping T_m fixed and varying T_g or, equivalently, keeping T_g fixed and varying T_m.
(ii) $\eta(T = T_m)$ was assumed to be a linear extrapolation of the Arrhenius plot for liquid Ni and T_g/T_m was varied by keeping T_g fixed and varying T_m.

It is not clear which of the two models is the more generally applicable; it is likely that the true picture is intermediate between the two models.

R_c has been estimated, using both of the models, for the rough ($f = 1$) and the smooth ($f = 0{\cdot}2\,\Delta T_r$) interface conditions, as a function of T_g/T_m. These calculations are plotted in figure 1 together with the theoretical cooling rates for some real alloys for which viscosity–temperature curves have been established or estimated from available data. (For each alloy the value of f was chosen depending upon the type of crystalline phase considered to be involved in the nucleation and growth process.) The agreement between the predicted cooling rates and experimentally estimated values has been found to be good over the gamut of glass formers, from about 10^2 K s^{-1} (Chen and Turnbull 1969) for a Pd–Si–Cu alloy to about 10^{10} K s^{-1} (Davies and Hull 1976) for Ni, with, for example, in the intermediate cooling rate range, R_c about 10^6 K s^{-1} for a Au–Ge–Si alloy and about 10^5 K s^{-1} for an Fe–P–C alloy.

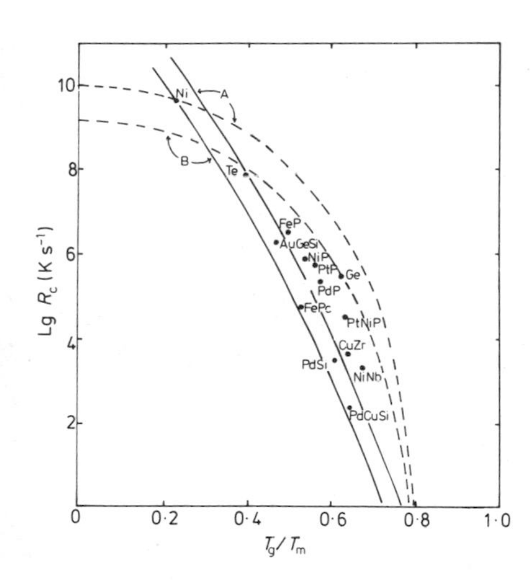

Figure 1. The variation of critical cooling rates for glass formation (R_c) with reduced glass temperature (T_g/T_m) for the model alloy, calculated using viscosity models (i) (broken lines) and (ii) (full lines) and for the two interface conditions $f = 1$ (curves A) and $f = 0{\cdot}2\,\Delta T_r$ (curves B) (see text). Cooling rates calculated for some real alloys are also shown. (Actual compositions of the alloys: $Au_{77{\cdot}8}Ge_{13{\cdot}8}Si_{8{\cdot}4}$, $Fe_{82{\cdot}5}P_{17{\cdot}5}$, $Ni_{81}P_{19}$, $Pt_{80}P_{20}$, $Pd_{81}P_{19}$, $Fe_{80}P_{13}C_7$, $Pt_{40}Ni_{40}P_{20}$, $Cu_{60}Zr_{40}$, $Ni_{60}Nb_{40}$, $Pd_{82}Si_{18}$, $Pd_{77{\cdot}5}Si_{16{\cdot}5}Cu_6$).

The curves in figure 1 are based on the liquid viscosity of nickel, and depict the dependence of R_c on T_g/T_m at a fixed viscosity. R_c is, however, sensitively dependent on viscosity ($R_c \propto 1/\eta$) and the close agreement between the model curves and the calculated R_c for individual alloys reflects the fact that most liquid metal viscosities, at T_m, fall within a narrow range of the order of a few centipoise.

The maximum thickness of glass (x_{max}) can also be estimated from the TTT diagram, assuming unidirectional cooling, no heat loss from the external surfaces, and using the relation

$$x_{max} \sim (D_t\, t_N)^{1/2} \tag{4.1}$$

where D_t is the thermal diffusivity. For Pd-based alloys and for Ni, $D_t \sim 10\ \mathrm{mm^2\,s^{-1}}$ so that x_{max} can be estimated as about 3·0 mm for $Pd_{77.5}\,S_{16.5}\,Cu_6$ and about 1 μm for Ni, in good agreement with observed maximum thicknesses.

Since only relatively small variations in the liquid viscosity are possible for metallic systems the magnitude of R_c (and hence x_{max}) is governed largely by T_g/T_m. In transition metal/metalloid systems T_g appears to be only slowly dependent on alloy composition (Lewis and Davies 1976), and can be roughly predicted theoretically (Davies 1976), so that the controlling variable in these systems is T_m. This emphasizes the importance of understanding the occurrence of very low temperature eutectics in these alloys around 20 at.%. The models of Bennett *et al* (1971), Chen and Park (1973), and Nagel and Tauc (1975) based respectively on considerations of atomic packing, chemical bonding and electron densities of states have all been proposed to account qualitatively for high glassy stability around this composition in such systems. They might all more properly be considered as explanations of the deep eutectics in these systems.

In metal/metal glass-forming alloys, for example Cu–Zr, Ni–Nb, T_g is enhanced substantially by the presence of the refractory metals, and the glass-forming ability depends rather less on the existence of very low melting temperatures.

5. The effect on R_c of modifications to the nucleation rate equation

5.1. *Heterogeneous nucleation*

We have previously considered homogeneous nucleation only, which was justified by the fact that in many easy glass-forming transition metal alloys, a large degree of bulk undercooling is possible. The effect of heterogeneous nuclei may, however, be significant, especially for alloys with a lower melting temperature. Onorato and Uhlmann (1976) have discussed this problem for a variety of glass-forming materials, including the hypothetical case of a pure metal. In this section we will consider the case of metallic glass formers in general.

If the spherical cap model of the heterogeneous nucleus is adopted, the heterogeneous nucleation rate is given by

$$I_v^{het} = A_v N_s^0 \frac{kT}{3\pi a_0^3 \eta} \exp\left[\left(\frac{-1{\cdot}024}{T_r^3 \Delta T_r^2}\right) f(\theta)\right] \tag{5.1}$$

where A_v is the area of nucleating substrate per unit volume of the melt, N_s^0 is the number of atoms on unit area of the substrate and $f(\theta)$ is given by

$$f(\theta) = \tfrac{1}{4}(2 + \cos\theta)(1 - \cos\theta)^2. \tag{5.2}$$

A_v may be estimated from the volume density and area of the nucleating heterogeneities as about $3 \times 10^{-3}\ \mathrm{cm^{-1}}$ (Onorato and Uhlmann 1976). This corresponds to an upper limit to the density of $2 \times 10^7\ \mathrm{cm^{-3}}$, deduced from studies of homogeneous nucleation, together with an assumed area per particle of $1{\cdot}5 \times 10^{-10}\ \mathrm{cm^2}$. The total nucleation rate is then the sum of I_v^{hom} (equation (2.2)) and I_v^{het} (equation (5.1)). For a given viscosity–temperature relation the temperature dependence of the growth velocity will be fixed

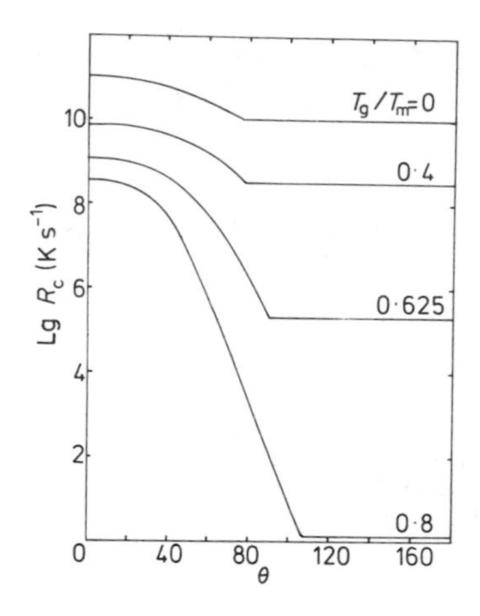

Figure 2. The theoretical variation of critical cooling rate (R_c) with contact angle (θ) for various T_g/T_m. (The calculations were performed for the model alloy with $T_m = 1000$ K and $f = 0{\cdot}2\,\Delta T_r$.)

so that any change in the temperature dependence of the nucleation frequency will be directly manifest in R_c.

R_c has been calculated for the model system as a function of contact angle θ, for various T_g/T_m, and these results are presented in figure 2. Heterogeneities with high angles of contact have little or no effect, R_c tending to the homogeneous case for $\theta \geqslant 100\,^\circ$C. Low contact angles, however, can modify R_c considerably, having an increasing effect with decreasing θ or increasing T_g/T_m.

The influence of nucleating heterogeneities on R_c in practice will obviously depend on the density and nature of the heterogeneities present. It is to be expected, however, that the density of nuclei in glass-forming alloys, typically transition metal based, would be much reduced as a result of the high temperatures involved in alloy preparation, and also the contact angle θ would generally be in the range of minimum effect ($\gtrsim 80°$). This is confirmed by the high undercoolings attainable in bulk quantities of such alloys. Nevertheless, for alloys based on low melting point metals heterogeneous nucleation could be an important factor unless the materials are very pure.

5.2. Temperature-dependent crystal/melt interfacial energy

In deriving equation (2.2) we follow Christian (1965) so that

$$\Delta G^* = \frac{16\pi}{3} \frac{\sigma^3}{(\Delta G_v)^3} \tag{5.3}$$

where σ is the crystal/melt interfacial energy and ΔG_v is the free energy difference between the crystal and liquid per unit volume and can be related to the molar heat of fusion by a model due to Hoffmann (1958) as

$$\Delta G_v = \frac{\Delta H_f^m}{V} \Delta T_r T_r . \tag{5.4}$$

In equation (5.3), σ was taken as temperature independent, but recent theoretical developments (Spaepen 1975, Spaepen and Meyer 1976) based on the dense random-packing model of the solid/liquid interface suggest that the origin of the term is largely entropic and thus temperature dependent, such that

$$\sigma = \sigma_m (T/T_m) = \sigma_m T_r \tag{5.5}$$

where σ_m is the interfacial energy at T_m.

Inclusion of this temperature dependence of the interfacial tension in equation (2.2) gives the homogeneous nucleation frequency as

$$I_v^{hom} = \frac{N_v^0}{a_0^3} \frac{kT}{3\pi\eta} \exp\left(\frac{-2{\cdot}00}{\Delta T_r^2}\right). \tag{5.6}$$

Turnbull and co-workers have carried out calculations of nucleation frequency for the two cases σ = constant (Turnbull 1969) and $\sigma = \sigma_m (T/T_m)$ (Spaepen and Turnbull 1977), in their discussions of metallic glass formation. It is difficult, however, to make comparisons between the two cases since, although the same viscosity was adopted in each case, the calculations were carried out for substantially different melt properties. We have calculated I_v^{hom} (equations (2.3) and (5.6)) for comparable melt properties and the nucleation frequencies in the σ = constant and $\sigma = \sigma(T)$ cases, plotted as a function of reduced temperature (T/T_m) for various T_g/T_m, are shown in figure 3. If we

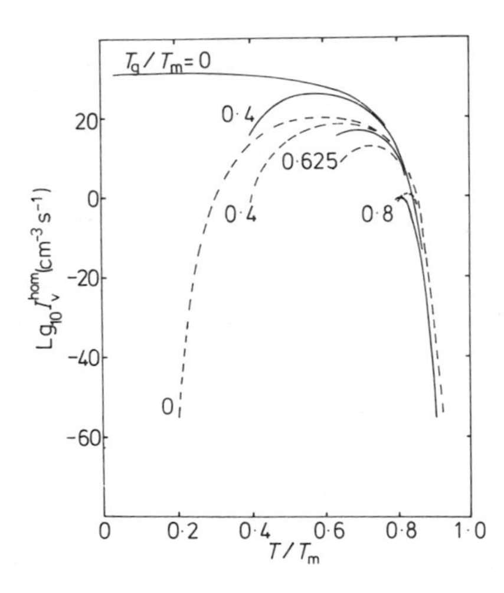

Figure 3. The variation of homogeneous nucleation frequency (I_v^{hom}) with reduced temperature (T/T_m) for various T_g/T_m. The solid lines are the nucleation rates with a temperature-dependent interfacial energy, the broken lines are the rates in the temperature-independent case.

include a temperature dependence of σ, then the nucleation rate at all temperatures of interest, and the peak nucleation rate in particular, is increased for all $T_g/T_m \lesssim 0{\cdot}7$. The difference between the temperature-independent σ nucleation rate and the temperature-dependent σ nucleation rate reduces with increasing T_g/T_m until at about $T_g/T_m = 0{\cdot}7$ the curves cross over. The exact details of the relative values of the two nucleation frequencies are of course determined by the form of the viscosity relation used. The corresponding calculated critical cooling rates for glass formation are shown in figure 4 as a function of T_g/T_m. As is expected, R_c ($\sigma = \sigma(T)$) is greater than R_c (σ = constant) by three orders of magnitude at $T_g/T_m = 0$ and this difference reduces with increasing T_g/T_m until at $T_g/T_m \sim 0{\cdot}67$, the cooling-rate curves intersect and R_c (σ = constant) > R_c ($\sigma = \sigma(T)$) for $T_g/T_m \gtrsim 0{\cdot}7$. Thus in the region of most practical

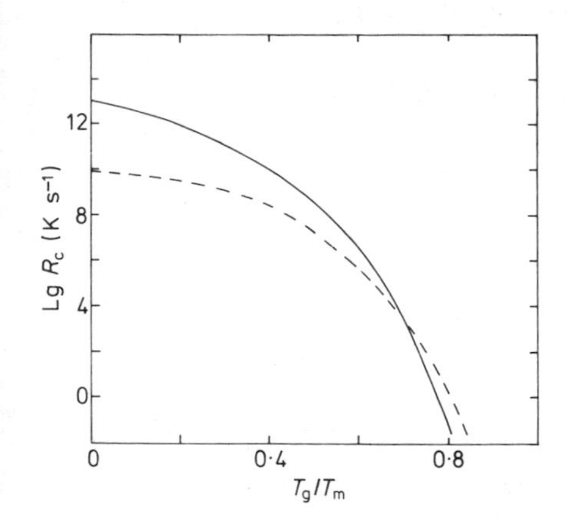

Figure 4. The variation of critical cooling rate (R_c) with reduced glass temperature (T_g/T_m) for the model alloy for a temperature-dependent interfacial energy (solid line) and a temperature-independent interfacial energy (broken line). (The calculations are for the model alloy with $T_m = 1000$ K and $f = 0{\cdot}2\,\Delta T_r$.)

interest for metallic glass formation ($0{\cdot}5 < T_g/T_m < 0{\cdot}75$), R_c is altered by only half an order of magnitude or less by the inclusion of a linear temperature dependence of the interfacial energy in the expression for I_v^{hom}.

Acknowledgments

The authors are grateful to the Science Research Council for financial support and to Professor B B Argent for the provision of laboratory and other research facilities.

References

Avrami M 1939 *J. Chem. Phys.* **7** 1130
—— 1940 *J. Chem. Phys.* **8** 212
—— 1941 *J. Chem. Phys.* **9** 177
Bennett C H, Polk D E and Turnbull D 1971 *Acta Metall.* **19** 1259
Buckle E R 1960 *Nature* **186** 875
Chen H S and Park B K 1973 *Acta Metall.* **21** 395
Chen H S and Turnbull D 1969 *Acta Metall.* **17** 1021
Christian J W 1965 *The Theory of Transformations in Metals and Alloys* (New York: Pergamon)
Davies H A 1975 *J. Non. Cryst. Solids* **17** 266
—— 1976 *Phys. Chem. Glasses* **17** in the press
Davies H A, Aucote J and Hull J B 1974 *Scripta Metall.* **8** 1179
Davies H A and Hull J B 1976 *J. Mater. Sci.* **11** 215
Davies H A and Lewis B G 1975 *Scripta Metall.* **9** 1107
Fulcher G S 1925 *J. Am. Ceram. Soc.* **6** 339
Hoffmann J D 1958 *J. Chem. Phys.* **29** 1192
Johnson W A and Mehl R F 1939 *Metall. Trans. AIME* **135** 416
Lewis B G and Davies H A 1976 *Mater. Sci. Eng.* **23** 179
Nagel S R and Tauc J 1975 *Phys. Rev. Lett.* **35** 380
Ofte D and Wittenberg L J 1963 *Trans. Metall. Soc. AIME* **227** 706
Onorato P I K and Uhlmann D R 1976 to be published
Spaepen F 1975 *Acta Metall.* **23** 729
Spaepen F and Meyer R B 1976 *Scripta Metall.* **10** 37
Spaepen F and Turnbull D 1977 *Proc. Int. Conf. on Rapidly Quenched Metals* (Cambridge, Mass.: MIT Press) in the press

Turnbull D 1962 *J. Phys. Chem.* **64** 609
—— 1969 *Contemp. Phys.* **10** 473
Uhlmann D R 1969 *Mater. Sci. Res.* **4** 172
—— 1972 *J. Non-Cryst. Solids* 7 337
Vogel H 1921 *Phys. Z.* **22** 645
Wilson J R 1965 *Metall. Rev.* **10** 381

Influence of electronic structure on glass-forming ability of alloys

S R Nagel† and J Tauc
Division of Engineering and Department of Physics, Brown University, Providence, Rhode Island 02912, USA

Abstract. In this paper a review is given of a model of metallic glass formation based on nearly-free-electron approximation. The model is extended to show how the eutectics in some binary alloys can be strongly influenced by these effects. Finally it is shown how the theory can predict certain correlations in metal–metalloid glasses and melts which have previously been ascribed to chemical bonds.

1. Introduction

Liquid metals, when cooled below their freezing temperatures, will usually form a polycrystalline solid. However, there do exist certain alloy compositions which, if cooled sufficiently rapidly, will form a glass. The reason for the enhanced glass-forming tendency exhibited by these alloys is not yet clearly understood, and there has been a variety of attempts to explain it. There is no doubt that geometrical considerations play an important role and one important approach to the problem focuses attention on the purely geometric aspects of randomly packing hard spheres (see Cargill 1975). However, it is certainly clear that these considerations are not entirely sufficient and the energy structure must also be taken into account. One possible way that this may be done is to hypothesize chemical bonding between the alloying elements (Chen and Park 1973, Bagley and DiSalvo 1976). However, no concrete proposal has yet been made as to how such chemical bonding arises in these materials. Another approach to understanding the electronic structure, which we will review and expand in this paper, is to start from a nearly-free-electron approximation (Nagel and Tauc 1975, 1976a). In principle at least, the electronic density of states and the cohesive energy can be calculated for a number of compositions and can be shown to favour glass formation in a certain concentration region of an alloy system. A complete calculation would, of course, include all chemical-bonding effects. However, we find that for one important class of glass formers this simple model, based only on second-order perturbation theory, gives very good qualitative agreement with the observed phenomena.

2. Nearly-free-electron model

The alloys which we consider are of the form $M_{1-x}X_x$ where M is a noble or a transition metal with a nearly filled d-band and X is a member of group IVa or Va. The glass-

† Present address: The James Franck Institute, The University of Chicago, Chicago, Illinois 60637, USA.

forming region is near the eutectic composition, that is, $x \simeq 0{\cdot}2$. An example of such a glass former is $Pd_{0.8}Si_{0.2}$ or $Au_{0.815}Si_{0.185}$. The addition of a third element such as copper to the palladium has been shown (Chen and Turnbull 1969) to increase the stability of some glasses. The photoemission spectra of one such particularly stable glass, $(Pd_{0.775}Cu_{0.06})Si_{0.165}$ was studied (Nagel *et al* 1976). When the valence band of the alloy was compared with that of pure palladium it was seen that the silicon in the alloy apparently acted to fill the palladium d-bands so that the d-bands lay completely below the Fermi level, E_F. Magnetic susceptibility measurements on this glass (Bagley and DiSalvo 1973) concluded that the states near the Fermi level were s- or p-like in character. Because of the similarity of the valence band of the alloy with those of the noble metals, it was suggested (Nagel and Tauc 1975) that the states near the Fermi level could be treated in terms of the nearly-free-electron approximation.

The second-order perturbation theory expression for the energy of an electron in a liquid or glass is:

$$E = E_k^0 + v(0) + \frac{\Omega}{8\pi^3} \int \frac{|v(q)|^2 S(q)}{E_{\mathbf{k}}^0 - E_{\mathbf{k}+\mathbf{q}}^0} \, d^3q \tag{1}$$

where $E_k^0 = \hbar^2 k^2/2m$, Ω is the volume, $v(q)$ is the pseudopotential which describes the interaction of the free electrons with the ionic cores and $S(q)$ is the (spherically symmetric) structure factor which gives information about the correlation between ionic positions. In general $S(q)$ will have a large first peak at $q = q_p$. This value of q is analogous to a reciprocal lattice vector in a crystal. When $|\mathbf{k} + \mathbf{q}_p| = |\mathbf{k}|$ this perturbation expansion will be invalid and, as in the crystal at a zone boundary, we expect a minimum in the density of states $D(E)$ at this point. This behaviour has actually been confirmed for a real liquid metal by a muffin-tin calculation for the density of states of liquid copper (Peterson *et al* 1976). A minimum in the density of states was found above the d-bands which was associated with the position of the first peak in $S(q)$. As a metal (with effective valence $\leqslant 1$) is alloyed with a metalloid (valence 4 or 5) the value of $2k_F$ is altered from less than q_p to greater than q_p. When $2k_F = q_p$, E_F lies in the minimum of $D(E)$. By using a rigid-band model for the states near E_F it was shown (Nagel and Tauc 1975) that the Fermi level lay in the minimum of the density of states, approximately in the concentration region where glass formation is observed in these metal–metalloid systems.

3. Glass-forming ability

Two arguments have been advanced to explain why the condition that E_F lie at a minimum of $D(E)$ should lead to enhanced glass-forming ability. The first dealt with a barrier against the nucleation of crystallites. If the liquid were to crystallize as a result of a fluctuation in which the system develops long-range order, $S(q)$ would be distorted and no longer spherically symmetric. Because $2k_F$ no longer coincides with the value of q_p in all directions, the total energy of the system is raised. This produces an energy barrier against nucleation.

The second argument involves the lowering of the eutectic in the alloy system (Nagel and Tauc 1976a, Tauc and Nagel 1976, Turnbull 1976). It was shown by Turnbull and

Cohen (1961) that the glass-forming tendency would be increased if the ratio of the glass temperature to the melting temperature, T_g/T_m, was increased. Thus at a eutectic, where T_m is low, glass formation is favoured especially if there is an interaction between the two elements of an alloy which abnormally depresses the eutectic temperature. Such an interaction apparently exists in the Au:Si system. Turnbull has calculated (Turnbull 1974) the liquidus curve for the Au:Si alloy system assuming that the liquid solution behaved ideally. The experimental curve is lowered significantly at the eutectic. One explanation of this fact could be based on the arguments presented above. Near the eutectic, where $2k_F = q_p$, the energy of the liquid is particularly lowered since occupied electron states have been shifted to lower energy. At either higher or lower concentrations of Si the Fermi level no longer lies at a minimum in the density of states. Thus there is a concentration-dependent interaction between Au and Si which is greatest when $2k_F = q_p$, that is, at the eutectic. This will help depress the eutectic temperature compared to the calculated value. Not only will this interaction act to lower the eutectic temperature but it can also raise the glass temperature T_g. Experimental evidence of this fact was found (Nagel and Tauc 1975) by comparing measurements of T_g (Chen 1974) and $S(q_p)$ (Sinha and Duwez 1971) for various alloys of the form $(Pt_{1-x}Ni_x)_{0\cdot75}P_{0\cdot25}$. It is found for these alloys that T_g depends in an almost linear way on the height of the first peak of the structure factor. Using an argument due to M H Cohen (1976, private communication) this variation is easily explicable. Cohen points out that in the free-volume theory of glass formation (Turnbull and Cohen 1961), T_g will scale with the heat of vaporization of the cohesive energy. Since the cohesive energy of the liquid is particularly large at the eutectic (which is why the eutectic temperature is abnormally low according to our model) T_g should also be large at the eutectic. The larger the value of $S(q_p)$, the greater will be the cohesive energy and the greater will be T_g.

The fact that the eutectics are influenced by these nearly-free-electron effects can be seen in yet another way. The eutectics in the Au:Si and Au:Ge occur at quite different compositions at 18·5 at.% Si and 27 at.% Ge respectively. These eutectic liquids were recently studied by x-ray diffraction (Waghorne *et al* 1976). The values of q_p obtained from this study were compared with calculated values of $2k_F$ (F J DiSalvo 1976, private communication). For both systems the eutectic was found to occur at the concentration for which $2k_F = q_p$. Although the composition at which the eutectic occurred varied significantly between the two systems, it was found that this may only be a manifestation of the fact that the Fermi surfaces of the two alloys have different diameters at the same composition. If (as argued above) the eutectic temperature can be appreciably lowered by the nearly-free-electron model, the eutectic composition can also be somewhat shifted from what it would be for a solution that behaved ideally. It should be noted that both Au:Si and Au:Ge have particularly simple phase diagrams with no intermetallic compounds. Thus there is no intermetallic compound formation, as in Cu:Si, which might also tend to displace the experimental eutectic composition from the ideal one.

4. Structural correlations in binary liquids

We now come to one final difficulty that has been raised concerning the validity of a nearly-free-electron approach to the stability of metallic glasses. There is some evidence that at the eutectic composition many transition metal–metalloid liquids and glasses have some partial ordering so that no metalloid atom has another metalloid atom as a nearest neighbour. The three partial-structure factors in a Co:P glass were determined by combined x-ray and neutron scattering (Blétry and Sadoc 1975). This showed that there were no P–P nearest neighbours. In the liquid Au:Si and Au:Ge alloys discussed above, the composition dependence of the x-ray interference function similarly indicated the absence of Si–Si and Ge–Ge neighbours (Waghorne *et al* 1976). The interpretation that has been given for these results is that there is a strong chemical bonding occurring between the unlike atoms which is much greater than that between either of the two pairs of similar atoms. Without such a strong chemical bond it is argued that the system would be completely disordered with all three partial-structure factors being similar. (If both atoms had the same radius one would expect the three partial-structure factors to be identical.) If such chemical bonding is responsible for the partial ordering of the liquid it would then appear very reasonable to assume that it plays a significant role in the formation of the glass (Bagley and DiSalvo 1976).

This effect can, however, be explained in terms of the nearly-free-electron approximation (Nagel and Tauc 1976b). The argument here is based upon the fact that the contribution of the band structure to the cohesive energy will favour a crystal structure which does not have a reciprocal lattice vector at a wavevector where the pseudopotential $v(q)$ is near zero (Heine and Weaire 1970). For a binary alloy, the perturbation expansion for the energy, equation (1), must be modified. In that equation, $|v(q)|^2 S(q)$ should be replaced by

$$|v_{\mathrm{M}}(q) - v_{\mathrm{X}}(q)|^2 x(1-x) + (1-x)^2 |v_{\mathrm{M}}(q)|^2 S_{\mathrm{MM}}(q) + x^2 |v_{\mathrm{X}}(q)|^2 S_{\mathrm{XX}}(q)$$
$$+ 2x(1-x)\, v_{\mathrm{M}}(q)\, v_{\mathrm{X}}(q)\, S_{\mathrm{MX}}(q).$$

It appears that for the three cases mentioned above, the x-ray structure factor (which is dominated by the partial-structure factor from the metal atoms $S_{\mathrm{MM}}(q)$) has its large first peak, at q_{p}, at a value of q where the metalloid pseudopotential is near zero, that is, $v_{\mathrm{X}}(q_{\mathrm{p}}) \simeq 0$. If $S_{\mathrm{XX}}(q)$ has its first peak also near q_{p}, which would be the case if there were metalloid–metalloid nearest neighbours, then this term would contribute very little to the cohesive energy. On the other hand, if there were no such neighbours and metalloid atoms were only second-nearest neighbours to each other, then the peak in $S_{\mathrm{XX}}(q)$ would move to a much smaller wavevector where $|v_{\mathrm{X}}(q)|^2$ is quite large. In this case the term involving $S_{\mathrm{XX}}(q)\ |v_{\mathrm{X}}(q)|^2$ will contribute significantly to the cohesive energy. From these simple considerations, it is easy to see why partial ordering occurs in these metal–metalloid systems. Far from being chemical bonding between the metal and metalloid atoms which is responsible for this behaviour, it is merely a property of the metalloid atoms. In the above we have considered $x^2|v_{\mathrm{X}}(q)|^2 S_{\mathrm{XX}}(q)$, the metal atom parameters have not entered. The metalloid atoms prefer to have a larger separation between them than the average separation of atoms in the alloy mixture.

Acknowledgments

This work was supported by NSF Grant DMR71-01814-A01, ARO Grant DAAG29-76-G-0221 and the NSF MRL program at Brown University.

References

Bagley B G and DiSalvo F J 1973 *Amorphous Magnetism* ed H O Hooper and A M deGraaf (New York: Plenum) p143
—— 1976 *Bull. Am. Phys. Soc.* **21** 385
Blétry J and Sadoc J F 1975 *J. Phys. F: Metal Phys.* **5** L110
Cargill G S III 1975 *Solid St. Phys.* **30** 227 (New York: Academic Press)
Chen H S 1974 *Acta Metall.* **22** 1505
Chen H S and Park B K 1973 *Acta Metall.* **21** 395
Chen H S and Turnbull D 1969 *Acta Metall.* **17** 1021
Heine V and Weaire D 1970 *Solid St. Phys.* **24** 249 (New York: Academic Press)
Nagel S R, Fisher G B, Tauc J and Bagley B G 1976 *Phys. Rev.* **B13** 3284
Nagel S R and Tauc J 1975 *Phys. Rev. Lett.* **35** 380
—— 1976a *Proc. 2nd Int. Conf. on Rapidly Quenched Metals, MIT, Cambridge, Mass. 1975* ed N J Grant and B C Giessen (Cambridge, Mass: MIT Press)
—— 1976b to be published
Peterson H K, Bansil A and Schwartz L 1976 *Bull. Am. Phys. Soc.* **21** 468
Sinha A K and Duwez P 1971 *J. Phys. Chem. Solids* **32** 267
Tauc J and Nagel S R 1976 *Comm. Solid St. Phys.* **7** 69
Turnbull D 1974 *J. Physique* **35** (Colloque 4) 1
—— 1976 *Proc. 2nd Int. Conf. on Rapidly Quenched Metals, MIT, Cambridge, Mass. 1975* ed N J Grant and B C Giessen (Cambridge, Mass.: MIT Press)
Turnbull D and Cohen M H 1961 *J. Chem. Phys.* **34** 120
Waghorne R M, Rivlin V G and Williams G I 1976 *J. Phys. F: Metal Phys.* **6** 147

Electronic properties of non-simple liquid metals

M Watabe
Department of Physics, Tohoku University, Sendai 980, Japan

Abstract. A review is given of recent progress in the theoretical work on electron states and electrical transport in liquid non-simple metals such as liquid transition metals and expanded metallic fluids near the critical point. The nearly-free electron approximation is not appropriate to such systems, and we have to employ some other approximation schemes for treating the strong electron–ion interaction. In this review, we discuss the tight-binding model, which has been studied extensively since the Tokyo conference (1972), and the muffin-tin potential model which forms the basis for the Ziman–Lloyd formalism.

1. Introduction

It is quite some time since the nearly-free electron (NFE) theories due to Ziman, Faber and Edwards were proposed for the study of the electronic structure and transport properties of liquid simple metals such as the alkali metals and some (non-transition) polyvalent metals (see, for example, Faber 1972). The essential features of the electronic properties of these liquid metals are now explained at least qualitatively and present research in this field is directed towards the refinement of quantitative treatments.

On the other hand, no such systematic theories have yet been established for liquid non-simple metals, such as liquid transition, noble and rare earth metals and expanded metallic fluids, for which the NFE theories are not valid. Although several attempts have previously been made to study the electronic structure and transport properties of liquid non-simple metals and some have been successful to a certain extent (see, for a review, Busch and Guntherodt 1974), it would still be highly desirable to construct a more systematic theoretical scheme. It is important to have a comprehensive understanding of the characteristic features of these systems where the scattering is strong and accordingly the short-range atomic correlations are expected to be more important than in the case of liquid simple metals. Furthermore, there is a need for a systematic interpretation of the considerable amount of experimental data which have been accumulated recently (Busch and Guntherodt 1974). With these purposes in mind, we discuss in this review the tight-binding model, which has been extensively investigated since the Tokyo conference, and the muffin-tin model which forms the basis of the Ziman–Lloyd formulation.

2. Tight-binding model

The tight-binding model is considered to be a rather simple starting point for studying the electronic properties of liquid non-simple metals. Although the model in its crudest forms (e.g., the single s-band model) is too simple to be applicable to a quantitative discussion of realistic systems (see below for some extensions and improvements), it is very useful for understanding some characteristic features of the electronic properties of liquid non-simple metals. Because it is so simple to manipulate, we can use it to test various approximation schemes for the electronic structure, discuss their features, compare them with each other, and hopefully select the most suitable methods for describing the electronic states in liquid non-simple metals.

In this connection it is interesting to quote the following remark from Ziman's (1973) opening lecture at the Tokyo conference. After mentioning the necessity for careful consideration of the resonances in atomic potentials, he says: 'It is surprising, also, to notice how little attention was paid to the other standard model of the electronic structure of transition metals. Only one paper at Brookhaven discussed the problem of setting up a tight-binding formalism for the d-electrons in a liquid transition metal. This, surely whould be the natural first step to a derivation of the transport properties in terms of conduction by s-electrons with scattering into empty states in the d-band, s–d hybridization, etc. It would be interesting to see whether a two-band model of a liquid transition metal could be made to work when each band must have its own spherical Fermi surface'.

2.1. *Density of states in the tight-binding model*

Extensive studies of the tight-binding approximation (TBA) for calculating the electronic density of states of liquid non-simple metals have actually been made in recent years, following the work of Ishida and Yonezawa (1973) and Roth (1972, 1973). The nature and the range of validity of various approximation schemes proposed up to now for this model, especially the Ishida–Yonezawa (IY) theory and the effective medium approximation (EMA) proposed by Roth (1974a), have been discussed by Roth (1975), Yonezawa and Watabe (1975), Watabe and Yonezawa (1975) and Yonezawa *et al* (1975) and will also be briefly reviewed here. We stress (i) that the EMA is the most natural extension of the well known coherent-potential approximation (CPA), which is known to be the best single-site approximation (see, for example, Yonezawa and Morigaki 1973) for a substitutionally disordered system to a structurally disordered system such as a liquid metal and (ii) that the IY theory is considered to be the first (and most practical) approximation to the EMA.

Most of the formal discussions of the TBA have been made for the single s-band model, in which it is assumed that the one-electron wavefunction, and hence the one-electron Green function, $\mathscr{G}(\mathbf{r}, \mathbf{r}')$, can be expanded in terms of atomic s-orbitals, $\phi_i(\mathbf{r}) = \phi(|\mathbf{r} - \mathbf{R}_i|)$, centred on the various atomic sites:

$$\mathscr{G}(\mathbf{r}, \mathbf{r}') = \sum_{i,j} \phi_i(\mathbf{r})\, \mathscr{G}_{ij}\, \phi_j^*(\mathbf{r}'). \tag{1}$$

For $\mathscr{G}_{ij}$, we have the equation

$$\sum_l (zS_{il} - H_{il})\, \mathscr{G}_{lj} = \delta_{ij} \tag{2}$$

where

$$S_{il} = S(\mathbf{R}_{il}) = \int \phi_i^*(\mathbf{r})\, \phi_l(\mathbf{r})\, d\mathbf{r}$$

is the overlap integral, and H_{il} is the matrix element of the one-electron Hamiltonian,

$$H_{il} = \int \phi_i^*(\mathbf{r})\, H\phi_l(\mathbf{r})\, \mathrm{d}\mathbf{r}.$$

The following simplifying assumptions are made for H_{ij}:

(i) For $i \neq j$, H_{ij} is a function only of the distance between atoms i and j, that is, $H_{ij} = H(\mathbf{R}_{ij})$.
(ii) For $i = j$, H_{ii} is a constant independent of the distribution of neighbouring atoms, that is, $H_{ii} = \epsilon_0$.

The equation for $\mathscr{G}_{ij}$ is given by

$$(z - \epsilon_0)\, \mathscr{G}_{ij} - \sum_{l \neq i} H'(\mathbf{R}_{il})\, \mathscr{G}_{lj} = \delta_{ij}, \tag{3}$$

where

$$H'(\mathbf{R}_{il}) = H(\mathbf{R}_{il}) - zS(\mathbf{R}_{il}). \tag{4}$$

In the usual two-centre approximation, we can write H' as

$$H'(\mathbf{R}_{il}) = t(\mathbf{R}_{il}) - (z - \epsilon_0)\, S(\mathbf{R}_{il}), \tag{5}$$

where $t_{il} = t(\mathbf{R}_{il})$ is the transfer integral. It is often further assumed that the atomic orbitals are orthogonal, so that $S_{ij} = \delta_{ij}$ and $S(\mathbf{R}_{il})$ in $H'(\mathbf{R}_{il})$ is neglected.

The ensemble-averaged density of states (per atom) is given in terms of the Green functions with $z = E + iO^+$ as

$$n(E) = -\frac{1}{\pi N}\, \mathrm{Im} \int \langle \mathscr{G}(\mathbf{r}, \mathbf{r}) \rangle\, \mathrm{d}\mathbf{r} \tag{6a}$$

$$= -\frac{1}{\pi}\, \mathrm{Im}\left[\left\langle \sum_i \mathscr{G}_{ii} \right\rangle + \left\langle \sum_{i \neq j} S_{ji}\, \mathscr{G}_{ij} \right\rangle\right] \tag{6b}$$

$$= -\frac{1}{\pi}\, \mathrm{Im}\left[\langle \mathscr{G}_{ii} \rangle_i + \rho \int S_{ji}\, \langle \mathscr{G}_{ij} \rangle_{ij}\, g_2(R_{ij})\, \mathrm{d}\mathbf{R}_{ij}\right] \tag{6c}$$

where N is the total number of atoms, $\rho = N/\Omega$ is the number density of atoms (where Ω is the volume of the system), $\langle \ldots \rangle$ denotes the configuration average over all atomic sites, and $\langle \ldots \rangle_i$ and $\langle \ldots \rangle_{ij}$ denote respectively the conditional average with the position of the ith atom held fixed and that with the positions of both the ith and jth atoms held fixed. To derive equation (6c) we have to use the fact that the system can be assumed to be statistically homogeneous so that $\langle \mathscr{G}_{ii} \rangle_i$ is independent of i (or $\mathbf{R}_i$) and $\langle \mathscr{G}_{ij} \rangle_{ij}$ is a function only of R_{ij}. The term, $g_2(R_{ij})$, in equation (6c) is the pair distribution function for ions. If we assume orthogonal orbitals, we have

$$n(E) = -\frac{1}{\pi}\, \mathrm{Im}\, \langle \mathscr{G}_i' \rangle_i. \tag{7}$$

The advantage of the formula in the form of equation (6c), where the correlation between the ith and jth atoms is expressed explicitly in terms of $g_2(R_{ij})$, for the discussion of non-orthogonality effects has been suggested by Yonezawa and Martino (1976).

In order to calculate $\langle \mathcal{G}_{ii} \rangle_i$ and $\langle \mathcal{G}_{ij} \rangle_{ij}$, it is convenient to expand $\mathcal{G}_{ij}$ in terms of H' as

$$\mathcal{G}_{ij} = G_0 \delta_{ij} + G_0(1 - \delta_{ij}) H'_{ij} G_0 + \sum_l{}' G_0 H'_{il} G_0 H'_{lj} G_0 + \sum_{l,m} G_0 H'_{il} G_0 H'_{lm} G_0 H'_{mj} G_0 + \ldots, \tag{8}$$

where $G_0 = (z - \epsilon_0)^{-1}$ and the prime on the summation indicates that any successive indices are different from each other. Then taking the configuration average of the expansion term by term, we have:

$$\langle \mathcal{G}_{ii} \rangle_i = G_0 + G_0 \rho \int \mathrm{d}\mathbf{R}_l g_2(\mathbf{R}_i, \mathbf{R}_l) H'_{il} G_0 H'_{li} G_0 + G_0 \rho^2 \int \mathrm{d}\mathbf{R}_l \, \mathrm{d}\mathbf{R}_m g_3(\mathbf{R}_i, \mathbf{R}_l, \mathbf{R}_m) H'_{il} G_0 H'_{lm} G_0 H'_{mi} G_0 + \ldots, \tag{9}$$

$$\langle \mathcal{G}_{ij} \rangle_{ij} = G_0 H'_{ij} G_0 + G_0 \rho \int \mathrm{d}\mathbf{R}_l [g_2(\mathbf{R}_i, \mathbf{R}_j)]^{-1} g_3(\mathbf{R}_i, \mathbf{R}_j, \mathbf{R}_l) \times H'_{il} G_0 H'_{lj} G_0 + \ldots, \tag{10}$$

where we need the s body distribution function, $g_s(\mathbf{R}_1, \ldots, \mathbf{R}_s)$, to express the average of the term containing s different atomic sites. Although these equations give the exact expansion for the ensemble-averaged Green functions, they are practically useless in these forms partly because they require full knowledge of many-body distribution functions and partly because, even if we have g_s for all s, the summation of all terms in the expansion is impossible. Therefore, an approximate theory for a liquid metal with atomic correlation is characterized by giving

(i) the criteria for choosing some important terms in the series expansion, and
(ii) an approximation for g_s.

Since we have very little knowledge of g_s for $s \geqslant 3$, it is necessary to approximate all higher-order distribution functions in terms of the pair distribution function g_2.

As for the approximation (i), almost all existing tight-binding theories employ the single-site approximation (SSA) (for some discussions of theories beyond SSA, see Katz and Rice (1972)), in which all multi-site terms describing the repeated scattering of an electron between two or more sites are neglected. Various SSA theories have been proposed by introducing different approximations for g_s. The best SSA theory known so far is the effective-medium approximation (EMA) due to Roth (1974b), which employs the Kirkwood superposition approximation for g_s. Here we give a brief derivation of the EMA. (See, for further details, Yonezawa *et al* 1975, Roth 1975, Yonezawa and Watabe 1975, Watabe and Yonezawa 1975.)

Figure 1 gives the diagrammatic representation of equations (9) and (10) with the Kirkwood approximation. The rules for counting diagrams are to associate $H'G_0$ with each directed solid line, g_2 with a broken line and $h_2 = g_2 - 1$ with a dotted line. The broken, g_2 lines are used to describe the correlation along the scattering chain, and the dotted, h_2 lines to represent out-of-chain correlations. The integration, $\rho \int \mathrm{d}\mathbf{R}$, is performed at each filled circle while unity is assigned to each open circle.

The h_2 lines correspond to the multiple occupancy correction in crystalline alloys and proper account of this is essential for the derivation of the coherent-potential approximation (CPA) which is known to be the best self-consistent SSA for the crystalline alloys. To be consistent with the SSA, we discard those correction terms which

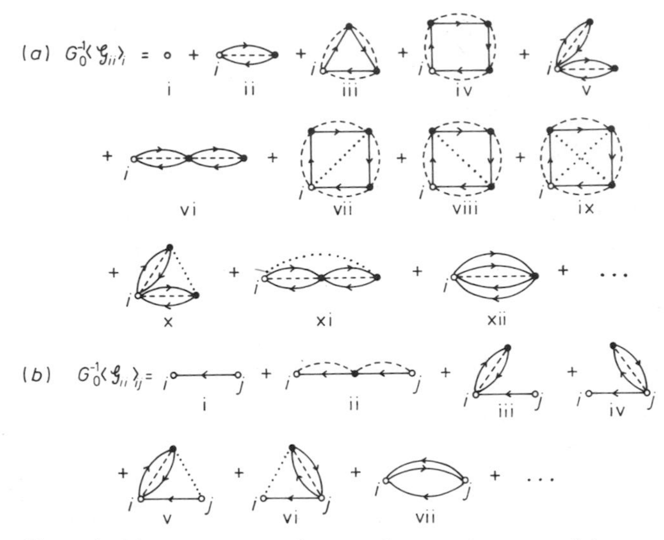

Figure 1. Diagrams representing some lower order terms of the perturbation expansion series of the conditionally averaged (*a*) site diagonal and (*b*) site off-diagonal Green functions in the tight-binding model.

have a structure analogous to the multi-site diagrams in the sense that they reduce the latter if the h_2 lines are replaced by an equivalence of vertices. For example, we discard in figure 1 the diagrams (*a*) ix, x and xi and diagrams (*b*) v and vi, since their structures are analogous respectively to diagram (*a*) xii and (*b*) vii which describe two-site effects and are to be discarded in the SSA.

By grouping some diagrams with common features, we renormalize locators and transfer integrals and obtain the equations determining $\langle \mathcal{G}_{ii} \rangle_i$ and $\langle \mathcal{G}_{ij} \rangle_{ij}$ expressed in compact forms by introducing the diagrams in figure 2. It is convenient to express the results, by means of the Fourier-transformed quantities, as

$$G_d = \langle \mathcal{G}_{ii} \rangle_i = G_0 \xi = G_0(1-\eta)^{-1} = (z - \epsilon_0 - \Sigma_d(z))^{-1}, \tag{11}$$

$$\begin{aligned} G_{nd}(k) &= \int \langle \mathcal{G}_{ij} \rangle_{ij} \exp(-i\mathbf{k} . \mathbf{R}_{ij}) \, d\mathbf{R}_{ij} \\ &= G_d \{H'(\mathbf{k}) + \rho M(\mathbf{k}) [1 - \rho G_d M(\mathbf{k})]^{-1} G_d M(\mathbf{k})\} G_d, \end{aligned} \tag{12}$$

where $M(\mathbf{k})$ is the Fourier transform of the modified transfer matrix M_{ij} represented by the heavy solid line in figure 2, and $\Sigma_d = G_0^{-1} \eta$ is a self-energy determined as the sum of all graphs which start from the position of the ith atom and return there only once. These quantities are given by the diagrams shown in figure 3. The self-consistent set of equations thus obtained is shown to be equivalent to the effective medium approximation derived originally by Roth (1973) in a rather intuitive way. The above derivation closely follows Yonezawa's (1968) theory of the self-contained SSA which is equivalent to the CPA and shows clearly that the EMA is the most natural extension of the CPA to a structurally disordered system.

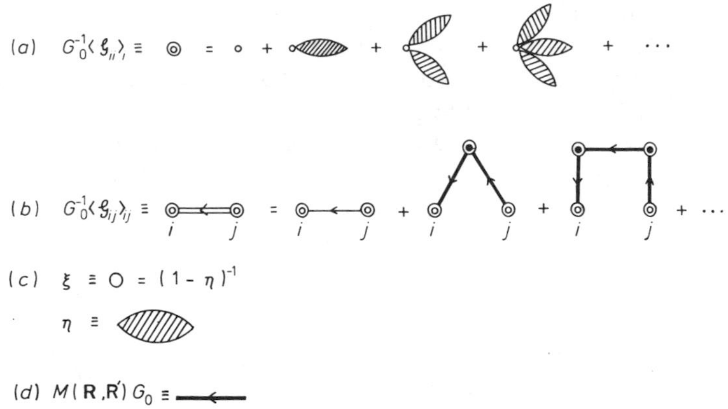

(a) $G_0^{-1}\langle \mathcal{G}_{ii}\rangle_i \equiv$ ◎ = ○ + ...

(b) $G_0^{-1}\langle \mathcal{G}_{ij}\rangle_{ij} \equiv$ ... + ...

(c) $\xi \equiv$ ○ $= (1-\eta)^{-1}$

$\eta \equiv$

(d) $M(\mathbf{R},\mathbf{R}')\,G_0 \equiv$

Figure 2. Diagrams describing the self-consistent single-site approximation.

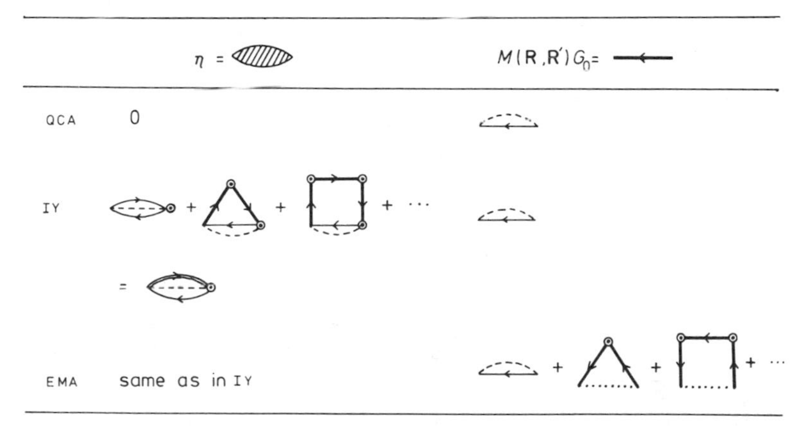

Figure 3. Various approximations for η and $M(\mathbf{R}, \mathbf{R}')$.

Other SSA theories only partially sum the self-contained single-site diagrams included in the EMA. These theories are specified by giving the approximations for Σ_d and $M(\mathbf{k})$. For example, the results for the quasi-crystalline approximation (QCA) and for the Ishida–Yonezawa (IY) theory are described in terms of diagrams in figure 3. (More detailed discussions are given in Yonezawa *et al* 1975.) For the QCA, $\eta = 0$, and hence, for orthogonal orbitals, we obtain only the atomic density of states (delta function at $E = \epsilon_0$) and this is completely unsatisfactory (for further discussion of the QCA, see Roth 1976). The IY theory can be viewed as a first approximation to the EMA, in which the renormalization of the transfer integral described by higher order terms in M_{ij} is not included. In other words, a $\mathbf{k}$-dependent part of the self-energy $\Sigma_1(\mathbf{k})$, defined by

$$\rho M(\mathbf{k}) = \rho \widetilde{H}'(\mathbf{k}) + \Sigma_1(\mathbf{k}) \text{ (where } \widetilde{H}'(\mathbf{k}) = \int \widetilde{H}'(\mathbf{R})\, g(R) \exp(-i\mathbf{k}.\mathbf{R})\, d\mathbf{R}),$$

is neglected in the IY.

The $\mathbf{k}$-independence of the self-energy is actually the great advantage of the IY theory for practical applications. Therefore it is very useful to know the range of validity of the IY theory compared to the EMA. This point has been investigated numerically for some particular choices of the transfer integral and of $g(R)$ and it has been found that the IY theory gives satisfactory results provided the densities are not too high. (Some numerical examples of comparison between the IY and the EMA are given in Roth (1976) for a single-band case. An example for a degenerate-band case is given below.) More specifically, for the so-called hard-core random case, Roth (1976) estimates that if $\bar{z} \ll 4$ then the IY theory is expected to be a good approximation. $\bar{z}$ is an effective coordination number defined by

$$\bar{t}\bar{z} = \rho \int g(R)\, t(R)\, \mathrm{d}\mathbf{R}, \tag{13a}$$

$$\bar{t}^2\bar{z} = \rho \int g(R)\, t^2(R)\, \mathrm{d}\mathbf{R}, \tag{13b}$$

with an effective hopping integral $\bar{t}$. It is also shown (Roth 1975) that for orthogonal orbitals the first three moments of the density of states are the same for the IY and EMA theories, and the two theories differ only in the fourth moment.

The range of validity of the EMA itself is rather difficult to assess since we do not know any exact solutions for three-dimensional cases of practical interest against which we can compare approximate solutions. Some general discussions have been made by Roth (1975, 1976), who showed that the EMA is in some sense an expansion in $\bar{z}^{-1}$, where $\bar{z}$ is the effective coordination number defined above, and hence should be best for large $\bar{z}$. As for the moments of the density of states, the EMA gives the first three moments exactly within the Kirkwood approximation employed for the higher-order distribution functions.

Some discussions of the analyticity of the solutions in the IY theory and the EMA have been given by Martino and Yonezawa (1975) and by Roth (1976). For the hard-core random case (square $g(R)$) the IY always gives results that are analytically correct while the EMA may not. For more general $g(R)$ the IY theory also may give a non-analytic solution. No definite condition for analyticity has been found.

Some extensions and improvements of the simple single s-band model treated above have been attempted recently. These are necessary if the model is to be applied to any realistic system. The nonorthogonality effects, originally studied by Roth (1973) and included in the above discussion, may be important since it is practically impossible to choose the mutually orthogonal atomic orbitals (different from site to site) as a basis for the expansion of the Green function. Numerical studies of these effects have been made by Yonezawa and her co-workers (Yonezawa *et al* 1976, 1977) in the IY approximation and by Roth (1976) in the EMA. It is worth pointing out that when the IY theory is employed some caution should be exercised in properly retaining the site exclusion property. The form of equation (6*c*) is preferred to the form expressed in terms of the continuum Green function (Roth 1973) since the site exclusion condition is explicit in the former. From numerical results it is concluded that for both the IY and EMA the effects of the nonorthogonality are such that (i) although the bandwidth is virtually unchanged, the change in band shape can be considerable, and (ii) the change in band shape becomes larger on decreasing the packing fraction (for the assumed random hard-core distribution function) at constant atomic density, and hence for an increase in the overlap of wavefunctions.

An improvement on the approximation (ii) assumed in the above discussions (described just before equation (3)) has been attempted by Yonezawa *et al* (1976).

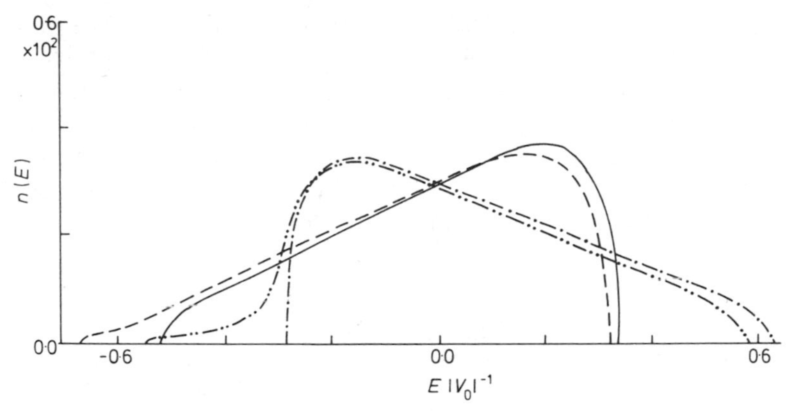

Figure 4. Curves of the density of states in a single s-band hard-core random liquid (——, – – –, with and without higher-order diagonal terms included but without non-orthogonality effects; —·—, —··—, with and without higher-order diagonal terms included but with non-orthogonality effects). $n(E)$ is given in units of $V_0^{-1}(p/32\pi)^{-1}$, where V_0 is the binding energy, and p is a dimensionless density parameter taken as 0·2. The packing fraction is taken as zero.

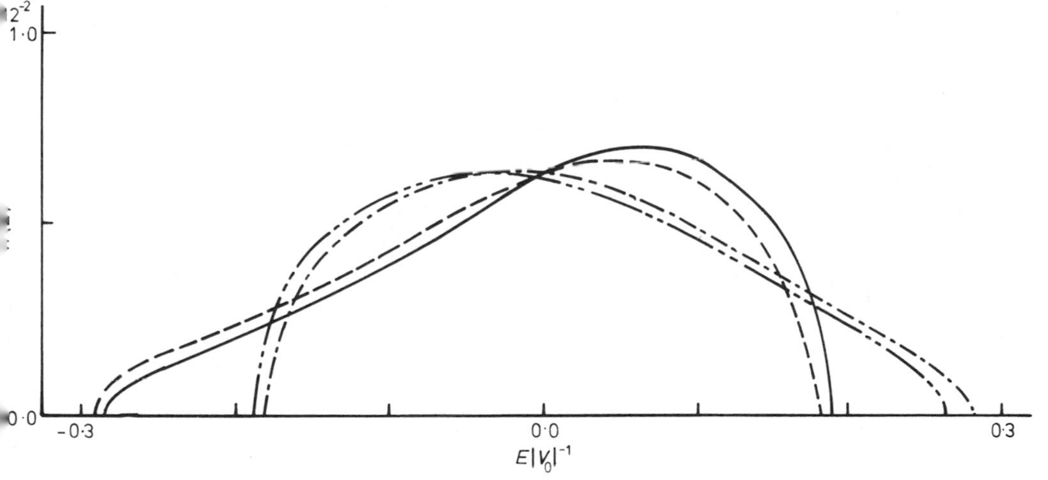

Figure 5. As figure 4 but with packing fraction equal to 0·02.

They show that the fluctuation of the diagonal elements, H_{ii}, due to neighbouring atoms coupled with the transfer process can give rise to a contribution of the same order as the nonorthogonality effects.

In figure 4 the density of states is given for the model of a hard-core random liquid with hydrogen-like 1s atomic orbitals of effective Bohr radius a^*. The dimensionless density, $p = 32\pi\rho(a^*)^3$ is taken as 0·2, and the packing fraction of the hard spheres as 0 (figure 4) and 0·02 (figure 5). For this choice of parameters the nonorthogonality effects are quite large. The above-mentioned behaviour of the effects due to the non-orthogonality and the higher-order terms with changing packing fraction is clearly seen in these figures. These effects due to the diagonal terms give rise to the tailing at the bottom of the band in addition to the downward shift of the band as a whole. As is the case for the nonorthogonality effects mentioned above, the effects are smaller for larger packing fractions at constant density.

In order that the model can be applied to real non-simple liquid metals where either several atomic states are degenerate or several different bands overlap in the same region of energy, the inclusion of the effects of more than one atomic state is essential. Such a multi-band formulation including effects of orbital degeneracy, angular-dependent hopping and hybridization can be done in a straightforward manner by replacing appropriate scalar quantities in the single-band equations with corresponding matrices with respect to the quantum numbers of relevant orbitals (Yonezawa and Martino 1976). The technique introduced by Yonezawa and Martino to reduce the three-dimensional integral in k-space involved in the resultant matrix equations to a one-dimensional integral is practically very important, since the self-consistent solution of the matrix equations is otherwise quite formidable. A similar technique has been developed independently by Niizeki (1976 private communication).

The application of the multi-band TBA has been carried out by Yonezawa and Martino (1976) for the 6s- and 6p-bands of fluid Hg in connection with the metal–nonmetal transition in this system under supercritical conditions (for more details, see Yonezawa *et al* 1977). Also, the first attempt to apply the TBA to the case of degenerate atomic states has been made by Ishida *et al* (1977) for the d-band of liquid Ni. They calculated the density of d-band states both in the IY approximation and in the EMA by using Herman–Skillman atomic wavefunctions and potentials and experimental values for $g(R)$. Hybridization effects with s- and p-bands are included approximately. The calculated density of states is rather featureless, having no fine structure and no pronounced tailing. As to the comparison between the IY and the EMA, the results are rather similar in both approximations; the calculated bandwidth is somewhat narrower in the EMA than in the IY, and the maximum of the density of states is near the centre of the band in the IY but shifts to a higher energy in the EMA (see Ishida *et al* 1977 for more details). Quantitatively their results should be taken as preliminary, but nevertheless they are expected to be useful for qualitative and, hopefully, semi-quantitative discussions. Actually Ishida *et al* (1976) have tried to investigate some general features of the effects of s–d hybridization based on these results, and also have been attempting to discuss systematic trends of the electronic properties in a liquid transition metal series.

As another interesting extension of the model we mention here the work by Yonezawa *et al* (1974). In connection with the metal–nonmetal transition (the so-called Mott transition) observed in supercritical alkali fluids, they tried to generalize the tight-binding theory of a liquid metal to include Coulomb interactions between electrons. That is, they extended the Hubbard model Hamiltonian, investigated extensively by Hubbard for pure solids, to the liquid case. One of the essential ingredients of Hubbard's approximation, the so-called scattering correction, is known to be equivalent to the alloy CPA. Therefore the extension of the treatment of this scattering correction to the liquid case is equivalent to the extension of the tight-binding theory to liquid alloys, which combines the CPA treatment of alloy disorder with the treatment of liquid disorder discussed above. Such an extension of the tight-binding theory to liquid alloys has also been made by Movaghar *et al* (1974) (see also Movaghar and Miller 1975) and by Roth (1975).

2.2. *Transport properties in the tight-binding model*

Extensions of the tight-binding theories to the calculation of electrical transport coefficients, such as electrical conductivity, thermopower and the Hall coefficient, is straightforward. Just in the same way as the IY and other liquid tight-binding theories extend the Matsubara–Toyozawa (1961) theory for the electronic structure of a completely random distribution to the case with atomic correlations, we can extend the theories for transport developed by Matsubara and Kaneyoshi (1966, 1968) for the random case to the liquid case.

The first such attempt was made by Ishida and Yonezawa (1973). They calculated the DC conductivity of the single s-band model using the Green function obtained in the IY approximation. A more systematic investigation of the problem has been performed by Itoh (1975). He derived expressions for the conductivity of the single s-band system which are consistent with both the IY and EMA calculations of the density of states. A formulation for the multi-band case has been given by Niizeki (1976 private communication). Yonezawa *et al* (1977) have also discussed the electrical conductivity of the multi-band system.

Here we explain briefly the calculation of the conductivity in the IY approximation. Following Matsubara and Toyozawa (1961), Matsubara and Kaneyoshi (1966) we can write the conductivity as:

$$\sigma_{\alpha\beta} = \int \mathrm{d}E \left(-\frac{\partial f(E)}{\partial E}\right) \sigma_{\alpha\beta}(E), \tag{14}$$

$$\sigma_{\alpha\beta}(E) = \sigma^{++}_{\alpha\beta}(E) - \sigma^{+-}_{\alpha\beta}(E) - \sigma^{-+}_{\alpha\beta}(E) + \sigma^{--}_{\alpha\beta}(E), \tag{15}$$

$$\sigma^{\pm\pm}_{\alpha\beta}(E) = \frac{\hbar}{\pi\Omega}\left\langle \sum_{i,j,l,m} \mathscr{J}_{\alpha,mi}\, \mathscr{G}_{ij}(E^{\pm})\, \mathscr{J}_{\beta,jl}\, \mathscr{G}_{lm}(E^{\pm})\right\rangle, \tag{16}$$

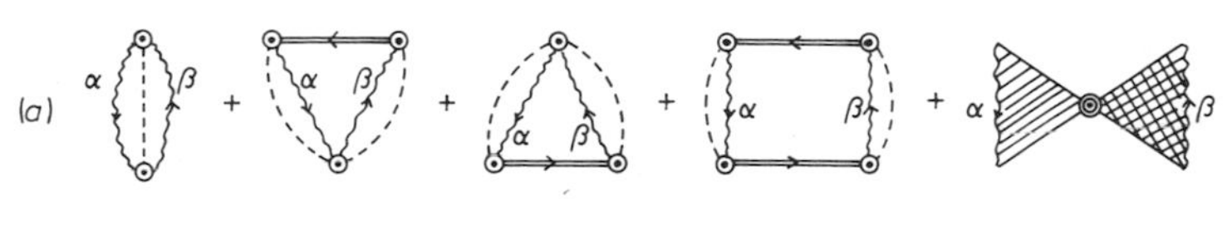

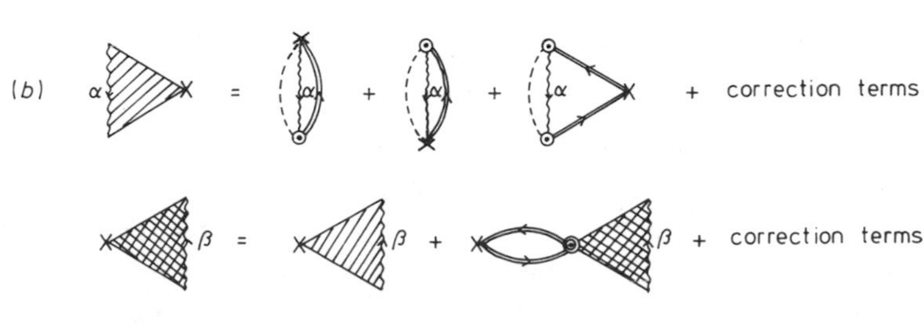

(c)

Figure 6. Diagrammatic equations for the calculation of the conductivity tensor $\sigma_{\alpha\beta}(\alpha, \beta = x, y, z)$ in the Ishida–Yonezawa approximation.

where $E^{\pm} = E + i0^{\pm}$, $\mathscr{J}_{ij}$ is the current operation in the site representation given by

$$\mathscr{J}_{ij} = \left(-\frac{e}{i\hbar}\right) \mathbf{R}_{ij}\ t(\mathbf{R}_{ij}) \tag{17}$$

and $f(E)$ is the usual Fermi distribution function. In the case of the multi-band system, $\mathscr{J}_{ij}$ and $\mathscr{G}_{ij}$ become matrices. Expanding the Green functions in equation (16) in terms of the transfer integral t_{ij}, we obtain the perturbation series which can be conveniently described by using the diagrammatic method. Retaining only the single-site diagram and employing the extended chain approximation, just in the same way as the calculation of the density of states in the IY theory, we obtain the equation for determining the conductivity in the IY approximation. This is described diagrammatically in figure 6. Wavy lines denote the current vertices. The vertex corrections described by the last diagram of figure 6(*a*) are determined by the equations described by the diagrams in figure 6(*b*). (The correction terms are needed to correct the improper treatment of the atomic correlation in the lowest-order terms.) It can be easily shown that these vertex corrections vanish for the single s-band model. The conductivity for this model is then given by

$$\sigma_{\alpha\beta}(E) = \delta_{\alpha\beta}\left(\frac{2\pi\hbar\rho}{3}\right)^2 \int \frac{\mathrm{dk}}{(2\pi)^3}\ |\mathbf{J}(\mathbf{k})|^2 \left[-\frac{1}{\pi}\,\mathrm{Im}\ G(\mathbf{k}, E^{+})\right]^2, \tag{18}$$

where $\mathbf{J}(\mathbf{k}) = (e/\hbar)\ (\partial\tilde{t}(\mathbf{k})/\partial\mathbf{k})$, with $\tilde{t}(\mathbf{k})$ the Fourier transform of $t(R)\,g(R)$, and

$$G(\mathbf{k}, z) = G_{\mathrm{d}}(z) + G_{\mathrm{d}}\rho\tilde{t}(\mathbf{k})\ G(\mathbf{k}, z) = [z - \Sigma_{\mathrm{d}}(\mathbf{k}) - \rho\tilde{t}(\mathbf{k})]^{-1}. \tag{19}$$

For the multi-band case, however, the vertex corrections do not vanish and a prescription for how to calculate their contributions has been given recently by Niizeki (1976 private communication).

For the single s-band case it is also easy to calculate the Hall coefficient in the IY approximation by extending the Matsubara–Kaneyoshi (1968) theory (see also Fukuyama *et al* 1970) to the liquid case. We obtain the antisymmetric part of the off-diagonal element of the conductivity tensor as

$$\begin{aligned}\sigma_{xy}^{(\mathrm{a})} &= \tfrac{1}{2}(\sigma_{xy} - \sigma_{yx}) \\ &= \left(\frac{2\pi^2}{9}\right)\left(\frac{e^3H}{\hbar^2c}\right)\int \mathrm{d}E\left(-\frac{\partial f}{\partial E}\right)\int \frac{\mathrm{dk}}{(2\pi)^3}\,\frac{1}{k}\left[\frac{\mathrm{d}\tilde{t}(k)}{\mathrm{d}k}\right]^3\left[-\frac{1}{\pi}\,\mathrm{Im}\ G(k, E^{+})\right]^3\end{aligned} \tag{20}$$

from which we can calculate the Hall coefficient by the well known formula

$$R = \sigma_{xy}^{(\mathrm{a})}/H\sigma_{xx}^2. \tag{21}$$

H in equations (20) and (21) denotes the magnetic field.

As an example of the application of the above formulae, we show in figure 7 the results for the hard-core random liquid with hydrogenic atomic orbitals as mentioned before (Itoh 1976). The parameters are taken as $p = 0{\cdot}25$ and $\eta = 0{\cdot}44$, which are very close to the values taken by Ishida and Yonezawa (1973) to simulate liquid Ni. Figure 7(*a*) shows the density of states in the IY approximation for this model, and figures 7(*b*) and (*c*) show respectively the usual DC conductivity (diagonal part of the conductivity tensor), σ_{xx}, and the Hall conductivity defined by $\sigma_H = \sigma_{xy}^{(\mathrm{a})}/H$. In these

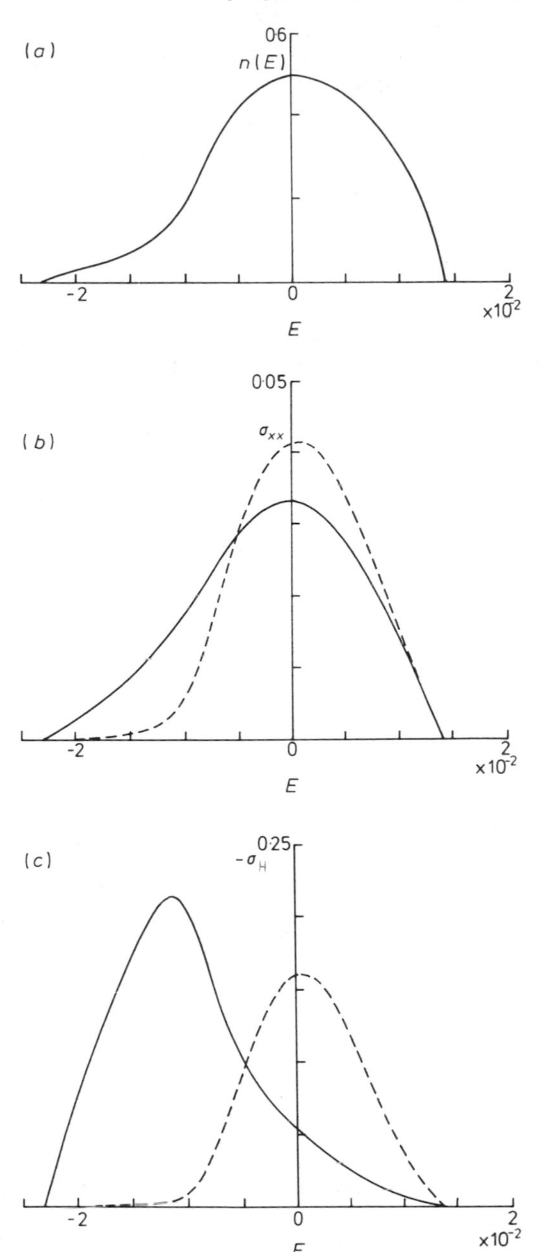

Figure 7. (*a*) Density of states, (*b*) DC conductivity and (*c*) Hall conductivity calculated in the Ishida–Yonezawa approximation for a single hydrogenic s-band hard-core random liquid with model parameters chosen to simulate liquid Ni at the melting point.

figures σ_{xx} and σ_H calculated in the IY approximation are compared with the values of the same quantities calculated in the random-phase model (RPM) due to Mott (1967), Hindley (1970) and Friedman (1971) (see also Kaneyoshi 1976). The RPM results correspond to retaining only the diagonal Green functions and hence can be obtained from equations (18) and (20) by replacing $G(k, E)$ with $G_d(E)$. Agreement between the IY and the RPM results is rather good for σ_{xx} except in the low-energy region, but for

σ_H, agreement is quite poor over the whole energy range. For both σ_{xx} and σ_H, the IY values in the low-energy region are very large compared with the small values of the RPM. Figure 8 shows the ratio of the calculated Hall coefficient in the IY approximation to the free-electron value defined by

$$R_{\mathrm{fe}}(E) = -\,(n_{\mathrm{el}}(E)\,ec)^{-1}, \tag{22}$$

$$n_{\mathrm{el}}(E) = \int_{-\infty}^{E} n(E')\,\mathrm{d}E', \tag{23}$$

where $n(E)$ is the density of states calculated in the IY approximation (shown in figure 7(*a*)). In the lower energy region, the ratio is rather close to unity (free-electron behaviour) but in the peak region $R(E)$ is very small compared with the free-electron value. These results can be understood from the behaviour of the spectral function

$$\begin{aligned}\rho(k, E) &= -\,(1/\pi)\ \mathrm{Im}\ G(k, E^{+}) \\ &= \frac{1}{\pi}\,\frac{|\mathrm{Im}\ \Sigma_d|}{(E - \mathrm{Re}\ \Sigma_d - \rho\tilde{t}(k))^2 + (\mathrm{Im}\ \Sigma_d)^2}\end{aligned} \tag{24}$$

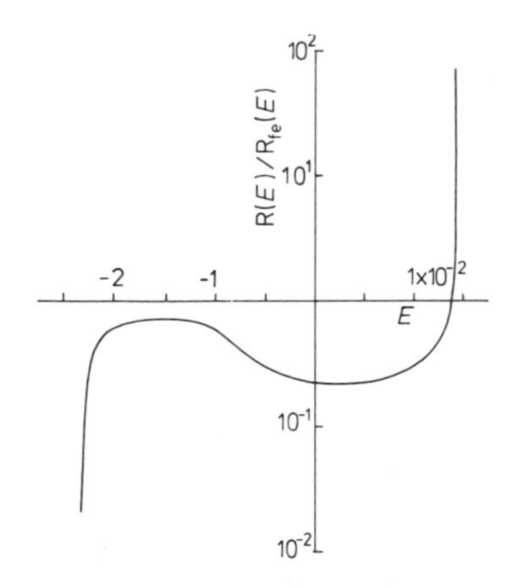

Figure 8. Ratio of the Hall coefficient in the Ishida–Yonezawa approximation to the free-electron value calculated for the same model as used in figure 7.

shown in figure 9. In the lower energy region the spectral function is sharply peaked, which means that we can treat the states there as being nearly the same as free-electron. Roth (1977) has made a similar calculation of the Hall coefficient.

In order to discuss the transport properties of realistic transition metals it is necessary to include the contribution of the s-band and the effect of s–d hybridization in some ways. This has been attempted recently by ten Bosch and Bennemann (1975). They investigated the s–d hybridization by treating the d-electrons in the tight-binding approximation and the s-electrons in the weak scattering approximation. From their numerical results they suggest that the d-electron contributions to the electrical conductivity in liquid transition and rare earth metals are important. Ishida *et al* (1977) also considered the s–d hybridization effects based on their multi-band theory for the d-electron Green function.

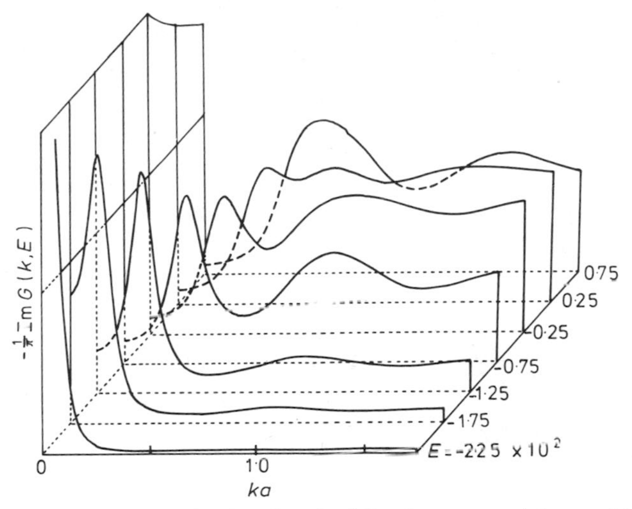

Figure 9. 'Spectral weight function' for the same model as used in figures 7 and 8.

3. Muffin-tin potential model (Ziman–Lloyd formalism)

As is well known, the muffin-tin model has been studied extensively and shown to be a very useful model for describing the electronic band structure of crystalline metallic systems. Investigation of the model for a liquid metal was initiated by Ziman (1966) and by Lloyd (1967a,b). In particular, Lloyd (1967a) derived a formula for the density of states of a system of non-overlapping spherical potentials which is valid for any configuration of scattering centres. The formalism is a natural generalization of the Green function method, or KKRZ method, of determining the band structure of crystalline systems (see, for example, Ziman 1972a) and reduces to it for a regular lattice. Lloyd (1967b) also considered the application of the formalism to a liquid metal.

The formula derived by Lloyd (1967a,b) for the integrated density of states for a muffin-tin liquid (i.e., the Lloyd formula), is given as

$$N(E) = N_0(E) + \frac{1}{\pi N} \langle \mathrm{Im\ Tr\ ln\ } \mathbf{T}(E) \rangle. \tag{25}$$

Here $N_0(E)$ is the integrated density of states for free electrons and $\mathbf{T}(E)$ is a generalized total t-matrix:

$$\mathbf{T}^{-1}(E) = \| \delta_{ij}\, \delta_{LL'}\, t_l^{-1}(E) - G_{0,LL'}(\mathbf{R}_i - \mathbf{R}_j, E) \|. \tag{26}$$

The labelling of the matrix specifies the positions of the individual atoms as well as the angular momentum subscripts $L = (l, m)$.

$$t_l(E) = -E^{-1/2} \exp\,[\mathrm{i}\delta_l(E)]\ \sin \delta_l(E)$$

is the t-matrix of the individual atom, $\delta_l(E)$ being the scattering phase shift at the energy E, and $G_{0,\,LL'}(\mathbf{R}_i - \mathbf{R}_j, E)$ is a quantity related to the free-particle propagator (see, for its precise definition, Lloyd 1967a,b, and a review article by Lloyd and Smith 1972).

From equation (25) the excess density of states is obtained as

$$\delta n(E) = n(E) - n_0(E)$$

$$= -\frac{1}{\pi N}\left\langle \mathrm{Im\ Tr}\ \mathbf{T}(E)\frac{\partial \mathbf{T}(E)^{-1}}{\partial E}\right\rangle$$

$$= -\frac{1}{\pi N}\ \mathrm{Im}\left[\left\langle \sum_i \mathrm{Tr}\ \langle \mathbf{T}_{ii}(E)\frac{\partial \mathbf{t}(E)^{-1}}{\partial E}\right\rangle - \sum_{i\neq j}\mathrm{Tr}\left\langle \mathbf{T}_{ij}(E)\frac{\partial \mathbf{G}_0(\mathbf{R}_j - \mathbf{R}_i, E)}{\partial E}\right\rangle\right] \quad (27)$$

where in the last expression the trace operation with respect to the atomic labels is explicitly written and the remaining trace denotes only that over the angular momentum subscripts. The matrices are defined as $\mathbf{t}(E) = \| \delta_{LL'} t_l(E) \|$, $\mathbf{G}_0(\mathbf{R}, E) = \| G_{0,\ LL'}(\mathbf{R}, E) \|$ and $\mathbf{T}_{ij}(E)$ is the (i, j) matrix element of $\mathbf{T}(E)$, which is given as a perturbation series expansion with respect to $\mathbf{t}(E)$ by

$$\mathbf{T}_{ij} = \mathbf{t}\delta_{ij} + \mathbf{t}(1 - \delta_{ij})\ \mathbf{G}_0(\mathbf{R}_i - \mathbf{R}_j)\,\mathbf{t} + \sum_l{}'\ \mathbf{t}\mathbf{G}_0(\mathbf{R}_i - \mathbf{R}_l)\ \mathbf{t}\mathbf{G}_0(\mathbf{R}_l - \mathbf{R}_j)\,\mathbf{t}$$

$$+ \sum_{l,m}{}'\ \mathbf{t}\mathbf{G}_0(\mathbf{R}_i - \mathbf{R}_l)\ \mathbf{t}\mathbf{G}_0(\mathbf{R}_l - \mathbf{R}_m)\ \mathbf{t}\mathbf{G}_0(\mathbf{R}_m - \mathbf{R}_j)\,\mathbf{t} + \ldots. \quad (28)$$

By comparing the above results with the equations in the multi-band tight-binding theory it is clear that the two formalisms are equivalent if we note the following correspondence:

Muffin-tin model		Tight-binding model
$\mathbf{t}(E)$	⟷	$\mathbf{G}_0(E)$
$\mathbf{G}_0(\mathbf{R}_{ij}, E)$	⟷	$\mathbf{H}'(\mathbf{R}_{ij})$
$\mathbf{T}_{ij}(E)$	⟷	$\mathbf{G}_{ij}(E)$
$\delta n(E)$	⟷	$n(E)$

It is also of some interest to note that if we introduce a matrix

$$\mathbf{G}(E) = \| \mathbf{G}_{ij}(E) \| = \| \delta_{ij}\delta_{\mu\nu}\, G_{0,\mu}(E)^{-1} - H'_{\mu\nu}(\mathbf{R}_{ij}, E) \|^{-1}$$

where ν and μ are the suffixes specifying the relevant atomic levels, we can write the integrated density of states in the tight-binding model as

$$N(E) = \frac{1}{\pi N}\langle \mathrm{Im\ Tr\ ln}\ \mathbf{G}(E)\rangle. \quad (29)$$

By noting the above correspondence the approximation schemes proposed for the tight-binding model can easily be extended to the muffin-tin model. In fact an approximation scheme equivalent to the IY theory was proposed by Lloyd (1967b) for the muffin-tin case before Ishida and Yonezawa (1973) developed their scheme for the tight-binding case. (The scheme might therefore be more accurately called the Lloyd–Ishida–Yonezawa (LIY) approximation.) The modified quasi-crystalline approximation (MQCA) proposed recently by Schwartz *et al* (1975) can be shown to be simply the non-self-consistent version of the Lloyd–Ishida–Yonezawa approximation.

The muffin-tin model has been applied to the calculation of the density of states of liquid noble and transition metals by several people including Peterson *et al* (1976) for liquid Cu in the MQCA, Honda (1976) for liquid Cu in the MQCA and the LIY approximation, Asano and Yonezawa (1977) for liquid Ni in the MQCA, the LIY, and the EMA, and papers presented at this conference). The results of these numerical studies have revealed that a more careful investigation of the properties (especially the analyticity property) of the solutions is required in this model than in the tight-binding model. In fact it has been shown by the authors mentioned above that the MQCA and the LIY theories lead to very unsatisfactory results, producing a negative density of states for certain energies within the d-bands. Only the EMA studied by Asano and Yonezawa (1977) for liquid Ni gives a physically acceptable result, and this seems to suggest that the requirement of the full self-consistency should be very important for obtaining a reliable description of the electronic spectrum in the muffin-tin model. For further discussions on the EMA results for liquid Ni, see Asano and Yonezawa (1977). Further work in this direction is desirable.

Acknowledgment
I should like to thank Dr Fumiko Yonezawa for many helpful discussions on the problems presented in this paper.

References

Asano S and Yonezawa I 1977 this volume
Busch G and Güntherodt H-J 1974 *Solid St. Phys.* **29** 235 (New York: Academic Press)
Faber T E 1972 *Introduction to the Theory of Liquid Metals* (London: Cambridge Univ. Press)
Friedman L 1971 *J. Non. Cryst. Solids* **6** 329–41
Fukuyama H, Saitoh M, Uemura Y and Shiba H 1970 *J. Phys. Soc. Japan* **28** 842–60
Hindley N K 1970 *J. Non. Cryst. Solids* **5** 17–30
Honda K 1976 *MSc Thesis* Tohoku University
Ishida Y and Yonezawa I 1973 *Prog. Theor. Phys.* **49** 731–53
Ishida Y, Martino F, Kushida K and Yonezawa F 1977 to be published
Itoh M 1975 *PhD Thesis* Tohoku University
—— 1976 to be published
Kaneyoshi T 1976 *Phil. Mag.* **33** 11–20
Katz I and Rice S A 1972 *J. Phys. C: Solid St. Phys.* **5** 1165–82
Lloyd P 1967a *Proc. Phys. Soc.* **90** 207–16
—— 1967b *Proc. Phys. Soc.* **90** 217–31
Lloyd P and Smith P V 1972 *Adv. Phys.* **21** 69–142
Martino F and Yonezawa F 1975 *J. Phys. F: Metal Phys.* **5** 1146–9
Matsubara T and Kaneyoshi T 1966 *Prog. Theor. Phys.* **36** 695–711
—— 1968 *Prog. Theor. Phys.* **40** 1257–72
Matsubara T and Toyozawa Y 1961 *Prog. Theor. Phys.* **26** 739–56
Mott N F 1967 *Adv. Phys.* **16** 49–144
Movaghar B and Miller D E 1975 *J. Phys. F: Metal Phys.* **5** 261–77
Movaghan B, Miller D E and Bennemann K H 1974 *J. Phys. F: Metal Phys.* **4** 687–702
Peterson H K, Bansil A and Schwartz L 1976 *Structure and Excitations of Amorphous Solids* (Am. Inst. Phys. Conf. Proc. 31) pp 378–83
Roth L M 1972 *Phys. Rev. Lett.* **28** 1570
—— 1973 *Phys. Rev.* **B7** 4321–37
—— 1974a *J. Physique* **35 C4** 317–23
—— 1974b *Phys. Rev.* **B9** 2476–84
—— 1975 *Phys. Rev.* **B11** 3769–79

—— 1976 *J. Phys. F: Metal Phys.* **6** 2267–88
—— 1977 this volume
Schwartz L, Peterson H K and Bansil A 1975 *Phys. Rev.* **B12** 3113–23
ten Bosch A and Bennemann K H 1975 *J. Phys. F: Metal Phys.* **5** 1333–41
Watabe M and Yonezawa F 1975 *Phys. Rev.* **B11** 4753–62
Yonezawa F 1968 *Prog. Theor. Phys.* **40** 734–57
Yonezawa F, Ishida Y, Martino F and Asano S 1977 this volume
Yonezawa F, Ishida Y and Martino F 1976 *J. Phys. F: Metal Phys.* **6** 1091–112
Yonezawa F and Martino F 1976 *J. Phys. F: Metal Phys.* **6** 739–47
Yonezawa F and Morigaki K 1973 *Prog. Theor. Phys. Suppl.* **53** 1–76
Yonezawa F, Roth L M and Watabe M 1975 *J. Phys. F: Metal Phys.* **5** 435–42
Yonezawa F and Watabe M 1975 *Phys. Rev.* **B11** 4746–52
Yonezawa F, Watabe M, Nakamura M and Ishida Y 1974 *Phys. Rev.* **B10** 2322–32
Ziman J M 1966 *Proc. Phys. Soc.* **88** 387–405
—— 1972 *Principles of the Theory of Solids* 2nd edn (London: Cambridge UP)
—— 1973 *The Properties of Liquid Metals* ed S Takeuchi (London: Taylor and Francis) pxiii

A muffin-tin potential model of a liquid transition or noble metal: formalisms and numerical results of the electronic density of states in various approximations

S Asano† and F Yonezawa‡
† Department of Physics, Tokyo University, Komaba, Meguro-ku, Tokyo, Japan
‡ Research Institute for Fundamental Physics, Kyoto University, Kyoto 606, Japan

Abstract. For the purpose of evaluating the electronic density of states of non-simple liquid metals such as the liquid transition and noble metals, we first study various approximation schemes in a muffin-tin potential model of a liquid metal. In this model, individual scattering centres are represented by non-overlapping, spherically symmetric atomic potentials. It is shown that the general formalism in the work of Ziman and Lloyd is analogous to that used in the renormalized perturbation theories in the tight-binding approximation (TBA). All the approximate treatments developed for the TBA therefore apply to the muffin-tin potential model. On the basis of this formalism we have performed numerical calculations of the density of states of liquid Ni in several single-site approximations (SSA) such as the Ishida–Yonezawa (IY) method (equivalent to the modified quasi-crystalline approximation), the self-consistent IY method (equivalent to the Lloyd approximation) and the effective medium theory (equivalent to the coherent potential approximation). We also discuss the analyticity of these SSA theories.

1. Introduction

There has been a sizeable amount of work carried out on the electronic structure and other electronic properties of liquid metals. As long as simple liquid metals are concerned, the nearly-free electron (NFE) theories have been successful in explaining characteristic features of the electronic properties. In the NFE model, electrons are scattered by weak scattering centres distributed randomly and interaction with these weak scatterers is described by a pseudopotential which is treated as a perturbation. It is widely accepted that the problems concerning simple liquid metals have been essentially solved. The current trend of research in this field therefore moves toward the refinement of theories and the attainment of quantitative results. On the other hand, we are still on the way towards a systematic understanding of non-simple liquid metals. By non-simple metals we mean those for which the NFE theories no longer hold, such as the transition, noble and rare earth metals, their alloys, expanded fluid metals, etc. From the point of view of studying the electronic structure and some other properties of non-simple liquid metals, the tight-binding approximations (TBA) have been extensively developed (for references, see, for instance, Watabe 1977). Although the TBA serves as a reasonable and yet simple first approach to the strong scattering limit, not only from the purely theoretical point of view but also from the practical point of view, it would be still desirable to have another model of a realistic non-simple liquid metal which could be complementary to the TBA. A muffin-tin potential model has been regarded as such and this is the theme of the present paper.

The definition of a muffin-tin potential model of a liquid metal and the basic ideas concerning the treatment of the model may be most concisely given by quoting the following paragraph from Ziman (1966). 'A large number of identical isotropic spheres are distributed throughout a large space, and bathed in a continuous uniform medium which would, in the absence of the spheres, allow the lossless propagation of waves. But these waves are scattered by the spheres, so that the excitations are dissipated and dispersed. The problem is to construct a formula for propagation in the compound system as if this were a macroscopically uniform continuum. Such a formula should depend only on the scattering properties of the individual spheres, and on their spatial distribution.' The statement in the last sentence is especially important. An additional advantageous aspect of the muffin-tin potential model is that it covers both the weak and the strong scattering limits.

Since the quantities of physical interest are given by an average over all possible spatial distributions of the spheres, we have to specify appropriate methods for the ensemble average. As we shall see later, the Ziman–Lloyd formalism has the same mathematical structure as that of the TBA for non-orthogonal orbitals. Therefore, the approximations that have been developed extensively for the TBA can also be applied to the case of the muffin-tin potential model. With this in mind, we will discuss in §3 the nature of various single-site approximations (SSA).

As an application of our formalism to a realistic system, we give the results of our calculation of the electronic structure of liquid Ni using various SSA theories. The modified quasi-crystalline approximation (MQCA) gives a negative density of states. The Ishida and Yonezawa (1973) theory (IY) works well and gives a reasonable density of states in the energy region near the resonance but seems to show a breakdown in analyticity at higher energies. On the other hand, the effective medium approximation (EMA) (Roth 1974) seems to work in the whole energy region.

2. Formalism

The general expression for the integrated density of states of the model system defined in the previous section has been given by Lloyd (1967) as

$$N(E) = N_0(E) - (\pi V)^{-1} \operatorname{Im} \ln [\det \| \delta_{LL'} \delta_{jj'} + t_L G_{LL'}(\mathbf{R}_j - \mathbf{R}_{j'}) \|] . \tag{1}$$

$N_0(E)$ is the free-electron integrated density of states, V is the volume of the system, t_L is related to the phase shift, $\eta_L(E)$, by $t_L = \tan \eta_L(E)/E^{1/2}$ and $G_{LL'}(\mathbf{R}_j - \mathbf{R}_{j'})$ is related to the free-particle propagator. The determinant is over both L and j, where L stands for the pair of angular momentum quantum numbers l and m. The formula (1) is defined for a given configuration of scattering centres. The quantities of physical interest, however, are given by an average over all possible configurations. What we need therefore is the ensemble-averaged density of states $n(E)$ which is related to $N(E)$ by

$$n(E) = \left\langle \frac{\mathrm{d}}{\mathrm{d}E} N(E) \right\rangle . \tag{2}$$

Since it is difficult to take directly the ensemble average of equation (1) or of its derivative, the general steps that are usually taken are described as follows:

(i) Expand equation (1) in powers of t_L and $G_{LL}'(\mathbf{R}_j - \mathbf{R}_j')$ and take the first derivative of each term in the expansion series with respect to the energy, E.
(ii) Take the ensemble average of each term in the expansion series.
(iii) Take the summation of the averaged terms.

For this purpose, Lloyd has given the expanded expression of $N(E)$ as:

$$N(E) = N_0(E) - (\pi V)^{-1} \operatorname{Im} \left\{ \sum_{n=1}^{\infty} \frac{1}{n} \operatorname{Tr} [-t_L G_{LL'}(\mathbf{R}_j - \mathbf{R}_{j'})]^n \right\}. \tag{3}$$

Here, the symbol for power represents a matrix power, with the matrix indices being (L, j). In order to reduce the matrix subscripts to only the angular momentum indices (L), Lloyd introduced diagrams to represent those parts of the above matrix power and trace which involve the atomic positions. Each atomic position is represented by a point (hereafter referred to as an atom) and each factor which contains two atomic positions is represented by a line joining those points (hereafter referred to as an electron line).

Following Lloyd (1967), we can now obtain a series for the usual density of states by differentiating equation (3) with respect to E. Differentiation using the chain rule shows that the factor $1/n$ which appears in equation (3) is removed and, accordingly, the summation of the diagrams at a later stage is made easier. The diagrammatic series for the density of states is then shown to be expressed by

$$\frac{\mathrm{d}N(E)}{\mathrm{d}E} = \frac{\mathrm{d}N_0(E)}{\mathrm{d}E} - \frac{1}{\pi V} \operatorname{Im} \left[\operatorname{Tr}\left(\frac{\mathrm{d}\mathbf{t}(E)}{\mathrm{d}E} \mathbf{D}_1 \right) + \operatorname{Tr}\left(\frac{\mathrm{d}\mathbf{G}}{\mathrm{d}E} \mathbf{D}_2 \right) \right], \tag{4}$$

where D_1 is the contribution of all possible topologically different diagrams which start from a specified atom and eventually return to that atom, while D_2 is the contribution from all diagrams which start from a specified atom and eventually return to a second specified atom. The matrix $\mathbf{t}(E)$ is defined by $\langle L | t(E) | L' \rangle \equiv \{-t_L \delta_{LL'}\}$. The contribution of each diagram is then counted by associating the free-particle (angular momentum matrix) propagator $\mathbf{G}(\mathbf{R}_i - \mathbf{R}_j) \equiv \{G_{LL'}(\mathbf{R}_i - \mathbf{R}_j)\}$ with each electron line and a $\mathbf{t}(E)$ matrix with each scattering atom.

3. Approximations

The next step is to take the ensemble average of each diagram in equation (4). It is easy to see that the formula is analogous to the perturbation theory employed in the tight-binding (TB) approaches. For the sake of explanation, it is convenient to use the formulation presented by Yonezawa *et al* 1976 (hereafter referred to as YIM). The averaged density of states per atom is given by

$$n(E) = \pi^{-1} \operatorname{Im} \operatorname{Tr} (\hat{\mathscr{G}}_{ii} + \rho \int \hat{S}^*_{ij} \hat{\mathscr{G}}_{ij} g(R_{ij}) (1 - \delta_{ij}) \, \mathrm{d}^3 \mathbf{R}_{ij}), \tag{5}$$

where ρ is the number density of atoms, $g(R_{ij})$ is the pair distribution function and S_{ij} is the overlap integral. $\hat{\mathscr{G}}_{ii}$ and $\hat{\mathscr{G}}_{ij}$ are the conditional averages of the Green's function as defined in YIM. Using the diagram representation, $\hat{\mathscr{G}}_{ii}$ is the ensemble average of the contribution of all possible topologically different diagrams which start from a specified atom and eventually return to that atom, while $\hat{\mathscr{G}}_{ij}$ is the conditional average, with $\mathbf{R}_{ij}$ being kept fixed, of all diagrams which start from a specified atom and eventually return to a second specified atom. In the case of the TB treatments, an unperturbed locator matrix, $\hat{G}^{(0)} = (E^+ - E_\mu)^{-1}$, is associated with a point in the diagrams, while a transfer matrix, $\hat{t}(\mathbf{R}_{ij}) = \{t^{\mu\nu}(\mathbf{R}_{ij})\}$, is assigned to a line joining sites i and j. Thus, the one-to-one correspondence between the second term of equations (4) and (5) is apparent and interpretation is straightforward. That is, $\hat{G}(0) \leftrightarrow \hat{\mathbf{t}}(E)$ and $\hat{t}(\mathbf{R}_{ij}) \leftrightarrow \mathbf{G}(\mathbf{R}_{ij}) = \{G_{LL'}(\mathbf{R}_{ij})\}$.

Both equation (5) and the average of equation (4), $n(E) = \langle \mathrm{d}N(E)/\mathrm{d}E \rangle$, are the exact expressions for the averaged density of states. In practice, however, the exact summation of all the averaged terms is usually impossible. In addition, taking the ensemble average of complicated terms in these expansion series is itself difficult since there we need many-point atomic distribution functions $g^{(n)}(\mathbf{R}_1, \mathbf{R}_2, \ldots \mathbf{R}_n)$. In actual calculations, approximations enter the formalism at the stages (ii) and (iii) described in §2. There have been quite a number of papers on various approximation schemes in the TB model concerning these two stages (for detailed references, see for instance, Watabe 1977). Existing approximations at stage (ii) are characterized by the way in which $g^{(n)}(\mathbf{R}_1, \mathbf{R}_2, \ldots \mathbf{R}_n)$ are decoupled in terms of $g(\mathbf{R}_{12})$ etc, while at stage (iii) we approximate the total summation by appropriate partial summations of dominant terms.

For our numerical calculation of the density of states of Ni on the basis of the muffin-tin potential model, we have picked out three single-site approximations (SSA): (i) the non-self-consistent Ishida–Yonezawa (NSCIY) method which is equivalent to the MQCA proposed by Schwartz *et al* (1975); (ii) the self-consistent Ishida–Yonezawa (SCIY) method which is equivalent to the Lloyd (1967) approximation; (iii) the EMA (Roth 1974) which is equivalent to the coherent potential approximation (Yonezawa *et al* 1975, Yonezawa and Watabe 1975, Watabe and Yonezawa 1975). The nature of these approximations is discussed fully in the above references.

The MQCA is the easier of the three with regard to actual calculations because the theory is not self-consistent. The SCIY introduces a renormalized scattering matrix $\tilde{\mathbf{t}}(E)$ and this must be determined in a self-consistent manner. For instance, the dispersion relation for either the MQCA or the SCIY is given by

$$\det \| \tilde{\mathbf{t}}^{-1}(E) - \mathbf{G}(E, \mathbf{k}) \| = 0, \tag{6}$$

where $\tilde{\mathbf{t}}(E)$ in the MQCA is defined directly in terms of $\mathbf{t}(E)$ and $\mathbf{G}(E, \mathbf{k})$, the Fourier transform of $G_{LL'}(\mathbf{R}_{ij})$, while in the SCIY, $\tilde{\mathbf{t}}(E)$ must be determined through a self-consistent equations which include $\mathbf{t}(E)$ and $\mathbf{G}(E, \mathbf{k})$. On the other hand, the EMA requires that both the scattering matrix and the propagator are defined in a self-consistent manner; this indicates that the renormalized propagator $\tilde{\mathbf{G}}(E, \mathbf{k})$ must be solved in a self-consistent manner for all values of $\mathbf{k}$ in the momentum space for each energy E, which needs a tremendous amount of time and money. The dispersion relation in the EMA is defined by

$$\det \| \tilde{\mathbf{t}}^{-1}(E) - \tilde{\mathbf{G}}(E, \mathbf{k}) \| = 0. \tag{7}$$

4. Numerical results

We now discuss some applications of the formula given above to realistic liquid metals. In this paper, we report the calculated density of states and the dispersion relation of liquid Ni and in a separate paper (Yonezawa *et al* 1977) we present the results for expanded liquid Hg. The present formula in the muffin-tin potential model is quite general, covering both the cases of simple and non-simple liquid metals, as explained

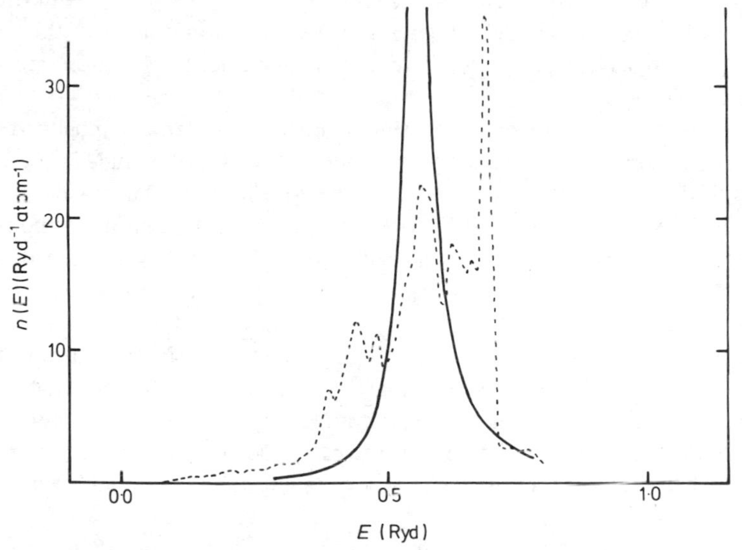

Figure 1. The density of states (solid curve) of a single muffin-tin potential of Ni placed in a free medium. The result is compared to the total density of states of crystal Ni (broken curve).

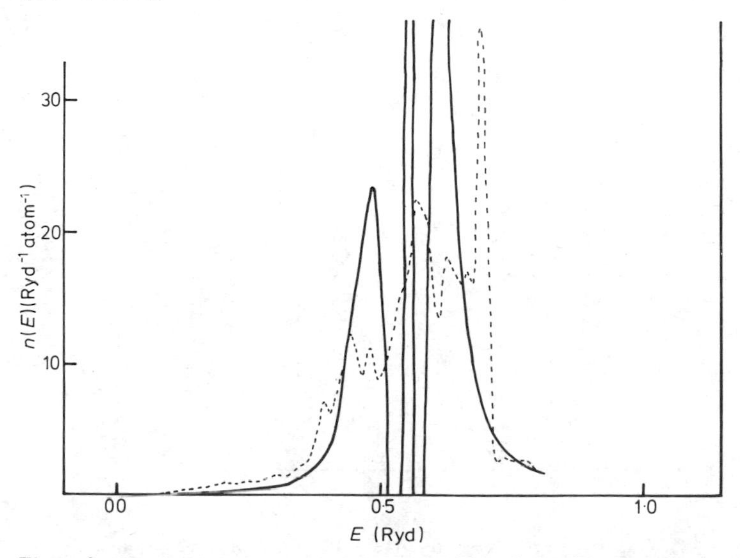

Figure 2. The density of states (solid curve) within the muffin-tin radius of liquid Ni in the MQCA compared to the total density of states of crystal Ni (broken curve).

in §1. Although in this paper we treat only liquid Ni, the detailed procedures used in our calculation remain the same for other liquid transition or noble metals. We even expect that the results obtained are similar, at least qualitatively, for all liquid transition and noble metals.

The muffin-tin potential is determined by using a self-consistent potential for a solid with a FCC structure, where the starting atomic wavefunctions are of the non-relativistic Hartree–Fock–Slater type with a Kohn–Sham value, $\lambda = \frac{2}{3}$. The muffin-tin radius r_i is chosen to be half the hard-core diameter σ; that is, $r_i = \frac{1}{2}\sigma = 2{\cdot}094$ a.u. The muffin-tin zero is given as $V_0 = -0{\cdot}0320$ Ryd. The Percus–Yevick solution is used for the pair-correlation function, $h(R) \equiv g(R) - 1$, where the packing fraction η is chosen to be 0·45.

Figure 1 gives the density of states of a single muffin-tin potential embedded in a free medium. The contribution from the free-electron part is not included. The result is compared to the total density of states of crystal Ni. Figures 2, 3 and 4 show the calculated densities of states of liquid Ni within the muffin-tin radius using the MQCA, the SCIY and the EMA. For comparison, in each of these figures the total density of states of crystal Ni is given by a broken curve. As is apparent from the figures, the MQCA yields a negative density of states near the resonance level, which means that the imaginary part of the self energy changes sign. The SCIY works over the energy region near the resonance level and gives a reasonable bandwidth, but at a higher energy, where the electrons are nearly free, the imaginary part of the self energy changes sign and analyticity seems to break down. The EMA on the other hand shows a reasonable behaviour over the whole energy region and the imaginary part of the self energy retains the same sign.

Therefore, our numerical work indicates that the EMA solution is analytic while analyticity does not hold in the MQCA and the SCIY. This fact concerning the various

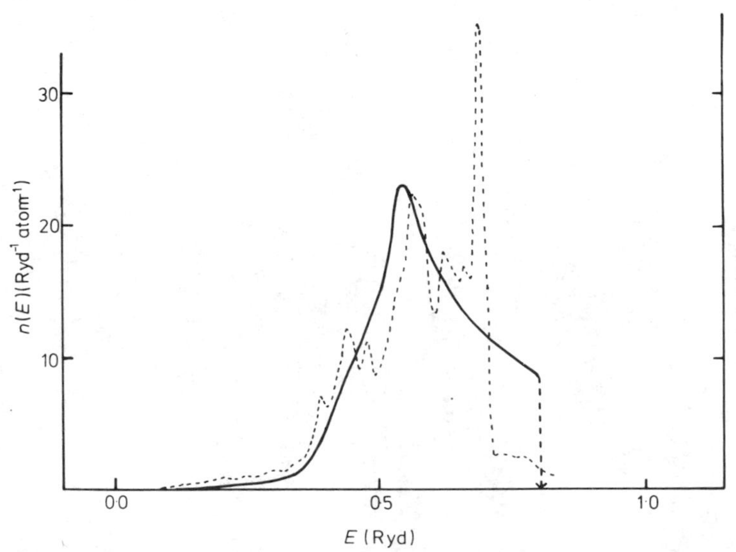

Figure 3. The density of states (solid curve) within the muffin-tin radius of liquid Ni in the SCIY approximation compared to the total density of states of crystal Ni (broken curve).

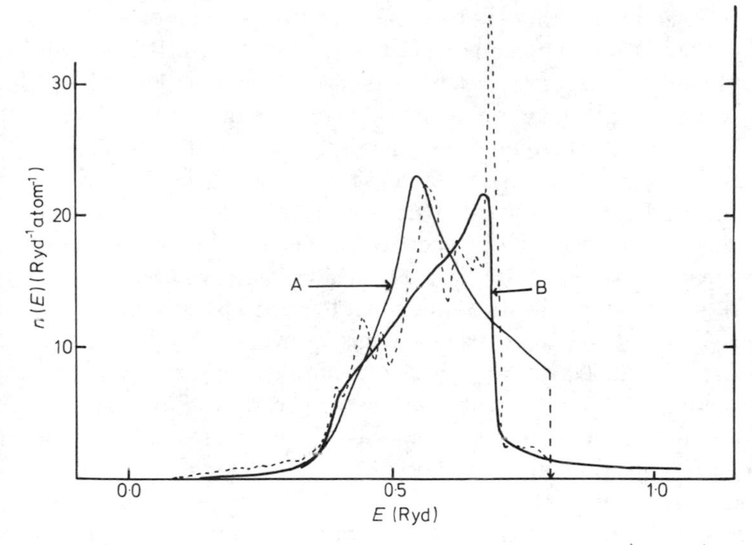

Figure 4. The densities of states within the muffin-tin radius of liquid Ni both in the SCIY (curve A) and in the EMA (curve B) compared to the total density of states of crystal Ni (broken curve).

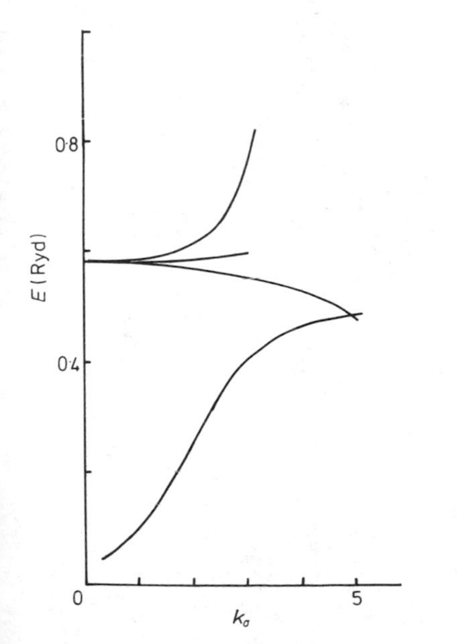

Figure 5. The dispersion relation, E versus Re k, of liquid Ni defined by equation (6) in the MQCA (Im k is not given in the figure).

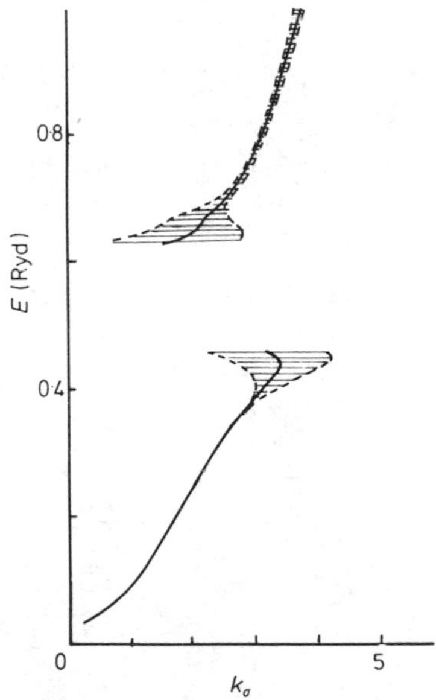

Figure 6. The dispersion relation of liquid Ni defined by equation (7) in the EMA; the solid curves in the centre represent E versus Re k while the imaginary part $|\text{Im}\, k|$ is shown by the hatched region.

approximations has not been observed in the TBA theories and therefore it seems that the present formula for the muffin-tin potential model more severely reveals the behaviour of the analyticity of the approximations. Although we have no detailed proof, we have some grounds to predict that the breakdown in the analyticity of the SCIY is related to their self-consistent requirements; that is, only the scattering matrix $\tilde{\mathbf{t}}(E)$ is renormalized in a self-consistent manner, as shown in equation (6). The EMA defines both the renormalized scattering matrix, $\tilde{\mathbf{t}}(E)$, and the renormalized propagator, $\tilde{\mathbf{G}}(E, \mathbf{k})$, by means of the simultaneous self-consistent equations. In other words, the self-consistent requirements for $\tilde{\mathbf{t}}(E)$ and $\tilde{\mathbf{G}}(E, \mathbf{k})$ are treated on the same basis.

The calculated density of states by the EMA shows that the sharp peak is maintained near the upper-band edge of the d-resonance band and this means that the characteristic feature of the crystal band for the FCC (close packed) structure remains, even for the case of a liquid, with no drastic change. Note that the Fermi level E_F for both crystal and liquid Ni falls at the sharp peak. These results for the EMA are compatible with the experimental result that the change of magnetic susceptibility of Ni is almost continuous at the melting point.

The dispersion curves are given in figures 5 and 6, figure 5 representing the relation E against Re k in the MQCA (defined by equation (6)) and figure 6 showing the relation E against Re k and a hatched region corresponding to $|\mathrm{Im}\, k|$ in the EMA. In the region near the resonance level, the imaginary part becomes so large that the dispersion relation no longer makes sense.

References

Ishida Y and Yonezawa F 1973 *Prog. Theor. Phys.* **49** 731–53
Lloyd P 1967 *Proc. Phys. Soc.* **90** 207–31
Roth L M 1974 *Phys. Rev.* **B9** 2476–84
Schwartz L, Peterson H K and Bansil A 1975 *Phys. Rev.* **B12** 3113–23
Yonezawa F, Roth L M and Watabe M 1975 *J. Phys. F: Metal Phys.* **5** 435–42
Yonezawa F, Ishida Y and Martino F 1976 *J. Phys. F: Metal Phys.* **6** 1091–101
Yonezawa F, Ishida Y, Martino F and Asano S 1977 this volume
Yonezawa F and Watabe M 1975 *Phys. Rev.* **B12** 4746–52
Watabe M 1977 this volume
Watabe M and Yonezawa M 1975 **B11** 4753–62
Ziman J M 1966 *Proc. Phys. Soc.* **88** 387–405

The electronic spectrum of d-band liquid metals

A Bansil†, H K Peterson‡ and L Schwartz‡

† Physics Department, Northeastern University, Boston, Massachusetts 02115, USA
§ Physics Department, Brandeis University, Waltham, Massachusetts 02154, USA

Abstract. Electronic states in liquid Cu are considered within the framework of the muffin-tin hamiltonian. Results for the complex energy bands, spectral functions and the average density of states are presented. Some aspects of the electronic structure of metal–metalloid glasses are also considered in light of the present calculations.

1. Introduction

Recent authors have devoted considerable attention to the one-electron properties of disordered metallic systems (Ehrenreich and Schwartz 1976). Of particular interest are materials which involve transition or noble metals. There the electrons are scattered by a d-resonance in the ionic potentials and methods based on perturbation theory are inherently unreliable. It is generally agreed therefore, that the electronic properties of this class of disordered materials are best treated within the framework of the muffin-tin model. This model is, of course, widely used in band structure calculations on the corresponding ordered solids. More recently, it has been applied to the study of random substitutional alloys. In particular, Bansil *et al* (1974, 1975) and Schwartz and Bansil (1974) have shown that the muffin-tin model together with the average t-matrix approximation (ATA) yields a satisfactory description of a variety of transition and noble metal alloys. While an analogous approach to the liquid metal problem, the quasicrystalline approximation (QCA) has been formulated by Lax (1951, 1952) and Ziman (1966) and has been shown by Peterson *et al* (1974) to give reliable results in the case of a simple one-dimensional model; calculations based on the muffin-tin hamiltonian have not yet been carried out.

The principal assumption underlying the muffin-tin model is that the total electronic potential can be written as a sum of spherically symmetric non-overlapping potentials $v(|\mathbf{r}-\mathbf{R}_\alpha|)$. (Here $\mathbf{R}_\alpha$ denotes the position of the αth atom.) Using the angular momentum representation, it can be shown quite generally that the ionic potentials $v(r)$ enter the calculation of the exact electronic density of states only through the energy shell matrix elements $\tau_l(E)$ of the corresponding scattering operators $t_l(k, k')$. (Physically, the fact that the potentials are non-overlapping implies that an electron propagates as a free particle between scattering events and, therefore, that only energy-shell matrix elements are necessary.) Because the density of states for each configuration of the liquid depends only on $\tau_l(E)$, the same is true for the ensemble-averaged spectrum. We emphasize, however, that this feature of the muffin-tin model may not be preserved by a given approximation scheme. Indeed, in an earlier publication (Schwartz *et al* 1975) the present authors have shown that the QCA density of states $\langle\rho(E)\rangle$ cannot be expressed in terms of just the quantities $\tau_l(E)$. (Physically, this aspect of the QCA is due to an approximate decoupling of the higher-order density correlation functions, which leads to a spurious overlap of the ionic spheres and, therefore, to the introduction of the off-shell matrix elements of the atomic scattering operators.) For this reason we proposed an alternative decoupling procedure, the modified QCA (MQCA) (Schwartz *et al* 1975), within which $\langle\rho(E)\rangle$ depends only on $\tau_l(E)$. We have found, however, that the MQCA does not lead to a reliable description of the electronic spectrum for a liquid noble or transition metal (Peterson *et al* 1976). Indeed, although the MQCA predicts a reasonable spectrum both above and below the d-bands, it can give a *negative* density of states within the d-bands. In view of this, we have reconsidered the implementation of the original QCA equations. We are encouraged in this attempt by the fact that one can prove, using arguments based on the optical theorem, that the QCA spectral functions (and hence the density of states) are positive definite. The equations relevant to the application of the QCA are summarized in §2 and the results of preliminary numerical calculations are described in §3.

2. QCA formalism

Within the QCA the average electronic density of states per atom $\langle\rho(E)\rangle$ is computed in terms of three input units: (i) the matrix elements $t_l(k, k')$ of the atomic scattering operators ($\tau_l(E) \equiv t_l(\kappa, \kappa)$ where $\kappa^2 = E$); (ii) the mean ionic density n; and (iii) the radial distribution function $g(R)$. The central equations are:

$$\langle\rho(E)\rangle = \frac{2}{n}\int \frac{\mathrm{d}^3k}{(2\pi)^3} A(\mathbf{k}, E) \tag{1}$$

where the spectral function $A(\mathbf{k}, E)$ is given by

$$A(\mathbf{k}, E) = -\pi^{-1}\,\mathrm{Im}\left\{(4\pi)^2 \sum_{LL'} F_L(\mathbf{k}, E)\,[\tau^{-1}(E) - B_{\mathbf{k}}(E)]^{-1}_{LL'}\, F^*_{L'}(\mathbf{k}, E)\right\}. \tag{2}$$

Here

$$F_L(\mathbf{k}, E) = \frac{Y_L(\mathbf{k})\, t_l(k, \kappa)}{(E - k^2)\, \tau_l(E)} \tag{3}$$

$$B^{LL'}_{\mathbf{k}}(E) = -(4\pi\mathrm{i}\kappa) \sum_{L_1} \mathrm{i}^{-(l-l')}\, C^{LL'}_{L_1}\, Y^*_{L_1}(\mathbf{k}) \int_0^\infty R^2\, \mathrm{d}R\, \mathrm{J}_l(kR)\, \mathrm{H}_l(\kappa R)\, g(R) \tag{4}$$

and the coefficients $C^{LL'}_{L_1}$ are given in terms of angular integrals of three spherical harmonics

$$C^{LL'}_{L_1} = \int \mathrm{d}\Omega_{\mathbf{x}}\, Y^*_L(\mathbf{x})\, Y_{L'}(\mathbf{x})\, Y_{L_1}(\mathbf{x}).$$

$Y_L(\mathbf{k})$ is the spherical harmonic of the angles associated with $\mathbf{k}$ and $\mathrm{J}_l(x)$ and $\mathrm{H}_l(x)$ are the spherical Bessel and Hankel functions respectively. In the work described below, $g(R)$ was taken from x-ray diffraction measurements (Waseda and Ohtani 1974) on molten Cu at 1150 °C and the ionic potentials were constructed by the renormalized atom procedure (Hodges *et al* 1972).

It can be shown quite generally that the spectral function $A(\mathbf{k}, E)$ is independent of the direction of the vector $\mathbf{k}$. (Physically, this reflects the fact that the liquid is, on average, an isotropic system.) The angular integration in equation (1) can then be performed by simply evaluating $A(\mathbf{k}, E)$ for $\mathbf{k}$ along the positive z axis. Having so oriented the vector $\mathbf{k}$, we find that the weighting factors $F_L(\mathbf{k}, E)$ and the structure functions $B_{\mathbf{k}}^{LL'}(E)$ reduce to (Schwartz *et al* 1975)

$$F_L(\mathbf{k}, E) \to \delta_{m,\,0} F_l(k, E) \equiv \delta_{m,\,0} \frac{[4\pi(2l+1)]^{1/2}\, t_l(k, \kappa)}{(E-k^2)\, \tau_l(E)} \tag{5}$$

and

$$B_{\mathbf{k}}^{LL'}(E) \to \delta_{m,\,m'} B_{|m|}^{ll'}(k, E) = \begin{cases} \delta_{m,\,m'}\, \delta_{l,\,l'}\, B_{|m|}^{l}(E), & k = 0 \\ \delta_{m,\,m'}\, B_{|m|}^{ll'}(k, E), & k \neq 0. \end{cases} \tag{6}$$

3. The electronic spectrum

It is possible to show rigorously that the free electron singularities present in equations (2) and (3) do not contribute to $A(k, E)$ (Mijnarends and Bansil 1976). In view of this fact the principal contributions to $\langle\rho(E)\rangle$ arise from the inverse matrix $[\tau^{-1} - B_{\mathbf{k}}]^{-1}$. Hence, the electronic spectrum may be characterized in terms of the complex energy bands determined by the secular equation

$$\|\tau^{-1}(E) - B_{\mathbf{k}}(E)\| = 0. \tag{7}$$

For a given real momentum k, solutions of equation (7) will be found at a sequence of complex energies $E_n(k)$. The real and imaginary parts of $E_n(k)$ describe, respectively, the mean energy and lifetime of the corresponding quasiparticle state.

In figure 1 we compare the complex spectrum of liquid Cu with the real energy bands of crystalline Cu. The symmetries exhibited by the liquid bands in the middle panel are related to the form of the structure functions as summarized in equation (6). The fact that the determinant $\|\tau^{-1} - B_{\mathbf{k}}\|$ is a function of just $|\mathbf{k}|$ implies that the complex bands in the liquid are spherically symmetric. In addition, equation (6) shows that at $k = 0$

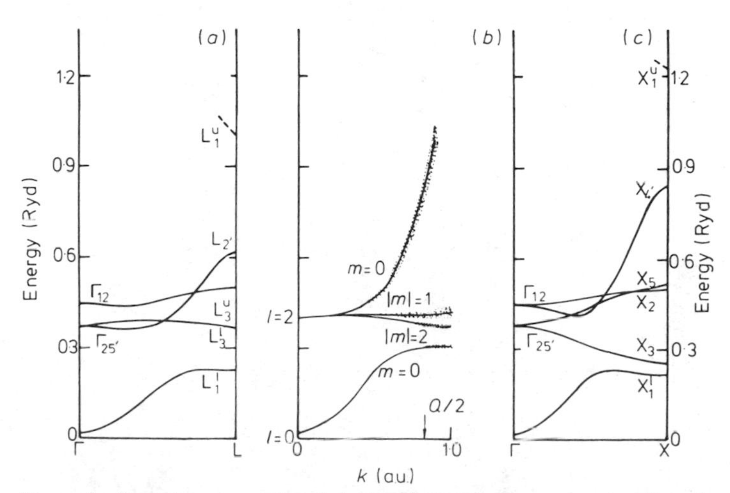

Figure 1. Complex energy bands in liquid Cu (*b*) are compared with real energy bands of crystalline Cu along two symmetry directions (*a* and *c*). (Note that, in principle, the liquid bands extend to $k = \infty$.)

both l and m are good quantum numbers and that the corresponding roots of equation (7) are $(2l+1)$-fold degenerate. For $k \neq 0$, the bands are labelled by $|m|$ and the five-fold degenerate d-level (at $k = 0$) gives rise to three bands (for $k \neq 0$): $m = 0$, $|m| = 1$ and $|m| = 2$. Note that only the $m = 0$ d-band hybridizes with the free-electron-like conduction band. It should be emphasized that the strength with which the complex bands contribute to the density of states is modulated by the weighting factors $F_L(\mathbf{k}, E)$. In particular, the fact that $F_L(\mathbf{k}, E)$ is proportional to $\delta_{m,0}$ (cf equation 5) implies that only bands with $m = 0$ will contribute to the final spectrum. The physical reasons for the vanishing of the spectral weight in $|m| \neq 0$ bands are not clear to us. It could possibly be a result peculiar to the QCA in the sense that, in a more accurate approximation scheme, the $|m| \neq 0$ bands might give a reduced but finite contribution to $\langle \rho(E) \rangle$. (We emphasize that there is no net loss of d-states; the spectral weight of the $|m| = 1$ and $|m| = 2$ bands is simply shifted to the d-components of the $m = 0$ bands.) Further work is required to clarify this effect.

Comparing the real parts of the liquid quasiparticle spectrum with the crystal energy bands shown in the outer panels of figure 1, we note that the large gaps at the Brillouin zone faces are absent in the liquid. These gaps are a consequence of the invariance of the crystal under lattice translations. In transition metals their magnitude is enhanced because symmetry allows just one of the levels (e.g. L_1^u or X_1^u) to mix with the d-states (thereby raising these levels significantly). Since no such requirements obtain in the liquid, it is not surprising that this feature of the crystal band structure is lost. Finally, we note that in the crystal the d-levels Γ_{12} and $\Gamma_{25'}$ are separated by approximately 1 eV due to crystal field effects which are, of course, absent in the liquid. In this connection figure 1(*a*) is of interest because for $k \neq 0$ the energy bands along $\Gamma \rightarrow L$ may be labelled by $|m|$ just as in the liquid. If the crystal field splitting could be discounted in figure 1(*a*) (so that Γ_{12} and $\Gamma_{25'}$ became degenerate) then the resulting crystal d-bands would in fact be quite similar to those of the liquid.

The damping of the quasiparticle states in the liquid is indicated by vertical shading of the bands in figure 1(*b*). Typical values of Im $[E_n(k)]$ are 0·025 Ryd at E_F and

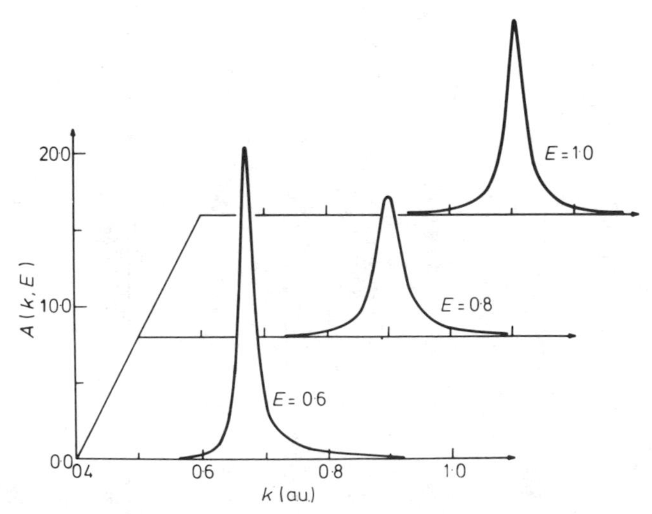

Figure 2. Spectral density $A(k, E)$ for the indicated values of energy E (given in Rydbergs).

0·01 Ryd for d-states at $k \simeq 1$. The damping is k dependent and is seen to approach zero as $k \rightarrow 0$. This can be understood on the basis of perturbation-theory arguments which show that Im $[E_n(k)]$ must vanish as $k \rightarrow 0$.

A complete description of the electronic spectrum involves the computation of the spectral density function $A(k, E)$ and the average density of states. Figure 2 shows a few typical $A(k, E)$ curves as a function of k, holding E fixed. The positions of the peaks in figure 2 are found to be in very good agreement with $k(E)$ values read from figure 1(*b*) at the appropriate energies. (Since the imaginary parts of the complex bands are related to the width of $A(k, E)$ as a function of E (for fixed k), they cannot be compared directly with the halfwidths of the peaks in figure 2.) Recall that in a perfect crystal the spectral density consists of a series of δ function peaks located at $[\mathbf{k}(E) + \mathbf{K}_i]$, where $\mathbf{K}_i$ denotes a reciprocal lattice vector. This feature which results from the periodicity of the crystal is destroyed in the liquid and, as seen in figure 2, $A(k, E)$ is essentially composed of a single peak. It is noteworthy that the tails of the peaks in figure 2 extend farther on the high-k side. This asymmetry is related to the fact that the complex d-bands possess a k-dependent damping which increases for larger values of k, and consequently the higher k d-states contribute to the spectral density in a broader range of values of E. As we go away from the d-bands this effect decreases and the peaks become more symmetric. (This is seen most clearly by comparing the $E = 0{\cdot}6$ and $E = 1{\cdot}0$ curves in figure 2.)

The density of states calculations on liquid Cu and Fe, based on equation (1) are currently in progress. Preliminary results outside the d-bands are considered below. As already noted, the QCA is expected to yield a reasonable density of states even in the energy region of the d-bands. The present calculations would thus allow us to delineate the characteristic changes in the d-band spectrum resulting from structural disorder in a liquid or an amorphous metal.

It is seen from figure 3 that $\langle\rho(E)\rangle$ for liquid Cu obtained within the QCA formalism exhibits a well defined minimum near 0·8 Ryd. (As expected, these results are in general agreement with our earlier calculations based on the MQCA (Peterson *et al* 1976), although the minimum is less pronounced.) Referring to figure 1(*b*), we find that the energy 0·8 Ryd corresponds to $k \simeq Q/2$, where Q is the position of the first peak in the

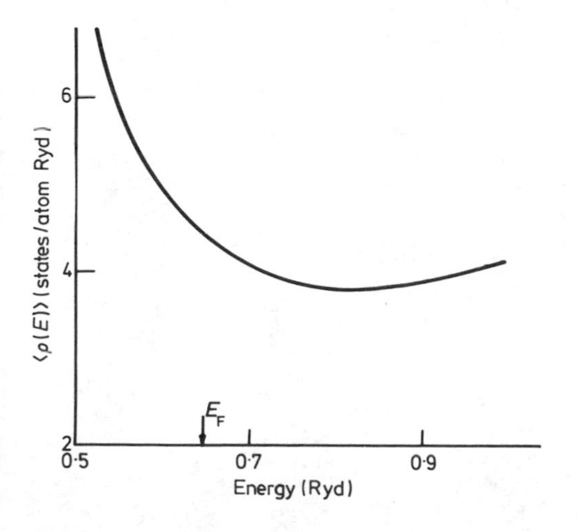

Figure 3. Electronic spectrum of liquid Cu in the vicinity of E_F using the QCA.

liquid Cu structure function. (The upper $m = 0$ band in figure 1(*b*) does, in fact, show an increase in slope near $k = Q/2$.) We conclude that the minimum in figure 3 is due to structural effects and may thus be viewed as an analogue of the Brillouin zone energy gaps in a crystal. The total number of states below 0·8 Ryd is found to be 11·6 which is in close agreement with the value 11·7 proposed by Nagel and Tauc (1975) in their suggestions of an electronic basis for the stability of metallic glasses. Although the existence of a minimum in the density of states appears to be confirmed by our calculations, we emphasize that its detailed shape is quite different from that predicted by the perturbation theory.

In conclusion, we note that in order to understand the electronic spectrum of metallic glasses involving two d-band metals, a multicomponent muffin-tin model will have to be used. Such calculations would also allow us to describe the effects of metalloid atoms in the metal–metalloid glasses, beyond the rigid band approximation.

Acknowledgments

We wish to thank Professors B C Giessen, Jan Tauc and S R Nagel for several important discussions. The work was supported by the National Science Foundation.

References

Bansil A, Ehrenreich H, Schwartz L and Watson R E 1974 *Phys. Rev.* **B9** 445
Bansil A, Schwartz L and Ehrenreich H 1975 *Phys. Rev.* **B12** 2893
Ehrenreich H and Schwartz L 1976 *Solid St. Phys.* **31** 149 (New York: Academic Press)
Hodges L, Watson R E and Ehrenreich H 1972 *Phys. Rev.* **B5** 3953
Lax M 1951 *Rev. Mod. Phys.* **23** 287
—— 1952 *Phys. Rev.* **85** 621
Mijnarends P E and Bansil A 1976 *Phys. Rev.* **B13** 2381
Nagel S R and Tauc J 1975 *Phys. Rev. Lett.* **35** 380
Peterson H K, Bansil A and Schwartz L 1976 *Am. Inst. Phys. Conf. Proc. 31* p378
Peterson H K, Schwartz L and Butler W H 1974 *Phys. Rev.* **B11** 3678
Schwartz L and Bansil A 1974 *Phys. Rev.* **B10** 3261
Schwartz L, Peterson H K and Bansil A 1975 *Phys. Rev.* **B12** 3113
Waseda Y and Ohtani M 1974 *Phys. Stat. Solidi* **B62** 535
Ziman J M 1966 *Proc. Phys. Soc.* **88** 387

Electronic structure and electrical resistivity of liquid transition metals

L E Ballentine
Department of Physics, Simon Fraser University, Burnaby, BC, Canada V5A 1S6

Abstract. The electrical resistivity of liquid transition metals and alloys has been treated with considerable success by means of a theory in which the scattering amplitudes of single atoms are calculated from phase shifts, and the sum of these amplitudes from all atoms determines the scattering probability to be used in the Boltzmann equation. Implicit in the theory is the assumption that propagating states possess a fairly well defined **k** vector, and that scattering may be described as a transition from **k** to **k**$'$. The same formula may be derived from more fundamental theory based on the Kubo formula if the spectral function $\rho(k, E_F)$ has a well defined sharp peak, that is if Γ/E_F is small, Γ being the imaginary part of the self-energy function. Simple estimates based on phase shifts yield $\Gamma \approx 1$ Ryd at the d-resonance, so that $\Gamma/E_F > 1$ in the middle of the 3d transition series and the simple theory of resistivity would be inapplicable. More sophisticated calculations of $\rho(k,E)$ suggest a smaller value of Γ, but definitive results have not yet been obtained.

1. Introduction

The purpose of this paper is to pose a problem and indicate the direction in which the solution is to be sought. However, the problem itself remains unsolved.

The electrical resistivity of liquid transition metals and alloys has been interpreted with considerable success (Evans *et al* 1971, Dreirach *et al* 1972) by means of a simple (in hindsight) extension of Ziman's theory for simple liquid metals. The scattering amplitude of a single atom is computed from phase shifts, rather than a pseudopotential, and a superposition of the amplitudes from all atoms determines the scattering probability to be used in the Boltzmann equation. Implicit in the theory is the assumption that propagating states possess a fairly well defined **k** vector, and that scattering may be described as a transition from **k** to **k**$'$.

The same formula can be derived from a more fundamental theory based on the Kubo formula if the spectral function has a well defined sharp peak. The spectral function is defined as (Ballentine 1975)

$$\rho(k,E) = \left\langle \sum_n |\langle \mathbf{k} | \psi_n \rangle|^2 \, \delta(E - E_n) \right\rangle \tag{1}$$

where $|\psi_n\rangle$ and E_n are eigenvectors and eigenvalues of the one-electron Hamiltonian for instantaneously fixed positions of the ions, and the outer brackets denote an average over the ensemble of arrangements of the ions. The spectral function describes the momentum distribution of electrons with energy E. It is related to the complex self-energy function $\Sigma(k, E)$ through the relations

$$G(k, E) = [E - k^2 - \Sigma(k, E)]^{-1} \tag{2}$$

$$\rho(k, E) = -\text{Im } G(k, E)/\pi \tag{3}$$

and the width of the peak of $\rho(k, E)$ is governed by the imaginary part of the self-energy function. If we denote $\Gamma = -\text{Im } \Sigma(k_0, E)$, where $k_0 = k_0(E)$ is the position of the peak, then the empirically successful theory of electrical resistivity is theoretically justifiable provided $\Gamma/\epsilon_F \ll 1$. We are thus motivated to calculate $\rho(k, E)$ for liquid transition metals.

2. Calculation of the spectral function

2.1. *Simple-minded estimate*

The self-energy function can be expressed as a cluster expansion involving the t-matrix of a single atom and the pair, triplet, etc correlation functions of the liquid. The reader should consult Ballentine (1975) for a systematic derivation and review of such expansions, which we shall use without derivation in this paper. The lowest order (single atom) term is just proportional to the diagonal matrix element of the t-matrix.

$$\Sigma^{(01)}(k, E) = N\langle \mathbf{k} | t(E) | \mathbf{k} \rangle \tag{4a}$$

$$= -\frac{N}{\Omega}\frac{4\pi}{k}\sum_{l=0}^{\infty}(2l+1)\sin\delta_l \exp(\mathrm{i}\delta_l). \tag{4b}$$

Here N is the number of atoms in the volume Ω over which the momentum eigenfunctions are normalized, $\delta_l = \delta_l(E)$ is the phase shift of the lth partial wave, and the wavevector is $k = E^{1/2}$ in our units ($\hbar = 2m = 1$).

At resonance ($\delta_2 = \pi/2$), neglecting all non-resonant partial waves, this yields $\Gamma = 20\pi/(k\Omega_0)$, where $\Omega_0 = \Omega/N$ is the volume per atom in Bohr units. Taking typical values of these parameters for a liquid transition metal (e.g. for Fe we have $\Omega_0 = 89{\cdot}1$, resonance energy $E_d \simeq 0{\cdot}64$ Ryd) yields $\Gamma \simeq 1$ Ryd, and hence $\Gamma/E > 1$ at resonance. In such a case $\rho(k, E)$ would be so broad (see figure 1) that the theory of electrical

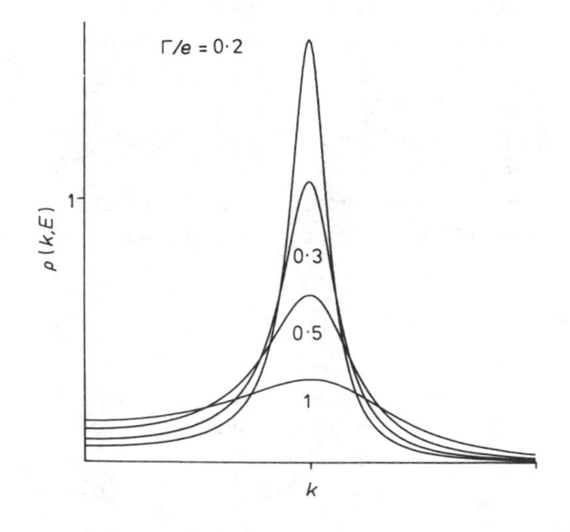

Figure 1. Spectral function $\rho(k,E) = -\pi^{-1}\,\text{Im}(E - k^2 + \mathrm{i}\Gamma)^{-1}$ for $E = 1$, to illustrate variation in peak width.

resistivity described in §1 could not be justified. Of course it is the value of Γ/ϵ_F at the Fermi energy that is important, and it will be smaller if the Fermi level is well above or well below the resonance energy. (For Ni we have $\Gamma/\epsilon_F \simeq 0{\cdot}14$.) But on the simple approximation (4) the conclusion $\Gamma/\epsilon_F > 1$ is inevitable for elements near the middle of the transition series.

2.2. *Cluster expansions of the Green function*

The ordinary t-matrix used in equation (4) describes the scattering of one atom (or ion) in free space, and its analytic forms for $E > 0$ and $E < 0$ are quite different, corresponding to the qualitative difference between complex and real exponential wavefunctions. But in a metal $E = 0$ (the muffin-tin zero) is not the dividing line between propagating and non-propagating states. For this reason Ballentine (1975) suggested that the cluster expansion should be reformulated in terms of the *Medium-adapted t-matrix*, which is defined by the operator equation

$$\hat{t}(E) = v + vG(E)\hat{t}(E). \tag{5}$$

Here $G(E)$ is the ensemble-average Green operator whose (diagonal) form in k-space is given by equation (2). The ordinary t-matrix satisfies a similar equation with the free-particle Green operator G_0 replacing G. In place of equation (4*a*) we now have

$$\Sigma^{(11)}(k,E) = N\langle \mathbf{k}\,|\,\hat{t}(E)\,|\,\mathbf{k}'\rangle. \tag{6}$$

Equations (5) and (6) must be solved self-consistently. The effect of the imaginary part of Σ on the solution of (5) is to reduce the strength of the scattering due to $\hat{t}$ compared with that due to t, thus the conclusions of the previous section are likely to be modified.

The computational labour of solving equation (5) is greatly reduced by taking the

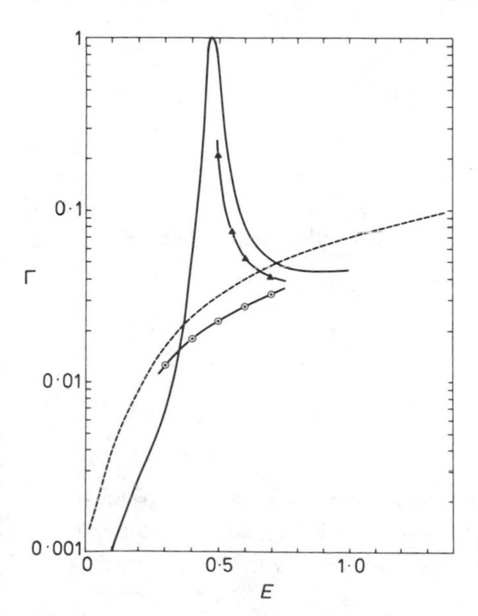

Figure 2. Imaginary part of self-energy function for liquid Ni in several approximations. Full curve: $\Sigma^{(01)}$, first order in bare t-matrix; broken curve: $\Sigma^{(11)}$, first order in medium-adapted t-matrix; triangles: $\Sigma^{(02)}$, second order in bare t-matrix; circles: $\Sigma^{(12)}$, second order in medium-adapted t-matrix.

one-atom potential v to be of separable form for each partial wave,

$$\langle \mathbf{r} | v | \mathbf{r}' \rangle = -(4\pi)^{-1} \sum_l u_l(r)\, u_l(r')\, \mathrm{P}_l(\cos\theta) \tag{7}$$

where $\mathrm{P}_l(\cos\theta)$ is a Legendre polynomial and θ is the angle between $\mathbf{r}$ and $\mathbf{r}'$. The functions $u_l(r)\,(l = 0, 1, 2)$ were determined so as to reproduce the phase shifts calculated by O Dreirach (unpublished) for liquid Ni over the range $0 < E \leqslant 0{\cdot}98$ Ryd.

The result for Γ shown in figure 2 is much smaller than that obtained from the bare t-matrix. But rather than vindicating the theory of resistivity, this actually indicates the inadequacy of equation (6) as an approximation. This one-atom approximation takes no account of correlations that prevent atoms from overlapping. Hence the d-band is spread out so broadly that it can no longer be considered a resonance.

The next and most obvious step is to go to a two-atom approximation,

Here (as in Ballentine 1975) the wavy line represents $\hat{t}$ (equation (5)), the double horizontal line represents G (equation (2)), and the small circle connecting the upper ends of the wavy lines represents $n^2h(\mathbf{R}_1 - \mathbf{R}_2)$ with $n = N/\Omega$ being the atomic density and $h(r) = g(r) - 1$ being the total correlation function for pairs of atoms. Integration over the atomic positions $\mathbf{R}_1$ and $\mathbf{R}_2$ is implied. As may be seen from figure 2, this improvement does not qualitatively change the result obtained from the one-atom approximation $\Sigma^{(11)}$, thus we must improve our partially summed cluster expansion in some other direction.

The nature of the needed improvement can be discerned by considering the diagrams that are included in the medium-adapted t-matrix. It is related to the ordinary t-matrix by

$$\hat{t} = t + t(G - G_0)\,\hat{t}. \tag{9}$$

Using the approximation $\Sigma^{(11)}$ to calculate G leads to the following expansion of equation (9):

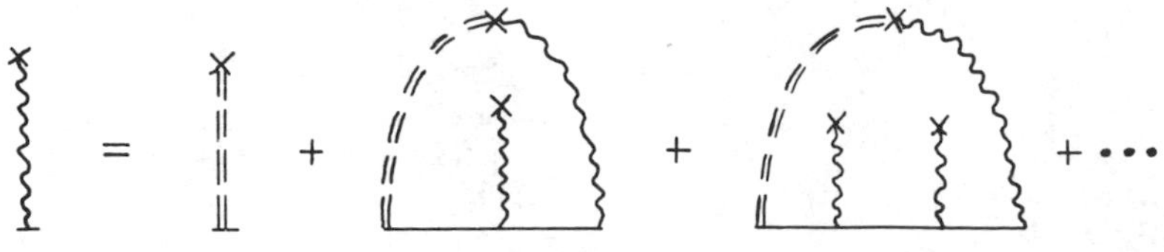

= + + + ⋯ (10)

Here the double dashed line represents t. Now in the second diagram on the right no account is taken of correlation between the two atoms, and hence their overlap may lead to large spurious contributions which would be cancelled by the following diagram that is not included in the series:

(11)

The correlation function $h(R_{12})$ in this diagram would combine with 1 (i.e. the absence of any such factor) in the previous diagram to yield $1 + h(R_{12}) = g(R_{12})$ as a factor in their sum, and there would then be no spurious contribution from overlapping of those two atoms. It is apparent that if this diagram and similar ones corresponding to the higher-order terms in (10) were included, the resultant elimination of spurious overlap contributions would considerably reduce the difference between $\hat{t}$ (so revised) and t. We have not yet carried out such a calculation, but it appears to be just within the realm of practical feasibility.

A qualitative indication of its effect is indicated by another approximation, $\Sigma^{(02)}$, which is similar to (8) except that t is used instead of $\hat{t}$. The result, shown in figure 2, is qualitatively similar to that of $\Sigma^{(01)}$. However, $\rho(k, E)$ is not positive definite in this approximation for $E < 0{\cdot}5$ Ryd, confirming our earlier belief that one should not base the cluster expansion on the bare-atom t-matrix.

3. Conclusions

The ultimate goal of the research programme, to evaluate the parameter Γ/E that controls the width of the spectral function and hence to determine the domain of applicability to transition metals of the simple theory of resistivity, has not yet been achieved. I would guess that the truth lies between the (very wide) limits of approximation $\Sigma^{(02)}$ and $\Sigma^{(12)}$, in which case the resistivity theory of Dreirach *et al* (1972) would be valid for Ni but very doubtful for Fe.

On the other hand, we have shown the relative importance of certain kinds of diagrams in the cluster expansion. The diagram (11), which corrects for overlap in a one-atom† term, appears to be more important than a two-atom† term like the last diagram of (8). A theory that includes both of these contributions and more has already been formulated by Watabe and Yonezawa (1975), but its technical complexity has so far prevented any quantitative use of it. However, more is learned by proceeding step by step, from the simplest theory to more complex ones, than by going immediately to the most sophisticated theory. For that reason we have presented this report of work still in progress.

References

Ballentine L E 1975 *Adv. Chem. Phys.* **31** 263–327
Dreirach O, Evans R, Güntherodt H-J and Künzi H-U 1972 *J. Phys. F: Metal Phys.* **2** 709–25
Evans R, Greenwood D A and Lloyd P 1971 *Phys. Lett.* **A35** 57–8
Watabe M and Yonezawa F 1975 *Phys. Rev.* **B11** 4753–62

† By a one-atom term we mean an irreducible diagram in which scattering begins and ends at the same atom. All such diagrams can be regarded as contributing to a 'renormalized t-matrix', of which $\hat{t}$ is an example. A two-atom term is one in which scattering begins at one atom and ends at a second atom. Unfortunately these definitions as well as terminology like 'single-site approximation' are not universally agreed upon.

Electrical resistivity of liquid rare earth metals and their alloys

H-J Güntherodt, E Hauser† and H U Künzi
Institut für Physik, Universität, Basel, Switzerland

Abstract. The electrical resistivity of the rare earth metals, including the heavy ones, has been measured in the liquid and solid state at high temperatures. In contrast to the results at room temperature the electrical resistivity increases nearly linearly from Ce to Er. At the melting point the resistivity changes very little and in the liquid state the resistivity increases slowly with increasing temperature except for Eu and Yb which show negative temperature coefficients of the electrical resistivity. The liquid alloys of the rare earth metals can be divided into different groups according to their behaviour as a function of temperature and concentration. The present experimental results give strong evidence that the spin disorder scattering contribution in the liquid state cannot be large.

1. Introduction

The determination and understanding of the physical properties of the liquid rare earth elements is only in its infancy. It seems that up to about five years ago almost no experimental investigations had been carried out on liquid rare earth metals, due to their corrosive properties and their high melting points. At low temperatures, on the other hand, where such problems are less severe, extensive research over the past thirty years has revealed a tremendous amount of interest. It is the purpose of this contribution to present new data on the electrical resistivity at high temperatures and in the liquid state. As up to now only a small part of the experimental data has been explained theoretically we would like to compare the new results with the results observed in liquid simple metals and in transition metals. Interesting results are also obtained by comparing the physical properties in the liquid state with those in the solid state. In particular, it seems that for the rare earth metals the transport properties show little change on melting and this distinguishes them from other groups of metals.

2. Experimental

Apart from the problems involved in the mechanical construction of high-temperature furnaces and measuring cells the main difficulties are encountered in the search for appropriate crucible materials to construct measuring cells. For mechanical construction one has to keep in mind that at high temperatures most materials lose some of their strength and may become soft. This is especially important for measurements that depend on the geometrical dimensions of the cell. The chemical problems due to the

† Present address: Division of Engineering and Department of Physics, Brown University, Providence, Rhode Island 02912, USA

corrosive nature of the rare earth metals and alloys are much harder to solve. For melting rare earth metals, crucibles made out of the refractory metals W, Mo or Ta are usually recommended, but as these materials are highly conductive their use is restricted to measurements where their contribution can easily be subtracted from the measured value. Measurement cells made out of Mo and Ta indeed proved to give reproducible results for the electrical resistivities and the magnetic susceptibilities of the pure rare earth metals and even for a number of alloys. For example, for resistivity measurements we used thin-walled tubes of either Ta or Mo. The resistivity of the liquid metal was calculated by assuming that the resistance of the cell is parallel to the resistance of the sample. At high temperatures the electrical resistance of the measuring cell is usually of the same magnitude as that of the liquid metal so that with this method good accuracy may still be achieved. A number of cross checks with liquid simple metals were made and the reproducibility of the resistivity measurements on the liquid rare earth metals proved to be better than ±3%.

3. Electrical resistivity of the rare earth metals

In figure 1 the electrical resistivities of liquid rare earth metals are compared with some typical resistivities for liquid simple and transition metals. From this figure it is clear that the resistivities of liquid rare earth metals are among the largest of any liquid metals. The value for liquid Er is more than ten times that of liquid Cu.

A comparison of the resistivity behaviour in the solid and liquid states of the rare earth metals shows that there are a number of general features which are observed for practically all the rare earth elements (Busch *et al* 1970, Güntherodt *et al* 1974a,b, Hauser 1976).

(i) At temperatures well above that of magnetic ordering, the electrical resistivity increases less strongly than linearly with temperature.
(ii) At the high-temperature phase transition the electrical resistivity increases by about 7% for the light rare earth metals. For the heavy rare earth metals this jump is less than 3% and may even be negative.

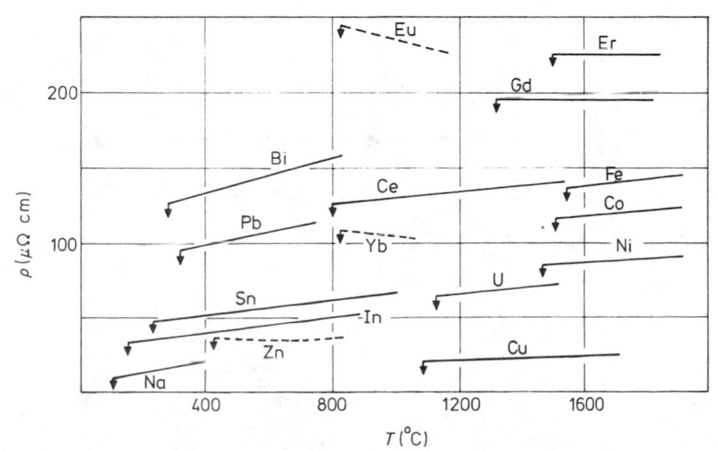

Figure 1. Typical values for the electrical resistivities of liquid metals

(iii) At the melting point the electrical resistivity increases for all the rare earth metals by only 3–6%.
(iv) The temperature coefficient of electrical resistivity in the liquid state is very small and positive or may even be zero for the heavy rare earth metals. Exceptions are Eu and Yb which have negative temperature coefficients of electrical resistivity.

Figure 2 shows the electrical resistivity of Dy as an example of a heavy rare earth element. The results illustrate points (i)–(iv) above. The values in the solid state have been obtained by directly cooling the melt in the measuring cells. The samples are therefore supposed to be polycrystalline. Our results are in reasonable agreement with those of Colvin *et al* (1960), Habermann and Daane (1964) and Zinov'ev *et al* (1974). The low-temperature behaviour up to the magnetic phase transition is characterized by a strong spin scattering contribution.

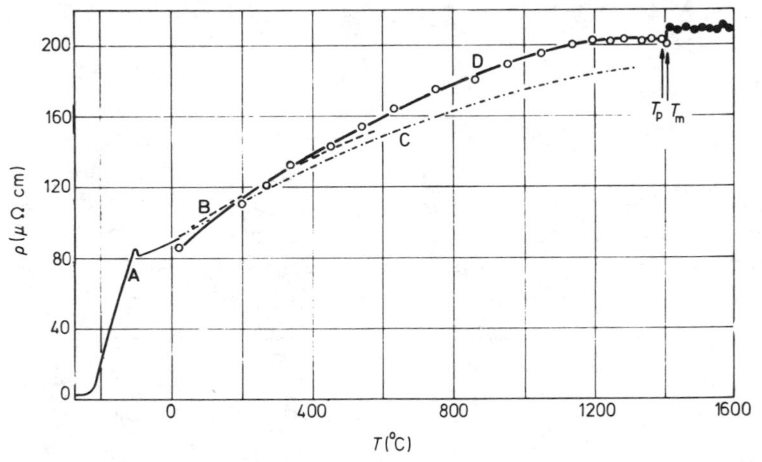

Figure 2. Temperature dependence of the electrical resistivity of Dy. Curve A, Colvin *et al* 1960); curve B, Habermann and Daane (1964); curve C, Zinov'ev *et al* (1974); curve D, present work.

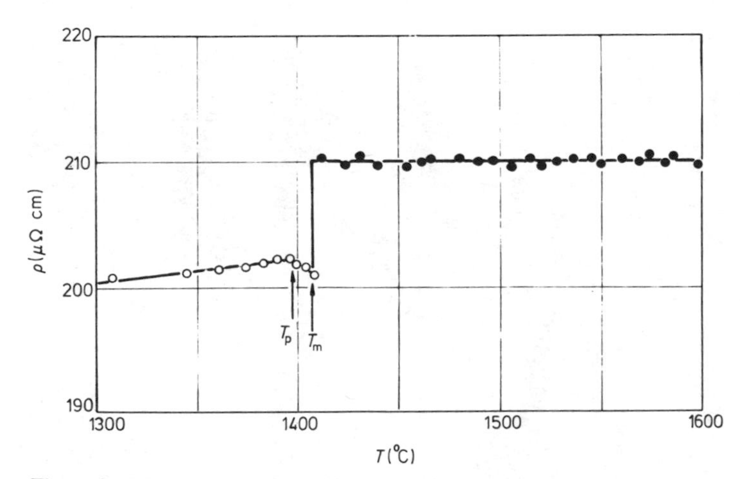

Figure 3. Electrical resistivity of Dy at high temperatures.

Figure 3 shows the resistivity of liquid Dy close to the melting point on an enlarged temperature scale. Similar temperature-independent resistivities have also been observed for liquid Gd, Ho and Er. Figure 4 shows the corresponding results for Nd as a typical example of the light rare earth metals.

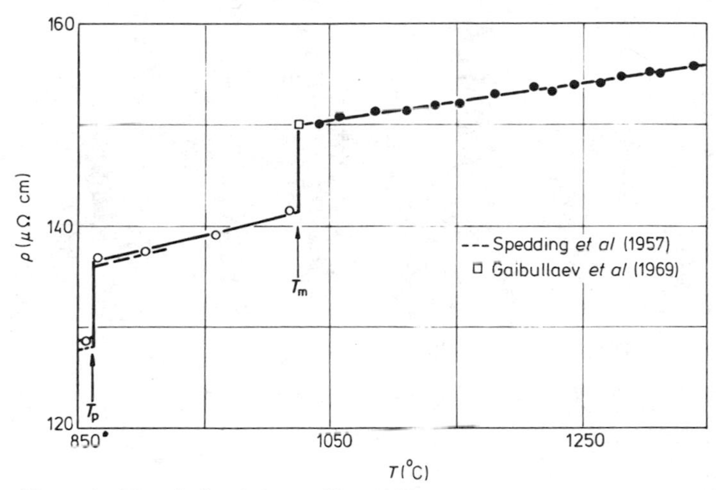

Figure 4. Electrical resistivity of liquid Nd.

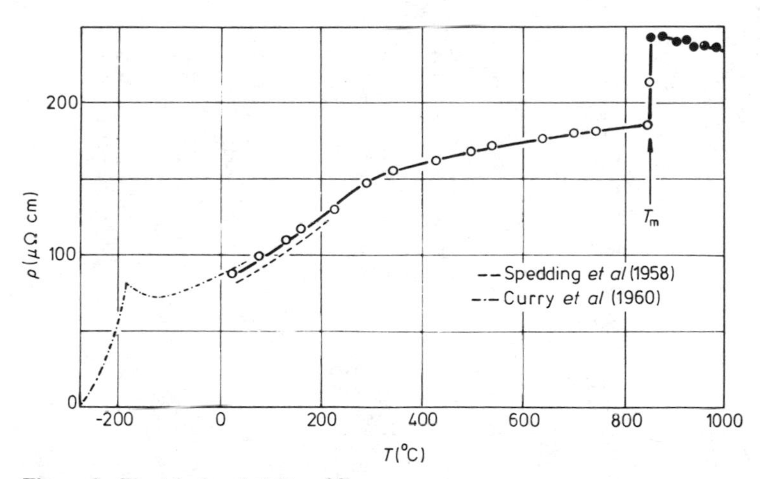

Figure 5. Electrical resistivity of Eu.

Figure 5 shows the electrical resistivity as a function of temperature for Eu. Above the melting point the resistivity decreases with increasing temperature as is characteristic for the liquid simple metals Zn and Ba. This behaviour has also been observed for liquid Yb (figure 6). The resistivity of liquid Eu is one of the largest for liquid metals and approaches that of liquid Ba for which $\rho = 314\,\mu\Omega\,\text{cm}$ (Güntherodt *et al* 1976). A compilation of values for the electrical resistivity and its temperature coefficient for the other liquid rare earth metals is presented in table 1.

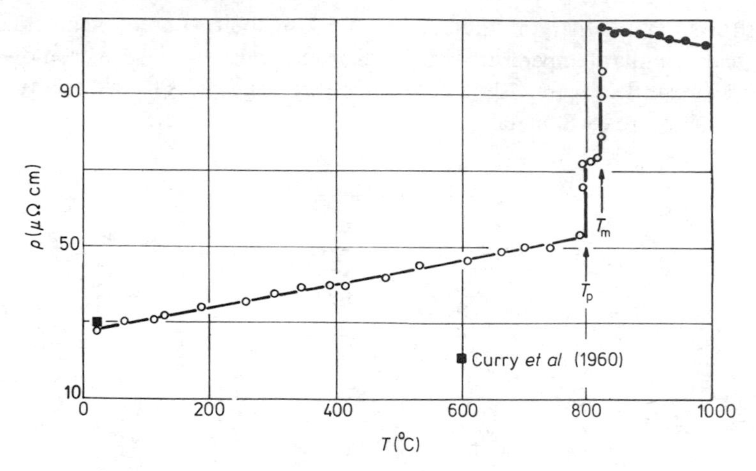

Figure 6. Electrical resistivity of Yb.

Table 1. Electrical resistivity of liquid rare earth metals at their melting point.

	ρ ($\mu\Omega$cm)	$d\rho/dT$ ($10^{-8}\,\Omega$cm/°C)	
La	118	—	a
	138	6·5	b
	135	1·8 ± 0·3	c
Ce	130	—	a
	127	1·9 ± 0·3	c
Pr	140	—	a
	140	1·2 ± 0·3	c
Nd	150	—	a
	150	1·8 ± 0·3	c
Eu	244	−5·5 ± 2·0	c
Gd	195	0 ± 1·0	c
Tb	193	1·6 ± 0·5	c
Dy	210	0 ± 0·5	c
Ho	221	0 ± 1·0	c
Er	226	0 ± 0·5	c
Yb	108	−2·5 ± 1·0	c

(a) Gaibullaev *et al* (1969)
(b) Krieg *et al* (1969)
(c) present work

Figure 7 shows the electrical resistivity of the rare earth metals as a function of the atomic number at three different temperatures. The uppermost line (black circles) represents the values in the liquid state. Since the temperature dependence of the resistivity in the liquid state is rather small we have plotted values corresponding to temperatures near the melting point. From this curve it follows that the electrical resistivity depends roughly linearly on the atomic number.

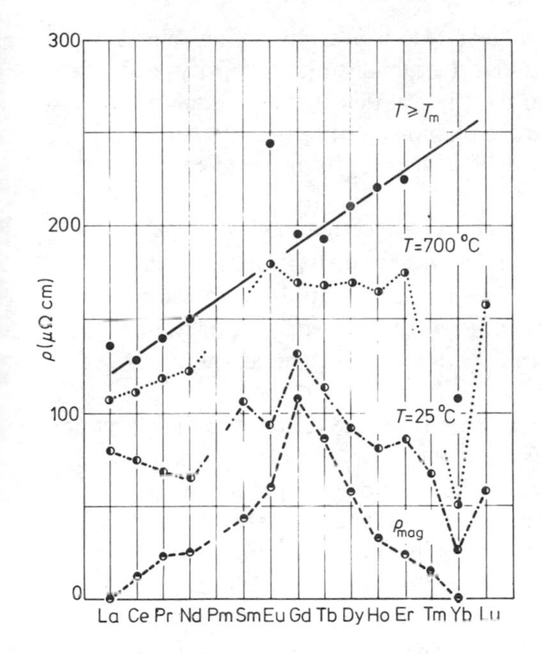

Figure 7. Electrical resistivities of the rare earth metals at different temperatures.

The value for liquid Er is almost twice as large as that of liquid Ce. Remarkable deviations from this linear plot are observed for Eu and Yb. Values for Pm and Tm have not yet been measured at high temperatures. In this connection it should be recalled that the rare earth series is mainly characterized by the continuous filling up of the 4f shell. This is only interrupted twice, namely for Eu and Yb where we observed exceptional behaviour in the resistivity.

In the solid state at 700 °C the resistivities of the heavy rare earth metals do not vary much from metal to metal so the behaviour is different from that in the liquid state. At room temperature the well known maximum of the resistivity occurs for Gd. This maximum has been explained in terms of the spin-disorder contribution to the resistivity. This is largest for metals with an approximately half-filled 4f shell. Values of this magnetic contribution which have been separated from the total resistivity are given for polycrystalline samples in the lower curve of figure 7.

At low temperatures the variation of the total resistivity ρ of the rare earth metals has been explained in terms of Mathiessen's rule as the sum of the residual resistivity ρ_r, the phonon contribution ρ_p and the magnetic (spin-disorder) contribution ρ_m due to the exchange interaction between the localized moments and the conduction electrons:

$$\rho(T) = \rho_r + \rho_p(T) + \rho_m(T).$$

At high temperatures, in the paramagnetic state, the magnetic contribution is predicted to be temperature independent and should occur with its maximum value (see for example Meaden 1971). We might therefore expect that the plot of resistivity versus atomic number should also show a peak at Gd for high temperatures.

The observed disappearance of this maximum in the electrical resistivity with increasing temperature raises questions concerning the magnitude of the spin-disorder scattering contribution in the liquid state and in the solid state at high temperatures. Whether the disappearance of the resistivity peak means that the spin-disorder resistivity really decreases at high temperatures (contrary to the classical theoretical prediction) or whether the spin-disorder contribution and the phonon contribution are not simply additive (Mathiessen's rule) cannot rigorously be concluded from measurements of the total resistivity.

Nevertheless Barabanov *et al* (1976) have recently suggested that in the paramagnetic region the spin-disorder scattering should give rise to a resistivity behaviour which is equivalent to that observed in dilute alloys (Kondo effect). Thus in the paramagnetic region they argue that the magnetic contribution to the resistivity is temperature dependent and should decrease logarithmically with increasing temperature provided the s–f coupling parameter, J_{sf}, is negative.

In order to analyse our data along these lines we first have to separate the magnetic contribution from the total resistivity. This is possible if we assume that Mathiessen's rule is applicable and that at high temperatures (near the melting point) the spin-disorder contribution is negligibly small. The latter assumption implies that at high temperatures the linear increase of the electrical resistivity across the rare earth series is due to a phonon contribution which increases with the atomic number or with the progressing lanthanide contraction. In figure 7 this linear increase at the melting point is only shown for the liquid state but as the change of the electrical resistivity on melting is rather small this remains essentially the same for the solid state.

Relevant values for the spin scattering contribution can now be deduced by normalizing all the resistivity temperature curves at a given high temperature, that is by dividing $\rho(T)$ by its value ρ_s at the given temperature.

Such reduced resistivities $\rho(T)/\rho_s$ are independent of the effects which give rise to the linear variation across the rare earth series at high temperatures and for a given crystal structure the phonon contribution ρ_p/ρ_s might be expected to be a 'universal' curve. Since it has a full f shell, Lu has no spin scattering contribution, and the reduced resistivity of Lu should represent this curve for the heavy rare earth metals which

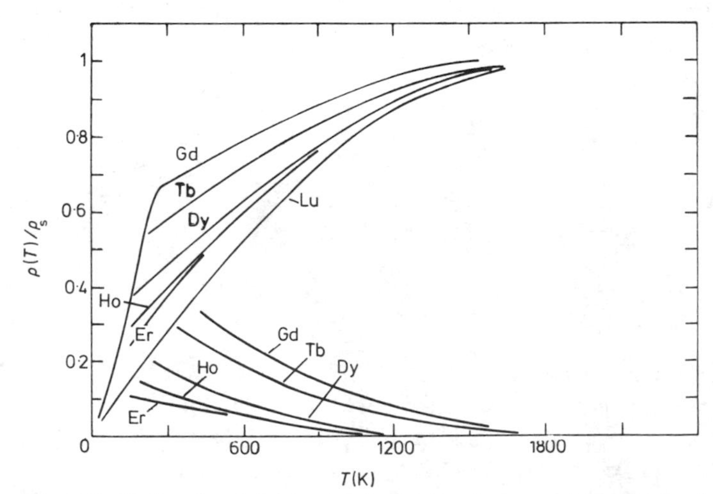

Figure 8. Reduced resistivities and the spin scattering contribution for the heavy rare earth metals.

together with Lu all crystallize in the HCP structure. Differences between the reduced curves of Lu and any of the heavy rare earth element are therefore proportional to the spin scattering contribution. Figure 8 shows the reduced resistivities and the spin scattering contribution (lower curves) for Gd, Tb, Dy, Ho, Er and Lu. The reduced resistivities are all normalized to unity at the melting temperature of Lu. For metals with lower melting points the resistivity curve of the HCP phase was extrapolated to this temperature. Figure 9 shows that this magnetic contribution indeed decreases as $\log T$. It certainly remains to show that this qualitative explanation also holds quantitatively.

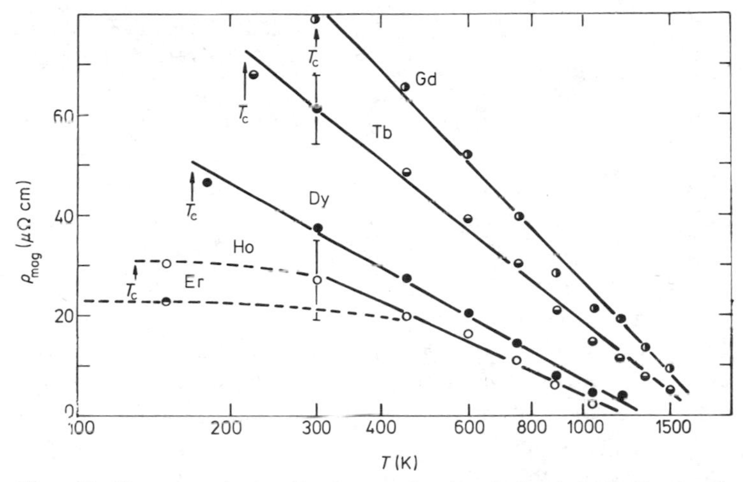

Figure 9. The magnetic contributions to the electrical resistivities for the heavy rare earth metals.

4. Electrical resistivity of the alloys

As has been observed in our earlier investigations (Busch and Güntherodt 1974) for alloys of simple and transition metals, the alloys of the rare earth metals with themselves or with other metals can be divided into different groups. The important criterion is again whether or not negative temperature coefficients of the electrical resistivity occur for certain concentration ranges. For alloys which do not show negative temperature coefficients we observe two characteristic concentration dependences: either a more parabolic behaviour in alloys of the rare earth metals with the noble and transition metals, or an almost linear variation with concentration for the alloys of the rare earth metals with themselves. The latter group evidently reproduces the linear variation with atomic number observed for the pure liquid rare earth elements. The third group of rare earth alloys shows negative temperature coefficients of the electrical resistivity for certain concentrations. These consist of alloys with polyvalent simple metals. In these alloys a pronounced maximum of the electrical resistivity as a function of concentration is observed.

Figure 10 shows the concentration dependence for three Cu alloys. Very similar results have been observed for Cu–Ce and Cu–Pr alloys. The resistivity increases quite

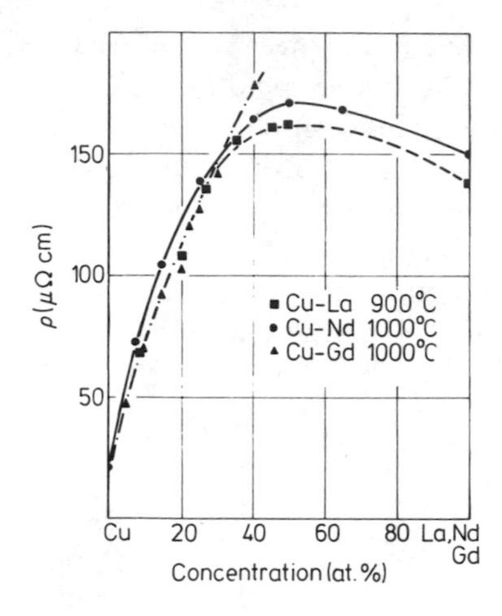

Figure 10. Concentration dependence of the electrical resistivities for liquid Cu rare earth alloys.

strongly on the Cu-rich side up to about 40 at.% rare earth and shows a maximum near 50 at.% rare earth. It is interesting that for these alloys within the range from pure Cu to a concentration of 30 to 40 at.% rare earth, the electrical resistivities follow practically the same curve. Deviations which are due to the different electrical resistivities of the pure rare earth metals occur only at higher concentrations. For all the measured Cu/rare earth systems the temperature coefficient of the electrical resistivity is quite small and positive. Recent measurements on Co–Ce alloys also indicate a parabolic maximum of the electrical resistivity as a function of concentration.

In contrast, a more or less linear concentration dependence has been observed for alloys of the rare earths with themselves namely in Ce–La, La–Gd and Ce–Er. Figure 11 shows the concentration dependence for La–Gd. The concentration dependence of this alloy follows closely the linear Z-dependence described in figure 7 except that it does not reproduce the large value of liquid Eu. The temperature coefficient of the electrical resistivity in these alloys is, as in the Cu alloys, small and positive.

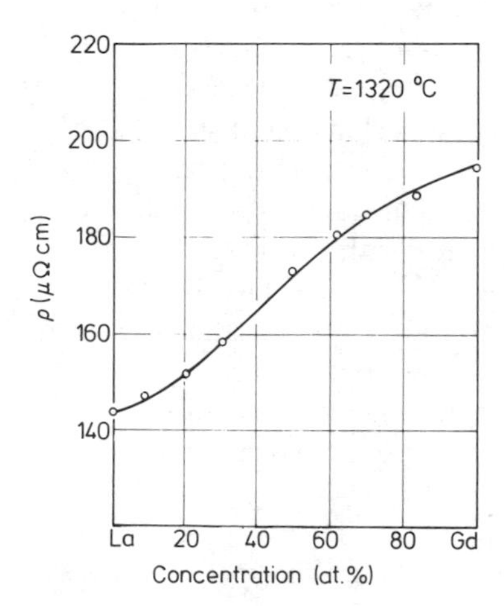

Figure 11. Concentration dependence of the electrical resistivity for liquid La–Gd alloys.

Figure 12 shows the concentration dependence of the electrical resistivity at 1500 °C and its temperature coefficient for Ce–Sn alloys. These, as well as La–Sn and Gd–Sn alloys, show a pronounced maximum in the middle of the concentration range and associated with this large maximum are negative temperature coefficients of the electrical resistivity. This is shown explicitly in figure 13. For comparison the results for liquid Eu and Yb are also shown. Both the magnitude and the temperature coefficient of resistivity in liquid Eu are very close to the corresponding values in Ce–Sn at concentrations of 51 at.% Sn. It follows that alloys of the rare earth metals with polyvalent metals reproduce the behaviour of Eu. We can contrast this with the La–Gd alloys mentioned above.

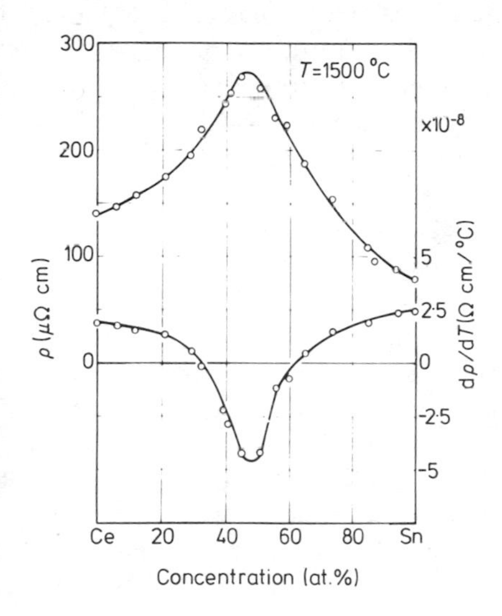

Figure 12. Concentration dependence of the electrical resistivity for liquid Ce–Sn alloys. Upper curve, ρ; lower curve, $d\rho/dT$.

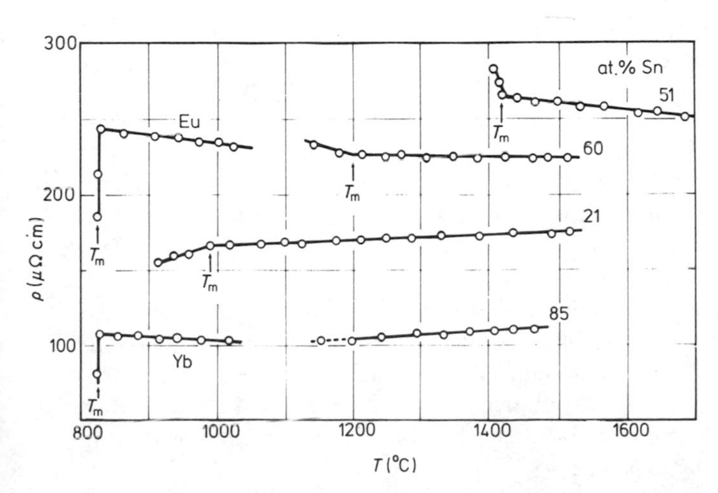

Figure 13. Temperature dependence of the electrical resistivities of liquid Ce–Sn alloys.

5. Discussion

At present the behaviour of the electrical resistivity of the rare earth metals at high temperatures is not well understood whereas at low temperatures in spite, or perhaps because of, the complex effects associated with the magnetic phase transition the resistivity has been studied in much greater detail. The main problems at high temperatures are concerned with the magnitude of the spin-disorder contribution, the nonlinear temperature dependence and the rather large values of the electrical resistivity itself.

The question of the spin-disorder resistivity has already been discussed in terms of the argument of Barabanov *et al* (1976). Apart from these arguments and the fact that the electrical resistivity increases linearly across the rare earth series, the alloys give further evidence that the spin-disorder contribution cannot be large in the liquid state. Cu-rich alloys of the light rare earth metals and Gd all show the same concentration dependence (figure 10) in spite of the different magnetic structure and hence different contributions of the spin-disorder resistivity. Also both La–Gd and Ce–Er alloys show much the same concentration dependence in spite of the fact that the spin-disorder contribution in Gd should be considerably larger, or in other words an alloy of about 50 at.% Ce and 50 at.% Er which both have small spin-disorder contributions has the same resistivity as Gd which should have the largest spin-disorder resistivity. In addition, recent measurements of liquid Gd–Sn alloys gave the same resistivity concentration curve as has been observed for liquid Ce–Sn.

At high temperatures in the solid state the resistivity shows significant deviations from the usual linear behaviour of the phonon resistivity. This nonlinearity even remains after subtraction of the spin-disorder contribution as can be seen from Lu which has no magnetic contribution. Kasuya (1966) has attributed this nonlinear behaviour of $\rho(T)$ in pure rare earth metals to a strongly temperature-dependent electron–phonon contribution. Within the Mott s–d model, this can occur if the electronic density of states has a pronounced structure in the neighbourhood of the Fermi energy so that contributions to ρ which arise from the expansion of the Fermi distribution function become important. Many transition metals (e.g. Ta, Pd, Pt) exhibit a similar but usually weaker nonlinear temperature dependence and this is usually attributed to this mechanism.

For the liquid rare earth metals we presume that we are confronted with the problem of structural (potential) disorder scattering. For liquid metals in general this problem seems best described within the Ziman type of theory which has had considerable success in predicting the electrical resistivity of liquid simple metals as well as transition metals. However, an attempted application of the Ziman theory to the rare earth metals, as given by Evans *et al* (1971) and Dreirach *et al* (1972) in the *t*-matrix formulation, has given rise to additional questions.

Firstly, there is the question whether this theory which has primarily been established to explain the resistivity of weak scatterers is still valid for strong scatterers as represented by the heavy rare earth metals. But as this theory has recently been applied to explain the large electrical resistivity of liquid Ba ($\rho = 314\ \mu\Omega$ cm) (Ratti and Evans 1973) it seems that this model is applicable in much stronger scattering regimes than was previously expected.

The fact that the 4f states are far below the Fermi energy and furthermore are atomic-like states suggests that the main contribution to the electrical resistivity should be due to 5d resonant scattering. A rough idea about the magnitude of d-phase shift associated with this resonant scattering can be gained from the concentration dependence of the electrical resistivity in Cu-rich rare earth alloys. An analysis in terms of the virtual bound d-states model as given by Friedel (1956, 1962) and the Friedel sum rule shows that the virtual bound d-band of the light rare earth metals is occupied by about two electrons per rare earth atom. This number may of course be somewhat different for the concentrated alloys and the pure rare earth metals.

Nevertheless this gives some indication that in the trivalent rare earth metals we have about one s-like electron per atom which acts as a conduction electron. In terms of the Faber–Ziman theory this explains the temperature and concentration dependence of the electrical resistivity in the alloys. In alloys of the rare earth metals with the noble metals or with themselves the conduction electron concentration is near one electron per atom and hence $2k_F$ is always smaller than K_p, the position of the first peak in the structure factor, and consequently the temperature coefficient of the resistivity remains positive. This corresponds exactly to the situation that has been encountered in the alloys of the noble metals with themselves.

On the other hand in alloys of rare earth with polyvalent simple metals such as tin, the number of conduction electrons per atom gradually increases from one to four with increasing tin concentration. In the concentration range where we have about two conduction electrons per atom, $2k_F$ is of the order of K_p and the resistivity shows a negative temperature coefficient since the structure factor decreases near K_p for increasing temperature.

The fact that in the rare earth/tin alloys the resistivity has its maximum value in exactly the concentration range where negative temperature coefficients occur, is a further strong indication that the negative temperature coefficients are caused by a structural effect. In this range the contribution of the structure factor greatly enhances the magnitude of the electrical resistivity.

For the divalent rare earth metals Eu and Yb we have the striking results that in the liquid state both metals show negative temperature coefficients of electrical resistivity. This is again readily explained in terms of structure factor arguments since in these two metals we certainly have two conduction electrons per atom. We note that the electronic configuration of Eu, apart from the 4f electrons, resembles Ba and that of Yb is close to that of Sr. Both Ba and Sr are divalent and show negative temperature coefficients of electrical resistivity in the liquid state. The resistivity of liquid Ba is large (314 $\mu\Omega$ cm) and is close to that measured in liquid Eu whereas Sr has a lower value which is close to that of Yb. The resistivities of the alkaline earth metals, including Ba and Sr, have been successfully calculated by Ratti and Evans (1973) and others using a Ziman-like theory.

In summary, we believe that the present experimental data gives a strong indication that the Ziman theory might be extended to the liquid rare earth metals and alloys. Quantitative calculations should be performed in order to test these ideas. Of course, such numerical calculations need to be based on potentials which give realistic band structures. Hopefully this approach should explain the linear increase of the electrical resistivity across the rare earth series and account for the detailed concentration dependence in alloys.

Acknowledgments

We are grateful to Dr R Evans for stimulating discussions. The authors are indebted to the 'Schweizerische Nationalfonds zur Förderung der wissenschaftlichen Forschung' and to the Research Center of Alusuisse for financial support.

References

Barabanov A F, Kikoin K A and Maksimov L A 1976 *Solid St. Commun.* **18** 1527–9
Busch G and Güntherodt H J 1974 *Solid St. Phys.* **29** 235–313 (New York: Academic Press)
Busch G, Güntherodt H J, Künzi H U and Schlapbach L 1970 *Phys. Lett.* **31A** 191–2
Colvin R V, Legvold S and Spedding F H 1960 *Phys. Rev.* **120** 741–5
Curry M A, Legvold S and Spedding F H 1960 *Phys. Rev.* **117** 953–4
Dreirach O, Evans R, Güntherodt H J and Künzi H U 1972 *J. Phys. F: Metal Phys.* **2** 109–25
Evans R, Greenwood D A and Lloyd P 1971 *Phys. Lett.* **34A** 51–8
Friedel J 1956 *Can. J. Phys.* **34** 1190–211
—— 1962 *J. Phys. Radium* **23** 692–700
Gaibullaev F, Regel A R and Khusanov Kh 1969 *Sov. Phys.–Solid St.* **11** 1138–9
Güntherodt H J, Hauser E and Künzi H U 1974a *Phys. Lett.* **47A** 189–90
—— 1974b *Phys. Lett.* **48A** 201–2
Güntherodt H J, Hauser E, Künzi H U, Evans R, Evers J and Kaldis E 1976 *J. Phys. F: Metal Phys.* **6** 1513–22
Habermann C E and Daane A H 1964 *J. Less Common Metals* **7** 31–6
Hauser E 1976 *Thesis* ETH Zurich
Kasuya T 1966 *Magnetism* vol 2B ed G T Rado and H Suhl (New York: Academic Press) pp215–94
Krieg G, Genter R B and Grosse A V 1969 *Inorg. Nucl. Chem. Lett.* **5** 819–23
Meaden G T 1971 *Contemp. Phys.* **12** 313–37
Ratti V K and Evans R 1973 *J. Phys. F: Metal Phys.* **3** L238–43
Spedding F H, Daane A H and Hermann K W 1957 *J. Metals* (July) **9** 895–97
Spedding F H, Hanak J J and Daane A H 1958 *Trans. Met. Soc. AIME* **212** 379–83
Zinov'ev, Gel'd L P, Chuprikov G E and Moreva N I 1974 *Sov. Phys.–Solid St.* **16** 235–36

The thermoelectric power and resisitivity of liquid Co–Ni

B C Dupree, J E Enderby, R J Newport and J B Van Zytveld
Department of Physics, University of Leicester, Leicester LE1 7RH

Abstract. The thermoelectric power, S, and electrical resistivity, ρ, have been measured as functions of concentration, c, for the liquid alloy system Co–Ni. It turns out that both $\rho(c)$ and $S(c)$ follow the theoretical predictions of Evans *et al* (1971 *Phys. Lett.* **35A** 57) if a low effective valence (of the order of 0·2) can be assumed. For valencies nearer unity, theory and experiment do not agree.

1. Introduction

In recent years two schemes have been proposed for the calculation of the electronic transport properties of transition metals and their alloys. The essence of the 'Bristol' method (see, for example, Evans *et al* 1971, Dreirach *et al* 1972) is to suppose that the nearly free electron formula for the resistivity ρ and the thermopower S (Ziman 1961) can be applied to liquid transition metals, provided the scattering from a single centre is calculated from the transition matrix:

$$t(k, k') = \frac{-2\pi\hbar^3}{m(2m\epsilon)^{1/2}}\left(\frac{1}{\Omega_0}\right)\sum_l (2l+1)\sin\eta_l(\epsilon)\exp(i\eta_l(\epsilon))\,P_l(\cos\theta).$$

The notation is standard, Ω_0 representing the atomic volume and $\eta_l(\epsilon)$ the lth partial-wave phase shift calculated at energy ϵ. In this model both ρ and S are governed by the proximity of the Fermi energy ϵ_F to a resonance in the $l = 2$ phase shift and to a good approximation are given by the relatively simple expressions

$$\rho = \frac{30\pi^3\hbar^3}{m\epsilon_F k_F{}^2\Omega_0 e^2}\,\bar{a}\sin^2\eta_2(\epsilon_F) = A\bar{a}\sin^2\eta_2(\epsilon_F) \tag{1}$$

$$S = \frac{\pi^2 k_0{}^2 T}{3|e|\epsilon_F}\left(-2 + \frac{\epsilon_F}{\bar{a}}\frac{\partial a}{\partial\epsilon}\bigg|_{\epsilon_F} + 2\epsilon_F\cot\eta_2(\epsilon_F)\frac{\partial\eta_2}{\partial\epsilon}\bigg|_{\epsilon_F}\right). \tag{2}$$

The methods used to calculate the Fermi energy ϵ_F, the radius of the Fermi sphere k_F and the phase shifts are discussed in detail by Dreirach *et al* (1972) and by Brown (1973). The liquid structure factor evaluated at a momentum transfer value, q, of $2k_F$ and represented by $\bar{a}$ can be obtained either from experiment (Waseda and Tamaki 1975, Waseda and Ohtani 1975) or from a solution of the Percus–Yevick equation for hard spheres (Ashcroft and Lekner 1966).

A basic objection to the method has, however, been formulated by Mott (1972). He argues that there are two mean free paths in the problem, one associated with the s and p electrons and the other with d electrons. Furthermore, the dominant contribution to the resistivity comes about, Mott argues, from s–d scattering and depends sensitively on $N_d(\epsilon)$, the d contribution to density of states. The effect of multiple scattering is to make $N_d(\epsilon)$ considerably broader than the single-site d-resonance from which it ultimately derives. Thus whilst the resonance theory in its elementary form may apply to liquid alloys involving an appreciable amount ($\gtrsim$25 at. %) of a non-transition metal, it might fail for those cases in which the d-band overlap integral is greater than a critical value (estimated by Mott as about 0·05 eV).

It was to throw further light on the theoretical problems posed by liquid transition metals that the series of experiments reported here were undertaken.

2. Experimental

The idea behind the present study was similar to that developed by Howe and Enderby (1967) who recognized that the difficulties (see §3 below) involved in making a direct comparison between theory and experiment could be avoided by focusing attention on the resistivity and the thermoelectric power of suitably chosen binary alloy systems. The method, which allows S for the alloy to be predicted from the properties of the pure components, works when the Fermi energy, atomic volume and liquid structure factors are concentration independent; it was applied successfully by Howe and Enderby (1967) to liquid Ag–Au.

The system chosen for this study was liquid Co–Ni because (i) the composition range and the evidence from the phase diagram (Hanson 1958) and the structure data (Waseda and Tamaki 1975) strongly favour a substitutional model for a; (ii) interpolating the data provided by Dreirach *et al* (1972) we find a change in ϵ_F across the composition range of less than 10%; (iii) according to Heine (1967), k_F for Co and Ni are essentially equal.

The experiments were carried out at temperatures of 1500 °C in a specially constructed apparatus, the principle of which was briefly described in an earlier communication (Dupree *et al* 1975; a full description of the method will appear elsewhere). The method ensured that the total sample contamination introduced by the tungsten counter electrodes did not exceed about 0·4 at.% in the most unfavourable case and

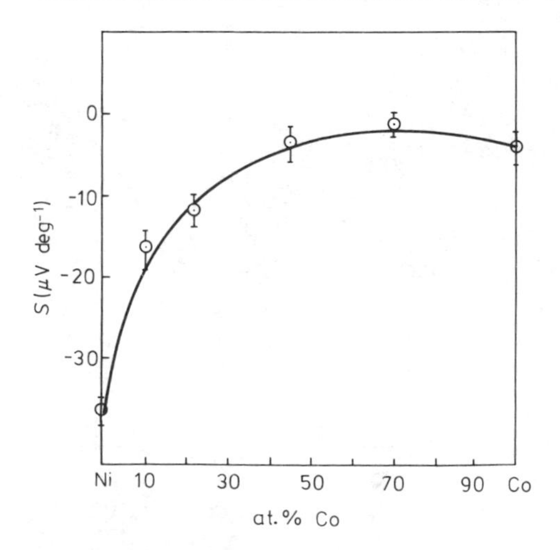

Figure 1. Thermoelectric power for liquid Ni–Co alloys at 1500°C.

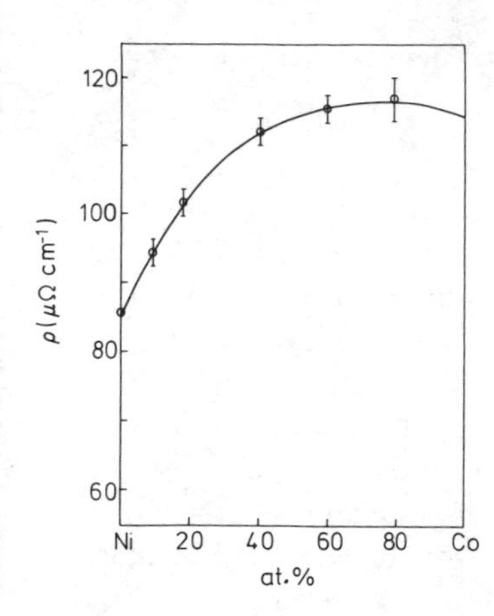

Figure 2. The resistivity for liquid Ni–Co alloys at 1500 °C

tests showed that this level of impurity did not, within experimental error, affect the observed thermoelectric power or resistivity. The basic technique was checked by measuring S and ρ for pure liquid Ni and comparing the values obtained with those reported by Howe and Enderby (1973). A satisfactory degree of agreement was achieved. The results for the liquid alloy system are presented in figures 1 and 2.

3. Discussion

The direct comparison of equation (1) with experimental observations on pure liquid transition metals is extremely difficult; this is because the value of A is somewhat arbitrary (depending as it does on the choice for E_F and k_F) and because $\sin^2 \eta_2(\epsilon_F)$ depending as it does on the choice for E_F and k_F) and because $\sin^2 \eta_2(\epsilon_F)$ depends sensitively on the position and the width of the d-resonance. Furthermore, in the low-momentum transfer region, the absolute value of $a(2k_F)$ is uncertain to perhaps as much as 50%.

One way round this difficulty is to replace $a(q)|t|^2$ in the expression for ρ and S by

$$\begin{aligned}\langle T_{\text{alloy}}\rangle^2 &= c_1|t_1|^2(1-c_1+c_1a_{11}(q)) + c_2|t_2|^2(1-c_2+c_2a_{22}(q))\\ &\quad + c_1c_2(t_1^*t_2+t_1t_2^*)(a_{12}(q)-1)\end{aligned}$$

where c_1 and c_2 are the concentration of the two types of ion and an obvious notation is used to label the transition matrices and the structure factors; we then assume that the parameters needed to evaluate T for the liquid alloy can be derived directly from the properties of the pure components. It then follows by straightforward extensions of equations (1) and (2) to the alloy case that the resistivity and the thermoelectric power of the alloy are given by

$$\rho_{\text{alloy}} = \alpha c^2 + \beta c + \gamma \tag{3}$$

$$S_{\text{alloy}} = \alpha' c^2 + \beta' c + \gamma' \tag{4}$$

where c is the atomic fraction of component 1. The full expression for $\alpha, \beta \ldots \gamma'$ will not be reproduced here. Apart from ϵ_F and k_F, which we treat as coupled parameters, $\alpha, \beta \ldots \gamma'$ are defined completely by the thermoelectric power and resistivity of the pure components 1 and 2. It therefore remains to decide a value of $\bar{a}$, $\partial\bar{a}/\partial\epsilon$ and ϵ_F; the following schemes were tried.

Scheme 1

The effective valency (Z) was chosen as unity and k_F fixed at a value of $1 \cdot 33$ Å^{-1}. To allow for uncertainties in the muffin-tin zero, ϵ_F was allowed to depart from the free-electron value by $\pm 0 \cdot 02$ Ryd. The structure factor used was the average of those appropriate to pure liquid Ni and pure liquid Co (Waseda and Ohtani 1975) and the error on $\bar{a}$ was assumed to be $\pm 0 \cdot 02$ subject to the condition that $\bar{a}$ must not be less than the long-wavelength limit. The error on $(\partial a/\partial q)|_{\epsilon_F}$ was taken as $\pm 10\%$. The results are shown in figure 3(a); the range of calculated values found when the input parameters are allowed to vary in the manner specified above fall within the thickness of the theoretical line. Thus the predicted variations of S and ρ with concentration, $S(c)$ and $\rho(c)$, do not agree with experiment.

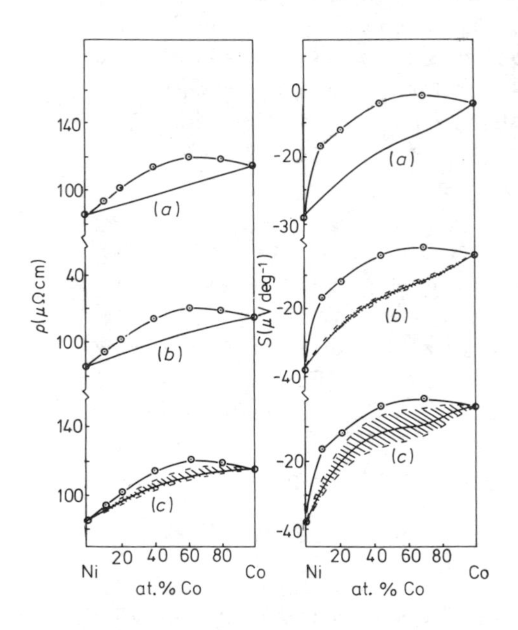

Figure 3. Resistivity and thermoelectric power of liquid Co–Ni alloys. Circles give experimental results; shaded areas give theory according to scheme 1(a), scheme 2(b) and scheme 3(c).

Scheme 2

Scheme 1 was followed except that Z was chosen as $0 \cdot 46$ (Brown 1973) so that k_F becomes $1 \cdot 02$ Å^{-1} and $\bar{a}$ is changed accordingly. Once again the calculated values of the transport coefficients (figure 3b) are not very sensitive to errors in the input parameters.

Scheme 3

Scheme 1 was again followed but this time Z was taken as $0 \cdot 2$. In this case (figure 3c) the calculated properties are very sensitive to the input parameters. It does appear, however, that if such a low value of Z is used, the theory can account qualitatively for the general form of $S(c)$ and $\rho(c)$, particularly when allowance is made for the rather crude approximations made in deriving equations (3) and (4).

4. Conclusions

The analysis we offer of our experimental data suggests that the single-site resonance method can be applied to liquid Ni–Co if low values of Z can be assumed. On the other hand, for Z values usually quoted in the literature, the predicted forms of $\rho(c)$ and $S(c)$ are essentially independent of the choice of ϵ_F, $\bar{a}$ and $(\partial a/\partial q)|_{\epsilon_F}$ and do not agree with experiment. Further work is clearly necessary if the general validity of small effective valences is to be established with confidence.

Acknowledgments

We are very grateful for helpful comments from Professor N F Mott and Dr R Evans. Dr J P D Hennessy is to be especially thanked for his major contribution to the resistivity apparatus. We wish to thank the Science Research Council for supporting the project.

References

Ashcroft N W and Leckner J 1965 *Phys. Rev.* **145** 83
Brown J S 1973 *J. Phys. F: Metal Phys.* **3** 1003
Dreirach O, Evans R, Güntherodt H-J and Kunzi H V 1972 *J. Phys. F: Metal Phys.* **2** 709
Dupree B C, Van Zytveld J B and Enderby J E 1975 *J. Phys. F: Metal Phys.* **5** L200
Evans R, Greenwood D A and Lloyd P 1971 *Phys. Lett.* **35 A** 57
Hanson M 1958 *Constitution of Binary Alloys* (New York: McGraw-Hill)
Heine V 1967 *Phys. Rev.* **153** 673
Howe R A and Enderby J E 1967 *Phil. Mag.* **16** 467
—— 1973 *J. Phys. F: Metal Phys.* **3** L12
Mott N F 1972 *Phil. Mag.* **26** 1249
Waseda Y and Ohtani M 1975 *Z. Phys.* **B21** 229
Waseda Y and Tamaki S 1975 *Phil. Mag.* **32** 275
Ziman J M 1961 *Phil. Mag.* **6** 1013

Electrical transport in amorphous and liquid transition metal alloys

H-J Güntherodt, H U Künzi, M Liard, R Müller, R Oberle and H Rudin
Institut für Physik, Universität, Basel, Switzerland

Abstract. Strong similarities have been observed between the electrical transport properties of transition metal alloys in the glassy and in the liquid state. In particular, the magnitudes of the electrical resistivity and its temperature coefficient in the glassy and liquid states are similar and this suggests that an extension of the Faber–Ziman theory in the t-matrix approximation to glassy alloys of transition metals may be very useful for interpreting the experimental data.

1. Introduction

Amorphous alloys can be prepared by a variety of methods from either the gas phase or from the liquid state. Glassy metals (or metallic glasses) obtained by rapid quenching of the melt are best suited for a comparison with the liquid state. In the last few years such metallic glasses have become technologically as well as scientifically interesting (Gilman 1975). Extensive structural studies of amorphous metals have been carried out. A comprehensive review (Cargill 1975) suggests that close-packed, continuous random structural models provide a realistic framework to analyse the structure. This approach is an extension of ideas developed for the liquid state of metals. Most experiments for studying the atomic arrangements in amorphous metals have employed x-ray and electron scattering. In this paper one of the first experiments on the structure using neutron scattering is reported. We propose new experiments to determine the partial interference functions of glassy binary alloys by using the isotope technique.

While the analysis of the structure of amorphous metals has used methods developed for the liquid state, the electrical transport properties of amorphous metals have been generally treated on the basis of concepts developed for the solid state at low temperatures. We attempt, as an alternative, to understand the electrical transport data of metallic glasses at high temperatures in terms of concepts derived during the last few years from studies of the liquid state of transition metal alloys (Busch and Güntherodt 1974). This viewpoint is supported by experimental data on electrical transport in a variety of metallic glasses. The data show strong similarities to results obtained in the corresponding liquid transition metal alloys.

2. Experimental

2.1. Total interference function by neutron scattering

Up to now, many amorphous alloys have been available only as foils or thin films produced by conventional splat cooling or evaporation techniques etc. Due to their small dimensions these samples were not well suited for neutron scattering experiments,

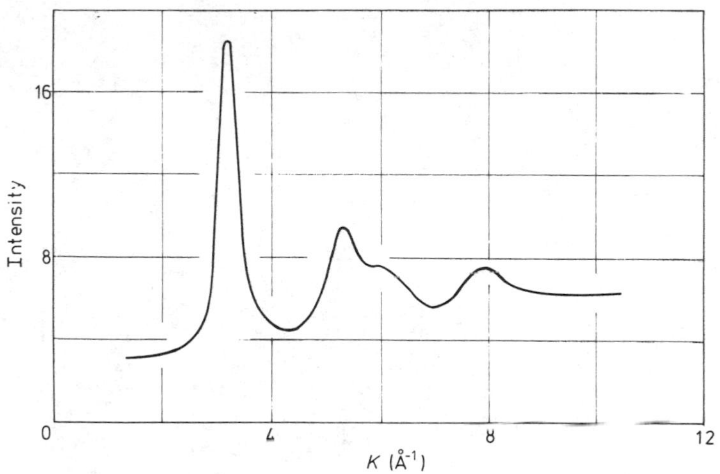

Figure 1. Neutron scattering intensity versus scattering vector of Metglas 2826A.

where much larger sample volumes are required. However, ribbons of glassy metals, produced by splat cooling from the melt using a variety of roller techniques, have recently become available. These are much more suitable for neutron scattering experiments. A ribbon of Metglas 2826A (Allied Chemical), of 15 m total length, was wound into a coil of 19 mm inner diameter, 20 mm outside diameter, and 50 mm height. This coil-shaped Metglas sample was supported by a thin-walled cylindrical vanadium container, which was mounted in the high-vacuum, high-temperature furnace of a two-axis spectrometer. The intensity of the scattered neutrons has been measured as a function of scattering vector and temperature by using thermal neutrons of $\lambda = 1{\cdot}05$ Å, in the same way as for liquid metals. Figure 1 shows the measured interference function versus scattering vector. We observe a liquid-like structure factor, but as is characteristic for amorphous samples, the second maximum is split into a double peak. Measurements at 300 °C do not differ significantly from room temperature data. For temperatures of 450 and 600 °C we observe Bragg peaks characteristic of crystalline samples. Since the structure of an amorphous alloy would have to be described in terms of the partial interference functions (for example, for a binary alloy AB in terms of $a_{AA}(K)$, $a_{AB}(K)$ and $a_{BB}(K)$), and this would be a difficult task for the five-component Metglas, we did not think it worthwhile to make the necessary detailed corrections for absorption, background, incoherent, magnetic and multiple scattering. The aim of our contribution is to show the advantage of using ribbons of metallic glasses to perform elastic and inelastic neutron scattering. It becomes possible to measure the partial interference functions of a binary metallic glass, by using three ribbons containing different isotopes. We believe further progress in understanding the structure and glass-forming ability of amorphous alloys can be made only by a systematic study of the partial interference functions.

2.2. *Electrical transport*

The aim of this section is to compare the electrical transport properties in the glassy and liquid states of corresponding transition metal alloys.

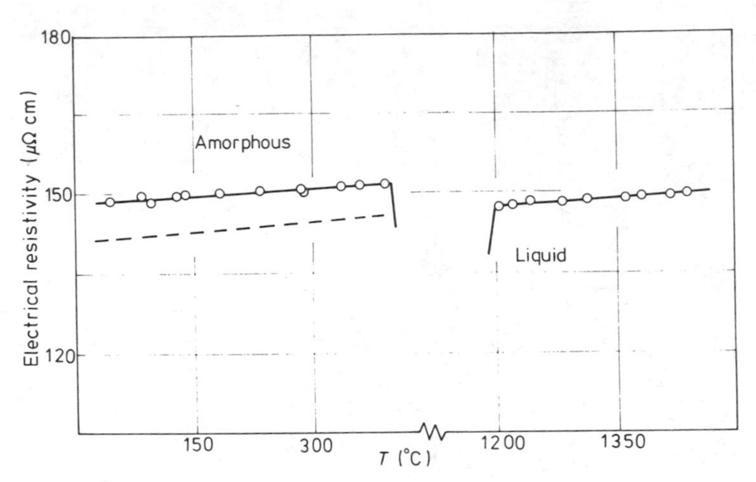

Figure 2. Electrical resistivity of Metglas 2605 ($Fe_{80}B_{20}$) in the glassy and liquid state. (Broken line, from Hasegawa 1976.)

For the Metglas 2826A we were able to show, on the basis of electrical transport data (Fischer *et al* 1976), that the five-component Metglas can be treated as a pseudo-binary alloy $Fe_{82}Ge_{18}$. The transition metal components Fe, Cr, Ni, act as a 3d configuration close to Fe and the metalloids B and P correspond to Ge with four conduction electrons. Due to the high vapour pressure of P it was not feasible to heat the Metglas 2826A sample up to the melting point in order to compare directly its amorphous and liquid states. However, the Metglas 2605 ($Fe_{80}B_{20}$) can be studied in the liquid state. Figure 2 shows the electrical resistivity of such a sample in the glassy and liquid states and strong similarities are apparent. For comparison, we show the resistivity data measured by R Hasegawa (private communication, 1976), these lying at smaller values. Note that the resistivity data in the glassy state do not show any change at the Curie point of 380 °C.

The slightly higher resistivity values of $Fe_{80}B_{20}$ in the glassy state compared to the liquid state might be due to the presence of magnetic scattering in the glassy state. We have good evidence that the magnetic scattering seems to be less important in the liquid state at high temperatures (Güntherodt *et al* 1977).

The classical examples of glassy alloys are the Pd–Si alloys at concentrations of around 80 at.% Pd. Figure 3 shows the electrical resistivity in the glassy, crystalline and liquid states of $Pd_{81}Si_{19}$. The electrical resistivity has a small positive temperature coefficient in the glassy state with a resistivity value of 80 $\mu\Omega$ cm at room temperature. At a temperature of 380 °C the transition into the crystalline state occurs and this state is characterized by a larger temperature coefficient and a smaller value of the resistivity. By heating up the crystalline sample the electrical resistivity changes at the melting point T_m. The resistivity values in the liquid state lie in a region which follows roughly from an extrapolation of the glassy state data and the temperature coefficient of the electrical resistivity is slightly larger than in the glassy state. This may be due to the difference in temperature dependence of the ionic structure in the glassy and in the liquid state. Furthermore, figure 3 shows that the amorphous, crystalline and liquid

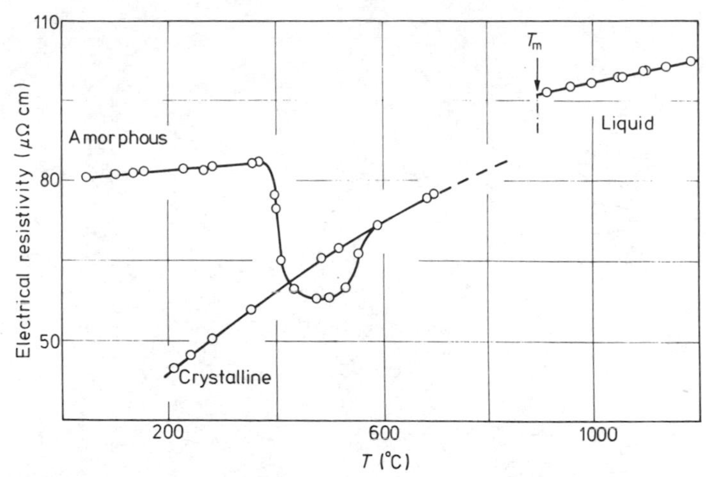

Figure 3. Electrical resistivity of glassy, crystalline and liquid $Pd_{81}Si_{19}$.

states tend to show the same resistivity values at high temperatures. In the liquid state, negative temperature coefficients of the electrical resistivity are found to occur when the Si concentration is increased to 30 and 40 at.% Si.

A comparison of the resistivity of liquid $Pd_{81}Si_{19}$ with that of pure liquid Pd (Güntherodt *et al* 1975) shows that the resistivity in the alloy is only slightly higher than the resistivity of liquid Pd. Consequently an explanation of the resistivity of pure liquid Pd should be useful in understanding the resistivity of Pd–Si alloys. In fact there are calculations of the resistivity of pure liquid Pd (Brown 1973) available and significantly, such calculations were carried out before the experimental values were measured.

Figure 4 shows the results of measurements of the Hall coefficient in glassy and liquid $Pd_{81}Si_{19}$: again the values in each state look very similar. We emphasize that such values of the Hall coefficient indicate that Si provides approximately four conduction electrons in the glassy and in the liquid state.

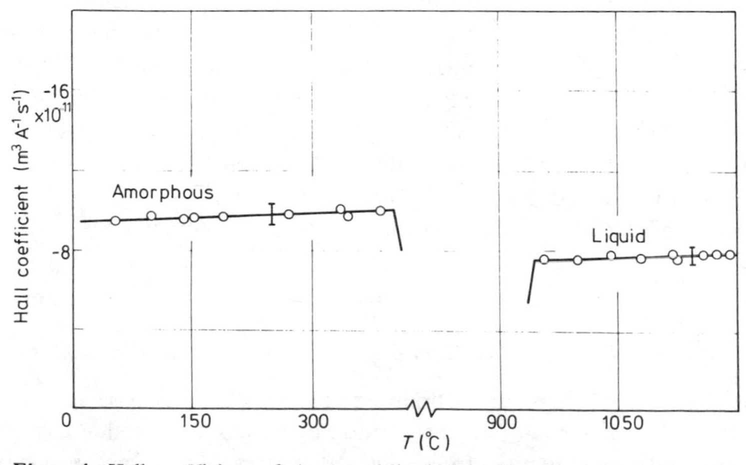

Figure 4. Hall coefficient of glassy and liquid $Pd_{81}Si_{19}$ (the error bar corresponds to ±5%).

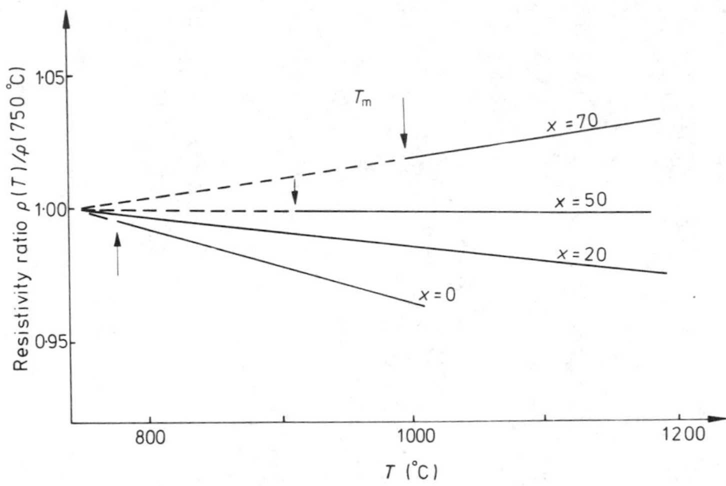

Figure 5. Electrical resistivity of liquid $(Pd_xCu_{100-x})_{80}Ge_{20}$ alloys.

The last examples which show strong similarities between electrical transport in the glassy and liquid states are some ternary alloys. In a recent letter (Tangonan 1975) the electrical resistivity of glassy $(Pd_xCu_{100-x})_{80}P_{20}$ alloys has been reported. The Cu-rich alloys show a negative temperature coefficient and the Pd-rich alloys show a positive temperature coefficient of the electrical resistivity. By changing the Pd to Cu concentration ratio, a continuous change of the temperature coefficient has been observed.

We have performed measurements of the electrical resistivity of liquid $(Pd_xCu_{100-x})_{80}Ge_{20}$ alloys because, with the high vapour pressure of P, it was not feasible to work with $(Pd_xCu_{100-x})_{80}P_{20}$ in the liquid state. The only difference between these systems is that P provides five and Ge four conduction electrons in the alloys. Figure 5 shows the resistivity ratio versus temperature plot for different concentrations of $(Pd_xCu_{100-x})_{80}Ge_{20}$ alloys and this is very similar to the plot presented by Tangonan (1975) for the glassy phosphorus alloys. Our Cu-rich alloys show a negative temperature coefficient while the Pd-rich alloys show a positive temperature coefficient of electrical resistivity. For comparison, the electrical resistivity of a liquid Cu–Ge alloy containing 20 at.% Ge is shown. This alloy has the largest negative temperature coefficient amongst the alloys presented in figure 5. The electrical resistivity of such liquid alloys has been qualitatively explained in terms of the Faber–Ziman theory (Busch and Güntherodt 1974) and numerical calculations (Dreirach *et al* 1972) are available which are in good agreement with experimental data.

3. Discussion

3.1. Theory of electrical resistivity of liquid transition metals and their alloys

The electrical resistivity of liquid transition metals and their alloys has been explained in terms of the following model. Liquid transition metals consist of randomly distributed ions and conduction electrons. The electrical resistivity can be calculated in a single-site approximation by considering the scattering of the conduction electrons by a non-overlapping muffin-tin potential. In this approach the electrical resistivity of pure liquid transition metals is given by (Evans *et al* 1971):

$$\rho = \frac{3\pi\Omega}{4e^2\hbar v_F^2 k_F^4} \int_0^{2k_F} a(K)\,|t(K)|^2 K^3\, dK \tag{1}$$

where Ω is the atomic volume, v_F the Fermi velocity, and $t(K)$ the single-site t-matrix. The interference function $a(K)$ of liquid transition metals (Waseda and Tamaki 1975) is not very different from that, say, of liquid copper. The value of $2k_F$ is determined by assuming that the liquid transition metals provide one or less conduction electrons per atom. The main experimental support for this choice comes from the alloying behaviour. Additional support follows from the interpretation of the Hall coefficient of liquid Ni and from a very recent determination of the normal and anomalous Hall coefficients of liquid Co in the paramagnetic range. The latter can be carried out by plotting the Hall coefficient R_H versus magnetic susceptibility χ (figure 6). We obtain in the solid state a normal Hall coefficient of $+4{\cdot}2 \times 10^{-11}\,m^3\,A^{-1}\,s^{-1}$ and in the liquid state, $-16{\cdot}2 \times 10^{-11}\,m^3\,A^{-1}\,s^{-1}$. This latter value corresponds roughly to 0·5 conduction electrons per Co atom in the liquid state.

The electrical resistivity can be written in the following approximation, if the d phase shift η_2 of transition metals is dominant at the Fermi energy E_F:

$$\rho = \frac{30\pi^3\hbar^3}{me^2k_F^2E_F\Omega}\, a(2k_F)\, \frac{\Gamma^2}{\Gamma^2 + 4(E_{res} - E_F)^2} \tag{2}$$

where Γ is the width and E_{res} is the energy of the scattering resonance, lying approximately at the centre of the 3d band. Figure 7 shows a schematic representation of E_{res} and E_F for some liquid transition metals. Since Mn has a half-filled d-band, the difference $(E_{res} - E_F)$ is very small in this metal. As the number of 3d electrons and the Fermi energy increase, the difference $(E_{res} - E_F)$ increases.

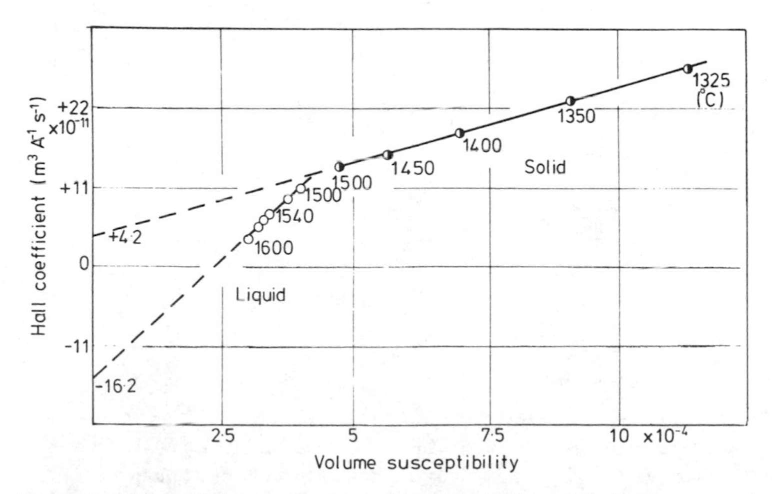

Figure 6. Hall coefficient versus magnetic susceptibility of liquid Co.

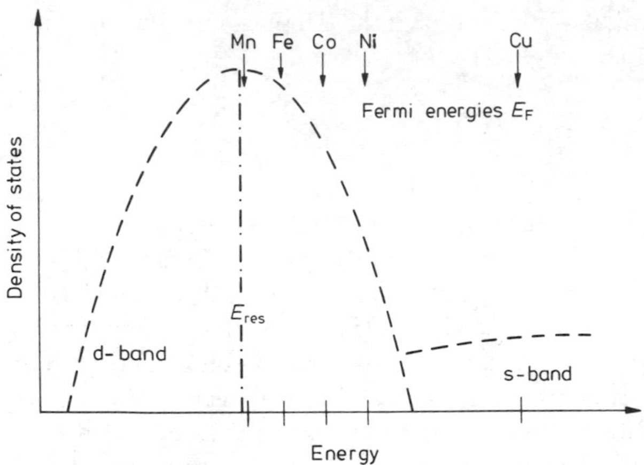

Figure 7. Schematic representation of resonance energy E_{res} and Fermi energy E_F of liquid transition metals.

From equation (2) it is clear that there are two separate factors which contribute to the electrical resistivity of liquid transition metals. The first contribution arises from the interference function. Since the value of $2k_F$ corresponding to one or less conduction electrons is small, then only a small contribution comes from the interference function. The main contribution arises from the resonant scattering of conduction electrons from the 3d states lying in the conduction band. The increasing value of $(E_{res}-E_F)^2$ going from Mn to Co, Ni and Cu should bring about a decrease in electrical resistivity. This is indeed observed, as is shown in figure 8 by the experimental values of electrical resistivity of liquid transition metals at their melting points. The temperature dependence of the electrical resistivity of liquid Fe, Co, Ni and Pd is shown in figure 9. We see that Ni and Pd have nearly the same resistivity values in the liquid state.

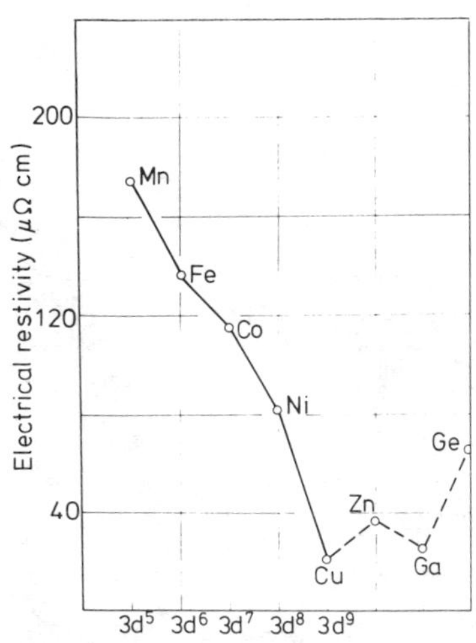

Figure 8. Electrical resistivity of liquid transition metals at their melting points.

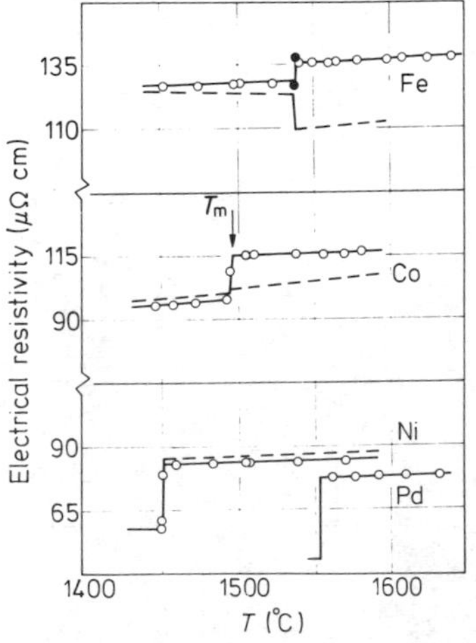

Figure 9. Electrical resistivity of liquid Fe, Co, Ni and Pd as a function of temperature.

At present it is still a matter for discussion as to whether there might be an additional contribution to the electrical resistivity of liquid transition metals from spin-disorder scattering.

The electrical resistivity of liquid alloys of transition metals is given in the single-site approximation by (Dreirach *et al* 1972):

$$\rho = \frac{3\pi\Omega}{4e^2 v_F^2 \hbar k_F^4} \int_0^{2k_F} |U(K)|^2 K^3 \, dK \tag{3}$$

with

$$|U(K)|^2 = c_A \, |t_A|^2 (1 - c_A + c_A a_{AA}) + c_B \, |t_B|^2 (1 - c_B + c_B a_{BB})$$
$$+ c_A c_B (t_A{}^* t_B + t_A t_B{}^*) (a_{AB} - 1),$$

where $c_{A,B}$ are the concentrations, $t_{A,B}$ the *t*-matrices of the components A and B, and a_{AA}, a_{BB} the partial structure factors of the alloy. In order to properly discuss binary alloys of transition metals the three partial interference functions are required. In general, these quantities are not yet known for liquid transition metal alloys. For numerical calculations of the electrical resistivity we refer to the paper by Dreirach *et al* (1972). In the present paper we restrict ourselves to qualitative discussions.

On alloying with polyvalent metals, the density of conduction electrons increases and so $2k_F$ increases giving rise to a larger contribution to the resistivity via the term $a(2k_F)$. For liquid transition metal alloys containing polyvalent metals such as B, Ge, Si, a maximum of the electrical resistivity and negative temperature coefficients are expected for concentrations where $K_p = 2k_F$, where K_p denotes the position of the first peaks in the partial interference functions. Measurements of the electrical resistivity of liquid alloys containing Ge and the transition metals Mn, Fe, Co, Ni and Cu have shown that the concentration range, over which negative temperature coefficients occur, exhibits a systematic variation which is correlated to the number of d-electrons of the transition metals. For more details see figure 29 in Busch and Güntherodt (1974). The different variations of $2k_F$ have been interpreted in terms of charge transfer (Bennemann 1976).

In view of the similarities between the structure, the magnitudes of the resistivity and its temperature coefficients we believe that a similar model should explain the observed high-temperature electrical transport data of glassy alloys.

3.2. Discussion of the electrical transport of glassy transition metal alloys

Here we attempt to explain the behaviour of the electrical resistivity of glassy alloys such as Metglas 2826A, 2605 ($Fe_{80}B_{20}$), $Pd_{81}Si_{19}$ and $(Pd_xCu_{100-x})_{80}P_{20}$ in terms of the model suggested above. Metglas 2826A contains metalloids with an average number of four conduction electrons per atom while Metglas 2605 contains B with three conduction electrons. These alloys have resistivities of 180 and 150 $\mu\Omega$ cm respectively and the difference can be understood in the model. Pure liquid Fe has a resistivity of 135 $\mu\Omega$ cm, and alloying with B increases the value of $a(2k_F)$ which, in turn, contributes more to the resistivity. This contribution is further increased by alloying with metalloids which provide four conduction electrons per atom.

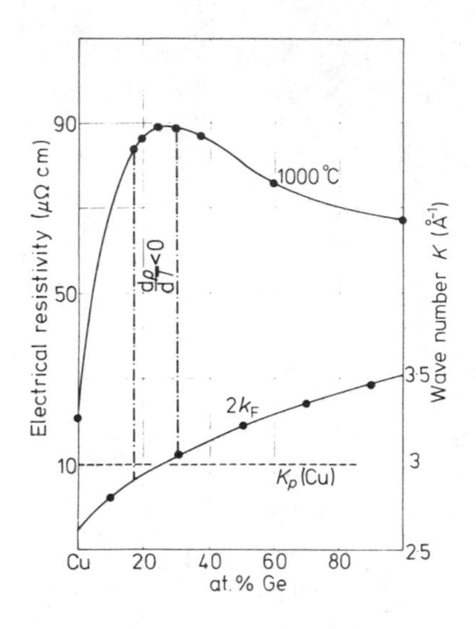

Figure 10. Electrical resistivity and wave numbers K_p and $2k_F$ of liquid Cu–Ge alloys.

The electrical resistivity of glassy $Pd_{81}Si_{19}$ can be explained in a similar manner. The value of $a(2k_F)$ in pure liquid Pd is not too large, but alloying with Si provides four conduction electrons per Si atom and the value moves closer to the value of K_p. However, to observe negative temperature coefficients we would need a concentration of 30 or 40 at.% Si. The different temperature coefficients in the glassy and liquid alloys are attributed to a different temperature dependence of the interference functions in the glassy and liquid states. Experimental confirmation of this speculation would be most useful.

The electrical resistivity of ternary $(Pd_xCu_{100-x})_{80}P_{20}$ alloys can be understood by comparison with data on liquid Cu–Ge alloys. Figure 10 shows the electrical resistivity and the wavenumbers K_p and $2k_F$ of liquid Cu–Ge alloys. Because of lack of structural data on liquid Cu–Ge, the K_p has been fixed at the value of pure liquid Cu. The $2k_F$ values are calculated according to the free electron model assuming one conduction electron for liquid Cu and four for liquid Ge, respectively. We see that the negative temperature coefficients occur in the concentration range where K_p is close to k_F. We assume that changing the Pd to Cu concentration ratio does not drastically affect the value of K_p. The structure data (Waseda and Ohtani 1975, Waseda and Tamaki 1975) show that the interference functions of Cu and Pd are very similar. However, for a fixed metalloid concentration, the value of $2k_F$ will be smaller for Pd-rich alloys since pure Cu has one conduction electron per atom whereas we assume Pd has less than one electron per atom. For Cu-rich alloys containing 20 at.% Ge or P, we have $K_p \sim 2k_F$, but for the corresponding Pd-rich alloys the value of $2k_F$ seems to be smaller than K_p. The fact that the negative temperature coefficients disappear at lower Pd concentrations in the Ge alloys than in the P alloys can be attributed to the difference in the number of conduction electrons between Ge and P.

These types of argument readily explain why $(Pd_xNi_{100-x})_{80}P_{20}$ alloys show positive temperature coefficients of the electrical resistivity whereas those of $(Pd_xNi_{100-x})_{75}P_{25}$ show negative temperature coefficients (Boucher 1972). The key to understanding this resistivity behaviour is provided by figure 29 in Busch and

Güntherodt (1974). It follows that for the first alloy, a change of the Pd to Ni ratio does not affect $2k_F$, K_p and hence the resistivity and its temperature coefficient. In the second alloy system the additional number of conduction electrons provided by the higher concentration of P increases the value of $2k_F$. In contrast to Pd–Cu–P, the Pd–Ni–P alloys do not show any significant variation of electrical resistivity when the ratio of Pd to Ni is changed. A very similar interpretation of the electrical resistivity of amorphous Ni–P alloys has been given by Cote (1976).

4. Conclusion

The strong similarities of electrical transport properties in glassy and liquid transition metal alloys have been demonstrated, but only for certain concentrations such as $Pd_{81}Si_{19}$ or $Pd_{77.5}Cu_6Si_{16.5}$. Certainly it would be more convincing to make comparisons over a larger concentration range where the glassy alloys are stable. It would be useful, as a next step, to look for similarities or differences between properties other than those of electrical transport – especially between the properties which depend significantly on the motion of the ions.

Acknowledgments

One of the authors (H-J G) is very much indebted to IBM for the opportunity of an introduction into the field of amorphous metals at the IBM Watson Research Center, Yorktown Heights. Many stimulating discussions with Dr P Chaudhari and Dr C C Tsuei are gratefully acknowledged. We are grateful for the $Pd_{81}Si_{19}$ sample prepared at IBM, Yorktown Heights by the technique of splat cooling. We would like to thank the 'Schweizerische Nationalfonds zur Förderung der wissenschaftlichen Forschung' and the Research Center of Alusuisse for financial support.

References

Bennemann K H 1976 *J. Phys. F: Metal Phys.* **6** 43
Boucher B Y 1972 *J. Non. Cryst. Solids* 7 277
Brown J S 1973 *J. Phys. F: Metal Phys.* **3** 1003
Busch G and Güntherodt H-J 1974 *Solid St. Phys.* **29** 235 (New York: Academic Press)
Cargill G S III 1975 *Solid St. Phys.* **30** 229 (New York: Academic Press)
Cote P J 1976 *Solid St. Commun.* **18** 1311
Dreirach O, Evans R, Güntherodt H-J and Künzi H U 1972 *J. Phys. F: Metal Phys.* **2** 709
Evans R, Greenwood D A and Lloyd P 1971 *Phys. Lett.* **A35** 57
Fischer M, Güntherodt H-J, Hauser E, Künzi H U, Liard M and Müller R 1976 *Proc. 2nd Int. Conf. on Rapidly Quenched Metals (Section I)* ed N J Grant and B C Giessen (Cambridge, Mass.: MIT Press)
Gilman J J 1975 *Phys. Today* **28** 46
Güntherodt H-J, Hauser E and Künzi H U 1977 this volume
Güntherodt H-J, Hauser E, Künzi H U and Müller R 1975 *Phys. Lett.* **54A** 291
Tangonan G L 1975 *Phys. Lett.* **54A** 307
Waseda Y and Ohtani M 1975 *Z. Phys.* **B21** 229
Waseda Y and Tamaki S 1975 *Phil. Mag.* **32** 273

Electrical resistivity of liquid alloys of manganese with antimony, tin and indium as a function of temperature and composition

J G Gasser and R Kleim

Laboratoire de Physique des Milieux Condensés, Faculté des Sciences, Ile du Saulcy, 57000 Metz, France

Abstract. The electrical resistivity of liquid binary alloys of manganese with antimony, tin and indium has been measured as a function of temperature and composition between the liquidus temperature and 1200 °C. Measurements were performed by a four-probe direct current method, using a quartz cell fitted with tungsten electrodes. We observed a negative temperature coefficient for atomic compositions greater than 55 at.% Mn for Mn–Sb, 45 at.% Mn for Mn–Sn and 36 at.% Mn for Mn–In and positive values elsewhere in the studied concentration range. For Mn–Sb no particular behaviour is observed near the eutectic composition. The resistivity as a function of temperature and composition is discussed in terms of the Faber–Ziman theory.

1. Introduction

The theory of Evans *et al* (1971, 1972) based on a *t*-matrix formulation of Ziman's nearly-free-electron model (Ziman 1961, Bradley *et al* 1962, Faber and Ziman 1965) has stimulated much experimental and theoretical work (for a review see Busch and Güntherodt 1974). Nevertheless, very little is known about the transport properties and more specifically the resistivity, for manganese with other normal or noble metals, except for Mn–Ge, Mn–In, Mn–Cu (Güntherodt and Künzi 1973) and Mn–Sb (Ohno *et al* 1973), the latter only for small Mn concentrations. We have measured the electrical resistivity of binary alloys of Mn with three polyvalent metals with increasing valency (In, Sn and Sb), as a function of temperature and concentration for Mn concentrations up to about 80 at.%. Some experimental details are given in §2, and in §3 and §4 we present and discuss our results over the full studied concentration range, with special attention for small Mn concentrations.

2. Experimental

The alloys were prepared using Sb and Mn with 99·99% purity supplied by Roc Ric and Johnson Matthey respectively. For Sn and In the purity, from the latter manufacturer, is 99·999%.

Resistivity measurements were performed by the four-probe method, using a quartz cell fitted with tungsten electrodes. A pressure of argon up to 7 bar could be applied over the liquid. The precision of the electrical measurements was better than 0·3% and the uncertainty in composition estimated to be less than 1%. Temperatures were measured by four chromel–alumel thermocouples giving an overall precision of 0·7%. Full experimental details will be given in a later paper.

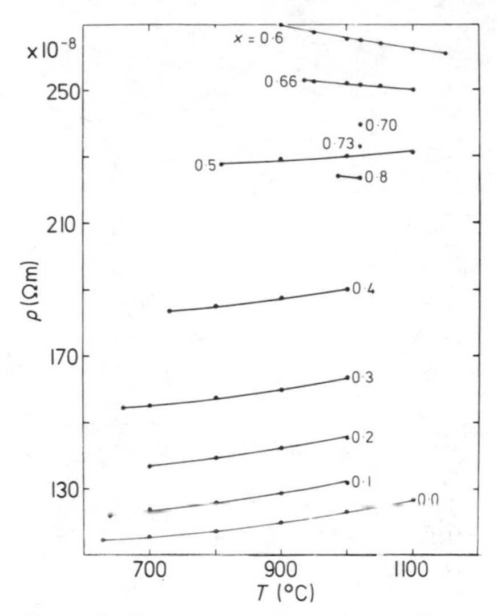

Figure 1. Temperature dependence of the electrical resistivity of liquid $Mn_x Sb_{1-x}$ alloys.

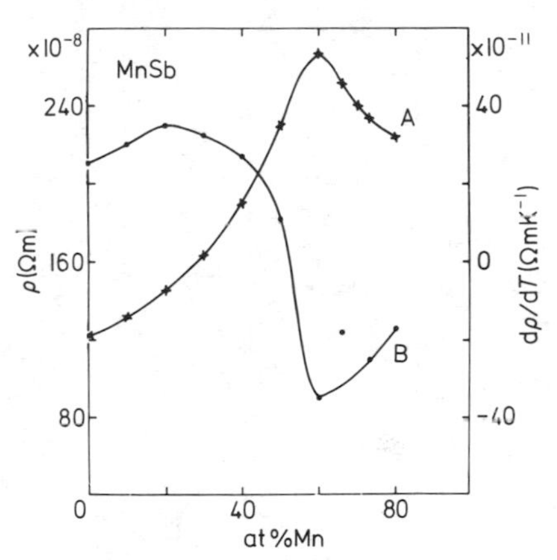

Figure 2. Electrical resistivity of liquid Mn–Sb alloys (A) and their temperature coefficient (B), as a function of concentration at 1000 °C.

3. Results

3.1. Mn–Sb alloys

The resistivity as a function of temperature is shown in figure 1, and as a function of concentration at 1000 °C in figure 2. The temperature coefficient at 1000 °C is also shown in figure 2. The resistivity of liquid Sb increases on alloying with Mn, and shows a maximum near 60 at.% Mn. No reliable measurements could be made after 80 at.% Mn, but the curve may be extrapolated to pure Mn giving a value of about 200 $\mu\Omega$ cm, consistent with the values extrapolated for Mn–In and Mn–Sn systems and with the value of 175 $\mu\Omega$ cm obtained in the same manner by Güntherodt and Künzi (1973). However, this extrapolation is not very significant because the temperature is below the melting point of pure Mn (1243 °C), but it may give a useful approximation for the resistivity of liquid Mn, for which large discrepancies exist (Grube and Speidel 1940, Mokrovsky and Regel 1953, Vostryakov *et al* 1964, Baym *et al* 1969). The temperature coefficient, after passing through a maximum at about 20 at.% Mn, becomes negative for Mn concentrations greater than 54 at.%. The minimum, located at 60 at.% Mn, is low (–35 $n\Omega$ cm K^{-1}) and comparable to that of Mn–Ge. The maximum of ρ and the minimum of $d\rho/dT$ lie at approximately the same concentration. This behaviour of the Mn–Sb system shows a great similarity with that of Ni–Ge, Co–Ge, Fe–Ge and Mn–Ge alloys studied by Güntherodt and Künzi (1973). The maximum in $d\rho/dT$ seems to be a peculiarity of this system that has not been observed elsewhere.

3.2. Mn–Sn alloys

For these alloys, the resistivity versus temperature and concentration are plotted in figures 3 and 4 respectively. There is no evidence of a resistivity maximum in the 0–80 at.% Mn concentration range, although it may be possible that a small hump exists

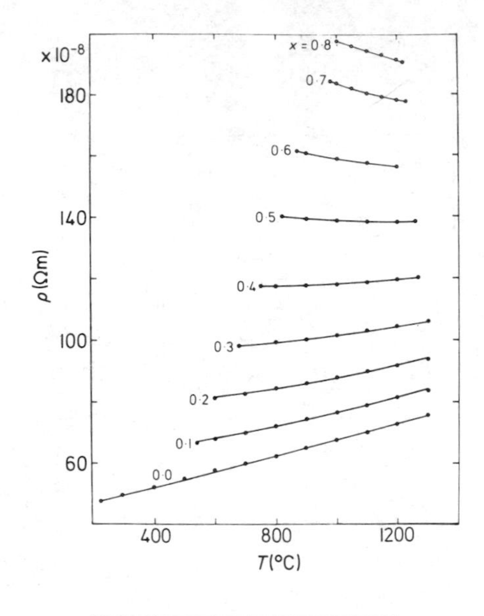

Figure 3. Temperature dependence of the electrical resistivity of liquid $Mn_x Sn_{1-x}$ alloys.

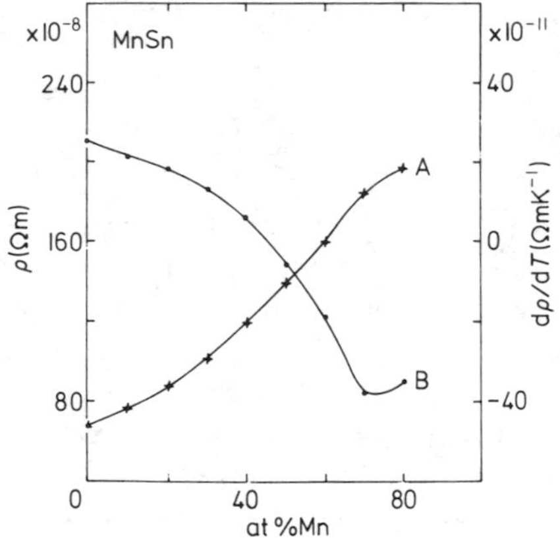

Figure 4. Electrical resistivity of liquid Mn–Sn alloys (A) and their temperature coefficient (B), as a function of concentration at 1000 °C.

for high Mn concentrations. To confirm this, it would be necessary to determine a resistivity isotherm at a higher temperature or, at least, to have a reliable measurement for pure Mn in the liquid state. In this respect, the Mn–Sn system differs from Mn–Sb and from the systems studied previously (Busch *et al* 1970, Güntherodt and Künzi 1973). Nevertheless, concerning the temperature coefficient, negative values are obtained for Mn concentrations greater than 45 at.%. The minimum value of about $-38\ n\Omega\,cm\,K^{-1}$ at 74 at.% Mn, is comparable to the values for Mn–Sb alloys.

3.3. Mn–In alloys

Figures 5 and 6 show the variation of resistivity with temperature and concentration, the latter together with the temperature coefficient. These figures look similar to those of Mn–Sn. No maximum in resistivity is observed over the whole concentration range and our results are in agreement with these of Güntherodt and Künzi (1973). Negative temperature coefficients occur after 36 at.% Mn. This system, as well, probably, as the

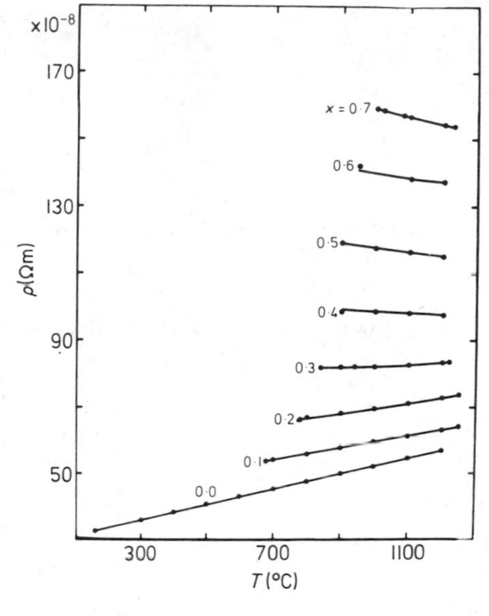

Figure 5. Temperature dependence of the electrical resistivity of $Mn_x In_{1-x}$ alloys.

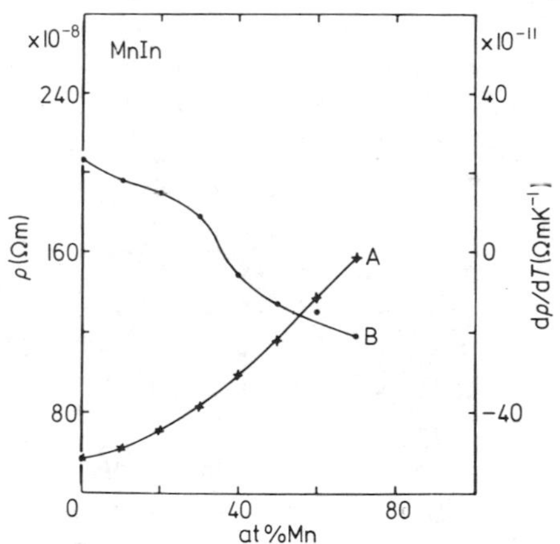

Figure 6. Electrical resistivity of liquid Mn–In alloys (A) and their temperature coefficient (B), as a function of concentration at 1100 °C.

Mn–Sn system, does not enter into any previously proposed classification (Busch and Güntherodt 1974).

4. Discussion

The temperature coefficient and the variation of the resistivity with concentration may be understood by an extension of the Faber–Ziman formula to transition alloys, given by Evans *et al* (1972) and Dreirach *et al* (1972). The general lines of a qualitative discussion, in terms of the relative positions of the Fermi diameter $2k_F$ of the alloy and those of the main peaks K_p in the partial interference functions, is well known (see, for example, Faber 1972, Busch and Güntherodt 1974). In table 1 we give the position K_p of the first peak of the interference function for the pure metals taken from experimental data on x-ray or neutron diffraction given by Waseda and Suzuki (1971), by Knoll and Steeb (1973) for Sb, by Waseda and Suzuki (1972) for In and Sn, and by

Table 1.

	Mn	Sb	Sn	In
$K_p(\text{Å}^{-1})$	2·83	2·15	2·21	2·30
$2k_F(\text{Å}^{-1})$	2·46	3·18	3·08	2·91

Waseda and Tamaki (1975) for Mn. To calculate k_F we used the measurements of the densities made by Crawley and Kiff (1972) for Sb, by Crawley (1968) for In, by Lucas (1972) for Sn and by Vatolin and Esin (1963) for Mn. For the latter element we assume an effective number of conduction electrons such that the valency could be taken equal to one (Dreirach *et al* 1972).

The true partial interference functions a_{ij} are not known for Mn alloys but, to a first approximation, a_{11} (index 1 for Mn) and a_{22} (index 2 for Sb, Sn or In) may be assumed to be identical with those of the pure metals, and a_{12} has its main peak located between those of a_{11} and a_{22}. On alloying Mn with a polyvalent metal, the small $2k_F$ value for pure Mn increases, but not necessarily as in a free electron picture because of the filling up of the 3d states of the transition metal (Dreirach *et al* 1972). The value of $2k_F$ may then become equal to the value of K_p for a_{11} and a_{12}, the latter obviously satisfying the necessary condition $2k_F(\text{Mn}) < K_p(a_{12}) < 2k_F(\text{Sb})$. The integrand in Ziman's formula becomes maximum if $2k_F(\text{alloy}) \simeq K_p(\text{alloy})$, and a maximum may occur in the resistivity with respect to concentration. For Mn–Sb alloys this maximum is observed for 60 at.% Mn, and a linear interpolation gives the crude value $2k_F(\text{alloy}) = 3{\cdot}03\ \text{Å}^{-1}$ for this concentration which, when compared with $K_p = 2{\cdot}83\ \text{Å}^{-1}$ of pure Mn, suggests that a_{11} may play the dominant role. The different terms in Ziman's formula are weighted by the concentrations, but this does not have a great influence near 60 at.% Mn. In other respects, it is by no means obvious that the above conditions, even when they are satisfied, necessarily involve such a maximum, because of the contribution from the other terms in Ziman's formula. The systems Mn–In and Mn–Sn may be representative examples for this latter case.

The occurrence of a relatively large concentration range where the temperature coefficient is negative, seems to be a more general property of transition or noble metals alloyed with polyvalent metals. The mechanism is basically the same as that in pure zinc. As the temperature increases, the main peaks in a_{ij} flatten, and if $2k_F$ of the alloy is located near such a peak, then the resistivity decreases. So we may say that a maximum in resistivity, with respect to concentration, involves a negative temperature coefficient in nearly the same concentration range about the maximum, but that the converse is not necessarily true. Negative temperature coefficients have also been observed for Mn–Cu (Güntherodt and Künzi 1973), and this alloy may perhaps be considered as a limiting case.

It may be worthwhile to discuss here the case of dilute alloys. As has been done for other systems (Ohno *et al* 1973) the increment of resistivity $\Delta\rho$ for dilute solutions of transition metals in polyvalent metals may be calculated by an extension to liquid metals of a formula strictly valid only in the solid state (Freidel 1956):

$$\Delta\rho = \frac{10\pi\hbar c}{ne^2k_F}(\sin^2\eta_2\uparrow + \sin^2\eta^2\downarrow)$$

Table 2.

	In	Sn	Sb
$\Delta\rho_{cal}$ ($\mu\Omega$ cm)	1·03	0·70	0·53
$\Delta\rho_{exp}$ ($\mu\Omega$ cm)	0·85	1·03	0·86

where c is the concentration of Mn, n the number of conduction electrons per atom in the solvent and k_F the Fermi radius of the solvent.

For Mn, the effective number of localized electrons with up and down spins are taken to be 5 and 1 respectively; this is consistent with the assumption of one effective valence electron per atom. The formula for $\Delta\rho$ has also been used by Güntherodt and Zimmermann (1973) for evaluating the number of 5d electrons in liquid rare earth metals, and is related to phase shift by means of Friedel's sum rule.

The calculated results, $\Delta\rho_{cal}$, for a dilute solution of 1 at.% Mn in solvent In, Sn and Sb are shown in table 2, together with the experimental values $\Delta\rho_{exp}$ at 700 °C. There is a reasonable agreement between the calculated and experimental values, but we think that this calculation is a relatively crude approximation and the discrepancies may not be meaningful. In particular, the results are very sensitive to the number of electrons with up and down spins, and this is still an open question for Mn. None the less, there is an overall agreement in favour of a nearly-free-electron picture in spite of the small ratio of mean free path to interatomic spacing (1·1 for Sb, 2·2 for Sn and 3·2 for In).

5. Conclusion

We have mainly discussed our results in the spirit of Ziman's model, but some alternative hypotheses may be interesting in explaining the behaviour of these systems. Mott (1972) suggests that it is likely that ρ decreases because the strong scattering by the transition atoms decreases with rising temperature, due to the broadening of the Fermi level. No numerical calculations are available, and, at least for normal metals, Faber (1972) thinks that this is a small effect.

On the other hand, the mean free path is small for pure Sb and, consequently, probably also for Mn–Sb alloys. This material may retain some covalent character: 'bonds' are broken and the resistivity decreases. This may perhaps be related to the subsidiary maximum in the interference function for pure Sb (Orton 1976). In other respects, in alloys of antimony with normal metals such as Sb–Cd, there is a strong evidence for clustering effects (Gasser and Kleim 1975). The band character of the d states which is not incorporated in the single-site model of Evans, has been taken into account by ten Bosch and Bennemann (1975), and their model leads to interesting qualitative conclusions for pure transition metals and alloys. Band structure effects and the role played by the density of states in determining the resistivity is also discussed by Evans *et al* (1973).

Concerning the variation of resistivity with concentration, we see that it is almost linear for Mn–Cu (Güntherodt and Künzi 1973), and has a strong maximum for

Mn–Sb; Mn–In and Mn–Sn are intermediate. These facts may be related to the increasing valency of the normal metal, but more experimental work is needed to draw a general inference.

References

Baym B A, Akhentsev Y N and Gel'd P V 1969 *Zh. Prikl. Khim* **42** 588
ten Bosch A and Bennemann K H 1975 *J. Phys. F: Metal Phys.* **5** 1333
Bradley C C, Faber T E, Wilson E G and Ziman J M 1962 *Phil. Mag.* **7** 865
Busch G and Güntherodt H-J 1974 *Solid St. Phys.* **29** 235 (New York: Academic Press)
Busch G, Güntherodt H-J, Künzi H U, Meier H A and Schlapbach L 1970 *Mater. Res. Bull.* **5** 567
Crawley A F 1968 *Trans. AIME* **242** 2237
Crawley A F and Kiff M R 1972 *Metall. Trans.* **3** 157
Dreirach O, Evans R, Güntherodt H-J and Künzi H U 1972 *J. Phys. F: Metal Phys.* **2** 709
Evans R, Greenwood D A and Lloyd P 1971 *Phys. Lett.* **35A** 57
Evans R, Güntherodt H-J, Künzi H U and Zimmermann A 1972 *Phys. Lett.* **38A** 151
Evans R, Gyorffy B L, Szabo N and Ziman J M 1973 *The Properties of Liquid Metals* ed S Takeuchi (London: Taylor and Francis) p319
Faber T E 1972 *An Introduction to the Theory of Liquid Metals* (London: Cambridge University Press)
Faber T E and Ziman J M 1965 *Phil. Mag.* **11** 153
Friedel J 1956 *Can. J. Phys.* **34** 1190
Gasser J G and Kleim R 1975 *J. Physique Lett.* **36** L93
Grube G and Speidel H 1940 *Z. Elektrochem.* **46** 233
Güntherodt H-J and Künzi H U 1973 *Phys. Kondens. Mater.* **16** 117
Güntherodt H-J and Zimmermann A 1973 *Phys. Kondens. Mater.* **16** 327
Knoll W and Steeb S 1973 *Phys. Chem. Liquids* **4** 39
Lucas L D 1972 *Mém. Sci. Rev. Metall.* **129** 395
Mokrovsky H P and Regel A R 1953 *Zh. Tekh. Fiz.* **23** 2121
Mott N F 1972 *Phil. Mag.* **26** 1249
Ohno S, Okazaki H and Tamaki S 1973 *J. Phys. Soc. Japan* **35** 1060
Orton B R 1976 *Z. Naturf.* **31A** 397
Vatolin N A and Esin D A 1963 *Fiz. Met. Metall.* **16** 936
Vostryakov A A, Vatolin N A and Esin O A 1964 *Russ. J. Inorg. Chem.* **9** 1034
Waseda Y and Suzuki K 1971 *Phys. Stat. Solidi* **B47** 581
—— 1972 *Phys. Stat. Solidi* **B49** 339
Waseda Y and Tamaki S 1975 *Phil. Mag.* **32** 273
Ziman J M 1961 *Phil. Mag.* **6** 1013

Separation of the normal and anomalous Hall effect in amorphous ferromagnetic Ni–Au alloys

G Bergmann
Institut für Festkörperforschung, Kernforschungsanlage Jülich, 5170 Jülich, Germany

Abstract. Amorphous $Ni_{0\cdot65}Au_{0\cdot35}$ alloys are prepared by quenched condensation and are ferromagnetic. The Hall resistivity is measured in the temperature range between 4 K and $2T_C$. In contrast to the liquid alloy one can separate the normal and the anomalous Hall effect. At helium temperature the anomalous Hall effect is saturated at low fields and the high-field Hall effect yields a Hall constant of $R_\infty = -12{\cdot}4 \times 10^{-11}\, m^3 A^{-1} s^{-1}$. Above the Curie temperature, $T_C = 155$ K, the anomalous Hall constant follows a Curie–Weiss law. The extrapolation of the measured Hall constant to $T \rightarrow \infty$ yields the Hall constant $R_0 = -11{\cdot}9 \times 10^{-11}\, m^3 A^{-1} s^{-1}$. The agreement between R_∞ and R_0 is quite good and we try to interpret them as the normal Hall constant. (The larger value of R_∞ is due to the fact that spin-up and spin-down electrons have slightly different mean free paths in the ferromagnetic state.) It corresponds to about 0·6 conduction electrons/atom. From the temperature dependence of the Hall constant in the amorphous phase we derive the value of $-12{\cdot}6 \times 10^{-11}\, m^3 A^{-1} s^{-1}$ for the liquid alloy, which is in good agreement with the measured value of the liquid of $-11{\cdot}5 \times 10^{-11}\, m^3 A^{-1} s^{-1}$.

A fuller account can be found in *Solid St. Commun.* **18** 897 (1976).

Soft x-ray band spectra of liquid metals

C F Hague
Laboratoire de Chimie Physique 'Matière et Rayonnement' Associé au CNRS,
Université Pierre et Marie Curie, 11 rue P et M Curie, 75231 Paris Cedex 05, France

Abstract. The L_3 emission and absorption spectra of liquid Fe, Co and Ni are reported for the first time. In view of the small changes in the band distributions expected between the solid and liquid phase, a method is proposed for testing the origin of the modifications observed relative to the solid x-ray spectra.

1. Introduction

Soft x-ray spectroscopy (SXS) has been little used in the study of liquid metals, yet it can provide a straightforward description of the densities of state of valence and conduction bands. An x-ray emission is obtained when an electron from the valence band drops down to fill a hole in an inner shell and x-ray absorption occurs when an electron is transferred from an inner shell to the conduction band. Partial densities of state only are observed because of the selection rules and the spectrum recorded is the convolution product of the valence or conduction band distributions with the inner level distribution. Despite this intrinsic limit to resolution many cases remain where valuable information can be obtained. Good examples are the L (valence band → 2p, 2p → conduction band) and K (valence band → 1s, 1s → conduction band) emission and absorption spectra of the 3d transition metals and the normal metals, since the width at half-maximum of the inner level remains fairly small ($\lesssim 1$ eV).

The contributions which have already been made in this field concern the L emission spectrum of liquid Al (Caterall and Trotter 1963), its K emission spectrum (Farineau 1938, Fischer and Baun 1965) and more recently the K emissions of liquid Cu and Fe (Garg and Källne 1975).

A number of other valence band results have been recorded by UV photoemission spectroscopy (UPS) and x-ray photoelectron spectroscopy (XPS). These techniques provide a better resolution of densities of state if a sufficiently monochromatic excitation radiation is employed (e.g. Koyama and Spicer 1971, Eastman 1971, Williams and Norris 1974, Petersen and Kunz 1974). They are more subject to surface contamination and secondary effects than SXS. Also, in SXS the thickness of the target involved in the emission process can be varied by adjusting the energy of the incident electron beam.

Here we report on the L_3 emission (valence band → $2p_{3/2}$) and absorption ($2p_{3/2}$ → conduction band) spectra of liquid Fe, Co and Ni. The absorptions are computed from the emission curves by a method suggested by Liefeld (1968) using the self-absorption effect of the radiation in the target. Comparisons are made with thin-film absorption results (Bonnelle 1966).

2. Experimental details

A 500 mm radius bent crystal spectrometer has recently been constructed for the study of liquid metals. Only the essential points need be mentioned here.

A high power work-accelerated electron gun heats the target and provides the necessary excitation energy for the ionization of the inner level. The electron beam is incident normally to the target surface and the emitted radiation is also viewed normally to it. The spectrometer is fitted with an entrance slit in such a way that metallic vapours are trapped before they can reach the crystal and differential pumping is established between the main tank and the source chamber. Well trapped diffusion pumps give a vacuum better than 5×10^{-8} Torr in the anode region. The pressure climbs to 5×10^{-7} Torr, as read on a Penning gauge placed some 10 cm from the target, when the liquid state is attained.

The electron beam is focused onto the top of a 6 mm diameter rod of the pure element to form a liquid drop. Thus, contamination from contact with foreign materials is avoided. Two thermocouple junctions are spaced down the target rod to provide a differential temperature reading. As an added precaution the liquid state is verified by eye by means of a port-hole. The electron gun high-voltage generator being well stabilized, no particular problems were encountered in maintaining the liquid state.

3. Results

The maximum depth at which atoms will be ionized in the target is a function of the energy of the incident electron beam. The photoabsorption cross section in the region of the Fermi level varies rapidly so the shape of the emission is bound to be modified as it traverses the target materials. The true bandshape would be observed for an incident electron beam hardly greater than the ionization threshold of the inner level since emission would then occur close to the surface. However, the intensity of the emitted radiation would then be low and surface contamination effects would be reinforced.

It was decided to take advantage of the self-absorption phenomenon to obtain an indication of the densities of state in the conduction band and also to assess the origin of the changes in emission bandshapes.

The relation between the true intensity I_0 and that of the observed emission I produced at a depth x in the target is given for each wavelength λ by

$$I = I_0 \exp(-\mu x)$$

where μ is the linear absorption coefficient of the target at λ. By taking the ratio of the emission spectra in the same phase at two different electron beam potentials we have a method of determining the self-absorption curve as suggested by Liefeld (1968):

$$\ln(I_2/I_1) = \mu(x_2 - x_1). \qquad (1)$$

where x_1 and x_2 are the effective penetrations at both potentials.

It should be noted that the penetration of the electron beam is inversely proportional to the density of the target material ρ so that, for a given element, the value $\mu x = m$ is independent of the density which in fact varies appreciably at the solid (s) to liquid (l) transition. It follows that the intensities of the liquid and solid phase spectra, both observed at the same excitation potential, are related by:

$$\ln \frac{I_1}{I_s} = \left[\left(\frac{\mu}{\rho} \right)_s - \left(\frac{\mu}{\rho} \right)_1 \right] m - \ln \frac{I_{0s}}{I_{01}}. \tag{2}$$

This reveals that the differences in intensity observed between the solid and liquid curves takes into account not only the true changes in the valence band but also any variation in the mass absorption coefficient, i.e. in the photoabsorption cross sections. In the most probable case where both absorption and emission vary, one should ideally use relation (1) to solve (2).

To obtain meaningful results it is necessary to have low statistical errors and the curves presented here were each obtained from step-by-step numerical accumulation for about ten successive scans.

3.1. L_3 emission spectra

The smoothed L_3 emission curves of liquid Ni, Co and Fe are represented in figure 1 at two excitation voltages. It will be noted that the spectra obtained at 5 kV are distinctly narrower than at 2·5 kV revealing the presence of self-absorption. The Ni band at 5 kV has a reinforced satellite emission on the high energy side due to multiple ionization. The $\ln (I_1/I_s)$ curves are plotted above each spectrum.

The two curves obtained for Ni are comparable and rise sharply near the emission maximum which reflects a displacement of the solid phase spectrum towards higher energies. However, surface oxidation is very likely present in the solid phase and in the solid phase only. Indeed it has been shown that the L_3 Ni band shifts to a higher energy in NiO (Bonnelle 1966).

The equivalent Co curves present a slightly different picture; there is a marked drop at 2·5 kV probably revealing an oxidized solid target but the effect is less marked at 5 kV. The curves flatten out to zero at the emission peak but on the other hand no shift is to be observed in the oxide emission (Bonnelle 1966). One can conclude that the influence of oxidation is small on the 5 kV data but the difference seen between solid and liquid spectra is still due to some contamination in the solid.

The Fe results are again a little different. At 2·5 kV $\ln (I_1/I_s)$ is only slightly negative. One can nevertheless expect oxidation to be responsible since the opposite result is encountered at 5 kV. Thus one might conclude that some modification in the density of states between the liquid and solid phase is detectable at 5 kV in this case.

3.2. L_3 absorption spectra

Our experimental results shown in figure 2 concern the liquid phase. From what has been said above the self-absorption curves for the solid phase were considered unreliable at the present stage. Instead the absorption spectra obtained from solid thin films (Bonnelle 1966) are given for comparison.

In Ni the resemblance between the self-absorption curve and the thin-film result is quite close. It has been shown that in the case of a sharp peak representing the empty d states in the conduction band, as here, the amplitude of the maximum is very dependent

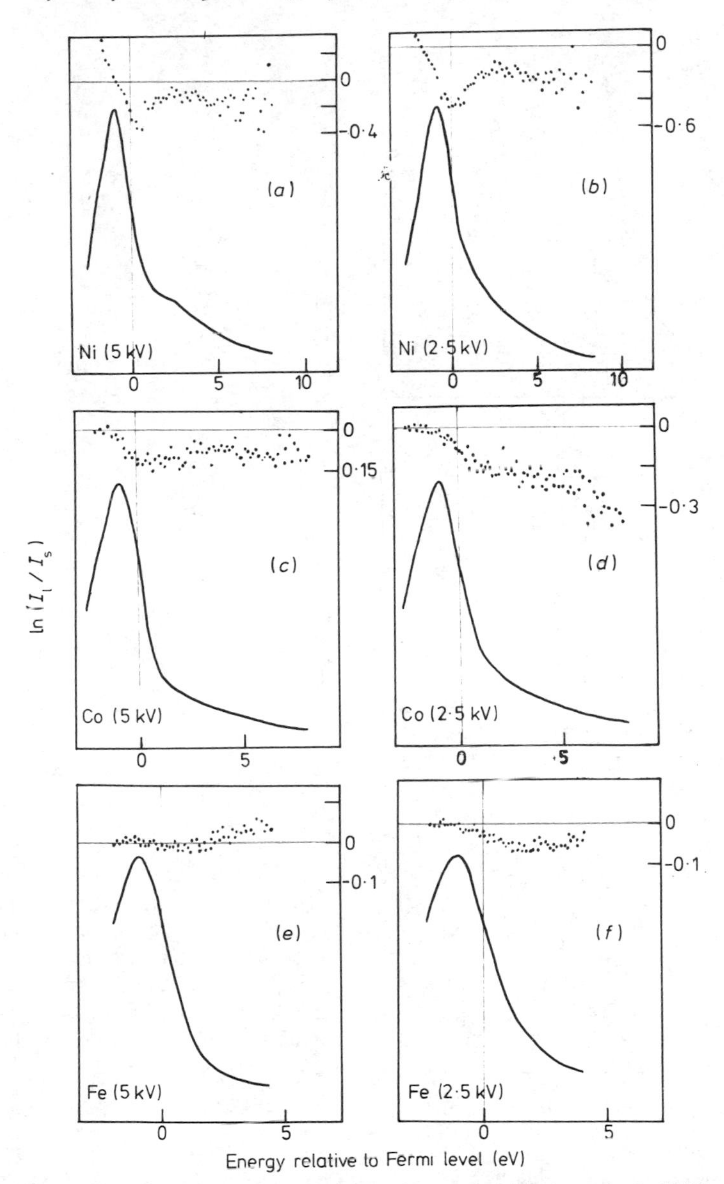

Figure 1. L_3 emission curves of the liquid metals with above the ln (I_l/I_s) plots: (*a*) Ni at 5 kV; (*b*) Ni at 2·5 kV; (*c*) Co at 5 kV; (*d*) Co at 2·5 kV; (*e*) Fe at 5 kV; (*f*) Fe at 2·5 kV.

on the thickness of the absorbant (Parratt 1959). This may explain the attenuation observed in the peak in the liquid spectrum.

Agreement is generally more convincing between the two absorption curves of Co. The normalization of such curves remains a fairly arbitrary process so that the small deviation observed at the high energy end may not be significant and must await confirmation from similar solid phase measurements.

The Fe results are considerably different. However, the thin-film spectrum is most likely perturbed, as pointed out by Bonnelle (1966), by an anomalous reflection in the

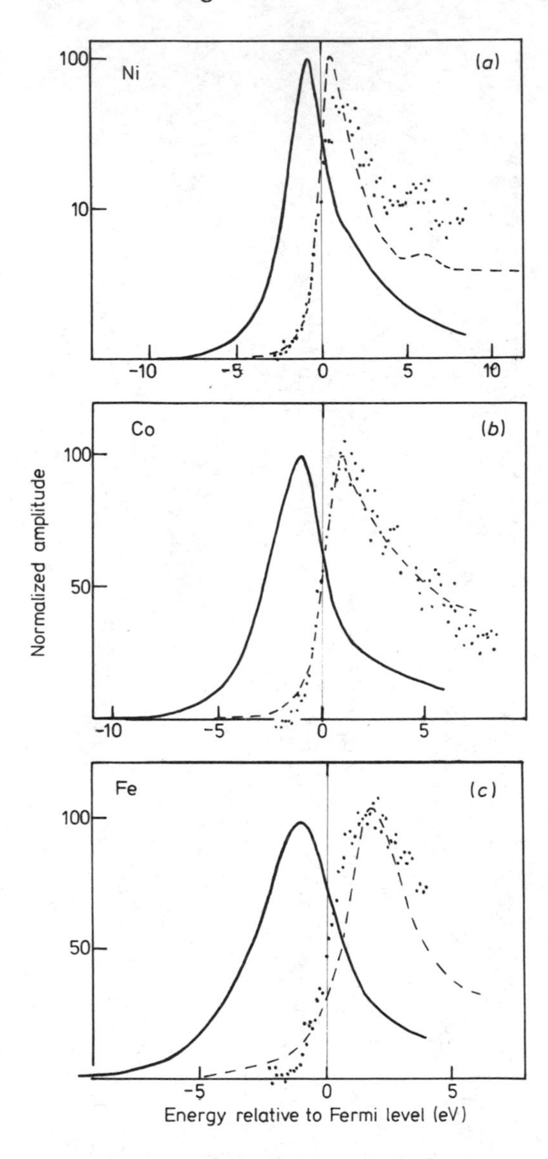

Figure 2. L_3 emission curves of the liquid metals at 2·5 kV (full curve), L_3 self-absorption curves of liquid metals (dotted curve), and thin solid film L_3 absorption curves according to Bonnelle (1966) (broken curve). (*a*) Ni, (*b*) Co, (*c*) Fe.

analysing crystal containing Fe which occurs close to the Fermi level. In fact our solid self-absorption curve (not presented here for the sake of clarity) follows exactly that of the liquid result except for a 20% increase in amplitude compatible with the 5 kV $\ln(I_l/I_s)$ data.

4. Discussion

It is generally admitted that the L emission spectra of the transition metals describe closely the $N(E)$ distribution of the d states in the valence band. Likewise the peak in the absorption curve close to the Fermi level is characteristic of the unfilled d states in the conduction band.

There is hardly any change in the interatomic distances or coordination number of the transition metals in the liquid phase so that any modification in the densities of state will be small. The more recent calculations using a cluster method (House and Smith 1973) or a hard sphere description (Gaspard 1976) bear this out.

Our SXS results do not suggest the presence of extra structure close to the Fermi level nor indeed do the K spectra obtained by Garg and Källne (1975). It may be added that d resonance states which had been defined in a calculation on Fe by Anderson and McMillan (1967) have since been refuted (Olson 1975).

Despite the blurring effect on fine structure due to inner level broadening the method of analysis presented here should be able to reveal quite small alterations in bandshape. Indeed some modification in the conduction band of Fe may be present. As we have seen, this method has essentially underlined, so far, the importance of oxidation effects in the solid spectra. This was not altogether a surprise as results on solid Ni_3Fe ordered and disordered alloys (Hague *et al* 1976) had to be obtained from polished targets protected by a thin film of Al to avoid contamination. Such a technique if employed here could not have provided a valid comparison with the liquid phase. It appears that the high vapour pressure encountered at the fusion (especially Fe) helps to break up the oxide film and reliable results can be obtained in the liquid phase.

5. Conclusion

These preliminary measurements on the L_3 emission and absorption curves of liquid transition metals seem encouraging despite the difficulties encountered with surface contamination present in the solid phase. In our opinion they justify further effort to improve experimental conditions. Moreover the method of analysis described here provides a reliable test as to the validity of the results. Confrontation with theoretical calculations will have to await new measurements.

Acknowledgments

The author is most grateful to Professor C Bonnelle for useful discussions. Thanks are also due to the Délégation Générale de la Recherche Scientifique et Technique for financial support.

References

Anderson P W and McMillan W L 1967 *Proc. Int. School of Physics 'Enrico Fermi' Course 37* ed W Marshall (New York: Academic Press)
Bonnelle C 1966 *Ann. Phys., Paris* **1** 439–81
Caterall J A and Trotter J 1963 *Phil. Mag.* **8** 897–902
Cauchois Y 1956 *C. R. Acad. Sci., Paris* **242** 100–2
Eastman D E 1971 *Phys. Rev. Lett.* **26** 1108–10
Farineau J 1938 *Ann. Phys., Paris* **10** 20–102
Fischer D W and Baun W L 1965 *Phys. Rev.* **A138** 1047–9
Garg K B and Källne E 1975 *Phys. Stat. Solidi* **B70** K 121–5
Gaspard J P 1976 *Williamsburg Conf. on Amorphous Substances, AIP Conf. Proc.* in the press
Hague C F, Källne E, Mariot J-M, Dufour G, Karnatak R C and Bonnelle C 1976 *J. Phys. F: Metal Phys.* **6** 899–907

House D and Smith P V 1973 *J. Phys. F: Metal Phys.* **3** 753–8
Koyama R Y and Spicer W E 1971 *Phys. Rev.* **B4** 4318–29
Liefeld R J 1968 *Soft x-ray Band Spectra and the Electronic Structure of Metals and Materials* ed D J Fabian (London: Academic Press) pp 163–72
Olson J J 1975 *Phys. Rev.* **B 12** 2908–16
Parratt L G 1959 *Rev. Mod. Phys.* **31** 616–45
Petersen H and Kunz C 1974 *Vacuum Ultraviolet Radiation Physics* ed E E Koch, R Haensel and C Kunz (Braunschweig: Pergamon) pp 587–9
Williams G P and Norris C 1974 *Vacuum Ultraviolet Radiation Physics* ed E E Koch, R Haensel and C Kunz (Braunschweig: Pergamon) pp 585–6

Positron annihilation in liquid binary alloys containing transition metals

K Tsuji†, H Endo†, Y Kita‡, M Ueda‡ and Z Morita‡
† Department of Physics, Faculty of Science, Kyoto University, Kyoto 606, Japan
‡ Department of Metallurgy, Faculty of Engineering, Osaka University, Suita, Osaka 565, Japan

Abstract. The angular distribution of positron annihilation has been measured in liquid binary alloys that contain a transition metal or a noble metal (i.e., Ni_xGe_{1-x}, Fe_xGe_{1-x} and Au_xGe_{1-x}). The experimental results for liquid Ni–Ge and Fe–Ge alloys showed that the high momentum component did not increase even for compositions of 40 at.% of Ni or Fe, while for liquid Au–Ge alloys the broad component appeared at small concentrations of Au. The variation in measured high momentum component for these alloys is discussed in relation to the extension of the d-wave function.

1. Introduction

In simple metals the total photon angular distribution of positron annihilation $N(\theta)$ consists of a parabolic-shaped distribution arising from annihilation with the conduction electrons, superimposed on a very small background arising from annihilation with the core electrons. In contrast with this, in the transition and noble metals there appears in the $N(\theta)$ curve a large and broad component resulting from annihilation with the d-electrons. It is interesting therefore to add transition or noble metals to simple metals and observe the change in the shape of the $N(\theta)$ curve, especially in the curvature near the end of the parabolic portion where it merges with the background. The positron annihilation rate strongly depends on the wavefunction of the outer-core electrons at the positron position since the positive ion core gives a repulsive force to the positron. In this paper we report the variation in high momentum component associated with the d-electrons, as the concentration of transition metals and noble metals in liquid Ge alloys is changed.

2. Experimental

The positron annihilation measurements were made using apparatus installed in the Oarai Branch of the Research Institute for Iron, Steel and Other Metals, Tohoku University. A small amount of ^{58}Co was obtained by irradiating pure Ni foil in the JMTR reactor of the Japan Atomic Energy Research Institute adjacent to the Oarai Branch of Tohoku University. The ^{58}Co was dispersed into the liquid specimen which was then put into a quartz tube with an inner diameter of 8 mm and sealed off under vacuum. The temperature was raised by a Mo heater surrounding the specimen container. Coincident photon pairs were counted with a NaI(Tl) detector of 20 cm length placed behind slits 280 cm away from the specimen. The angular resolution was 1 mrad.

3. Results

Figure 1 shows $N(\theta)$ obtained for liquid Ge–Ni alloys. The data were corrected for random coincidence and the small amount of annihilation in the quartz cell. Also shown in figure 1 are the results for $N(\theta)$ in polycrystalline Ge obtained by M Hasegawa (1975, private communication) and in polycrystalline Ni irradiated by fast neutrons at 6×10^{20} neutrons/cm^2. All the curves are normalized to the same total area.

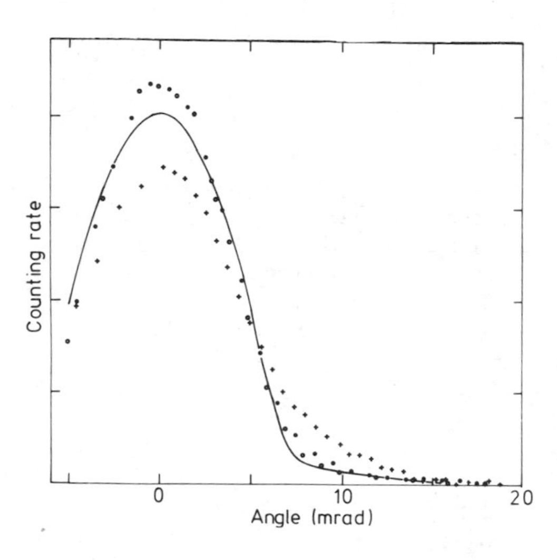

Figure 1. $N(\theta)$ for liquid $Ge_{67}Ni_{33}$ (○) and $Ge_{50}Ni_{50}$ (•). $N(\theta)$ for polycrystalline Ge (full curve) and polycrystalline Ni (+) irradiated by fast neutrons are also shown for reference.

The core-annihilation part of $N(\theta)$ in liquid $Ge_{67}Ni_{33}$ and in liquid $Ge_{50}Ni_{50}$ is very different from that in Ni, which has a broad component, and is quite similar to that in polycrystalline Ge. It should be noted that the core-annihilation part is scarcely affected by the addition of Ni, even up to concentrations of 50 at.%. In figure 2 the results for liquid $Ge_{63}Fe_{37}$ are shown and the data for polycrystalline Fe (Berko and Zuckerman 1964) and Ge given for reference. In the angular distribution curve for Fe there is a large and broad component of $N(\theta)$ resulting from annihilations with the d-electrons, but this is not observed for liquid $Ge_{63}Fe_{37}$.

Figure 3 shows $N(\theta)$ for liquid Au–Ge alloys. The broad component, associated with d-electrons in the high momentum region, increases with the addition of Au to Ge, in contrast with Ni–Ge and Fe–Ge alloys.

4. Discussion

Many theoretical investigations have been carried out recently on the physical properties of liquid transition and noble metals. Although the nearly-free electron approximation is quite successful in understanding many physical properties of liquid simple metals, it is not applicable for these metals. It is well known that the d-electrons play an important role in the physical properties of transition and noble metals and it has been reported that variations in the d-electron state have a large effect on the values of resistivity, magnetic susceptibility and thermoelectric power etc when transition metals are added to liquid Ge.

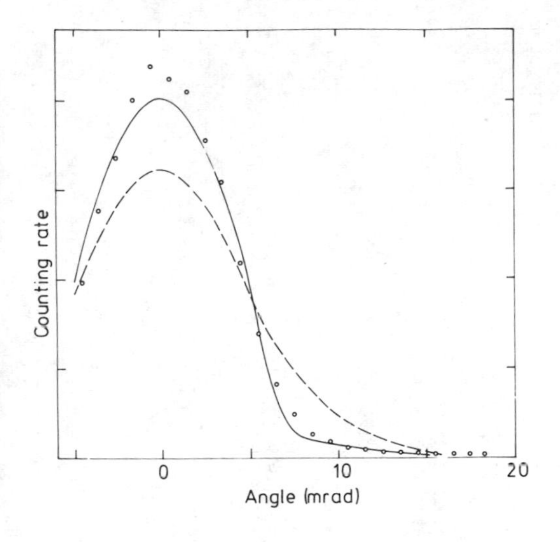

Figure 2. $N(\theta)$ for liquid $Ge_{63}Fe_{37}$ alloy (○). $N(\theta)$ for polycrystalline Ge (full curve) and Fe (broken curve) are also shown for reference.

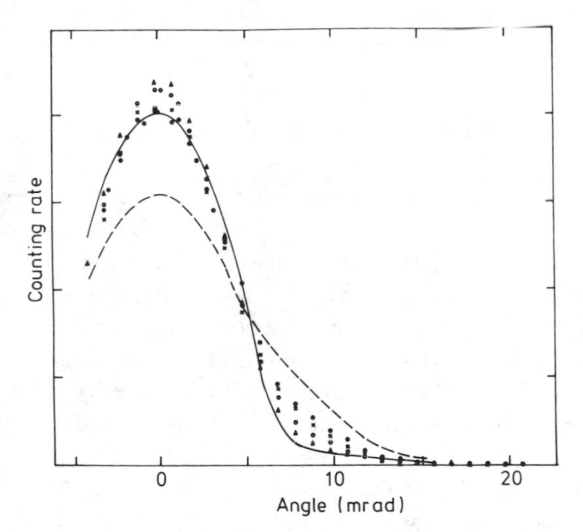

Figure 3. $N(\theta)$ for liquid $Au_{73}Ge_{27}$ at 900 °C (●), $Au_{60}Ge_{40}$ at 940 °C (×), $Au_{40}Ge_{60}$ at 900 °C (○) and $Au_{20}Ge_{80}$ at 940 °C (△). $N(\theta)$ for polycrystalline Au (broken curve) and Ge (full curve) are also shown for reference.

Since the positron annihilation experiment gives information on the aspect of the wavefunction which concerns the outer-shell electrons, the extent of overlap of the d-electron wavefunction affects the profile of $N(\theta)$ and the broad component in $N(\theta)$ increases as the wavefunction of the d-electrons extends over a wider range. Figure 4 shows the curve of $N(\theta)$ for liquid Ni–Ge and Fe–Ge alloys estimated from the free-electron model and convoluted by the angular resolution function of the system and the momentum distribution of thermalized positrons. The effective mass of a positron was assumed to be $2m_e$. The values of the Fermi cut-off for liquid $Ge_{67}Ni_{33}$, $Ni_{50}Ge_{50}$ and $Ge_{63}Fe_{37}$ were deduced to be 6·70, 6·65 and 6·63 mrad. A slight change in the profile of $N(\theta)$ with alloying could possibly be attributed to the contribution of the d-electrons. The atomic volume of Ge is about twice that of Ni and Fe and therefore the mean atomic distances of Ni atoms in liquid Ni–Ge and Fe atoms in liquid Fe–Ge alloys, in the concentration ranges investigated in the present work, are still too large

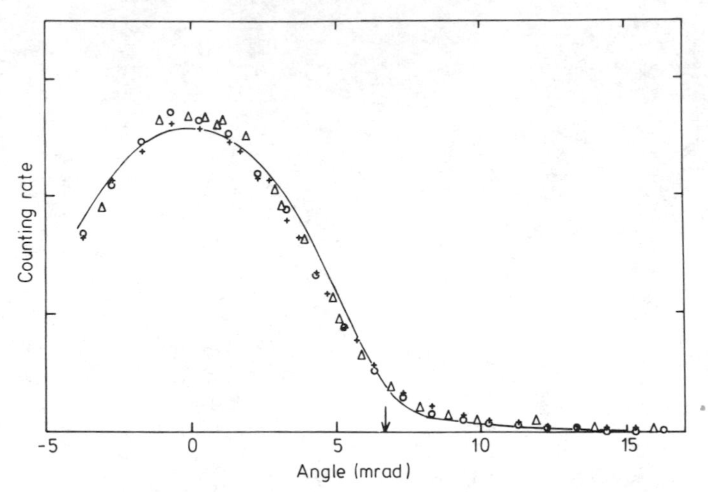

Figure 4. $N(\theta)$ for liquid $Ge_{67}Ni_{33}$ at 840 °C (△), $Ge_{50}Ni_{50}$ at 960 °C (+) and $Ge_{63}Fe_{37}$ at 960 °C (○). The solid curve indicates $N(\theta)$ for liquid $Ge_{50}Ni_{50}$ calculated from the free-electron model assuming the number of free electrons per atom to be four for Ge and two for Ni.

for the wavefunctions of the d-electrons in the atoms of Ni or Fe to overlap in a wide range.

The transition from the paramagnetic to the ferromagnetic state occurs at around 30 at.% Fe for amorphous Fe–Ge and at over 50 at.% Ni for amorphous Ni–Ge alloys. The ionic structures in these concentration regions are quite similar to those of the corresponding liquid alloys (Fukushima *et al* 1974). The magnetic susceptibility for liquid Ni–Ge alloy, measured by Güntherodt and Meier (1973), increases rapidly at concentrations over 50 at.% Ni. Measurements of x-ray photoemission for amorphous Ni–Ge and Fe–Ge systems (Tamura *et al* 1974) suggest that the positions of peaks associated with the d-state of Ni and Fe in the density of states, are still far from the Fermi level in the concentration range around 40 at.% Ni and Fe and that the widths of the peaks are small compared with pure Ni and Fe. For concentrations over 40 at.% Ni and Fe, however, the positions of the peaks shift towards the Fermi level and the widths of the peaks become broad. This experimental evidence is consistent with the picture built up around our results from the positron annihilation experiment. Since the amplitude of the positron wavefunction is large around the central position between the ions, the broad component arising from annihilation with the d-electrons would not be apparent when the wavefunction of the d-electrons is strongly localized at the Ni or Fe atoms.

In liquid Au–Ge alloys, where the atomic volume of Au is similar to that of Ge, the mean atomic distances between the Au are 4·7 Å, 3·7 Å, 3·2 Å and 3·0 Å for $Au_{20}Ge_{80}$, $Au_{40}Ge_{60}$, $Au_{60}Ge_{40}$ and $Au_{73}Ge_{27}$ alloys respectively. In the low concentration range the broad components in $N(\theta)$ appear for liquid Au–Ge alloys in contrast with those for liquid Ni–Ge and Fe–Ge alloys. The wavefunctions of the d-electrons of Au in liquid Au–Ge alloys start to overlap at around 20 at.% Au.

Many investigators have pointed out that positrons are easily trapped at vacant sites. As shown in figure 4, the deviation from the free-electron parabola in $N(\theta)$ is small for liquid Ni–Ge and Fe–Ge. The broad components are observed in the angular distribution curve for polycrystalline Ni irradiated by fast neutrons and for Cu in the liquid state (Itoh *et al* 1972). The small change in the core annihilation in the angular distribution curve, which is observed for liquid $Ni_{33}Ge_{67}$, $Ni_{50}Ge_{50}$ and $Fe_{37}Ge_{63}$ alloys, could not be related to the vacancy trapping. Lock and West (1974) discussed the possibility that the positrons select the annihilation sites in some binary alloys. One could not expect any charge transfer between Ge and Ni atoms because chemical shifts are not found in photoelectron emission experiments on the Ge–Ni system. X-ray diffraction data reveal that the ionic structures for these alloys are understood by the random packing model and that the concentration fluctuation is very small. Therefore the suggestion that the particular species attract the positrons in these alloys does not provide a satisfactory explanation for the present experimental results.

References

Berko S and Zuckerman J 1964 *Phys. Rev. Lett.* **13** 339–41
Fukushima J, Tamura K, Endo H, Kishi K, Ikeda S and Minomura S 1974 *AIP Conf. Proc. No.* 20 pp108–11
Güntherodt H J and Meier H A 1973 *Phys. Kondens. Mater.* **16** 25–51
Itoh F, Kuroha K, Kai K and Takeuchi S 1972 *J. Phys. Soc. Japan* **33** 567
Lock D G and West R N 1974 *J. Phys. F: Metal Phys.* **4** 2179–85
Tamura K, Fukushima J, Endo H, Kishi K, Ikeda S and Minomura S 1974 *J. Phys. Soc. Japan* **36** 565–71

Low-density metallic fluids and the metal–nonmetal transition

F Hensel
Institute of Physical Chemistry, D-3550 Marburg/L., Lahnberge, West Germany

Abstract. A brief review is presented of the continuous metal–nonmetal transition observed in expanded fluid metals and fluid selenium. Recent experimental results are summarized and discussed.

1. Introduction

During recent years there has been a growing interest in the electronic and thermodynamic properties of those one-component and two-component liquids in which a continuous transition from a metallic to a semiconducting to an insulating state can be experimentally induced by a simple change in one or two of the intensive variables of temperature, pressure, density and mole fraction. Examples that have been known for a long time are the solutions of alkali metals in liquid ammonia (Schindewolf 1968, Thompson 1968), solutions of electropositive metals in their molten halides (Arendt and Nachtrieb 1970, Bronstein and Bredig 1958) and semiconducting liquid alloy systems (Ilschner and Wagner 1953, Enderby 1974, Schmutzler *et al* 1976). All these systems have the disadvantage that they are multicomponent systems. It is therefore necessary to take into account the influence of the solvent, that is, the main component, on the minority component which exhibits the effect to be investigated.

During the past 10 years metal–nonmetal transitions have also been studied on single component systems including liquid metals and the liquid semiconductor, selenium. A number of experiments at high temperatures and pressures have demonstrated that metallic liquids such as mercury (Hensel and Franck 1966, Kikoin and Sechenkov 1967), caesium and potassium (Freyland and Hensel 1972) can be transformed to a semiconducting or insulating state by expanding the metal above its critical point into a low-density fluid. Because of the large interatomic bond strength of metals, the temperatures and pressures at the critical point are high, and therefore experimental work is very difficult. Thus mercury, caesium, rubidium and potassium are the only metals whose critical data have been directly measured. Table 1 gives these data together with the liquid chalcogenide elements sulphur, selenium and tellurium. The latter elements are of particular interest because they exhibit another very interesting progressive variation of the electrical properties from insulating to metallic with increasing atomic weight. Liquid sulphur is an insulator, selenium is a liquid semiconductor, tellurium exhibits both metallic and semiconducting behaviour, whereas polonium is a liquid metal. Selenium is the only example of an elemental 'liquid semiconductor'. Its liquid structure near the melting point is assumed to be dominated by relatively large

Table 1.

Substance	Critical temperature T_c (K)	Critical pressure p_c (bar)	Reference
Hg	1760	1510	(Hensel and Franck 1966)
Cs	2020	110	(Renkert *et al* 1971)
Rb	2100	130	(Chung and Bonilla 1973)
K	2200	155	(Freyland and Hensel 1972)
Te	2500 (estimated)	270 (estimated)	(Endo *et al* 1977)
Se	1760	380	(Hoshino *et al* 1976b)
S	1315	180	(Baker 1971)

linear chains (Eisenberg and Tobolsky 1960), that is, it is assumed that the covalent bonds between atoms in two-fold coordination of crystalline selenium persist to a large extent on melting. Theoretically this assumption leads to well developed energy gaps and semiconducting properties (Mott and Davis 1971). However, if the temperature is increased to very large values, the chains rupture and the coordination number increases; the decrease of the chain length goes in the direction of a completely random atomic configuration which is typical for metals. Therefore selenium becomes metallic at very high temperatures (Hoshino *et al* 1975a). Again these very high temperatures can be achieved provided the heating up to the liquid takes place at pressures greater than the liquid–vapour critical pressure. However, as can be seen from table 1, the unavoidable critical pressures of the metals and the chalcogens are relatively high. Experiments for obtaining data under such extreme conditions are chiefly limited by the reduced strength and increased chemical reactivity of the sample containers at the high temperatures. This is surely the reason why little attention was paid to the liquid–vapour phase behaviour of metals until a few years ago. It was only through the development of a special experimental technique favoured by the rapid advances in the production of high-temperature construction materials during the past few years, that a number of experiments have been possible on Hg, Cs, Rb, K and Se, all of which have critical temperatures lower than 2000 °C. The fundamental thermodynamic and electronic properties of density, vapour pressure, critical data, viscosity, electrical conductivity, thermoelectric power, Hall effect, Knight shift, optical absorption and structure factor have been measured in some cases. The aim of this paper is to review the available experimental results and to discuss some recent selected data.

2. The liquid–vapour phase transition and the metal–nonmetal transition

The metal that has been most intensively studied is mercury. Figure 1(*a*) shows the DC electrical conductivity measured by various groups of workers (Hensel and Franck 1966 Kikoin and Sechenkov 1967, Schmutzler and Hensel 1972, Postill *et al* 1967) as a function of the pressure for a number of subcritical and supercritical temperatures. At

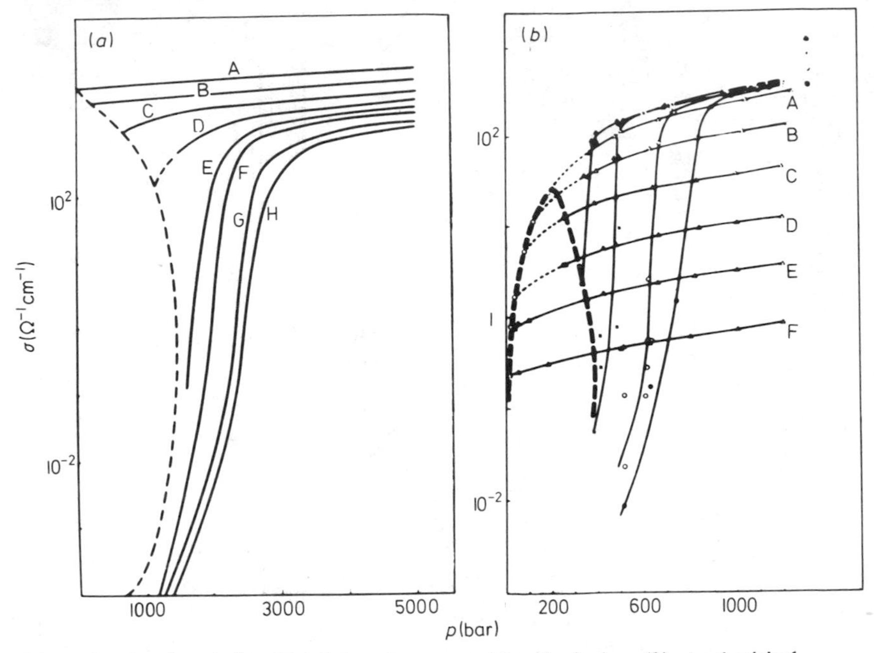

Figure 1. The electrical conductivity of mercury (*a*) and selenium (*b*) at subcritical temperatures as a function of pressure. In (*a*) curve A = 600 °C, B = 900 °C, C = 1200 °C, D = 1400 °C, E = 1500 °C, F = 1550 °C, G = 1650 °C, H = 1700 °C. In (*b*) curve A = 1400 °C, B = 1300 °C, C = 1200 °C, D = 1100 °C, E = 1000 °C, F = 900 °C; ▲ 1600 °C, . 1700 °C, ○ 1500 °C, ● 1750 °C.

subcritical temperatures in the liquid phase, the conductivity, σ, changes only slightly with pressure, whereas a strong dependence on pressure is observed at supercritical temperatures. A pressure decrease of only 500 bar in the vicinity of the critical point can decrease the conductivity by up to five orders of magnitude to values that are certainly nonmetallic. This strong pressure dependence, which is also found for caesium (Pfeifer *et al* 1973, Hensel 1973) and potassium (Freyland and Hensel 1972) is undoubtedly due to the great increase in the mean interatomic distance as the pressure is decreased in this range. The large decrease of the conductivity indicates a gradual transition from a fluid metal to a fluid nonmetal. In other words, the electron distribution changes from a nearly-free electron gas, with long-range screening of ionic charges, to localized electrons on the atoms or molecules. Since there is an interdependence of electron–electron screening and interatomic forces, it was speculated (Krumhansl 1966) that the liquid–vapour phase transition across the boiling curve will coincide with the transition in the electronic structure from a metallic liquid to an insulating vapour.

As far as the experimental evidence is concerned, the electrical conductivity of mercury, caesium and potassium decreases steeply as the critical point is approached along the saturation curve. Quantitatively, however, a striking difference is observed for the conductivity of divalent mercury and monovalent caesium or potassium near the critical point. This is demonstrated by figure 2, which gives a plot of the conductivity of mercury and caesium at a slightly supercritical temperature (T/T_c = 1·025) versus the

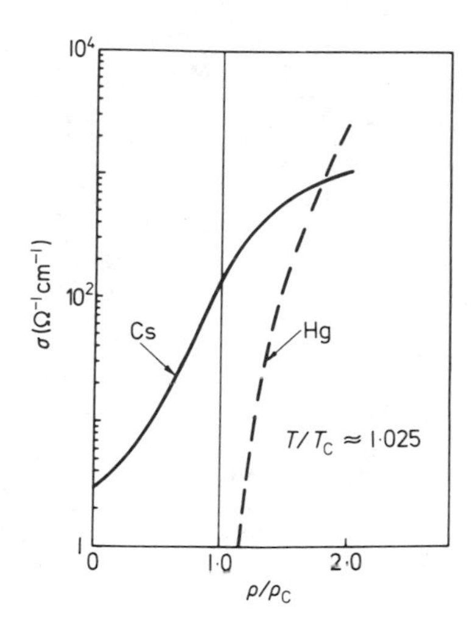

Figure 2. Conductivity isotherms at a supercritical temperature against reduced density for caesium and mercury.

reduced density ρ/ρ_c (Hensel 1974, Freyland *et al* 1974). At the critical density the conductivity of mercury (σ_{Hg} is about $10^{-1}\,\Omega^{-1}\,cm^{-1}$) is three orders of magnitude smaller than the conductivity of caesium (σ_{Cs} is about $200\,\Omega^{-1}\,cm^{-1}$). In addition, the value of the absolute thermoelectric power of mercury, which is about $-500\,\mu V K^{-1}$ (Schmutzler and Hensel 1972) at the critical point, is very large and certainly nonmetallic, whereas the value for caesium is $-60\,\mu V K^{-1}$ at the critical point (Freyland *et al* 1974). The latter is relatively small and not unreasonable for a metal. Thus, on the basis of this argument it is tempting to speculate that one observes two different types of critical point. For mercury close to the critical point a type of liquid–vapour phase transition similar to that in nonmetallic liquids is observed, that is, the liquid mercury exists in a high-density metallic form and a low-density nonmetallic form. For caesium (the same is true for potassium) it cannot be excluded that the liquid–vapour phase transition is associated with the metal–nonmetal transition. It must be pointed out, however, that it is not clear whether the existence of a relatively high conductivity at the critical point of caesium and potassium means that the state of these two elements can really be considered as metallic in the critical region (i.e., that the forces remain long-range compared with an insulating liquid such as argon). An additional difficulty typical of all fluid systems is obvious from figure 2. The metal–nonmetal transition is not indicated by sharp breaks or jumps in the transport properties. Therefore it is necessary to inquire how the transition density can be specified in terms of the experimental data. An attempt to specify the transition density for expanded mercury is shown in figure 3. In figures 3(*a*) and 3(*b*) are shown the temperature coefficient of the conductivity, σ, at constant volume, $(\partial \ln \sigma/\partial T)_V$ (Schmutzler and Hensel 1972) and the volume coefficient of σ at constant temperature, $(\partial \ln \sigma/\partial \ln V)_T$ (Schmutzler and Hensel 1972) as a function of the density. In figures 3(*c*) and 3(*d*), plots of the Knight shift (El-Hanany and Warren 1975) and the Hall mobility (Even and Jortner 1972) versus the density are shown. All these quantities change in a characteristic manner around $9\,g\,cm^{-3}$, but it is obvious that it is difficult to identify a particular density; a gradual change of the slopes of the curves in figure 3 is observed in the relatively large density range between $10\,g\,cm^{-3}$ and $8{\cdot}5\,g\,cm^{-3}$. It is only for

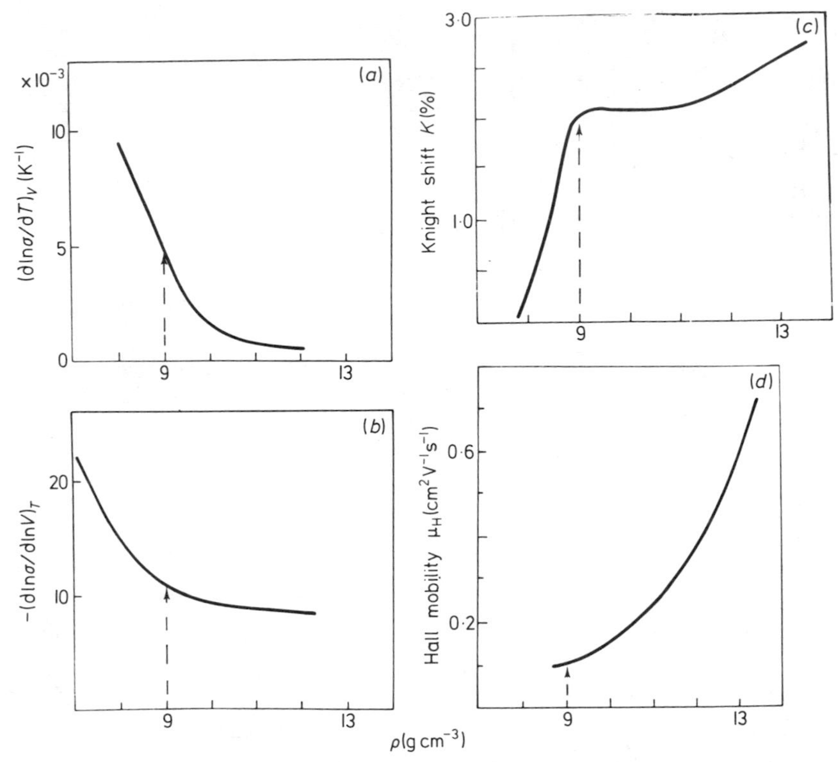

Figure 3. Temperature and volume coefficient of the conductivity, the Knight shift and the Hall mobility of fluid mercury as a function of density.

$(\partial \ln \sigma / \partial \ln V)_T$ and for the Knight shift that two ranges with different slopes can clearly be distinguished. Both properties exhibit a relatively sharp increase or decrease at densities smaller than 9 g cm^{-3}. It was suggested by Mott (1971) that the turnover of the volume coefficient of σ at 9 g cm^{-3} indicates that the corresponding conductivity, $\sigma = 200\ \Omega^{-1}\ \text{cm}^{-1}$ at 9 g cm^{-3}, represents the minimum metallic conductivity occurring when localization is about to set in. The density of 9 g cm^{-3} considerably exceeds the thermodynamic critical density of 5·3 g cm^{-3} (Hensel and Franck 1968). For mercury there is no correlation between the metal–nonmental transition and the thermodynamical critical point, in contrast to the above-mentioned theoretical predictions.

The conduction behaviour of selenium sharply differs from the behaviour of the metals. This is demonstrated in figure 1(b), which shows a plot of conductivity isotherms versus pressure at subcritical and supercritical temperatures. It is only at temperatures close to the critical temperature that the conductivity becomes strongly dependent on pressure. This is surely related to the strong pressure dependence of the density in this region. More pronounced is the effect of the temperature on the conductivity at constant pressure. The conductivity isobars show a very large positive slope at subcritical conditions, giving a maximum conductivity near the critical temperature that is many orders of magnitude higher than the conductivity near the melting point. It was suggested (Hoshino *et al* 1976a) that the temperature effect in the liquid results from the combined effect of the breaking up of the selenium chains (i.e., the coordination number is increased), the thermal activation of current carriers and the decreasing density from thermal expansion. The conductivity of molten selenium throughout the entire liquid range at equilibrium pressure is given in figure 1(*b*) by the broken curve in heavy type. It reaches a maximum value of 25 $\Omega^{-1}\,cm^{-1}$ and decreases to a quite small value near the critical point. This small conductivity can be considered consistent with the result of an analysis of the vapour pressure curve of selenium in terms of 'correspondence state theory' which leads to the suggestion that selenium is essentially a molecular fluid in the vicinity of the critical point rather than a metallic one (Hoshino *et al* 1976b).

3. The optical properties of compressed mercury vapour and the metal–nonmetal transition

For any divalent metal, such as fluid mercury, a separation of the valence and the conduction band is expected to occur at sufficient expansion, whether the substance is crystalline or not. In the latter case, however, the situation is more complicated, because in disordered, especially in fluid systems, the singularities in the density-of-states function at the band edges are smeared out; that is, the band edges exhibit tails within the region which is forbidden in the corresponding crystalline solid (Mott and Davis 1971). This is clearly demonstrated by the results of an optical absorption experiment. The spectral dependence of the absorption coefficient of mercury at the constant supercritical temperature of 1640 °C and densities between 1 $g\,cm^{-3}$ and 4·8 $g\,cm^{-3}$ is shown in figure 4 (Uchtmann and Hensel 1975). The absorption edges exhibit the expected red shift with increasing density and are found to be exponential in form at values of K up to about $2 \times 10^3\,cm^{-1}$. The exponential shape of the absorption edge,

$$K = A \exp\,[(\hbar\omega)^{1/2}], \tag{1}$$

reminds us of the Urbach rule which is obeyed in many materials, for example, in alkali halides (Urbach 1953), many semiconductors and in amorphous semiconductors and glasses (Tauc 1974). It is supposed that the similar absorption behaviour in all these different substances appears for the same physical reason. As a general physical model, Dow and Redfield (1972) propose that the Urbach rule arises from an electric-field

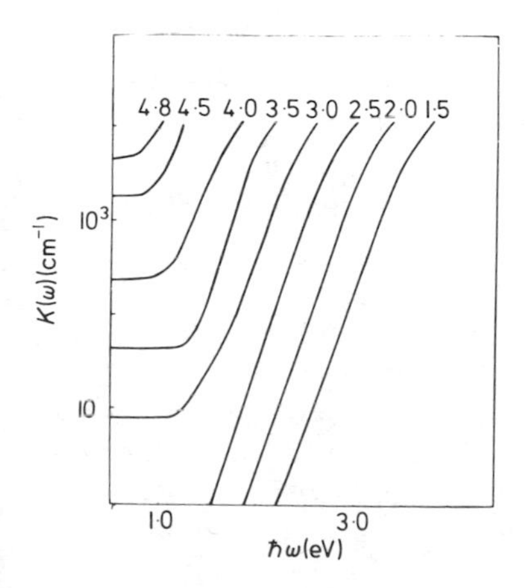

Figure 4. Optical absorption edges in fluid mercury at different densities and constant temperature of 1640 °C.

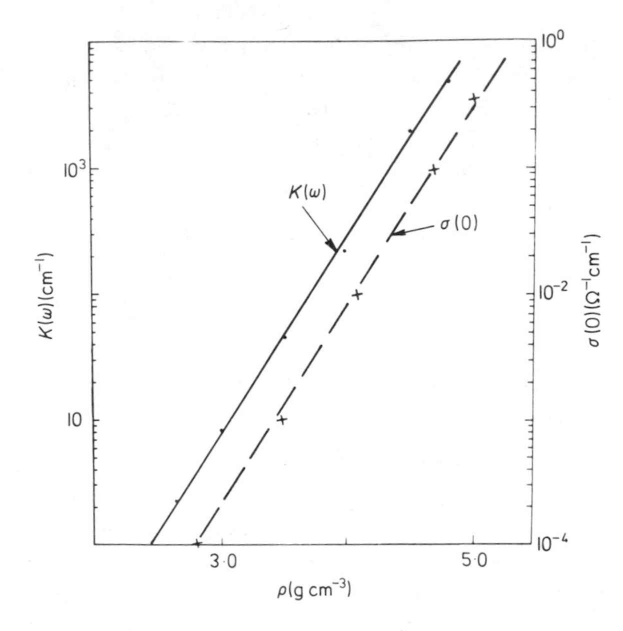

Figure 5. Density dependence of the DC conductivity and the optical frequency absorption constant (at $\hbar\omega = 0{\cdot}5$ eV) of supercritical mercury. $T = 1640\,°\mathrm{C}$.

broadening of an exciton line. In amorphous and fluid materials one must consider the fluctuations in composition or density as possible sources of local fields. However, there is an additional possibility, which has been suggested for disordered semiconductors by Tauc (1974), namely that the Urbach tail arises from electronic transitions between states in the band-edge tails, the density of which is assumed to be an exponential function of energy.

At densities larger than $3\ \mathrm{g\ cm}^{-3}$ the absorption curves flatten out to approximately constant values of the absorption coefficient at long wavelengths up to the infrared region. The level of these long wavelength tails increases very rapidly with increasing density and the density dependence is the same as observed for the DC conductivity. This is shown in figure 5, which shows a plot of K at a constant photon energy of 0·5 eV, together with a plot of the DC conductivity, both as a function of the density. This equivalence of the density dependence for the DC conductivity and the optical frequency absorption constant is strong evidence that both mechanisms have the same physical origin. If both involve carriers that are thermally activated to extended states in the conduction and valence bands, then the relative frequency independence of the infrared conductivity implies extremely short relaxation times ($\tau \sim 10^{-15}$ s).

In the strong absorption region of the edge ($K > 2 \times 10^3\ \mathrm{cm}^{-1}$) the absorption constant K increases less steeply with increasing $\hbar\omega$ and can be described by the relation

$$K\hbar\omega = \text{constant} \times (\hbar\omega - E_0)^2. \tag{2}$$

This relation is of a form similar to that for direct inter-band electronic transitions in crystalline semiconductors, and was derived by Tauc under the assumption that transitions take place between pairs of states in parabolic valence and parabolic conduction bands separated by an energy $\hbar\omega$, and that the optical matrix element is constant in the energy range under consideration (Tauc 1974). The constant E_0 is used to define an optical gap ΔE, which is shown as a function of the density in figure 6. It should be

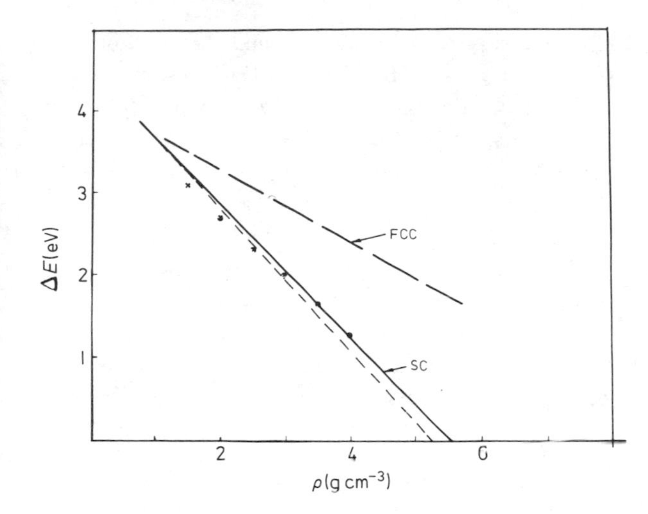

Figure 6. Optical gap in mercury as a function at the constant temperature of 1640 °C; the points are the experimental data. The curves give the calculated band edges for SC and FCC crystal structures (Overhof *et al* 1976).

emphasized that the range in which the dependence (2) is observed is too small to be sure of its validity and that ΔE is an extrapolated rather than a real zero in the density of states. The author feels that not too much physical significance should be given to ΔE. Despite this fact it is surely tempting to compare the density dependence of the optical gap defined in this way with the gap calculated for expanded mercury crystals (Yonezawa *et al* 1977, Overhof *et al* 1976, Devillers and Ross 1975). The main result of these calculations is that the energy gap is very sensitive to changes in structure. This is in accordance with the old empirical rule of chemistry that both density and coordination environment determine whether a fluid or amorphous material is metallic or not. In figure 6 the calculated band gaps as a function of density for FCC and SC structures are included. The absolute values, as well as the slopes of the curves, are different for each structure. The agreement between the measured gap and that calculated for the SC structure, evident from figure 6, is extremely close and should be regarded to be partly accidental. The plot in figure 6 demonstrates that the valence band (the 6s-band) and the conduction band (the 6p-band) cross at a density between 5 and 6 g cm^{-3}. At this density the DC conductivity is about $10^{-1}\,\Omega^{-1}\,\mathrm{cm}^{-1}$ and the thermoelectric power is about $-500\,\mu\mathrm{V\,K}^{-1}$; these are certainly nonmetallic values.

This is quite consistent with the current picture of disordered semiconductors. In this picture the band gap of the crystalline solid is replaced by a mobility gap and all the states within the mobility gap are localized. Conduction at sufficiently high temperatures proceeds by thermal excitation of carriers across the mobility gap. The last mechanism leads for thermal excitation of electrons (Mott and Davis 1971) to the following correlation between the thermoelectric power S and the electrical conductivity σ:

$$\ln (\sigma/\sigma_0) = \left(1 - \frac{e}{K} S\right). \tag{3}$$

Schmutzler and Hensel (1972) have shown that σ and S for mercury in the density range between 8 g cm^{-3} and 5 g cm^{-3} are consistent with equation (3); that is, with an excitation of electrons from localized states at the Fermi energy to free states, across the critical energy which divides ranges in which states are localized from those in which they are not. On the other hand, $(\partial \ln \sigma/\partial \ln V)_T$, $(\partial \ln \sigma/\partial T)_V$, the Hall mobility and the Knight shift indicate a change of the electronic state of mercury for densities around $9\,g\,cm^{-3}$. Therefore, for the density range between $8\,g\,cm^{-3}$ and $9\,g\,cm^{-3}$, Cohen and Jortner (1974) have proposed an alternative explanation. They suggested the existence of an inhomogeneous transport regime in mercury between $9\,g\,cm^{-3}$ and $8\,g\,cm^{-3}$. They assumed that the microscopic inhomogeneities arise from density fluctuations. The metal–nonmetal transition in such a system of metallic and semiconducting clusters was then discussed within the framework of percolation theory.

However, it is necessary to point out that from existing experimental results it cannot be decided whether the Cohen–Jortner proposal is right. More accurate experimental work on expanded mercury is necessary to make it possible to formulate a theory that can be tested experimentally.

4. Atomic transport properties of expanded fluid metals

We turn next to the existing experimental results of the viscosity in low-density metallic fluids. Our knowledge of these properties lags far behind that of the electrical or thermodynamical properties. This surprises no one, because the experimental problems associated with such measurements are much greater. However, the knowledge of the viscosity coefficient in these systems is of particular interest. The immediate interest in the viscosity stems in part from the long-range considerations of a possible application of such highly conducting low-density fluids or vapours to MHD converters. In addition, viscosity data should provide valuable information on the interaction of the particles. In principle, the viscosity can be determined from the pair potential function, although this is quite difficult. And indeed, as mentioned already above, for a metal the knowledge of the density dependence of the interatomic forces is certainly of interest, because however we may describe the forces in a fluid metal, the description must change with density. As the distance between the atoms increases the metallic description must change to a nonmetallic description. Despite these facts the viscosities for only two metals are measured up to high temperatures and pressures. Tsai and Orlander (1974) measured the viscosity of liquid caesium up to 1600 °C at the equilibrium vapour pressures, that is, about 150 °C beyond the critical temperature. For these conditions caesium is surely metallic: thus the measurements extend only into the liquid subcritical range. The same is true for the viscosity measurements on liquid and gaseous mercury measured by von Tippelskirch *et al* (1975). Figure 7 shows a plot of the data obtained for the coexisting liquid and vapour. Unfortunately, the experimental results do not extend to densities smaller than $9\,g\,cm^{-3}$ where the transition from a metallic to a nonmetallic state is expected to occur. The curves, including the dotted lines, are calculated with the aid of the modified Enskog theory of Dymond and Alder (1966). They have used the Enskog theory with the concept of a van der Waals fluid in molecular dynamics computer-simulated calculations to calculate the viscosity of dense rare gases. The two main assumptions made in their calculation are that the collisions between atoms are similar to those between hard spheres and that the core diameters depend on temperature. The temperature dependence can be calculated from experimental equation of state data. Correlation at high densities arises from back-scattering

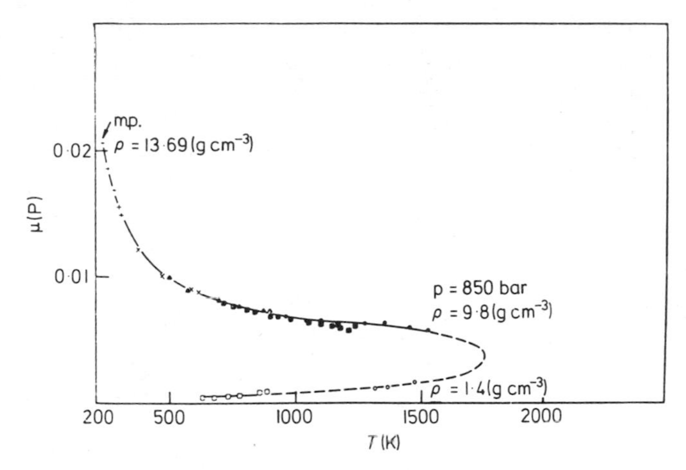

Figure 7. Viscosity of liquid and gaseous mercury (von Tippelskirch *et al* 1975).

after collisions and this correlation can be estimated from molecular dynamics computer-simulated experiments. The calculation for mercury was made using the following liquid metal hard-sphere parameters obtained from the equation of state data (Hensel and Franck 1966):

$$\sigma = 2{\cdot}81 \times 10^{-8} \left(\frac{234}{T}\right)^{0{\cdot}0912} \tag{4}$$

for the metallic range and

$$\sigma = 4{\cdot}93 \times 10^{-8} \left(\frac{234}{T}\right)^{0{\cdot}283} \tag{5}$$

for the nonmetallic diluted gas range. The melting temperature of 234 K was chosen as the reference temperature. The agreement between calculated and measured viscosities is surprisingly good.

5. Expanded monovalent fluid metals

For an array of monovalent atoms, whether they are crystalline or not, a metal–nonmetal transition can only occur as a consequence of the interaction between the electrons (Mott 1961). The formal treatment is due to Hubbard (1964). He finds that if the distances between the atoms exceed a critical value, a splitting of the conduction band into two bands occurs. The lower band is occupied and the upper band is empty and the material is a nonmetal with an activation energy for conduction. In several papers it has been pointed out by Mott (1967, 1972a,b), that in a disordered expanded array of one-electron atoms with a mean interatomic distance near that for the metal–nonmetal transition, the density of states can be represented by two overlapping Hubbard bands, so that the position of the Fermi energy is in the minimum of the

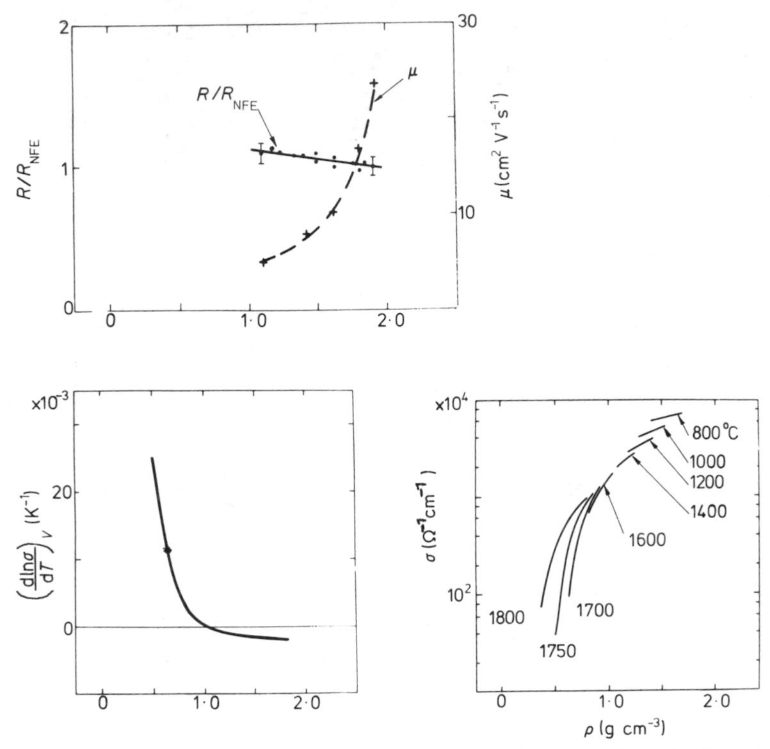

Figure 8. Conductivity, temperature coefficient of the conductivity, Hall constant and Hall mobility of fluid caesium as a function of density.

density of states. If the depth of the minimum is large enough, the states around the Fermi energy should become localized, that is, a mobility gap should be formed. This gives a gradual transition to a nonmetallic state which experimentally should be very similar to the transition observed for divalent mercury.

Figure 8 shows, for fluid caesium, the results of the electrical conductivity σ, the temperature coefficient $(\partial \ln \sigma/\partial T)_V$ at constant volume (Freyland *et al* 1974), the Hall constant and the Hall mobility (Even and Freyland 1975) as a function of the density. Two different ranges can be distinguished. At densities larger than 1 g cm^{-3} the temperature coefficient is negative and its absolute value is quite small. These results can certainly be explained in terms of Ziman's nearly-free electron model, because up to this point the electron mean free path is larger than the mean interatomic distance (Mott and Davis 1971), and the Hall constant R is close to the free-electron value, whereas the Hall mobility $\mu = \sigma R$ is larger than 2 cm^2 V^{-1} s^{-1}. The latter hypothesis was recently proved experimentally by Block *et al* (1977). They showed that the density dependence of the electrical conductivity, which was recently measured by Pfeifer *et al* (1976) and is shown in figure 9, can be well represented by the nearly-free electron theory even for an expansion of the fluid metal to a volume of 65% of the volume at the melting point. The experimental results agree satisfactorily with those calculated within the NFE model using the measured structure factors of expanded rubidium (Block *et al* 1976).

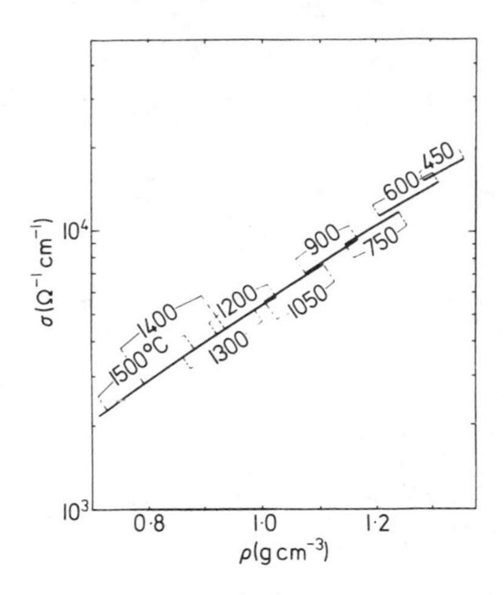

Figure 9. Conductivity of expanded liquid rubidium as a function of density at different temperatures.

Returning to figure 8 it is obvious that at 1 g cm^{-3} and a temperature of about 1600 °C, $(\partial \ln \sigma / \partial T)_V$ changes sign. At this point the conductivity, σ, is about 1500 Ω^{-1} cm^{-1} and the electron mean free path is nearly equal to the mean inter-atomic distance. According to Mott (1971), for smaller values of σ a strong-scattering metallic conductivity region is expected. Over about one order of magnitude to around 300 Ω^{-1} cm^{-1}, σ should depend on the depth of the minimum in the density of states at the Fermi energy. It has also been predicted by Mott that for conductivities smaller than 300 Ω^{-1} cm^{-1} the states in the minimum of the density of states around the Fermi energy should be localized, that is, a mobility gap should be formed. This prediction is consistent with the observed linear dependence of the logarithm of the conductivity on the absolute thermoelectric power with a slope of e/k (equation 3), which is obtained at conductivities smaller than 300 Ω^{-1} cm^{-1} (Freyland *et al* 1974).

However, the author believes that the situation is more complicated and that one has to prove whether a monovalent vapour, such as caesium, is indeed monovalent. The diatomic molecules Cs_2 and K_2 are quite stable; they have binding energies of about 0·5 eV. If, however, diatomic molecules are formed in the dense metal vapour, then the metal–nonmetal transition will occur by an overlap of bands as in the case of mercury. The stability and formation of covalent chemical bonds in highly and electronically conducting liquids and their role for localization and mobility gap formation is an unsolved theoretical question. The most obvious evidence that the formation of chemical bonds is of fundamental importance for the metal–nonmetal transition, in at least some systems, stems from the temperature-induced transition in fluid selenium and from semiconducting alloys (see, for example, the papers of Hoshino *et al* 1977, Krüger *et al* 1977, Freyland and Steinleitner 1977).

References

Arendt R H and Nachtrieb N H 1970 *J. Chem. Phys.* **53** 3085
Baker E H 1971 *Trans. AIMME* **C80, C93**
Block R, Suck J B, Freyland W, Hensel F and Gläser W 1977 this volume
Block R, Suck J B, Gläser W, Freyland W and Hensel F 1976 *Ber. Bunsenges. Phys. Chem.* **80** 718
Bronstein H R and Bredig M 1958 *J. Am. Chem. Soc.* **80** 2077

Chung J W and Bonilla C F 1973 *Proc. 6th Symp. on Thermophysical Properties, Atlanta, Georgia* (ASME) p397
Cohen M H and Jortner J 1974 *Proc. 5th Int Conf. Amorphous and Liquid Semiconductors* ed J Stuke and W Bredig (London: Taylor and Francis)
Devillers M A C and Ross R G 1975 *J. Phys. F: Metal Phys.* **5** 73
Dow J D and Redfield D 1972 *Phys. Rev.* **5B** 594
Dymond J H and Alder B J 1966 *J. Chem. Phys.* **45** 2061
Eisenberg A and Tobolsky A 1960 *J. Polym. Sci.* **46** 19
El-Hanany V and Warren W W 1975 *Phys. Rev. Lett.* **34** 1276
Enderby J E 1974 *Amorphous and Liquid Semiconductors* ed J Tauc (London: Plenum Press)
Endo H, Hoshino H, Schmutzler R W and Hensel F 1977 this volume
Even U and Freyland W 1975 *J. Phys. F: Metal Phys.* **5** L104
Even U and Jortner J 1972 *Phys. Rev. Lett.* **28** 31
Freyland W and Hensel F 1972 *Ber. Bunsenges. Phys. Chem.* **76** 347
Freyland W, Pfeifer H P and Hensel F 1974 *Proc. 5th Int. Conf. Amorphous and Liquid Semiconductors* ed J Stuke and W Bredig (London: Taylor and Francis)
Freyland W and Steinleitner G 1977 this volume
Hensel F 1973 *The Properties of Liquid Metals* ed S Takeuchi (London: Taylor and Francis) p357
—— 1974 *Angew. Chem.* **13** 446
Hensel F and Franck E U 1966 *Ber. Bunsenges. Phys. Chem.* **70** 1154
—— 1968 *Rev. Mod. Phys.* **40** 697
Hoshino H, Schmutzler R W and Hensel F 1976b *Ber. Bunsenges. Phys. Chem.* **79** 1186
—— 1977 this volume
Hoshino H, Schmutzler R W, Warren W W and Hensel F 1976a *Phil. Mag.* **33** 255
Hubbard J 1964 *Proc. R. Soc.* **A281** 401
Ilschner B R and Wagner C N 1953 *Acta Metall.* **6** 712
Kikoin I K and Sechenkov A R 1967 *Phys. Met. Metallogr.* **24** 5
Krüger K D, Fischer R and Schmutzler R W 1977 this volume
Krumhansl J A 1966 *Physics of Solids at High Pressures* ed C Tomizuka (New York: John Wiley)
Mott N F 1961 *Phil. Mag.* **6** 287
—— 1971 *Phil. Mag.* **24** 1
—— 1972a *Phil. Mag.* **26** 1
—— 1972b *J. Non. Cryst. Solids* **8–10** 1
Mott N F and Davis E A 1971 *Electronic Processes in Noncrystalline Materials* (Oxford: Clarendon Press)
Overhof H, Uchtmann H and Hensel F 1976 *J. Phys. F: Metal Phys.* **6** 523
Pfeifer H P, Freyland W and Hensel F 1973 *Phys. Lett.* **43A** 111
—— 1976 *Ber. Bunsenges. Phys. Chem.* **80** 716
Postill D R, Ross R G and Cusack N E 1967 *Adv. Phys.* **16** 493
Renkert H, Hensel F and Franck E U 1971 *Ber. Bunsenges. Phys. Chem.* **75** 507
Schindewolf U 1968 *Angew. Chem.* **80** 165
Schmutzler R W and Hensel F 1972 *Ber. Bunsenges. Phys. Chem.* **76** 531
Schmutzler R W, Hoshino H, Fischer R and Hensel F 1976 *Ber. Bunsenges. Phys. Chem.* **80** 107
Tauc J 1974 *Amorphous and Liquid Semiconductors* (New York: Plenum Press)
Thompson J C 1968 *Rev. Mod. Phys.* **40** 704
von Tippelskirch H, Franck E U, Hensel F and Kestin J 1975 *Ber. Bunsenges. Phys. Chem.* **79** 889
Tsai H C and Orlander D R 1974 *High Temp. Sci.* **6** 142
Uchtmann H and Hensel F 1975 *Phys. Lett.* **53A** 239
Urbach F 1953 *Phys. Rev.* **92** 1324
Yonezawa F, Ishida Y, Martino F and Asano S 1977 this volume

Theoretical approaches to the electronic properties of expanded liquid mercury

F Yonezawa†, Y Ishida‡, F Martino§ and S Asano¶

† Research Institute for Fundamental Physics, Kyoto University, Kyoto 606, Japan
‡ Department of Applied Physics, Tokyo Institute of Technology, Ohokayama, Tokyo 152, Japan
§ Department of Physics, The City College of New York, New York, NY 10031, USA
¶ Department of Physics, Tokyo University, Komaba, Tokyo, Japan

Abstract. We present the first calculation of the electronic structure and conductivity of expanded liquid mercury. All the calculations are carried out from first principles. It is shown that the density of states $n(E)$, at the normal density $\rho = 13{\cdot}6\ \mathrm{g\ cm^{-3}}$, has a dip at the Fermi level E_F. When the density ρ is decreased, this dip grows until the band opens up. We put an emphasis on the investigation of the dependence of the electronic structure and the 'band open-up' density ρ_0 on various methods, potentials and structures. For this purpose we have performed systematic calculations through several methods: (i) a multi-band tight-binding model of a liquid metal; (ii) a muffin-tin potential model of a liquid metal; and in addition, for the sake of comparison, (iii) the KKR calculation for artificially expanded crystal mercury. We conclude that there is an appreciable potential dependence and a strong structure dependence of ρ_0, etc. We also give a discussion of Mott's g value defined by $g = n(E_F)/n_{\mathrm{free}}(E_F)$, the conductivity versus g, and the fractional s-character of the density of states in connection with the Knight shift.

1. Introduction

We report the first calculation of the density of states and the conductivity of expanded liquid Hg. Our purposes are: (i) to show that theoretical treatments of liquid metals based on first principles do lead to a separation of the 6s and 6p bands of Hg at low enough densities, which is somehow related to transformation from a metallic state to an insulating state; (ii) to elucidate the behaviour of electronic properties of fluid mercury at reduced densities; and (iii) to give some comments on previous work in the light of the first-principle calculations with a view to examining the predictions proposed so far.

In practice, the plan of our work is described in table 1. We are interested in working out what input data are necessary to explain the most essential features of the output physical properties and to show how the former affects the latter. Before explaining details of our flow chart in table 1, we would like to mention that our philosophy for the present work has partially been motivated by a review talk due to Cusack (1973) at the Tokyo conference four years ago, from which we quote: 'What would be interesting . . . but perhaps a theoretician would regard it as unreasonably time consuming . . . is a calculation of the density of states of liquid via several different theoretical objects with several different approximations to the higher-order correlation functions, but starting with the same potential and the same pair distribution in all cases. Often on moving into the present literature from one calculation to another, every aspect changes at once and this makes comparison difficult'.

We did not find it very time consuming although it was rather money consuming. However, it was interesting to take this line of research. Besides performing calculations via various different methods and approximations starting with the same potential and the same pair-distribution function $g(R)$, we have also made an attempt the other way around; that is, carrying out calculations through the same method and the same approximation starting with different potentials and structures.

2. Input data, methods and output quantities

Now, let us turn to our flow chart in table 1. As our input data, we need information about the atomic nature and the atomic configurations. The atomic nature can be described by an atomic potential. On the other hand, the present stage of experiments and theories generally allows us to discuss atomic configurations only in terms of the pair-distribution function $g(R)$. As illustrated in table 2, we take both nonrelativistic and

Table 1. Flow chart of the calculation.

Input data		Black box		Output information
				(i) $n(E)$
				(ii) ρ_0
(i) Atomic information –potential–				(iii) ρ_0 on {potential, structure, methods}
	→	?	→	
(ii) Information about structure $-g(R)-$				(iv) $g = n(E_F)/n_{free}(E_F) \rightarrow R_{free}/R$
				(v) dispersion curve → localization
				(vi) σ (conductivity) → $\sigma \propto g^2$?
				(vii) $n_s(E)/n(E)$ → Knight shift

relativistic Hartree–Fock–Slater (HFS) atomic potentials and wavefunctions. As for $g(R)$, we use two kinds of model pair-distribution function: (i) a step-function $g(R)$ for a hard-core random liquid as defined in table 2, and (ii) a Percus–Yevick pair distribution.

Our black box is characterized by the methods employed for the calculations: (i) a multi-band tight-binding approximation (TBA) of a liquid metal (Yonezawa and Martino 1976a,b); (ii) a muffin-tin potential model of a liquid metal on the basis of the Ziman–Lloyd formalism (Ziman 1966, Lloyd 1967a,b). The approximations we have used are two typical self-consistent single-site approximations (SSSA) due to Ishida and Yonezawa (1973) (IY) and Roth (1974) (the effective medium approximation to be abbreviated as EMA). For the sake of comparison, we have also evaluated (iii) the band structure of artificially expanded crystal mercury by the nonrelativistic KKR method. We shall also refer to previous work by Overhof *et al* (1975) and by Fritzson and Berggren (1976).

More explicit input data are listed in table 2 for both the TBA and the muffin-tin potential model. For the former, our formulation requires atomic levels E_μ(atomic) and

Table 2. Input data and explanation of the black box.

Input data		
(i) Atomic information	Potential { (*a*) nonrelativistic { (*b*) relativistic } →	(i) TBA { $E^{\mu}_{\rm at}$ (atomic level); $t^{\mu\nu}(R_{ij})$ (transfer integral) (ii) muffin-tin potential model { t_L (phase shift); $G_{LL'}(R_{ij})$ (Green's function)
(ii) Information about structure	$g(R)$ { (*a*) hard-core random liquid { (*b*) Percus–Yevick	$g(R) = \begin{cases} 0 & R < \sigma \\ 1 & R > \sigma \end{cases}$
Black box		
(i) As liquid	{ (*a*) TBA (*b*) muffin-tin potential model }	Approximations (i) Ishida–Yonezawa (ii) EMA
(ii) As expanded crystal	{ (*a*) KKR (*b*) OPW	{ Overhof *et al* 1975 { the present authors Fritzson and Berggren 1976

transfer integrals $t^{\mu\nu}_{ij} = t^{\mu\nu}(R_{ij})$ where μ, ν indicate atomic orbitals and i, j atomic sites. For the latter, phase shifts t_L and Green's functions $G_{LL'}(R_{ij})$ are necessary, where the label L stands for the pair of angular momentum quantum numbers l and m (where $-l \leqslant m \leqslant l$ etc). All these quantities are of course derived from the atomic potential, the most essential input parameter.

As our output results, we have evaluated (see table 1):

(i) the density of states $n(E)$ for densities from $\rho = 13{\cdot}6$ to $2\,{\rm g\,cm^{-3}}$;
(ii) the density ρ_0 at which the 'band open-up' takes place;
(iii) the dependence of ρ_0 on input information, that is the potential and structure, and on the methods used;
(iv) Mott's g value defined as the ratio of the density of states at the Fermi level $E_{\rm F}$ to that of free electrons, that is $g = n(E_{\rm F})/n_{\rm free}(E_{\rm F})$;
(v) the dispersion curve, E versus ${\rm Re}(k)$, as well as $|{\rm Im}(k)|$ related to the magnitude of damping and, accordingly, to the possibility of electronic localization;
(vi) the conductivity $\sigma(E)$ for showing the behaviour of $\sigma(E)$ as a function of energy and comparing our results with the proposed relation, $\sigma(E) \propto g^2$;
(vii) the fractional s-character of the density of states, $n_s(E)/n(E)$, in order to give a few words on the Knight shift.

3. Density of states in TBA

There exist accumulated experimental data about liquid mercury such as the anomalous impurity effect of the electrical conductivity over a wide density range including the normal liquid density 13·6 g cm^{-3} (Hensel 1973), a continuous metal–nonmetal (MNM) transition in the DC conductivity of expanded fluid Hg at about 9 g cm^{-3} (Schmutzler and Hensel 1972, Duckers and Ross 1972), etc. It is also inferred that there would probably be an optical gap at 5 g cm^{-3} if the band tailing did not mask it. A number of the existing experimental results are generally regarded to be explained qualitatively by supposing that there exists a minimum in the density of states at 13·6 g cm^{-3}, from which a pseudogap evolves at $\rho \simeq 9$ g cm^{-3} and separation of the bands takes place at $\rho \simeq 5$ g cm^{-3}, although the detailed mechanism of the MNM transition is still controversial.

In order to see if the density of states does really have this tendency, we calculate from first principles the mean density of states $n(E)$ as determined by a generalization of the ensemble-averaged one-band Green's function to a mixture of s- and p-bands (Yonezawa and Martino 1976a,b). Our model Hamiltonian is then

$$H = \sum_{i,\mu} E_\mu \, |i\mu\rangle\langle i\mu| + \sum_{i \neq j} \sum_{\mu,\nu} t_{ij}^{\mu\nu} \, |i\mu\rangle\langle j\nu|, \tag{1}$$

where μ, ν denote orbitals s, p_x, p_y and p_z and i, j designate disordered atomic sites. To calculate the averaged Green's function matrix which is related to the density of states per atom by

$$n(E) = -\pi^{-1} \, \mathrm{Im} \, \mathrm{Tr} \, \langle \hat{G}_{ii} \rangle \equiv -\pi^{-1} \, \mathrm{Im} \sum_\mu \left\langle \langle i\mu | \frac{1}{E-H} | i\mu \rangle \right\rangle \tag{2}$$

we use the IY method. Note that the hat on G denotes that $\hat{G}$ is a matrix. The angular brackets are used to express the ensemble average. The calculated density of states is

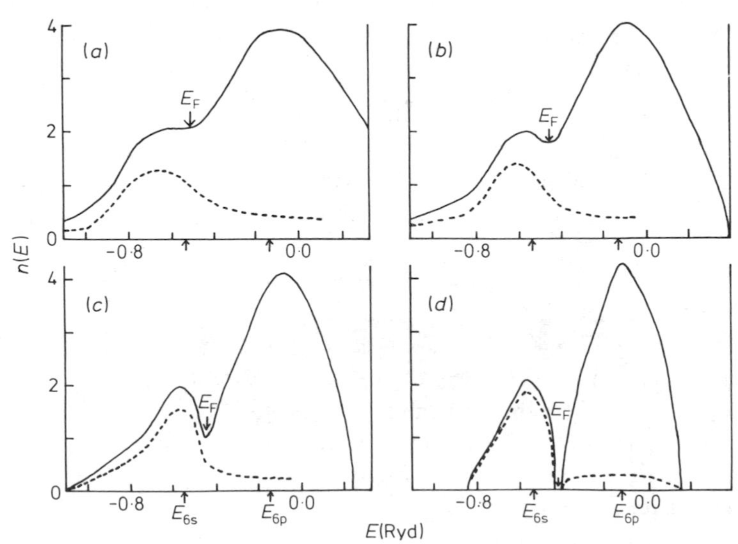

Figure 1. Density of states for: (*a*) ρ = 13·6, (*b*) ρ = 7·0, (*c*) ρ = 4·0 and (*d*) ρ = 2·0 g cm^{-3}, for a hard-core random liquid with σ = 5·4 a.u. (by the TBA and the IY method). Density of states $n(E)$ per atom is scaled by $(\rho/13{\cdot}6)^{1/3}$. Both $n(E)$ and E are expressed in atomic units. The broken line indicates the s-character of the density of states, $(n_s E)$, and the Fermi energy, E_F, and the atomic levels, E_{6s} and E_{6p}, are also shown.

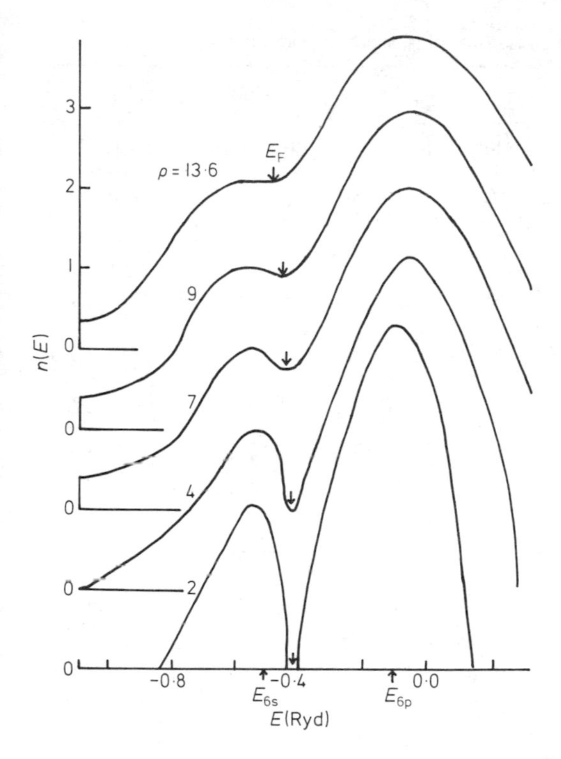

Figure 2. The same density of states as illustrated in figure 1 for the densities (in g cm^{-3}) as labelled on each curve.

shown by the full curves in figure 1 for various densities from 13·6 to 2 g cm^{-3}. An inverted arrow shows the Fermi energy E_F while upright arrows mark the atomic levels E_{6s} and E_{6p}. The relativistic HFS atomic wavefunctions and potentials and a step-function $g(R)$ are used. The results show that the density of states has the tendency expected from the analysis of experimental data, which is more clearly seen in figure 2. The band open-up occurs at $\rho_0 \simeq 2{\cdot}5$ g cm^{-3} for this approximation. We do not take the precise value of ρ_0 seriously, as will be explained in a succeeding section.

4. The band open-up density ρ_0

Now, let us show how the band open-up density ρ_0 is influenced by the input data, that is, the potential and $g(R)$, and by the method used for the calculation. Our concern is by no means to hit the experimental value for ρ_0 on the nose, since this is rather meaningless when there is considerable ambiguity about input data. What is important is to make clear the qualitative tendency of the obtained results when parameters are changed systematically, and this is nothing but the idea as stated in §1.

To start with, let us study the results of various band calculations for artificially expanded crystal Hg. In table 3, we list the calculated values for ρ_0. The self-consistent nonrelativistic KKR band calculation is performed for the various crystal structures, FCC, BCC, WS (Wigner–Seitz), and SC starting with both nonrelativistic and relativistic atomic HFS wavefunctions and potentials. In contrast to the conclusion due to Devillers and Ross (1975), the structure dependence of ρ_0 is found to be strong. The band open-up density ρ_0 is lower for structures with smaller coordination numbers. Similar

Table 3. The band open-up density ρ_0 for various potentials, structures and methods.

	Liquid Hg			Expanded crystal Hg			
Method	TBA		Muffin-tin	Nonrelativistic KKR			
Structure	$g(R)$: step function (i)	$g(R)$: step function (ii)	$g(R)$: PY	FCC	BCC	WS	SC
Potential							
(i) Nonrelativistic wavefunction	3·2			6·5	5·5		4
(ii) Relativistic wavefunction	4	2·5	$\gtrsim 5$	10	8	7·5	5·5
(iii) Relativistic KKR (Overhof *et al* 1975)				8·2			5·5
(iv) OPW (Fritzson and Berggren 1976)				6·5	5·5		4

results have also been obtained by Overhof *et al* (1975) through the relativistic KKR method and by Fritzson and Berggren (1976) via the OPW method. Their results are also shown in table 3. Talking about the potential dependence, we have found that relativistic potentials lead to larger densities ρ_0 since the 6s band shrinks considerably in this case. The calculated results therefore suggest that, for Hg which is comparatively heavy, the relativistic effects are not negligible.

Now, let us turn to our TBA model for a hard-core random liquid with the step-function $g(R)$. The calculation is carried out on the basis of the Ishida–Yonezawa (IY) theory. In case (i), an atomic potential without screening is used. Here again, the relativistic effects are appreciable, as can be seen from table 3. Case (ii) corresponds to an atomic potential with screening. In all these examples, the hard-core diameter σ is taken to be 5·4 a.u., which is the value calculated for Hg at the normal liquid density 13·6 g cm^{-3} (Schiff 1969). The relation of ρ_0 with the hard-core diameter σ is shown in figure 3, where the results using both (i) and (ii) are given. In both cases, the density ρ_0 is extremely sensitive to the hard-core diameter. As is seen from figure 3, the density

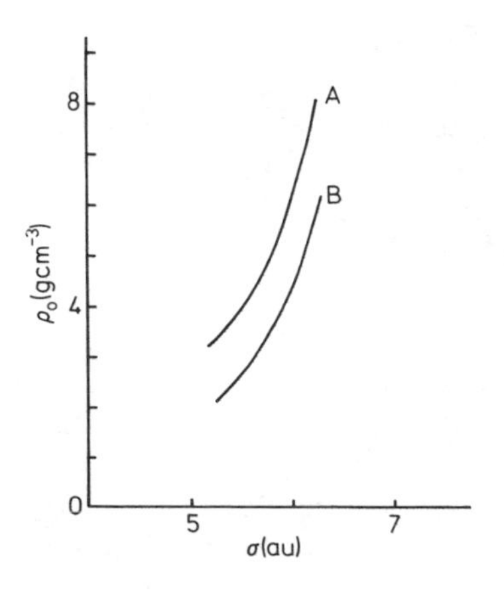

Figure 3. The band open-up density ρ_0 plotted against the hard-core diameter σ. Cases (A) and (B) respectively correspond to the relativistic HFS potential without and with screening.

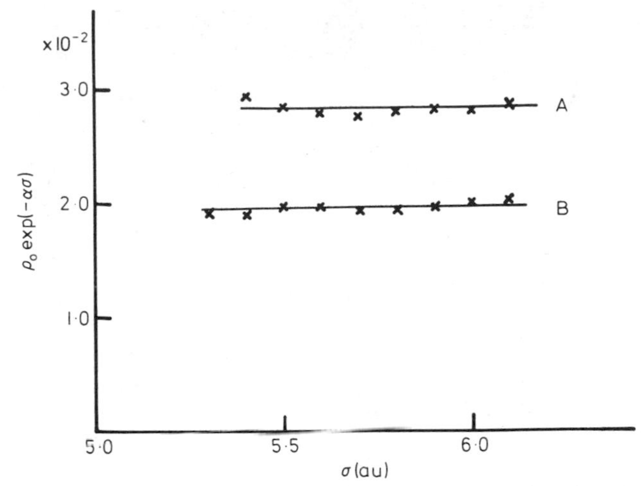

Figure 4. $\rho_0 \exp(-\alpha\sigma)$ versus σ where A, B correspond to the relativistic HFS potential without and with screening, respectively. The parameter α is determined from the effective Bohr radius.

ρ_0 increases by a factor of two when the hard-core diameter is increased by 10 to 15%. Since it is expected that the band open-up density is determined by the bandwidth, it is interesting to analyse the results in figure 3 somehow or other in terms of the bandwidth. Roughly speaking, the bandwidth is proportional to the density ρ and to the hopping integral. To a first approximation, the hopping integral is defined by $\exp(-\alpha\sigma)$, where α^{-1} is related to the effective Bohr radius, that is, to the size of the atomic wavefunction. Figure 4 gives the relation of $\rho_0 \exp(-\alpha\sigma)$ with σ. It is surprising that this quantity is almost constant for a given potential. This indicates that the band open-up density ρ_0 can be estimated from the value of ρ which satisfies the equation $\rho \exp(-\alpha\sigma) = \text{constant}$.

Before we proceed to the next section, it is interesting to show the behaviour of Mott's g value. The solid curve in figure 5 shows the calculated g value for the hard-core diameter $\sigma = 5{\cdot}9$ a.u. The result is compared to the inverse of the experimental Hall coefficient R_{exp} by Even and Jortner (1972). It is proposed that $R_F/R = g$ where R_F is the Hall coefficient for free electrons. As is apparent from the figure, the experimental g value shows a sharper drop around 9 g cm^{-3} than the calculated g value. Since our calculation is based on the SSSA theory, cluster effects are not fully taken into account. This is considered to be one of the main reasons why the calculated g value fails to drop sharply around 8–9 g cm^{-3} where the localization is regarded as setting in and the cluster effects are extremely important.

5. The density of states and the dispersion relation in a muffin-tin potential model

A formalism based on a muffin-tin potential model has been developed by Ziman (1966) and Lloyd (1967a,b) to calculate the density of states of liquid metals. Their theoretical scheme is shown to have the same mathematical structure as the tight binding (TB) formalism in terms of nonorthogonal atomic orbitals. A formal corres-

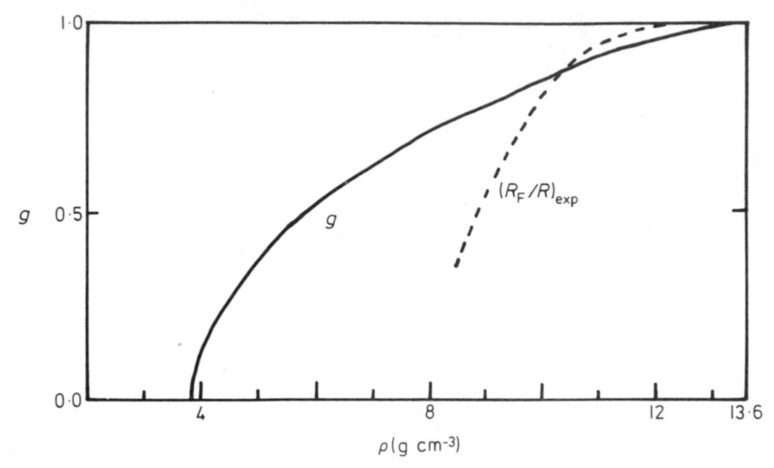

Figure 5. Full curve: $g = n(E_F)/n_{free}(E_F)$ versus ρ (g cm^{-3}). Dashed curve: (R_{free}/R_{exp}) from Even and Jortner (1972).

pondence is attained by replacing the unperturbed locator $G^{(0)} = (E_\mu)^{-1}$ and the transfer integral $t^{\mu\nu}(\mathbf{R}_{ij})$ in the TB theories by a phase shift t_L and the Green's function $G_{LL}'(\mathbf{R}_{ij})$ respectively (Asano and Yonezawa 1977, Watabe 1977). The correspondence is not only in a formal mathematical language but also in the physical meanings that each of these quantities stands for. As explained in §2, all these quantities are of an atomic nature and can be evaluated from atomic potentials. Note that, because of the above-described formal correspondence, all the approximations proposed for the TB theories fully hold in the muffin-tin potential model.

We have calculated the density of states and the dispersion relation in this muffin-tin potential scheme. We have started with the relativistic HFS atomic potential and used Percus–Yevick (PY) pair distribution. The calculation is performed for the EMA and

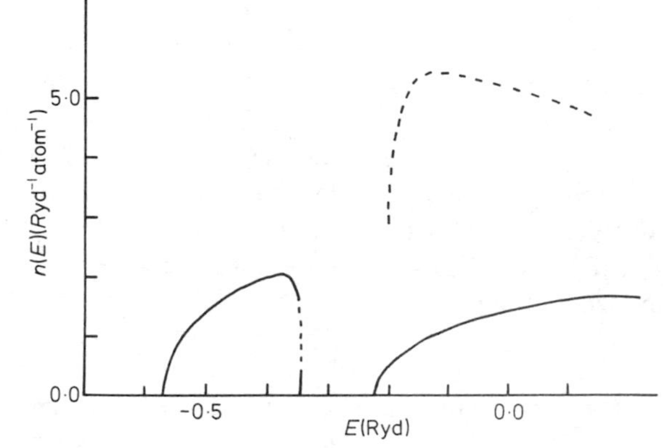

Figure 6. The density of states for ρ = 5 g cm^{-3} for a Percus–Yevick liquid with σ = 5·4 a.u. by the muffin-tin potential model and the EMA. Full curves denote the density of states in the muffin-tin radius while the dashed curve corresponds to the total density of states. A general formulation for the density of states in the muffin-tin radius and for the total density of states was given by Hamazaki *et al* (1976).

the obtained density of states is shown in figure 6. There is a definite bandgap for $5\,g\,cm^{-3}$. Therefore ρ_0 by this method becomes larger than the results of the TBA on the basis of the IY. The source of this difference may be twofold. First the TBA tends to give somewhat broader bandwidths than the muffin-tin potential formalism. Secondly, the EMA generally yields a little broader bandwidth than the IY. In this connection, it is appropriate to mention here why we have restricted ourselves to the IY theory in studying the TBA. The IY is much easier than the EMA in actual calculation. More importantly, the IY and the EMA give similar results not only qualitatively but also quantitatively as far as the TBA is concerned. This is not the case for the muffin-tin

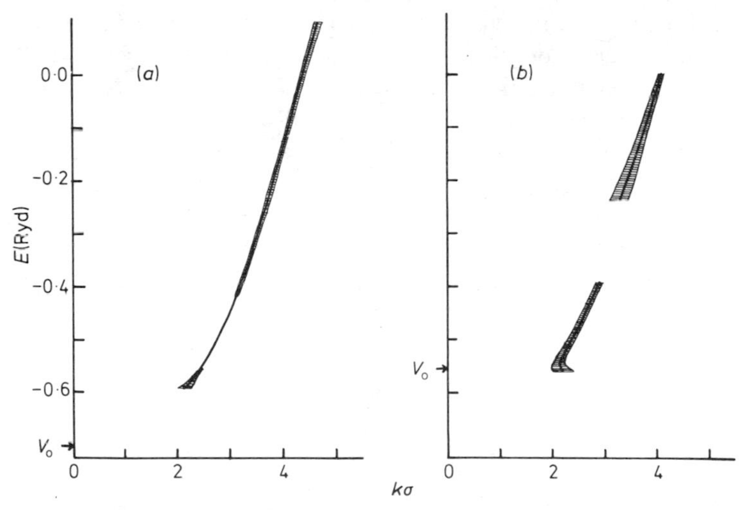

Figure 7. The dispersion curve for (*a*) ρ = 13·6 and (*b*) $\rho = 10\,g\,cm^{-3}$ showing E versus $k\sigma$ where σ is twice the muffin-tin radius. The calculation is based on the same model and approximations as described in figure 5. See text for details.

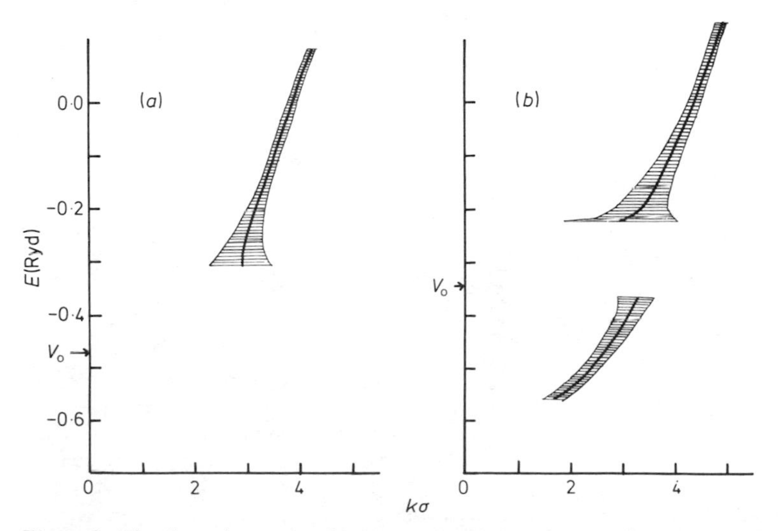

Figure 8. The dispersion curves for (*a*) ρ = 8 and (*b*) $\rho = 5\,g\,cm^{-3}$. Details are the same as given in figure 7.

potential model. In the latter model, the IY yields a breakdown in analyticity under certain conditions whilst the EMA does not. Therefore, the muffin-tin potential model must be studied in the framework of the EMA.

The dispersion relation $E = E(k)$ is determined from the solution of the equation $\det |\tilde{t}_L^{-1}\,\delta_{LL'} - \tilde{G}_{LL'}(\mathbf{R}_{ij})| = 0$ where the tilde indicates that both the phase shift and the Green's function are renormalized in the sense as defined by the EMA. Since the calculation is self-consistent, we no longer have a solution with both E and k real and we have searched for solutions where k is complex for a given real energy E. Figures 7 and 8 show the dispersion curves for various densities where the central solid curve represents the relation E against $\mathrm{Re}(k)$, and the magnitude of $|\mathrm{Im}(k)|$ is expressed by the size of the shadowed region. Since $|\mathrm{Im}(k)|$ is related to the damping, or the inverted lifetime of the state, it gives some idea of the localization of electronic states. The result indicates that mercury at $13{\cdot}6\,\mathrm{g\,cm^{-3}}$ is a good metal. When the density is decreased to $10\,\mathrm{g\,cm^{-3}}$, the damping starts to grow near the middle of the band and we could not find the solution in this region. This is exactly the region where the density of states has a dip, which is expected to become a pseudogap for reduced densities. The damping becomes much more serious for $8\,\mathrm{g\,cm^{-3}}$ and we understand by this that electrons in the states in the middle of the band have a very short mean free path. For $5\,\mathrm{g\,cm^{-3}}$, the gap sets in and at this density the states in the band tail region are considered to be almost localized because the damping there is remarkably large.

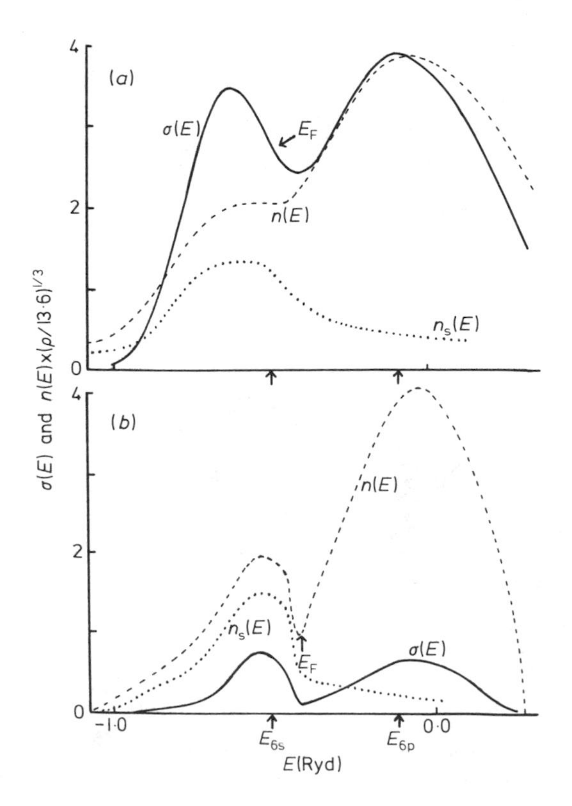

Figure 9. Conductivity $\sigma(E)$ as a function of energy for (*a*) $g = 1{\cdot}0$ and (*b*) $g = 0{\cdot}33$, for the same model and approximations described in figure 1. See text for details.

6. Electrical conductivity

In order to obtain the conductivity by means of the Kubo formula for instance, we need to evaluate the ensemble average $\langle GG \rangle$, which is usually much more difficult to obtain than $\langle G \rangle$. Although the relation $\langle GG \rangle = \langle G \rangle \langle G \rangle$ is correct only for limited cases, here we use this relation as a first approximation. The calculated conductivity by the IY method for the TBA is given in figure 9. The graph of $\sigma(E)$ against g is given in figure 10. Except for a region just below the normal liquid density ($\leqslant 13{\cdot}6\ \mathrm{g\ cm^{-3}}$), $\sigma(E)$ may be expressed as $\sigma(E) \propto g^s$ where $s \geqslant 2$. Note that $\sigma(E) \propto g^2$ has been proposed (Mott and Davis 1971).

7. The fractional s-character and the Knight shift

Recently, the Knight shift in liquid Hg has been measured by El-Hanany and Warren (1975) for densities from the normal liquid density to less than $8\ \mathrm{g\ cm^{-3}}$. Since their experimental results do not seem to be explained in the framework of previously proposed predictions, it is interesting to see the fractional s-character of the density of states at E_F since this is expected to give the most dominant contribution in determining the Knight shift. The s-part of the density of states, $n_s(E)$, is given in figure 9 by broken curves. The fractional s-character, $n_s(E)/n(E)$, as a function of E, is given in figure 11 for various values of g (which is related to density ρ in the way given in figure 5 for instance). Arrows on the curves indicate the Fermi energy E_F. It is readily seen that the fractional s-character at E_F remains almost constant ($n_s(E_F)/n(E_F) \simeq 0{\cdot}5$) for all values of g or ρ. This does not agree with the conclusion of the band calculation due to Devillers and Ross (1975) that there is a substantial increase in the fractional s-character of states at the Fermi level when the density ρ of Hg is decreased. Our result also seems to contradict Mott's proposal that the s-band density of states varies much more slowly with energy near E_F than does the p-band.

We believe that our conclusion concerning the almost constant value of $n_s(E_F)/n(E_F)$ will not be altered drastically even if the calculation is performed through different

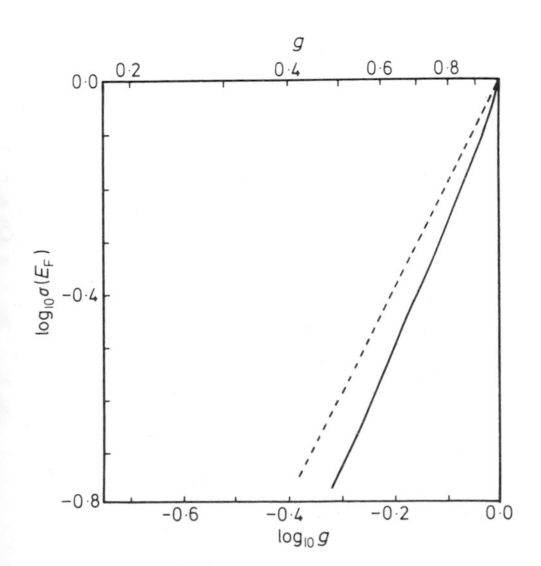

Figure 10. Full curve, log(σ) versus log(g). Dashed curve, log(g^2) versus log(g).

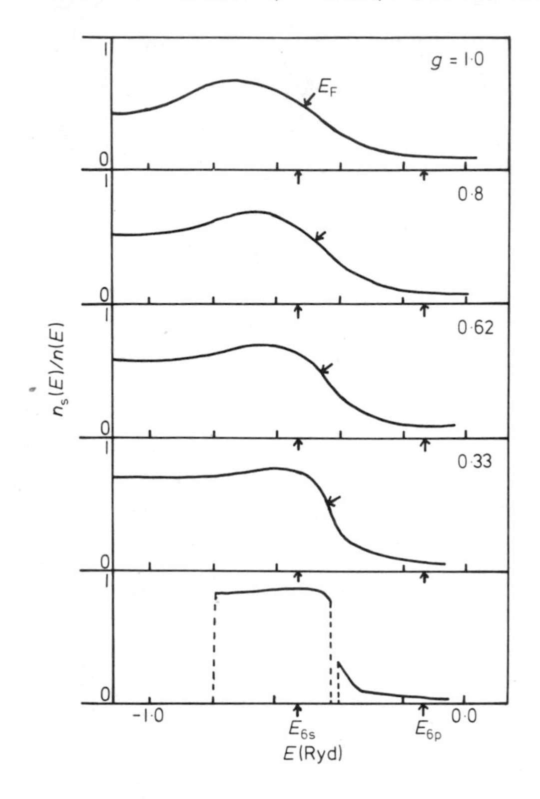

Figure 11. The fractional s-character of the density of states $n_s(E)/n(E)$ as a function of energy for various values of g determined from figure 1.

methods or approximations. Whether this is true or not, we may at least propose the following. Since the fractional s-character varies quite sharply near E_F for $g \lesssim 0{\cdot}33$, it would be interesting from an experimental point of view to see what happens to the Knight shift if one can realize the situation in which only the position of the Fermi level is shifted either upwards or downwards without changing other quantities; for example by adding impurities which do not cause too much volume change.

References

Cusack N 1973 *The Properties of Liquid Metals* ed S Takeuchi (London: Taylor and Francis) pp157–71
Devillers M A C and Ross R G 1975 *J. Phys. F: Metal Phys.* **5** 73–80
El-Hanany U and Warren W W Jr 1975 *Phys. Rev. Lett.* **34** 1276–9
Even U and Jortner J 1972 *Phil. Mag.* **25** 712–20
Duckers L J and Ross R G 1972 *Phys. Lett.* **30A** 715
Fritzson P and Berggren K F 1976 *Solid St. Commun.* to be published
Hamazaki M, Asano S and Yamashita J 1976 *J. Phys. Soc. Japan* **41** 378–87
Hensel F 1973 *The Properties of Liquid Metals* ed S Takeuchi (London: Taylor and Francis) pp357–63
Ishida Y and Yonezawa F 1973 *Prog. Theor. Phys.* **4** 731–53
Lloyd P 1967a *Proc. Phys. Soc.* **90** 207–16
—— 1967b *Proc. Phys. Soc.* **90** 217–31
Mott N F 1968 *Adv. Phys.* **16** 49
Mott N F and Davis E A 1971 *Electronic Processes in Non-crystalline Materials* (Oxford: Clarendon Press) ch 3

Overhof H, Uchitman H and Hensel F 1975 preprint
Roth L M 1974 *Phys. Rev.* **B9** 2476–84
Schiff D 1969 *Phys. Rev.* **186** 151–9
Schmutzler R W and Hensel F 1972 *Ber. Bunsenges. Phys. Chem.* **76** 531
Watabe M 1977 this volume
Yonezawa F and Martino F 1976a *Solid St. Commun.* **18** 1471–4
—— 1976b *J. Phys. F: Metal Phys.* **6** 739–47
Ziman J M 1966 *Proc. Phys. Soc.* **88** 387–405

Electrical conductivity and thermoelectric power of expanded mercury and dilute amalgams

K Tsuji†, M Yao†, H Endo†, S Fujiwaka‡ and M Inutake‡
† Department of Physics, Kyoto University, Kyoto 606, Japan
‡ Institute of Plasma Physics, Nagoya University, Nagoya 464, Japan

Abstract. The electrical conductivity σ and thermoelectric power S of liquid Hg and dilute amalgams (Hg–Au, Hg–Cd, Hg–Tl and Hg–Bi) have been measured in the temperature range 20–1600 °C, and in the pressure range 1–1700 bar. Addition of an element having a different number of valence electrons from that of Hg gives rise to a change of the density of states near the Fermi level, and affects the metal–nonmetal transition observed in expanded Hg.

Experimental results for σ and S for Hg were in fairly good agreement with previous values (Schmutzler 1971). The thermoelectric power near the critical point showed a large negative value and tended abruptly to zero with increasing temperature, as Duckers and Ross had observed.

The variations of σ and S with temperature at constant pressure in dilute Bi amalgams were smaller than those of pure Hg. The critical point of Hg–0·5 at.% Cd was determined to be at T_c = 1500 ± 20 °C and P_c = 1680 ± 30 bar.

1. Introduction

The overlap between the valence and conduction bands, and the density of states $N(E)$ at the Fermi level E_F, both decrease when the mean atomic distance between Hg atoms is continuously increased, giving finally a metal–nonmetal transition. A large and continuous reduction of density can be achieved fairly easily for liquid Hg at high temperatures and high pressures.

Recently Zillgit *et al* (1972) studied the effect of excess electrons on the metal–nonmetal transition in liquid Hg by measuring the conductivity of liquid dilute In amalgams at high temperatures. They found that, at a given pressure, the concentration dependence of the conductivity changed remarkably with increasing temperature.

The simultaneous measurements of the density and conductivity were made by Even and Jortner (1974). The large excess volumes and large differences in the conductivity (relative to pure Hg) were observed for In amalgams at constant temperature and pressure. They suggested that the large difference in the conductivity between In amalgam and pure Hg at constant temperature and pressure originated mainly from the density variation between them.

In the present paper we report the data on the electrical conductivity σ and thermoelectric power S of Hg and its amalgams (containing impurities with various valencies) over an extensive range of temperature and pressure. The paper also describes the vapour pressure curve of liquid amalgams, which is terminated by the critical point.

2. Experimental

The liquid specimen was contained in an electrically insulating tube of pure sintered alumina. All other parts in direct contact with the specimen were made of molybdenum, which is compatible with Hg up to 1700 °C. At the closed end of the cell, which was located in a furnace, a Pt–30% Rh:Pt–6% Rh thermocouple was connected to the molybdenum cell by a screw. The test size of the specimen was about 6 mm in axial length and 1·7 mm in diameter. A four-terminal technique was used for the conductivity measurement. The absolute thermoelectric power was obtained from the difference between the Seebeck voltages for the Pt:Hg and Hg:Pt–6% Rh couples. The temperature was measured using the Pt–30% Rh:Pt–6% Rh thermocouple, and the pressure was measured using a Heise gauge.

Simultaneous temperature–pressure conditions up to 1600 °C and 1700 bar were achieved using a water-cooled steel vessel pressurized with argon gas and containing an internal molybdenum resistance furnace. The vessel was filled with alumina powder in order to obtain good thermal insulation and to prevent compressed-argon convection.

To determine the vapour pressure curve, the electrical conductance of the liquid specimen was measured with increasing temperature at fixed pressures. At the boiling temperature the conductance of the specimen abruptly decreased by an order of magnitude 4 or even more. Both the temperature and pressure values of the liquid–vapour equilibrium mixture were determined from observation of the discontinuous change in conductance when the liquid specimen was evaporated.

At the critical point the discontinuous change in conductance was no longer observed since, at this point, any distinction between liquid and vapour disappeared. Thus the critical point was determined.

3. Results and discussion

Figure 1 shows the vapour pressure curves of liquid pure Hg and dilute amalgams,

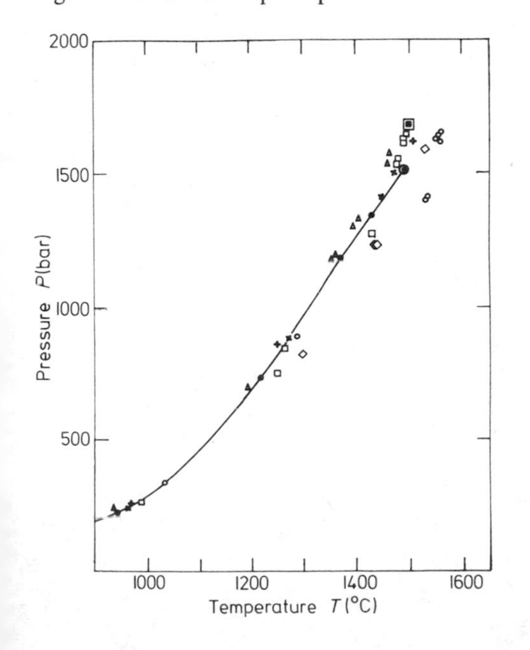

Figure 1. The vapour pressure curves of Hg and dilute amalgams. ● Hg; △ Hg–0·5 at.% Au; □ Hg–0·5 at.% Cd; ◇ Hg–2 at.% Cd; ○ Hg–0·5 at.% Tl; + Hg–0·5 at.% Bi; × Hg–2 at.% Bi.

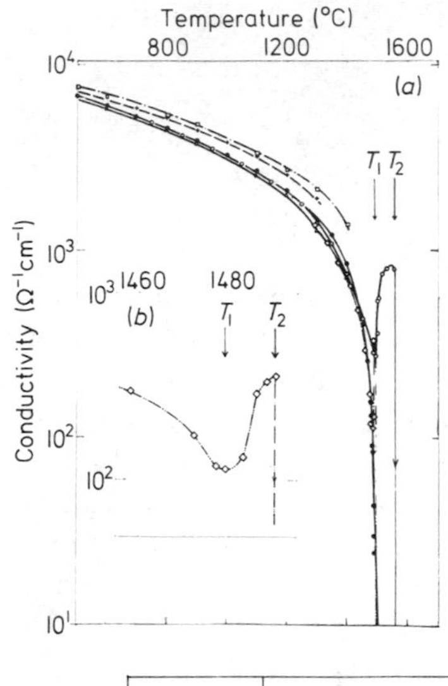

Figure 2. (*a*) Temperature dependence of conductivity of Hg and dilute amalgams at 1600 bar. ● Hg; △ Hg–0·5 at.% Au; ◇ Hg–0·5 at.% Cd; □ Hg–2 at.% Cd; ○ Hg–0·5 at.% Tl; + Hg–2 at.% Bi. The arrows indicate T_1 and T_2 of Hg–0·5 at.% Tl. (*b*) Temperature dependence of conductivity of Hg–0·5 at.% Cd at 1620 bar plotted in enlarged scale near T_1 and T_2.

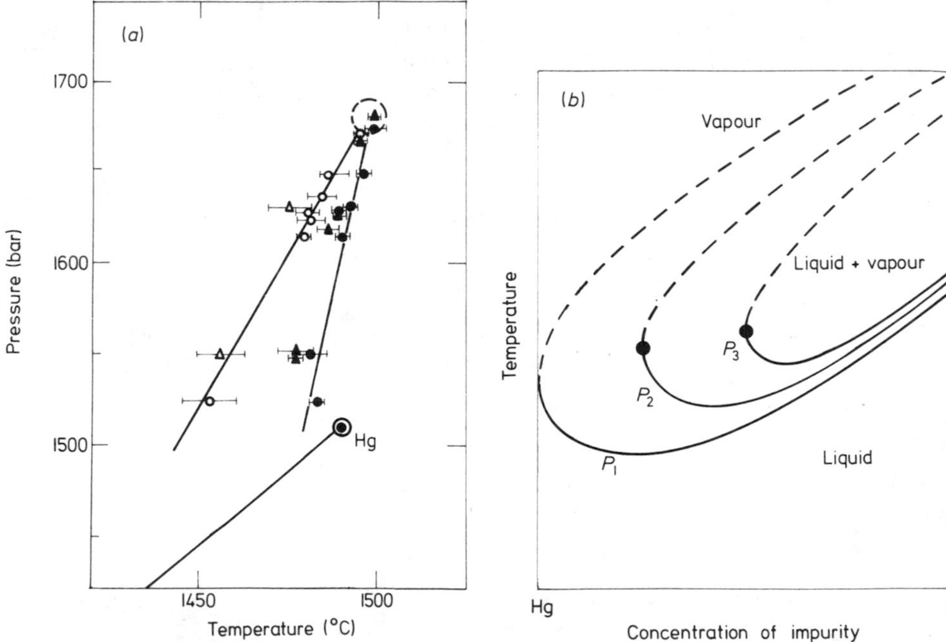

Figure 3. (*a*) The estimation of the critical point of Hg–0·5 at.% Cd. The critical point of pure Hg is also shown for reference. (*b*) Schematic phase diagram of dilute amalgams ($P_1 < P_2 < P_3$). The full lines indicate the temperatures of the beginning of two phases of dilute amalgams and the broken lines indicate the temperature of the end of the two phases. The full circles indicate the critical points of amalgams.

Hg–0·5 at.% Au, Hg–0·5 at.% and 2 at.% Cd, Hg–0·5 at.% Tl and Hg–0·5 at.% and 2 at.% Bi. All the plotted points were reproducible within ±5 °C and ±5 bar. The locus of points in the P–T plane for pure Hg agrees with that established by Hensel and Frank (1968).

It is interesting that the locus in the P–T plane changes by the addition of a small amount of the other elements. The addition of Cd and Tl moves the locus to a higher temperature at a given pressure (below 1300 bar), while Au moves the locus to a lower temperature.

Our data for the change of the conductivity with temperature of the amalgams at 1600 bar are shown in figure 2(*a*). The conductivities of the amalgams along the isobars decreased with increasing temperature, but began to increase at the temperature T_1 (lower than the temperature T_2, where an abrupt jump occurred). The arrows in figure 2(*a*) indicate T_1 and T_2 at 1600 bar for Hg–0·5 at.% Tl and in figure 2(*b*) for Hg–0·5 at.% Cd at 1620 bar.

In figure 3(*a*), open and full circles denote the values of T_1 and T_2 on heating, and open and full triangles the values on cooling. With increasing pressure, T_2–T_1 decreases, and becomes zero where the conductivity drops continuously by raising temperature. Such phenomena were reproducible and reversible for several trials.

The schematic phase diagram for dilute amalgams at various pressures is shown in figure 3(*b*). When the specimen is heated at a pressure below the critical pressure, it begins to separate into two phases at the temperature denoted by the full line corresponding to T_1. With further heating, the vapour phase becomes dominant and an abrupt jump occurs at a certain temperature corresponding to T_2. With increasing pressure the phase boundary should move towards higher concentrations of impurity, as seen in figure 3(*b*). At the pressure where the two-phase region of the amalgam begins to disappear at a given concentration, T_2–T_1 becomes zero. Thus the critical point of the amalgam was assumed to be located in this region. In this manner the critical point for Hg–0·5 at.% Cd was determined to be at $T_c = 1500 \pm 20\ ^\circ C$ and $P_c = 1680 \pm 30$ bar, though the supercritical region was not achieved for the other amalgams.

Figure 4 shows the thermoelectric power of Hg and amalgams against temperature along the isobar 1600 bar. It was found that the thermoelectric power of liquid Hg near

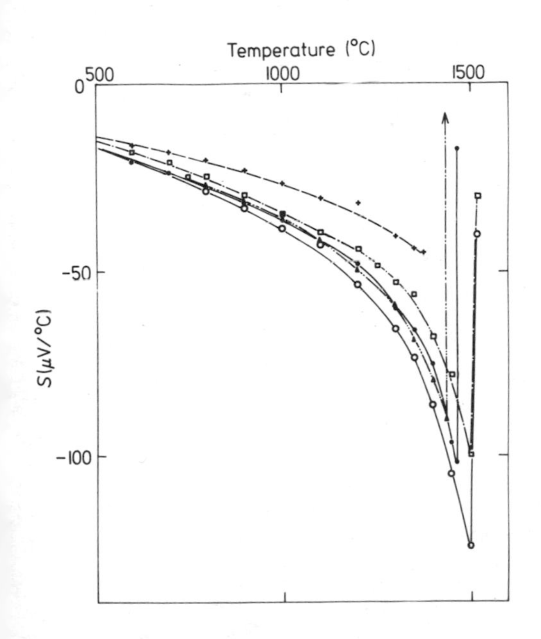

Figure 4. Temperature dependences of thermoelectric power of Hg and dilute amalgams at 1600 bar. ● Hg; ▲ Hg–0·5 at.% Au; □ Hg–2 at.% Cd; + Hg–2 at.% Bi; ○ Hg–0·5 at.% Tl.

the critical point changed rapidly from large-negative to near-zero values, as previously reported by Duckers and Ross (1972).

Figures 5 and 6 show respectively the electrical conductivity and thermoelectric power of Hg–Cd as a function of Cd concentration at different temperatures and a pressure of 1600 bar.

Figure 7 shows the logarithmic plot of conductivity against absolute thermoelectric power for pure Hg and amalgams. It is interesting that $\log\sigma$–S curves for liquid amalgams (except for Hg–Bi amalgams) are on almost the same line as pure Hg. These facts indicate that the difference in conductivity between Hg and Hg–Cd amalgams at a given temperature and pressure is assigned to the density variation between them, because Cd has the same valency as Hg. However, for Hg–Bi amalgams the deviation of $\log\sigma$–S curves from that of pure Hg arises from the shift of E_F due to addition of the element (with valency 5) as well as from volume effects.

As long as accurate measurements of the excess volume have not been done, it cannot be excluded that the whole effect may be due to change in the structures, the inter-

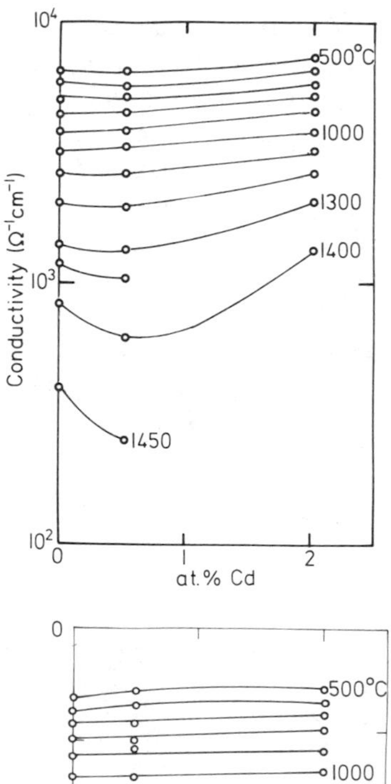

Figure 5. Concentration dependence of conductivity for dilute Hg–Cd amalgams at various temperatures at 1600 bar.

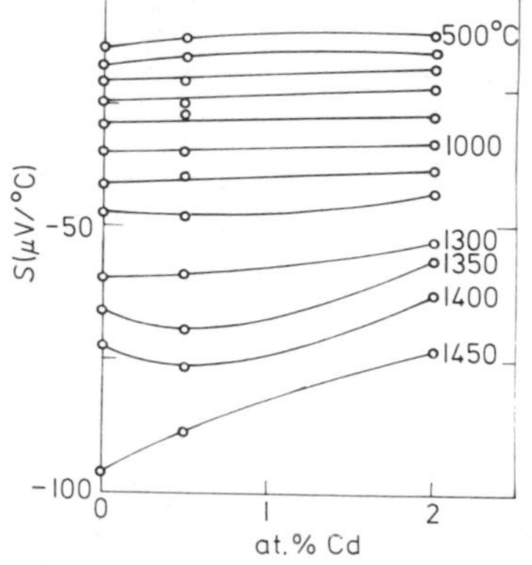

Figure 6. Concentration dependences of thermoelectric power for dilute Hg–Cd amalgams at various temperatures at 1600 bar.

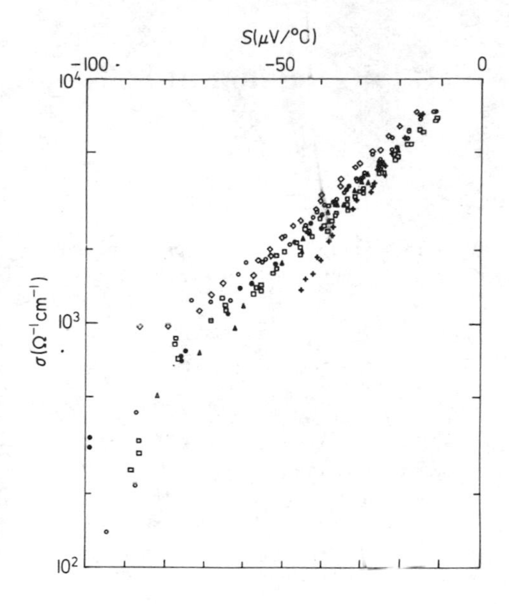

Figure 7. The logarithmic plot of conductivity against thermoelectric power for Hg and dilute amalgams. ● Hg; △ Hg–0·5 at.% Au; □ Hg–0·5 at.% Cd; ◇ Hg–2 at.% Cd; ○ Hg–0·5 at.% Tl; + Hg–2 at.% Bi.

atomic distances and the density of states of Hg caused by the addition of the other elements.

It is necessary to note that a temperature gradient in liquid alloys might cause thermal diffusion and produce a concentration gradient. A complete discussion for σ and S requires the knowledge of density data for the mixture.

Acknowledgment

We are very grateful to Professor F Hensel for helpful advice concerning the design of the apparatus.

References

Duckers L J and Ross R G 1972 *Phys. Lett.* **38A** 291–2
Even U and Jortner J 1974 *Phil. Mag.* **30** 325–34
Hensel F and Frank V 1968 *Rev. Mod. Phys.* **40** 697–703
Schmutzler R W 1971 *Disertation* Universität Karlsruhe
Zillgitt M, Schmutzler R W and Hensel F 1972 *Phys. Lett.* **39A** 419–21

Electronic transport properties of fluid selenium at sub- and supercritical conditions

H Hoshino†, R W Schmutzler‡ and F Hensel‡

† Department of Chemistry, Faculty of Science, Hokkaido University, Sapporo 068, Japan
‡ Fachbereich Physikalische Chemie der Universität Marburg, 3550 Marburg/Lahn, Biegenstrasse 12, West Germany

Abstract. The paper describes new data on the critical point, and the new results of measurements of the electrical conductivity and the thermoelectric power of selenium as a function of temperature and pressure up to supercritical conditions.

The conductivity isobars as well as the thermoelectric power (with negative sign) show marked changes from nearly metallic behaviour to more insulating behaviour depending on the pressures and temperatures applied. Special considerations are given to the transport properties in terms of the variation of the structure from the liquid to the dense vapour.

1. Introduction

Recently, the properties of various liquid semiconductors have been discussed in terms of partly covalent and partly ionic bonding between the component atoms (Enderby 1974). Therefore, knowledge of the behaviour of purely covalent bonding, as a limiting example, is of special interest.

Selenium, whose liquid structure is dominated by the presence of rings and relatively large chain-like units (Eisenberg and Tobolsky 1960, Massen *et al* 1964, Keezer and Bailey 1967), is the only example of an elemental 'liquid semiconductor' which shows semiconducting properties over a very extended temperature range. Its gradual transformation to more metallic behaviour is expected in terms of increasing local atomic coordination number, if the covalent bonds are ruptured, for example, by thermal agitation (Andreev *et al* 1975, Baker 1968, Gobrecht *et al* 1971, Rabit and Perron 1974). Moreover, many previous studies suggest a tendency towards more metallic behaviour as the bonding arrangement is altered by application of pressure (Drickamer and Frank 1973) or by alloying (Perron 1967,1972). It is interesting to study liquid selenium under very high temperatures and pressures, since appreciable changes of density can occur near the liquid–gas critical point and such volume changes may produce significantly large changes in electronic structure. Similar studies of some metallic elements, mercury, caesium and potassium, have demonstrated drastic variation of physical properties in the supercritical fluid state (Hensel 1974, Ross and Greenwood 1969). Therefore, interesting effects can be expected in selenium when the conditions are changed from the normal, subcritical liquid to the dense supercritical vapour whose physical properties may be determined by the presence of molecular species such as Se_2 (Rau 1967).

In this study experimental results are presented for the vapour pressure curve up to the the critical point and the DC electrical conductivity, together with some preliminary results of the thermoelectric power of selenium up to supercritical conditions.

2. Experimental procedures

In order to study these properties, the measurements were extended to temperatures up to 1750°C and pressures up to 1200 bar. Because of the highly corrosive nature of liquid selenium at these high temperatures high-purity sintered alumina and high-purity graphite electrodes were employed with great success. The experiments were performed in an internally heated autoclave with argon as pressure medium. Pairs of temperature and pressure values on the liquid–vapour curve have been determined by the observation of the discontinuous change in resistance when the liquid selenium is evaporated at constant pressure or constant temperature. The DC electrical conductivity was measured by means of the usual four-probe method.

The absolute thermoelectric power of selenium was determined by measuring the thermal EMF of a solid graphite/fluid selenium thermocouple. The experimental details are given elsewhere (Hoshino *et al* 1976a,b).

3. Experimental results and discussion

As a fundamental thermodynamic property, the vapour pressure curve up to the critical point is of importance for any liquid, but for semiconducting liquids it has some additional interest. The reason is that the thermodynamic transition from liquid to vapour across the saturation curve incorporating the transition in the electronic structure from a semiconducting liquid to a low-density non-conducting vapour. Therefore, it is reasonable to consider vapour pressure data together with the electrical transport properties for a fluid like selenium.

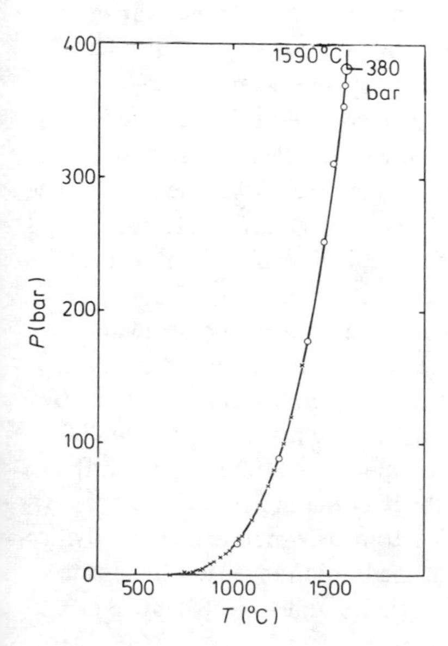

Figure 1. The vapour pressure of liquid selenium as a function of temperature. o: results of this work. X: results of Baker (1968).

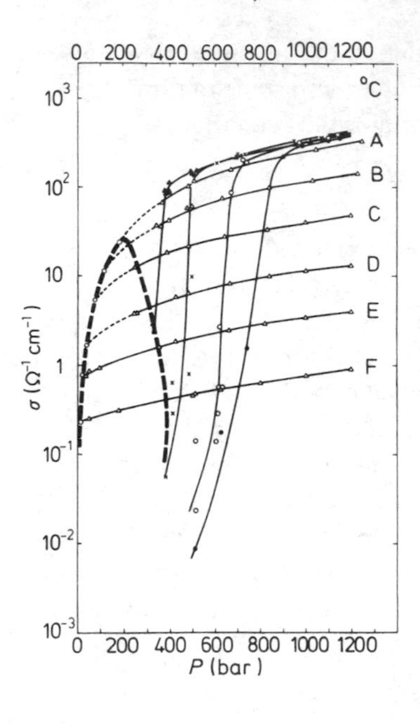

Figure 2. The conductivity isotherms of fluid selenium as a function of pressure. The bold dotted curve gives the conductivities of liquid selenium at saturation pressure. • 1750 °C; ○ 1700 °C; × 1600 °C; ▲ 1500 °C; Δ, A, 1400 °C; B, 1300 °C; C, 1200 °C; D, 1100 °C; E, 1000 °C; F, 900 °C.

Figure 1 shows the vapour pressure curve of liquid selenium up to the critical point, which was found at T_c=1590°C and P_c=380 bar (Hoshino *et al* 1976a). As indicated in figure 1 the vapour pressure data of the present work agree well with the most recent data of Baker (1968) up to 1380° C.

Figure 2 displays the specific electrical conductivity of sub- and supercritical selenium as a function of pressure (Hoshino *et al* 1976a,b). Each of the points in figures 1 and 2 is an average of several independent determinations of temperature, pressure and conductivity. In the range investigated here the conductivity changes over four orders of magnitude. At subcritical temperatures the increase of the conductivity with rising pressure is relatively small. It is only close to the critical temperature that the conductivity becomes strongly pressure dependent, changing from nearly metallic to insulating. This is surely related to the strong pressure dependence of the density in this region. The bold dotted curve shows the partly extrapolated electrical conductivity of liquid selenium at equilibrium pressure. The conductivity increases along the coexistence curve with increasing temperature to a still nonmetallic value of about 25 Ω^{-1} cm^{-1} at about 1400° C and about 180 bar, and then decreases to a quite small value near the critical point.

In figure 3 the electrical conductivity is plotted as a function of temperature at constant pressure. The dominant feature of the conductivity isobars is the sharp rise from 'semiconducting' conductivity less than about 1 Ω^{-1} cm^{-1} at low temperatures to maxima in the nearly metallic range of 100–500 Ω^{-1} cm^{-1}. The maximum in the 400 bar isobar lies at about 100° C lower than the critical temperature. The maxima shift to higher temperatures with increasing pressure. Above the maxima the conductivity drops rapidly at the highest temperatures. The transformation to nearly metallic conductivities can be induced either by increasing the temperature in the liquid range (left hand side of figure 3) or by increasing the pressure in the vapour (right hand side of figure 3).

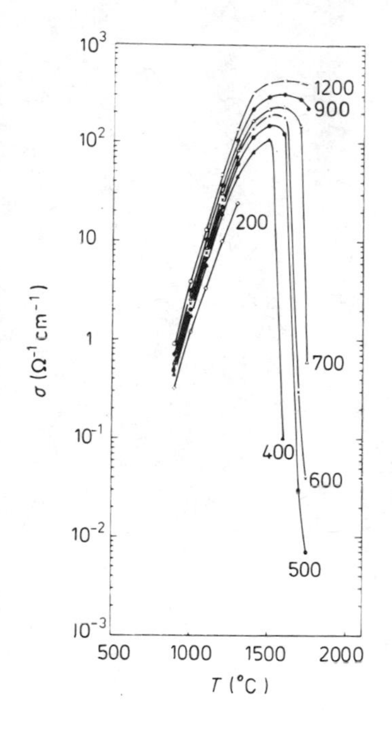

Figure 3. The conductivity isobars as a function of temperature for fluid selenium. The numbers on each curve denote pressure in bars.

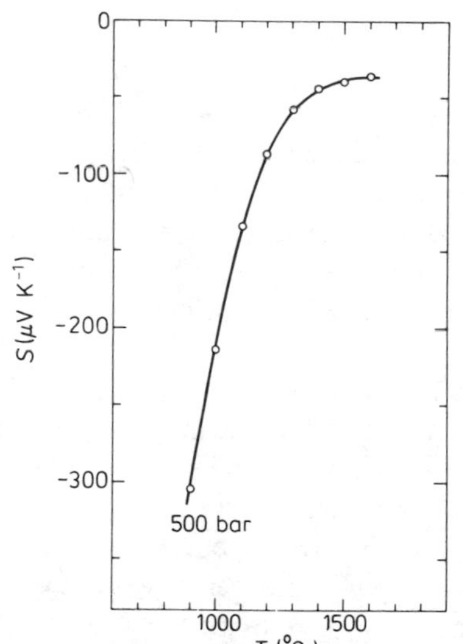

Figure 4. The absolute thermoelectric power of fluid selenium at 500 bar as a function of temperature.

Approaching the critical point the conductivity decreases to a value of about 0·1 Ω^{-1} cm^{-1}. This indicates that fluid selenium is clearly nonmetallic at the critical point.

Moreover, not only the conductivity, but also other properties of the fluid are affected by the transformation due to temperature and density changes. Thus, as shown in figure 4, the thermoelectric power at 500 bar also indicates the steep but continuous change from semiconducting to nearly metallic behaviour. It changes from about -300 $\mu V\ K^{-1}$ at 900 °C to $-36\ \mu V\ K^{-1}$ at 1600 °C. These preliminary results include a

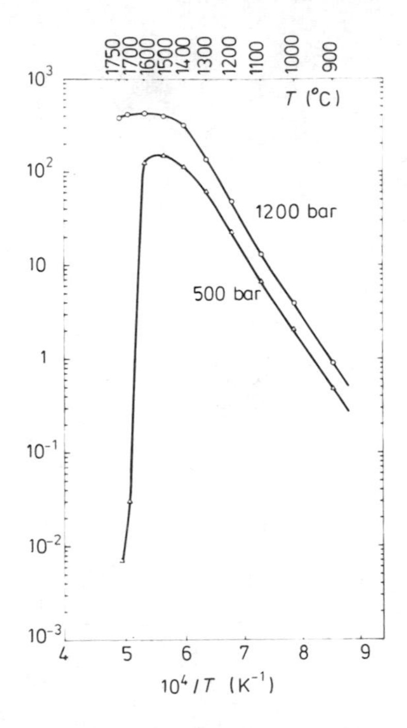

Figure 5. The logarithmic plot of conductivity isobars as a function of reciprocal temperature.

maximum possible error of about ± 20%, stemming partly from uncertainties in the absolute thermoelectric power of the graphite electrodes used (Kazandzhan *et al* 1970), and partly from possible errors of temperature and pressure measurements.

In figure 5 the electrical conductivity of fluid selenium is shown as a function of reciprocal temperature at constant pressure. As long as no density data of selenium up to the experimental conditions applied here are known, any quantitative interpretation of the conductivity and thermoelectric power is difficult. However, the transformation to nearly metallic transport properties probably due to electrons in liquid selenium may be qualitatively interpreted from the following points of view.

As the structural studies (Waseda *et al* 1974, Tourand 1973) and viscosity data (Keezer and Bailey 1967) have indicated, selenium in the subcritical region is a highly associated liquid and has a large number of covalent bonds between atoms in two-fold coordination. Then bonding models analogous to those of the crystal give for liquid selenium a well developed energy gap and semiconducting electronic properties (Mott and Davis 1971). This causes an increase in the conductivity with temperature due to the effects of thermal activation of electrons across the energy gap. In this range of temperature up to about 1100°C an apparent activation energy may be extracted from the slope of the curves in figure 5, which is only slightly pressure and temperature dependent. However, it must be emphasized that these activation energies may be very different from those estimated from an isochoric representation, which is physically more meaningful.

If the temperature is raised further, the limitation of such a picture for considering the activation energy becomes apparent; that is, the slope increases and then decreases as the maximum approaches. Here, a possible explanation can be given in terms of smearing of the energy gap into a 'pseudogap' (Mott 1971) as selenium chain- and ring-

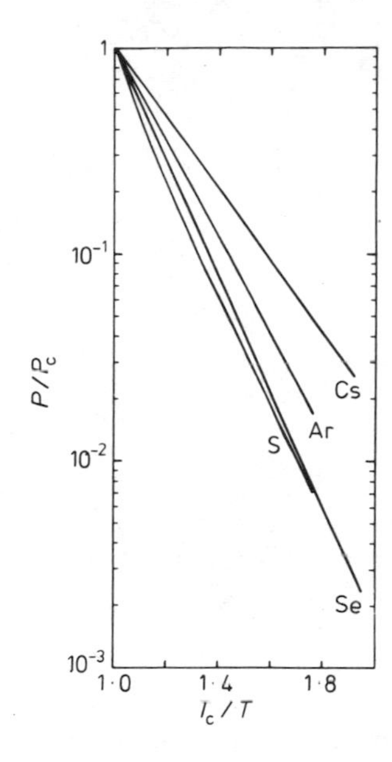

Figure 6. The logarithmic plot of reduced pressure, P/P_c, as a function of reciprocal reduced temperature, T/T_c.

structures rupture and the coordination number increases. The electronic conductivity values in the range of maxima can possibly be described by the so-called strong-scattering or diffusive model of metallic conduction (Cohen 1970).

At the highest temperatures the decrease in the conductivity should appear by a decrease in density due to thermal expansion. Such a decrease in the density should favour the formation of small molecular species, like Se_2 or Se_3. Although there is no direct evidence yet for these species in the dense vapour at high temperatures, some indication of their existence near the critical point can be obtained from the only available thermodynamic property, the vapour pressure curve up to the critical point.

In figure 6 reduced vapour pressure data of selenium are shown as a function of reciprocal reduced temperature together with those for the metal caesium, for the inert gas argon and for the molecular liquid sulphur (Baker 1971). The critical point is used as a reference state. It was demonstrated by Ross and Greenwood (1969) that the reduced vapour pressures of the metals potassium, sodium, caesium and mercury agree within the experimental error. The curve for inert gases which obey a principle of corresponding states within their group (Pitzer 1955) is clearly distinguishable from that for the metals and that for the molecular liquid sulphur. The poor correlation between the reduced pressures of metals and selenium, and the relatively small deviation of the selenium data from the sulphur data suggest that selenium is essentially a molecular fluid like sulphur. The latter is found to be a mixture of different molecular species such as S_2 or S_4 near the critical point (Baker 1971, Weser 1975). Thus the drops in the conductivity at the highest temperatures are at least consistent with such a molecular model.

In the dense selenium vapour the strong decrease of the conductivity with further increase of temperature may be caused by a marked increase in the mean intermolecular distances. Similar effects are supposed to be present in the metallic fluid under supercritical conditions (Hensel 1973, 1974).

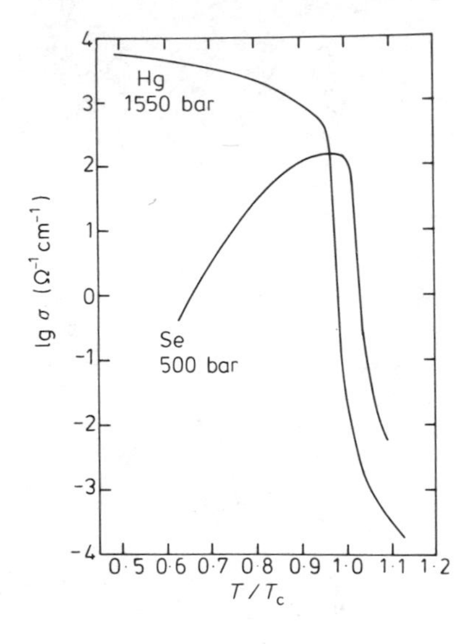

Figure 7. The electrical conductivity of fluid selenium and fluid mercury at supercritical pressures as a function of reduced temperature, T/T_c.

This is demonstrated in figure 7, where the logarithms of the electrical conductivity of fluid selenium and fluid mercury at supercritical pressures is plotted as a function of reduced temperature. There is a clear resemblance in the behaviour of the conductivity above the critical temperature, although the conduction mechanism well below the critical temperature is largely different between two systems.

4. Conclusions

Although the qualitative nature of the interpretation given above must be emphasized, we can conclude that the changes in volume and temperature play very important roles for the changes in the electronic transport properties of fluid selenium. Moreover, our understanding will become clearer as data become available for the density and other physical properties up to the supercritical state of selenium.

Acknowledgments

This work was partly supported by the Deutsche Forschungsgemeinschaft. One of the authors (H H) is thankful to the Alexander von Humboldt-Stiftung for a grant during this work.

References

Andreev A, Turgunov T and Alekseev V A 1975 *Sov. Phys.–Solid St.* **16** 2376
Baker E H 1968 *J. Chem. Soc.* **A** 1089
—— 1971 *Trans. Inst. Mining Metall.* **C80** C93
Cohen M H 1970 *J. Non. Cryst. Solids* **4** 391
Drickamer H G and Frank C W 1973 *Electronic Transitions and the High Pressure Chemistry and Physics of Solids* (Chapman and Hall)
Eisenberg A and Tobolsky A 1960 *J. Polym. Sci.* **46** 19
Enderby J E 1974 *Amorphous and Liquid Semiconductors* ed J Tauc (New York: Plenum) p 361

Gobrecht H, Gawlik D and Mahdjuri F 1971 *Phys. Kondens. Mater.* **13** 156
Hensel F 1973 *The Properties of Liquid Metals* ed S Takeuchi (London: Taylor and Francis) p 357
—— 1974 *Angew. Chem.* **86** 459
Hoshino H, Schmutzler R W and Hensel F 1976a *Ber. Bunsenges. Phys. Chem.* **80** 27
Hoshino H, Schmutzler R W, Warren W W and Hensel F 1976b *Phil. Mag.* **33** 255
Kazandzhan B I, Lobanov A A, Selin Yu I and Tsurikov A A 1970 *Zav. Lab.* **36** 1100
Keezer R C and Bailey M W 1967 *Mater. Res. Bull.* **2** 185
Massen C H, Weijts A G L M and Poulis J A 1964 *Trans. Faraday Soc.* **60** 317
Mott N F 1971 *Phil. Mag.* **24** 1
Mott N F and Davis E A 1971 *Electronic Processes in Non-crystalline Materials* (Oxford: Clarendon)
Perron J C 1967 *Adv. Phys.* **16** 659
—— 1972 *J. Non. Cryst. Solids* **8-10** 272
Pitzer K S 1955 *J. Am. Chem. Soc.* **77** 3427
Rabit J and Perron J C 1974 *Phys. Stat. Solidi* **B65** 255
Rau H 1967 *Ber. Bunsenges. Phys. Chem.* **71** 711
Ross R G and Greenwood D A 1969 *Prog. Mater. Sci.* **14** 173
Tourand G 1973 *J. Physique* **34** 937
Waseda Y, Yokoyama K and Suzuki K 1974 *Phys. Condens. Matter* **18** 293
Weser G 1975 *Thesis* Marburg University

Electrical conductivity of liquid tellurium at high temperatures and pressures

H Endo†, H Hoshino‡ §, R W Schmutzler‡ and F Hensel‡

† Department of Physics, Faculty of Science, Kyoto University, Kyoto, Japan

‡ Fachbereich Physikalische Chemie, Philipps-Universität Marburg, D 3550 Marburg/Lahn, West Germany

Abstract. The paper describes new results for the electrical conductivity of liquid tellurium as a function of temperature and pressure up to 1600°C and 1000 bar. The conductivity isobars exhibit broad maxima which can be understood within the structural model of liquid tellurium developed by Cabane and Friedel.

1. Introduction

The liquid chalcogen tellurium provides an example of an electronically conducting liquid which exhibits some features of both metallic and semiconducting behaviour (Perron 1967, Busch and Güntherodt 1967, Warren 1972). Its conductivity of about 1000 $\Omega^{-1}\,cm^{-1}$ at the melting point is relatively large and its Knight shift value is in a range typically observed for liquid metals. On the other hand, the temperature coefficients of the conductivity and of the Knight shift are positive.

Neutron scattering data (Tourand *et al* 1972) indicate that the number of atoms in the first coordination shell is about two near the melting point and changes to about three near 900°C. The two-fold (binary) coordination is typical of the helical chain-like structure of selenium, while the three-fold (ternary) coordination is similar to that of the arsenic crystal structure. Such a change of coordination was found to be consistent with the Knight shift measurements (Warren 1972) and velocity of sound measurements (Gitis and Mikhailov 1966). Between the melting point and 900°C, Cabane and Friedel (1971) have suggested the existence of regions of two-fold coordination which are semiconducting, mixed with regions of three-fold coordination which are metallic. They assume that the number of free carriers increases with temperature due to the increase in ternary sites.

Since the first coordination number increases to a value larger than five at 1800°C (Tourand *et al* 1972), Cabane and Friedel (1971) proposed that the local structure of liquid tellurium tends to a more metallic cubic structure similar to α-polonium with metallic character at high temperature (about 1700°C) or high pressure. In order to test this proposal, measurements of the electrical conductivity as a function of pressure and temperature have been performed up to 1000 bar and 1600°C.

§ Present address: Department of Chemistry, Faculty of Science, Hokkaido University, Sapporo, Japan.

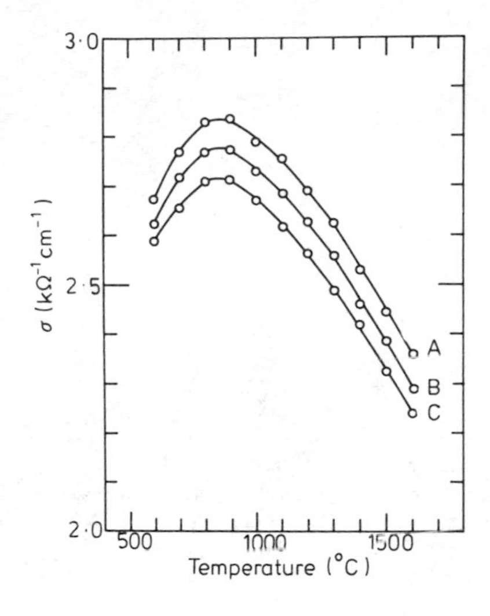

Figure 1. The electrical conductivity of liquid tellurium as a function of temperature at a pressure of: A, 1000 bar; B, 750 bar; C, 500 bar.

2. Experimental results and discussion

The electrical conductivity has been measured by means of a DC four-probe method in a cell of high purity alumina with tungsten or graphite electrodes. The required temperatures and pressures were obtained in an internally heated autoclave. The applied experimental arrangements and procedures were the same as those used for liquid selenium (Hoshino *et al* 1976a). The purity of the sample material was 99·999%. The reproducibility of the data was checked over several experimental runs and was found to be better than ± 2%.

Figure 1 shows the electrical conductivity of liquid tellurium as a function of temperature at various pressures. Up to about 900 °C the conductivity values along isobars increase and then decrease through broad maxima at about 1000 °C. Above 1000 °C the values of $(d\sigma/dT)_p$ are almost constant irrespective of pressure. The maxima shift slightly to higher temperatures with increasing pressure.

According to Cabane and Friedel (1971), the number of free carriers estimated from Hall effect data repidly increases up to 850 °C and almost saturates at 900 °C. This is expected even under pressure since the local atomic structure of liquid tellurium might not be markedly modified at the relatively small pressures up to 1000 bar. Therefore the observed increase in the conductivity below 900 °C may be explained by the increase in the number of free carriers.

It seems reasonable to suppose that the effect of increasing temperature results from the combined effects of increasing coordination number and decreasing density from thermal expansion. The former effect should increase the conductivity while the latter should work in the opposite direction (Hensel 1974). In figure 2 conductivities at different constant temperatures are plotted as a function of pressure. The values of the conductivity change almost linearly with pressure. The pressure coefficients of the conductivity below 900 °C are rather small compared with those above 900 °C.

A direct comparison of our results with recent data of Alekseev *et al* (1975) for the conductivity of liquid tellurium shows deviations up to a factor of two for the conductivity data and up to a factor of 3 to 5 for the pressure coefficients of the

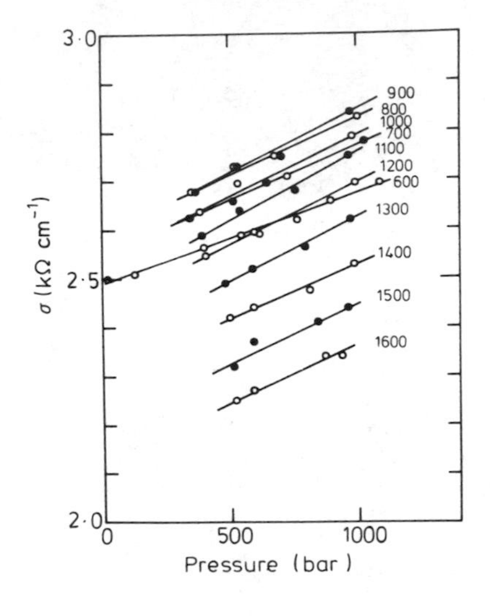

Figure 2. The electrical conductivity of liquid tellurium as a function of pressure. The numbers on the curves denote temperature in °C.

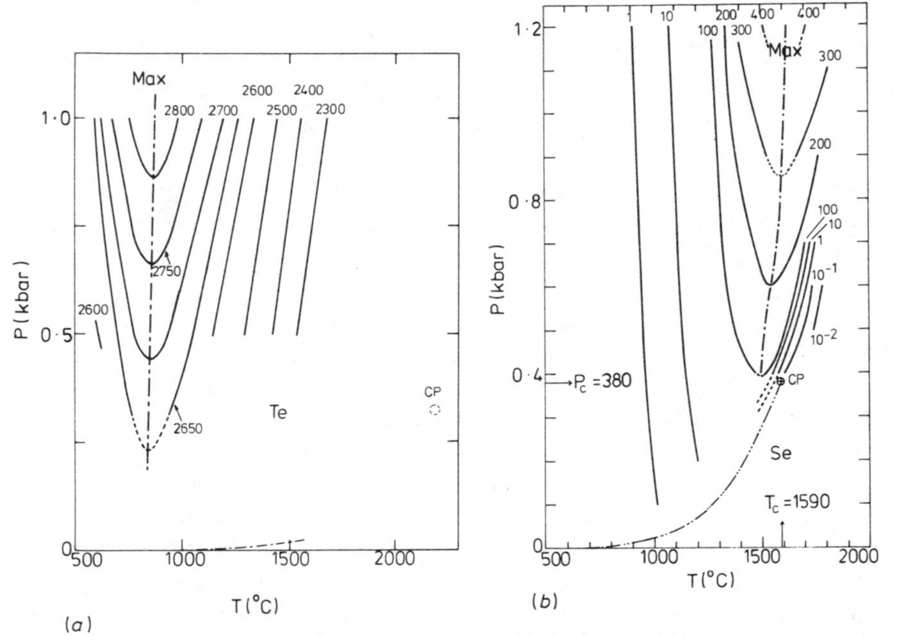

Figure 3. The constant conductivity curves in $P-T$ diagrams for (*a*) liquid tellurium and (*b*) liquid selenium. The numbers on the curves denote the value of the conductivity in Ω^{-1} cm^{-1}. The point, CP denotes the critical point and – - - - - - – indicates the liquid–vapour coexistence curve.

conductivity. The data of Alekseev *et al* are always larger than ours. A possible explanation is that there occurs a considerable diffusion of tellurium into the ceramics used as electrical insulators and this may have a marked effect on the conductivity measurements. This error is certainly avoided in our experiment.

A comparison of the temperature dependence of the electrical conductivity of liquid tellurium and liquid selenium (Hoshino *et al* 1976b), both at moderate pressures, shows that under the present experimental conditions the conductivity of liquid tellurium is always larger than that of liquid selenium. It is obvious that in both elements the conductivity varies through maxima and the change found in liquid tellurium is much smaller than that in selenium. This is certainly related to the larger stability of the bonds within the chain molecules in selenium. The difference in the pressure and temperature dependence of the conductivity between selenium and tellurium is given in figure 3, which shows plots of temperature against pressure for constant conductivities. It is obvious from this figure that the maxima in the conductivities, that is the nearly metallic conductivities, occur at much higher temperatures for selenium than for tellurium. For selenium the maximum conductivity is observed near the critical temperature of the liquid–vapour phase equilibrium, whereas for tellurium the maxima occur far beyond the critical temperature. The latter was determined by means of the corresponding states principle. In accordance with the rule (Pitzer 1955) that any group of substances with effectively similar intermolecular forces should conform themselves to the principle of corresponding states, the correspondence between the vapour pressures of sulphur, selenium and tellurium was used to determine the critical point of tellurium (Hoshino *et al* 1976a, Baker 1967).

3. Conclusion

The reported results of the conductivity of liquid tellurium are in accord with the structural and bonding model proposed by Cabane and Friedel. The temperature dependence of the conductivity can be explained mainly by a variation of the number of free carriers and a decrease in density due to thermal expansion.

Acknowledgments

One of the authors (HH) is grateful to the Alexander von Humboldt-Stiftung for its support during this work. Likewise one of us (HE) is indebted to Deutscher Akademischer Austauschdienst for its support to perform this work in Marburg University. The authors are thankful to Mr H Schüssler for his technical support.

References

Alekseev V A *et al* 1975 *Fiz. Tekh. Poluprov.* 9 139
Baker E H 1967 *J.Chem. Soc.* **A1** 558
Busch G and Güntherodt H J 1967 *Physik Kondens. Mater.* **6** 325
Cabane B and Friedel J 1971 *J. Physique* **32** 73
Gitis M B and Mikhailov I G 1966 *Sov. Phys.–Acoust.* **12** 17 and 131
Hensel F 1974 *Angew. Chem.* **86** 459
Hoshino H, Schmutzler R W and Hensel F 1976a *Ber. Bunsenges. Phys. Chem.* **80** 27
Hoshino H, Schmutzler R W, Warren W W and Hensel F 1976b *Phil. Mag.* **33** 255
Perron J C 1967 *Adv. Phys.* **16** 657
Pitzer K S 1955 *J. Am. Chem. Soc.* **77** 3427
Tourand G, Cabane B and Breuil M 1972 *J. Non. Cryst. Solids* **8-10** 676
Warren W W 1972 *Phys. Rev.* **B6** 2522

An ultrasonic investigation of the metal–nonmetal transition in lithium–ammonia solutions

D E Bowen
Physics Department, The University of Texas, El Paso, Texas 79968, USA

Abstract. Measurements of the absorption coefficient for ultrasound (α/f^2) have been made as a function of metal concentration, temperature and frequency in $Li-NH_3$ solutions. In the concentrated region (greater than 8 mole % metal) the absorption has a behaviour similar to that in liquid metals. The metal–nonmetal transition is marked by a change in the ultrasonic absorption: α/f^2 becomes frequency dependent with a relaxation frequency in the 10–20 MHz range. Strong temperature dependence is also noted in the transition region. The absorption as a function of concentration reaches a maximum at the consolute or critical concentration yet the absorption does not fit critical theory expressions. The implications of this data for models of the solutions will be discussed.

1. Introduction

Solutions of metals in ammonia are known to undergo a metal–nonmetal transition as the metal concentration is varied between 10 and 1 mole %. Two theoretical models for this transition have been presented recently.

Cohen and Jortner (1975) have presented an inhomogeneous model for metal–ammonia solutions in the intermediate concentration range (1–9 mole % metal (mpm)). In this model the solutions are microscopically inhomogeneous, with a volume fraction C of the material occupied by metallic clusters of mean concentration about 9 mpm, and the remaining consisting of more dilute clusters of mean concentration of about 2·3 mpm. Using a modified effective-medium theory they have presented reasonable explanations of the Hall effect, the Hall mobility, thermal conductivity, thermoelectric power, and the electrical conductivity.

Another model has been proposed by Mott (1974, 1975); this is not an inhomogeneous model but discusses the metal–nonmetal transition in terms of band overlap or band crossing. In this model the concentrated solutions are treated as single-band conductors while the solutions in the range 0·3–3 mpm are intrinsic semiconductors; in the region 3–6 mpm there is band overlap. Using these ideas good qualitative explanations of the electrical properties as well as the existence of the phase region have been made.

The ideas presented in the above two models are somewhat controversial and even though the metal–ammonia solutions have been extensively investigated it appears that more experimental data is necessary. In this paper we present the results of ultrasonic absorption measurements in lithium–ammonia solutions in the concentration region of the metal–nonmetal transition. Comparisons between these results and those seen in normal liquid metals will then be presented.

2. Experimental

The measurements were made using a pulse-echo technique with a variable-path cell that has been previously described (Bowen and Priesand 1974). This apparatus can determine the absorption coefficient (α/f^2) to within 4%. For these experiments solutions of known concentration were prepared and the absorption measured as a function of temperature and frequency.

3. Results

Absorption was obtained for 17 different solutions with concentrations between 1 and 14 mpm; frequencies between 10 and 90 MHz were used. The measurements were made as functions of temperature between -63 °C and -45 °C.

Figure 1 shows the absorption (plotted as α/f^2) as a function of the metal concentration at two frequencies, 10 and 30 MHz. Figure 2 shows the frequency dependence at

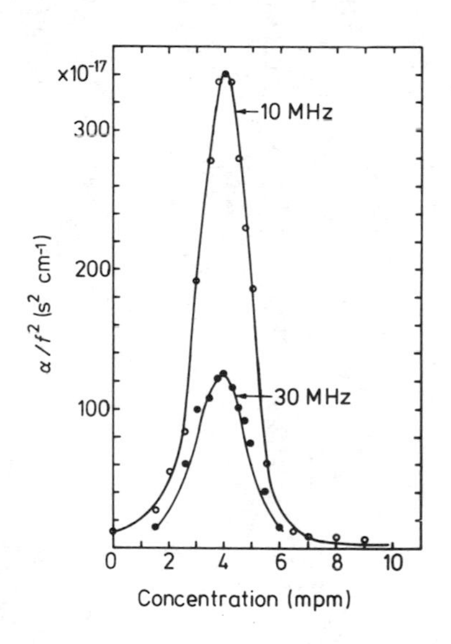

Figure 1. α/f^2 versus concentration for lithium–ammonia solutions at -63 °C.

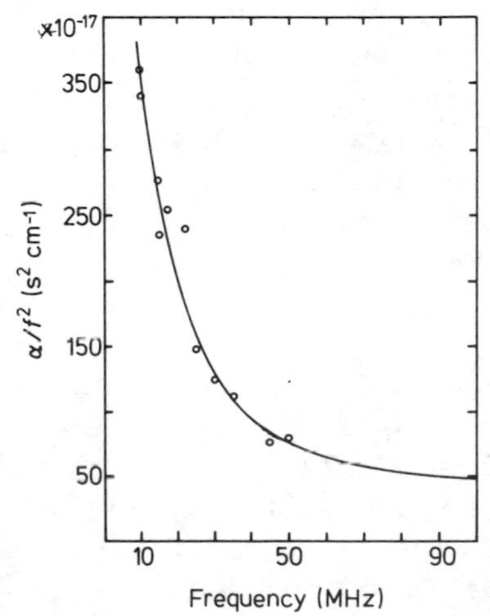

Figure 2. α/f^2 versus frequency for lithium–ammonia solutions at -63 °C.

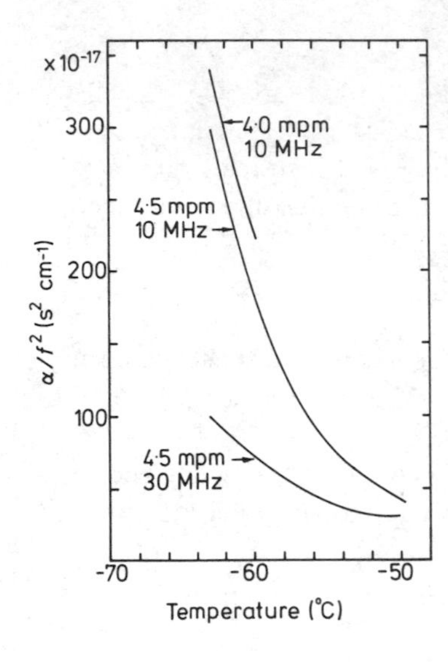

Figure 3. α/f^2 versus temperature for lithium–ammonia solutions.

Table 1. Results in lithium–ammonia solutions

Concentration region	Absorption coefficient (α/f^2)
$7 < x < 20$ mpm	Frequency independent Temperature independent Weak concentration dependence
$1 < x < 7$ mpm	Frequency dependent Temperature dependent Maximum at 4 mpm

$T = -63$ °C while the temperature dependence is shown in figure 3. The peak in the absorption and the nature of the frequency dependence of some of the data have been previously discussed (Bowen 1975) and will not be discussed here.

The frequency behaviour and temperature dependence shown in these figures were not obtained for all concentrations. Indeed, using these measurements it is possible to separate the solutions into two distinct concentration regions. Table 1 gives these regions and the characteristics of α/f^2 in each region. Concentrated solutions are thus seen to have temperature- and frequency-independent absorption coefficients with dependence occurring as the solutions are diluted below 7 mpm.

4. Discussion

Concentrated metal–ammonia solutions are considered to be liquid metals (Cohen and Thompson 1968). This conclusion is based upon considerable experimental evidence such as the electrical conductivity and the Hall coefficient. Figure 4 shows the electrical conductivity for lithium–ammonia solutions as a function of concentration (Nasby and Thompson 1968). We see here that the conductivity reaches metallic values at about 8 mpm (Allgaier 1969). At saturation (20 mpm) the conductivity reaches a value of

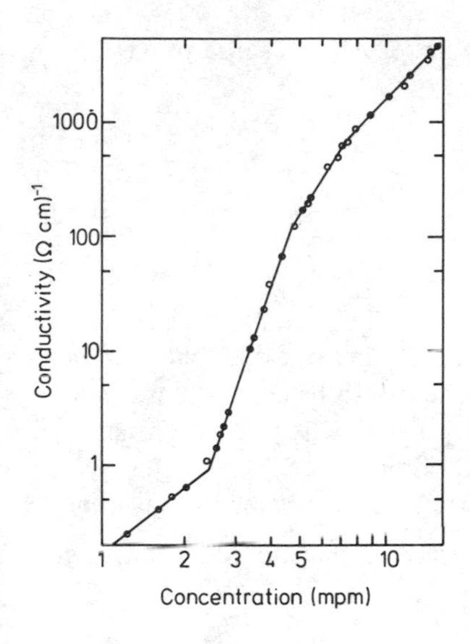

Figure 4. Electrical conductivity versus concentration for lithium–ammonia solutions at –62 °C (from Nasby and Thompson 1968).

15 000 Ω^{-1} cm^{-1} which is greater than that of liquid mercury at room temperature (Morgan *et al* 1965). Further evidence for the liquid metal nature of these solutions comes from measurements of the Hall coefficient, an example of which is shown in figure 5 (Nasby and Thompson 1968). These measurements indicate that the carriers are electrons with a density equal to that of the metal valence electrons. Deviations begin as the solutions are diluted below about 8 mpm.

We now turn to the results obtained for normal liquid metals; these have been summarized by Webber and Stephens (1968). If a plane sound wave propagates through a material the amplitude of the pressure wave will vary according to

$$P = P_0 \exp(-\alpha x) \tag{1}$$

where α is the absorption coefficient and x is the distance traversed by the wave. From

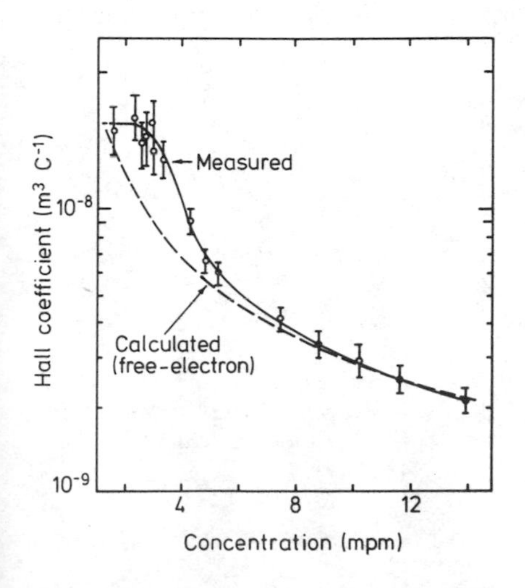

Figure 5. Hall coefficient versus concentration for lithium–ammonia solutions at –50°C (from Nasby and Thompson 1968).

classical hydrodynamics one can obtain the Stokes–Kirchoff classical absorption coefficient α_c (Beyer and Letcher 1969):

$$\frac{\alpha_c}{f^2} = \frac{8\pi^2}{3\rho c^3}\left[\eta_s + 3(\gamma - 1)\frac{K}{4C_p}\right] \tag{2}$$

$$= \frac{\alpha_s}{f^2} + \frac{\alpha_T}{f^2}. \tag{3}$$

This expression indicates that the absorption is due to mechanisms associated with the shear viscosity (η_s) and the thermal conductivity (K). Here ρ is the density, c is the sound velocity, γ is the ratio of specific heats and C_p is the specific heat at constant pressure. It should be noted that α_c/f^2 is independent of the frequency f.

Measurements of α/f^2 show two major deviations from that predicted by equation (2);

(i) α/f^2 is greater than that predicted by equation (2) and is frequency independent or
(ii) α/f^2 is greater and depends on frequency.

If the deviation from α_c/f^2 is frequency independent the excess absorption can be formally attributed to a volume viscosity η_B, defined by

$$\eta_B = \frac{4}{3}\eta_s\left(\frac{\alpha - \alpha_c}{\alpha_s}\right) \tag{4}$$

where α_s is the contribution to α_c due to the shear viscosity.

Measurements of the absorption in liquid metals (Webber and Stephens 1968) indicate that α/f^2 is frequency independent and slightly greater than α_c/f^2 (i.e., $\eta_B/\eta_s \simeq 2$). Figure 6 shows the results of Kim *et al* (1971) for sodium. These results are typical of those seen in liquid metals.

Figure 7 shows the results of our measurement in the lithium–ammonia solutions along with our calculations of α_c/f^2 using data from the literature. It is seen that α/f^2 does not change until a concentration of around 8 mpm is reached.

Table 2 gives comparison between the results seen in a variety of liquid metals (Webber and Stephens 1968) and in the metal–ammonia solutions. For the solutions it

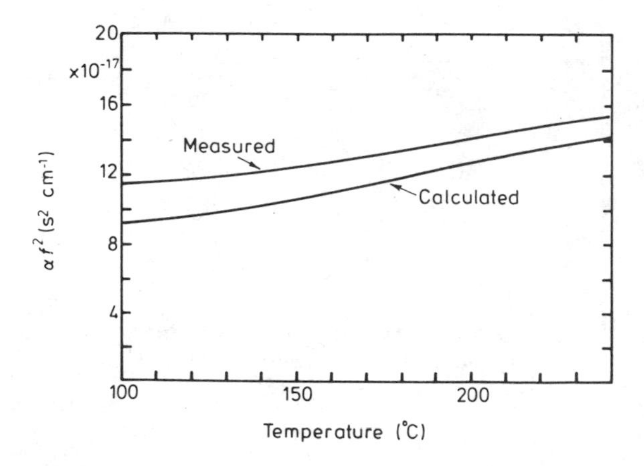

Figure 6. α/f^2 versus temperature for liquid sodium (from Kim *et al* 1971).

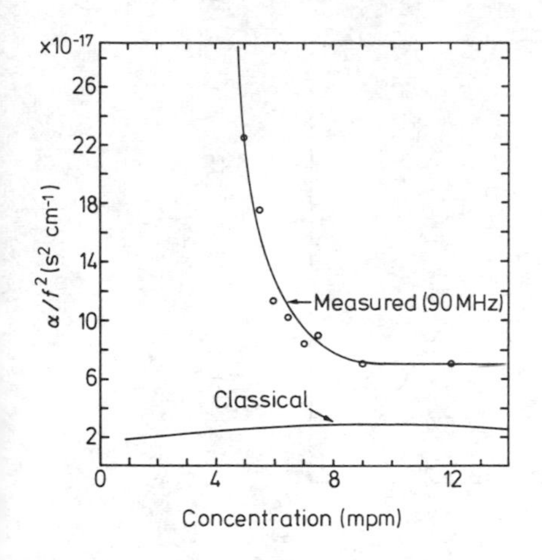

Figure 7. α/f^2 versus concentration for lithium–ammonia solutions at −60 °C and 90 MHz.

Table 2. Sound absorption and ratio of bulk to shear viscosities in liquid metals.

Metal	T (K)	$\alpha_S/f^2 \times 10^{17}$ (cm^{-1} s^2)	$\alpha_T/f^2 \times 10^{17}$ (cm^{-1} s^2)	$\alpha/f^2 \times 10^{17}$ (cm^{-1} s^2)	η_B/η_S
Na	373	1·24	8·31	11·5	2·0
K	343	2·64	24·56	31·9	2·40
Zn	723	0·75	3·48	3·7	−0·9
Cd	633	0·48	10·8	14·5	9·0
Hg	298	0·98	4·29	5·71	0·6
Ga	303	0·37	1·04	1·58	0·6
Sn	513	0·49	3·8	5·63	3·5
Pb	613	1·13	6·44	9·4	2·1
Bi	553	1·07	4·66	8·05	2·9
Rb	316	6·10	52·30	75·5	3·73
Cs	308	9·73	77·35	110·3	3·18
In lithium–ammonia solutions					
12 mpm	213	2·73	–	7·0	2·40
7 mpm	213	2·79	–	8·5	2·72
6 mpm	213	2·72	–	11·5	4·30
5 mpm	213	2·62	–	22·5	10·12

is noted that the contribution to α_c/f^2 from the thermal conductivity is less than 0·1% and is not shown; this is in contrast to the liquid metals where the absorption due to thermal conductivity is dominant, being 5 to 12 times that due to shear viscosity. However, the ratio η_B/η_S for the solutions is of the same order as that in the liquid metals down to a concentration of 5 mpm.

We thus see that an absorption mechanism different from those described by either viscosity or thermal conductivity occurs in the solutions as they are diluted below 8 mpm. First it should be noted that this increase in α/f^2 (or in the ratio η_B/η_S) is most likely not due to critical effects associated with the two-phase region. Figure 8 shows the coexistence curve for the lithium–ammonia solutions (Schettler and Patterson

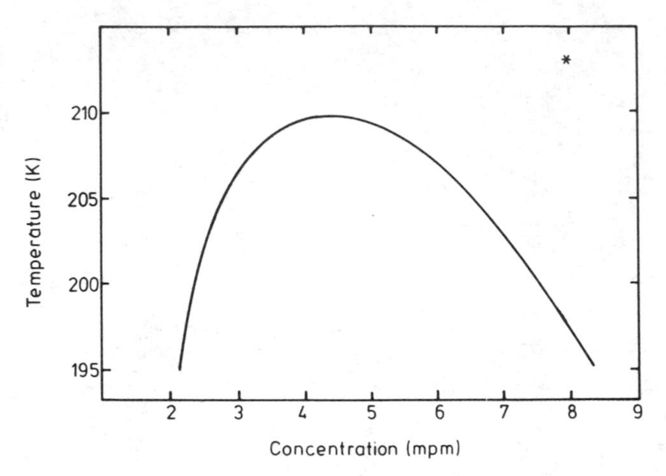

Figure 8. Mixed-phase region for lithium–ammonia solutions (from Schettler and Patterson 1964). The asterisk indicates the location of the solutions for the data shown in figure 7.

1964). The measurements shown in figure 7 are at a point 15 K above the phase transition and at a concentration well away from the critical point. Chieux and Sienko (1970) have pointed out that critical effects at the critical concentration occur only within 1·8 K. This point has been further discussed by Mott (1974). The explanation of the increase in α/f^2 must thus lie elsewhere.

It is interesting to interpret these data in terms of the percolation model of Cohen and Jortner (1975). According to this model the solutions become microscopically inhomogeneous as the concentration is diluted below 9 mpm. Between 2 and 9 mpm the solution consists of a volume concentration C of metallic clusters of concentration 9 mpm and a volume concentration $1-C$ of clusters of concentration 2·3 mpm. This model then says that the solutions are 'mixtures' of two solutions and one is directed to see if the absorption here parallels that seen in other binary mixtures (Blandamer 1973). Since both 2 and 9 mpm lithium–ammonia solutions would be classified as associated liquids we are interested in the general category of mixtures of associated liquids. A characteristic of these systems is that α/f^2 for some mixtures is often many times larger than that of either component and is frequency dependent. As the concentration is varied between the two components, α/f^2 reaches a maximum. These effects can be explained in most instances by a model involving the equilibria between the two components (Andreae *et al* 1965). The frequency dependence is then related using a relaxation approach to the rate constants involved. The trends in the data for the lithium–ammonia solutions, in other words the onset of a frequency dependence at 9 mpm and a maximum in α/f^2 at 4 mpm, certainly indicate that we are looking at a mixture and thus lend considerable support to the Cohen–Jortner model.

References

Allgaier R S 1969 *Phys. Rev.* **185** 227
Andrea J H, Edmonds P D and McKellar J F 1965 *Acustica* **15** 74
Beyer R T and Letcher S V 1969 *Physical Acoustics* (New York: Academic Press)

Blandamer M J 1973 *Introduction to Chemical Ultrasonics* (New York: Academic Press)
Bowen D E 1975 *J. Phys. Chem.* 79 2895
Bowen D E and Priesand M A 1974 *Proc. Ultrasonic Symp., IEEE Cat. No.* 74, CHO 896-1SU 540
Chieux P and Sienko M J 1970 *J. Chem. Phys.* **53** 566
Cohen M H and Jortner J 1975 *J. Phys. Chem.* 79 2900
Cohen M H and Thompson J C 1968 *Adv. Phys.* **17** 857
Kim M G, Kemp K A and Letcher S V 1971 *J. Acoust. Soc. Am.* **49** 706
Morgan J A, Schroeder R L and Thompson J C 1965 *J. Chem. Phys.* **43** 4494
Mott N F 1974 *Metal–Insulator Transitions* (London: Taylor and Francis)
—— 1975 *J. Phys. Chem.* 79 2915
Nasby R D and Thompson J C 1968 *J. Chem. Phys.* **49** 969
Schettler P D and Patterson A 1964 *J. Phys. Chem.* **68** 2865
Webber G M B and Stephens R W B 1968 *Physical Acoustics IVB* ed W P Mason (New York: Academic Press) p 53

The metal–nonmetal transition in M–NH_3 solutions †

U Even‡, R D Swenumson§ and J C Thompson
The University of Texas at Austin, Austin, Texas 78712, USA

Abstract. The metal–nonmetal (MNM) transition in M–NH_3 solutions has been described in terms of an inhomogeneous regime by which the transition occurs continuously as the metallic concentration is changed. The presence of a phase separation between metallic and nonmetallic components in some M–NH_3 solutions provides an indication of the presence of inhomogeneities in those solutions. The apparent absence of a phase separation in Cs–NH_3 solutions suggests that a different mechanism for the MNM transition may exist there. Therefore, the trends of conductivity or other transport parameters through the transition in Cs–NH_3 may not be the same as in the other M–NH_3 solutions. We have measured the conductivity, density and Hall coefficient across the MNM transition in Cs–NH_3 and other solutions so that a comparison can be made and the role of inhomogeneities associated with phase separations determined. There are distinct differences between the solutions which may be correlated with the presence or absence of fluctuations.

1. Introduction

If one looks at the phase diagrams of M–NH_3 solutions it is immediately apparent that Cs–NH_3 solutions differ from the other alkalis in that there is no phase separation (Thompson 1976). This impression is confirmed by an analysis of the activities. One can calculate the mean square composition fluctuations from $(\partial(\ln a_M)/\partial x_M)^{-1}$, and there is no sign of the large fluctuations one commonly associates with phase separations. Note that there are stronger fluctuations in Na–NH_3 than in Li–NH_3.

Recent theories of electrical conduction by Cohen and Jortner (1975) have given a central role to composition fluctuations. They suggest that such fluctuations dominate the process by which the solution takes itself from the metallic phase which is found for metal concentrations above 10 MPM (mole % metal) to the solvated electron which we all know to exist below 1·0 MPM. If, indeed, fluctuations are important then Cs–NH_3 solutions clearly should not exhibit the same electrical transport properties in the 1–10 MPM range as do solutions of the lighter alkalis.

We have undertaken to attempt to evaluate this prediction (Mott 1975) and have measured conductivities and Hall coefficients in Li–, Na–, and Cs–NH_3 solutions at concentrations above 1·0 MPM and at temperatures above the consolute or freezing points.

† Supported in part by the US National Science Foundation and the R A Welch Foundation of Texas.
‡ R A Welch Foundation Post-doctoral fellow.
§ R A Welch Foundation Pre-doctoral fellow.

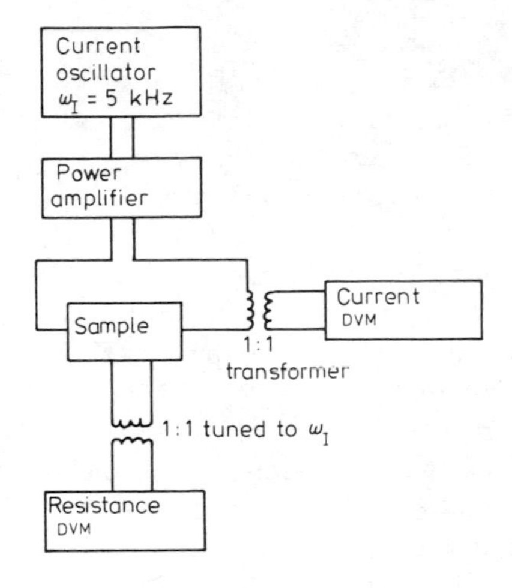

Figure 1. The conductivity measurement system.

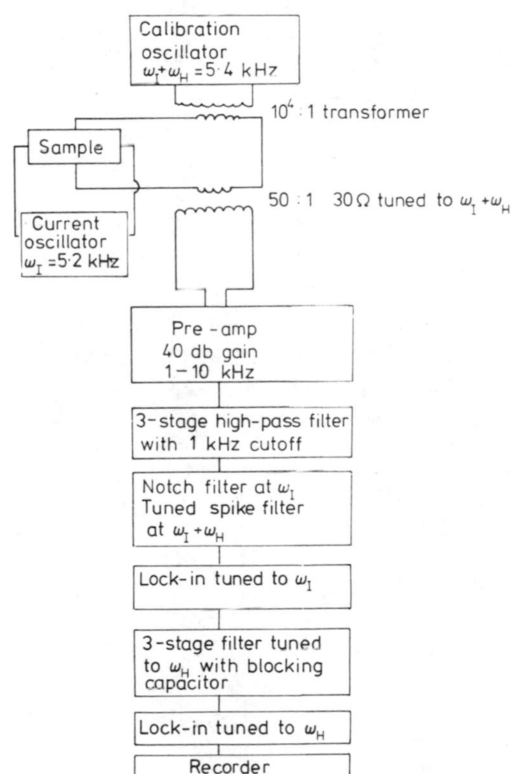

Figure 2. The Hall voltage measurement system.

2. Experimental

Our conductivity measurements employed conventional four-probe cells operating at 5·2 kHz (figure 1). The Hall effect was measured (figure 2) in a double-AC experiment with the magnetic field at 200 Hz and the current at 5·2 kHz. The Hall voltage appeared at 5·0 and 5·4 kHz. The Hall voltage at 5·4 kHz was measured using a pair of lock-in amplifiers in series, with one tuned to 5·2 kHz and the other to 200 Hz. The lock-in tuned to 5·2 kHz rejects the misalignment voltage at 5·2 kHz. A series of filters is employed to diminish other unwanted harmonics of ω_I and ω_H.

The importance of reducing IR drop between Hall electrodes cannot be overemphasized. Remarkable improvement was seen in one misaligned cell when the Hall electrodes were mechanically bent into alignment.

Severe electrode problems have plagued previous investigators. We have employed a variety of electrodes Au, W, Ag-plated W, Au-plated W, Cs-plated W (the latter for solutions containing no Cs). We find the bare W electrodes to yield large apparent Hall voltages and even sign reversals. The Au was dissolved from the electrodes in Cs solutions and the Cs was dissolved from the electrodes in Li solutions. In Hall effect measurements the fundamental problem is that any nonlinearity in either electrode or detection system can introduce a 'mixer' effect and produce a signal at the same frequency (though possibly different phase) as the Hall voltage.

Each solute shows electrode effects in R_H but none in conductivity measurements. In Li–NH_3 the value of R_H decreases by two orders of magnitude as the electrode is changed from 'clean' W to W coated with Cs, Ag or Au. Furthermore, solutions of the lighter alkalis exhibit a reversal in the sign of R_H as the concentration is reduced below 2·5 MPM with noble metal coated W electrodes. With Ag-plated W electrodes, the application of a steady state (DC) current between one Hall (+) and one current electrode (–), of several mA for periods of several minutes would cause a further irreversible reduction of the Hall voltage. We took this as an indication of a film coating (possibly oxide) on the Hall electrode and rejected data obtained with such electrodes. It should be emphasized that our success with a given electrode depended upon the solute. Clean W was sufficient in Cs solutions; we presume that Cs is sufficiently reactive to clean the electrode. Ag- or Au-plated W was sufficient in Li–NH_3 solutions, but only pure Au electrodes were totally satisfactory in Na–NH_3 solutions, as well as in Li solutions.

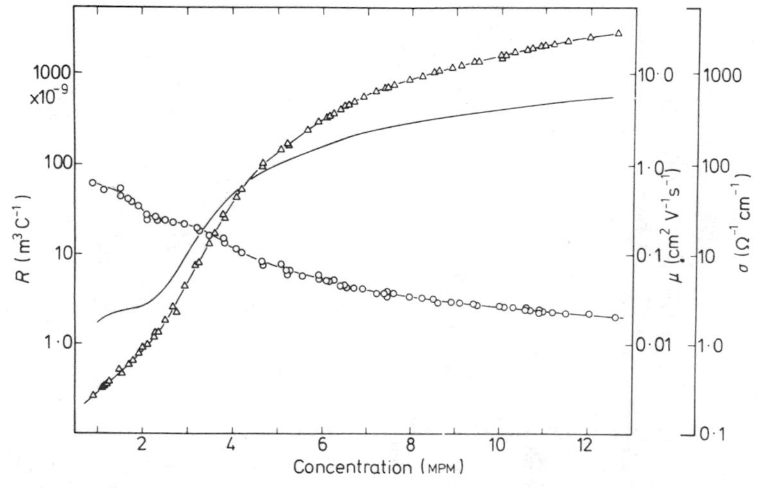

Figure 3. The transport properties at –55 °C for Li–NH_3 solutions with triangles for conductivity, open circles for Hall coefficient, and the solid line for mobility.

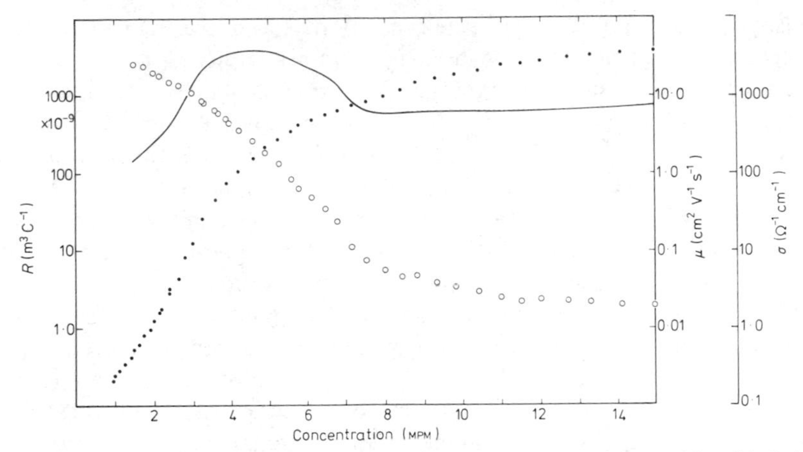

Figure 4. The transport properties at −33 °C for Na–NH_3 solutions with solid circles for conductivity, open circles for Hall coefficient, and the solid line for mobility.

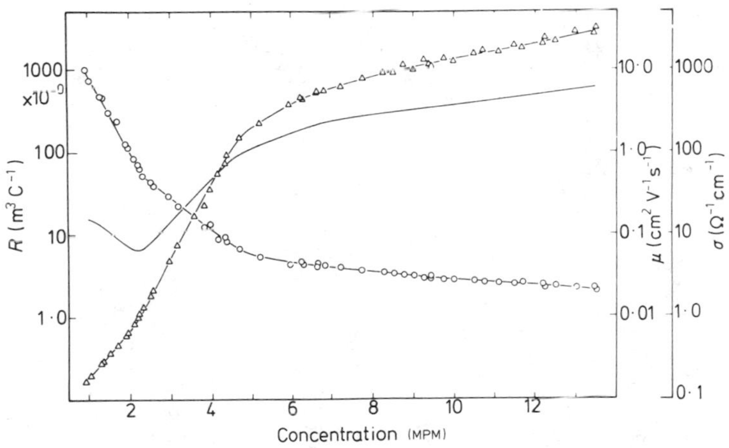

Figure 5. Shown are the transport properties at −55 °C for Cs–NH_3 solutions with triangles for conductivity, open circles for Hall coefficient, and the solid line for mobility.

We have chosen to regard the noble metal electrode results as 'real' on the basis of the following criteria. (i) The results are more reproducible from run-to-run and cell-to-cell with Ag or particularly Au electrodes. (ii) Values of V_H are linear in current and field. (iii) Values of V_H do *not* change following prolonged operation of the cell with large DC currents between the Hall electrodes. (iv) There is no variation of V_H during the course of a run, if a given concentration is reproduced. (v) Microscopic examination of the electrodes following an experiment failed to reveal any contamination or etching.

All data were taken at the consolute point plus 10 degrees, except the Cs–NH_3 which was taken at −55 °C.

3. Results and discussion

We have obtained conductivity and Hall coefficient data on solutions of Li, Na, or Cs in NH_3 as well as density data for Cs–NH_3 solutions. For the present, the importance of the density data is that differences in concentration dependence among the solutions are not altered if one uses mole fraction instead of number density.

Figure 3 shows the results for Li (we have perhaps three times as much data as is shown). We see that R_H is at $10^{-6}\,m^3\,C^{-1}$ at the lowest concentration, while the mobility is at $0{\cdot}1\ cm^2\,V^{-2}\,s^{-1}$. Our conductivity results are consistent with previous workers. Our R_H is well above Nasby's lone data point (Thompson 1976) below 2 MPM. Note the minimum in μ_H at 2·3 MPM. Figure 4 shows comparable results for Na–NH_3 solutions. In this system R_H is higher and there is a hump in μ_H centred at 4 MPM. Finally in figure 5 we display the results for Cs–NH_3 solutions. The observed R_H is only $100 \times 10^{-9}\,m^3\,C^{-1}$ and the limiting mobility is $0{\cdot}01\,cm^2\,V^{-1}\,s^{-1}$. For both Na and Cs our conductivity data are in complete agreement with previous workers.

These results are compared in the next three figures. In figure 6 we see the conductivities are essentially the same. Note that the conductivity falls below $100\ \Omega^{-1}\ cm^{-1}$ near 4 MPM and since this is the minimum metallic conductivity, the MNM transition would be placed near 4 MPM on such a model. The conductivity activation energy is

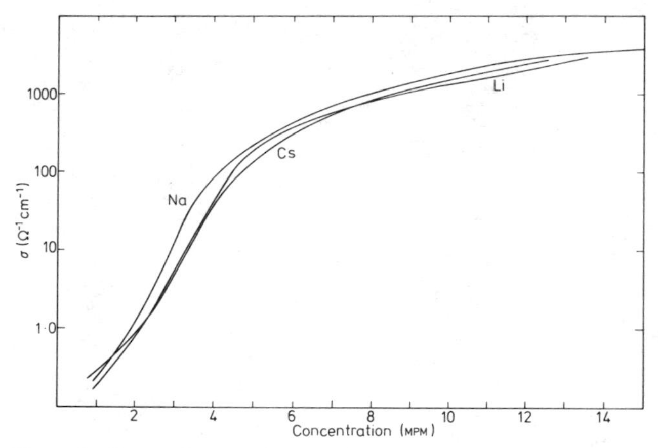

Figure 6. Comparison of the conductivities for Li, Na and Cs systems.

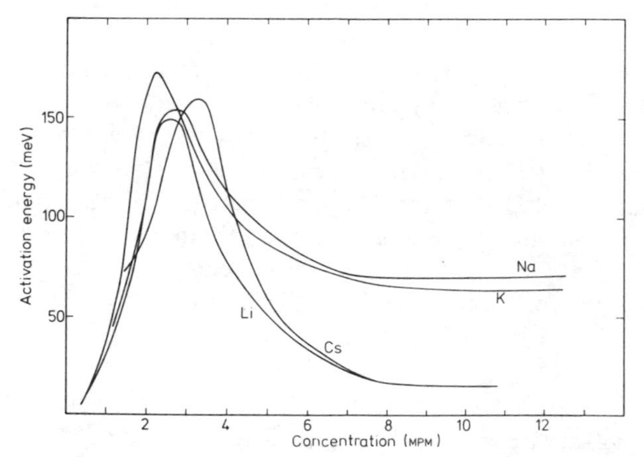

Figure 7. Activation energies between −30 °C and −50 °C for Li, Na, K and Cs systems.

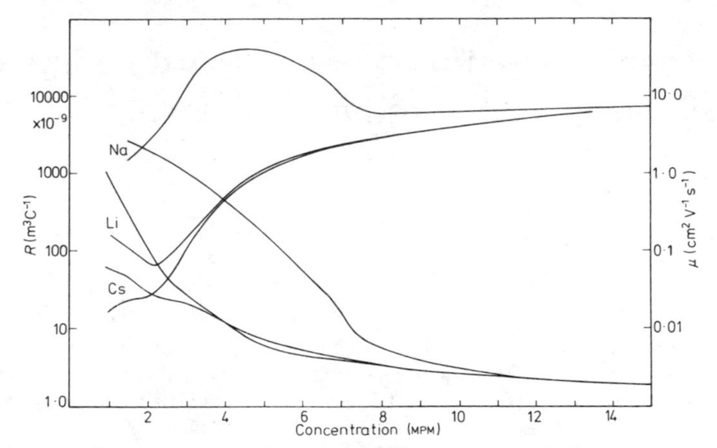

Figure 8. Comparisons of the Hall coefficient and mobilities for Li, Na and Cs systems.

shown in figure 7, where some differences may be seen for both K and Cs solutions. (Some of these latter data are not ours.)

The Hall coefficient and mobilities are shown in figure 8. First, let us look at the general features. We find the Hall coefficient moves above the free-electron value near 7 MPM for Li, 11 MPM for Na, and 10 MPM for Cs though the trends in R_H are similar at lower concentrations. The mobility curve for the Na–NH_3 solutions is reminiscent of the observation of a peak in the mobility of excess electrons in Ar near the critical point (Jahnke *et al* 1970). There are clear differences between Na–NH_3 solutions and the others. It is precisely in the Na–NH_3 solutions that the chemical potentials may be used to predict the presence of substantial composition fluctuations.

The efficacy of percolation processes is difficult to assess (Cohen and Jortner 1975). We have performed a computer simulation of R_H in an inhomogeneous material (Swenumson and Thompson 1976) and such a simulation will fit the Li–NH_3 results but not the Na–NH_3. We also note that there are changes in behaviour near 2 MPM, below which both Cohen and Jortner (1975) and Mott (1974, 1975) suggest that a new conduction process, perhaps semiconducting, sets in. Further measurements are in progress.

References

Cohen M H and Jortner J 1975 *J. Phys. Chem.* **79** 2900
Jahnke J A, Meyer L and Rice S A 1970 *Phys. Rev.* **A3** 734
Mott N F 1974 *Metal–Insulator Transitions* (London: Taylor and Francis)
—— 1975 *J. Phys. Chem.* **79** 2915
Swenumson R D and Thompson J C 1976 *Phys. Rev.* **B14** in press
Thompson J C 1976 *Electrons in Liquid Ammonia* (Oxford: Clarendon Press)

On the structure of metal/inert element binary alloys in which a metal–nonmetal transition is observed

J J Quinn and J G Wright
School of Mathematics and Physics, University of East Anglia, Norwich, NR4 7TJ, England

Abstract. Thin films of lead–xenon, lead–argon and cobalt–argon having various compositions have been prepared in high vacuum at 4 K and their structures examined by *in situ* scanning electron diffraction. The lead–xenon and cobalt–argon systems have been found to be two phase, which is in agreement with the observed percolation-type resistivity against composition curve. The lead–argon system appears to be single phase and exhibits a structural change from amorphous, throughout the conducting regime, to FCC, throughout the insulating regime. The abrupt change in structure is coincident with the metal–nonmetal transition and is further evidence that the transition is Mott-like. Some general observations about the nature of metal/inert element systems are made.

1. Introduction

During the past six years, a number of experiments have been reported on alloy systems composed of an inert element and a metal, where a metal–nonmetal transition has been observed. So far, transitions have been observed in Na–Ar (Cate *et al* 1970), Hg–Xe (Raz *et al* 1972), Cu–Ar (Endo *et al* 1973), Rb–Kr and Cs–Xe (Phelps *et al* 1975), Pb–Ar (Eatah *et al* 1975), Fe–Xe (Shanfield *et al* 1975), Pb–Xe (Eatah 1975, Hilder 1976 private communication), In–Ar (G Hilder 1976 private communication) and Ga–Xe, Cu–Xe and Ag–Xe (R Ryberg and O Hunderi 1976 private communication). The results from resistivity studies show that two forms of transition appear to occur as shown schematically in figure 1.

The resistivity against composition curve of the type A in figure 1 is consistent with percolation theory (Phelps *et al* 1975) and is observed in Pb–Xe, Cs–Xe, Rb–Kr and Fe–Xe whilst that of type B which gives a sharper boundary between the conducting and non-conducting regimes is observed in Na–Ar, Cu–Ar, Pb–Ar, In–Ar and possibly Hg–Xe. This latter type of transition has been discussed from the metallic point of view, in terms of a Mott–Hubbard transition (Berggren *et al* 1974, Mott 1974) and from the dielectric point of view in terms of a classical dielectric catastrophe (Berggren 1974). In order to understand more fully the nature of the transitions a programme of structure studies has been commenced, the preliminary results of which are discussed here.

2. Experimental procedure

The experiments have been performed using a UHV scanning electron diffraction system described previously by Leung and Wright (1974). In the present studies it has been

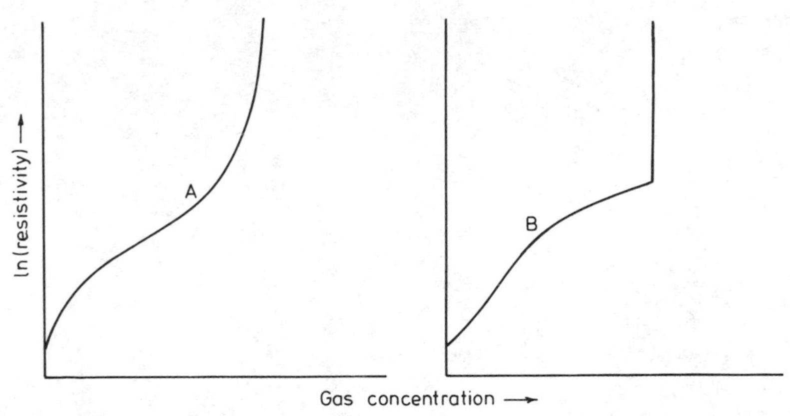

Figure 1. Schematic representation of the two basic types of metal–nonmetal transition observed in metal/inert element systems.

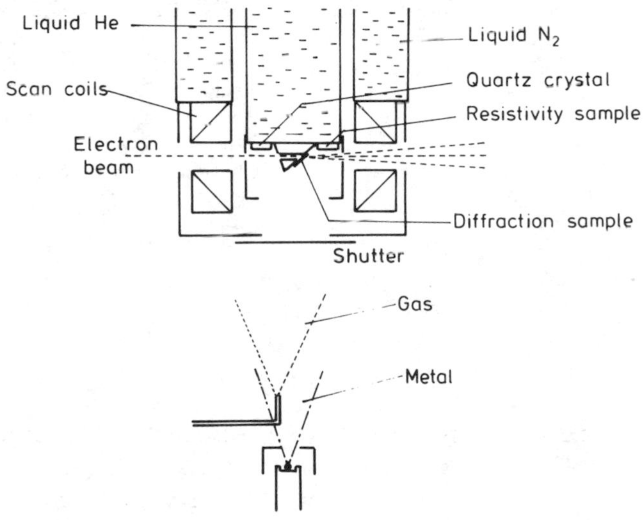

Figure 2. The deposition and sample area of the scanning electron diffraction system.

operated in the HV region giving a trapped background impurity level in the samples of no more than a few per cent. The system has been modified to allow codeposition of the gas from an injection nozzle and the metal from an electron-beam evaporator. The diffraction substrates (of formvar) are mounted on the bottom of a cryostat that is cooled by liquid helium, together with a sapphire substrate for resistivity measurements and an oscillating quartz crystal for measurement of deposition rates (and consequently composition). The essential features of the sample and deposition areas of the apparatus are shown in figure 2. Diffraction samples were typically 150 Å thick. The work to date has been mainly concerned with Pb–Xe and Pb–Ar systems which, as far as the resistivity measurements are concerned, represent the two classes of metal–nonmetal transitions that are observed in metal/inert element alloys. Some studies on the Co–Ar system have also been made.

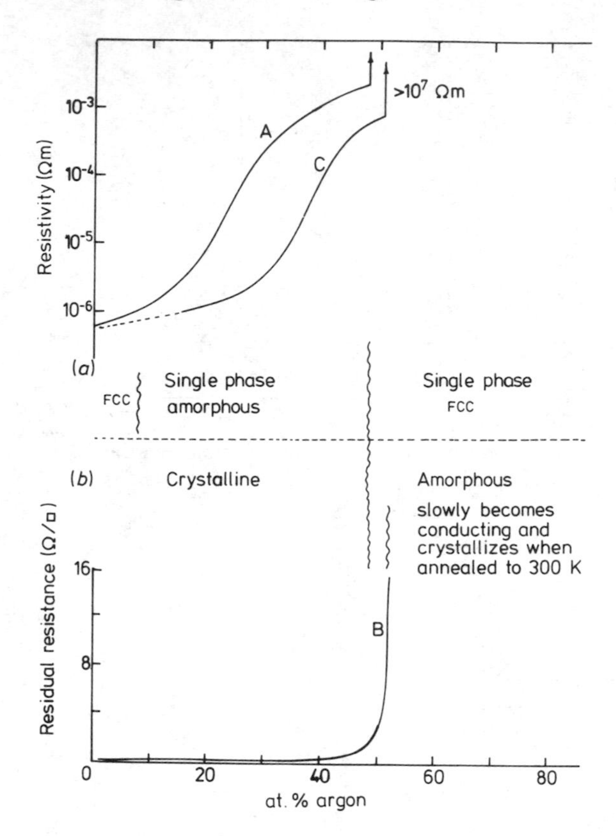

Figure 3. The correlation between resistivity in the as-deposited samples (*a*), resistance of the residual lead deposit after re-evaporation of the gas (*b*), and the structural observations in lead–argon: A, from Eatah *et al* (1975); B, from Eatah 1975; C, from G Hilder (1976 private communication).

3. Results

3.1. Lead–argon

Figure 3 shows the nature of the structural results obtained from Pb–Ar in comparison with resistivity and residual (after the gas has re-evaporated) resistivity measurements made by Eatah *et al* 1975, Eatah 1975 and G Hilder (1976 private communication). The following are the major points:

(i) As the concentration of argon is increased from zero, the superconducting transition temperature in the lead is quickly reduced (Eatah 1975) and the samples lose their crystalline structure and become amorphous. The amorphous structure persists in all samples until the metal–nonmetal transition composition is reached.
(ii) On the nonmetal side of the transition a single-phase FCC diffraction pattern is observed (figure 4, curve A). Since the lattice parameters of lead and argon are very close, this could be thought to be due to a two-phase mixture. However, near to the transition, the lattice parameter is measurably less than either lead or argon and it is certainly the case that a two-phase mixture could readily be detected, as shown in figure 4, curve B, which shows the diffraction profile obtained from a layer of argon on top of a lead layer.
(iii) On heating the insulating samples, the argon re-evaporated at about 35 K. During this process the crystalline diffraction profile disappears. Once the re-evaporation is

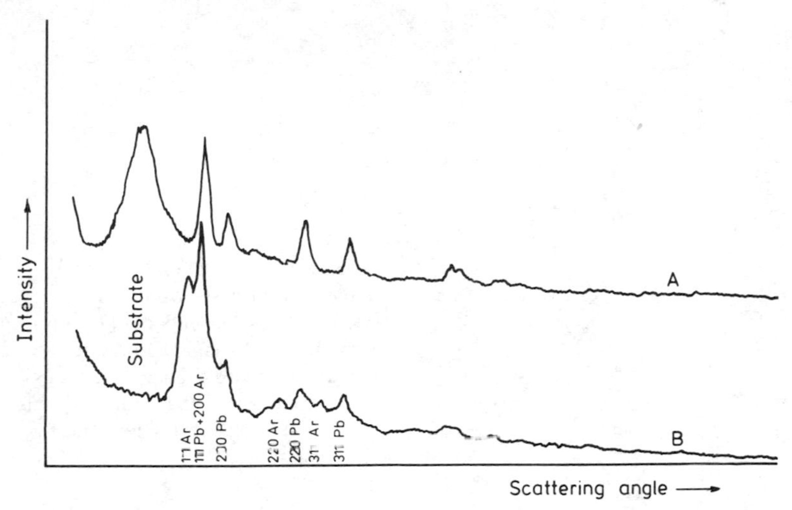

Figure 4. Diffraction profiles due to 40 kV elastically scattered electrons from: A, a film of 42 at.% Pb/58 at.% Ar co-deposited to form a single-phase FCC lattice and B, a film of argon formed on a film of lead (32 at.% Pb). Both profiles contain a contribution from the form varsubstrate which is larger in A due to a thinner deposit.

complete the observed profile is that of an amorphous residue and indicates a mean interatomic separation slightly larger than expected for lead. The samples are non-conducting in agreement with the earlier observations of Eatah (1975). On further annealing to room temperature and above, the mean interatomic separation becomes that which is expected for lead and the samples slowly become conducting. Only at temperatures above room temperature does the amorphous lead crystallize.
(iv) When the gas evaporates from the samples which are initially conducting, that is those which have a lead concentration greater than 50 at.%, the lead immediately forms a crystalline deposit with a proper metallic conductivity.

3.2. Lead–xenon

In the case of lead–xenon the deposit appears to consist of very small crystallites of both materials in all compositions. This deduction is based upon the appearance of only two main peaks characteristic of Xe–Xe and Pb–Pb and the absence of any Pb–Xe peak (which would be expected if the sample were single-phase amorphous), together with signs of peak formation for scattering from the 220 and 113 + 222 planes of the expected FCC structures. The general features of the system are shown in figure 5.

3.3. Cobalt–argon

The cobalt–argon system shows a very high degree of phase separation forming FCC argon and amorphous cobalt. The resistivity measurements indicate a percolation-like transition similar to that observed in lead–xenon, but with the sample becoming nonmetallic at about 75 at.% argon.

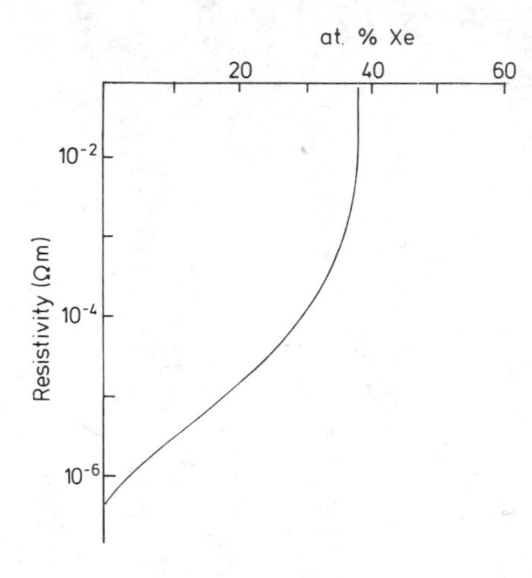

Figure 5. The correlation between resistivity in the as-deposited samples, from Eatah (1975) and G Hilder (1976 private communication), and the structural observations in lead–xenon. The lead–xenon system is two-phase microcrystalline, with a conducting lead residue after the xenon evaporates.

4. Discussion

4.1. Lead–argon

The structural results observed for lead–argon are interpreted as being due to a single-phase system at all compositions. The amorphous structure observed on the conducting side of the transition is consistent with the view that metallic films formed by low-temperature condensation form an amorphous structure when significant levels of impurities are included, provided that phase separation does not take place. When the gas leaves the metal, the observed crystallization is consistent with a single-phase system but inconsistent with a two-phase system. If the system had been two-phase it is not likely that the lead would undergo atomic rearrangement when the argon re-evaporated. Furthermore, it is unlikely that if phase separation had occurred, anything other than crystalline lead would have been observed in the as-deposited state.

On the non-conducting side of the transition the observation (iii) in §3.1 is consistent with a single-phase structure which disintegrates when the argon leaves. The lead atoms are released, from what must be a Van der Waals' bound solid, but appear to be unable to form a metal until temperatures greater than 250 K are reached. This would suggest that as metal atoms at low temperatures are moved together, they are unable to form a metal until some activation energy is overcome.

The nature of the structure change in lead–argon as the composition is varied suggests that the metal–nonmetal transition is not inconsistent with the Mott–Hubbard type. Once the electron density falls below a critical value, the system consists of a collection of neutral atoms of nearly similar size, with the consequence that a Van der Waals' FCC solid forms. If the electron density is sufficiently high to allow conductivity, the samples should be regarded as consisting of amorphous metallic lead containing a very high level of impurity in the form of argon atoms.

4.2. Lead–xenon and cobalt–argon

The structure of these systems indicate two-phase systems which would be entirely consistent with the percolation-like resistivity observations.

5. Conclusions

The most exciting aspects of metal–nonmetal transitions in metal/inert element alloys are concerned with systems where the transition is discontinuous. The preliminary results reported in this paper would indicate that the transition could be of a Mott–Hubbard type in keeping with the theoretical deductions of Berggren *et al* (1974). It seems probable that this type of transition is seen in systems where the inert element is argon or possibly neon, but not the heavier inert elements which are more likely to cluster on account of their deeper pair potentials and consequently more likely to cause phase separation. However, the cobalt results, which are the first to report a well defined percolation type of transition in argon, suggest that the range of metals which may be used may also be limited to perhaps those whose energy of condensation is not much more than that of copper.

Acknowledgments

The authors would like to thank the SRC for financial support for this work. They would also like to thank Professor N E Cusack for valuable discussions.

References

Berggren K F 1974 *J. Chem. Phys.* 9 3399–402
Berggren K F, Martino F and Lindell G 1974 *Phys. Rev.* **B9** 4092–102
Cate R C, Wright J G and Cusack N E 1970 *Phys. Lett.* **A32** 467–8
Eatah A I 1975 *PhD Thesis* University of East Anglia
Eatah A I, Cusack N E and Wright J G 1975 *Phys. Lett.* **51A** 149–50
Endo H, Eatah A I, Wright J G and Cusack N E 1973 *J. Phys. Soc. Japan* **34** 666–71
Leung P K and Wright J G 1974 *Phil. Mag.* **30** 185–94
Mott N F 1974 *Metal–Insulator Transitions* (London: Taylor and Francis)
Phelps D J, Avci R and Flynn C P 1975 *Phys. Rev. Lett.* **34** 23–6
Raz B, Gedanken A, Even U and Jortner J 1972 *Phys. Rev. Lett.* **26** 1643–6
Shanfield Z, Montano P A and Barrett P H 1975 *Phys. Rev. Lett.* **35** 1789–92

NMR studies of liquid semiconductors and the metal–nonmetal transition

W W Warren Jr
Bell Laboratories, Inc, Murray Hill, New Jersey 07974, USA

Abstract. Several classes of metallic liquids can transform to a semiconducting or insulating state under changes of an appropriate thermodynamic variable. Familiar examples are the semiconducting liquid alloy systems, for which composition and temperature are the relevant parameters, and certain elemental liquids whose properties are radically altered with changes in density or temperature. Nuclear magnetic resonance (NMR) provides one of the few *microscopic* probes which is widely applicable and compatible with the difficult experimental conditions under which these transitions often occur. In this paper selected experimental results are reviewed with special emphasis on their relationship with some issues of importance for current theoretical models: covalency and ionicity in semiconducting liquid alloys; the relationship between chemical bonding and liquid structure; microscopic inhomogeneity and fluctuations; the strong-scattering description of electron transport ('random-phase model'); and localization and the 'mobility edge'. The evidence presently available tends to support some current theoretical concepts in certain cases, e.g. the Mott pseudogap model for liquid Te alloys, while demonstrating that no generally valid model of the metal–nonmetal transition is yet at hand.

1. Introduction

The properties of liquid semiconductors are intimately related to the existence and nature of the metal–nonmetal (MNM) transition. Most liquids exhibiting semiconductor-like electronic properties are found to transform continuously to a metallic state under changes of an appropriate thermodynamic variable such as temperature or composition. Conversely, it is obvious that, at least in principle, any metallic liquid can be induced into a semiconducting or insulating state by expansion to very low density or by alloying with a nonmetal. Consequently, a successful theory or model of semiconducting liquids must explain not only the properties of the semiconducting state, but must also account for the transition to a metallic condition. Similarly, the most illuminating experiments seem to be those which extend beyond the semiconducting state and through the MNM transition.

There are several distinct groups of liquid systems which exhibit semiconducting properties and a MNM transition:

(i) semiconducting liquid alloys (Tl–Te, Ga–Te, Mg–Bi, etc)
(ii) metal–molten-salt solutions (Na–NaCl, Bi–BiI_3, etc)
(iii) metal–ammonia solutions (Na–NH_3, Cs–NH_3, etc)
(iv) elemental liquid semiconductor (Se)
(v) expanded liquid metals (Hg, Cs, Se).

Much recent activity on these systems has centred on development and testing of simple physical models of the electronic structure and properties. Major efforts at model development have taken such diverse starting points as conventional semiconductor band theory (Cutler 1971), the nearly free-electron (NFE) approximation (Enderby and Collings 1970, Enderby 1974), the disorder-induced pseudogap and electronic localization (Mott 1971), and classical percolation theory (Cohen and Jortner 1973a, b, 1974). While it is inappropriate to discuss these models in detail here, it is possible to pose certain broad questions whose answers bear on the validity of the theoretical models. For example, how do the properties of semiconducting liquid alloys depend on formation of chemical bonds and on the relative ionicity and covalency of those bonds? How does chemical bonding affect the liquid structure? Are microscopic inhomogeneity and fluctuation phenomena of predominant importance for some systems? What are the limits of validity for the NFE description and what is the nature of electronic transport outside this regime? Finally what is the mechanism of electron localization and how valid is the notion of a mobility edge?

Nuclear magnetic resonance (NMR) has proven a useful experimental tool for providing at least partial answers to these questions. It has now been applied to all of the classes of MNM transition listed above, although only very recent and incomplete data are available for metal–molten-salt solutions (Dupree and Warren 1977), liquid Se (Seymour and Brown 1973, W W Warren and R Dupree unpublished) and expanded metals (El-Hanany and Warren 1975). The advantages of the NMR method are that it is a microscopic probe sensitive to the local atomic environment, it is selective with respect to atomic species, it can be applied to a wide variety of elemental constituents, and NMR experiments are feasible at high temperatures and pressures with chemically reactive sample materials.

This paper will review the available NMR data with particular emphasis on their interpretation and relevance to the questions posed above. Selected data will be discussed for all the system classes listed except metal–ammonia solutions. Recent summaries of results for those systems have been given, for example, by Thompson (1973) and Lelieur (1973). Following a brief review of the NMR method and experimental principles, the main part of this paper is divided into two sections dealing, first, with liquid structure and chemical bonding and then with electron transport. A short summary of the principal conclusions is presented in the final section.

2. Basic principles and interpretation of NMR

Nuclear magnetic resonance is the spectroscopy of the Zeeman energy levels of an assembly of nuclear magnetic moments in a static magnetic field H_0. For typical laboratory fields ($H_0 \sim 10$ kOe) the magnetic splittings fall roughly in the range 1–40 MHz and are thus accessible with radio-frequency radiation. In a standard experiment the sample is placed in a tuned coil oriented at right angles to the static field. When the coil is excited by RF current one can detect either an absorption of RF energy or a signal induced by the precessing nuclear magnetization. The response is maximum when the field and applied RF frequency ω are related by the resonance condition $\omega = \omega_0 = \gamma_n H_0$ where γ_n is the nuclear gyromagnetic ratio†.

There are several parameters characterizing the resonance effect whose measurement provides the raw experimental data. In the case of liquid metals and semiconductors, the most important of these are the centre frequency of the resonance, which is usually shifted slightly from the value $\gamma_n H_0$, and the spin-lattice relaxation time T_1. The latter is the time constant for approach to equilibrium of the nuclear spin system and its thermal environment. In liquids it is often possible to determine T_1 from the absorption (or dispersion) lineshape but, in general, this quantity must be measured directly using transient NMR techniques (Abragam 1961).

The value of NMR in the study of condensed matter derives from the sensitivity of the measurable parameters to the local electrical and magnetic environment. In the case of metals and semiconductors, two specific interactions are usually of predominant importance. The first of these, the magnetic-contact hyperfine interaction, couples the nuclear and electronic magnetic dipole moments (μ_n and μ_e) according to

$$H_c = \frac{8\pi}{3} \mu_n \cdot \mu_e \delta(\mathbf{r}_n) \tag{1}$$

where $\mathbf{r}_n$ is the nuclear position. The static or time-averaged effect of this interaction is a shift of the resonant frequency proportional to the electronic magnetization $\langle \mu_e^z \rangle$. In metals this is the usual Knight shift and $\langle \mu_e^z \rangle / H_0$ is the Pauli paramagnetic susceptibility. In addition to the shift, time-dependent fluctuations in the local magnetic field due, say, to translational movement of the electrons, can induce mutual nuclear–electronic spin flips. This provides a mechanism for spin-lattice relaxation.

The second important interaction is the nuclear–electric quadrupole interaction which couples the nuclear quadrupole moment to the gradient of the local electric field. This interaction vanishes if the local environment has cubic or higher symmetry. Normally the rapid atomic motion in a liquid leads to time-averaged spherical symmetry and hence to the absence of a quadrupolar shift. However, fluctuations of the electric field gradient can cause nuclear spin flips giving a quadrupolar spin-lattice relaxation process. In general it is difficult to establish whether an observed relaxation time is determined by a magnetic or a quadrupolar process. However, there are special cases for which the spin-lattice relaxation time can be measured separately for two isotopes of the same element, for example ^{69}Ga and ^{71}Ga. Since the ratio of the relaxation times for the two isotopes depends on the type of process, these measurements can be used to identify the relaxation mechanism or to determine the separate contributions if both are present (Narath and Alderman 1966, Cornell 1967). The availability of two isotopes is an important consideration in choosing appropriate systems for NMR studies.

3. Chemical bonding and structure

Formation of chemical bonds and the associated structural changes are of fundamental importance for the transition to a semiconducting state in many systems. The most

† These remarks can serve as only the briefest introduction to the subject of NMR. For more complete background the reader is referred to one of several general references such as Abragam (1961) or Slichter (1963).

familiar evidence for this is probably provided by the semiconducting liquid alloys for which plots of conductivity against composition exhibit sharp minima at simple stoichiometries (see, for example, Enderby 1974). These and related anomalies in other electronic properties normally occur at the concentration corresponding to the normal ionic valence states of the constituents, for instance Mg_3Bi_2, Tl_2Te, In_2Te_3, etc. Large negative heats of mixing (Nakamura and Shimoji 1971, Maekawa *et al* 1973) and maxima in the viscosity (Glazov *et al* 1969) attest to the presence of strong attractive interactions between unlike atoms while small values of the concentration–concentration correlation function at $k = 0$, $S_{cc}(0)$, indicate that fluctuations from the average composition are minimized at the stoichiometric compositions (Bhatia and Hargrove 1974, Thompson *et al* 1976).

The transition to semiconducting-like electronic properties in these alloy systems is best understood as a tendency for a gap to open in the electronic density of states as a result of chemical bond formation. Similarly, liquid Se and S are nonmetallic near the melting points because of well developed molecular species such as Se_8 and polymeric chains. One possible indication of the degree of perfection of this bonding is provided by the density of states at the Fermi level, $N(E_F)$, since this should be essentially zero in the gap for completely saturated bonds while approaching the free-electron value in the metallic regime. In this connection, NMR measurements are of interest since the Knight shift provides a measure of $N(E_F)$ and, hence, the bond perfection.

3.1. *The Knight shift*

The Knight shift for nuclei of element A can be expressed quite generally in the following form (Slichter 1963):

$$K_A = \frac{1}{N_A}\frac{8\pi}{3}\sum_{i=1}^{N_A}\sum_{\nu} |\psi_\nu(r_i^A)|^2 \chi_\nu \tag{2}$$

where N_A is the number of atoms of element A in the system, $|\psi_\nu(r_i^A)|^2$ is the probability amplitude for the νth eigenstate at atomic position r_i^A, and χ_ν is the susceptibility of the state ν. The shift can be nonzero only for singly occupied states ($\chi_\nu \neq 0$) having s-character at the nuclear positions. Closed-shell ionic configurations or covalent bonds do not give a Knight-shift contribution. It should be noted that there may be additional shift contributions due to p-electrons and the orbital effects of both filled and unfilled states. In most cases not involving transition metals these terms are substantially smaller than equation (2). Only the filled-state orbital contribution (chemical shift) is present in fully bonded insulators.

For a system of non-interacting, degenerate electrons equation (2) reduces to the standard expression

$$K_A = \frac{16\pi}{3}\mu_\beta^2 \overline{\langle|\psi(r^A)|^2\rangle_F} N_s^A(E_F) \tag{3}$$

where μ_β is the Bohr magneton, $\langle\ \rangle_F$ denotes an average over all states in an energy interval kT at the Fermi level, the bar denotes an average over all A-atom sites and $N_s^A(E_F)$ is the density of states at the Fermi level having s-character at A-atom sites.

The usual derivation of equation (3) tacitly assumes microscopic homogeneity so that $|\psi_\nu(r_i^A)|^2$ and χ_ν can be separately averaged over all states and sites. The situation is quite different for an inhomogeneous system. As an extreme case, consider a liquid consisting of small regions of which some are metallic and others nonmetallic. If these regions are sufficiently small, a given nucleus can diffuse rapidly from one region to another and the average shift becomes

$$K_A = CK_A^{metal} \tag{4}$$

where C is the fraction of A atoms located in metallic regions (Cohen and Jortner (1974).

For elemental liquid semiconductors and stoichiometric binary alloys near the melting point, thermodynamic evidence favours microscopic homogeneity (Thompson *et al* 1976). The range of Knight shifts and nature of bonding for such systems can be conveniently summarized by a plot such as that shown in figure 1. Here the Knight shifts, normalized to their values in the pure elemental liquid metals, are plotted against the electronegativity difference ΔE (Pauling 1960). The philosophy of this plot is that essentially two parameters characterize the chemical bonding in a liquid semiconductor. One is the degree of bond perfection, or conversely, the degree of metallization which is roughly characterized by K/K_{metal}. The second parameter must describe the relative ionic or covalent character of the bond; the electronegativity difference is one of a number of ways of representing the ionicity.

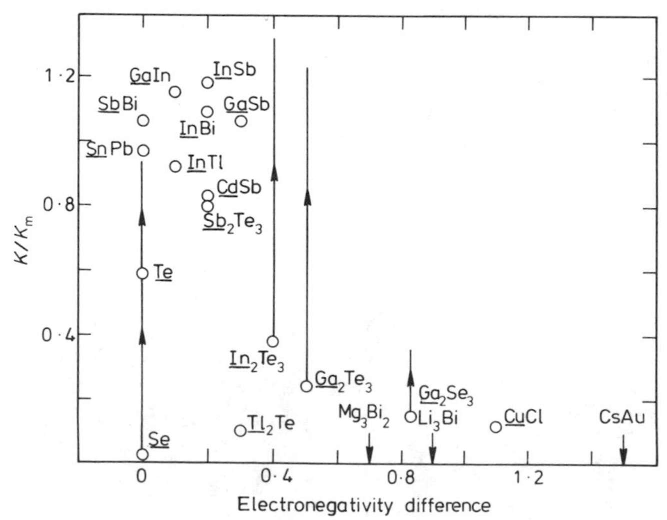

Figure 1. Relative Knight shifts K/K_m near the melting point against electronegativity difference ΔE for the more electropositive element (underlined) in stoichiometric binary liquid alloys. The normalization factor K_m is the shift in the pure liquid metal. Also shown are shifts for elemental liquid Te and Se. Vertical lines indicate the range of K/K_m achieved on heating. For reference, the ΔE values are indicated for three important liquids whose shifts have not yet been measured (Mg_3Bi_2, Li_3Bi, and CsAu). Knight-shift data are taken from Moulson and Seymour (1967), Kerlin and Clark (1975), Warren and Clark (1969), Warren (1971, 1972a, b, 1973 unpublished).

A plot such as figure 1 emphasizes the fact that the nonmetallic liquids span a continuous range of ionicity from fully covalent Se to the molten halides. Most of the well studied Te alloys lie in the highly covalent region. A second important point is that the more covalent liquids can be driven into the metallic region with sufficient increase in temperature whereas this becomes increasingly difficult with greater ionicity. In other words, covalent bonds and conduction electrons can coexist as in the Cabane and Friedel (1971) model of liquid Te. Well developed negative ions, on the other hand, locally tend to exclude conduction electrons (Flynn 1974) leading to severe modulation of the conduction-electron charge density on the scale of interatomic distances. It is interesting that attempts to bridge the 'forbidden region' of figure 1 (i.e. large shift and large ΔE) by alloying a metal and a molten salt usually produce large regions of liquid–liquid phase separation (Bredig 1964). Even outside the region of macroscopic phase separation there is evidence of severe microscopic inhomogeneity in more ionic alloys (Dupree and Warren 1977).

We turn now to consider the Knight-shift behaviour of some specific systems in somewhat more detail.

3.1.1. Selenium. Se is a purely covalent liquid semiconductor with a high degree of bond satisfaction near the melting point. The structure is believed to consist of linear chains containing up to about 10^3 atoms (Eisenberg and Tobolsky 1960). The chains are terminated by broken bonds which provide well localized paramagnetic centres (Koningsberger *et al* 1971). The semiconducting properties result from the gap between the filled non-bonding π-states and the empty antibonding states, analogous to crystalline Se. As the temperature is raised, the chains rupture and increasing numbers of dangling bonds are created until eventually Se becomes metallic as the critical temperature is approached (Hoshino *et al* 1976).

There is no observable Knight shift in liquid Se near the melting point. A small negative chemical shift was reported by Seymour and Brown (1973). The main effect of the paramagnetic centres is to provide a source of spin-lattice relaxation (Brown *et al* 1972, Seymour and Brown 1973). Recent experiments to much higher temperatures (Warren W W and Dupree R unpublished) show the gradual development of a Knight shift (figure 2). At the highest temperature reached, the shift value is roughly one-half those observed for other liquid metals in the same part of the periodic table. Thus the high degree of bonding in two-fold coordination is largely destroyed on heating from T_m to temperatures greater than about 1300 °C. The situation in this temperature range is therefore similar to that of liquid Te in which an appreciable number of free carriers apparently coexist with residual two-fold covalent bonding (Cabane and Friedel 1971). The high-temperature Knight shift in Se is comparable with that of Te (Warren 1972b) when one takes into account the difference in s-electron hyperfine coupling.

3.1.2. Gallium–tellurium. Of the more covalent binary alloys, the most complete Knight-shift data are available for Ga–Te (Warren 1972a). The covalent character of this system is emphasized by a relatively small value of ΔE and by the crystal structure of the stoichiometric compound Ga_2Te_3. This material and the closely related In_2Te_3 crystallize in a defect ZnS structure having one-third of the metal sites unoccupied. These crystals are therefore very similar to those of the III–V compounds such as InSb or GaAs.

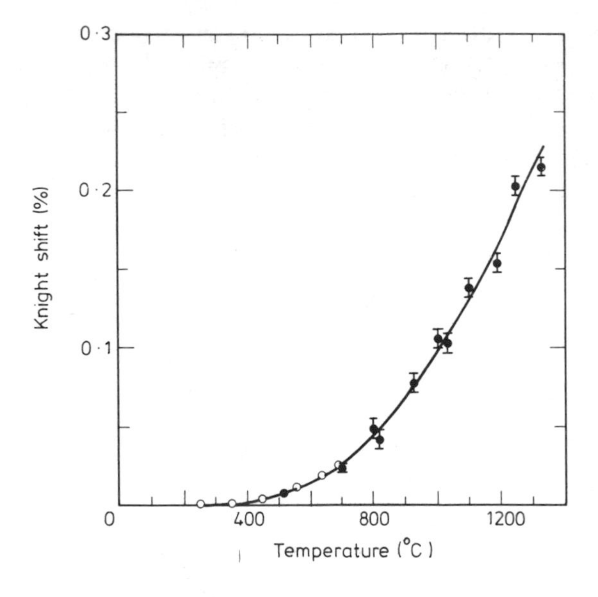

Figure 2. Knight shift against temperature for elemental liquid ^{77}Se (W W Warren and R Dupree unpublished). Data below the normal boiling point (open points) were obtained with a sample sealed in a quartz ampoule; higher temperature data (full points) were obtained with an internally heated autoclave.

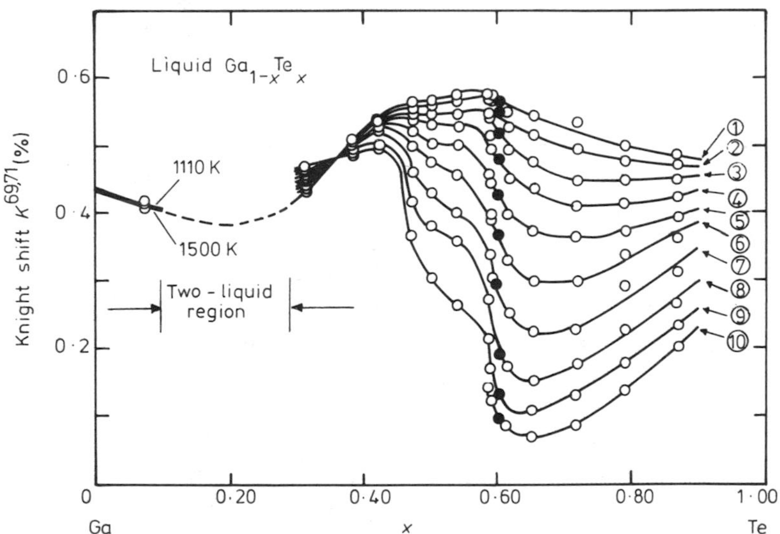

Figure 3. Knight shift $K^{69,71}$ against composition for $^{69,71}Ga$ in Ga–Te liquid alloys (Warren 1972a). The temperatures ① through to ⑩ are 1500 K, 1450 K, 1400 K, 1350 K, 1300 K, 1250 K, 1200 K, 1150 K, 1110 K and 1070 K.

The $^{69,71}Ga$ Knight shifts in Ga–Te are reproduced in figure 3. At low temperatures, the Ga shift reaches a minimum near, but slightly to the Te-rich side, of the stoichiometric composition Ga_2Te_3. There is also a weak feature near GaTe. The value of K at the minimum is about 20% of the value in pure Ga indicating that there are always many more broken bonds in this system than in Se at low temperatures. The shift is strongly temperature dependent in the vicinity of the minimum. The Te shifts show a similar temperature and composition dependence except that the asymmetry of the minimum is reversed relative to the composition of Ga_2Te_3. Comparison of the temperature dependences of K_{Ga} and K_{Te} with the magnetic susceptibility (Glazov *et al* 1969, Fischer and Guntherodt 1977) suggests that the temperature dependence of the shift results mainly from the density of states $N_s(E_F)$. Thus the temperature-induced MNM transition can be best understood as a filling of the gap between the sp^3 hybrid valence states and the conduction states covalent bonds rupture.

3.1.3. Copper–tellurium. Bonding in the Cu–Te system is probably considerably more ionic than in systems such as Tl–Te or Ga–Te although this point remains unsettled. The Pauling-scale electronegativity difference, $\Delta E = 0{\cdot}2$, implies much more covalent character than the Phillips scale (Phillips 1973) which gives $\Delta E = 0{\cdot}68$ for bonding in tetrahedral symmetry. The latter value is considerably larger than for Ga–Te on this scale ($\Delta E = 0.34$) being comparable with Ga–Se ($\Delta E = 0{\cdot}66$). Measurements of the partial liquid structure factors by neutron diffraction (Hawker *et al* 1974) suggest that the more ionic description is appropriate for these liquid alloys.

The Knight-shift data in Cu–Te, shown in figure 4, contrast sharply with those in Ga–Te and also suggest relatively more ionic character. Both the Cu and Te shifts do tend to small values near the composition of Cu_2Te in analogy with Ga_2Te_3. However, the temperature and composition dependences are very different in Cu–Te. This is especially evident for the Cu shifts in highly Te-rich alloys where K_{Cu}/K_m approaches a constant value of about 0·3. Such behaviour suggests that the dilute impurity is significantly ionized. If the bonding were covalent we should expect the bond perfection to be at least as sensitive to temperature as the covalent bonds in the Te host. This does appear to be the case for Ga–Te where, extrapolated to the dilute limit, $K_{Ga}/K_m \simeq 1$ and the temperature dependence is similar to that of K_{Te} in pure Te.

3.1.4. Bismuth–bismuth triodide. NMR data for metal–molten-salt mixtures have only recently been obtained. Data for the $Bi–BiI_3$ system are reported in Dupree and Warren (1977) and need only be summarized here. The essential observation is that at temperatures near the liquidus, $K_{Bi}/K_m \simeq 1$, even in alloys for which the electrical conductivity has dropped to ionic values ($\sigma \lesssim 10\,(\Omega\,\mathrm{cm})^{-1}$). As the temperature is raised, K_{Bi} *decreases* rapidly, in contrast with the behaviour of the Te alloys described previously and contrary

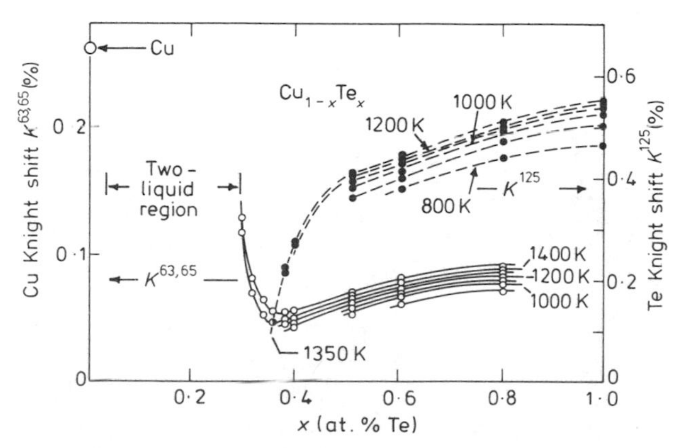

Figure 4. Knight shifts $K^{63,65}$ and K^{125} against composition for, respectively, ^{63,65}Cu and ^{125}Te in Cu–Te liquid alloys (Warren 1973).

to simple expectation based on the behaviour of the conductivity and magnetic susceptibility (Grantham and Yosim 1963, Topol and Ransom 1963). These Knight-shift results suggest that the mixtures are sufficiently inhomogeneous that equation (3) is inapplicable and only the more metallic regions contribute to the observed shift.

3.2. Quadrupolar relaxation

Additional information about bonding and structure can be inferred from the quadrupolar rates in liquid alloys. The quadrupolar contribution R_q to the total relaxation rate $1/T_1$ can be generally expressed for a liquid in the form

$$R_q = \omega_q{}^2 \tau_i \qquad (5)$$

where ω_q is the average coupling between the quadrupole moment and the instantaneous electric field gradient and τ_i is a correlation time characterizing the lifetime of a typical arrangement of near neighbours. The value of τ_i is clearly sensitive to the formation of chemical bonds and will tend to increase the relaxation rate if neighbouring atoms remain associated longer than the typical diffusional times in liquid metals ($\tau_i \lesssim 10^{-12}$ s). The NMN transition can also change R_q through changes in ω_q. This effect is discussed for In–Te alloys by von Hartrott *et al* (1977).

Quadrupolar rates for two systems Cu–Te and Ga–Te are shown in figure 5. As with the Knight shifts, the data in these two systems are very different. The relaxation rate for ^{69}Ga exhibits a sharp maximum relative to both the pure metal and the dilute (Te-rich) alloys. Similar results have been observed for In–Te (von Hartrott *et al* 1977). For Cu–Te, on the other hand, R_q is nearly independent of composition and exhibits only a very weak feature near Cu_2Te. The peak value of R_q in Ga–Te is much greater than could be expected from the range of ω_q values measured for ^{69}Ga in various crystals (Biryukov *et al* 1969). This peak can, however, be explained easily by formation of bonded structures of composition Ga_2Te_3 with a bonding lifetime $\tau_i \sim 10^{-11}$ s. This time is sufficiently short that the

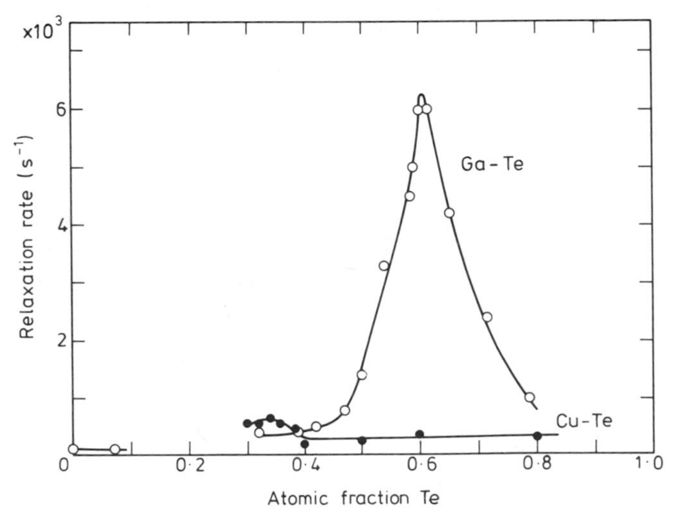

Figure 5. Quadrupolar relaxation rates against composition for ^{69}Ga and ^{63}Cu, respectively, in Ga–Te and Cu–Te liquid alloys (Warren 1972a and unpublished).

liquid cannot be considered molecular in the usual sense, but bond lifetimes of this order can strongly affect electronic properties since the structure is essentially static on the electronic timescale. The absence of a similar strong peak in Cu–Te is consistent with the notion of more ionic bonding in that system.

4. Electron transport

4.1. Magnetic relaxation enhancement: limit of the NFE *regime*

The magnetic relaxation rate obeys a general expression analogous to equation (5):

$$R_m = {\omega_m}^2 \tau_e \tag{6}$$

where ω_m is the average strength of the magnetic interaction and τ_c is the correlation time for local magnetic field fluctuations. This form is valid provided that $\omega_0\tau_e \ll 1$. In metals and semiconductors, the local magnetic field is provided mainly by electron spins.

If the electrons are itinerant and degenerate, equation (6) yields (Warren 1971) the approximate expression

$$R_m \simeq \frac{16{\gamma_n}^2 K^2 kT}{{\gamma_e}^2 \hbar} \frac{\tau_e}{\hbar N(E_F)}. \tag{7}$$

Now for a metal in the NFE regime, the electronic mean free path λ exceeds the average interatomic separation a so that the correlation time is roughly the time required for an electron moving at the Fermi velocity v_F to traverse one atomic volume. Since $a/v_F \simeq \hbar N(E_F)$, the final factor in equation (7) reduces to unity and the relaxation rate depends only on the Knight shift and temperature. A more exact derivation for metals gives the Korringa (1950) formula

$$(R_m)_{Korr} = \frac{4\pi {\gamma_n}^2 K^2 kT}{{\gamma_e}^2 \hbar}. \tag{8}$$

Thus so long as $\lambda > a$, the local field fluctuations are independent of the scattering and equation (8) should apply. In actual practice there are corrections to equation (8) from electron–electron interactions and p- and d-electron effects, but these are normally less than roughly ±50% in non-transition metals.

The situation is very different if the scattering becomes sufficiently strong that $\lambda \sim a$ or if the electrons enter localized states. In this case the local field will depend on the 'residence time' of an electron near a nucleus or, if the electron is well localized on a resonant nucleus, the correlation time will be the spin relaxation time of the local moment (Walstedt and Narath 1972). At the limit of the NFE regime the ratio $\tau_e/\hbar N(E_F)$ begins to exceed unity and the relaxation rate is enhanced relative to the Korringa rate by a factor η:

$$\eta \equiv \frac{(R_m)_{expt}}{(R_m)_{Korr}} \simeq \frac{\tau_e}{\hbar N(E_F)}. \tag{9}$$

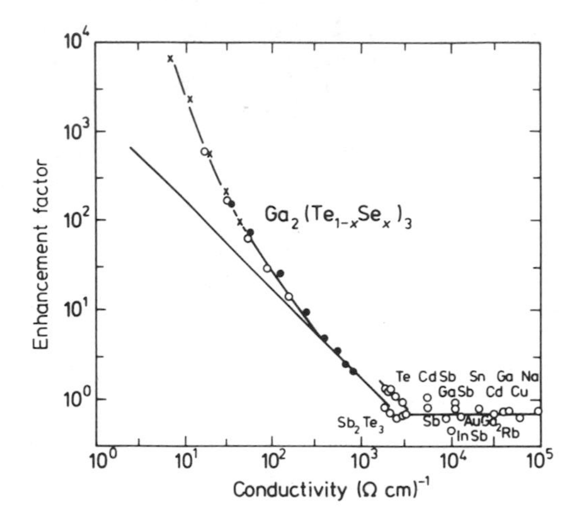

Figure 6. Log–log plot of magnetic relaxation enhancement factor η against electrical conductivity σ for elemental liquid metals and stoichiometric liquid alloys (Warren 1971, 1973).

In the range of diffusive transport (Cohen 1970) the enhancement is related in a simple way to the DC conductivity (Warren 1971)

$$\sigma = \frac{1}{3}\left(\frac{e^2a^2}{\Omega\hbar}\right)\eta^{-1} \equiv \sigma_0\eta^{-1} \tag{10}$$

where Ω is the atomic volume. For characteristic values of a and Ω in liquid alloys ($a \simeq 2{\cdot}5$ Å, $\Omega \simeq 35$ Å^3) we find $\sigma_0 \simeq 1500\ (\Omega\ \text{cm})^{-1}$. Thus the onset of enhancement when $\sigma \simeq \sigma_0$ gives a direct experimental indication of the limit of the NFE transport regime.

A log–log plot of η against σ for a number of metals and stoichiometric alloys demonstrates the enhancement onset (figure 6). For alloys in the pseudobinary system $Ga_2(Se_xTe_{1-x})_3$, σ_0 lies between 1500 and 2000 $(\Omega\ \text{cm})^{-1}$ while for Te and Sb_2Te_3, σ_0 may be a bit higher. The essential point is that the enhancement appears where theory predicts the limit of the NFE regime (Mott and Davis 1971) and, initially at least, η increases with decreasing σ according to equation (10). The deviation for $\sigma \lesssim 200$ $(\Omega\ \text{cm})^{-1}$ reflects the onset of conductivity by electrons excited to higher mobility states away from E_F. This will be discussed in more detail in §4.3.

4.2. The strong-scattering regime

A transport regime in which the electronic states are extended but $\lambda \sim a$ is known variously as the strong-scattering regime, the diffusion regime, the Brownian motion regime, etc. Electronic transport in this situation has been extensively discussed by Cohen (1970), Mott (1971), Hindley (1970), Friedman (1971, 1973) and others. An essential feature of these theoretical treatments is the prediction that as the density of states changes with some thermodynamic variable, the conductivity should change in proportion to the square of $N(E_F)$. More precisely,

$$\sigma \propto [a^3JN(E_F)]^2 \tag{11}$$

where J is the interatomic transfer integral.

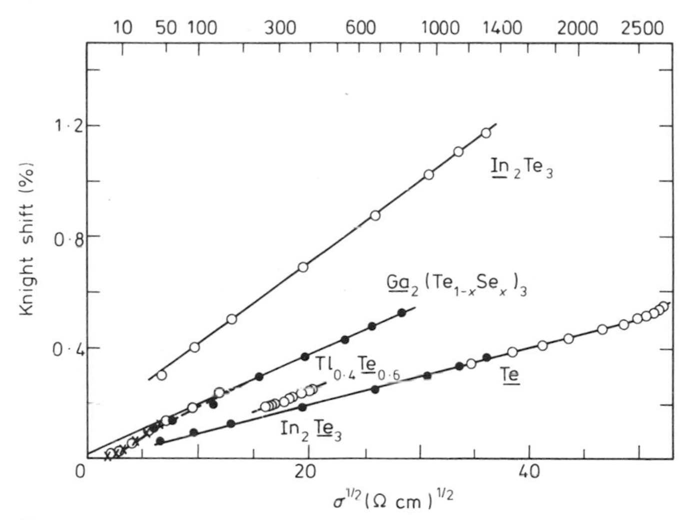

Figure 7. Knight shifts against the square root of electrical conductivity in liquid chalcogen alloys. In each case the shifts correspond to the underlined elemental component. Data are those of Warren (1971, 1972b, 1973) except for $Tl_{0\cdot4}Te_{0\cdot6}$ (Brown *et al* 1971).

In a stoichiometric alloy or elemental liquid semiconductor where $N(E_F)$ changes with temperature, we expect a^3J to be roughly constant and $K_A \propto N_s^A(E_F) \propto N(E_F)$. This proportionality would hold if the fractional s-character remains roughly constant so that K_A varies in proportion to the susceptibility. Since the shift can contain a small temperature-independent term K_0 due, say, to the chemical shift, the best test is a plot of K against $\sigma^{1/2}$ as is shown in figure 7 for several Te alloys and Te. These plots yield very good straight lines for conductivity values above about 100 $(\Omega\ \text{cm})^{-1}$. The Te data begin to deviate at high conductivities near 2500 $(\Omega\ \text{cm})^{-1}$.

An inhomogeneous model has been proposed by Cohen and Jortner (1973b) for liquid Te and Te alloys. According to this model the observed shift is related to the shift in the metallic fluctuations by equation (4). Classical percolation theory applied to Te predicts a nearly linear K versus σ relationship and the data were found to obey this nearly as well as the K versus $\sigma^{1/2}$ plot in figure 7. The explanation for this is that the Te data cover only a variation in conductivity of about a factor of two and, when the experimental error is included, it is not possible to distinguish between these two alternative dependences. As we have seen, however, the K versus $\sigma^{1/2}$ plot is linear over a full order of magnitude variation in σ for In_2Te_3. A system such as this would provide a much more stringent test of the inhomogeneous model. Although it has been argued that the model should apply to the Te alloy systems (Cohen and Jortner 1973b), these have not yet been analysed in detail from this point of view.

The predictions of the strong-scattering model for the Hall effect are far less satisfactory than for the conductivity. Friedman (1971, 1973) has advanced a theory which predicts $R_H \propto [a^3JN(E_F)]^{-1}$ and hence $1/R_H \propto \sigma^{1/2}$. It has been frequently noted that this prediction fails for Te (Tieche and Zareba 1963) and has also been found to be incorrect for In_2Te_3 (Tschirner *et al* 1975). Cohen and Jortner (1974) have taken this as evidence for breakdown of microscopic homogeneity although one can argue as

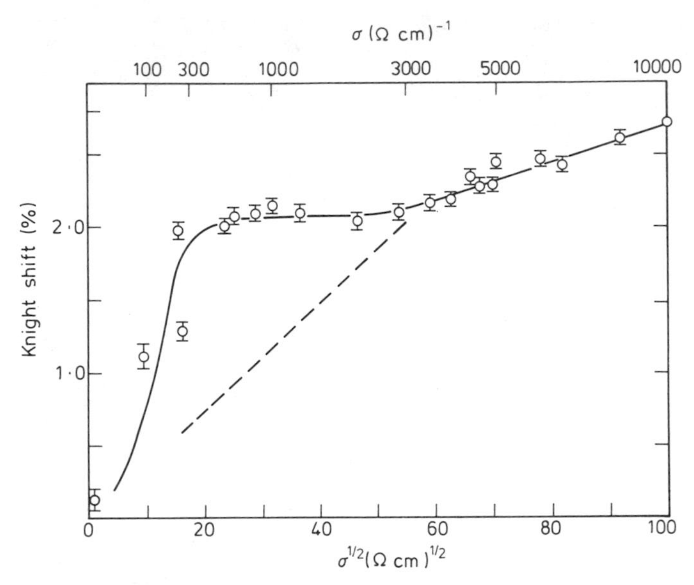

Figure 8. Knight shifts against the square root of electrical conductivity for expanded liquid Hg (El-Hanany and Warren 1975).

well that these systems are homogeneous, as suggested by other experimental data, and that the difficulty lies with this particular treatment of the Hall coefficient.

One case where the Friedman theory and strong-scattering model have looked promising is expanded liquid Hg. As the density is decreased, the value of R_H maintains its free-electron value until $\sigma \simeq 3000\,(\Omega\,\text{cm})^{-1}$ and then $1/R_H$ begins to decrease as $\sigma^{1/2}$ (Even and Jortner 1972a,b, 1973). Recent measurements of the Knight shift, however, seem inconsistent with this model (El-Hanany and Warren 1975). In the range $300 \lesssim \sigma \lesssim 3000\,(\Omega\,\text{cm})^{-1}$ where the strong-scattering model implies a change of a factor $10^{1/2}$ in $a^3JN(E_F)$, the shift is essentially constant (figure 8). Thus either $N(E_F)$ is constant or the factors a^3J, $\langle|\psi(r_i)|^2\rangle_F$, and $N_s(E_F)/N(E_F)$ vary in such a way as to compensate the change in $N(E_F)$ and thereby cancel the density dependence of the Knight shift.

Although one cannot at this stage exclude the possibility of accidental cancellation in the 'strong-scattering' range, there is reason to suspect that $N_s(E_F)$ is, in fact, independent of density in this range. The argument is based on structural considerations. Available evidence on the structure of expanding liquids suggests that a increases much less rapidly than $\rho^{-1/3}$ due to generation of free volume (Ocken and Wagner 1966, Waseda *et al* 1974, Block *et al* 1977). In liquid Ar, a is essentially constant from the triple point to the critical point while the mean number of nearest neighbours decreases from nearly 12 to about 4 (Pings 1968). We have recently explored the effect of volume expansion on $N(E_F)$ and $N_s(E_F)$ in Hg with a series of band-structure calculations for different structures and constant interatomic spacing (L F Mathiess and W W Warren unpublished). According to this very simple model, $N_s(E_F)$ is nearly independent of coordination number in the density range where the Knight shift is constant. The total density of states, however, actually *increases* as the density decreases in this region, in complete contradiction to the strong-scattering model. (Note that a^3J in equation (11) remains constant for constant a.) Finally, it is probably still necessary to include density fluctuations to obtain a proper description of the MNM transition in Hg. At a density of 9·0 g cm^{-3}, for example, the value of the isothermal compressibility implies a width $(\overline{\Delta\rho^2})^{1/2} \simeq 5$ g cm^{-3} for the probability distribution of fluctuations involving 20 atomic volumes. This means that significant portions of the material have local densities below that at which the MNM transition occurs. This problem is clearly deserving of more experimental and theoretical consideration.

4.3. Localization in a pseudogap: the 'mobility edge'

Mott (1969) has extended the basic model of Anderson (1958) to argue that a sharp transition to localized electronic states should occur when the density of states falls below a critical value. The conductivity of states at this 'mobility edge' (Cohen *et al* 1969) should lie in the range 100–300 (Ω cm)$^{-1}$. This is the so-called 'minimum metallic conductivity'.

The relaxation-rate enhancement η provides a sensitive measure of the tendency toward localization since it depends on the interaction time of an electron with a particular nucleus. The very large enhancements ($\eta \sim 10^4$) observed for $Ga_2(Se_xTe_{1-x})_3$ (figure 6) indicate increasingly strong localization as the density of states decreases. This observation provides qualitative support for the Mott model.

Since the nuclear relaxation rate is dominated by electrons at the Fermi level, the conductivity given by equation (10) should be lower than the DC conductivity when the Fermi level states in a pseudogap are localized. This is observed for the alloys shown in figure 6 when $\sigma \lesssim 200$ (Ω cm)$^{-1}$. This effect can be used to estimate the sharpness of the mobility edge, that is the rapidity with which $\sigma(E)$ drops as $N(E)$ decreases. It was found (Warren and Brennert 1974) that for the $Ga_2(Se_xTe_{1-x})_3$ alloys, η is a unique function of the Knight shift K, irrespective of the temperature or composition. Since $\sigma(E_F) = \sigma_0/\eta$ and K = constant × $N(E_F)$, these data provide an empirical relationship

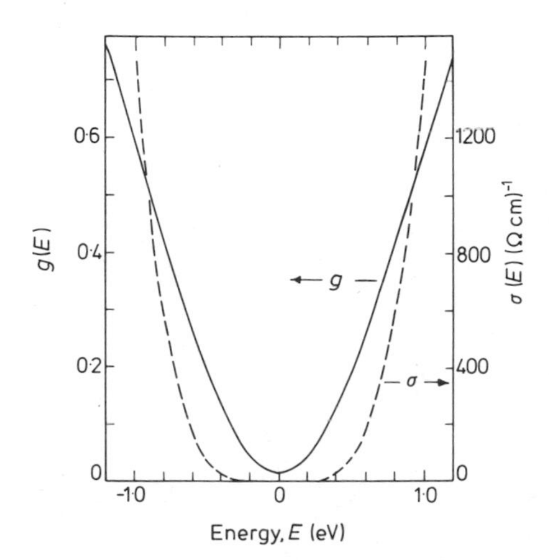

Figure 9. Model pseudogap $g(E)$ and variation of conductivity $\sigma(E)$ with energy for $Ga_2(Te_{1-x}Se_x)_3$ pseudobinary liquid alloys (Warren and Brennert 1974).

between $\sigma(E_F)$ and $N(E_F)$ if suitable values are estimated for the proportionality constant†.

In figure 9 the variation of $\sigma(E)$ is shown for an assumed pseudogap form of $N(E)$. The conductivity in the centre of the pseudogap is sufficiently small that if the Fermi level were placed in the centre of the gap these states would make a negligible contribution to the conductivity. On the other hand, the drop in $\sigma(E)$ occurs over an energy range much wider than kT. There is no identifiable energy E_c at which a sharp edge appears. For this reason it may be better to speak of a 'mobility shoulder'. A simple model of this kind was successful in explaining the magnitude and temperature dependence of the conductivity in $Ga_2(Se_xTe_{1-x})_3$ alloys (Warren and Brennert 1974).

4.4. Microscopic conductivity in non-stoichiometric alloys

We have been discussing the changes in electronic transport in elements and stoichiometric alloys where temperature and density are the variables. We now consider briefly the variation in conductivity as the composition of a liquid alloy is varied. This situation is more complex, but very interesting, since the conducting states may be associated preferentially with one of the component species. The nuclear relaxation enhancement and the inferred microscopic conductivity σ_0/η reflect the mobility of electrons with respect to the resonant nuclear species and therefore probe the distribution of conducting states.

The composition dependence of σ_0/η for ^{69}Ga in Ga–Te is compared with recent data of Valiant and Faber (1974) in figure 10. The constant σ_0 was chosen to yield the correct conductivity for Ga_2Te_3 in the (high-conductivity) strong-scattering regime. For alloys which are Ga rich relative to Ga_2Te_3, the variation of η gives a reasonably good fit to the composition dependence σ, including a weak feature observed near the composition of GaTe. On the other hand, in Te-rich alloys the conductivity associated with the Ga sites is substantially smaller than the DC conductivity. Thus the principal conducting states must be mainly associated with the Te atoms and the mobile electrons, on the average, spend more time near a Ga nucleus than near a Te nucleus.

5. Summary

We conclude with a brief summary of the contributions of NMR data toward answering the fundamental questions raised in the introductory section.

The Knight shift provides a convenient measure of the degree of chemical-bond satisfaction in liquid chalcogens and their alloys. This information is particularly valuable in non-stoichiometric alloys where the ability to discriminate between component species provides qualitative indications of the ionicity of the bonding. It is evident that liquid semiconductors occur with widely varying ionicities ranging from covalent Se and Te to the molten salts. Similarly there is a continuous range of metallic character which often can be strongly affected by changes in temperature. However,

† Extension of the expression $\sigma(E_F) = \sigma_0/\eta$ beyond the strong-scattering range and into a regime of possibly weak localization is based on the notion (Mott 1969) that localized states extend over many atoms near the onset of localization. Thus the nuclei would continue to sense a short-range diffusive transport mode until the localized states were confined to about one atomic volume.

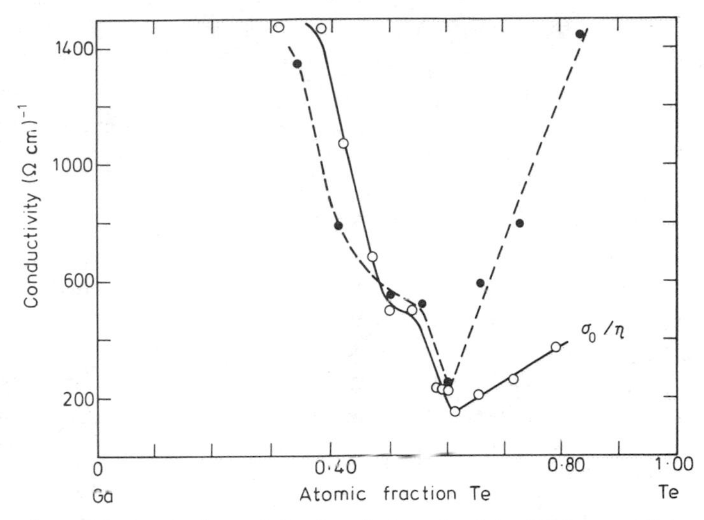

Figure 10. Comparison of the electrical conductivity (Valiant and Faber 1974) (broken curve) with the microscopic conductivity σ_0/η (full curve) determined from ^{69,71}Ga NMR (Warren 1972a) in Ga–Te liquid alloys at 950 °C. For alloys that are Ga-rich relative to Ga_2Te_3 the macroscopic and microscopic conductivities are in reasonable agreement while for Te-rich alloys the macroscopic conductivity is dominated by states associated with Te atoms and the Ga states have much lower microscopic mobilities.

these temperature-induced nonmetal–metal transformations can only be achieved in the more covalent systems, presumably by thermal dissociation of large 'molecular' units or network structures. The structural units are easily identified as linear chains in liquid Se while their form in alloys such as Ga–Te remains obscure. The stronger tendency toward structural association in covalent liquids is demonstrated by the much stronger quadrapolar relaxation enhancement in Ga_2Te_3 compared with Cu_2Te.

Many questions remain unanswered concerning fluctuations and microscopic inhomogeneity. In stoichiometric liquid chalcogen alloys, correspondence of the microscopic and macroscopic conductivities suggests that the homogeneous strong-scattering description is appropriate. Most features of the NMR data for these alloys are in quantitative agreement with calculations based on the Mott pseudogap model. Liquid Te has frequently served as a test case for comparison of different models but, unfortunately, Te exhibits such a small variation of its electronic properties with temperature that such comparisons are inconclusive. The most striking evidence of microscopic inhomogeneity is found for Hg and Bi–BiI_3 which exhibit microscopic electronic properties distinctly different from their macroscopic properties. It is noteworthy that these systems are characterized by the presence of second-order phase transitions and, hence, large thermodynamic fluctuations near the regions of the MNM transitions.

Magnetic relaxation enhancement gives a direct experimental indication of the NFE transport regime. Beyond this breakdown, the strong-scattering picture gives a good correlation of the NMR properties and the conductivity, but there remain major difficulties with the Hall coefficient. The validity of the strong-scattering model for expanded Hg remains open to question.

Formation and properties of highly localized states such as those at the chain ends in liquid Se are reasonably well understood. However, the nature of the more weakly localized states in alloys such as Ga_2Te_3 remains obscure. Although relatively delocalized in comparison with those in Se, these states are sufficiently well localized as to produce dramatic relaxation enhancement and strongly affect the conductivity. There is evidence in Ga–Te for preferential localization on one species in non-stoichiometric alloys.

Interpretation of Knight shift and relaxation data in $Ga_2(Se_xTe_{1-x})_3$ indicates that the concept of a mobility edge caused by localization in the pseudogap has qualitative validity for this system. However, there is no evidence for a sharp edge at a critical energy E_c. Rather the conductivity drops smoothly over several multiples of kT as the energy moves into the region of low density of states. This mobility shoulder is sufficiently 'soft' that states at E_F make a non-negligible contribution to the total conductivity even when the conductivity is well below the so-called 'minimum metallic conductivity'.

Acknowledgments

Major contributions to the experimental work reported in this paper have been made by my collaborators Dr U El-Hanany and Dr R Dupree. Expert technical assistance was provided throughout the experimental programme by G F Brennert. Discussions with many persons have contributed greatly to my understanding of the problems discussed in this paper. Among these are Profs M Cutler, J F Enderby, F Hensel, J Jortner and N F Mott to whom I am deeply grateful.

References

Abragam A 1961 *Principles of Nuclear Magnetism* (Oxford: Clarendon Press)
Anderson P W 1958 *Phys. Rev.* **109** 1492
Bhatia A B and Hargrove W H 1974 *Phys. Rev.* **B10** 3186
Biryukov I P, Voronkov M G and Safin I A 1969 *Tables of Nuclear Quadrapole Resonance Frequencies* (Jerusalem: Israel Program for Scientific Translations)
Block R *et al* 1977 this volume
Bredig M A 1964 *Molten Salt Chemistry* ed M Blander (New York: Interscience)
Brown D, Moore D S and Seymour E F W 1971 *Phil. Mag.* **23** 1249
—— 1972 *J. Non. Cryst. Solids* **8–10** 256
Cabane B and Friedel J 1971 *J. Physique* **32** 73
Cohen M H 1970 *J. Non. Cryst. Solids* **4** 391
Cohen M H, Fritzsche H and Ovshinsky S R 1969 Phys. Rev. Lett. **22** 1065
Cohen M H and Jortner J 1973a *Phys. Rev. Lett.* **30** 696
—— 1973b *Phys. Rev. Lett.* **30** 696
—— 1974 *J. Physique* **35** C4–345
—— 1976 *Phys. Rev.* **B13** 5255
Cornell D A 1967 *Phys. Rev.* **153** 208
Cutler M 1971 *Phil. Mag.* **24** 381
Dupree R and Warren W W 1977 this volume
Eisenberg A and Tobolsky A V 1960 *J. Polymer Sci.* **46** 19
El-Hanany U and Warren W W 1975 *Phys. Rev. Lett.* **34** 1276
Enderby J E 1974 *Amorphous and Liquid Semiconductors* ed J Tauc (New York: Plenum) p 361
Enderby J E and Collings E W 1970 *J. Non. Cryst. Solids* **4** 161
Even U and Jortner J 1972a *Phys. Rev. Lett.* **28** 31
—— 1972b *Phil. Mag.* **25** 715
—— 1973 *Phys. Rev.* **B8** 2536

Fischer M and Guntherodt H J 1977 this volume
Flynn C R 1974 *Phys. Rev.* **B9** 1984
Friedman L 1971 *J. Non. Cryst. Solids* **6** 329
—— 1973 *Phil. Mag.* **28** 145
Glazov V M, Chizhevskaya S N and Glagoleva N N 1969 *Liquid Semiconductors* (New York: Plenum)
Grantham L F and Yosim S J 1963 *J. Chem. Phys.* **38** 1671
von Hartrott M *et al* 1977 this volume
Hawker I, Howe R A and Enderby J E 1974 *Amorphous and Liquid Semiconductors* ed J Stuke and W Brenig (London: Taylor and Francis) p 85
Hindley N K 1970 *J. Non. Cryst. Solids* **5** 17
Hoshino H, Schmutzler R W, Warren W W and Hensel F 1976 *Phil. Mag.* **33** 255
Kerlin A L and Clark W G 1975 *Phys. Rev.* **B12** 3533
Koningsberger D C, van Wolput J H M C and Rieter P C U 1971 *Chem. Phys. Lett.* **8** 145
Korringa J 1950 *Physica* **16** 601
Lelieur J P 1973 *Electrons in Fluids* ed J Jortner and N R Kestner (Berlin: Springer-Verlag) p 305
Maekawa T, Yokokawa T and Niwa K 1973 *The Properties of Liquid Metals* ed S Takeuchi (London: Taylor and Francis) p 501
Mott N F 1969 *Phil. Mag.* **19** 835
—— 1971 *Phil. Mag.* **24** 1
Mott N F and Davis E A 1971 *Electronic Processes in Non-Crystalline Materials* (Oxford: Clarendon Press) p 81
Moulson D J and Seymour E F W 1967 *Adv. Phys.* **16** 449
Nakamura Y and Shimoji M 1971 *Trans. Faraday Soc.* **67** 1270
Narath A and Alderman D W 1966 *Phys. Rev.* **143** 328
Ocken H and Wagner C N J 1966 *Phys. Rev.* **149** 122
Pauling L 1960 *The Nature of the Chemical Bond* (New York: Cornell University Press) 3rd edn
Phillips J C 1973 *Bonds and Bands in Semiconductors* (New York: Academic Press)
Pings C J 1968 *Physics of Simple Liquids* ed H N V Temberley, J S Rowlinson and G S Rushbrooke (Amsterdam: North-Holland) p 387
Seymour E F W and Brown D 1973 *The Properties of Liquid Metals* ed S Takeuchi (London: Taylor and Francis) p 399
Slichter C P 1963 *Principles of Magnetic Resonance* (New York: Harper and Row)
Thompson J C 1973 *Electrons in Fluids* ed J Jortner and N R Kestner (Berlin: Springer-Verlag) p 287
Thompson J C, Ichikawa K and Granstaff S M Jr 1976 *J. Phys. Chem. Liquids* in the press
Tieche Y and Zareba A 1963 *Phys. Kondens. Mater.* **1** 402
Topol L E and Ransom L D 1963 *J. Chem. Phys.* **38** 1663
Tschirner H U, Wolf R and Wobst M 1975 *Phil. Mag.* **31** 237
Valiant J C and Faber T E 1974 *Phil. Mag.* **29** 571
Walstedt R E and Narath A 1972 *Phys. Rev.* **B6** 4118
Warren W W 1971 *Phys. Rev.* **B3** 3708
—— 1972a *J. Non. Cryst. Solids* **8–10** 241
—— 1972b *Phys. Rev.* **B6** 2522
—— 1973 *The Properties of Liquid Metals* ed S Takeuchi (London: Taylor and Francis) p 395
Warren W W and Brennert G F 1974 *Amorphous and Liquid Semiconductors* ed J Stuke and W Brenig p 1047
Warren W W and Clark W G 1969 *Phys. Rev.* **177** 600
Waseda Y, Yokoyama K and Suzuki K 1974 *Phil. Mag.* **29** 1427

NMR studies of molten Bi–BiI_3 mixtures

R Dupree† and W W Warren Jr
Bell Laboratories, Murray Hill, New Jersey 07974, USA

Abstract. NMR results for the system Bi–BiI_3 over the concentration range 15–100 mole % Bi metal at temperatures from 400 °C to 750 °C are reported. Although the apparent Knight shift, K, decreases monotonically with decreasing metal concentration at constant temperature, the magnitude of the shift appears anomalously large when compared with the conductivity in the salt-rich mixtures; for example, $K \approx 1{\cdot}1$ % at 400 °C in a mixture containing 30% Bi with a conductivity about 1 $\Omega^{-1}\,cm^{-1}$. Equally surprising, K decreases rapidly with increasing temperature, the temperature coefficient increasing by a factor of nearly 20 in going from pure Bi to a mixture containing 30% Bi. In contrast, the conductivity has been observed to increase with T while the magnetic susceptibility becomes less diamagnetic. The behaviour of the Knight shift in these highly ionic mixtures thus differs sharply with that in more covalent systems such as Tl–Te and Ga–Te, or in expanded Hg or metal–NH_3 solutions where the Knight shift tends to a high (i.e., more metallic) value as the conductivity increases and the susceptibility becomes more paramagnetic. These results suggest the possible presence of metallic concentration fluctuations which, however, seem difficult to reconcile with existing models of the salt-rich mixtures based on the chemical reaction $2Bi + BiI_3 \rightleftharpoons 3BiI$. Attempts to calculate the concentration dependence of the Knight shift using simple fluctuation theories were unsuccessful. The nuclear spin–lattice and spin–phase memory relaxation rates have been measured and increase strongly with decreasing Bi–metal concentration. This increase undoubtedly reflects much stronger contributions from electric quadrupolar relaxation in salt-rich melts although enhanced magnetic relaxation cannot be excluded.

1. Introduction

Although metal–molten salt mixtures are a class of materials that have been investigated for over a hundred years, and the metal–nonmetal transition is quite well documented for some of these systems, there have been very few studies which give information at the microscopic level on the behaviour in such mixtures. NMR is a technique which is capable of giving such information, so we have therefore undertaken an NMR study of the Bi–BiI_3 system. Because of the more ionic nature of the solvent this system's behaviour should be rather different from that of, say, the metal–ammonia solutions or liquid tellurium alloys. The Bi–BiI_3 system has been well studied previously. In particular both the magnetic susceptibility (Topol and Ransom 1963) and the electrical conductivity (Grantham and Yosim 1963) have been measured across the entire concentration range, and the thermodynamic properties are quite well documented (Predel and Rothacker 1971). Furthermore it has the lowest consolute temperature (730 K, Yosim *et al* 1962) of any molten salt–metal system, it does not react with the quartz containers used in NMR and both ^{209}Bi and ^{127}I nuclei are, in principle, suitable for investigation.

† Present address: Physics Department, University of Warwick, Coventry CV4 7AL, England.

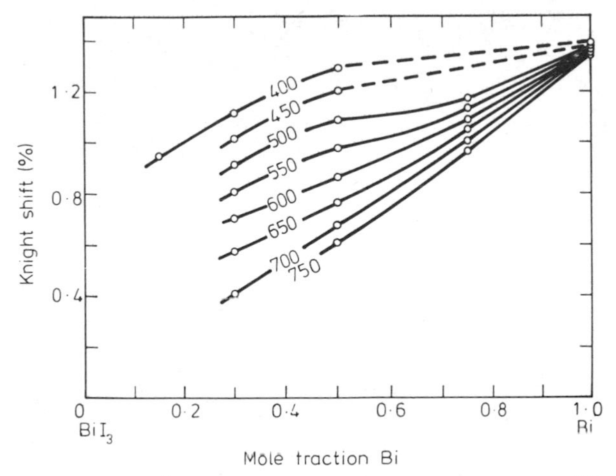

Figure 1. The ^{209}Bi Knight shift against concentration at different temperatures in Bi–BiI$_3$ mixtures.

2. Experimental

Our samples were prepared *in situ* in sealed quartz ampoules using 5N pure bismuth (Asarco) and Fisher 'purified' BiI$_3$. NMR measurements were made at 13 MHz using a pulsed coherent spectrometer with boxcar integration and a Nicolet 1040 for signal enhancement. No dispersing medium was used and the resonance signal was observed within the RF skin depth of the more highly conducting samples. The Knight shift of the ^{209}Bi was measured using the Knight shift of pure bismuth, $K = 1\cdot39\%$ at 400 °C (Rossini and Knight 1969) as reference.

3. Results

The bismuth shifts are shown as a function of concentration at various temperatures in figure 1. The bismuth shift was reversible with temperature for all concentrations, except the 75% Bi mixture close to the temperature for phase separation, indicating that the samples were in equilibrium. It can be seen that the shift is large and strongly temperature dependent in all the mixtures. No resonance was observed for iodine in any sample, nor was resonance observed in molten pure BiI$_3$, presumably due to very strong quadrupolar relaxation in both cases. The ^{209}Bi shift in pure molten BiBr$_3$ and BiCl$_3$ was therefore measured in an effort to obtain some idea of its likely value in the iodide. In BiBr$_3$ the shift was 0·36% at the melting point and increased slightly with temperature. A preliminary measurement of the shift in liquid BiCl$_3$ yielded a somewhat larger value in the range 0·4–0·5%.

Less extensive measurements of the nuclear spin–lattice relaxation time, T_1, were made; the relaxation rate is shown as a function of concentration at 400 °C in figure 2. The relaxation rate increased strongly with decreasing concentration of bismuth such that once the metal concentration was less than 50%, T_1 was shorter than 20 μs and only T_2^* could be measured. In the 30% sample T_2^* decreased rapidly with increasing temperature such that at 700 °C, where $T_2^* \simeq 1\cdot4\,\mu$s, the resonance was barely observable.

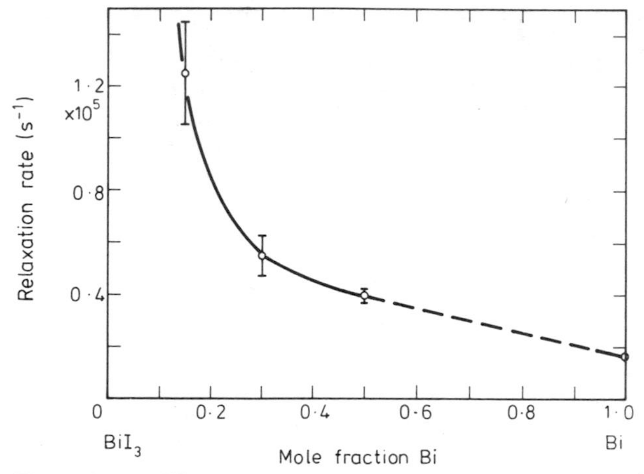

Figure 2. The ^{209}Bi relaxation rate against concentration at 400°C.

4. Discussion

The conductivity of the Bi–BiI_3 system increases by four orders of magnitude in going from pure BiI_3 to the pure metal. Although it increases almost exponentially as bismuth is added to BiI_3 (in contrast to $BiCl_3$ where there is initially very little change of conductivity as bismuth is added), the conductivity does not attain a 'metallic' value ($\sim 1000\ \Omega^{-1}\ cm^{-1}$) until a concentration of about 75% bismuth is reached. Similarly there is an exponential increase of conductivity with temperature for concentrations of bismuth less than about 70%. The generally accepted interpretation of this behaviour (Raleigh 1963, Ichikawa and Shimoji 1966, 1969), is that in the bismuth-rich region the conductivity mechanism is that of a normal liquid metal containing a high concentration of ionized scattering sites, whilst in the salt-rich mixtures the reaction $2Bi + BiI_3 \rightleftharpoons 3BiI$ is postulated to occur. Thus both Bi^+ and Bi^{3+} ions are present and conduction occurs by electron pair exchange (or hopping) between these ions. Evidence for the existence of BiI is somewhat tentative. For instance, Predel and Rothacker (1971) argue against the formation of BiI on thermodynamic grounds; however the above picture can be supported by the susceptibility measurements of Topol and Ransom (1963) who could fit the susceptibility for concentrations up to approximately 40% Bi by using the reaction above. Since the susceptibility becomes more paramagnetic as the temperature increases, this implies that the reaction must shift to the left, that is, more free bismuth atoms are formed.

Although the large chemical shift in the pure salt complicates the interpretation, our results clearly show a striking difference between the Knight shift and conductivity data. First, the magnitude of the shift is completely different from that expected since, in the alkali metal–ammonia systems, for instance, the Knight shift has decreased by an order of magnitude by the time the conductivity has dropped below $1000\ \Omega^{-1}\ cm^{-1}$ and in semiconducting tellurium alloys $K \propto \sigma^{1/2}$ and has also decreased by an order of magnitude when $\sigma \sim 300\ \Omega^{-1}\ cm^{-1}$. At these conductivities the Knight shift of bismuth is still large and, for instance, for 50% Bi at 400 °C where $\sigma < 30\ \Omega^{-1}\ cm^{-1}$, the shift is still 90% of its metallic value.

Secondly, whereas the conductivity (and the paramagnetism) increases as the temperature increases, the Knight shift decreases rapidly. (Note that the chemical shift in pure salts *increases* with temperature.) The large Knight shifts observed for Bi–BiI$_3$ mixtures also appear to be incompatible with the chemical reaction invoked above since Bi^{3+} and Bi^{+} should be diamagnetic and the concentration of unreacted Bi is much too small to give the required shift. Furthermore the temperature dependence is opposite to that predicted.

Since both the conductivity and susceptibility measurements imply that the average density of free carriers is much too small to give the observed shift, the implication is that concentration fluctuations to a locally more metallic mixture are present in the salt-rich mixtures. As the temperature is raised the fluctuations become less important and the Knight shift decreases. There are considerable problems with this description however, since a homogeneous solution presumably would have shifts at least as small as those given by the highest temperature data and the distribution of local concentrations is clearly required to extend to about 75% bismuth. If the fluctuations are weak (i.e., only small changes in concentration occur and the number of particles involved is large), then the mean square fluctuation in concentration is given by

$$\langle \Delta x^2 \rangle = \frac{k_B T}{N(\partial^2 g/\partial x^2)_{T,P,x}}, \quad (1)$$

where g is the Gibbs free energy per atom and x is the concentration.

The vapour pressure measurements of Predel and Rothacker (1971) show that the free energy against concentration curve has a very small curvature over a considerable proportion of the concentration range corresponding roughly to the miscibility gap. In this range $\langle \Delta x^2 \rangle$ will be relatively large. However from their data we find that $\partial^2 G/\partial x^2 \sim 10$ kcal/mole at 815 K and a concentration of 50%, thus for $N = 10$ atoms (the smallest number for which we could reasonably expect any metallic character) $(\langle \Delta x^2 \rangle)^{1/2} = 0{\cdot}13$. The width of the distribution is much too small to explain our data. Also, this number of atoms is probably too small for equation (1) to be valid and use of a larger value of N leads to still sharper distributions.

Recently, Cutler (1976) has derived an expression for the probability of a fluctuation in composition for strong (i.e., large) changes in composition. His theory yields a probability distribution:

$$P_N(x;\bar{x}) = \exp(-N\delta g/kT), \quad (2)$$

where $N\delta g$ is the change in free energy determined by a construction analogous to Maxwell's construction when N atoms with composition x separate from an infinite number of atoms with average composition $\bar{x}$. Because of the shape of the free energy curve, this expression predicts a much higher probability of a fluctuation to metal-rich than to salt-rich regions. Although this expression is not, strictly speaking, a correct description of the distribution of local concentrations in a finite system, we have used it to estimate the Knight shift since it is the only available function for describing strong changes in composition. It predicts a significant probability for a concentration fluctuation to 75% when the mean concentration is 50%, but the magnitude of the calculated shift in the salt-rich region is too small. For example, if we assume that a homogeneous solution is characterized by a chemical shift of 0·30% up to 30% bismuth and by the observed shifts at 750 °C for higher concentrations of bismuth, then a calculation of the average shift in an inhomogeneous mixture using equation (2) yields a shift value of 0·67% for a 50% bismuth sample at 542 °C. The experimental value is 1·0%. Attempts to improve the fit by assuming a larger chemical shift and a larger Knight shift for a homogeneous solution of 75% bismuth at this temperature naturally increase the calculated shift. However, all calculations predict a very rapid decrease in shift between 50% and 30% bismuth in disagreement with experiment. Thus there appears to be no fluctuation theory that is adequate to explain our data. Of course, as noted by Cutler, these theories have neglected the surface free energy which could well have a significant effect.

Another problem with a fluctuation model is that the volume of mixing measurements of Keneshea and Cubicciotti (1959) suggest that for $x \leq 0{\cdot}67$, bismuth goes into interstitial spaces between the iodine atoms so that there is little volume available for 'cluster' formation.

So far we have not commented on the relaxation data. If fluctuations occurred then there would be a contribution to the relaxation from the exchange of bismuth atoms between the different environments. However this turns out to be very small compared with the observed relaxation rate. In covalent alloys there is a strong enhancement of the ordinary Korringa relaxation as the electrons start to become localized (Warren 1971). In some cases this can be as much as 1000 times. Whilst localization of the electrons could enhance the relaxation rate in this system as well, our inability to observe resonance in pure BiI_3 means that very strong quadrupolar relaxation is also occurring. Thus we are unable to determine how important localization effects are.

5. Conclusions

The behaviour of this molten salt–metal mixture is very different from both the more covalent liquid metal–tellurium alloys and the alkali metal–ammonia systems. In both the latter, the behaviour of the conductivity and Knight shift can be well correlated. For instance, Seymour and Brown (1973) found that $K(\mathrm{Te}) \propto \sigma^{1/2}$ in Se–Te alloys. In the Bi–BiI_3 system there is no corresponding correlation between the conductivity and Knight shift since the former increases with temperature while the latter decreases. Furthermore 'metallic' type shifts are observed even when the conductivity is below $1\,\Omega^{-1}\,\mathrm{cm}^{-1}$.

There has been much recent discussion of the importance of microscopic inhomogeneities for the metal–nonmetal transformations of liquids (see, for example, Cohen and Jortner 1974). This matter remains highly controversial and there is little evidence for microscopic inhomogeneities in a number of these systems. However, our Knight shift data indicate that there are microscopic inhomogeneities in Bi–BiI_3 mixtures which are large enough to produce a locally metallic susceptibility. The present theory of concentration fluctuations is not able however to predict these inhomogeneities, perhaps because of its neglect of the surface free energy of the fluctuations. Clearly, in view of the unexpected nature of these results, further measurements on other systems are required to see whether the behaviour is peculiar to the Bi–BiI_3 mixture. Such measurements are currently under way.

References

Cohen M H and Jortner J 1974 *J. Physique* **35** C4–345
Cutler M 1976 to be published
Grantham L F and Yosim S J 1963 *J. Chem. Phys.* **38** 1671
Ichikawa K and Shimoji M 1966 *Trans. Faraday Soc.* **62** 3543
—— 1969 *Ber. Bunsenges. Phys. Chem.* **73** 302
Keneshea F J Jr and Cubicciotti D J 1959 *Phys. Chem.* **63** 1472
Predel B and Rothacker D 1971 *Thermochim. Acta* **2** 25
Raleigh D O 1963 *J. Chem. Phys.* **38** 1677
Rossini F A and Knight W D 1969 *Phys. Rev.* **178** 641
Seymour E F W and Brown D 1973 *The Properties of Liquid Metals* ed S Takeuchi (London: Taylor and Francis) p 399
Topol L E and Ransom L D 1963 *J. Chem. Phys.* **38** 1663
Warren W W Jr 1971 *Phys. Rev.* **B3** 3708
Yosim S J, Ransom L D, Sallach R A and Topol L E 1962 *J. Phys. Chem.* **66** 28

Nuclear spin relaxation in liquid In–Te alloys

M von Hartrott, K Nishiyama, J Rossbach, E Weihreter and D Quitmann
Institut für Atom- und Festkörperphysik, Freie Universität Berlin, D1 Berlin 33, Boltzmannstrasse 20, Federal Republic of Germany

Abstract. The nuclear spin relaxation rate of the 340 μs isomer $^{117}Sb^{m}$ was measured in liquid In_cTe_{1-c} for $c = 1 \ldots 0{\cdot}2$ at $T = 1000$ K using the TDPAD technique. For this impurity atom the quadrupolar relaxation rate R_Q is dominant. Its concentration dependence is discussed by starting from the known results for metallic alloys. It is argued that the reduction factor for the density of states at E_F, g, has a large influence on R_Q.

1. Introduction

The liquid In–Te alloy is of particular interest since near the 2:3 composition it becomes a liquid semiconductor which may be regarded as typical for the Te compounds. The drop in electrical conductivity is usually discussed by introducing the gap parameter (Mott 1971):

$$g = N(E_F)/N_{FE}(E_F) \tag{1}$$

that is, the reduction of the density of states at the Fermi energy below the free electron value. Studies of the nuclear spin relaxation rates have been reported for several liquid semiconductors, especially Ga_2Te_3 and In_2Te_3 (Warren 1971); however, one is essentially restricted to the magnetic rate R_M there. Additional information about the conduction electrons and the charge state of the constituents can be expected from that part of the hyperfine interactions which is determined by the electrostatic forces, that is, the quadrupolar relaxation rate R_Q.

The 340 μs isomeric state of $^{117}Sb^{m}$ is a very useful probe, since the magnetic moment is small and R_Q is dominant, as was already found in liquid In_cSb_{1-c} (von Hartrott *et al* 1976). The strong quadrupole interaction of Sb (compared for example with In) may be ascribed to the fact that the conduction electrons have mainly p-character at the Sb site.

2. Experiment and results

At the cyclotron of the Gesellschaft für Kernforschung, Karlsruhe, the isomeric state $^{117}Sb^{m}$ is produced by the reaction $^{115}In\,(\alpha, 2n)$ using short time pulses (3 μs). The anisotropy of the γ radiation (1000 keV and 1323 keV) from the decay of the isomer is observed by the TDPAD technique using a perpendicular magnetic field of $B \approx 10^2$ gauss, and the relaxation time of the alignment of the nuclear spin is extracted; for details see for example Recknagel (1974). The compositions of In_cTe_{1-c} and the temperatures used were c = 1430–1000 K; c = 0·9–0·2, 1000 K. The samples were

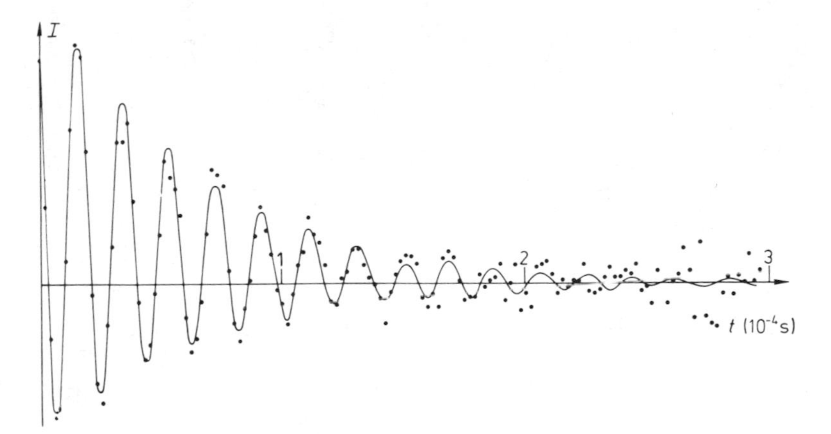

Figure 1. Time spectrum (counts per time channel against time) for $In_{0.64}Te_{0.36}$ at 1000 K. Correction for lifetime of the isomer and for background is applied. The line is the fitted curve.

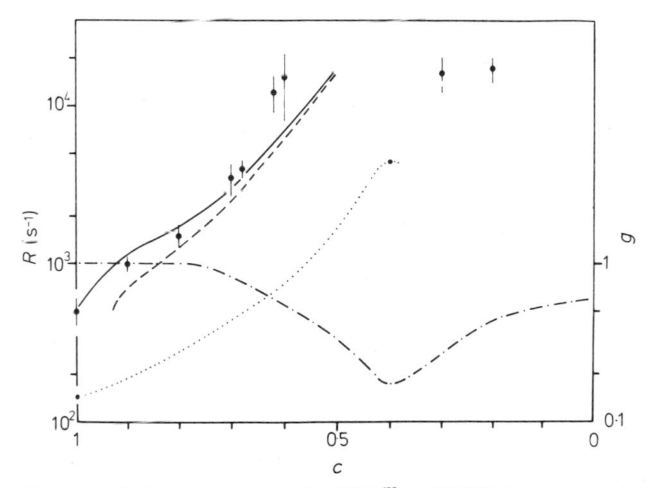

Figure 2. Relaxation rates R for $^{117}Sb^m$ at 1000 K. Experimental results (points) and analysis according to § 3.1 and 3.3 (full curve); incoherent contribution (broken curve) and magnetic contribution (dotted line) are given separately. The parameter g of equation (1) is included (chain curve, right-hand scale).

melted from pure metals (In 99·999%; Te 99%) under Ar (99·999%) atmosphere. Graphite crucibles with a mica window for the impinging α particle beam, Ar atmosphere and a calibrated thermocouple were used. The targets were checked for loss of Te; the uncertainty $\Delta(1-c)/(1-c)$ did not exceed 5%. The uncertainty of temperature was $\Delta T = ^{+10}_{-30}$ K.

Figure 1 shows a time spectrum after correcting for the nuclear lifetime and background. From such spectra, the relaxation rates presented in figure 2 were obtained. Each point represents at least two measurements. The errors quoted result from the statistics of the measurements and include estimates of the uncertainties due to the background and to the correlation of the fit parameters. For $c = 0{\cdot}5$ and $c = 0{\cdot}6$, no anisotropy was observed. By inspection of the spectra it was estimated that $R \gtrsim 2 \times 10^4 \text{s}^{-1}$.

3. Discussion

In a TDPAD experiment with γ rays, one measures the relaxation of the $k = 2$ orientation parameter, whereas NMR measures the $k = 1$ term (nuclear magnetization). The quadrupolar rate R_Q as well as the magnetic rate R_M is the product of a purely nuclear factor and the field fluctuation J_Q or J_M. With the known moments $|Q| = 0{\cdot}5$ barn (von Hartrott *et al* 1976), $\mu = 1{\cdot}487$ nuclear magnetons (Fromm *et al* 1975), and $I = 25/2$, one has $J_M/R_M = 1{\cdot}59 \times 10^{-6}$ erg s^2 cm^{-3} and $J_Q/R_Q = 1{\cdot}94 \times 10^{16}$ erg s^2 cm^{-5} for the present experiment on ^{117}Sbm. In the following we will try to outline an analysis of the dependence of J_M and J_Q on concentration for the region $c = 1 \ldots 0{\cdot}5$.

3.1. Magnetic rate R_M

For $^{117}\text{Sb}^m$, $R_M \ll R_Q$ in liquid In, InSb, and Sb_2Te_3 (see von Hartrott *et al* 1976, Warren 1971). We estimate that this will be the case in In_cTe_{1-c} also: comparing In in In with In in In_2Te_3 at high temperatures ($g \to 1$), J_M is increased by one order of magnitude; the same is true for Ga in Ga and Ga in Ga_2Te_3; at lower temperatures, R_M rises further by a factor of about 2 in In_2Te_3 and Ga_2Te_3 (Warren 1971). Sb as a probe atom is presumably less disturbed in In_2Te_3 than In is (see §3.3 below). Therefore, an increase in J_M by a factor of 30 for Sb in In_2Te_3 over Sb in In will be used as an upper limit. These points and a smooth interpolation curve are included in figure 2.

3.2. The quadrupolar rate R_Q in alloys

Of the various theoretical expressions proposed for R_Q in liquid metallic two-component alloys, we use here the expression of Gabriel (1974) which extends the calculations performed in k-space to the simplest case of a substitutional alloy. It is instructive to introduce the average electric field gradient (EFG) U_{coh} and its mean square scatter U_{inc}^2

$$U_{coh}(k) \equiv \overline{U(k)} = cU_1(k) + (1-c)\, U_2(k) \tag{2}$$

$$U_{inc}(k)^2 \equiv \overline{U^2} - \overline{U}^2 = c(1-c)\,(U_1(k) - U_2(k))^2. \tag{3}$$

This separation into a coherent and an incoherent contribution is analogous to the one used in neutron scattering. In equations (2) and (3) $U_1(k)$ is the Fourier transform of the effective EFG, $U_1(r)$, produced at the probe nucleus by a neighbour of type 1 at distance r†. With

$$a(k) = 1 + \frac{4\pi N}{V} \int_0^\infty (g(r) - 1) \frac{\sin kr}{kr} r^2 \, dr \tag{4}$$

where $g(r)$ is the pair correlation function, and with $S_s(k, \omega)$ the Fourier transform of

† We consider U only as a function of $|\mathbf{r}|$ or $|\mathbf{k}|$; for a full discussion see the references quoted.

the self part of the van Hove correlation function $G_s(r,t)$, the expression for J_Q becomes

$$J_Q = \int_0^\infty [a(k)\, U_{coh}(k)^2 + U_{inc}(k)^2]\, S_s(k,0)\, d^3k. \tag{5}$$

The first term ('coherent' contribution) describes the situation in the 'average' liquid†. Only this term occurs for an isolated probe atom in a homogeneous liquid. The second, 'incoherent' term is the fluctuation of the EFG due to the difference between U_1 and U_2.

The EFGS $U_1(k)$ and $U_2(k)$ will be considered schematically within the linear screening approximation (see for example Sholl 1976, Schirmacher 1976 and references therein). First, the Coulomb potential of the neighbouring ions, Ze/r, is shielded by the dielectric constant ϵ; the EFG is to be calculated from this reduced potential $\mu_{1,2}(k) = Z_{1,2}4\pi e/k^2\epsilon(k)$. The k-dependence of $\epsilon(k)$ is strongly influenced by the Fermi wave-number k_F. Second, the electrons within the atomic cell of the probe atom increase the resulting EFG by the Sternheimer effect which we take into account by a factor $(1-\gamma_{probe})$. Let us separate $U_0(k)$, the common k-dependent factor of $U_{1,2}(k)$:

$$U_{1,2}(k) = Z_{1,2}(1-\gamma_{probe})\, U_0(k). \tag{6}$$

$U_0(k)$ is independent of concentration or type of atom 1 and 2‡, except for the dependence of ϵ on g and k_F. Equation (5) becomes

$$J_Q = (1-\gamma_{probe})^2 \left[Z^2_{coh} \int U_0(k)^2 a(k)\, S_s(k,0)\, d^3k + Z^2_{inc} \int U_0(k)^2\, S_s(k,0)\, d^3k \right] \tag{7}$$

$$Z_{coh} = cZ_1 + (1-c)\, Z_2$$

$$Z^2_{inc} = c(1-c)(Z_1 - Z_2)^2. \tag{8}$$

Equation (7) leads back to the formulation used in Claridge *et al* (1972) and Titman and Jolly (1972) if we further introduce

$$S_s(k, \omega \approx 0) = \frac{Dk^2/\pi}{\omega^2 + (Dk^2)^2} \approx \frac{1}{\pi D k^2} \tag{9}$$

where D is the diffusion constant and the divergence at $k = 0$ is cancelled by the k^2 factor in d^3k. The second term of equation (7) is essentially the integral I_1 of Sholl (Claridge *et al* 1972). The first integral corresponds to $(I_1 + I_2)$ of that paper; the strong reduction $(I_1 + I_2) \ll I_1$ found in alloys (e.g. Seymour 1974, von Hartrott *et al* 1976) is seen to be due to the additional weighting function $a(k)$ which suppresses the contributions from the small k-region where U_0^2 is largest.

As a function of k, U_0^2 oscillates faster than the associated weight functions aS_s and S_s (see for example Warren 1974). The 'coherent' term may, due to the small overlap, depend sensitively on small changes of U_0 and/or $a(k)S_s(k, \omega \approx 0)$. The incoherent term is independent of $a(k)$.

† In this connection it might be better to give up the Vineyard approximation and to introduce $S(k, \omega)$ here.

‡ The same cutoff function is assumed for both U_1 and U_2.

3.3. Analysis of R_Q in In–Te

Experimental evidence has been given for essentially ionic conductance in the liquid semiconductor Cs–Au (Schmutzler *et al* 1977). The electronegativity difference is considerable between In and Te also; it equals that between Cs and Au on the Allred–Rochow scale. We take the view here that a disproportionate distribution of the conduction electron charge takes place between In and Te cells; at this point, we deviate once (and strongly) from the linear screening-picture. Roughly, the Te atoms will gain electrons to become Te^{2-} at the expense of In. One then has effective charges

$$Z_{In} = +3; \quad Z_{Te} = -2 \qquad \text{for} \qquad c = 1 \ldots 0{\cdot}4.$$

Upon alloying Te to In the number of electrons able to shield these ionic fields is depressed. The reduction factor g from equation (1) as a function of concentration is included in figure 2. It was derived using the following data: the measured conductivity σ at 1000 K (Ninomiya *et al* 1972); the value of σ expected for $g = 1$ was extrapolated from (Tschirner *et al* 1975): $\sigma(g = 1) = 6 \times 10^3\ \Omega^{-1}\ \text{cm}^{-1}$, which corresponds to a mean free path of 3·1 Å; and the average number of electrons per atom. The change in density was neglected.

In the In–Te alloys, the 'incoherent' term of equation (7) is dominant. It is proportional to $\int U_0(k)^2\ dk$ which has its largest contributions at small $k (k \lesssim k_F)$. In this k-region, the dielectric constant is essentially proportional to $N(E_F)$, since $4\pi e^2 \chi/k^2 \gg 1$ and the susceptibility $\chi(k)$ is proportional to $N(E_F)$. On the other hand, in the model underlying equation (1), E_F and k_F do not change as g falls somewhat below 1, and therefore the k-dependence of $\chi(k)$ does not change with concentration. Thus we can introduce $1/g^2$ as a factor in the second term of equation (7). The first term gets its largest contribution from the region around the maximum of $a(k)$, which is not far from $k = 2k_F$ where $\epsilon \approx 1$; therefore it appears a reasonable approximation to neglect the influence of g on the first term. Thus approximately

$$J_Q \propto Z_{coh}^2 \int a(k)\ U_0(k)^2\ dk + \frac{Z_{inc}^2}{g^2} \int U_0(k)^2\ dk. \tag{10}$$

The large difference in ionic charges, $\Delta Z = 5$ and the explicit appearance of g introduce strong dependencies of R_Q on c. We want to test here whether these effects may be sufficiently strong to explain the variation of R_Q with c. We therefore neglect the differences in $a(k)$ for In–In, In–Te, and Te–Te pairs as well as the change of $S_s(k, \omega)$ with c. As to the effective Sternheimer factor $(1 - \gamma_{probe})$ in In–Te, the approach to the ionic state implies a sensible reduction of the number of conduction electrons in an In-cell†. For Sb in In–Te, however, we note that the electronegativity of Sb falls right in between that of In and Te, and that InSb and Sb_2Te_3 are both almost metallic in the liquid state. We therefore neglect the change of the Sb Sternheimer factor with c. It also

† According to the calculations of Feiock and Johnson (1969) the Sternheimer antishielding factor $(1 - \gamma_\infty)$ increases noticeably with the number of outer electrons: $-\gamma_\infty = 20; 25; 29; 36; 51$; for a 5+; 3+; 2+; 1+; 0-charged atom, respectively, near $Z = 50$. It is therefore not surprising that R_Q is small for In (and Ga) at the 2:3 composition (Warren 1971). The increase of electronic correlation time which shows up in the magnetic rate is probably without influence on R_Q which is governed by the much slower movement of the ions.

appears improbable that the magnetic interaction strength increases beyond what is observed for In (this was used in § 3.1).

With the assumptions discussed, evaluation of equation (5) becomes very simple. We finally obtain for the concentration dependence for $c = 1 \ldots 0{\cdot}5$

$$R_Q \propto [(Z_2 + c\Delta Z)^2 \, \alpha + c(1-c) \, \Delta Z^2/g^2] \tag{11}$$

where the factor $\alpha < 1$ reflects the reduction of the coherent term due to the additional weighting factor $a(k)$. From the region $c = 1 \ldots 0{\cdot}7$ one gets the proportionality constant and

$$\alpha \approx \tfrac{1}{8}.$$

The resulting curve is included in figure 2. We have thus arrived at a satisfactory agreement based on a few, reasonable parameters.

Near the conductivity minimum, very high rates are predicted in accord with the experimental limit (§ 2). The results for $c = 0{\cdot}3$ and $0{\cdot}2$ (figure 2) conform qualitatively to the symmetry about $c = 0{\cdot}4$ expected from the picture discussed here and from the behaviour of the transport properties. We have also measured R for the isomer $^{132}Xe^m$ in liquid In_cTe_{1-c} for $c \leqslant 0{\cdot}3$; again, R_Q shows a rapid increase as one proceeds from $c = 0$ to $c = 0{\cdot}2$. This concentration region will be discussed separately.

Acknowledgments

We are obliged to the Gesellschaft für Kernforschung, Karlsruhe, for cyclotron time and technical assistance. The financial support of the Deutsche Forschungsgemeinschaft, Sfb 161 is gratefully acknowledged. One of us (JR) is supported by the Hahn–Meitner Institut, Berlin.

References

Claridge E, Moore D S, Seymour E F W and Sholl C A 1972 *J. Phys. F: Metal Phys.* **2** 1162

Feiock F D and Johnson W R 1969 *Phys. Rev.* **187** 39

Fromm W D, Brinckmann H F, Doenau F, Heiser C, May F R, Pashkevich V V and Rotter H 1975 *Nucl. Phys.* **A243** 9

Gabriel H 1974 *Phys. Stat. Solidi* **B64** K63

von Hartrott M, Hadijuana J, Nishiyama K and Quitmann D 1976 *Hyperfine Interactions* **3** in the press

Krüger K P, Fischer R and Schmutzler R W 1977 this volume

Mott N F 1971 *Phil. Mag.* **24** 1

Ninomiya Y, Nakamura Y and Shimoji M 1972 *Phil. Mag.* **26** 953

Recknagel E 1974 *Nuclear Spectroscopy and Reactions* ed J Cerny (New York: Academic Press) section VII C

Schirmacher W 1976 *J. Phys. F: Metal Phys.* **6** L157

Seymour E F W 1974 *Pure Appl. Chem.* **40** 41

Sholl C A 1976 *J. Phys. F: Metal Phys.* **6** L161

Titman J M and Jolly R I 1972 *Phys. Lett.* **A39** 213

Tschirner H U, Wolf R and Wobst M 1975 *Phil. Mag.* **31** 237

Warren W W Jr 1971 *Phys. Rev.* **B3** 3708

—— 1974 *Phys. Rev.* **A10** 657

Magnetic susceptibility of liquid Tl–Te alloys†

J A Gardner and M Cutler
Physics Department, Oregon State University, Corvallis, Oregon 97331, USA

Abstract. Measurements of the magnetic susceptibility χ of Tl_xTe_{1-x} in the range $x < 2/3$ have been analysed in relation to the electrical conductivity σ in terms of their mutual dependence on the density of states predicted by the Pauli–Landau theory for the susceptibility and the diffusive mechanism for electronic transport. It is found that the predicted relation $\chi = c_1 + c_2\sigma^{1/2}$ is obeyed in the range $\sigma \gtrsim 500\ \Omega^{-1}\ cm^{-1}$ where the metallic approximation is valid. The constant $c_1 = -4{\cdot}2 \times 10^{-5}\ cm^3$/mole is independent of x, indicating that the core susceptibilities of Tl and Te are both equal to this value. The constant c_2 is also independent of x except for $x = 0$, and its value provides insight about the electronic behaviour of the valence band. In alloys with 30–50 at.% Tl there is evidence of a positive deviation from the above relation between χ and σ which indicates that the paramagnetic susceptibility is enhanced when the Fermi energy approaches the top of the valence band. This seems to be evidence of spin correlation, and a possible mechanism is proposed in terms of electronic states due to dangling bonds.

1. Introduction

The liquid binary alloys Tl_xTe_{1-x} have large changes in electrical conductivity σ with composition x, and for $x \lesssim 2/3$, σ is also strongly dependent on temperature T (Cutler and Mallon 1966). This implies that the density of states $N(E)$ near the Fermi energy E_F is changing with x or T. Since the Pauli–Landau susceptibility reflects the behaviour of $N(E)$ near E_F, measurements of the magnetic susceptibility χ can be expected to yield information about the electronic structure. Measurements of χ have been reported previously for this alloy (Brown *et al* 1971). The present study is more extensive and detailed, and makes possible a useful analysis of the behaviour of χ in relation to σ. Our measurements will be reported more completely in another paper (Gardner and Cutler 1976). We discuss here the results in the composition range $x < 2/3$, where the positive thermopower indicates that E_F is in the valence band. The fact that σ varies with T makes it possible to analyse the relative behaviour of χ and σ in terms of the diffusive model for electronic transport, and deduce values of parameters describing the electronic behaviour. We also find that when E_F is near the valence band edge, there is an interesting deviation from the normal relationship between χ and σ which seems to indicate an enhancement of the paramagnetic susceptibility.

2. Relationship between χ and σ

In the experimental range $100 \lesssim \sigma \lesssim 2500\ \Omega^{-1}\ cm^{-1}$, it is expected that the electrical conductivity is determined by the diffusive mechanism with the result that

$$\sigma(E) = A\,[N(E)]^2, \qquad (1)$$

† Supported in part by the National Science Foundation and the Oregon State General Research Fund.

where A is a constant which depends on the momentum matrix element between wavefunctions on adjoining atomic sites (Mott 1969, Hindley 1970). The electrical conductivity is obtained from

$$\sigma = \int \sigma(E) \left(\frac{-\partial f}{\partial E}\right) \mathrm{d}E, \tag{2}$$

where f is the Fermi–Dirac distribution function (Mott and Davis 1971). In the metallic approximation, which becomes accurate when $\sigma \gtrsim 500\ \Omega^{-1}\,\mathrm{cm}^{-1}$,

$$\sigma \simeq \sigma(E_\mathrm{F}) = A\,[N(E_\mathrm{F})]^2. \tag{3}$$

In the same metallic approximation,

$$\chi = \chi_\mathrm{d} + \mu_\mathrm{B}^2\,(\alpha - \beta) N(E_\mathrm{F}), \tag{4}$$

where χ_d is the diamagnetic contribution of the ion cores, μ_B is the Bohr magneton, and α and β are factors describing respectively the electron–electron enhancement and the diamagnetic response of the band electrons. In the free electron model $\alpha = 1$ and $\beta = 1/3$, and in normal metals $\alpha - \beta \sim 1$.

When equation (3) is substituted into equation (4), a simple relation between χ and σ is obtained:

$$\chi = \chi_\mathrm{d} + \mu_\mathrm{B}^2\,(\alpha - \beta)\,\sigma^{1/2}/A^{1/2}. \tag{5}$$

Since χ_d is expected to depend only on composition, plots of χ against $\sigma^{1/2}$ at constant x should yield straight lines. The experimental results are shown in figure 1 for compositions x between 0·1 and 0·55. Parallel straight lines are obtained, and somewhat unexpectedly the intercepts are nearly the same. The fact that

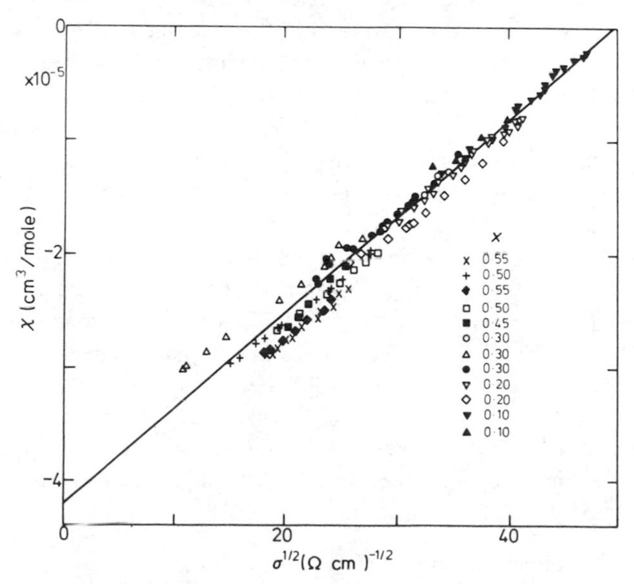

Figure 1. Plots of χ versus $\sigma^{1/2}$ for various compositions of Tl_xTe_{1-x} between $x = 0{\cdot}1$ and 0·55. Repeated compositions refer to different sources of data for σ.

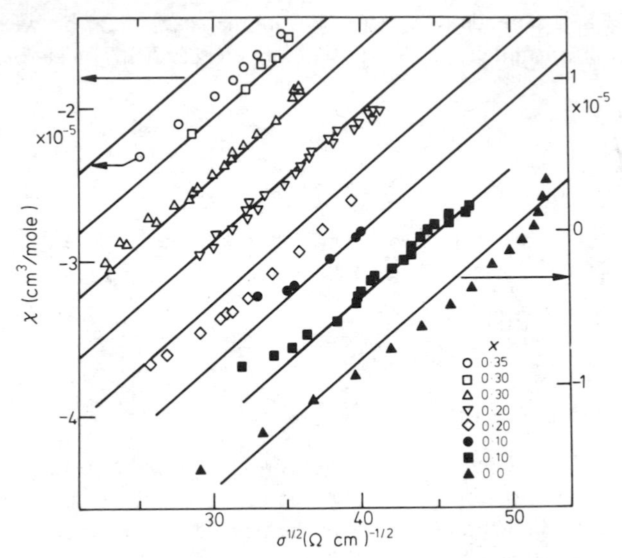

Figure 2. Plots of χ versus $\sigma^{1/2}$ with successive displacements $\Delta\chi = 0{\cdot}4 \times 10^{-5}$ cm^3/mole. The solid line in each graph is the same as the one in figure 1. Repeated compositions have different sources of data for σ.

χ_d ($= -4{\cdot}2 \times 10^{-5}$ cm^3/mol) is independent of composition implies that the core susceptibilities of Tl and Te are equal. The observed value of χ_d is close to theoretical estimates for the two elements (Mendelsohn *et al* 1970).

A clearer representation of this result is shown in figure 2 where the vertical axis is displaced for various compositions between pure Te and 35 at.% Tl, and the same straight line is drawn for each composition. The deviations of the experimental curves from the common line for the Tl–Te alloys are within the experimental error $\Delta\chi \lesssim 0{\cdot}2 \times 10^{-5}$ cm^3/mol. This error is the result of uncertainty in the susceptibility of the fused quartz sample capsule, and it is constant for a given composition.

The slope of the curves yields the value $A = 1200$ eV^2 atom^2 Ω^{-1} cm^{-1}, assuming that $\alpha - \beta = 1$. The corresponding value of $N(E_F)$ at high Te concentrations is around 1 eV/atom, and since the valence band is expected to contain two electrons per atom, the result suggests that the valence band is narrow, with a width of 1–2 eV.

Figure 2 also shows the result for pure Te. Except at large σ, the points fall on a straight line with a smaller slope, indicating a larger value of A. This implies a larger coordination number and supports the view that Tl–Te and pure Te have different molecular structures (Cutler 1976a). There is evidence that Te has a three-fold coordinated structure (Tourand *et al* 1972), and it has been proposed that Tl–Te contains chain-like molecules with two-fold coordination (Cutler 1971). The curved region in figure 2 occurs at $\sigma \gtrsim 2500\ \Omega^{-1}$ cm^{-1}, where the diffusive model has questionable validity. It may also reflect a change in structure of the liquid, since it encompasses a large temperature range (700–900 °C).

3. Evidence of enhanced paramagnetism

Careful examination of the curves in figure 1 for $x = 0{\cdot}30$ and $0{\cdot}50$, which cover a wider range of σ than other compositions, discloses that a positive deviation from a straight line occurs at low σ. Since the metallic approximation is poor in this range, it is necessary to use Fermi–Dirac integrals in evaluating σ in equation (2) and the Pauli susceptibility, which in turn requires a knowledge of $\sigma(E)$ and $N(E)$. In a recent study (Cutler 1976b) it has been found that $\sigma(E) = -BE$ in this experimental range, where $B = 2960\ \Omega^{-1}\ \text{cm}^{-1}\ \text{eV}^{-1}$ and E is measured from the top of the valence band. With this and equations (1) and (2), the expression for χ becomes

$$\chi = \chi_d + \mu_B{}^2(\alpha-\beta)\int N(E)\left(\frac{-\partial f}{\partial E}\right)\mathrm{d}E$$

$$= \chi_d + \mu_B{}^2(\alpha-\beta)\,(\sigma^*)^{1/2}/A^{1/2}, \tag{6}$$

where

$$(\sigma^*)^{1/2} = (BkT/4)^{1/2}\int_0^\infty y^{-1/2}\,[\exp(y-\xi)+1]^{-1}\,\mathrm{d}y, \tag{7}$$

and

$$\xi = -E_F/kT = \ln\,[1+\exp(\sigma/BkT)]\,. \tag{8}$$

In the limit of $\sigma/BkT \gg 1$, equations (6), (7) and (8) reduce to equation (5), but plots of χ against $(\sigma^*)^{1/2}$ should be linear for any value of σ. This curve is plotted in figure 3 for $x = 0{\cdot}30$ and $0{\cdot}50$, and it is seen that the paramagnetic deviation at small σ^* is even more pronounced. Allowing for experimental errors, the results for $x = 0{\cdot}55$, 0·45, and 0·40, which have shorter ranges of σ, fall in a pattern consistent with the two curves in figure 3. They imply that there is an enhanced paramagnetic susceptibility when E_F approaches the valence band edge. Because of the relative effects of x and T on σ^*, the larger deviation for $x = 0{\cdot}30$ indicates that the enhancement decreases with increasing T.

We believe that this enhanced susceptibility may be related to similar phenomena observed in heavily doped crystalline silicon and some other materials (Quirt and Marko 1972, 1973, Brown and Holcomb 1974). In this case, there seems to be transition from Pauli to Curie behaviour when the carrier density is just above the value where a metal–insulator transition occurs. It has been proposed that this is

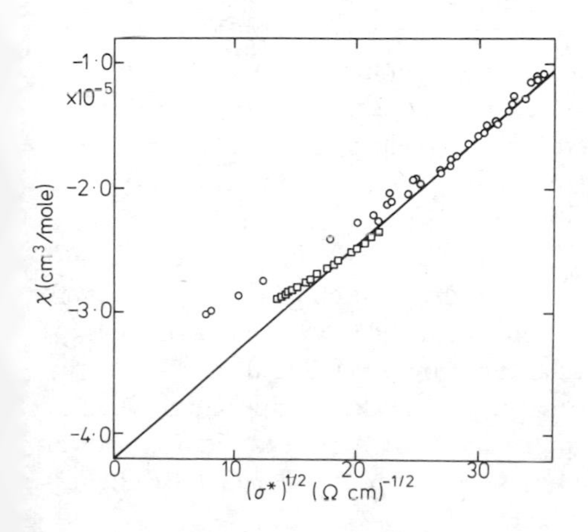

Figure 3. Plots of χ versus $(\sigma^*)^{1/2}$ for $x = 0{\cdot}30$ (circles) and $0{\cdot}50$ (squares).

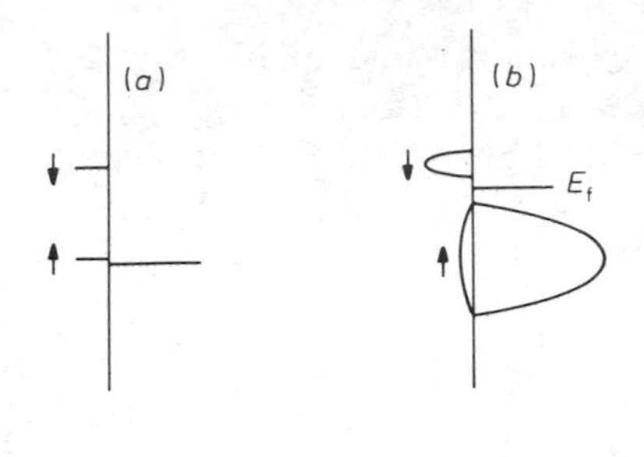

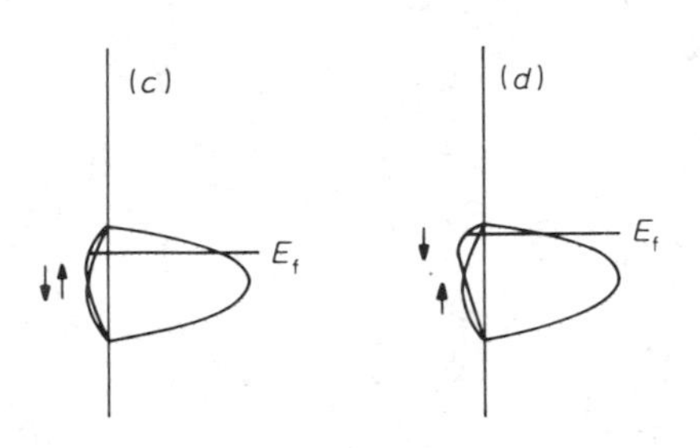

Figure 4. Diagrams illustrating the expected magnetic properties of electronic states due to dangling bonds. The density of states due to dangling bonds is indicated to the left of the vertical line, and those due to the non-bonding electrons are to the right. The one-electron state is indicated by an arrow pointing up, and the two-electron state by an arrow pointing down. (*a*) Discrete molecules. (*b*) Condensed phase with a low density of dangling bonds. (*c*) Metallic limit with a high density of dangling bonds. (*d*) Intermediate case where E_F and the virtual acceptor states are close to the band edge.

associated with the development of the Hubbard gap in the impurity band as the result of an increased correlation energy. The magnetic behaviour has been interpreted in terms of the theory for a highly correlated electron gas (Mott 1972, Berggren 1974).

In Te-rich Tl–Te alloys, the carriers are believed to be generated by broken Te–Te bonds, but unlike crystalline semiconductors, the impurity bands due to the acceptor states are absorbed into the valence band (Cutler 1971). The energy level structure is illustrated in figure 4. Figure 4(*a*) shows the energy levels of the last electron for non-bonding Te orbitals and dangling bond orbitals containing one or two electrons for the case where the molecules are isolated. The one-electron dangling bond state (arrow pointing up) has approximately the same energy as the non-bonding state, but the two-electron state (arrow pointing down) is appreciably higher because of Coulomb repulsion. In a condensed phase the non-bonding states broaden into the valence band. At low densities of dangling bonds, the two-electron states are localized or form an impurity band as indicated in figure 4(*b*), but it seems likely that the one-electron states will be absorbed into the valence band because of their great similarity. Since the two-electron states are empty, the dangling bond states would have Curie-type behaviour. (There is reason to believe that further interaction between dangling bond states and non-bonding states on neighbouring atoms may occur in amorphous solids at low temperatures, with the result that there are no paramagnetic states at thermal equilibrium (Mott *et al* 1975).) At high densities of dangling bonds the Coulomb energy of the two-electron states is reduced by screening, and the two-electron band is absorbed into the valence band. At low Tl concentrations the Fermi energy is well into the valence band, as illustrated in figure 4(*c*), and the distribution of the enrgy levels for the virtual one- and two-electron states is believed to be similar, with the result that Pauli susceptibility occurs. As the density of dangling bonds is reduced and E_F approaches the band edge, an intermediate situation can be expected where the energy distribution of the two-electron levels rises appreciably with respect to the one-electron levels, as illustrated in figure 4(*d*). This would cause a temperature-dependent enhancement of the paramagnetic susceptibility such as the one observed, and it constitutes our tentative explanation for the phenomenon.

4. Conclusions

The study of the magnetic susceptibility in relation to the electrical conductivity confirms the relation between the two which is implied by the theory for transport by the diffusive mechanism. The resulting analysis leads to quantitative information about the pertinent parameters (χ_d and A). In addition, a deviation from the above relationship has been found which does not seem to be explainable by deficiencies in the transport theory, and indicates that the paramagnetic susceptibility is enhanced when the Fermi energy approaches the valence band edge. It seems likely that this is the result of an increased electron–electron correlation, and we have proposed a qualitative explanation in terms of the electronic states due to dangling bonds.

References

Berggren K F 1974 *Phil. Mag.* **30** 1
Brown D, Moore D S and Seymour E F W 1971 *Phil. Mag.* **23** 1244
Brown G C and Holcomb D F 1974 *Phys. Rev.* **B10** 2394
Cutler M 1971 *Phil. Mag.* **24** 381
—— 1976a *J. Non. Cryst. Solids* **21** 137
—— 1976b *Phys. Rev.* to be published
Cutler M and Mallon C E 1966 *Phys. Rev.* **144** 642
Gardner J A and Cutler M 1976 *Phys. Rev.* to be published
Hindley N K 1970 *J. Non. Cryst. Solids* **5** 17
Mendelsohn L B, Biggs F and Mann J B 1970 *Phys. Rev.* **A2** 1130
Mott N F 1969 *Phil. Mag.* **19** 835
—— 1972 *Adv. Phys.* **21** 785
Mott N F and Davis E A 1971 *Electronic Processes in Non-Crystalline Materials* (Oxford: Clarendon Press)
Mott N F, Davis E A and Street R A 1975 *Phil. Mag.* **32** 961
Quirt J D and Marko J R 1972 *Phys. Rev.* **B5** 1716
—— 1973 *Phys. Rev.* **B7** 3842
Tourand G, Cabane B and Breuil M 1972 *J. Non. Cryst. Solids* **8–10** 676

The magnetic susceptibility of liquid Te alloys

M Fischer and H-J Güntherodt
Institut für Physik, Universität, Basel, Switzerland

Abstract. The magnetic susceptibility of liquid Ga–Te and In–Te alloys has been measured as a function of concentration and temperature up to 1000 °C. Strong diamagnetic deviations are found from the ideal linear behaviour of the susceptibility as a function of concentration. These diamagnetic deviations decrease with increasing temperature. Preliminary calculations of the magnetic susceptibility of these alloys favour an interpretation in terms of covalent bonding rather than an ionic description.

1. Introduction

In this paper we present new measurements of the magnetic susceptibilities over the entire concentration range of In–Te and Ga–Te. Previous measurements have been published only for the stoichiometric compositions Ga_2Te_3 and In_2Te_3 (Glazov *et al* 1969). These older results are in poor agreement with our present data. They show a large change of the susceptibility at the melting point and a quite different temperature dependence.

2. Experimental method and results

The magnetic susceptibility has been determined by the Faraday method. The samples were fused in quartz capsules under high vacuum. After measuring the susceptibility of the sample, the empty capsule was also measured and its contribution subtracted. The capsule was so constructed that its contribution was less than 10% of the total measured value. We estimate that our accuracy is 1%.

The experimental results for the magnetic susceptibilities of liquid Ga–Te and In–Te are shown in figures 1 and 2 a function of concentration for different temperatures. In both cases large diamagnetic deviations are observed at the stoichiometric compositions. These diamagnetic deviations diminish very rapidly with increasing temperature.

3. Discussion

The aim of this work was to investigate whether the ionic or the covalent bonding model is more appropriate for describing these alloy systems. The magnetic susceptibility should be a good probe for studying the chemical bonding in such alloys. A full discussion and interpretation of our results will appear elsewhere but preliminary calculations indicate that the magnitude of the susceptibilities is better explained in a covalent bonding model.

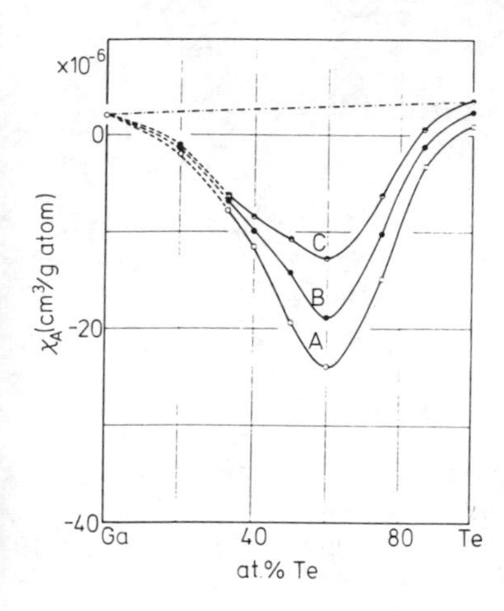

Figure 1. Magnetic susceptibility, χ, versus concentration of liquid Ga–Te at 800 °C (curve A), 900 °C (curve B) and 1000 °C (curve C).

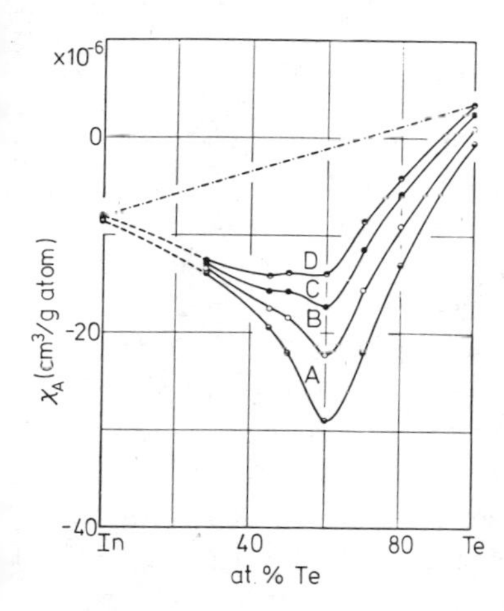

Figure 2. Magnetic susceptibility, χ, versus concentration of liquid In–Te at 700 °C (curve A), 800 °C (curve B), 900 °C (curve C) and 1000 °C (curve D).

Acknowledgment

We would like to thank Professor M Cutler and Dr W Freyland for helpful discussions. We are indebted to the Schweizerische Nationalfonds zur Förderung der wissenschaftlichen Forschung for financial support.

References

Glazov V M, Chizhevskaya S N and Glagoleva N N 1969 *Liquid Semiconductors* (New York: Plenum Press) p144

Localized impurity states in liquid semiconductors

S Tamaki and S Ohno
Department of Physics, Faculty of Science, Niigata University, Niigata 950–21, Japan

Abstract. The magnetic susceptibilities of cobalt and manganese solutes in liquid Se–Te have been measured. The cobalt solute proves to be non-magnetic in all compositions of the solvent alloys whereas the manganese solute is magnetic. In the case of the manganese solute, however, the Weiss temperature is large and positive and the mechanism of s–d interaction in the semiconductor solvent might be different from that of liquid metals.

1. Introduction

A series of experimental studies of the localized impurity states in liquid metals has been reported by the authors (Tamaki 1967, 1968a,b, Tamaki and Takeuchi 1967, Ohno *et al* 1973). These results are satisfactorily described by using the concept of the virtual bound state proposed by Friedel (1956) and developed by Anderson (1961). As a next step, our attention is concentrated on the 3d-transition solutes in liquid semiconductors and in fact an attempt has been made to study the localized impurity states in liquid tellurium (Ohno *et al* 1976). Liquid Se–Te alloys are of particular interest because the two components represent opposite extremes of semiconductor behaviour. The electrical resistivity of liquid Te shows a typical semiconducting behaviour but its magnitude is only a few times larger than that of liquid Sb which seems to be one of the ill-conditioned liquid metals (Epstein *et al* 1957, Takeda *et al* 1976). On the other hand, the electrical resistivity of liquid Se shows the obvious existence of an energy gap between the conduction and valence bands (see, for example, Mott and Davies 1971). In this paper, we report an examination of the magnetic properties of cobalt and manganese solutes in liquid Se–Te alloys.

2. Experimental procedure

The magnetic susceptibility apparatus used in this investigation was a torque balance based on the Faraday method. Details of the apparatus were given in a previous paper (Tamaki 1968b). As standard samples, we used $MnSO_4 \cdot 5H_2O$ ($\chi_g = 30{\cdot}4 \times 10^{-6}$ CGS emu at room temperature) and Cr ($\chi_g = 3{\cdot}17 \times 10^{-6}$ CGS emu). Measurements of the susceptibility were carried out with $H = 9500$ Oe and $H\,dH/dx = 6{\cdot}0 \pm 0{\cdot}3$ kOe^2 cm^{-1}.

An alloy sample of about 3 g was put into the vacuum-sealed quartz tube. This sample was kept at high temperatures for several hours during the measurement. Each measurement was carried out from the melting temperature of the specimen to about 900 °C. Purities of the materials were 99·99 at.% for Te and Se and 99·9 at.% for Co and Mn.

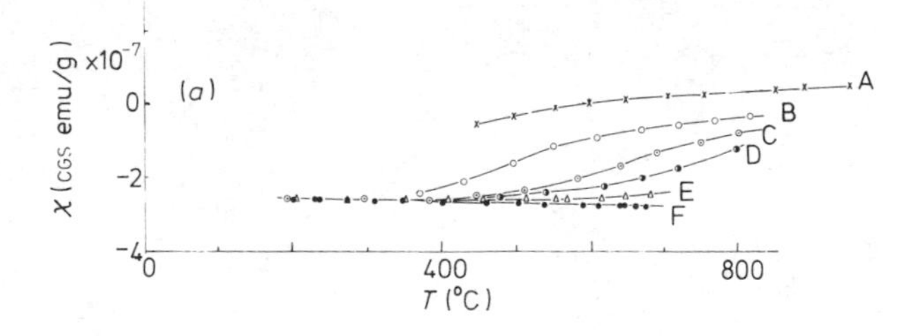

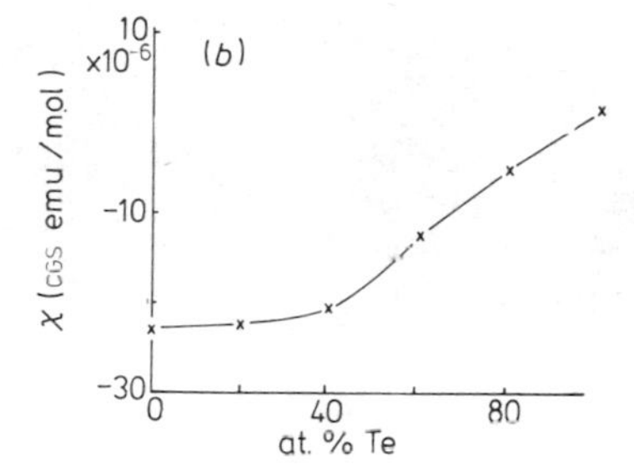

Figure 1. (*a*) Temperature dependence of the susceptibility of liquid Se–Te alloys: A, Te; B, $Se_{0\cdot2}Te_{0\cdot8}$; C, $Se_{0\cdot4}Te_{0\cdot6}$; D, $Se_{0\cdot6}Te_{0\cdot4}$; E, $Se_{0\cdot8}Te_{0\cdot2}$; F, Se. (*b*) The concentration dependence of the molar susceptibility at 650 °C.

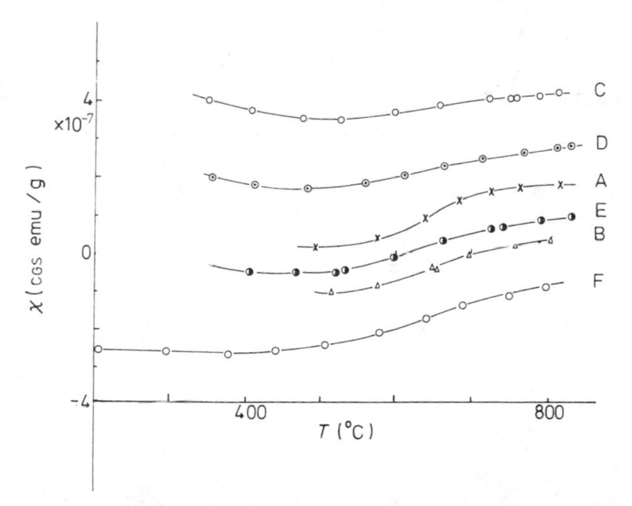

Figure 2. Temperature dependence of the susceptibility of cobalt (curve A, 2·0 at.%; curve B, 1·1 at.%) and manganese solute (curve C, 2·0 at.%; curve D, 1·4 at.%; curve E, 0·7 at.%) in $Se_{0\cdot4}Te_{0\cdot6}$ solvent (curve F).

3. Results and discussion

The susceptibility of liquid tellurium increases rapidly with an increase in temperature, while the temperature dependence of the susceptibility of liquid Se is very small. Consequently, the susceptibility of liquid Se–Te alloys has a complicated temperature dependence with variation of composition: this is shown in figure 1. Using these

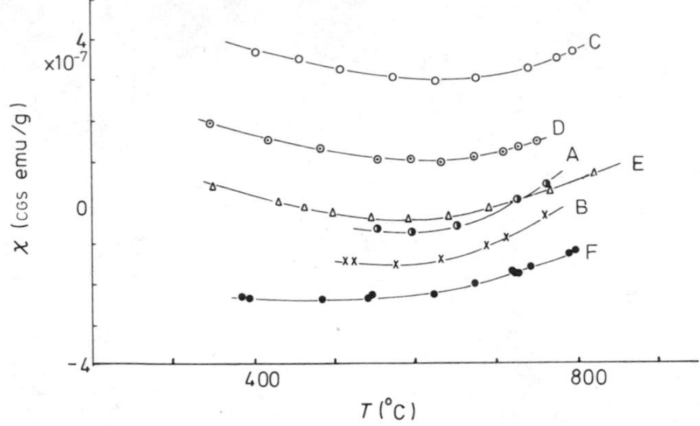

Figure 3. Temperature dependence of the susceptibility of cobalt (curve A, 2·0 at.%; curve B, 1·0 at.%) and manganese solute (curve C, 1·9 at.%; curve D, 1·2 at.%; curve E, 0·6 at.%) in $Se_{0\cdot6}Te_{0\cdot4}$ solvent (curve F).

results, the concentration dependence of the susceptibility of Se–Te at 650 °C is obtained and is also given in figure 1.

In figures 2 and 3, typical results for the temperature dependence of the susceptibilities of alloys with cobalt and manganese solutes are shown. In the case of the cobalt solute, the temperature dependence of the susceptibility is nearly parallel to that of the solvent alloy and the increment in the susceptibility is proportional to the concentration of cobalt. The increment depends linearly on the composition of the solvent alloys, ranging from $\Delta\chi = 0{\cdot}25 \times 10^{-7}$ for Se to $\Delta\chi = 1{\cdot}75 \times 10^{-7}$ CGS emu for the Te solvent. These magnitudes are of the same order as a paramagnetic susceptibility increment due to the virtual bound state for cobalt in liquid metals.

In the case of the manganese solute, the susceptibility is experimentally expressed in the form:

$$\chi = (1-x)\chi_0 + x\Delta\chi + x\frac{C}{T-\theta}, \tag{1}$$

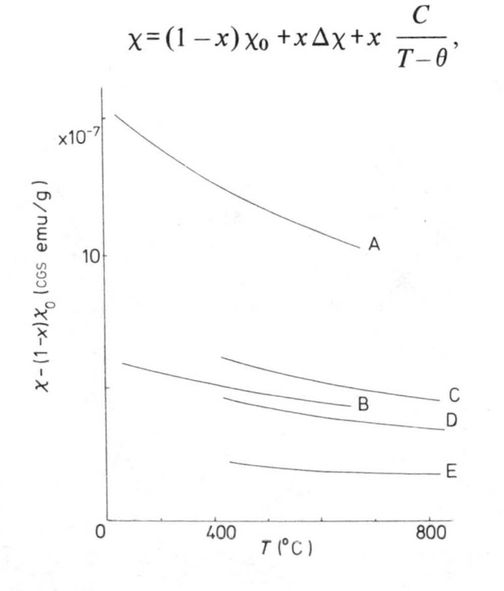

Figure 4. Temperature dependence of the additional susceptibility. Curve A, 2·8 at.% Mn and curve B, 0·9 at.% Mn in Se. Curve C, 2·1 at.% Mn, curve D, 1·4 at.% Mn and curve E, 0·7 at.% Mn in $Se_{0\cdot2}Te_{0\cdot8}$.

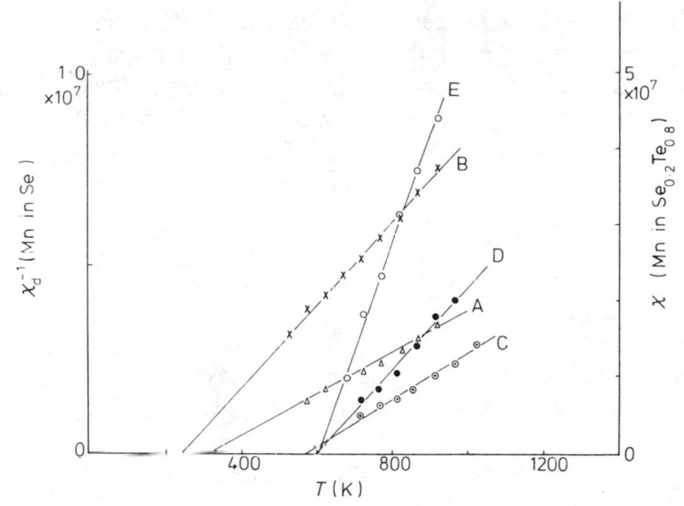

Figure 5. Temperature dependence of the inverse of the Curie–Weiss susceptibility term. Curve A, 2·8 at.% Mn and curve B, 0·9 at.% Mn in Se. Curve C, 2·1 at.% Mn, curve D, 1·4 at.% Mn and curve E, 0·7 at.% Mn in $Se_{0.2}Te_{0.8}$.

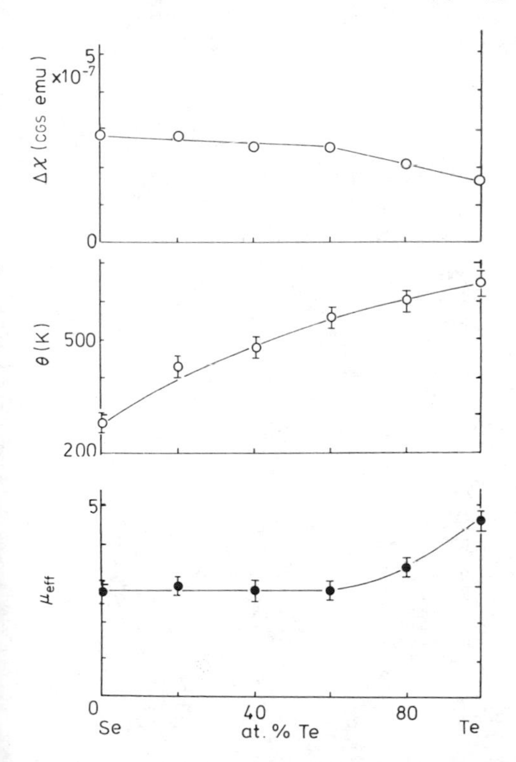

Figure 6. Concentration dependence of the temperature independent term in the additional susceptibility of 1·0 at.% manganese solute, $\Delta\chi$, of the Weiss temperature θ and of the effective Bohr magneton number μ_{eff}.

where x is the atomic fraction of the solute, χ_0 is the susceptibility of the solvent, $\Delta\chi$ is a temperature-independent term of the additional susceptibility and C is the Curie–Weiss term, $N\mu_{eff}^2/3k$, where μ_{eff} is the effective Bohr magneton number. In figure 4, typical results are given of the temperature dependence of the additional susceptibilities, $\chi-(1-x)\chi_0$. In figure 5, the temperature dependence of the inverse of the last term, $1/\chi_d=(T-\theta)/xC$, is shown. From these figures, $\Delta\chi$, θ and μ_{eff} are obtained and are presented in figure 6. Since the changes in these values are gradual, on going from Se to Te, the essential feature of the 3d-localized electrons should not be drastically changed for any composition of the solvent. This fact suggests that the magnetic behaviour of the 3d-localized electrons does not depend to any great extent on whether or not the solvent has degenerated conduction electrons. As seen in figure 6, the Weiss temperatures are positive, of the order of a few hundred degrees, in contrast to the liquid metal solvents (which are nearly zero). In a liquid metal solvent, the mixture of localized and conduction electrons leads to an antiferromagnetic effect and this is explained by the so-called compensation theorem (Anderson 1961). In other words, the negative sign of the effective s–d exchange integral results from the resonance between localized and degenerated conduction electrons.

It seems, therefore, that the resonance in the Se–Te alloys is not large enough to make the sign of the effective integral to be negative, but the 3d-localized electrons may resonate with the electrons of the valence band of the solvent alloys to give a different type of s–d exchange integral than that of the metals. Thus the present situation essentially is beyond the Friedel–Anderson scheme or the Kondo problem because of the large positive sign of the effective integral.

Acknowledgments

The authors express their cordial thanks to Dr Y Waseda for his valuable advice concerning the structure. Thanks are also due to Dr Y Tsuchiya for useful discussions.

References

Anderson P W 1961 *Phys. Rev.* **124** 41
Epstein A S, Fritzshe H and Lark-Horovitz K 1957 *Phys. Rev.* **107** 412
Friedel J 1956 *Can. J. Phys.* **34** 1190
Mott N F and Davies E A 1971 *Electronic Processes in Noncrystalline Materials* (Oxford: Clarendon Press)
Ohno S, Nomoto T and Tamaki S 1976 *J. Phys. Soc. Japan* **40** 72
Ohno S, Okazaki H and Tamaki S 1973 *J. Phys. Soc. Japan* **35** 1060
Takeda S, Ohno S and Tamaki S 1976 *J. Phys. Soc. Japan* **40** 113
Tamaki S 1967 *J. Phys. Soc. Japan* **22** 865
—— 1968a *J. Phys. Soc. Japan* **25** 379
—— 1968b *J. Phys. Soc. Japan* **25** 1602
Tamaki S and Takeuchi S 1967 *J. Phys. Soc. Japan* **22** 1042

The electronic structure of lithium-based liquid semiconducting alloys

V T Nguyen and J E Enderby
Department of Physics, University of Leicester, Leicester LE1 7RH

Abstract. The resistivity, ρ, and the thermoelectric power S, have been measured for liquid lithium alloys of the type Li_xM_{1-x} where $0 < x < 1$ and M represents a metal whose valence, Z, is in the range $2 \leqslant Z \leqslant 5$. It is found that for $Z \geqslant 3$, $\rho(x)$ and $S(x)$ become increasingly different from the predictions of the nearly free electron theory. These data, together with results for the spin-flip scattering cross section for small amounts of M in liquid Li, are interpreted within the framework of impurity ionicity.

Experimental evidence for ionic electrical transport in liquid Cs–Au alloys

K D Krüger, R Fischer and R W Schmutzler
Fachbereich Physikalische Chemie, Philipps-Universität Marburg, D-3550 Marburg (Lahn), Biegenstr. 12, West Germany

Abstract. The paper presents experimental data for the absolute thermoelectric power S of liquid Cs–Au alloys at 650 °C. It is shown that the thermopower is strongly concentration dependent and reaches typically nonmetallic values of $-1400\,\mu V\,K^{-1}$ near the equiatomic composition. As the thermopower of the alloys is always negative, normal liquid semiconductor theories cannot be applied for the interpretation. It is suggested that in the molten CsAu compound the electrical transport properties are dominated by Cs^+ and Au^- ions. Experimental evidence is given for this suggestion by electromigration experiments, the results of which can be explained by an ordinary electrolysis according to the Faraday law. It is concluded that the liquid CsAu compound consists of Cs^+ and Au^- ions.

1. Introduction

There are some liquid binary alloy systems of essentially metallic constituents (e.g. Mg–Bi, Li–Bi or Tl–Te) which exhibit at certain compositions clearly nonmetallic properties, as revealed by the large anomalies in their electrical transport properties (Ilschner *et al* 1953, Enderby and Collings 1970, Steinleitner *et al* 1975, Cutler and Mallon 1966, Cutler and Field 1968). The compositions, where these anomalies occur, correspond to the natural valence states of the atoms involved, which led to concepts to explain this nonmetallic behaviour by the formation of stable covalent or ionic bonds saturated within the short-range order of these melts (Ilschner *et al* 1953, Cutler 1971). As purely covalent bonds are restricted by definition to nonmetallic bonds between like atoms, nonmetallic bonding between unlike atoms always incorporates some degree of ionicity (Suchet 1965, Phillips 1973), which clearly is not a directly measurable physical quantity. It depends basically on the definition accepted and on the procedure applied for its evaluation (Hinze and Jaffé 1962, Hinze *et al* 1963). As a relative measure of the 'electronegative character' of the atoms, different 'electronegativity scales' have been proposed (Hinze and Jaffé 1962, Hinze *et al* 1963, Phillips 1973), which enable a degree of ionicity to be evaluated for a particular compound. The results based on the different scales are very similar, so that the choice of any preferred scale is somewhat arbitrary.

For alloy semiconductors of the type mentioned above, the question now arises as to when there is a large degree of ionicity (i.e. a large but still partial charge transfer) whether really ion-like species are formed, which must become mobile when going from the solid to the liquid phase, eventually giving rise to a transition from semiconductor to ionic conductor on melting. The purpose of this paper is to present the first example of

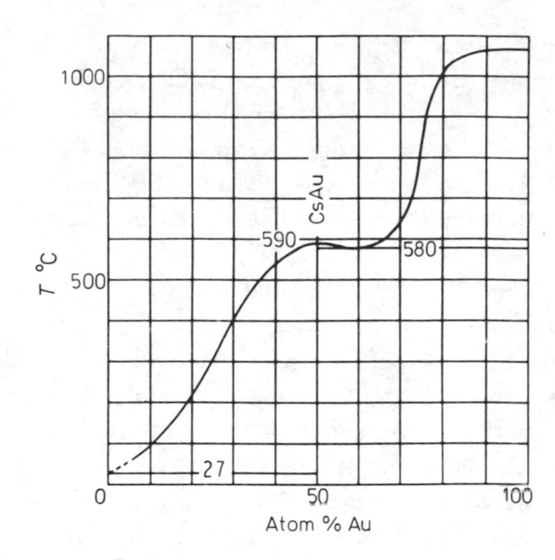

Figure 1. Phase diagram of the Cs–Au system according to Elliot (1965).

a compound semiconductor showing such a transition, namely the compound CsAu, and to discuss the experimental evidence for this transition.

The phase diagram of the Cs–Au system (figure 1) (Elliot 1965) shows the existence of the congruent melting stoichiometric CsAu phase with a very narrow range of homogeneity. This compound with a melting point of 590 °C is known to crystallize with the typically ionic CsCl structure and to be a semiconductor with an energy gap of 2·6 eV (Spicer *et al* 1959). The electrical conductivity of the solid compound given in the literature ($5\,\Omega^{-1}\,cm^{-1}$ to $100\,\Omega^{-1}\,cm^{-1}$) (Spicer *et al* 1959, Wooten and Condas 1963) seems to depend strongly on the purity of the samples, due perhaps to the effect of doping, or perhaps to the effect of phase separation on crystallizing because of the very narrow range of homogeneity of the compound. Taking the Pauling scale of electronegativity (Pauling 1960) as a means of comparison the resulting ionicity of CsAu is 51%, which is comparable to the ionicity of CsI (55%).

Measurements of the electrical conductivity as a function of the composition of liquid Cs–Au alloys have revealed a metal–nonmetal transition at about 44 at.% Au, and a minimum conductivity of about $5\,\Omega^{-1}\,cm^{-1}$ at the stoichiometric composition (Hoshino *et al* 1975) at 650 °C. Its low positive temperature coefficient ($d\ln\sigma/dT = 3{\cdot}5\times10^{-3}\,K^{-1}$) corresponding to an apparent activation energy of about 5 kcal mol^{-1}, together with the weak concentration dependence of the conductivity in the immediate vicinity of the stoichiometric composition gave a first hint that in this system the electrical transport properties are possibly dominated by Cs^{+} and Au^{-} ions. To test this suggestion, measurements of the absolute thermoelectric power and of the electromigration have been performed on this system.

2. Experimental method

For these investigations experimental arrangements were developed which allowed these measurements to be made at temperatures of about 600 °C to 700 °C, that is, at least some ten degrees above the melting point of the CsAu compound. Here a continuous variation in the homogeneous liquid is possible from pure Cs to alloys with about 65 at.% Au. All the materials in contact with the samples had to withstand corrosion by these highly reactive alloys. Furthermore, as the vapour pressure of pure Cs is about 1 atm at the temperatures applied in the experiments, the Cs partial pressure above the liquid alloys, unknown as yet, is probably not negligible. Therefore gas-tight measuring cells had to be used to avoid Cs losses during the experiments. Due to the very strong reaction of Cs and the Cs–Au alloys with oxygen or humidity even at room temperature, any handling of the samples had to be done in a high-purity inert gas atmosphere.

The measuring cell for the absolute thermoelectric power consisted essentially of a gas-tight sintered alumina capillary filled with the sample. At both ends, Pt 13% Rh/Pt thermocouples were fixed to the 0·5 mm thick molybdenum walls which closed the capillary and provided the electrical contact between the Pt wires and the sample. This also avoided any corrosion of the thermocouples by the Cs–Au alloys. Alumina and molybdenum were found to be compatible with the Cs–Au alloys at the experimental conditions. Two separate furnaces allowed us to impose temperature gradients up to 50 °C in either direction on the capillary. The whole assembly was inside a vacuum vessel to protect the molybdenum parts of the cell from corrosion by air at the high temperatures used. For the sample preparation, correctly weighed amounts of Cs (purity 99·98%, Merck, Darmstadt) and Au (purity 99·999%, Cerac pure, Menomonee Falls, Wisconsin) were filled under a high-purity argon atmosphere into a reservoir connected with the cell. After closing the cell together with the reservoir under vacuum, the whole assembly was heated in vacuum to 650 °C for about 1 to 2 h, so ensuring homogenization of the alloys. After this time the alloy was poured directly from the reservoir into the capillary at this high temperature, and the measurements of the Seebeck EMF as a function of small values of ΔT were made. Further details of the experiment have been published elsewhere (Schmutzler *et al* 1976).

The experimental setup for the electromigration experiment (i.e. the determination of concentration shifts in the stoichiometric CsAu alloy due to the passage of a large DC current) has to meet the same constraints concerning temperature, corrosion and tightness of the cell as described for the thermopower measurements. Among the different techniques described in the literature for the observation of the electromigration (Epstein 1972), the capillary method was most easily adapted to the needs imposed by the Cs–Au system: the measuring cell consisted in principle of a reservoir made from pure molybdenum containing the main part of the alloy and an electrode compartment defined by a very small sintered alumina capillary of 0·6 to 0·7 mm inner diameter and about 8 mm length. It contains less than about 1% of the total amount of the alloy placed in the cell. The electrode inside this cathode compartment was made from a 0·3 mm thick niobium wire, cemented into position by a high-temperature gas-tight frit; this provided good electrical insulation against the cell body. Sample preparation was accomplished in the same way as described for the thermopower measurements. Taking the cell body as anode and the electrode in the capillary as cathode, the enrichment of Cs within the capillary due to the passage of a DC current can be taken as a measure of the effect of electromigration. The electrical circuit was designed so that the entire circuit resistance is determined essentially by the cathode compartment, represented by the capillary. As the conductivity of the Cs–Au alloys is strongly concentration-dependent especially at excess concentrations of Cs (X'_{Cs}) greater than 0·15 (figure 2), it is possible to determine the amount of excess Cs, transported into the capillary by the

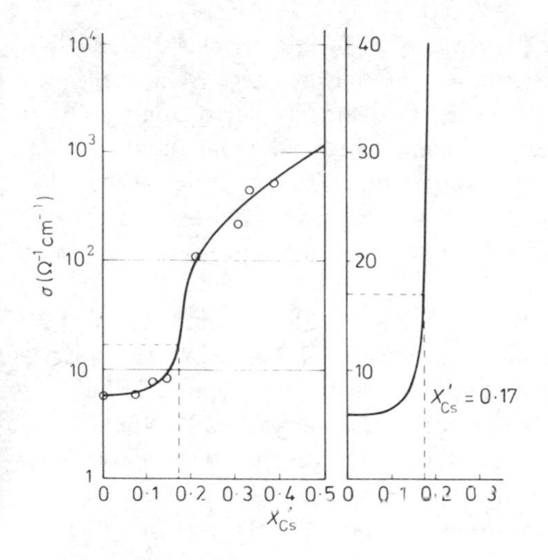

Figure 2. Electrical conductivity σ as a function of the excess concentration of Cs, X'_{Cs}, of the liquid CsAu–Cs system at 650 °C

DC current, by a simple resistance measurement. An AC technique must be used for this purpose to eliminate uncertainties due to large Seebeck voltages caused by small temperature gradients along the cell. A detailed description of the experiment will be published elsewhere (Krüger and Schmutzler 1976).

3. Results and discussion

The results for the absolute thermoelectric power of the Cs–Au alloys as a function of composition are given as the strong line with the experimental points in figure 3. They are deduced from the measured Seebeck coefficients of the Pt/Cs–Au alloy/Pt thermo-

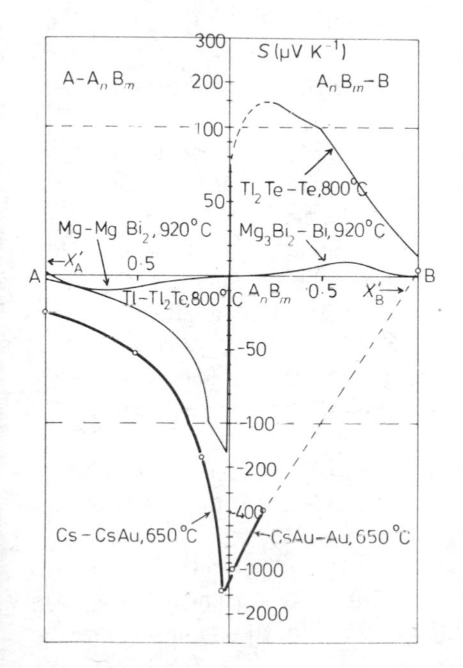

Figure 3. Absolute thermoelectric power S of the liquid Cs–Au system. For the comparison with 'normal' liquid semiconductor systems, the curves for the Mg–Bi and Tl–Te systems are given according to Enderby and Collings (1970) and Cutler (1971). The kinks in the curves at $\pm 100\ \mu V\ K^{-1}$ are only due to the change from a linear scale for $|S| < 100\ \mu V\ K^{-1}$ to a logarithmic scale for $|S| > 100\ \mu V\ K^{-1}$.

couple by correcting for the known absolute thermopower of Pt (Cusack and Kendall 1958). As can be seen from the figure, the thermopower is always negative in the vicinity of the stoichiometric compound and reaches very large, typically nonmetallic values (about $-1400\,\mu V\,K^{-1}$) near the equiatomic composition. The maximum possible error in these results due to the effect of thermal diffusion, to uncertainties of the overall concentration and to the determination of the temperature gradient along the cell is estimated to be smaller than about ±50%. Despite this fairly large possible error the above statements are still valid.

This behaviour of the absolute thermopower of the Cs–Au alloys cannot be explained by simple models of liquid semiconductors, which predict an n–p transition near the stoichiometric composition, causing a change of sign of the thermopower. This 'normal' behaviour is visualized in figure 3 by the curves for the systems Mg–Bi and Tl–Te, taken from the literature (Enderby and Collings 1970, Cutler 1971). However, there are some liquid semiconductors preserving the same sign of the thermopower on both sides of the stoichiometric composition, for example the Ge–Te or the Ga–Te systems (Valiant and Faber 1974). As an explanation of this behaviour, bond rearrangement between like and unlike atoms in these essentially covalently bonded systems is assumed in order to satisfy all the valencies of the atoms within the short-range order. So the same kind of excess state, due to for example vacancies within the structure, is always created, independent of the composition. This explanation can hardly be applied to a system like Cs–Au, as both components are unable to form stable covalent bonds within their respective pure condensed phases, so that predominantly covalent bonds in the CsAu compound cannot be expected. Furthermore, the very large value of the thermoelectric power, more than $1000\,\mu V K^{-1}$, would correspond to an energy gap or a mobility gap ΔE of at least 2 eV according to Mott and Davis (1971).

$$S = -\frac{k}{e}\left(\frac{\Delta E}{kT}+1\right)$$

when only electrons play an important role as current carriers, as must be assumed because of the negative sign of S.

In the above equation k, e and T are Boltzmann's constant, the elementary charge and the absolute temperature respectively. This large energy gap disagrees drastically with the observed activation energy of about 0·2 eV for the electrical conductivity. Therefore an alternative explanation may be given by the assumption of stable Cs^+ and Au^- ions, dominating the electrical transport properties of the stoichiometric liquid. The observed absolute value compares well with values observed in molten salt systems (Shimoji and Ichikawa 1967, 1969). However, as principally in these systems the effect of the electrode–ionic liquid interface cannot be separated from the bulk effect in the liquid, it is difficult to give any meaningful interpretation of the observed thermopowers.

Further information can be drawn from the electromigration experiments, which must result in an ordinary electrolysis according to Faraday's law in the case of purely ionic transport. Figure 4 shows a typical experimental curve (A) for the cell resistance as a function of time while passing a constant DC current of 10 mA, corresponding to a current density of about $3\,A\,cm^{-2}$, through the sample. Initially a strong linear decrease in resistance with time is observed, which bends over after about 1 min into a slowly

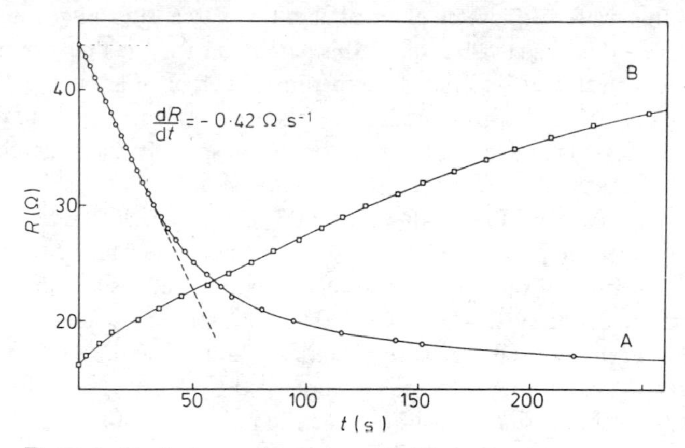

Figure 4. Typical experimental curve of the cell resistance as a function of time: A, during the passage of a constant DC current of 10 mA which is started at $t = 0$; B, after switching off the DC current at $t = 0$ having reached the stationary state previously.

varying value, becoming practically constant after 3–5 min. The total change of resistance is by about a factor of 2·7, which is much too large to be explained by a simple temperature increase, as can be seen from the temperature coefficient given above as $d \ln \sigma / dT = 3{\cdot}5 \times 10^{-3}\,K^{-1}$. With the assumption of a homogeneous concentration within the capillary in the stationary state, a limiting excess concentration of Cs, $X'_{Cs} = 0{\cdot}17$, can be evaluated from the stationary value of the cell resistance with the aid of figure 2.

To test the above suggestion of ion-dominated electrical transport, a correlation must be established between the total charge passed through the specimen and the mass transported into the cathode compartment by the DC current. For this purpose the ratio n_{Cs}/n_e (n_{Cs} = number of Cs atoms accumulated in the capillary, n_e = number of elementary charges crossing the orifice of the capillary) must be evaluated. It is about 10^{-6} for dilute metallic alloys (Epstein 1972), and Z^{-1} for liquid electrolytes according to Faraday's law of electrolysis, where Z is the valency of the ion involved. The difference of many orders of magnitude in this parameter thus provides a clear distinction between the two mechanisms.

For the evaluation of this parameter, density data for the liquid Cs–Au alloys are needed: these are not known from experiment as yet. They may be estimated by correcting the known density of the solid stoichiometric compound (Spicer *et al* 1959) for the volume increase on melting and thermal expansion, by the assumption of some reasonable numbers for these effects. The values used here are roughly those of CsBr, that is,

$$\alpha = \frac{1}{v}\left(\frac{\partial v}{\partial T}\right) = 2 \times 10^{-4}\,K^{-1}$$

and $(\Delta v/v)_{melting} = 0{\cdot}15$, so giving a molar volume of 61 $cm^3\,mol^{-1}$ for liquid CsAu at 650 °C. To estimate the concentration dependence of the mean molar volume of liquid Cs–CsAu mixtures, an ideal solution of excess Cs in liquid CsAu was assumed, taking the atomic volume of pure Cs at 650 °C as 88·6 cm^3/g-atom (Achener 1968). The molar volumes obtained in this way may have an error of about ±15% causing an equally large error in the parameter n_{Cs}/n_e. The evaluation of this parameter for the linear part of curve A in figure 4 gives a value of 1 within the experimental error of about ±25% including the uncertainty of the densities used. As this corresponds to the natural valency of the Cs ions, it is concluded, that the electrical transport in the stoichiometric liquid is essentially ionic.

To see if the fairly high value of the conductivity of $5\ \Omega^{-1}\ cm^{-1}$ is at least compatible with a purely ionic conduction, a rough estimate of the diffusion coefficient would be valuable. Then the application of the Nernst–Einstein equation enables an upper limit for the ionic conductivity of the system under interest to be estimated. As the capillary–reservoir arrangement used here is very similar to normal experimental arrangements for diffusion experiments the observation of the capillary resistance after switching off the electrolysing current should provide the possibility of such an estimate.

The curve labelled B in figure 4 shows the behaviour of the cell resistance as a function of time after switching off the DC current. As can be seen from the figure the resistance increases slowly and reaches the value it had before starting the electromigration experiment after some 20 min. With the assumptions of a homogeneous excess concentration $X'_{Cs} = 0{\cdot}17$ within the capillary as an initial condition and a vanishing excess concentration $X'_{Cs} = 0$ at the orifice of the capillary as a boundary condition, Fick's second law of instationary diffusion can be solved, giving a rough estimate of the diffusion coefficient of Cs in CsAu. So a value of $D_{Cs} = 2 \times 10^{-4}\ cm^2\ s^{-1}$ is obtained. The application of the Nernst–Einstein equation gives an upper limit for the contribution of the Cs ions to the electrical conductivity of about $4\ \Omega^{-1}\ cm^{-1}$ which is consistent with the suggestion of essentially ionic conduction in the liquid CsAu compound.

As all the experimental data available for the liquid CsAu system support the suggestion of the existence of stable Cs^+ and Au^- ions in the molten CsAu compound, where especially the electromigration experiment provides direct experimental evidence for this suggestion, it is concluded that the electrical transport properties are dominated in this system in the liquid state by the ions. So the compound CsAu is the first example of an ionic semiconductor showing a transition to ionic conduction on melting.

Acknowledgments

This work was supported by the Deutsche Forschungsgemeinschaft. The authors are grateful to Professor Dr F Hensel for many valuable discussions.

References

Achener P Y 1968 *Aerojet General Corp., Nuclear Div. San Ramon, California, Res. Rep.* AGN – 8195

Cusack N E and Kendall P 1958 *Proc. Phys. Soc.* **72** 898

Cutler M 1971 *Phil. Mag.* **24** 381, 401

Cutler M and Field M F 1968 *Phys. Rev.* **169** 632

Cutler M and Mallon C E 1966 *Phys. Rev.* **144** 642

Elliot R P 1965 *Constitution of Binary Alloys* 1st Suppl. (New York: McGraw-Hill)

Enderby J E and Collings E W 1970 *J. Non-Cryst. Solids* **4** 161

Epstein S G 1972 *Liquid Metals* ed S Z Beer (New York: Marcel Dekker)
Hinze J and Jaffé H H 1962 *J. Am. Chem. Soc.* **84** 540
Hinze J, Whitehead M A and Jaffé H H 1963 *J. Am. Chem. Soc.* **85** 148
Hoshino H, Schmutzler R W and Hensel F 1975 *Phys. Lett.* **A51** 7
Ilschner B R and Wagner C N 1953 *Acta Metall.* **6** 712
Krüger K D and Schmutzler R W 1976 *Ber. Bunsenges. Phys. Chem.* **80** 816
Mott N F and Davis E A 1971 *Electronic Processes in Non-Crystalline Materials* (Oxford: Clarendon Press)
Pauling L 1960 *The Nature of the Chemical Bond* (Ithaca, New York: Academic Press)
Phillips J C 1973 *Bonds and Bands in Semiconductors* (New York: Academic Press)
Schmutzler R W, Hoshino H, Fischer R and Hensel F 1976 *Ber. Bunsenges. Phys. Chem.* **80** 107
Shimoji M and Ichikawa K 1967 *Ber. Bunsenges. Phys. Chem.* **71** 1149
—— 1969 *Ber. Bunsenges. Phys. Chem.* **73** 302
Spicer W E, Sommer A M and White J G 1959 *Phys. Rev.* **115** 57
Steinleitner G, Freyland W and Hensel F 1975 *Ber. Bunsenges. Phys. Chem.* **79** 1186
Suchet J P 1965 *Chemical Physics of Semiconductors* (London: van Nostrand)
Valiant J C and Faber T E 1974 *Phil. Mag.* **29** 571
Wooten F and Condas G A 1963 *Phys. Rev.* **131** 657

Metal–nonmetal transition and change in the electronic and magnetic properties of liquid Cs–Sb and Cs–Au alloys

W Freyland and G Steinleitner
Fachbereich Physikalische Chemie, Philipps-Universität Marburg, 3550 Marburg (Lahn), Biegenstr. 12, West Germany

Abstract. The paper reports new experimental results of the electrical conductivity, absolute thermoelectric power, and magnetic susceptibility of liquid Cs–Sb alloys. Measurements have been performed for different compositions ranging from 0 to 25 at.% Sb and for temperatures up to 1000 °C. Near the compound composition Cs_3Sb a steep metal–nonmetal transition is observed with a minimum conductivity of $2\Omega^{-1}\,cm^{-1}$. A detailed comparison of these results with the corresponding data of the liquid Cs–Au system is given. From the discussion of the electronic and magnetic properties of these two systems, it is suggested that a high amount of covalent bonding exists in liquid Cs_3Sb, whereas liquid CsAu exhibits predominantly ionic bonding.

1. Introduction

The majority of liquid binary alloy systems investigated so far, which show clear deviations from metallic behaviour at some compositions, consist of a metallic component and a chalcogenide such as S, Se or Te (see, for example, the review by Enderby 1974). According to their electronic properties, these liquids have been described as 'liquid semiconductors'. A comparatively small number of liquid alloys are known which are composed of two metallic elements in the pure liquid state and which become nonmetallic at definite stoichiometric compositions. Up to now, the most prominent example of this type of alloy, with undoubtedly nonmetallic characteristics, is the liquid Cs–Au system (Schmutzler and Krüger 1976, Schmutzler *et al* 1976). Certainly the large difference in electronegativities between caesium and gold is one favourable requirement for the occurrence of nonmetallic electronic states in this system. However, the question arises whether this is a generally necessary condition.

Our knowledge about this kind of alloy is not yet sufficient to enable us to decide which of the physical and chemical properties of the metallic constituents are most essential to stabilize nonmetallic electronic states in the liquid. In view of this problem, it is of special importance to find out which type of chemical bonding mostly prevails for these nonmetallic liquid alloys. As for the type of the metal–nonmetal transition in these liquids, a second question of interest concerns the change in nature of the electronic states in relation to the bonding of the nonmetallic compound. In order to obtain, experimentally, some information about these questions it is necessary to look systematically for further examples of these alloys. This is not a straightforward task and the success of finding new systems may be just a little fortuitous: we found that liquid Cs_3Sb is one.

In this paper we report new measurements of the temperature dependence of the electrical conductivity of liquid Cs–Sb alloys in the concentration range between pure Cs and the compound composition Cs_3Sb, where a minimum conductivity of $2\Omega^{-1}\,cm^{-1}$ has been observed. According to the phase diagram (Elliott 1965) the solid compound Cs_3Sb exists and melts at 725 °C. It is semiconducting with a gap energy of 1·6 eV and crystallizes in a compensated interstitial lattice (Suchet 1965). At and near 25 at.% Sb, experimental values of the absolute thermoelectric power have also been obtained. New measurements of the magnetic susceptibility of this alloy are compared with previous susceptibility results of the Cs–Au system (Freyland and Steinleitner 1976). It is shown that the magnetic susceptibility can yield some direct information about the chemical bonding in liquid CsAu and Cs_3Sb. Special emphasis is given to the discussion of characteristic differences in the electronic and magnetic properties of these two alloys.

2. Experimental methods

The high corrosivity of the liquid metallic components, and the relatively high vapour pressure and chemical reactivity of liquid caesium at elevated temperatures, make special constructions of the liquid sample containers necessary. The cell for conductivity and thermopower measurements is shown schematically in figure 1. The samples used in these experiments were prepared from 99·999% pure Sb and 99·95% pure Cs. The reservoir was filled with these elemental materials under a pure argon atmosphere and then the cell was closed under vacuum. For a homogeneous mixing of the alloys, the whole cell was kept at 700 °C to 800 °C for several hours. Afterwards the cell, together with the surrounding furnaces, was inverted to pour the liquid alloy into the measuring compartment, which was a capillary tube made from Lucalox, a polycrystalline sintered α-alumina. The temperature gradient along this tube was controlled by two Pt–PtRh

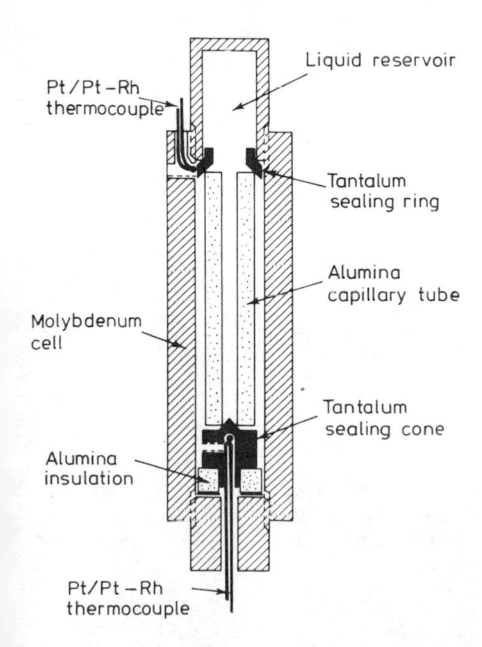

Figure 1. Construction of the cell used for conductivity and thermopower measurements of liquid Cs–Sb alloys.

thermocouples, which at the same time served as voltage and current probes for the conductivity and thermopower measurements. For this purpose, the tips of the thermocouples were mechanically clamped to the tantalum sealing ring and sealing cone at the top and bottom end of the cell respectively. For thermopower measurements, definite temperature gradients were produced by two separate resistance furnaces.

This technique has been applied to alloys with low values of electrical conductivity. The measurements of alloys with high conductivities have been performed in thin-walled molybdenum cells, the construction of which has been described previously (Steinleitner *et al* 1975). The sample preparation with these cells was the same as above. With both methods, the reproducibility of the experimental points was checked on several heating and cooling cycles and proved to be very satisfactory.

The Faraday method was used for the determination of magnetic susceptibility. An approximate volume of 1·5 cm^3 of sample was contained in a thin-walled molybdenum cylinder, which was tightly vacuum-sealed by a niobium cone. Measurements have been made at different magnetic fields ranging from 7 to 10 kG and under a vacuum of about 10^{-5} Torr. At the highest temperatures an absolute weight resolution of ±10 μg was achieved. For some concentrations of the Cs–Sb system, the tightness of the cell was not perfect, in which case a main error source was induced.

3. Experimental results

The electrical conductivity σ of liquid Cs–Sb has been measured in the composition range from 0 to 25·0 at.% Sb and at temperatures up to 980 °C. Figure 2 shows the concentration dependence of σ and of the temperature coefficient at constant composition, $d(\ln \sigma)/dT$, at 800 °C. With increasing antimony content the conductivity

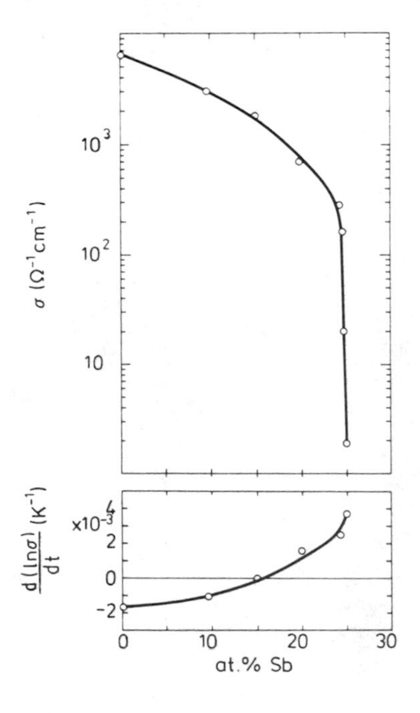

Figure 2. (*a*) Electrical conductivity, σ, and (*b*) temperature coefficient of σ at constant composition, $d(\ln \sigma)/dT$, of liquid Cs–Sb alloys at 800 °C.

decreases continuously down to about $200\,\Omega^{-1}\,cm^{-1}$ at 24·5 at.% Sb. A very sharp drop in σ by two orders of magnitude is observed in the narrow concentration range between 24·5 to 25·0 at.% Sb. The temperature coefficient changes from negative to positive values at about 15 at.% Sb, where the conductivity is $2 \times 10^3\,\Omega^{-1}\,cm^{-1}$. As will be seen later in figure 6, the logarithm of σ follows a reciprocal temperature dependence over a range of about 200 °C at 25 at.% Sb. For three concentrations between 24·5 and 25 at.% Sb, experimental values of the absolute thermoelectric power S have been determined. Within experimental error no clear variation of S with composition has been observed. The mean value was found to be $-20 \pm 10\,\mu V\,K^{-1}$.

Because of the strong concentration dependence of σ near the compound composition Cs_3Sb, it is very difficult to give exact numbers for the absolute accuracy of these data. The uncertainty in the composition itself, determined by simple weighing, was about ±0·1 at.% Sb. On the other hand a small leak in the cell during the measurement, which caused a shift in composition, could be noticed immediately by a time-dependent drift of the measured voltage and by a corresponding irreproducibility of the data on heating and cooling. An impression of the relative accuracy of the data is given by the scattering of the plotted experimental points in figure 6.

In figure 3, the data of the magnetic susceptibility per gram, χ_g of liquid Cs–Sb and Cs–Au and the respective temperature coefficients $d\chi_g/dT$, are plotted for different compositions. Referred to the broken line, which is a simple interpolation between the susceptibility values of the pure components, a pronounced increase of diamagnetism is found in both systems as the compound compositions are approached. We call particular attention to two features of the data. The concentration dependence is different in these two alloys near the compound compositions Cs_3Sb and CsAu. Related

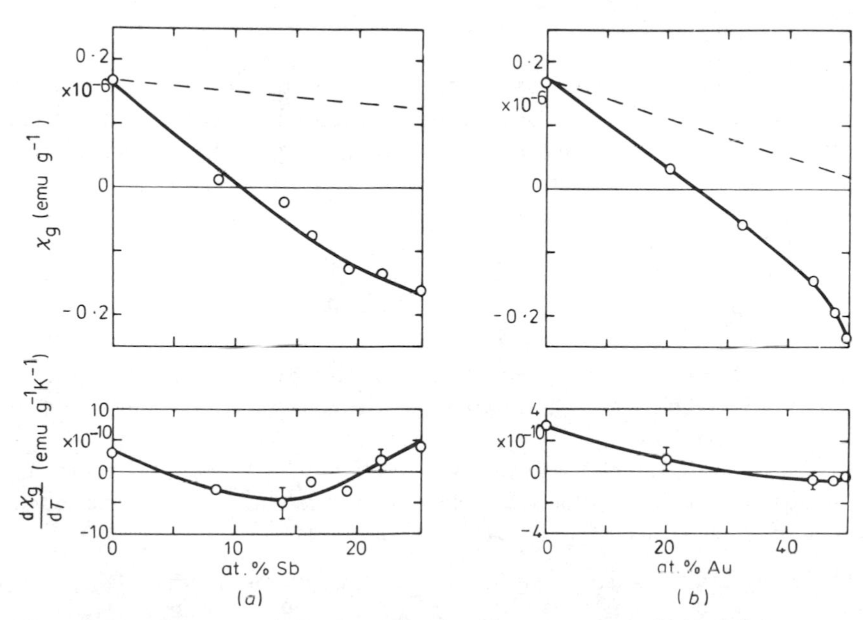

Figure 3. Magnetic susceptibility per gram, χ_g, and temperature coefficient of χ_g at constant composition, $d\chi_g/dT$, of (*a*) liquid Cs–Sb at 750 °C and (*b*) liquid Cs–Au at 650 °C.

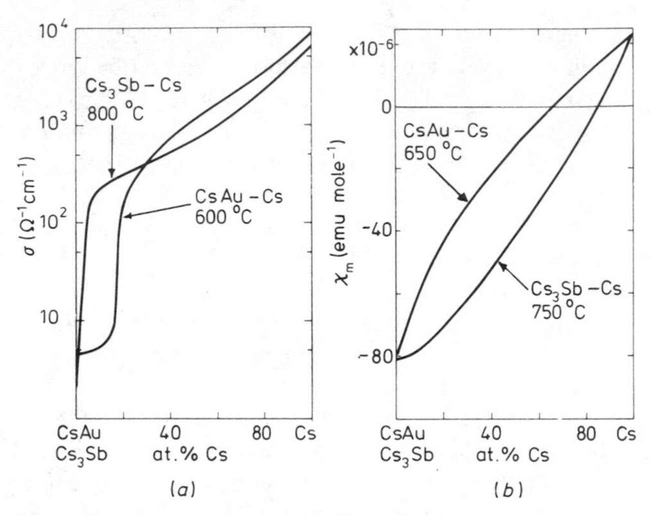

Figure 4. Comparison of (*a*) the electrical conductivity σ and (*b*) the molar magnetic susceptibility χ_m, of Cs–Au and Cs–Sb alloys; data of σ for Cs–Au have been taken from the literature (Schmutzler *et al* 1976).

to this fact a clear positive temperature dependence of χ_g is observed near Cs_3Sb, whereas near CsAu it is comparatively small and negative. The total behaviour of $d\chi_g/dT$ is not simple. For some concentrations, the uncertainty in $d\chi_g/dT$ is relatively large, and no correct concentration dependence of these data can be discussed. The absolute error of our susceptibility data is estimated to be ±5 to ±10% for values not close to zero.

4. Discussion

To understand the change of the electronic structure from metallic to nonmetallic states, a direct comparison of the concentration dependence of the electrical conductivity and the magnetic susceptibility of the two systems is very informative: this is illustrated in figure 4. In order to compare both alloys, the concentration scale has been converted, varying between the compound composition on the left and pure Cs on the right hand. In this plot, the conductivity of Cs_3Sb–Cs rises abruptly within a narrow concentration range of less than 5 at.% excess Cs. In contrast to this behaviour, σ of the nonmetallic CsAu–Cs remains approximately constant up to about 15 at.% access Cs, where the nonmetal–metal transition sets in. The molar magnetic susceptibilities χ_m are plotted in figure 4(*b*). The peculiar behaviour of χ_m is that in Cs–Au, the slope of the curve becomes steeper as the compound composition is approached, whereas in Cs–Sb it reduces.

This plot clearly demonstrates that a distinction exists between the nature of the electronic states in both systems. Starting from this result, the following discussion shall be concentrated particularly on the following aspects:

(i) Influence of Sb or Au on the charge carrier susceptibility in the metallic transport regime;
(ii) magnitude of the atomic susceptibility of CsAu and Cs_3Sb in relation to the chemical bonding;

(iii) interpretation of the difference in the temperature and concentration dependence of conductivity and susceptibility in the nonmetallic transport regime.

Consider first the metallic region. If, for the moment, we neglect structural effects (Timbie and White 1970), the charge carrier contribution χ^c to the total susceptibility χ should be mainly determined by the density of states at the Fermi level, $n(E_F)$, and according to the free electron model, χ_v^c per unit volume is given by:

$$\chi_v^c = \tfrac{4}{3}\mu_B{}^2\, n(E_F)_{\text{free}}, \tag{1}$$

where μ_B is the Bohr magneton. If we approximate the atomic contribution, χ^a, by the diamagnetism of the corresponding ions Cs^+, Au^+ and Sb^{5+} (Weiss and Witte 1973), it is possible to separate the experimental $\chi_v^c(\text{exp})$ from the total measured susceptibility. In this way some information may be gained about the reduction of $n(E_F)$, which is defined by the g-factor

$$g = n(E_F)/n(E_F)_{\text{free}} \simeq \chi_v^c(\text{exp})/\chi_v^c(\text{calc}). \tag{2}$$

Here $\chi_v^c(\text{calc})$ has been calculated according to equation (1).

The result of this evaluation is plotted in figure 5. As can be seen, the g-factor starts to decrease rapidly for high Cs concentrations which lie well inside the metallic conduction regime. Mott has given several arguments (see, for example, Mott and Davies 1971) that g should decrease by about a factor of three for a corresponding decrease of the conductivity from about $10^3\,\Omega^{-1}\,cm^{-1}$ to $10^2\,\Omega^{-1}\,cm^{-1}$. These two conductivity values define the approximate limits of Mott's 'strong scattering transport regime', which should lie between Mott's minimum metallic conductivity of about $10^2\,\Omega^{-1}\,cm^{-1}$ and $10^3\,\Omega^{-1}\,cm^{-1}$, where the electron mean free path and the interatomic distance are of the same magnitude. In this region we observe a reduction of g by about a factor of two for both Cs–Au and Cs–Sb. A similar reduction in g has been determined by Knight-shift measurements in the metal–nonmetal transition region of liquid In_2Te_3 and Ga_2Te_3 (Warren 1971). However, a critical point must be made about our evaluation of χ_v^c. A reduction of the electronic paramagnetism should at the same time lead to an increase of the atomic diamagnetism and thus an even steeper decrease of g with

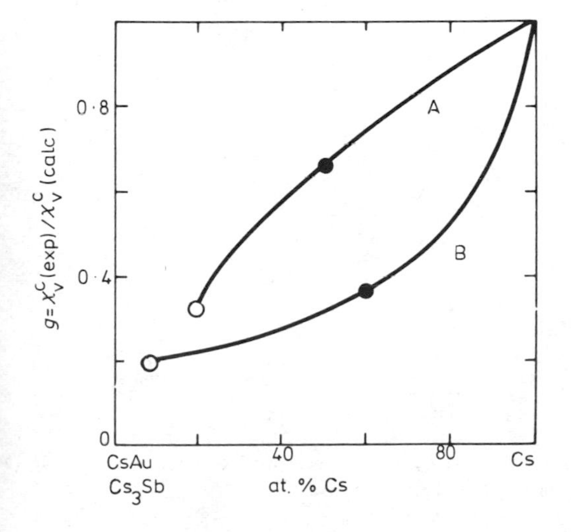

Figure 5. Concentration dependence of the g-factor of the two alloy systems: A, Cs–Au and B, Cs–Sb. The points ○ are for $\sigma = 10^2\,\Omega^{-1}\,cm^{-1}$ and ● are for $\sigma = 10^3\,\Omega^{-1}\,cm^{-1}$.

composition may result. Yet, in order to account for this effect, a separate measurement, for example of the liquid structure, is necessary. With the aid of such measurements an estimate of the concentration of new atomic or ionic species which may form in the metallic liquid should be possible.

We turn next to the interpretation of the observed electronic and magnetic properties in the nonmetallic transport regime and especially to the discussion of the chemical bonding in liquid CsAu and Cs_3Sb. Considering first the Cs–Au system, in a recent paper Freyland and Steinleitner (1976) give a detailed description of the temperature and composition dependence of the magnetic susceptibility of liquid Cs–Au. So we will restrict ourselves here to a summary of the main results. There is good evidence that charge carriers do not contribute to the susceptibility at the compound composition CsAu. Based on this consideration, the observed molar susceptibility of -81×10^{-6} emu mol^{-1} at 50 at.% Au can be described by a predominantly ionic bonding of Cs^+ and Au^- ions. This conclusion is strongly supported by the observed ionic migration in liquid CsAu (Schmutzler and Krüger 1976). The slope of the concentration dependence of χ_m and the sign and magnitude of $d\chi_m/dT$ for mixtures of CsAu with excess Cs, are consistent with a description of the dissolved Cs atoms by non-interacting impurity atoms. Here the Cs valence electron should be trapped at the Cs impurity centre. For the diameter of the orbital of the trapped electron, we found a value of about 10 Å. At the onset of the nonmetal–metal transition the mean interatomic distance can be estimated to be 10 Å. Thus the metal–nonmetal transition in the Cs–Au system possibly may be explained by the overlap of states on neighbouring Cs impurity centres leading to a delocalization of the trapped electrons.

In the case of liquid Cs_3Sb, a thermally activated dependence of σ on temperature has been observed over a temperature range of about 200 °C, as is shown in figure 6. From this plot an activation energy of $\Delta E = 0.44$ eV is determined. This magnitude of

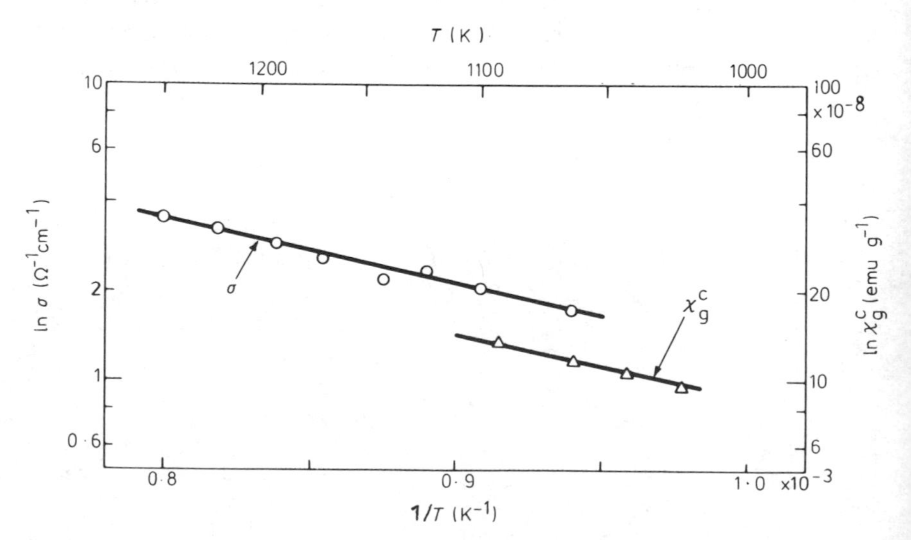

Figure 6. Logarithm of conductivity σ and of charge carrier susceptibility per gram, χ_g^c, of liquid Cs_3Sb as a function of the reciprocal temperature.

ΔE, together with the small absolute thermoelectric power of approximately $-20\,\mu V\,K^{-1}$, gives a good indication that electronic conduction should prevail in liquid Cs_3Sb. For purely intrinsic semiconduction the conductivity σ is given by

$$\sigma = \sigma_0 \exp(-\Delta E/kT). \tag{3}$$

To describe the σ data of Cs_3Sb with this relation, a value of $\sigma_0 \simeq 5 \times 10^2\,\Omega^{-1}\,cm^{-1}$ has to be taken. This is a reasonable value of σ_0, typically found in amorphous semiconductors (Mott and Davis 1971). The positive temperature dependence of χ_g can be explained in the same way, that is, by thermally excited electrons and holes. Therefore the charge carrier contribution χ_g^c should be given by

$$\chi_g^c = \frac{\mu_B{}^2}{\rho}\, n_0 (kT)^{1/2} \exp(-\Delta E/kT), \tag{4}$$

where ρ is the mass density and n_0 depends on the effective mass of electrons and holes. With the above $\Delta E = 0{\cdot}44$ eV, the experimental data fit this relation very well as is shown in figure 6. From this fit, a molar value of the atomic contribution to χ of $\chi_m^a = -136 \times 10^{-6}$ emu mol^{-1} is obtained. This value is equal to the molar susceptibility of solid Cs_3Sb measured at room temperature (Freyland and Steinleitner 1976). As the electrical gap energy of solid Cs_3Sb is 1·6 eV, the charge carrier susceptibility of the solid semiconductor at room temperature is negligibly small. So the coincidence of χ_m^a in the liquid state with χ_m in the solid gives good evidence that the type of bonding in liquid Cs_3Sb is the same as in the solid, that is, sp^3-hybridized bonding (Suchet 1965). A predominantly ionic bonding of $3Cs^+$ and Sb^{3-} can certainly be excluded. This would yield a molar susceptibility of $\chi_m^a = -180 \pm 15 \times 10^{-6}$ emu mol^{-1}, which is calculated from the literature value of Cs^+ and from an estimate of Sb^{3-} by the isoelectronic sequence Xe, I^-, Te^{2-} (Weiss and Witte 1973).

In conclusion it may be stated that magnetic susceptibilities give a valuable direct insight into the different chemical bonding in CsAu and Cs_3Sb, which also shows up in a distinct behaviour of the magnetic and electronic properties of these two alloys. In addition, some indications about the type of metal–nonmetal transition have been obtained. Possibly, the type of abrupt transition in liquid Cs–Sb is more similar to a change of the electronic states in highly degenerated semiconductors.

Acknowledgments

We are indebted to F Hensel for stimulating discussions. We would like to thank C Karaschinski for valuable technical assistance in performing part of the conductivity experiments.

References

Elliott R P 1965 *Constitution of Binary Alloys: First Supplement* (New York: McGraw-Hill)
Enderby J E 1974 *Amorphous and Liquid Semiconductors* (London: Plenum Press)
Freyland W and Steinleitner G 1976 *Ber. Bunsenges. Phys. Chem.* **80** (8) 810

Krüger K D, Fischer R and Schmutzler R W 1977 this volume
Mott N F and Davis E A 1971 *Electronic Processes in Non-Crystalline Materials* (Oxford: Clarendon Press)
Schmutzler R W, Hoshino H, Fischer R and Hensel F 1976 *Ber. Bunsenges. Phys. Chem.* **80** 107
Steinleitner G, Freyland W and Hensel F 1975 *Ber. Bunsenges. Phys. Chem.* **79** 1186
Suchet J P 1965 *Chemical Physics of Semiconductors* (London: Van Nostrand)
Timbie J P and White R M 1970 *Phys. Rev.* **B1** 2409
Warren W W Jr 1971 *Phys. Rev.* **B3** 3708
Weiss A and Witte H 1973 *Magnetochemie* (Weinheim: Verlag-Chemie)

The far-infrared reflectivity of liquid Mg–Bi alloys

E D Crozier, R W Ward and B P Clayman
Department of Physics, Simon Fraser University, Burnaby, BC, Canada V5A 1S6

Abstract. Electron transport measurements of liquid Mg–Bi alloys suggest marked local ordering of the atoms near the stoichiometric composition Mg_3Bi_2. Thermodynamic data and calculations have suggested that molecules of Mg_3Bi_2 may exist in the liquid state. Using high-resolution Fourier transform spectroscopy the reflectivity of Mg–Bi alloys has been measured as a function of temperature and composition over the frequency range 3 cm^{-1} to 300 cm^{-1}. Molecular vibrational levels were not observed. The reflectivity displayed an anomalous increase over a narrow range about the 60 at.% Mg composition. The data, which can be fitted to the Drude model, indicate that the Drude optical parameter $\sigma(0)_{opt}$ of Mg_3Bi_2 is enhanced by a factor of 100 relative to the DC value of $\sigma(0)$. The implications of the results for the electronic density of states are discussed.

1. Introduction

The fluid and electronic structures of liquid binary alloys which displays semiconducting behaviour are not completely understood. The bulk of the experimental work has concentrated on measurement of thermodynamic and DC electron transport properties. Most of the alloys are characterized by a composition-dependent DC conductivity $\sigma(0)$ which displays a pronounced minimum near a stoichiometric composition. Taken together, the experimental results indicate that marked short-range order exists near these critical stoichiometric compositions. The Mott pseudogap model (for review see Mott and Davis 1971) has provided a convenient basis for discussion of the electronic structure. In this model a dip develops in the density of states $N(E)$ near the Fermi energy E_F. When $\sigma(0) \lesssim 200\,\Omega^{-1}\,cm^{-1}$ Mott argues that states near E_F become localized and that electronic conduction occurs via thermally activated hopping processes. The NMR work of Warren (1977) provides the strongest experimental evidence for the existence of the pseudogap and the onset of localization.

Better understanding of the electronic processes requires knowledge of the local spatial arrangement of atoms. There are three ways of obtaining the required structural information. Neutron scattering (Hawker *et al* 1974) has been used to determine the partial structure factors in binary systems for which $\sigma(0) > 200\,\Omega^{-1}\,cm^{-1}$ and the extended x-ray absorption fine structure technique has been used to investigate the local structure in systems for which $\sigma(0) \ll 200\,\Omega^{-1}\,cm^{-1}$ (Crozier *et al* 1976). Thirdly, for those systems in which it is suspected that the local structure results from the presence of molecules, appropriate optical experiments provide a definitive structural probe.

In this paper we report measurements of the far-infrared reflectivity of liquid Mg–Bi alloys which display semiconducting behaviour for compositions near 60 at.% Mg. In calculating the composition dependence of thermodynamic data and the concentration–concentration correlation function in the long wavelength limit, McAlister and Crozier (1974) assumed that molecules of Mg_3Bi_2 existed with finite lifetimes. Using a similar model Ratti and Bhatia (1975) calculated with some success the DC conductivity and thermoelectric power of liquid Mg–Bi alloys. Our reflectivity measurements were initiated: (i) to test specifically for vibrational modes that could be associated with molecules of Mg_3Bi_2, (ii) to measure the reflectivity in a frequency range where hopping processes may be important, and (iii) to demonstrate the feasibility of making optical measurements in the frequency range 9×10^{10} Hz to 9×10^{12} Hz on chemically reactive liquids at elevated temperatures.

2. Experimental details and results

Far-infrared reflectivity measurements were taken over the range 20–300 cm^{-1} using as energy source and analyser a modified Beckmann RIIC FS-720 Michelson interferometer operating in the asymmetric mode. Typical instrumental resolution was 2·5 cm^{-1}. On some runs a lamellar interferometer was used to cover the range 3–30 cm^{-1}.

Samples of Mg–Bi alloys were made from Mg and Bi of 99·99% and 99·999% purity respectively supplied by Ventron Corporation, Beverly, Massachusetts, USA. They were contained in a 316 stainless steel crucible inside an evacuable, resistance-heated furnace which could be kept at a pressure of 10^{-5} Torr at 800 °C. The furnace was maintained at a slight overpressure of pure He during all runs. This was necessary to reduce the evaporation of Mg and consequent composition changes of the alloys. A stirring rod was built into the top of the furnace and could be lowered into the crucible or fully withdrawn at any time. The stirring was required to ensure homogeneity of the alloys which were thermally cycled between the liquid and solid states. The crucible could be moved vertically by an externally mounted micrometer in order to optimize optical alignment. The far-infrared radiation from an output light-pipe fitted to the interferometer was incident on the sample at a 16° angle. After one specular reflection from the surface of the sample, the radiation was collected and transmitted through a light-pipe to the detector.

The detector cryostat is similar to that described by Clayman *et al* (1971). It houses a doped germanium bolometer operating at 0·3 K. The bolometer's output signal was amplified and the resultant interferograms recorded digitally. The interferograms were Fourier-transformed in the laboratory on an HP2115A minicomputer immediately after recording. To obtain a reference spectrum the reflectivity of an aluminium mirror, placed in the furnace, was measured prior to each run. In addition, the dependence of bolometer output upon sample temperature was included by measuring the reflection spectrum of a polished copper mirror at room temperature and at several elevated temperatures between 700 °C and 850 °C. There was a 45% loss of signal on heating the Cu mirror from room temperature to 820 °C.

The normalized reflectivity was determined for liquid and solid alloys containing 55 to 70 at.% Mg. The high partial pressure of Mg placed an upper limit on the temperature range of practical measurements on the liquids. For example, at the composition 60 at.% Mg, evaporation losses became excessive 20 °C above the liquidus temperature. Typical reflectivity results obtained with a resolution of 2·6 cm^{-1} and normalized to the reflectivity of Cu at room temperature are shown in figure 1 for different compositions and different temperatures. The compositions are correct to within 1 at.% Mg and the temperatures within 1%. The normalized reflectivity is accurate to within 10%. At

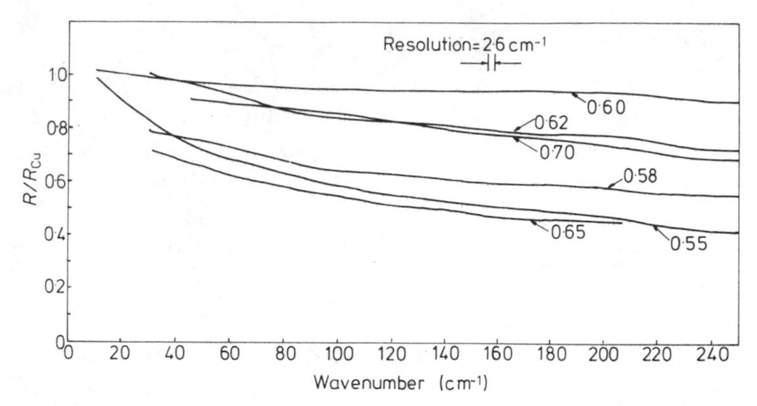

Figure 1. The reflectivity of liquid Mg–Bi alloys. The reflectivity is normalized to the reflectivity of Cu at room temperature. The resolution is 2·6 cm^{-1}. The numbers represent the composition in at.% Mg. The temperature of the reflectivity curves are 832 °C (60), 813 °C (62), 790 °C (58), 783 °C (65), 770 °C (70), 687 °C (55).

all compositions the reflectivity of the liquid was greater than that of the solid. The solid results will be published elsewhere.

There is no evidence in the reflectivity spectra of absorption due to molecular vibrational modes. Within experimental error, for all compositions investigated the relectivity decreases monotonically with increasing frequency.

The composition-dependence of the reflectivity is anomalous. To illustrate this the data of figure 1 are used to plot in figure 2 the reflectivity at the fixed frequency of 100 cm^{-1} as a function of at.% Mg. The reflectivity increases rapidly in a narrow composition range centred about Mg_3Bi_2. Similar behaviour is observed at other frequencies. Figure 2 was derived from data taken near the liquidus temperatures of the different compositions. However, it is anticipated that similar behaviour would be observed in an isothermal plot.

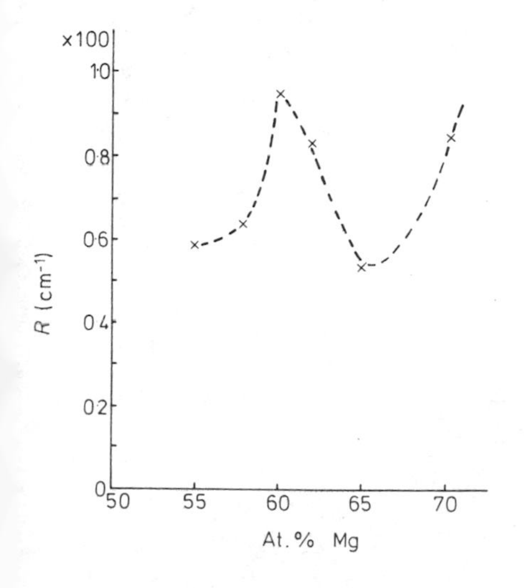

Figure 2. Reflectivity of liquid Mg–Bi alloys at temperatures near the liquidus temperatures. The data was taken from figure 1 for the fixed frequency of 100 cm^{-1}.

3. Model calculations

In the absence of observable molecular modes, the liquid behaviour is in sharp contrast with free-carrier absorption expected from the DC conductivity $\sigma(0)$ which decreases near 60 at.% Mg (Ilschner and Wagner 1958, Enderby and Collings 1970); thus on the nearly-free electron model the reflectivity would decrease. However, the frequency-dependence of the reflectivity can be fitted by a free-carrier absorption model.

At normal incidence the reflectivity R can be calculated from $R = r^*r$ where the reflection amplitude

$$r = \frac{1 - \hat{\epsilon}^{1/2}}{1 + \hat{\epsilon}^{1/2}}$$

and $\hat{\epsilon} = \epsilon_1 + i\epsilon_2$ is the complex dielectric constant which is related to the complex AC conductivity $\hat{\sigma}(\omega)$ by

$$\hat{\epsilon} = 1 + i(4\pi\hat{\sigma}(\omega)/\omega).$$

The reflectivity data can be fitted using the Drude nearly-free electron model wherein

$$\hat{\sigma}(\omega) = \sigma(0)_{opt}/1 - i\omega\tau$$

provided the relaxation time τ and $\sigma(0)_{opt}$ are treated as two adjustable parameters. The best fit parameters are tabulated in table 1 which also gives χ^2, a cumulative measure of the deviation of the calculated reflectivity from the experimental results. On this Drude fit to the data $\omega\tau \ll 1$ and $\omega_p\tau < 1$ where ω_p is the plasma frequency. Under these conditions the reflectivity is dominated by Re $\sigma(\omega)$ or Im $\epsilon(\omega)$. The rapid increase of the reflectivity near Mg_3Bi_2 is associated with a large increase in $\sigma(0)_{opt}$. At the stoichiometric compositions Mg_3Bi_2, $\sigma(0)_{opt}$ is enhanced by a factor of about 100 relative to the DC value of about 50 Ω^{-1} cm^{-1} found at 900 °C by Enderby and Collings (1970).

The composition dependence of $\sigma(0)_{opt}$ is shown in figure 3. This is compared with N_e, the number of free electrons per atom of alloy, and N_m, the number of Mg_3Bi_2 molecules per atom of alloy as calculated from the molecular association model of McAlister and Crozier (1974). This model led to reasonable agreement with the experimental concentration–concentration correlation function in the long wavelength limit $S_{cc}(0)$ for liquid Mg–Bi alloys near the composition 60 at.% Mg at a temperature 1 °C above the liquidus temperature of Mg_3Bi_2. It was assumed that the atoms were associated 90%

Table 1. Liquid alloy optical parameters

At.% Mg	$\sigma(0)_{opt}$ (Ω^{-1} cm^{-1})	τ (10^{-15} s)	χ^2
55	88	1	0·00072
58	133	1·2	0·0004
60	5333	2·3	0·00009
62	522	2·1	0·00160
65	79	0·94	0·00086
70	460	1·4	0·00200

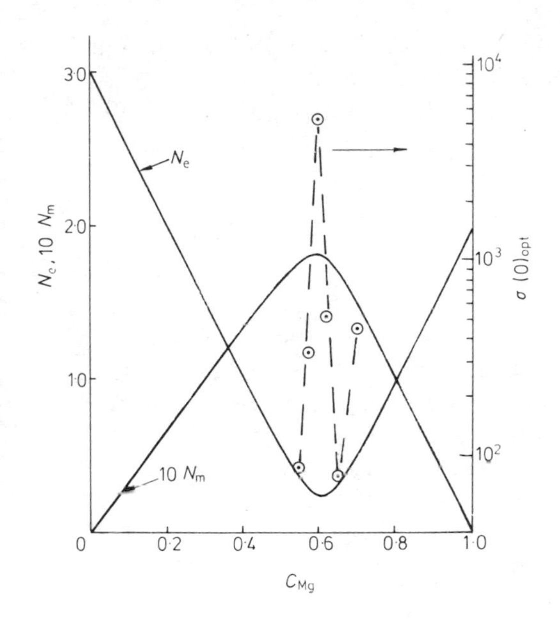

Figure 3. The composition dependence of particle densities in liquid Mg–Bi alloys. The dashed curve represents $\sigma(0)_{opt}$ which on the Drude model would be proportional to the free electron density. The solid lines represent N_e and N_m, the number of free electrons and molecules per atom of alloy derived using the molecular model of the text. Note the change of scale for N_m.

of the time as weakly bound molecules of Mg_3Bi_2 with an excess Gibbs free energy of formation of -15 kT per molecule. In calculating N_e it was assumed that fully ionized Mg and Bi would donate two and three electrons respectively to the electron gas. The observed increase in $\sigma(0)_{opt}$, which by the Drude model is proportional to the number of free electrons, is clearly inconsistent with the decrease in N_e predicted by the molecular formation model.

It is not realistic to attribute the enhancement of the experimental reflectivity to the presence of molecules. We calculated the reflectivity with the assumption that the molecules of Mg_3Bi_2 could be represented by single Lorentz oscillators with a frequency response specified by

$$\hat{\epsilon} = 1 + \left(\frac{4\pi Ne^2}{m}\right)\left(\frac{1}{(\omega_0{}^2 - \omega^2) - \mathrm{i}\Gamma\omega}\right)$$

where N is the number of oscillators per unit volume of mass m, natural frequency ω_0 and lifetime Γ^{-1}. The reflectivity of the alloys could be fitted with this model. But the fit was insensitive to ω_0; any value in the frequency range spanned by the measurements was adequate. In addition the other parameters were non-physical. For example, at 60 at.% Mg the fit parameters were: $N/m \sim 2 \times 10^6$ times the same ratio deduced from the dissociation model, $\omega_0 = 1{\cdot}89 \times 10^{13}$ Hz (100 cm^{-1}) and $\Gamma^{-1} = 1{\cdot}1 \times 10^{-15}$ s. With a lifetime of $1{\cdot}1 \times 10^{-15}$ s (or a linewidth of 3×10^4 cm^{-1}) it is not meaningful to talk about molecules.

On the bais of the Mott classification of liquid semiconductors (Mott and Davis 1971), when $\sigma(0) < 200\,\Omega^{-1}\,\mathrm{cm}^{-1}$, electronic states near the Fermi energy become localized. The recent Hall effect results of Even (unpublished) indicate that the Hall constant differs from the free-electron value only within a narrow range of about ± 2 at.% of Mg_3Bi_2. This suggests that it is only within this range that significant localization of states would occur and not the larger range implied by earlier measurements of $\sigma(0)$

(Enderby and Collings 1970, Ilschner and Wagner 1958). As indicated in figure 2 and figure 3 the infrared reflectivity is enhanced in an equally narrow range. This enhancement can be attributed to an electronic hopping between localized states. The reflectivity was calculated assuming the simple Debye model with a single hopping time τ:

$$\hat{\epsilon}(\omega) = \epsilon_\infty + \frac{\epsilon_0 - \epsilon_\infty}{1 - i\omega\tau}.$$

In the calculations ϵ_∞ was set equal to 1·0 and ϵ_0 and τ were treated as adjustable parameters. The reflectivity of all the alloys could be fitted by the model when $\omega\tau > 1$. In this regime the reflectivity is dominated by Im $\hat{\epsilon}$ which approximates the value $(\epsilon_0 - \epsilon_\infty)/\omega\tau$. Thus only this ratio can be determined from the fit. To illustrate the typical magnitudes involved, we find with an arbitrary choice of $\tau = 2 \times 10^{-12}$ s that $\epsilon_0 - \epsilon_\infty = 1{\cdot}1 \times 10^5$ for 60 at.% Mg. The hopping frequency is physically plausible but the value of $\epsilon_0 - \epsilon_\infty$ is anomalously high. The values of χ^2 obtained when fitting the data with the hopping model are comparable to those obtained with the Drude model at 58, 60 and 62 at.% Mg. A marginally poorer fit was obtained at 55, 65 and 70 at.% with the hopping model compared to the Drude model.

4. Conclusions

Earlier calculations of thermodynamic properties of liquid Mg–Bi alloys suggested that weakly bound molecules (excess Gibbs free energy of formation about $-15\,kT$/molecule) of Mg_3Bi_2 may exist (McAlister and Crozier 1974, Bhatia and Hargrove 1975) which affect the electronic structure of the alloys. Vibrational frequencies in such molecules would be low. Within our experimental uncertainty, no infrared active modes were observed over the frequency range 3 cm^{-1} to 300 cm^{-1} for alloys containing 55 to 70 at.% Mg. The enhanced reflectivity which approaches unity at 60 at.% Mg, cannot be attributed to a molecular model.

The composition and frequency dependence of the reflectivity may be taken as evidence of the existence of localized states near the Fermi energy over the restricted composition range 55 to 65 at.% Mg. The reflectivity implies an absorption maximum at frequencies less than 30 GHz. Consequently, we have been unable to fit the data with a Mott pseudogap model. The reflectivity data can be approximated by a simple Debye model with a single hopping frequency between localized states. However, the parameters obtained in fitting the data are questionable. The large value of $\epsilon_0 - \epsilon_\infty$ may reflect the fact that experimental data are not yet available for frequencies less than 30 GHz. It may also indicate the inadequacies of hopping as the only mechanism operative in the far-infrared regime. Experiments at lower frequencies (and higher than 9×10^{12} Hz) are clearly necessary.

Mg–Bi is only one of numerous liquid alloys which display semiconducting behaviour in the region of stoichiometric compositions. Optical measurements in the far infrared should be undertaken in these systems, particularly in the composition range where $\sigma(0) \sim 200\,\Omega^{-1}\,cm^{-1}$, to probe for the existence of molecules and to improve understanding of electron transport mechanisms.

Acknowledgments

The authors would like to thank the National Research Council of Canada for grants in support of this work.

References

Bhatia A B and Hargrove 1975 *Phys. Rev.* **B10** 3186–96
Clayman B P, Kirby R D and Sievers A J 1971 *Phys. Rev.* **B3** 1351–64
Crozier E D, Lytle F W, Sayers D E, Stern E A 1976 *Proc. Int. Conf. on Electrons in Fluids* ed G R Freeman
Enderby J E and Collings E W 1970 *J. Non. Cryst. Solids* **4** 161–7
Hawker I, Howe R A and Enderby J E 1974 *Proc. 5th Int. Conf. on Amorphous and Liquid Semiconductors* ed J Stuke and W Brenig (London: Taylor and Francis) pp 85–90
Ilschner B R and Wagner C 1958 *Acta Metall.* **6** 712–14
McAlister S P and Crozier E D 1974 *J. Phys. C: Solid St. Phys.* 7 3509–19
McAlister S P, Crozier E D and Cochran J F 1973 *J. Phys. C: Solid St. Phys.* **6** 2269–78
Mott N F and Davis E A 1971 *Electronic Processes in Non-crystalline Materials* (Oxford: Clarendon)
Ratti V K and Bhatia A B 1975 *J. Phys. F: Metal Phys.* **5** 893–902
Warren W W 1977 this volume

Liquid semiconductors under high pressure

K Tamura, M Misonou and H Endo
Department of Physics, Faculty of Science, Kyoto University, Kyoto, Japan

Abstract. The resistances of liquid semiconductors (Te–Se and Tl–Te alloys) were measured under high pressure by developing the Drickamer method. A sharp nonmetal–metal transition was observed for liquid $Te_{0.6}Se_{0.4}$ and $Te_{0.5}Se_{0.5}$ alloys. The resistances of liquid Tl_2Te and Tl_3Te_2 alloys gradually decreased with increasing pressure and a difference in pressure dependence of the resistance was found between them. The observed changes in the resistances of these liquid semiconductors under high pressure are discussed in connection with the variation of the coordination number, density of states and atomic configuration.

1. Introduction

It is well known that a nonmetal–metal transition can be induced by varying the temperature, pressure and concentration in various random systems. This transition has been observed for In_2Te_3, Ga_2Te_3 (Warren 1970a,b, 1971, 1972) and liquid Te (Busch and Güntherodt 1967), all of which make a gradual transition from semiconducting to metallic behaviour as the temperature is raised. Similar behaviour is shown by the liquid Te–Se system (Perron 1967, Alekseev *et al* 1975). Another example of a metal–nonmetal transition has been observed in the conductivity and the thermoelectric power of liquid Hg under high pressure and temperature (Schmutzler and Hensel 1972), this being due to separation of the conduction and valence bands at low density. In amorphous Mg–Bi alloys (Ferrier and Herrell 1969, Sik and Ferrier 1974), liquid Mg–Bi (Enderby and Collings 1970) and liquid Tl–Te alloys (Cutler and Mallon 1965, 1966), the conductivity goes through a deep minimum at compositions satisfying simple chemical valence requirements (e.g., Mg_3Bi_2, Tl_2Te). Recently, Hoshino *et al* (1976) have found metal–nonmetal transition in a liquid Cs–Au alloy near the equiatomic concentration, where ionic conduction plays an important role. A sharp nonmetal–metal transition is found at high pressures in amorphous Ge and Ge–Si alloys (Fukushima *et al* 1973) with the transition pressure, about 60 kbar for amorphous Ge, increasing with concentration of Si content. Mott (1975) comments that the observed sharp transition, characteristic of the amorphous state, was related to the formation of the conducting electron–hole gas. However, the nature of this transition is not yet clarified since it has still not been visualized experimentally.

It is interesting to investigate whether there is any discontinuous transition under high pressure in the liquid semiconductors, which are regarded as good testing materials since liquids are thermodynamically stable in comparison to amorphous solids. In the present paper we report the electrical resistivity of liquid Te, Te–Se and Te–Tl alloys under high pressure.

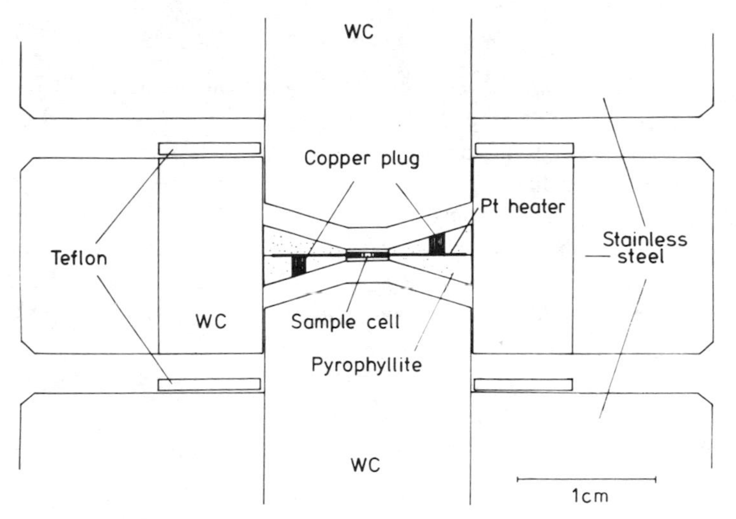

Figure 1. Schematic diagram of the opposite-anvil-type apparatus for the measurement of resistance of the liquid semiconductors under high pressure and high temperature.

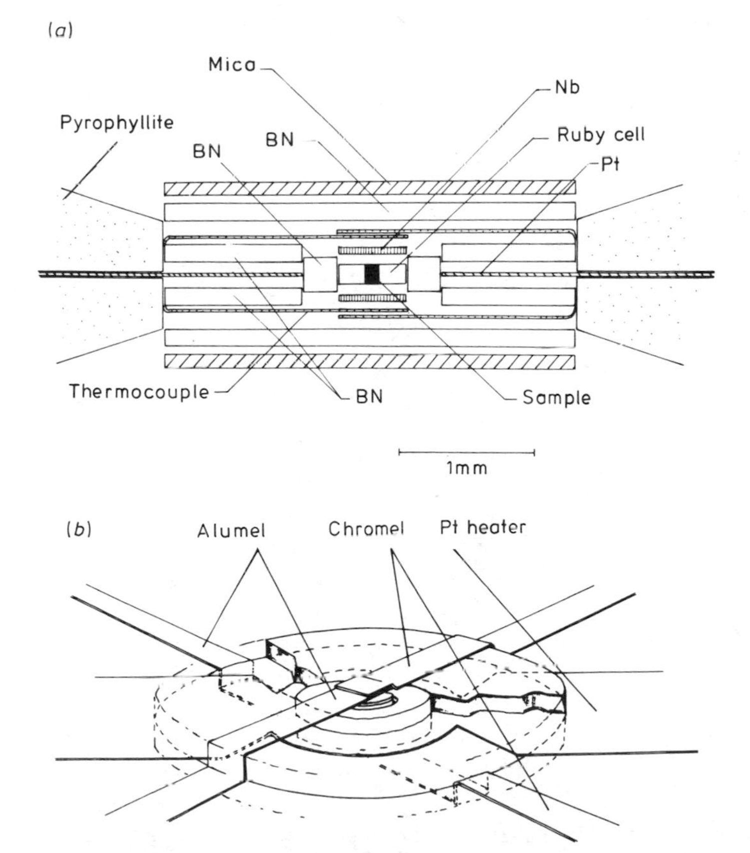

Figure 2. (*a*) The central assembly of the high-pressure cell and (*b*) the construction of the Pt heater and the alumel–chromel thermocouples.

2. Experimental

Pressure was applied by an external hydraulic press to a tungsten carbide (WC) opposite-anvil-type apparatus (figure 1) developed by Balchan and Drickamer (1961). Thin Pt ribbon (30 μm thick), with a 1·1 mm diameter hole, was used for the heater which was surrounded by pyrophyllite pellets. The heater current terminals were attached to the WC anvils, and current was passed through copper plugs situated in two holes drilled through the pyrophyllite. A heater current of about 30 A at 5V was used to raise the temperature of the specimens to 700 °C. The side jacket was electrically insulated from the anvils by paper sheets and the four electrode wires were fed through drilled holes into the inner cell.

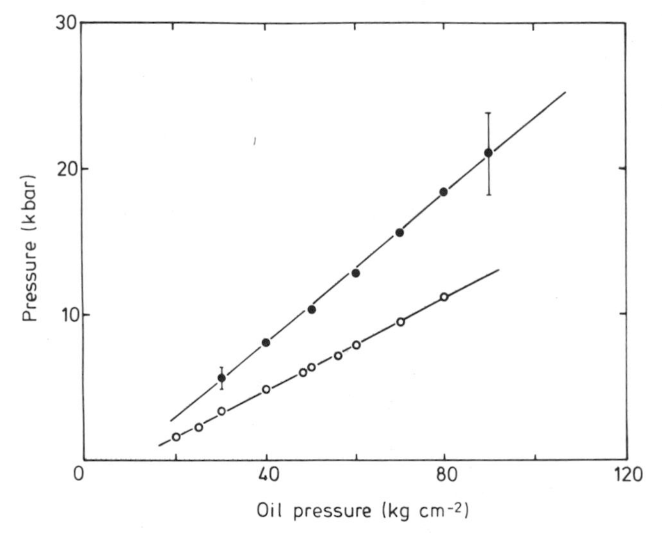

Figure 3. Pressure calibration curves for the primary oil pressure. Pressures were determined by measuring the change in resistance of indium along the melting curve. The calibration curves I (open circles) and II (full circles) were obtained by using the two kinds of ruby cell.

Figure 2 shows the assembly of the inner cell and the construction of the heater and electrodes. The size of the inner cell was 3·00 mm in diameter and 0·70 mm in thickness and 0·1 mm thick mica discs were used to prevent heat flow into the anvils. Boron nitride (BN) was used as a pressure medium since pyrophyllite attacked the Pt heater. Two pairs of alumel–chromel thermocouples 0·1 mm in diameter were used for the measurements of electrical resistance and temperature. Two different ruby cells were used for the specimen holder, the dimension of the first (I) being 0·5 mm × 0·1 mm × 0·15 mm thick, and the second (II) 0·3 mm × 0·05 mm × 0·10 mm thick. The liquid specimen was sealed by 0·05 mm thick niobium foil to which the electrodes were attached.

Pressure was determined by measuring the change of resistance along the melting curve of indium given by Kennedy and Newton (1963). The pressure calibration curves are shown in figure 3 for cells I and II respectively.

3. Results and discussion

Figure 4 shows the pressure dependence of the resistances of liquid Te and Te–Se alloys near the melting point. The resistance of liquid Te decreases gradually with increasing pressure (by a factor of 4 at a pressure of 10 kbar) and the 30 at.% Se liquid alloy shows similar behaviour. In contrast a sharp drop of resistance is found at a lower pressure of about 1·5 kbar for both the 40 and 50 at.% Se alloys, though slightly less steep for the 50 at.% Se alloy. The resistance of these two alloys then gradually decreases with pressure to approach that of liquid Te. For the 60 at.% Se alloy, the change of resistance is not so sharp. It should be noticed that with an increasing concentration of Se, the resistance drops at higher values of pressure and the value of resistance after the transition increases with Se content.

Neutron scattering experiments on liquid Te (Tourand *et al* 1972) suggest that the number of atoms in the first coordination shell is about two near the melting point and reaches nearly three towards 900 °C. The two-fold (binary) coordination is not far removed from the helical chain-like structure, while the three-fold (ternary) coordination is similar to that of an arsenic crystal structure. Such a conversion of coordination was found to be consistent with Knight-shift measurements (Warren 1972) and the velocity of sound (Gitis and Mikhailov 1966). It is suggested (Cabane and Friedel 1971, Cutler 1971a,b, Cohen and Jortner 1974) that the number of free carriers increases with the generation of ternary sites. This is responsible for the increasing metallic character of liquid Te, although the transport properties have been explained by the strong scattering or diffusive model (Warren 1972, Mott 1971, Cohen and Jortner 1974).

In the model of Cabane and Friedel (1971) the electron is promoted into the anti-bonding band if the three-fold bond is formed and as a result the Fermi level E_F is in

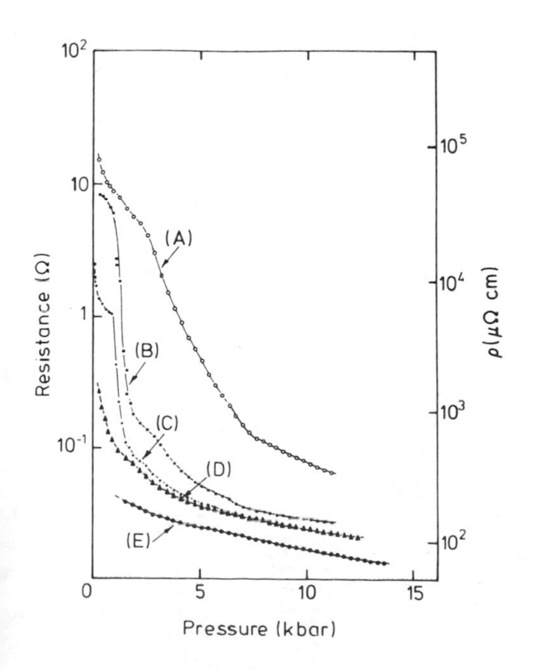

Figure 4. Electrical resistance against pressure for liquid Te–Se alloys at 500 °C: A, $Te_{0\cdot4}Se_{0\cdot6}$; B, $Te_{0\cdot5}Se_{0\cdot5}$; C, $Te_{0\cdot6}Se_{0\cdot4}$; D, $Te_{0\cdot7}Se_{0\cdot3}$; E, pure Te.

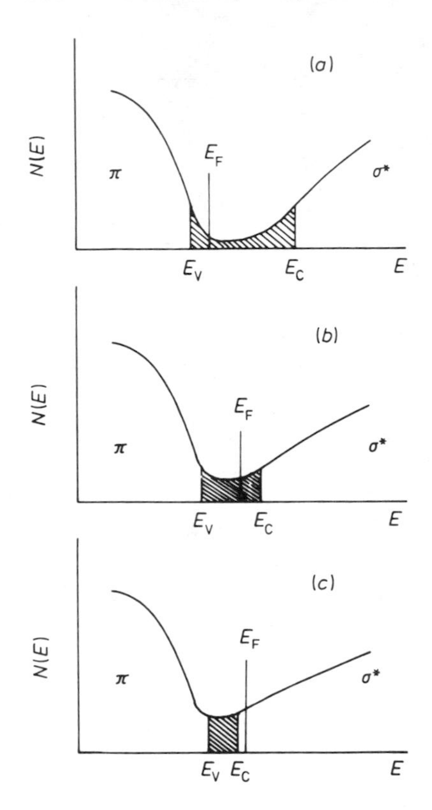

Figure 5. The density of states for liquid Te–Se alloys in the concentration of 40 and 50 at.% Se. With increasing pressure, the density of state changes from (*a*), through (*b*) to (*c*); E_F shifts to higher energy and then crosses over the mobility edge E_C.

the conduction band arising from the antibonding orbital. E_F moves to higher energy as the coordination number increases. The coordination number might be expected to increase with pressure. The valence band, being due to π-orbitals which do not take part in bonding along the chains, should not be unduly affected by a change in the coordination number, but the conduction band, due to antibonding orbitals between nearest neighbours, should be affected. With increasing pressure, E_F might shift remarkably and then cross over the mobility edge E_C since the density of states near E_F is small in the concentrations of 40 and 50 at.% Se (see figure 5); this is a reason why the sharp transition occurs for these alloys. The values of the resistances after the sharp transitions are still high, compared to that of liquid Te, because the density of states just above E_C is small. The fact that the gradual transition occurs in 60 at.% Se alloy can be attributed to the existence of an inhomogeneous mixture of chains with different lengths or rings (Hoyer *et al* 1975).

There is a sharp peak in the resistivity isotherms for Tl_2Te, and the isotherms for the thermoelectric power change at this composition from negative to positive as the Te concentration increases in liquid Tl–Te alloys. In addition, there is a significant asymmetry in the rate of change of the thermoelectric power with temperature and with composition of Te. Enderby and Simmons (1969) assumed that the Tl_2Te molecules formed a binary solution with Tl or Te atoms and gave rise to a band of bound states just below the conduction band. Cutler (1971) discussed that in the Te-rich side of liquid Tl–Te alloys the behaviour is to be explained in terms of mixtures of molecules of the type Tl–Te_n–Tl (n = 1,2,3, . . .), with an increasing number of Te–Te bonds

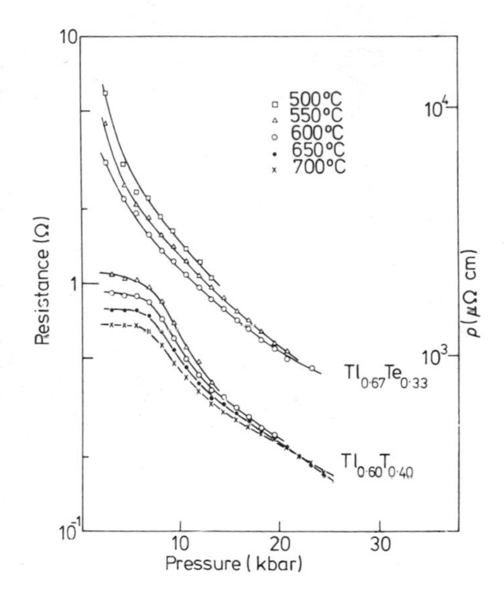

Figure 6. Curves of electrical resistance against pressure for liquid Tl_2Te and Tl_3Te_2.

breaking as the temperature is raised. Since such a liquid semiconductor may be microscopically inhomogeneous, it is expected that the continuous nonmetal–metal transition under pressure could be observed.

Figure 6 shows the pressure dependence of the resistance of liquid Tl_2Te and Tl_3Te_2 alloys. The resistance of liquid Tl_2Te alloy drops with increasing pressure by a factor of almost 10 at a pressure of 20 kbar and the resistance of liquid Tl_3Te_2 alloy decreases by a factor of about 3 at a pressure of 20 kbar, approaching the metallic value (about 400 $\mu\Omega$ cm) at 26 kbar. A graph of the logarithm of resistance of Tl_3Te_2 alloy against pressure exhibits two straight line regions, with a break in slope at about 10 kbar. It is interesting that the initial slope in the resistance against pressure curve for Tl_2Te is larger than that of Tl_3Te_2.

Further measurements of the resistance, the thermoelectric power and x-ray diffraction at much higher pressure and temperature are in progress.

Acknowledgment

We are indebted to Mr M Moriyasu for his kind support in the high-pressure measurements.

References

Alekseev V A, Turgunov T and Andreev A A 1975 *Sov. Phys.–Solid St.* **16** 2376
Balchan A S and Drickamer H G 1961 *Rev. Sci. Instrum.* **32** 308
Busch G and Güntherodt H J 1967 *Phys. Kondens. Mater.* **6** 525
Cabane B and Friedel J 1971 *J. Physique* **32** 73
Cohen M H and Jortner J 1974 *Proc. 5th Int. Conf. on Amorphous and Liquid Semiconductors* (London: Taylor and Francis) p 167
Cutler M 1971 *Phil. Mag.* **24** 381
—— *Phil. Mag.* **24** 401

Cutler M and Mallon C E 1965 *J. Appl. Phys.* **36** 201
—— 1966 *Phys. Rev.* **144** 642
Enderby J E and Collings E W 1970 *J. Non. Cryst. Solids* **4** 161
Enderby J E and Simmons C J 1969 *Phil. Mag.* **20** 125
Ferrier R P and Herrell D J 1969 *Phil. Mag.* **19** 853
Fukushima J, Tamura K, Endo H, Minomura S, Shimomura O, Asaumi K and Sakai N 1973 *Proc. 5th Int. Conf. on Amorphous and Liquid Semiconductors* (London: Taylor and Francis) p 1149
Gitis M B and Mikhailov I G 1966 *Sov. Phys.–Acoust.* **12** 17, 131
Hoshino H, Schmutzler R W and Hensel F 1975 *Phys. Lett.* **A51** 7
Hoyer W, Thomas E and Wobst M 1975 *Z. Naturf.* **30a** 1633
Kennedy G C and Newton R C 1963 *Solids under Pressure* (New York: McGraw-Hill) p 163
Mott N F 1971 *Phil. Mag.* **24** 1
—— 1975 *Phil. Mag.* **32** 159
Perron J C 1967 *Adv. Phys.* **16** 657
Schmutzler R W and Hensel F 1972 *Ber. Bunsenges. Phys. Chem.* **76** 531
Sik M J and Ferrier R P 1974 *Phil. Mag.* **29** 877
Tourand G, Cabane B and Breuil M 1972 *J. Non. Cryst. Solids* **8–10** 676
Warren W W 1970a *J. Non. Cryst. Solids* **4** 168
—— 1970b *Solid St. Commun.* 8 1269
—— 1971 *Phys. Rev.* **B3** 3708
—— 1972 *Phys. Rev.* **B6** 2522

Chemical potentials and resistivities of liquid Na–Te alloys

S M Granstaff Jr and J C Thompson
Department of Physics, University of Texas, Austin, Texas 78712, USA

Abstract. Data on the temperature and composition dependence of the chemical potential of Na in liquid Na–Te alloys, obtained from an electrochemical cell, are reported. A variety of thermodynamic parameters are calculated. In particular the long-wavelength limit of the concentration correlation function is calculated. It is shown that the complex Na_2Te has a large effect on the entire liquid Na–Te system, even to the extent of suggesting a 'pseudo-binary' alloy system Na_2Te–Te for the composition range $0{\cdot}33 < x_{Te} < 1{\cdot}0$. In addition the complex $NaTe_3$ exhibits some influence in a narrow composition range around this stoichiometry. The conductivity has been determined over the $0{\cdot}50 < x_{Te} < 1{\cdot}0$ range. There is a local maximum in the activation energy of $NaTe_3$.

1. Introduction

Many recent investigations have directed attention toward compound-forming molten metal alloys. These include a number of discussions of models of concentration fluctuations in these mixtures (Bhatia and Thornton 1970, Bhatia *et al* 1974, Bhatia and Hargrove 1974, McAlister and Crozier 1974, Ichikawa and Thompson 1973, Thompson 1974, Thompson and Granstaff 1976) and experimental investigations showing anomalies in transport and thermodynamic properties (Enderby and Collings 1970, Busch and Guntherodt 1967, Cutler and Mallon 1966, Rubin *et al* 1974, Valiant and Faber 1974, Popp *et al* 1974, Tschirner *et al* 1975). The alkali–chalcogen system, which falls into the compound-forming alloy class, has had attention drawn to it because of the liquid Na–S battery (Weber and Kummer 1967, Gupta and Tischer 1972) under development by a number of companies. Of the alloys in this system, Na with Te has the fewest compounds and, unlike the others, has no phase separation (Weber and Kummer 1967, Gupta and Tischer 1972, Hansen and Anderko 1958). Previous conductivity studies in the Te-rich range show no conductivity anomalies (Kraus and Glass 1929, Cutler and Leavy 1963). To the authors' knowledge none of the alloys in this system has been investigated over the complete concentration range for any property.

Equipped with a ceramic electrolyte in which Na^+ is the only carrier (Wittingham and Huggins 1971) and hoping to find useful information on both compound-forming alloys and the alkali–chalcogen system we set out to obtain the mean square composition fluctuations $\langle(\Delta x)^2\rangle$ by measuring the cell EMFS E of the simplest alloys, Na–Te (Ichikawa and Thompson 1973). This alloy has compounds in the solid at Na_2Te and $NaTe_3$ (Hansen and Anderko 1958). We have also measured the conductivity near $NaTe_3$.

The relevant thermodynamics is straightforward (Prigogine and Defay 1965). The partial molar thermodynamic parameters of Na may be calculated from EMFS E using the relations

$$\Delta\bar{G}_{\mathrm{Na}} = -FE, \tag{1}$$

$$\Delta\bar{S}_{\mathrm{Na}} = F(\partial E/\partial T), \tag{2}$$

$$\Delta\bar{H}_{\mathrm{Na}} = F[T(\partial E/\partial T) - E] \tag{3}$$

where $\Delta\bar{G}_{\mathrm{Na}}$, $\Delta\bar{S}_{\mathrm{Na}}$ and $\Delta\bar{H}_{\mathrm{Na}}$ are the partial molar free energy, entropy and enthalpy, respectively; F is the Faraday, and T is the absolute temperature. The partial molar parameters for Te were computed by numerical integration (using Simpson's rule) of the Gibbs–Duhem equation. The partial quantities were then combined to calculate the corresponding integral quantities. Since the concentration correlation function (Bhatia and Thornton 1970, Bhatia *et al* 1974, Bhatia and Hargrove 1974) $S_{\mathrm{cc}}(0)$ is given by

$$\begin{aligned} S_{\mathrm{cc}}(0) &= \frac{NkT}{(\partial^2 G_{\mathrm{Na}}/\partial x^2_{\mathrm{Na}})_{T,p}} \\ &= NkT\left(\frac{(1-x_{\mathrm{Na}})}{(\partial\Delta\bar{G}_{\mathrm{Na}}/\partial x_{\mathrm{Na}})_{T,p}}\right) \end{aligned} \tag{4}$$

it may also be computed from the observed E.

2. Experimental

The concentration cell may be schematically represented as

Na(liquid) | Na β-alumina | Na–Te alloy (liquid).

A sketch is given in figure 1. The Na and Na–Te are contained within a stainless steel (A) and boron nitride (C, D) vessel which has a Na β-alumina disc at B. The volumes of the reference side and the alloy side were approximately 0·5 and 0·7 cm^3, respectively. A stainless steel clamp insulated from A by boron nitride, E, is used to hold the vessel together. Stainless steel leads, G, connect the electrodes, A, to the external circuit. Insulated thermocouple leads, H, extend from a chromel–alumel thermocouple junction

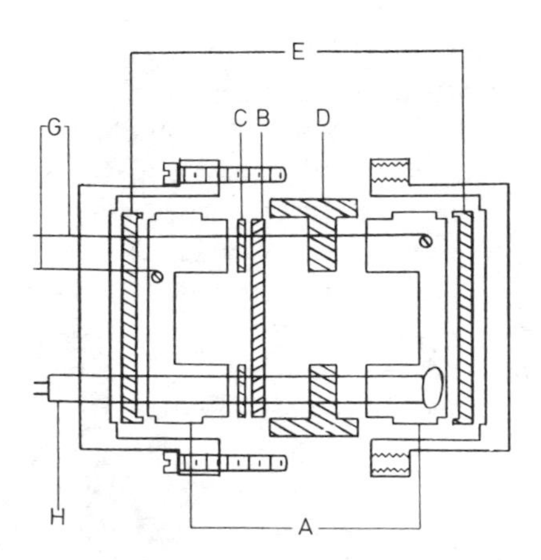

Figure 1. Na–Te experimental vessel: A, stainless steel container and electrodes; B, Na β-alumina disc; C, boron nitride sealing washer; D, boron nitride insulator; E, boron nitride insulators; G, stainless steel leads; H, insulated chromel–alumel thermocouple leads.

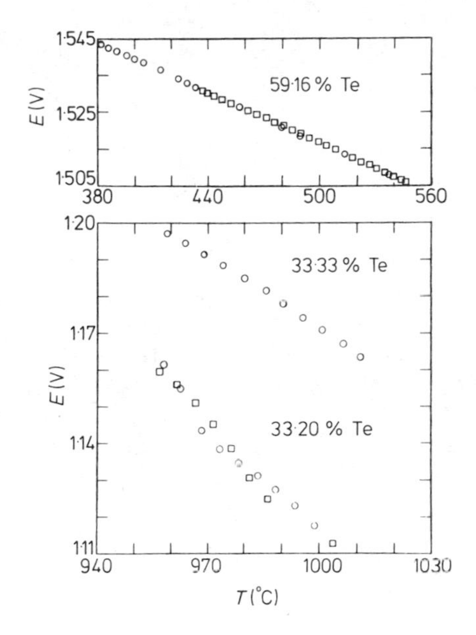

Figure 2. EMF versus temperature for 59·16, 33·33 and 32·20% Te for the Na–Te system. The squares indicate heating data and circles indicate cooling data.

cemented into the Na–Te container. E was measured on a Keithley 171 digital multimeter having greater than $10^9\,\Omega$ input impedance.

Samples were made from 99·99% Na and 99·995% Te metal. The samples were weighed and the cells assembled in a He-atmosphere glove box. The ceramic electrolytes were sintered sodium β-alumina discs obtained from TRW, Systems Group. These discs had a resistivity with sodium electrodes of $11{\cdot}7\,\Omega$ cm at 300 °C and uniform grain size of 3 μm.

Typical data for three concentrations are shown in figure 2. A data point was taken every 2–4 minutes at 5 °C intervals. Conductivity data were taken using conventional four-probe techniques in a Pyrex cell with tungsten leads; mixing of the Na and Te was done in a separate quartz vessel.

3. Results

The EMF values at 527 and 1000 °C are shown in figure 3; good agreement is shown with the 527 °C Morachevskii *et al* (1970) data for the concentration range $0{\cdot}60 < x_{\mathrm{Te}} < 1{\cdot}00$. Slope and EMF values for 1000 °C and the temperature ranges over which data were taken are listed in table 1. The values marked with a dagger were extrapolated to 1000 °C by a least squares linear regression fit of the data taken in the temperature ranges indicated. The errors in the EMF values, including the extrapolations, are less than 5%.

We turn next to the presentation of the relevant thermodynamic quantities calculated from the data. The integral values of the Gibbs free energy, entropy and enthalpy at 1000 °C are shown in figure 4 in the forms $\Delta G_{\mathrm{m}}/RT$, $\Delta S_{\mathrm{m}}/R$ and $\Delta H_{\mathrm{m}}/RT$ as functions of x_{Te}. The error in these functions includes, in addition to the original error of the data, the arbitrariness introduced in drawing smooth curves through the appropriate graphs of the sodium partial thermodynamic parameters in order to obtain the

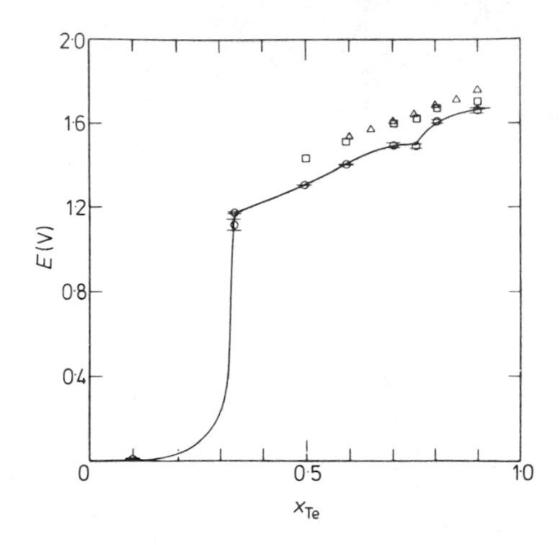

Figure 3. EMF versus concentration of tellurium for the Na–Te system: circles, 1000 °C data; squares, 527 °C data; triangles, Morachevskii *et al* (1970) data for 527 °C.

Table 1. EMF data for sodium–tellurium

x_{Te}	x_{Na}	Temperature (°C)	E (V) at 1000 °C	$\partial E/\partial T$ (mV K^{-1})
0·9000	0·1000	455–510	1·6609†	−0·0885
0·8043	0·1957	450–530	1·6070†	−0·1293
0·7545	0·2455	450–540	1·4926†	−0·2777
0·7027	0·2973	440–560	1·4956†	−0·2091
0·5916	0·4084	430–550	1·4029†	−0·2276
0·4976	0·5024	450–570	1·3033†	−0·2723
0·3333	0·6667	955–1015	1·1713	−0·6448
0·3320	0·6680	955–1005	1·1143	−1·0384
0·1000	0·9000	570–710	0·000152†	+0·0028

† Extrapolated from indicated temperature range.

tellurium parameters by Simpson's rule integration. This should be particularly kept in mind in the meagre data region at low Te concentration.

Our conductivity data together with those of Kraus and Glass (1929) and Cutler and Leavy (1963) are shown in figure 5. If the conductivity is represented in a conventional activated form then the pre-exponentials and activation energies are shown in figure 6.

Hansen and Anderko (1958) show a liquidus curve with a large peak at the solid compound Na_2Te and a much smaller peak at the solid compound $NaTe_3$. On the expectation that these two compounds may influence the behaviour of the liquid, the data will be examined by comparison with other compound-forming systems, such as Mg–Bi, and models for such systems.

Figure 4 shows a large negative entropy in the region near Na_2Te as do other compound-forming systems such as Tl–Te (Nakamura and Shimoji 1971). The large negative value of ΔG_m at the composition of Na_2Te shown in figure 4 indicates a strongly attractive interaction energy between the constituents of the system. The ΔH_m curve of figure 4, while having a similar shape, has a more negative value than the systems Tl–Te and Bi–Te examined by Maekawa *et al* (1971). Since ΔH_m is propor-

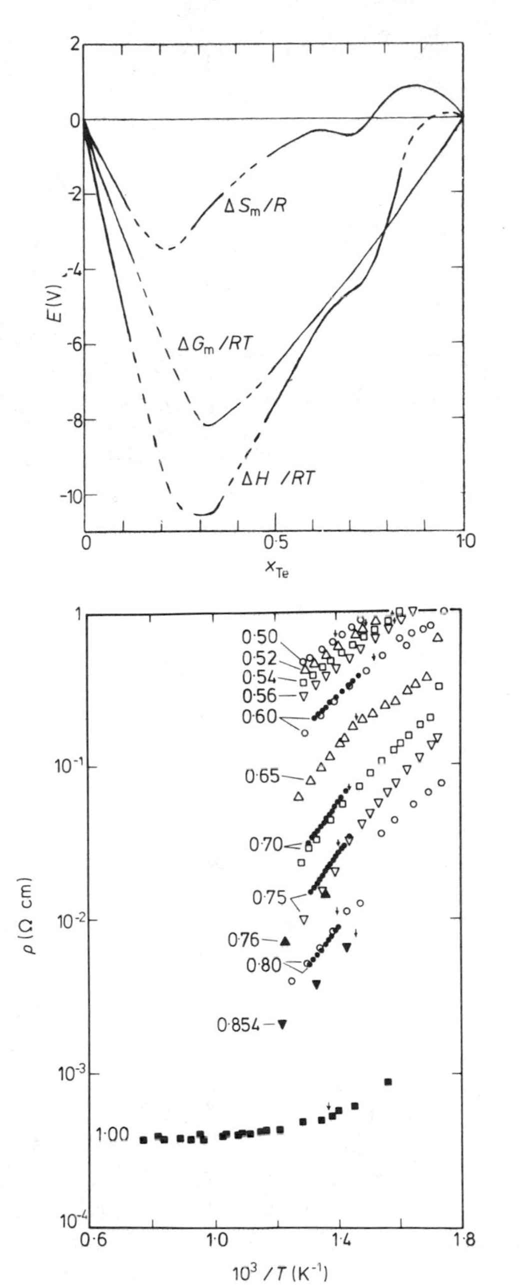

Figure 4. Thermodynamic functions of mixing versus concentration of tellurium for the Na–Te system.

Figure 5. Resistivity versus inverse temperature for the Na–Te system: solid circles, present data; open circles, squares and triangles, Kraus and Glass (1929); solid triangles, Cutler and Leavy (1963); numbers at left are concentration of tellurium; arrows indicate alloy melting temperature.

tional to ω/kT (where ω is the interchange energy) in the simple conformal model (Bhatia and Hargrove 1974, Guggenheim 1952) we may infer that the Na_2Te is more tightly bound than the others. However, the dependence of ΔH_m on x is *not* that appropriate to the *simple* conformal model. Nevertheless, this confirms our expectations that the Na_2Te complex is important even in the liquid.

The concentration correlation function $S_{cc}(0)$, equation 4, gives the next level of

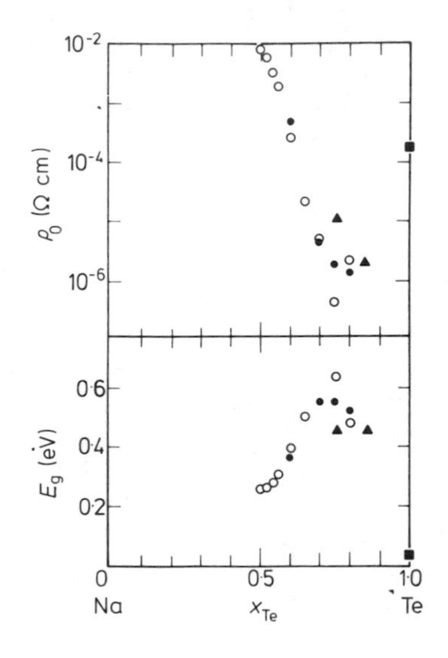

Figure 6. Pre-exponentials and activation energies for the Na–Te system: solid circles, present data; open circles, Kraus and Glass (1929); closed triangles, Cutler and Leavy (1963).

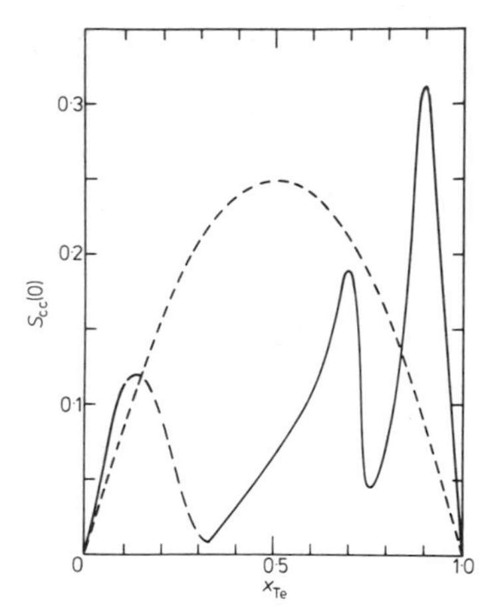

Figure 7. $S_{cc}(0)$ versus concentration of tellurium for the Na–Te system. The broken curve shows ideal solution values for $S_{cc}(0)$.

differentiation of the thermodynamic quantities. The relationship of S_{cc} (0) to the question of compound formation has already been discussed (Bhatia *et al* 1974, Bhatia and Hargrove 1974, McAlister and Crozier 1974, Ichikawa and Thompson 1973, Thompson 1974, Thompson and Granstaff 1976). Figure 7 shows $S_{cc}(0)$ for the Na–Te system. Again, additional errors have been introduced by the arbitrariness involved in drawing a smooth curve through the points of the EMF versus x_{Te} graph used in obtaining the function $S_{cc}(0)$. The main region of concern for these errors is that of low Te concentration as indicated by the dashed region of the $S_{cc}(0)$ curve in figure 7. The solid compounds Na_2Te and $NaTe_3$ are strongly reflected in the liquid by the deep notches occurring in $S_{cc}(0)$ at these compositions, especially the Na_2Te composition. Similar deep notches occur in $S_{cc}(0)$ for other systems such as Mg–Bi or Tl–Te (Ichikawa and Thompson 1970) which may be described by the Bhatia and Hargrove (1974) model.

The peak in $S_{cc}(0)$ at $x_{Te} = 0{\cdot}90$ may be attributed to the breaking-up of Te chains by the addition of Na. Alternatively, this peak could imply a size effect as discussed by Bhatia and March (1975). Their requirement for such an effect is a volume ratio greater than two. Although the Na and Te volume ratio is smaller than two, the size ratio of Na_2Te and Te is 2·93. This suggests that the concentration range $0{\cdot}33 < x_{Te} < 1{\cdot}0$ might be described as a 'pseudo-binary' alloy of Na_2Te with Te. $NaTe_3$ cannot be considered for the pseudo-binary description since the cohesion of that complex is known to be weak by the shallow dip in $S_{cc}(0)$. It would instead fit into this description as $(Na_2Te)\,Te_5$. Thus it appears that the complex Na_2Te, which may exist as some electron-sharing association, exerts a great influence over the entire liquid Na–Te alloy system.

Either description, Na–Te or Na_2Te–Te, indicates that anomalies in the transport properties may be expected at the compositions of Na_2Te and $NaTe_3$. Again, the Te-rich complex should show a weaker effect as the notch in $S_{cc}(0)$ is shallower and narrower. The incomplete conductivity data is consistent with the existence of an effect at Na_2Te. At $NaTe_3$ the anomaly is visible only as a slight shift in the pre-exponential and in the activation energy. In the absence of any real theory of transport in such strongly interacting systems we are unable to provide any meaningful interpretation of this increase in the activation energy beyond the simple observation that the existence of $NaTe_3$ requires tighter binding.

Acknowledgments

The authors wish to thank Dr L S Marcoux and Dr H Silverman of TRW, Systems Group for supplying the Na β-alumina discs so vital to this work. This work was supported in part by the US National Science Foundation under grant DMR 75-07828 and the R A Welch Foundation of Texas.

References

Bhatia A B and Hargrove W H 1974 *Phys. Rev.* **B10** 3186
Bhatia A B, Hargrove W H and Thornton D E 1974 *Phys. Rev.* **B9** 435
Bhatia A B and March N H 1975 *J. Phys. F: Metal Phys.* **5** 1100
Bhatia A B and Thornton D E 1970 *Phys. Rev.* **B2** 3004
Busch G and Guntherodt H-J 1967 *Phys. Kondens. Mater.* **6** 325
Cutler M and Leavy J F 1963 private communication
Cutler M and Mallon C E 1966 *Phys. Rev.* **144** 642
Enderby J E and Collings E N 1970 *J. Non. Cryst. Solids* **4** 161
Guggenheim E A 1952 *Mixtures* (Oxford: Clarendon Press)
Gupta N K and Tischer R P 1972 *J. Electrochem. Soc.* **119** 1033
Hansen M and Anderko K 1958 *Constitution of Binary Alloys* (New York: McGraw-Hill)
Ichikawa K and Thompson J C 1973 *J. Chem. Phys.* **59** 367
Kraus C A and Glass S W 1929 *J. Phys. Chem.* **33** 984
Maekawa T, Yokokawa T and Niwa K 1971 *J. Chem. Thermodyn.* **3** 143
McAlister S P and Crozier E D 1974 *J. Phys. C: Solid St. Phys.* **7** 3509
Morachevskii A G, Bykova M A and Rozova T T 1970 *Elektrokhimiya* **6** 1065
Nakamura Y and Shimoji M 1971 *Trans. Faraday Soc.* **67** 1270
Popp K, Tschirner H -U and Wobst M 1974 *Phil. Mag.* **29** 571
Prigogine I and Defay R 1965 *Chemical Thermodynamics* (London: Longmans)

Rubin I B, Komarek K L and Miller E 1974 *Z. Metallk.* **65** 191
Thompson J C 1974 *J. Physique Colloque* C4 **35** 367
Thompson J C and Granstaff Jr S M 1976 *Proc. Int. Conf. on the Electronic and Magnetic Properties of Liquid Metals* ed J Keller to be published
Tschirner H-U, Wolf R and Wobst M 1975 *Phil. Mag.* **31** 237
Valiant J C and Faber T E 1974 *Phil. Mag.* **29** 571
Weber N and Kummer J T 1967 *Proc. Ann. Power Sources Conf.* **21** 37
Wittingham M S and Huggins R A 1971 *J. Chem. Phys.* **54** 414

Thermodynamic properties of molten In–Te alloys

M Naoi, Y Nakamura and M Shimoji
Department of Chemistry, Hokkaido University, Sapporo 060, Japan

Abstract. The free energy and entropy of mixing of the molten In–Te system have been determined by means of the electromotive force method. Both the free energy and enthalpy of mixing show deep minima around the stoichiometric composition In_2Te_3 and a marked reduction of the entropy of mixing was found at the same composition. These results suggest that strong nonmetallic bonds with local ordering are present around the composition In_2Te_3 where the electrical properties are of the semiconducting type.

1. Introduction

In a previous paper (Ninomiya *et al* 1972), we have reported that the molten In–Te system shows marked semiconducting behaviour around the stoichiometric composition In_2Te_3. The electrical conductivity has a minimum and the thermoelectric power varies sharply at this composition. These results were confirmed later by Popp *et al* (1974). The Hall coefficient of the molten In–Te system also shows a marked maximum at the composition In_2Te_3 (Tschirner *et al* 1974). The calorimetric study of this system by Maekawa *et al* (1972) suggests the formation of very stable alloys around this composition. The purpose of the present study was to determine the free energy and entropy of mixing of the molten In–Te system as a function of alloy composition by use of the electromotive force (EMF) method and to correlate them to the electrical properties of the system.

2. Experimental

The constitution of the EMF cell used in the present work was:

$$\text{W, In(l)} \mid \text{InI (in NaI+KI eutectic mixture)} \mid \text{In–Te(l), W.} \qquad \text{(I)}$$

The purity of the metals used was 99·999% for In and 99·9999% for Te. Indium mono-iodide (InI) was prepared by direct reaction of indium metal and indium tri-iodide (InI_3) at 500 °C (Rolsten 1961). A two-legged H-type cell made of Vycor glass was used. Tungsten wires of 1 mm diameter were used as electrodes. The cell was evacuated and heated at 200 °C for several hours in order to dehydrate the electrolyte mixture and then filled with argon gas. The cell was placed in an alumina tube which was sealed under an argon atmosphere, then heated in an electric furnace and held at 800 °C for more than eight hours to allow the metals and electrolytes to attain equilibrium. The EMF measurements were made at different temperatures from 620 to 860 °C during repeated heating and cooling runs. The EMF values obtained were reproducible to within ± 0·5 mV for the alloys with $X_{In} \geqslant 0{\cdot}4$ (X_{In} is the atomic fraction of In).

The EMF data for the alloys with $X_{In} < 0{\cdot}4$ were not used for the determination of the thermodynamic functions because of the ambiguity of the valence of In ions in the electrolyte mixture. A considerable decrease of EMF values was observed with increasing InI concentration (0·02 to 5 wt%) in the electrolyte mixture, when X_{In} was smaller than 0·4 or $a_{In} < 10^{-3}$ (where a_{In} is the activity of In). This decrease of EMF should correspond to the relative increase of InI_3 concentration in the electrolyte mixture. The EMF values were practically independent of InI concentration for the alloys $X_{In} \geqslant 0{\cdot}4$. It may be noted that in the EMF cell similar to the cell (I) used in the study of molten In–Sb alloys (Hoshino *et al* 1965), the assumption of the monovalent In ions for the whole concentration range of the measurements ($a_{In} \geqslant 0{\cdot}05$) has been verified by Chatterji and Smith (1973), who used solid In_2O_3 as electrolyte.

Table 1. EMF data for molten In–Te alloys

X_{In}	E (mV) at 800 °C	dE/dT (μV/deg)	Temperature range (°C)
0·400	497·7	151	770–830
0·425	433·4	204	730–840
0·450	378·5	258	710–830
0·500	283·3	274	620–860
0·550	220·5	314	730–820
0·700	35·2	83·1	660–850
0·800	14·7	41·5	660–830
0·900	9·76	14·1	700–860

3. Results

From the EMF data for the alloys with $X_{In} \geqslant 0{\cdot}4$ (table 1), we have determined the partial molar free energy and entropy of In, $\bar{G}_{In}$ and $\Delta\bar{S}_{In}$, using the relations:

$$\Delta\bar{G}_{In} = -EF = RT \ln (a_{In}), \tag{1}$$

and

$$\Delta S_{In} = F(dE/dT), \tag{2}$$

where E is the EMF and F the Faraday constant. We have extrapolated the present data of $\Delta\bar{G}_{In}$ to the Te-rich side of the composition In_2Te_3, taking account of the phase diagram data (Grochowski *et al* 1964).

Using these results, we calculated the partial molar free energy of Te, $\Delta\bar{G}_{Te}$, by the graphical integration of the Gibbs–Duhem relation, and then determined the integral molar free energy of mixing, ΔG_m. The integral molar entropy of mixing, ΔS_m, can be calculated by combining the present results of ΔG_m with the calorimetric data of the enthalpy of mixing, ΔH_m, reported by Maekawa *et al* (1972). The values of partial molar entropy of In, $\Delta\bar{S}_{In}$, determined from the gradient of the curve of ΔS_m thus obtained, are consistent with the values of ΔS_{In} obtained directly from the EMF data in the In-rich region. This may indicate that the extrapolation was reasonable. The integral molar quantities obtained at 800 °C are shown in figure 1. The present data of ΔG_m are in poor agreement with those determined from the vapour pressure measurements

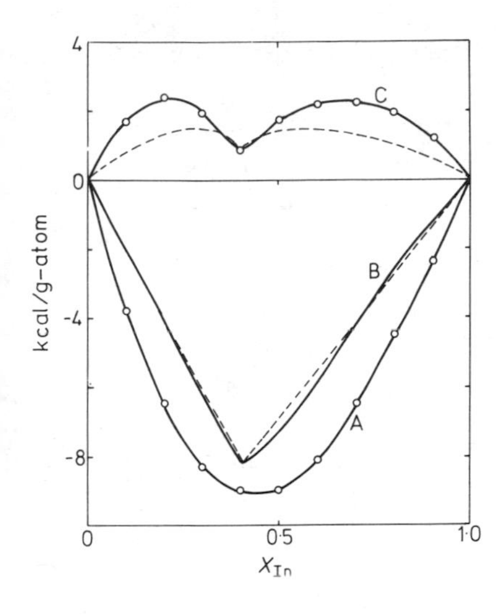

Figure 1. A, integral molar free energy ΔG_m; B, enthalpy ΔH_m (from Maekawa *et al* 1972); C, entropy $T\Delta S_m$ of molten In–Te alloys at 800 °C. The broken curves are explained in the text.

(Predel *et al* 1975). Their calorimetric data of ΔH_m also differ greatly from those of Maekawa *et al* (1972). Consequently, their values of ΔS_m, which are negative almost in the whole concentration range, are entirely different from ours. We believe that the present data, as well as the calorimetric data of Maekawa *et al*, are more reliable and some discussion will be given on the basis of these data.

4. Discussion

The thermodynamic properties of the molten In–Te system are characterized by deep minima in ΔG_m and ΔH_m and a marked depression of ΔS_m at the stoichiometric composition In_2Te_3 or $X_{In} = 0{\cdot}4$. These are very similar to those of the molten Tl–Te (Nakamura and Shimoji 1971) and Mg–Bi (Egan 1959) systems which show the characteristic changes at the stoichiometric composition Tl_2Te and Mg_3Bi_2 respectively.

In these 'compound' forming systems, two models have been suggested. The first of these involves the existence of fairly well defined molecular groups like Tl_2Te or Mg_3Bi_2 (Enderby and Simmons 1969, Nakamura and Shimoji 1971, Bhatia and Hargrove 1973, 1974). The cluster model of Hodgkinson (1976) is essentially an extension of the assumption of molecular assembly. In the second model the liquid is assumed to consist of ions such as Tl^+ and Te^{2-} (Nakamura and Shimoji 1969) or Mg^{2+} and Bi^{3-}. The 'molecular' and 'ionic' models thus represent the two extremes expressing the local ordering around the stoichiometric composition, which is strongly suggested from the thermodynamic data. A more plausible description, particularly for molten chalcogenides, should be one in which the bonds are partially ionic and partially covalent (Ninomiya *et al* 1974).

In the ionic model, we can assume a fused salt-like arrangement of the constituent atoms in molten In_2Te_3: indium atoms may be surrounded preferentially by tellurium atoms and *vice versa.* This preferential neighbouring of atoms may be represented by a quasi-chemical approximation (Guggenheim 1952) in a quasi-crystalline model. However, a quantitative estimation of the thermodynamic functions in terms of the quasi-chemical approach appears very difficult, since the coordination number of the quasi-lattice should vary drastically as we proceed from pure liquid In to pure liquid Te through In_2Te_3.

In view of the fact that $|\Delta G_m| \simeq 5RT$ at the composition In_2Te_3, it would appear reasonable to assume that the formation of the 'compound' In_2Te_3 is completed and the system can be treated conveniently by considering two pseudobinary systems: $In-In_2Te_3$ and $Te-In_2Te_3$. The 'compound' does not necessarily mean well defined molecular species; that is, it may be approximated to be ionic or molecular. In such a pseudobinary mixture model, any thermodynamic functions of mixing, ΔX_m, can be given by two terms: the term due to the formation of the $A_\mu B_\nu$ compound from the pure elements A and $B(\Delta X_f)$, and the term due to the mixing of $A_\mu B_\nu$ with excess A or B in each binary system (ΔX_p).

In the previous work on the molten Tl–Te system (Nakamura and Shimoji 1971), we have employed Flory's model for the estimation of ΔX_p, regarding Tl_2Te as trimer. If we apply this simple model to the present system, assuming In_2Te_3 molecules, we obtain the curves for ΔS_m and ΔH_m, which are shown in figure 1 by broken lines and compared with experiment. The only parameters used in this plot are $\Delta S_f = 0{\cdot}83$ cal/deg.g-atom and $\Delta H_f = -8{\cdot}11$ kcal/g-atom. The mixing in the respective pseudobinary system is assumed as athermal, though the data of ΔH_m indicate that ΔH_p of the $In-In_2Te_3$ system is negative in the In_2Te_3-rich region and positive in the In-rich region. This positive ΔH_p may give rise to the liquid–liquid phase separation in the In-rich region of the system. It is seen that the features of experimental curves can be reproduced fairly well in terms of this pseudobinary model. It should be noted, however, that apart from the validity of the lattice model we have had no experimental evidence for the existence of molecular entities such as In_2Te_3 yet. Detailed structure experiments would resolve this uncertainty.

We emphasize here that the local ordering around the stoichiometric composition is strongly related to the occurrence of the semiconductivity. According to the pseudogap model proposed by Mott (Mott and Davis 1971) the pseudogap appears to be formed around the composition In_2Te_3, since there the conductivity is of the order of 10^2 $\Omega^{-1}\,cm^{-1}$. The existence of the pseudogap in molten In_2Te_3 has also been suggested from the NMR study by Warren (1970). In the In-rich region the system can be treated as liquid metallic alloys to which the nearly-free electron model is appropriate. In the intermediate region in the $In-In_2Te_3$ pseudobinary system and Te-rich side of In_2Te_3, the electrical conduction may be interpreted in terms of the strong scattering model (Mott and Davis 1971) where the mean free path of electrons is short and independent of composition, probably of the order of atomic separation. It should also be noted that the concentration fluctuation, $S_{cc}(0)$, which is related to the thermodynamic data (Bhatia and Thornton 1971), shows a marked peak around $X_{In} \simeq 0{\cdot}8$, where the liquid–liquid phase separation takes place below the critical temperature. We cannot exclude therefore the possibility of the appearance of microscopic inhomogeneity in the solutions of the composition around $X_{In} \simeq 0{\cdot}8$, as often used in the explanation of the electrical properties in 'compound' forming systems (Cohen and Sak 1972, Hodgkinson 1976).

5. Conclusions

In this paper we have reported the experimental results of the free energy and entropy of mixing of the molten In–Te system. From these results together with the enthalpy data, it has been suggested that very stable liquid alloys with local ordering are formed around the stoichiometric composition In_2Te_3. The system can be considered as composed of two pseudobinary systems: In–In_2Te_3 and Te–In_2Te_3. The marked semiconductivity around the composition In_2Te_3 appears to be related to the formation of a pseudogap due to this local ordering.

Acknowledgments

The authors should like to acknowledge discussions with Professor T Yokokawa and Dr T Maekawa.

References

Bhatia A B and Hargrove W H 1973 *Nuovo Cim. Lett.* 8 1025–30
—— 1974 *Phys. Rev.* **B10** 3186–96
Bhatia A B and Thornton D E 1970 *Phys. Rev.* **B2** 3004–12
Chatterji D and Smith J V 1973 *J. Electrochem. Soc.* **120** 770–2
Cohen M H and Sak H 1972 *J. Non. Cryst. Solids* **8–10** 696–701
Egan J J 1959 *Acta Metall.* **7** 560–4
Grochowski E G, Mason D R, Schmitt G A and Smith P H 1964 *J. Phys. Chem. Solids* **25** 551–8
Guggenheim E A 1952 *Mixtures* (Oxford: Clarendon) pp 38–46
Hodgkinson R J 1976 *J. Phys. C: Solid St. Phys.* **9** 1467–82
Hoshino H, Nakamura Y, Shimoji M and Niwa K 1965 *Ber. Bunsenges. Phys. Chem.* **69** 114–8
Maekawa T, Yokokawa T and Niwa K 1972 *J. Chem. Thermodyn.* **4** 153–7
Mott N F and Davis E A 1971 *Electronic Processes in Non-Crystalline Materials* (Oxford: Clarendon) pp 6–101
Nakamura Y and Shimoji M 1971 *Trans. Faraday Soc.* **67** 1270–7
Ninomiya Y, Nakamura Y and Shimoji M 1972 *Phil. Mag.* **26** 953–60
—— 1974 *J. Non. Cryst. Solids* **17** 231–40
Popp K, Tschirner H U and Wobst M 1974 *Phil. Mag.* **30** 685–90
Predel B, Peihl J and Pool M J 1975 *Z. Metallkd.* **66** 268–74
Rolsten R F 1961 *Iodide Metals and Metal Iodides* (New York: Wiley) pp 259–60
Tschirner H U, Wolf R and Wobst M 1974 *Phil. Mag.* **30** 237–42
Warren W W Jr 1970 *J. Non. Cryst. Solids* **4** 168–77

The compressibility of some liquid gallium–tellurium alloys

S P McAlister† and E D Crozier
Physics Department, Simon Fraser University, Burnaby, BC, Canada V5A 1S6

Abstract. We report measurements of the ultrasonic velocity as a function of temperature for liquid Ga_3T_2, GaTe, Ga_2Te_3 and $GaTe_3$ and deduce the adiabatic compressibility using available density data. In liquid Ga_2Te_3 the sound velocity decreases nonlinearly with temperature. The observed compressibility deviates from the ideal behaviour for the system and shows possible evidence for compound formation.

1. Introduction

The fluid and electronic structures of binary liquid alloys that display semiconducting behaviour are not well understood. It has been suggested that molecules exist in such systems and affect the electronic properties, particularly in those alloys whose DC conductivity is less than about 200 Ω^{-1} cm^{-1} in which localization of electron states is likely. One alloy system which has attracted considerable attention is Mg–Bi for which properties such as the DC conductivity (Enderby and Collings 1970), Hall coefficient (U Even, unpublished) and thermodynamic properties (see for example McAlister and Crozier 1974) have been studied.

Ratti and Bhatia (1975) have endeavoured to explain quantitatively the electrical properties of Mg–Bi on the assumption that molecules of Mg_3Bi_2 do exist. The sharp peak observed in the compressibility (McAlister *et al* 1973) has also been considered indicative of molecular formation. However, measurements of the reflectivity in the far infrared failed to detect the existence of molecules (Crozier *et al* 1977). We have now undertaken measurements on liquid Ga–Te to determine if the anomaly in the compressibility is restricted to the Mg–Bi system. Ga–Te exhibits a number of intermetallic compounds in the solid state (Hansen 1958) including two congruently melting compounds GaTe and Ga_2Te_3. It is a suitable system for investigation since a variety of measurements suggest that molecular associations of Ga_2Te_3 may exist in the liquid. Measurements of the electrical resistivity and thermoelectric power have shown that there is a sharp maximum at the Ga_2Te_3 composition (Glazov *et al* 1969; Valiant and Faber 1974). Properties such as the viscosity (Glazov *et al* 1969) and the surface tension (Wobst 1970) also show extrema in this region. To date, the most direct evidence for compound formation comes from the work of Warren (1971, 1972) who has measured the Knight shift and the magnetic and quadrupolar relaxation. He concludes that his results are consistent with molecular associations with the composition Ga_2Te_3 existing over a concentration range of 10% on either side of the Ga_2Te_3 composition, and which appear to be stable for about 10^{-11} s.

† Present address: Division of Chemistry, National Research Council, Ottawa, Ontario, Canada K1A OR9.

In this paper we present results for the sound velocity and compressibility of four alloys in the Ga–Te system, having compositions of 40, 50, 60 and 75% Te. The compressibility shows marked deviations from ideal behaviour, although there does not appear to be a sharp peak at the Ga_2Te_3 composition.

2. Experimental details and results

Our experiments to determine the sound velocity were performed at 20·2 MHz using a conventional pulse-echo technique, the reflector being the flat end of a silica sheath normally used in our electromagnetic detection method (Turner *et al* 1972). In the conventional mode up to 150 fringes (spacing one half an acoustic wavelength) were used for each measurement, which was repeated at least three times at each temperature. We estimate our sound velocity results to be accurate to ±0·1%.

The alloys were prepared by melting the appropriate amounts of Ga (6N grade†) and Te (6N grade ‡) in the integral silica delay line/container assembly in a separate apparatus in about 0·5 atm of Ar. All the experiments were performed in a similar inert atmosphere to inhibit any composition changes during the course of an experiment due to evaporation of Te. The compositions of the alloys were thus assumed to be those given by the amounts of metal initially alloyed. Before measurements were made the alloys were gently stirred to ensure homogeneity. No difficulties were experienced in obtaining good acoustic coupling through the silica/liquid metal interface.

Table 1. Sound velocity and compressibility results. The Ga and Te results are taken from Gitis and Mikhailov (1966).

Comp. (at.% Te)	v_S at 835 °C ($m\,s^{-1}$)	$-(dv_S/dT)_p$ ($m s^{-1}\,°C^{-1}$)	ρ ($g\,cm^{-3}$)	κ_S ($10^{-11}\,m^2\,N^{-1}$)
40·0	1608 ± 1	0·16 ± 0·01	5·2	7·43
50·0	1333 ± 1	0·36 ± 0·01	5·16	10·9
60·0	1148 ± 1	see figure 1	5·09	14·9
75·0	1045 ± 1·5	0·27 ± 0·01	5·3	17·3
Ga	2663 ± 2	0·275	5·59	2·52
Te	1137 ± 5	–ve	5·53	14·1

Experiments were performed on four alloys having compositions corresponding to Ga_3Te_2, GaTe, Ga_2Te_3 and $GaTe_3$. The v_S results are shown in table 1 where we have also included the values for pure Ga and pure Te (Gitis and Mikhailov 1966). The sound velocity decreased linearly with temperature except for the alloy Ga_2Te_3. The results for this alloy are depicted in figure 1. The points in the figure were obtained by cycling the temperature a number of times: for example the first measurement was made at about 850 °C, the seventh at about 820 °C and the fourteenth at about 930 °C. Thus we are convinced that the observed curvature is real and not the result of any composition changes.

† Aluminum Company of America.
‡ Alpha Inorganics (Ventron Corporation).

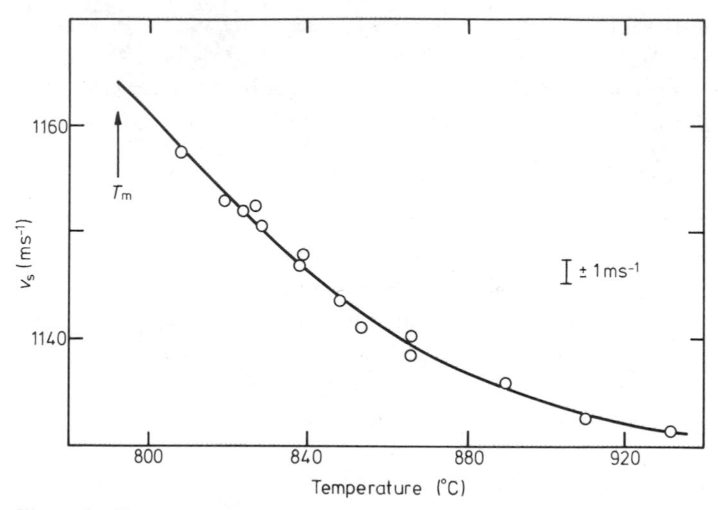

Figure 1. Temperature variation of the sound velocity (v_S) in liquid Ga_2Te_3. T_m indicates the melting point.

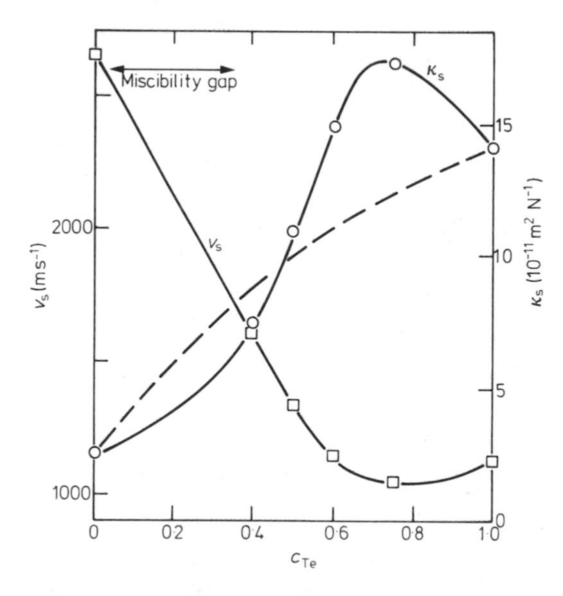

Figure 2. Variation of the sound velocity (v_S) and the adiabatic compressibility (κ_S) for Ga–Te at 835 °C. The broken curve is the ideal variation for κ_S. The curves drawn for the composition dependence of v_S and κ_S must be regarded as approximate because of the small number of points.

In figure 2 the variation with composition of the sound velocity at 835 °C is shown. To determine the adiabatic compressibility $\kappa_S = 1/\rho v_S^2$ density values were required. Density data for GaTe and Ga_2Te_3 (Glazov *et al* 1969) were used to estimate the volume change on alloying for the other two compositions, and hence obtain the density. The density value for pure Te was also taken from Glazov *et al* (1969) and that for Ga from Basin and Solov'ev (1967). This enabled us to calculate the adiabatic compressibility as a function of concentration, which is also shown in figure 2. Because of the small number of points in the sound velocity and density data the curves drawn are only approximate.

3. Discussion

In this section we give a qualitative analysis of our results, in particular focusing on the deviations from ideality observed in the system. Ideal behaviour of the compressibility in a liquid alloy follows from assuming that the molar volumes are additive and leads to the ideal compressibility

$$\kappa_s^0 = \frac{c_i V_{m_i} \kappa_{s_i} + c_j V_{m_j} \kappa_{s_j}}{c_i V_{m_i} + c_j V_{m_j}}$$

c_i being the mole fraction of species i whose molar volume is V_{m_i} and adiabatic compressibility κ_{s_i}. In most liquid alloy systems the deviation of the compressibility from ideal behaviour is small and of the same sign for all compositions, as in Cu–Sn for example (Turner *et al* 1973). The compressibilities of binary semiconducting systems such as Mg–Bi (McAlister *et al* 1973) Tl–Te (Turner 1974) and Ga–Te show pronounced deviations from ideal behaviour. In Mg–Bi and Ga–Te there is a rapid increase in κ_s, and a change in sign of the deviation from ideality, as the composition is approached at which extrema are found in the electrical properties. In the Tl–Te system the deviation is of the same sign and there is no rapid increase in κ_s near the composition of the extrema in the electron transport properties. The composition dependance of κ_s for liquid Ga–Te is compared with the ideal behaviour (broken curve) in figure 2. The variation of κ_s for $c_{Te} > 0{\cdot}6$ is subject to uncertainty because of inadequate density data.

In figure 3 we show the useful quantity $S_{cc}(0)$, or just S_{cc}, the long wavelength limit of the concentration–concentration correlation function. This function is related to the mean square fluctuation in the concentration $\langle(\Delta c)^2\rangle$ by

$$S_{cc} = N\langle(\Delta c)^2\rangle = NkT\left(\frac{\partial^2 G}{\partial c_i}\right)^{-1}_{T,P,N}$$

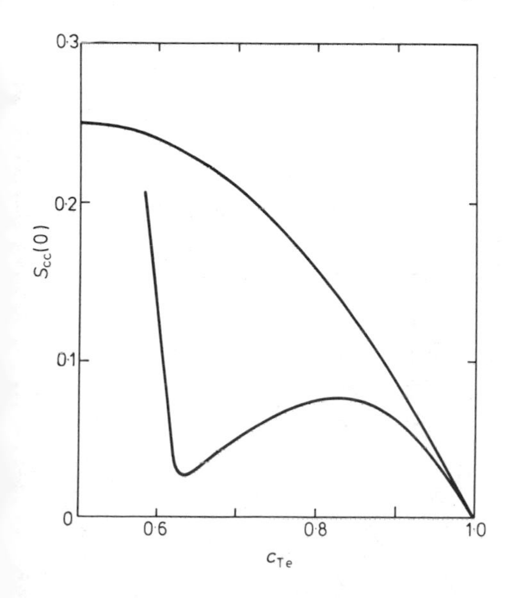

Figure 3. $S_{cc}(0)$ for Ga–Te at 841 °C. The upper parabolic curve is the ideal $S_{cc}(0)$, $c(1-c)$.

G being the Gibbs free energy of N atoms (Bhatia and Thornton 1970). The thermodynamic activity (a_i) data of Predel *et al* (1975) was used to determine S_{cc} for Ga–Te at 841 °C, using the relationship

$$NkT\left(\frac{\partial^2 G}{\partial c_i^2}\right)^{-1}_{T,P,N} = c_j\left(\frac{\partial \ln a_i}{\partial c_i}\right)^{-1}_{T,P,N}$$

where c_j is the mole fraction of the species j. In the figure we have not included S_{cc} for $c_{Te} < \sim 0{\cdot}6$ since there is very little data in this concentration range. The absolute values of S_{cc} shown are not that critical: for our present purposes we are interested only in the overall shape of the curve. The most noticeable features of the S_{cc} curve are the appearance of a minimum and the fact that S_{cc} is less than the curve $c_i c_j$ which applies to an ideal mixture. McAlister and Crozier (1974) showed that a minimum in S_{cc} is produced when the species form molecules in the liquid state, with the depth and position of the minimum dependent on the degree of dissociation of the molecules. For a one-component system S_{cc} is identically zero, but with dissociation at the compound concentration the value of S_{cc} becomes finite. These predictions are similar to the observations of S_{cc} for Ga–Te. The minimum does not occur exactly at the composition of Ga_2Te_3, but this may not be significant since very accurate thermodynamic data is really needed in the vicinity of the compound composition and the data of Predel *et al* (1975) show some scatter in this important region. Nevertheless the minimum is sufficiently close to $c_{Te} = 0{\cdot}6$ to suggest the formation of molecules of Ga_2Te_3. At the very least S_{cc} indicates that some non-ideal, short-range order exists in Ga–Te at and near $c_{Te} = 0{\cdot}6$. For the region $c_{Te} < 0{\cdot}6$ the rapid rise in S_{cc} is most probably due to the shift towards the immiscibility gap region at low Te compositions. Turner (1973) found the same behaviour in the similar system Tl–Te.

It is worth noting that the data of Predel *et al* (1975) was taken at 841 °C which is nearly 50 °C above the melting point of Ga_2Te_3, and it would have been very useful to

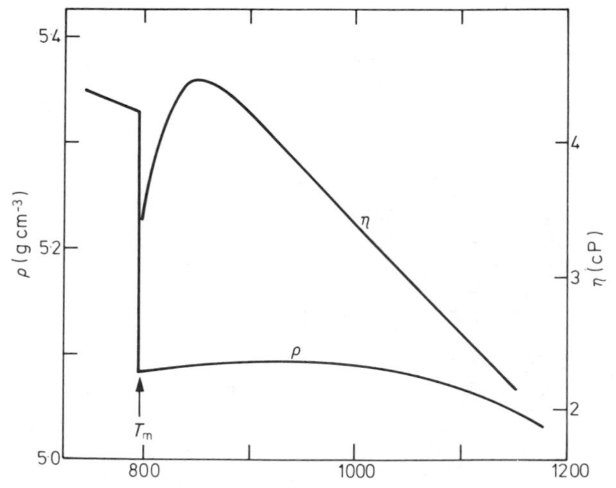

Figure 4. Temperature variation of the density (ρ) and the viscosity (η) for liquid Ga_2Te_3 (data taken from Glazov *et al* 1969). T_m indicates the melting point.

have had thermodynamic measurements at temperatures closer to the melting point of the compound. The effect of increasing the temperature would presumably be to cause dissociation of the compound molecules which would lead to an increase in the value of S_{cc} at the minimum. Thus near the liquidus temperature we would expect that the minimum is closer to zero and closer to $c_{Te} = 0{\cdot}6$.

In Mg–Bi the minimum in S_{cc} and the maximum in κ_s occur at the stoichiometric composition of Mg_3Bi_2. However, in Ga–Te the extrema in S_{cc} and κ_s are displaced from the characteristic composition Ga_2Te_3. This latter behaviour is consistent with Warren's NMR work which revealed that the Ga Knight shift has a minimum at $c_{Te} > 0{\cdot}6$. The relationship between S_{cc} and the Knight shift has been treated recently by Thompson *et al* (1976).

The negative values of $(dv_s/dT)_p$ for the alloys is normal and when combined with the thermal expansion coefficient α_p this gives a positive value for

$$\Delta_p = \frac{1}{\kappa_s}\left(\frac{d\kappa_s}{dT}\right)_p = \alpha_p + 2\beta_p,$$

where

$$\beta_p = -\frac{1}{v_s}\left(\frac{dv_s}{dT}\right)_p$$

which corresponds to a 'loosening' of the structure as the temperature is increased. It is usual for v_s to decrease linearly with temperature, hence the temperature variation shown by liquid Ga_2Te_3 (figure 1) deserves some attention. In figure 4 we show density and viscosity data (taken from Glazov *et al* 1969) for Ga_2Te_3 as a function of temperature. The very broad maximum observed for the density variation leads to α_p being very small. As a result β_p dominates the temperature variation of κ_s, and the compressibility does not increase rapidly for temperatures above $T \simeq 850\,^\circ C$. This leads to the deviation of the experimental compressibility from the ideal value decreasing with temperature, in other words the alloy becomes more ideal at the composition Ga_2Te_3. This is just the situation if increasing dissociation occurs as the temperature is raised. The unusual viscosity results depicted in figure 4 may also be reflecting this fact. Furthermore, Warren performed his NMR experiments as a function of temperature and concluded that they were consistent with the dissociation of the molecules with rising temperature. As already noted a nonzero minimum in S_{cc} (like that observed for Ga–Te) exists if dissociation of the compound molecules occurs.

We have found large deviations in the composition dependence of the compressibility for liquid Ga–Te from ideal behaviour which suggest that molecular associations of Ga_2Te_3 may exist. Furthermore the temperature dependence of the sound velocity and compressibility of Ga–Te alloys is consistent with the breakup of the molecules with increasing temperature. These anomalies differ from those observed in the other semiconducting systems Mg–Bi and Tl–Te.

Acknowledgments

We would like to thank the National Research Council of Canada for grants in support of the work reported in this paper.

References

Basin A S and Solov'ev A N 1967 *Zh. Prikl. Mekh. Tekh. Fiz.* **6** 83.
Bhatia A B and Thornton D E 1970 *Phys. Rev.* **B2** 3004
Crozier E D, Ward R W and Clayman B P 1977 this volume
Enderby J E and Collings E W 1970 *J. Non. Cryst. Solids* **4** 161
Gitis M B and Mikhailov I G 1966 *Sov. Phys.–Acoust.* **12** 14
Glazov V M, Chizhevskaya S N and Glagoleva N N 1969 *Liquid Semiconductors* (New York: Plenum)
Hansen M 1958 *Constitution of Binary Alloys* (New York: McGraw-Hill)
McAlister S P and Crozier E D 1974 *J. Phys. C: Solid St. Phys.* **7** 3509
McAlister S P, Crozier E D and Cochran J F 1973 *The Properties of Liquid Metals* ed S Takeuchi (London: Taylor and Francis) p 445
Predel B, Piehl J and Pool M J 1975 *Z. Metallkd.* **66** 268
Ratti V K and Bhatia A B 1975 *J. Phys. F: Metal Phys.* **5** 893
Thompson J C, Ichikawa K and Granstaff S M 1976 *J. Phys. Chem. Liq.* in the press
Turner R 1973 *J. Phys. F: Metal Phys.* **3** L57
—— 1974 *J. Phys. C: Solid St. Phys.* **7** 3686
Turner R, Crozier E D and Cochran J F 1972 *Can. J. Phys.* **50** 2735
—— 1973 *J. Phys. C: Solid St. Phys.* **6** 3359
Valiant J C and Faber T E 1974 *Phil. Mag.* **29** 571
Warren W W 1971 *Phys. Rev.* **B3** 3708
—— 1972 *J. Non. Cryst. Solids* **8–10** 241
Wobst M 1970 *Scripta Metall.* **4** 239

Structure of liquid and amorphous selenium by pulsed neutron diffraction using an electron LINAC

Kenji Suzuki and Masakatsu Misawa
The Research Institute for Iron, Steel and Other Metals, Tohoku University, Sendai 980, Japan

Abstract. Structure factors of liquid and amorphous selenium were measured over a range of the scattering vector as wide as 0·5 to 30 $Å^{-1}$ by means of time-of-flight neutron diffraction using hot pulsed neutrons generated by an electron LINAC. Based on experimental observation of the structure factor in the high-scattering-vector region which is mainly attributed to the intramolecular correlation, a disordered-chain model which has both helical- and ring-type arrangements in a single molecule has been proposed for the structure of the molecule in liquid and amorphous selenium.

1. Introduction

Liquid selenium has been considered to consist of a ring molecule having eight Se atoms and a chain molecule in which about 10^4 Se atoms link to one another near the melting point. It has been estimated from the measurement of viscosity (Glazov *et al* 1969) and a special solvent extraction (Briegleb 1929) that the fraction of ring molecules decreases and the average length of chain molecules becomes shorter with increasing temperature (Eisenberg and Tobolsky 1960). Optical spectroscopy measurements have been put forward as evidence that amorphous selenium is also a mixture of both the ring and chain molecules by Kawarada and Nishina (1975).

However, conventional x-ray (Waseda *et al* 1974) and neutron (Hansen and Knudsen 1975) diffraction experiments have not provided a clear picture of the structure of liquid and amorphous selenium which is compatible with the experimental observations described above. In order to obtain high-resolution structural information on a complicated system which includes a high degree of atomic ordering, such as liquid and amorphous selenium, the measurement of their structure factors has to be extended up to quite a wide range of values of the scattering vector.

In this work, we measure the structure factor of liquid and amorphous selenium over a range of the scattering vector as wide as 0·5–30 $Å^{-1}$ by means of time-of-flight neutron diffraction using hot pulsed neutrons produced by an electron LINAC. Based on the experimental observation of the structure factor in the high-scattering-vector region which is mainly attributed to intramolecular correlations, new structural models of the chain molecule in liquid and amorphous selenium are discussed.

2. Experimental

Measurements of the total structure factor $S(Q)$ were made using a time-of-flight pulsed neutron diffractometer installed on the Tohoku University electron LINAC. The LINAC was operated at an electron acceleration energy of 250 MeV with a peak beam current of 60 mA and at a pulse duration of 3 μs with a repetition frequency of 115 pulses/s, to generate pulsed neutrons, of which a range of wavelength from 0·3 to 1·5 Å with burst width of less than 40 μs was used in this work. The average power of the LINAC was only about 5 kW, which was enough to keep a statistical error of counts within less than 1% for a reasonable measuring duration.

Neutrons scattered from a sample were counted simultaneously at four scattering angles, $2\theta = 15, 30, 60$ and $150°$, by ^{3}He counters and their counting rates were stored in the memory of a computer as a function of the time-of-flight of neutrons through ordinary electronic amplifiers and pulse height analysers. Performance and operating experiences of the diffractometer have been described in previous papers (Misawa *et al* 1972, Suzuki *et al* 1975), together with procedures for correction and derivation of $S(Q)$ from the observed data.

The liquid selenium sample (purity 99·99%) was sealed in vacuum in a silica tube with a wall 0·30 mm thick and 10 mm inner diameter. An electric furnace, specially designed so that the heating elements were not located in the way of the neutron beams, was used to keep the liquid selenium at a desired temperature between the melting point and the boiling point during the measurement. The amorphous selenium sample was made by quenching the melt in water at 0 °C. The amorphous sample is 10 mm in diameter and 50 mm long.

3. Results

Figure 1 shows experimental structure factors $S(Q)$ for liquid and amorphous selenium. In liquid selenium, $S(Q)$ is found to be very temperature-dependent when $Q < 3$ Å^{-1}, while for $Q > 5$ Å^{-1} it shows oscillating behaviour which is essentially independent of temperature and persists up to values of Q as high as 25 Å^{-1}. More pronounced

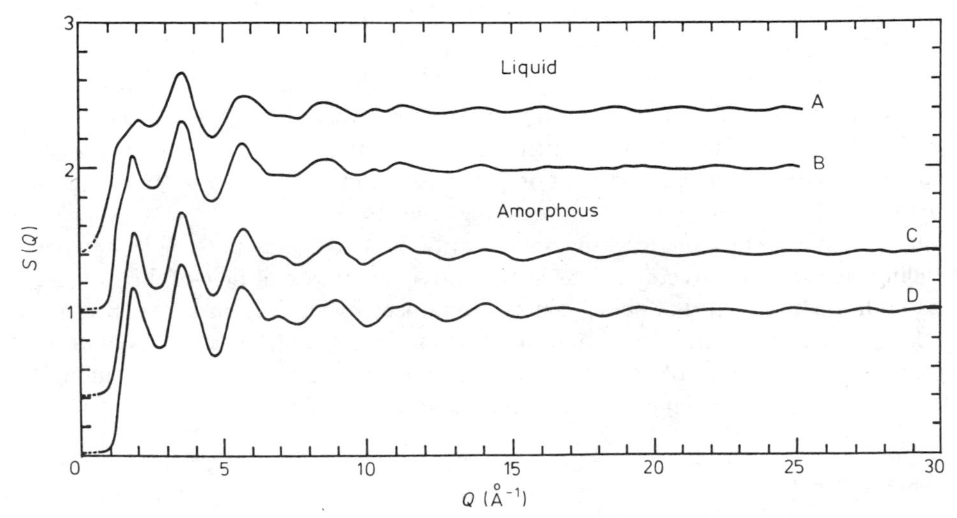

Figure 1. Observed structure factors of liquid selenium (A, at 650 °C; B, at 265 °C) and amorphous selenium (quenched from: C, 680 °C; D, 265 °C).

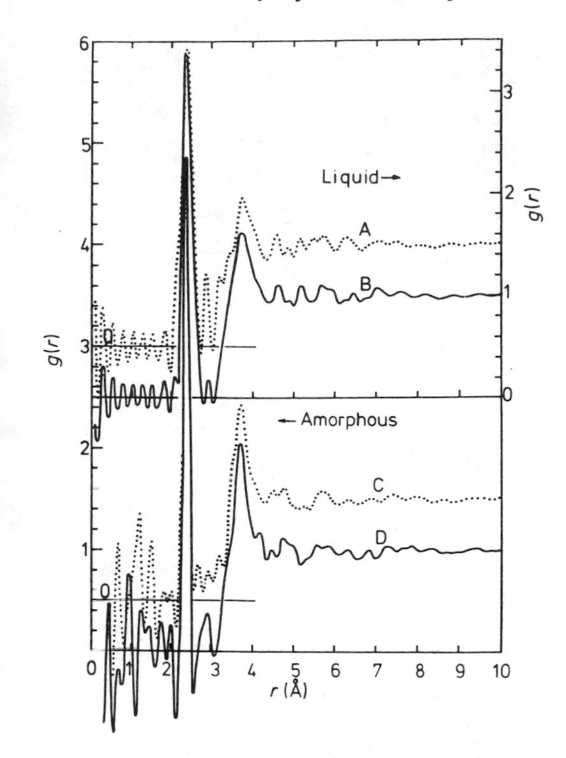

Figure 2. Pair distribution functions of liquid and amorphous selenium. A, liquid at 680°C; B, liquid at 265°C; C, amorphous, quenched from 680°C; D, amorphous, quenched from 265°C. $g(r)$ for liquid selenium is the Fourier transform of the $S(Q)$ corrected for the long-wavelength ripples in Q-space which contribute to $g(r)$ only inside the first peak. $g(r)$ for amorphous selenium is the Fourier transform of $S(Q)$ without any corrections.

oscillations of $S(Q)$ are found for $Q > 30\,\text{Å}^{-1}$ in the case of amorphous selenium. These observations suggest that atomic configurations among nearest-neighbour Se atoms are little influenced by temperature in the liquid state, and relatively long-range structures vary drastically with temperature. In the quenched amorphous samples we did not find much variation in $S(Q)$ with the temperature from which the melt was quenched (figure 1).

The pair distribution functions $g(r)$ of liquid and amorphous selenium are shown in figure 2. Since the Fourier transformation of $S(Q)$ into $g(r)$ has been truncated at $Q_{max} = 25\ \text{Å}^{-1}$, many fine peaks are well resolved but spurious ripples due to the truncation effect and small noises in $S(Q)$ in the high-Q region are inevitably superposed on $g(r)$ in figure 2. Therefore, the discussion in §4 will be given in terms of $S(Q)$ using momentum space, rather than $g(r)$ using real space. Temperature variations in $g(r)$ for liquid selenium are mainly found over a restricted region of $r = 2{\cdot}8$ to $4{\cdot}0$ Å, particularly including the lower side of the second peak. The function $g(r)$ for amorphous selenium indicates little temperature dependence in this experiment in contrast to liquid selenium.

4. Discussion

It is unlikely that long helical chain molecules with a straight symmetrical axis are conserved in liquid and amorphous selenium as well as in trigonal crystalline selenium. In fact we have pointed out that the position of peaks in the $g(r)$ of liquid selenium can be satisfactorily assigned by using the average position of Se atoms calculated in a free-rotating-chain model (Misawa 1976, Suzuki 1976).

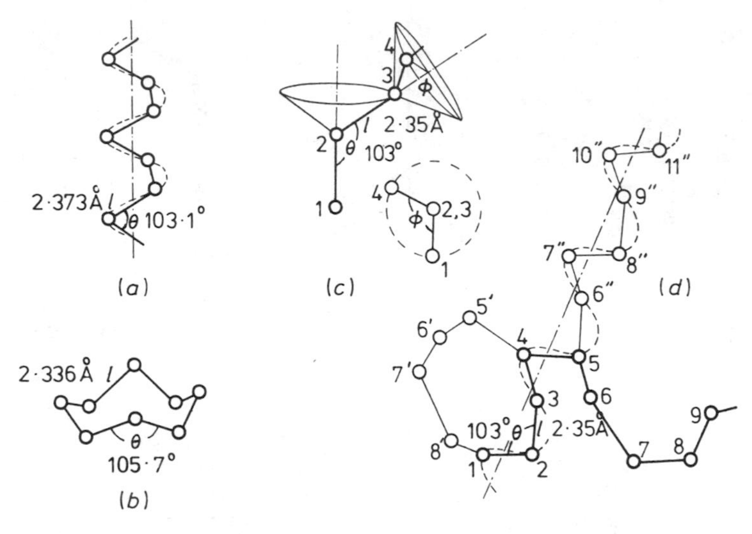

Figure 3. Structure models of molecules in liquid and amorphous selenium. (*a*) Helical chain, (*b*) ring, (*c*) free-rotating chain and (*d*) disordered chain.

If a liquid consists of molecules, $S(Q)$ asymptotically approaches the form factor of a single molecule in the liquid state with increasing Q, because the intermolecular correlations disappear in the high-Q region. Therefore, the structure of molecules existing in liquid and amorphous selenium can be directly verified by comparing the observed $S(Q)$ with the form factors calculated for various molecular structure models. To simplify discussions here, thermal fluctuations of Se atoms in the chemical bond are neglected, because configurational fluctuations of Se atoms are expected to be larger than thermal fluctuations particularly in the free-rotating-chain model and the disordered-chain model described below.

Molecular form factors are calculated for four molecular structure models as shown in figure 3. Since the bond lengths and bond angles between Se atoms in molecules are very similar in every model, atomic configurations in the range from the first atom (an arbitrary origin) to the third atom are on the same plane and cannot be distinguished among the four models.

In the free-rotating-chain model the fourth atom can rotate freely around the axis through the second and third atoms (figure 3(*c*)). Hence, there are no restrictions on the dihedral angle ϕ in the free-rotating-chain model.

In both the Se_8 ring (figure 3(*b*)) and helical-chain (figure 3(*a*)) models, the fourth atom is located at dihedral angle $\phi = 102°$ at which the rotational potential has minimum energy. Therefore, there are no differences in atomic configuration between these two models up to the fourth atom. According to whether the fifth atom is located at the cis-position or trans-position with respect to the first atom, the Se_8 ring structure and helical-chain structure can be distinguished. If the cis-type arrangement continues from the fifth to eighth atoms, the first and eighth atoms are connected with each other by a chemical bond, and a closed ring structure having eight Se atoms is realized. On the other hand if only the trans-type arrangement occurs, the helical-chain structure, as in trigonal crystalline selenium, appears. Furthermore, we can surely expect to find mixing of the cis- and trans-arrangements in a single molecule, which is hereafter called a disordered chain.

The form factors of the four molecular structure models described above are written as

$$F_1^{\alpha}(Q) = 1 + \sum_{n=2} \int C_n^{\alpha}(r) \frac{\sin Qr}{Qr} \mathrm{d}r,$$

where $C_n^{\alpha}(r)$ is the distribution function of the nth neighbour averaged over all atoms in a molecule. The four molecular structure models are labelled by a superior letter α, where α = r denotes the Se_8 ring, hc the helical chain, frc the free rotating chain and dc the disordered-chain molecules. Values of $C_n^{\alpha}(r)$ for the Se_8 ring and helical-chain molecules can be obtained from structural data in their crystalline states. The $C_n^{\alpha}(r)$ for the free-rotating-chain and disordered-chain molecules are calculated using the zero and finite rotational potentials (Misawa and Suzuki 1976).

The form factors, $F_1^{\alpha}(Q)$, calculated up to the eighth atom are shown for each molecular structure model in figure 4, together with the observed $S(Q)$ of amorphous selenium. In the high-Q region the $F_1^{\alpha}(Q)$ should be mainly contributed from the correlations between adjacent Se atoms. Since the atomic configurations between the nearest-neighbour Se atoms are almost same in the four models, the basic behaviour of the $F_1^{\alpha}(Q)$ when Q is higher than 20 $Å^{-1}$ is very similar. However, significant differences between the $F_1^{\alpha}(Q)$ can be found, as shown in figure 4, when Q is between 6 and 20 $Å^{-1}$. These differences may correspond to variations in the arrangement of atoms of a higher order than the fourth atom.

Since the thermal fluctuations of Se atoms are neglected in the simple models shown in figure 3, the calculated $F_1^{\alpha}(Q)$ include more pronounced oscillations even in the high-Q region compared with the observed $S(Q)$. However, the overall behaviour of the observed $S(Q)$ in the intermediate Q region from 6 to 20 $Å^{-1}$ has been reproduced fairly

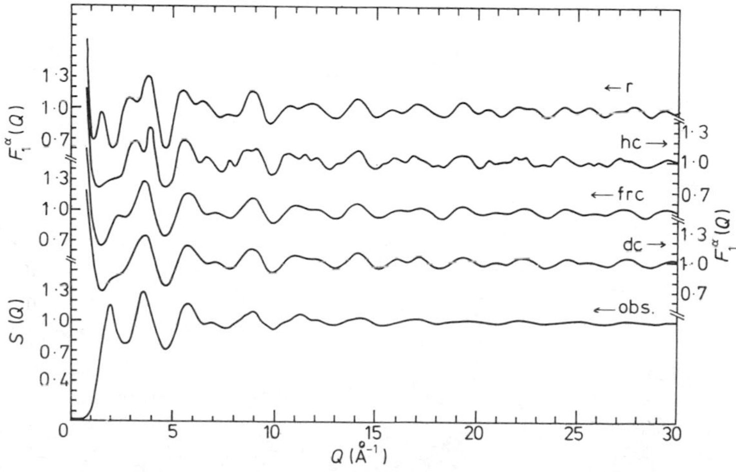

Figure 4. Molecular form factors calculated for structure models shown in figure 3.

well by F_1^{frc} and $F_1^{\mathrm{dc}}(Q)$. In contrast to this, $F_1^{\mathrm{r}}(Q)$ and $F_1^{\mathrm{hc}}(Q)$ show oscillatory behaviour which is different from the observed $S(Q)$.

As shown in figure 4, the structure of molecules in liquid and amorphous selenium should be considered to be a disordered chain including parts corresponding to both the helical and ring arrangements, in the range of temperature studied in this work. Pure helical and ring molecules are extreme examples of this structure. The structure and properties of liquid and amorphous selenium are expected to be systematically interpreted in terms of the disordered-chain model. In particular the structure of liquid selenium at high temperature can be simulated by the free-rotating-chain model. The fraction of the cis-type arrangement and the average number of Se atoms in a disordered chain must be determined as a function of temperature through the difference in potential energy between the cis- and trans-type arrangements. The statistical mechanics of the cis–trans transition in liquid selenium will be discussed in another paper (Misawa and Suzuki 1976).

Acknowledgments

The authors would like to thank Professor Igaki and Miss Ito for the preparation of amorphous selenium. One of us (KS) gratefully acknowledges partial support for this study by the 1975 RCA Grant Programme.

References

Briegleb G 1929 *Z. Phys. Chem.* **A144** 321
Eisenberg A and Tobolsky A V 1960 *J. Polym. Sci.* **46** 19
Glazov V M, Chizhevskaya S N and Glagoleva N N 1969 *Liquid Semiconductors* (New York: Plenum)
Hansen F Y and Knudsen T S 1975 *J. Chem. Phys.* **62** 1556
Kawarada M and Nishina Y 1975 *Jap. J. Appl. Phys.* **14** 1519
Misawa M 1976 *PhD Thesis* Tohoku University
Misawa M, Kai K, Suzuki K and Takeuchi S 1972 *Res. Rep. Lab. Nucl. Sci., Tohoku Univ.* **5** 73
Misawa M and Suzuki K 1976 *Proc. 79th Fall Meeting of Japan Institute of Metals* pp 65–67 (in Japanese)
Suzuki K 1976 *Ber. Bunsenges. Phys. Chem.* **80** (8) in the press
Suzuki K, Misawa M and Fukushima Y 1975 *Trans. Japan Inst. Metals* **16** 297
Waseda Y, Yokoyama K and Suzuki K 1974 *Phys. Condens. Matter* **18** 293

Solutions of nonmetals in liquid metals

C C Addison
Department of Chemistry, University of Nottingham, Nottingham, England

Abstract. The lecture was concerned with the nature of solutions of nonmetals in the liquid alkali metals (Addison 1967, 1971a,b, 1972, 1974), and reviewed recent work in the Nottingham Laboratories. Factors which control solubilities (Addison *et al* 1975) and the nature of the dissolved species (Addison *et al* 1976) were discussed; particular reference was made to the concept of solvation (Thompson 1972) in liquid metals, the relation between lattice energy and solvation energy (Adams *et al* 1975), and the similarity between solvation energies in the liquid alkali metals and in other solvents (Addison 1974). Interactions between dissolved nonmetals (Pulham and Simm 1971, Adams *et al* 1976) were considered in terms of solvation of the species involved. There are also peculiar features about precipitation from liquid metal media which are best interpreted in terms of solvation (Addison *et al* 1975, 1976). The paper considered reasons for the persistence of covalent bonds in the presence of an excess of free electrons. A detailed discussion of the topics referred to in the lecture is to be published by the Chemical Society shortly in *Chemical Society Reviews,* under the title 'The Liquid Alkali Metals as Solvents and Reaction Media'.

References

Adams P F, Down M G, Hubberstey P and Pulham R J 1975 *J. Less-Common Metals* **42** 325
Adams P F, Hubberstey P and Pulham R J 1976 *J. Less-Common Metals* **49** 253
Addison C C 1967 *Endeavour* **26** 91
—— 1971a *Proc. R. Inst.* **44** 317
—— 1971b *Presidential Address, British Association for the Advancement of Science* (Swansea)
—— 1972 *Sci. Prog. Oxford* **60** 385
—— 1974 *Chemistry in Britain* **10** 331
Addison C C, Creffield G K, Hubberstey P and Pulham R J 1976 *J. Chem. Soc. (Dalton)* 1105
Addison C C, Pulham R J and Trevillion E A 1975 *J. Chem. Soc. (Dalton)* 2082
Pulham R J and Simm P 1971 *Proc. Conf. Chemical Aspects of Corrosion and Mass Transfer in Liquid Sodium* (New York: Metallurgical Society)
Thompson R 1972 *J. Inorg. Nucl. Chem.* **34** 2513

Solubilities of rare gases in liquid metals

S Fukase and T Satoh
Department of Mathematics and Physics, National Defense Academy, Yokosuka 239, Japan

Abstract. To our knowledge there has been no article in which the solubilities of rare gases into liquid metals are discussed based on the electron theory. An interesting point for this subject is that the solubilities are considerably smaller than those in other normal liquids and that a conventional calculation based on the scaled particle theory yields rather poor results (Veleckis *et al* 1971).

In a very recent paper (Fukase and Satoh 1976), we have suggested that the change in the interaction energies among solvent molecules caused by adding a solute molecule (Neff and McQuarrie 1973), plays an important role for this problem and is closely related to the vacancy formation energy (or point-defect energy) in metals. Application of an approximate formula for vacancy formation energy recently proposed by Faber (1972) has led to encouraging results for liquid sodium.

In this report we introduce portions of this work and additional results of similar calculations for metals other than Na.

References

Faber T E 1972 *An Introduction to the Theory of Liquid Metals* (London: Cambridge University Press)
Fukase S and Satoh T 1976 *J. Phys. F: Metal Phys.* **6** 1233
Neff R O and McQuarrie D A 1973 *J. Phys. Chem.* 77 413
Veleckis E, Dhar S K, Cafasso F A and Feder H M 1971 *J. Phys. Chem.* 75 2832

Aspects of the solution chemistry of liquid alkali metals as elucidated from electrical resistivity studies

Peter Hubberstey
Inorganic Chemistry Department, Nottingham University, Nottingham NG7 2RD

Abstract. Resistivity techniques have been applied to a study of the behaviour of liquid alkali metals as solvents and reaction media; phase equilibria, solute reaction chemistry and the nature of the solute species have been investigated. Lithium and sodium are generally chosen as solvents and both non-metals and metals are studied as solutes.

Phase relationships in alkali metal systems have been investigated using both resistivity–concentration and resistivity–temperature methods; these techniques are particularly effective for the determination of solubility data (i.e. the hypereutectic liquidus) and of liquid–liquid miscibility gaps, phase boundaries which are difficult to examine using conventional thermal methods.

The stoichiometry and kinetics of chemical reactions in alkali metal solutions have been studied using resistivity techniques to monitor solute concentration. Thus, resistivity changes have shown the role of the solvent to be important when considering solute interaction in liquid metals; whereas association between H^- and O^{2-} occurs in sodium at 723 K according to the equilibrium

$$H^- + O^{2-} \rightleftharpoons OH^- + 2e.$$

H^- and N^{3-} act independently in lithium at 693 K. Furthermore, resistivity changes have been used to identify solvated solute species; dissolution of nitrogen in solutions of barium in liquid sodium at 573 K gives rise to a strongly bonded solvated unit of stoichiometry close to Ba_4N.

1. Introduction

Continuing interest is being shown in liquid alkali metal solutions since a fundamental understanding of their chemistry as solvent media is of both academic and technological importance (Addison 1974). The intention of this paper is to show how investigation of the electrical resistivities of these solutions may be used to derive information on their chemistry, and not to give an account, *per se*, of their electrical resistivities. Thus, resistivity studies of phase equilibria, solubility data, reaction stoichiometry and kinetics, and the question of solvation in liquid alkali metals, are considered. Lithium and sodium are generally chosen as solvents and both nonmetals and metals are studied as solutes.

2. Experimental

The stainless steel resistivity vessel is shown in figure 1. The bulk of the liquid metal (*ca* 50 g) was contained in a cylindrical reservoir A. It could be circulated independently through two loops using miniature electromagnetic pumps developed by Pulham (1971).

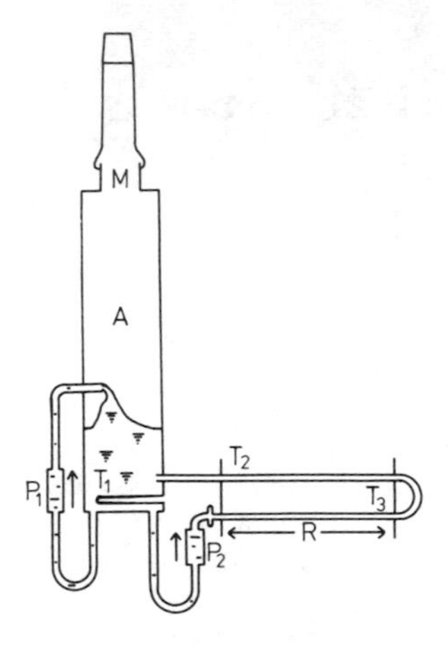

Figure 1. Resistivity vessel; A, main reservoir; M, glass-to-metal seal; P_1, P_2, pump ducts; R, resistivity capillary; T_1, T_2, T_3, thermocouples.

The main pump, P_1, provided a continuously regenerated clean metal surface for (i) rapid dissolution of the solute, and (ii) effective mixing of the solution. The subsidiary pump, P_2, fed the solution into a capillary, R, for continuous resistance measurement. Resistivities were calculated from calibration and sample resistance data and the capillary dimensions.

The resistivities of the solutions, and hence their chemistry, were investigated under both isothermal (ρ–c method) and constant concentration (ρ–T method) conditions; those solutions formed from solid solutes could only be studied using the latter method, whereas those produced from gaseous solutes could be studied using both methods. Details of the resistivity vessel and of both methods have been reported previously (Adams *et al* 1975).

3. Phase equilibria and solubility studies

Although resistivity techniques can be used to investigate many phase relationships, their use is particularly effective for the delineation of hypereutectic liquidi and of miscibility gaps, phase boundaries which are difficult to examine using conventional thermal methods (Down *et al* 1975a).

The ρ–c method for the determination of the solubility of metal salts (LiH) in liquid alkali metals (Li) is illustrated in figure 2; breaks in typical $\Delta\rho$–c isotherms (upper diagram) are correlated by broken lines with the hypereutectic liquidus of the Li–LiH phase diagram (lower diagram). ($\Delta\rho$ is the resistivity increase over that of the pure solvent at temperature, T.) Using this technique, the solubility of several lithium salts in liquid lithium has been investigated (Adams *et al* 1975, 1976); the data are of importance in the contexts of purification of and solvation in liquid alkali metals.

The ρ–T method is illustrated in figure 3 for the sodium–lithium system. The left hand diagram shows ρ–T curves for a number of typical solutions; breaks in these curves are correlated by broken lines with boundaries in the schematic phase diagram on the right hand side. For the more concentrated solutions, 44·5, 30·4 and 15·9 mol % Li, the immiscibility boundary, monotectic horizontal (444 K) and eutectic horizontal (365 K) were observed. For the more dilute solution, 7·5 mol % Li, the hypereutectic

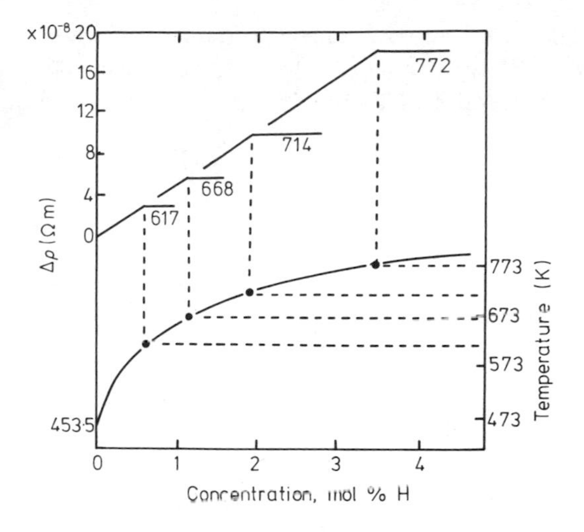

Figure 2. Increase in resistivity, $\Delta\rho$, over that of lithium for Li–H solutions, as a function of hydrogen concentration (temperatures (K) given against the curves), and the solubility of hydrogen in liquid lithium (0·0–4·5 mol % H).

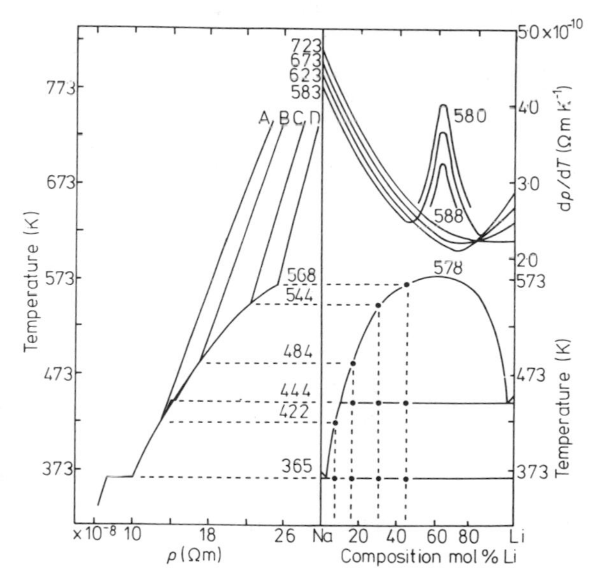

Figure 3. Resistivity–temperature curves (concentrations: A, 7·5; B, 15·9; C, 30·4; D, 44·5 mol % Li), and $(d\rho/dT)_T$–concentration isotherms (temperatures (K) given against the curves), for Na–Li solutions and the schematic Na–Li phase diagram.

liquidus and eutectic horizontal (365 K) were observed. Using this technique, we have defined the miscibility gap in the Na–Li system (Down *et al* 1975a), the hypereutectic liquidus in the Na–Li (Down *et al* 1975a), Na–Ba (Addison *et al* 1971), Na–Sr (Bussey *et al* 1976), Na–Ge, Na–Sn (Hubberstey and Pulham 1974) and Na–Pb systems (Hubberstey and Pulham 1972) and confirmed monotectic, peritectic, eutectic and peritectoid reactions, where present, in these systems.

An interesting feature of the resistivity of Na–Li solutions is the fact that $(d\rho/dT)_T$ is very sensitive to incipient immiscibility for several degrees above the miscibility boundary (Down *et al* 1975b). This phenomenon is illustrated in figure 3 within a series of $(d\rho/dT)_T$–c isotherms. Low-temperature isotherms (580, 583 and 588 K) pass through a peak, absent in the higher temperature isotherms (623, 673 and 723 K). The lowest temperature at which the entire peak can be observed is 578 K, the consolute temperature; the peak maximum occurs near to the consolute composition, 63 mol % Li. These observations have also been noted by Feitsma *et al* (1975) and by Schurmann and Parks (1971) and have been attributed to concentration fluctuations peculiar to systems exhibiting miscibility gaps.

4. Solution reaction studies

Our studies have shown that resistivity changes provide an elegant method of monitoring solute concentrations, and hence of studying both the stoichiometry and kinetics of chemical reactions in alkali metal solutions. Furthermore, we have obtained direct evidence for the occurrence of solvation in these solutions.

4.1. Reaction stoichiometry

Association between solutes is reflected in a deviation from additivity in the resistivity; a composite solute species will increase the resistivity of the solvent to an extent which differs from the sum of those of the separate components. When there is no association, however, the resistivity of the solution is expected to be additive. Solutions formed by reaction of NH_3 and of H_2O with liquid alkali metals exhibit these two patterns of behaviour. Thus, reaction of NH_3 could occur according to either equation (1) or (2):

$$6M + NH_3 \rightarrow M_3N + 3MH \tag{1}$$

$$2M + NH_3 \rightarrow MNH_2 + MH. \tag{2}$$

Similarly for H_2O,

$$4M + H_2O \rightarrow M_2O + 2MH \tag{3}$$

or

$$2M + H_2O \rightarrow MOH + MH. \tag{4}$$

Subsequent solution of these salts in the liquid metal will give rise to resistivity changes from which the solute species can be inferred.

The results for the reaction of NH_3 with lithium at 693 K are shown in figure 4(*a*) which includes those resistivity increases experimentally determined for H^- (line OA) and N^{3-} (line OB) and that derived for $3H^- + N^{3-}$ (line OC) in liquid lithium. A corresponding line for $H^- + NH_2^-$ cannot be included since resistivity data for NH_2^- in lithium are not available. The data points fall almost exactly on line OC, indicating complete breakdown of the NH_3 molecule on reaction with and dissolution in liquid lithium.

Reaction of H_2O with sodium (Pulham and Simm 1973) at 723 K, however, leads to a more complex situation as shown in figure 4(*b*), which includes those resistivity increases experimentally determined for H^- (line OD) and O^{2-} (line OE) and that derived for $2H^- + O^{2-}$ (line OF) in liquid sodium. A corresponding line for $H^- + OH^-$ cannot be included since resistivity data for OH^- in sodium are not available. At very low solute concentrations, the data points follow line OF indicating complete dissociation of the H_2O molecule. At higher concentrations, however, they deviate from this

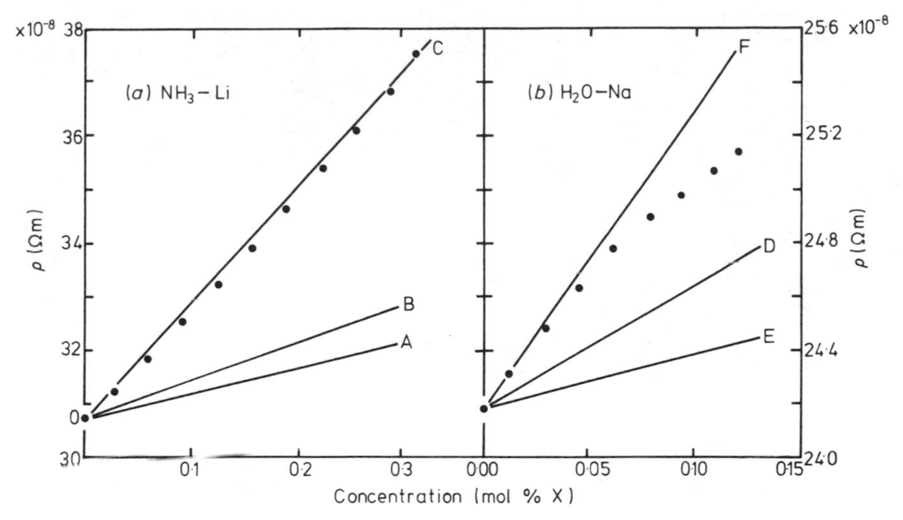

Figure 4. (*a*) Comparison of the resistivity data observed on dissolving the products of the reaction of ammonia with lithium at 693 K with those experimentally observed for H^- (line OA) and for N^{3-} (line OB) and that derived for $(3H^- + N^{3-})$ (line OC) in liquid lithium. (*b*) Comparison of the resistivity data observed on dissolving the products of the reaction of water vapour with sodium at 723 K with those experimentally observed for H^- (line OD) and for O^{2-} (line OE) and that derived for $(2H^- + O^{2-})$ (line OF) in liquid sodium. (All data are corrected, where necessary, for the presence of hydrogen equilibration pressures in the vessel.)

line correlating with an increase in OH^- concentration according to the equilibrium (5), the proportion of OH^- increasing with increasing solute concentration:

$$O^{2-} + H^- \rightleftharpoons OH^- + 2e. \qquad (5)$$

These results and others obtained on M–O–H solutions (M = Li, K, Rb, Cs) show that the solvent properties of sodium lie between those of lithium and those of the three heavier alkali metals; thus, O^{2-} and H^- act independently in lithium, O^{2-}, H^- and OH^- are in equilibrium in sodium and OH^- does not dissociate in potassium, rubidium and caesium (Hubberstey *et al* 1976). This gradation is in accord with the enthalpy changes, ΔH_{reac}, calculated for reactions (3) and (4), and depicted in figure 5(*b*) as a function of atomic number. With lithium, H_2O reacts to form M_2O and MH, whereas with potassium, rubidium and caesium, the formation of MOH and MH is favoured. This analysis is mirrored by the corresponding one, figure 5(*a*), for the reaction of NH_3 with alkali metals, which indicates that whereas reaction with lithium yields M_3N and MH, reaction with the heavier alkali metals will give MNH_2 and MH.

4.2. Solvation in solution

Very little direct evidence for solvation has been reported despite the growing interest in theoretical solvation models (Thompson 1972, Gellings *et al* 1972, 1974, Mainwood and Stoneham 1976). Perhaps the most elegant use of resistivity techniques in the study of liquid alkali metal solutions is in the positive identification of solvated solute species

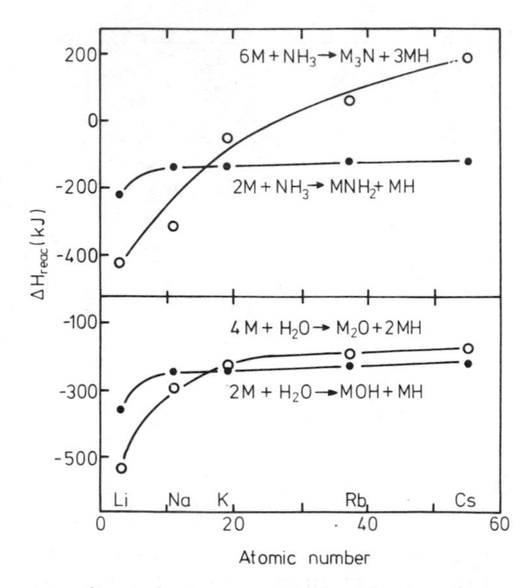

Figure 5. Enthalpy changes calculated for the reactions of ammonia and of water vapour with alkali metals.

in sodium–barium–nitrogen solutions (Addison *et al* 1976). Typical resistivity changes are shown in figure 6. Addition of barium to sodium gives an almost linear resistivity increase (section AB). Subsequent reaction with nitrogen causes a linear resistivity decrease (section BC); previous work (Addison *et al* 1975) has shown that this solution stage extends up to a Ba : N ratio near Ba_4N. At C, precipitation of Ba_2N commences until at D the resistivity returns to that of pure sodium. Further absorption of nitrogen (section DE) converts Ba_2N into Ba_3N_2. The initial resistivity decrease can only be interpreted by assuming that Ba atoms are incorporated with N^{3-} ions in solution to form a strongly bonded, solvated, Ba_4N unit, since independent behaviour of the two solutes would give rise to a resistivity increase (cf. that for nitrogen in lithium (figure 4(*a*) line OB). The decrease corresponds, therefore, to the sum of the loss of barium from (line BF), and the gain of Ba_4N units to (line GC) the solution; it is thus possible to

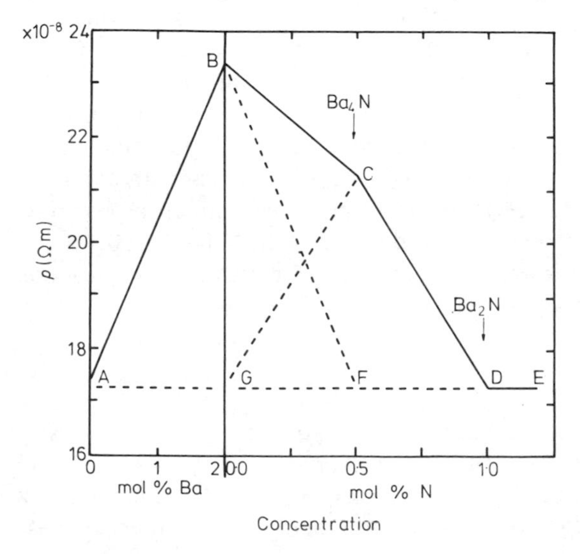

Figure 6. Resistivity changes observed in the sodium–barium–nitrogen system at 573 K, as a function of barium and nitrogen concentrations.

calculate the resistivity increase for Ba_4N units in sodium. This value, $8{\cdot}8 \times 10^{-8}$ Ωm (mol % $Ba_4N)^{-1}$, is similar to that for nitrogen in lithium, $6{\cdot}6 \times 10^{-8}$ Ωm (mol % N$)^{-1}$, indicating that solvated solute species are formed in all solutions of nonmetals.

4.3. Reaction kinetics

We have also utilized the linear interdependence between resistivity and solute concentration to monitor reaction kinetics in a study of lithium purification (Hubberstey *et al* 1975). The rate of hydrogen removal by reaction with yttrium was monitored by observing the resistivity decrease as a function of time. The results of a study of the variation in reaction rate as a function of hydrogen content of the yttrium, all other variables (e.g. temperature (673 K), hydrogen content of the lithium (0·25 mol % H)) being kept constant, are shown in figure 7. The main diagram shows the decrease in

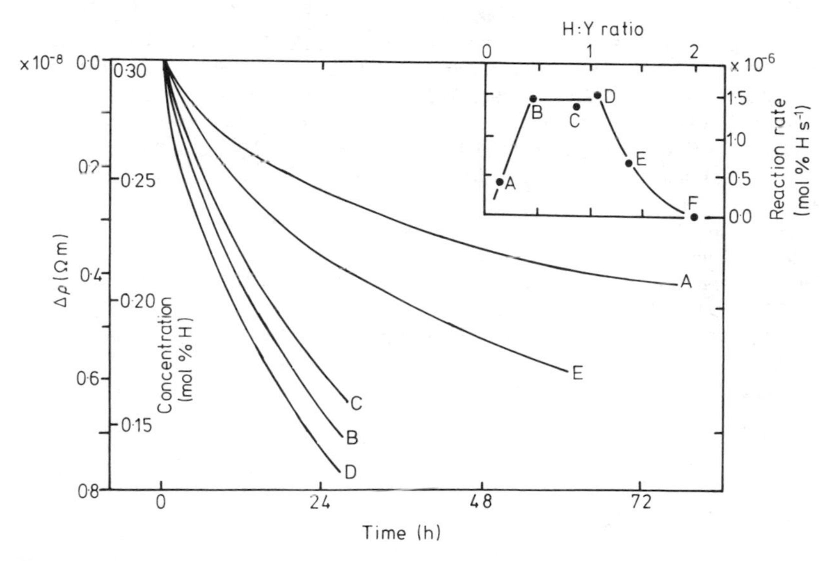

Figure 7. Resistivity decrease observed as a function of time for consecutive additions (A to E) of an yttrium sample to Li–H solutions (initial concentration, 0·30 mol % H) at 673 K, and (inset) the variation in reaction rate as a function of hydrogen content of the yttrium.

resistivity, and hence in hydrogen content of the lithium, for successive reactions (A–E) commencing at a constant Li–H concentration (0·30 mol % H). The reaction rates at a Li–H concentration of 0·25 mol % H are shown as a function of H : Y ratio in the inset diagram. Although a detailed discussion of the results is not appropriate in the present context, it is pertinent to note that they show that, after an initial increase, the rate is independent of H : Y ratio over a substantial concentration range before decreasing to zero when all the yttrium has been used to produce YH_2.

5. Conclusions

Resistivity measurements provide the basis of perhaps the most effective method for the elucidation of the chemistry of liquid alkali metals, the linear interdependence between resistivity and solute concentration being particularly valuable. Thus, many different phase boundaries have been readily delineated and both interactions between, and solvation of, solutes have been successfully examined.

References

Adams P F, Down M G, Hubberstey P and Pulham R J 1975 *J. Less-Common Metals* **42** 325
Adams P F, Hubberstey P, Pulham R J and Thunder A E 1976 *J. Less-Common Metals* **46** 285
Addison C C 1974 *Chem. in Britain* **10** 331
Addison C C, Creffield G K, Hubberstey P and Pulham R J 1971 *J. Chem. Soc.* **A** 2688
—— 1976 *J. Chem. Soc., Dalton* 1105
Addison C C, Pulham R J and Trevillion E A 1975 *J. Chem. Soc., Dalton* 2082
Bussey P R, Hubberstey P and Pulham R J 1976 *J. Chem. Soc., Dalton* in the press
Down M G, Hubberstey P and Pulham R J 1975a *J. Chem. Soc., Dalton* 1490
—— 1975b *J. Chem. Soc. Faraday Trans. I* **71** 1387
Feitsma P D, Hallers J J, Werff F V D and van der Lugt W 1975 *Physica* **B79** 35
Gellings P J, Huiscamp G B and van den Broek E G 1972 *J. Chem. Soc., Dalton* 151
Gellings P J, van der Scheer A and Caspers W J 1974 *J. Chem. Soc., Faraday Trans. II* **70** 531
Hubberstey P, Adams P F and Pulham R J 1975 *Proc. Int. Conf. on Radiation Effects and Tritium Technol. for Fusion Reactors, Gatlinburg, Tennessee* (USERDA Conf. 750989) vol 3 p 270
Hubberstey P, Adams P F, Pulham R J, Down M G and Thunder A E 1976 *J. Less-Common Metals* **49** 253
Hubberstey P and Pulham R J 1972 *J. Chem. Soc., Dalton* 819
—— 1974 *J. Chem. Soc., Dalton* 1541
Mainwood A and Stoneham M 1976 *J. Less-Common Metals* **49** 271
Pulham R J 1971 *J. Chem. Soc* **A** 1389
Pulham R J and Simm P A 1973 *Liquid Alkali Metals, Proc. BNES Conf., Nottingham* p 1
Schurmann H K and Parks R D 1971 *Phys. Rev. Lett.* **27** 1790
Thompson R 1972 *J. Inorg. Nucl. Chem.* **34** 2513

Improved model for solutions of nonmetals in liquid alkali metals: calculation of enthalpy of solution and electrical resistivity

M Oosterbroek, H P van de Braak and P J Gellings
Twente University of Technology, Enschede, The Netherlands

Abstract. A new model is proposed for the calculation of the enthalpy of solution of electronegative elements and noble gases in liquid alkali metals. This is based on the difference between the internal energies of the pure metal and of the solution. Contributions to those energies arise from free-electron terms and the Madelung energy, expressed in terms of simple model potentials and partial structure factors. Screening is accounted for by the band-structure energy.

Satisfactory agreement between experimental and calculated values is obtained using pure metal empty-core radii and crystal radii for the dissolved atoms in the model potentials and adjusting hard-sphere diameters in Percus–Yevick calculations for the structure factors. The model is able to explain the positive enthalpy of solution of the noble gases. Volume effects give an important contribution to corrections to the Madelung energy. The volume dependence of the electron-gas and band-structure terms are relatively small.

The electrical resistivity is obtained from Ziman's formula using identical structure factors and form factors based on the model potentials as used in the enthalpy calculations. Reasonable values for the resistivity change on dissolving the nonmetal are obtained but the nearly complete lack of experimental data makes verification practically impossible at this moment.

1. Introduction

Several models have been proposed (Gellings *et al* 1972, 1974, Thompson 1972, Feder and Schnijders 1970) to calculate the enthalpy of solution of nonmetals in liquid metals. These are all based on the assumption that the nonmetal dissolves as negative ions, in agreement with the calculations of Greenwood and Ratti (1972). Only the noble gases are assumed to dissolve as neutral atoms (Feder and Schnijders 1970).

All models have more or less serious defects, as shown for some cases by Abriata and Tarchitzky (1975). The main objections are:

(i) Volume effects are not taken into account.
(ii) The collection of metal atoms is usually considered as a uniform background, so that the coordination of guest ions by metal atoms cannot be calculated properly.
(iii) Screening effects are calculated and used in dubious ways: Gellings *et al* (1974) incorrectly use screened potentials for both interacting ions (see Abriata and Tarchitzky 1975); Thompson (1972) uses screening by ions; Feder and Schnijders (1970) have several terms in which screening plays a role.
(iv) Other physical properties, such as resistivity changes, cannot be calculated from these models.

In this paper we present a new method for the calculation of the enthalpy of solution of electronegative elements and noble gases in liquid alkali metals. This is again done on the basis of the Born–Haber cycle:

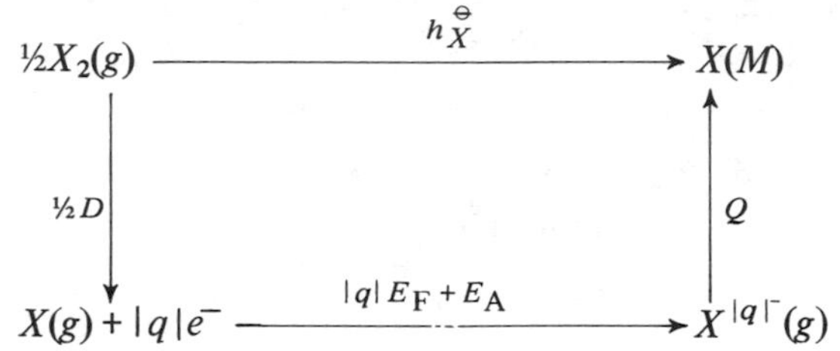

$$\begin{array}{ccc} \tfrac{1}{2}X_2(g) & \xrightarrow{\;h_X^{\ominus}\;} & X(M) \\ \big\downarrow \tfrac{1}{2}D & & \big\uparrow Q \\ X(g) + |q|e^- & \xrightarrow{\;|q|E_F + E_A\;} & X^{|q|^-}(g) \end{array}$$

Using Hess's law this gives for the partial molar enthalpy of solution,

$$h_X^{\ominus} = \tfrac{1}{2}D + |q|E_F + E_A + Q, \tag{1}$$

where D is the dissociation enthalpy of X_2, E_F the work function of the metal, E_A the electron affinity of X, and Q the energy of interaction of $X^{|q|^-}$ with the metal. For noble gases, $|q| = 0$ and only Q remains in equation (1). In this paper Q is calculated as the difference between the internal energies of the metal and the solution. Contributions to these energies arise from free-electron terms, the Madelung energy and the band-structure energy (which accounts for screening), all expressed in terms of model potentials and partial structure factors (Ashcroft and Langreth 1967a,b, Ashcroft 1972).

The electrical resistivity is obtained from the expression derived by Ziman (1972) using identical structure factors and form factors based on the same model potentials as used in the enthalpy calculations.

2. Energy of pure metal

The energy of the metal is calculated as proposed by Ashcroft (1972). The metal ions are supposed to be present as discrete entities in an originally uniform background of electrons. The ions cause local disturbances in the uniform density of the electron gas which is the so-called 'screening'.

The total energy U of a metal is then written as:

$$U = U_{el} + U_{Mad} + U_{bs}, \tag{2}$$

where U_{el} is the energy of the electron gas including exchange and correlation effects, U_{Mad} is the potential energy due to the non-uniform charge distribution of metal ions and U_{bs} is the band-structure energy.

A more detailed discussion and derivation of the terms is given by Ashcroft (1972). Using the formulation of Pines and Nozières (1966) for the exchange and correlation energies, the energy of the electron gas of a Z-valent metal consisting of N atoms is found to be:

$$U_{el} = ZN\left(\frac{2{\cdot}21}{r_s^2} - \frac{0{\cdot}916}{r_s} - 0{\cdot}115 + 0{\cdot}031 \ln r_s\right), \tag{3}$$

where the effective radius r_s is given by

$$r_s = (3\Omega/4\pi ZN)^{1/3} \tag{4}$$

with Ω = total volume of metal.

In order to calculate U_{Mad} the structure factor S_k and the model potential $V^M(r)$ are introduced. The structure factor is obtained as a Fourier transform of the pair distribution function $g(r)$:

$$S_k = 1 + \frac{N}{\Omega}\int [g(r)-1] \exp(-\mathbf{ik.r})\, d\mathbf{r}. \tag{5}$$

The structure factor can be approximated using the theory of a hard-sphere liquid (Ashcroft and Lekner 1966) with the hard-sphere diameter σ and the packing fraction η as the relevant parameters.

A model potenial is an approximation of the real potential. An often used form is a Coulomb potential outside a sphere with radius R_M and a constant value A inside that sphere. If $A = 0$ this is the potential of a charge distributed over the surface of the sphere which is called an empty-core or Ashcroft potential. If $A = Ze^2/R_M$ this is the potential of a sphere with a uniform charge and this is called a Shaw potential.

From these definitions the Fourier transforms of these potentials are found to be for monovalent ions:

$$\text{Coulomb potential, } V_k^{c} = 4\pi e^2/k^2; \tag{6}$$

$$\text{Ashcroft potential, } V_k^{a} = 4\pi e^2 \cos k\, R_M/k^2; \tag{7}$$

$$\text{Shaw potential, } V_k^{b} = 4\pi e^2 \sin k\, R_M/k^3. \tag{8}$$

The Madelung energy is obtained as the sum of the potential energies of the electron–electron interaction (for which a Coulomb potential is used), of the electron–ion interaction and of the ion–ion interaction, with a model potential $ZV^M(r)$ used for the metal ions.

The final result is:

$$U_{Mad} = \frac{Z^2N}{2\Omega}\sum_{k\neq 0} V_k^{c}(S_k - 1) + ZN\rho(V_0^{c} - V_0^{M}). \tag{9}$$

Here we have introduced the electron density $\rho = ZN/\Omega$ and

$$V_0^i = \lim_{k\to 0} V_k^i = \int V^i(r)\, d^3r. \tag{10}$$

The first term in equation (9) is an improved Ewald energy and the second term a correction for the introduction of a model potential.

The band-structure energy of a Z-valent metal, again with a model potential $ZV^M(r)$ is found to be

$$U_{bs} = \frac{Z^2N}{2\Omega}\sum_{k\neq 0}\frac{k^2}{4\pi e^2}\left(\frac{1}{\epsilon_k} - 1\right)|V_k^M|^2 S_k. \tag{11}$$

From equations (2), (3), (9) and (11) we obtain for the total energy:

$$U = ZN\left(\frac{2\cdot21}{r_s^2} - \frac{0\cdot196}{r_s} - 0\cdot115 + 0\cdot031 \ln r_s\right) + \frac{Z^2N}{2\Omega}\sum_{k\neq0} V_k^c(S_k - 1)$$

$$+ ZN\rho(V_0^c - V_0^M) + \frac{Z^2N}{2\Omega}\sum_{k\neq0}\frac{k^2}{4\pi e^2}\left(\frac{1}{\epsilon_k} - 1\right) |V_k^M|^2 S_k. \qquad (12)$$

3. Change of energy upon dissolving a nonmetal

We consider an assembly of N_M atoms of a liquid metal to which N_X atoms of a noble gas or electronegative element are added. The values of the different quantities occurring in equation (12) after dissolution are denoted by an asterisk. The model potential of a noble gas atom is taken to be zero. Electronegative elements are supposed to dissolve as negative ions with a model potential $Z'V^X(r)$, where $Z' = -|q|$.

Upon dissolution the total volume is changed by $\Delta\Omega$ and we put:

$$\frac{\Delta\Omega}{\Omega} = \frac{N_X}{N_M}\frac{\Omega_X}{\Omega_M}\frac{N_X}{N_M}\left(\frac{\sigma_X}{\sigma_M}\right)^3 = c\delta, \qquad (13)$$

with the concentration $c = N_X/N_M$ and where σ_X and σ_M are the hard-sphere diameters of X and M. When a noble gas is dissolved the total number of free electrons is unchanged and we obtain from equation (4), to first order in $\Delta\Omega$:

$$\Delta r_s/r_s = \Delta\Omega/3\Omega = c\delta/3. \qquad (14)$$

When an electronegative element is dissolved, $|q|N_X = \Delta N$ electrons are taken from the electron gas and this gives, to first order in ΔN and $\Delta\Omega$:

$$\frac{\Delta r_s}{r_s} = \frac{1}{3}\left|\frac{\Delta\Omega}{\Omega} + \frac{\Delta N}{N}\right| = \frac{c}{3}(\delta + |q|). \qquad (15)$$

For the change in energy of the electron gas upon dissolving a noble gas we obtain:

$$(\Delta U_{el})_{ng} = U_{el}^* - U_{el} = \frac{\partial U}{\partial r_s}\Delta r_s,$$

which gives, with equations (3) and (14),

$$\frac{(\Delta U_{el})_{ng}}{N_X} = \frac{Z\delta}{3}\left(-\frac{4\cdot42}{r_s^2} + \frac{0\cdot916}{r_s} + 0\cdot031\right). \qquad (16)$$

Similarly when an electronegative element is dissolved

$$(\Delta U_{el})_{ene} = U_{el}^* - U_{el} = \frac{\partial U}{\partial r_s}\Delta r_s + \frac{\partial U}{\partial N}\Delta N$$

or, together with equations (3) and (15):

$$\frac{(\Delta U_{\mathrm{el}})_{\mathrm{ene}}}{N_X} = \frac{\delta + |q|}{3}\left(-\frac{4\cdot 42}{r_s^2} + \frac{0\cdot 916}{r_s} + 0\cdot 031\right)$$

$$+ |q|\left(-\frac{2\cdot 21}{r_s^2} + \frac{0\cdot 916}{r_s} + 0\cdot 115 - 0\cdot 031 \ln r_s\right). \tag{17}$$

This second term is the energy needed to take $|q|$ electrons out of the metal; it is the quantity $|q|E_{\mathrm{F}}$ of equation (1).

To calculate the change in Madelung energy we have to use the partial structure factors S_{MM}, S_{MX} and S_{XX} (Enderby *et al* 1966, Ashcroft and Langreth 1967b). For solutions of noble gases the term $S_k - 1$ simply becomes $S_{MM} - 1$. For solutions of electronegative elements the term $S_k - 1$ becomes $[(S_{MM} - 1) - 2S_{MX}c^{1/2} + c(S_{XX} - 1)]$ as long as $c \ll 1$. Furthermore the change in the electron density ρ is, to first order:

$$\Delta\rho = \rho\left(-\frac{\Delta\Omega}{\Omega} + \frac{\Delta N}{N}\right).$$

For noble gas solutions $\Delta N = 0$ and for electronegative elements $\Delta N = |q|N_X$, so that:

$$\begin{aligned}&\text{Noble gas: } \Delta\rho = \rho c\delta;\\ &\text{Electronegative element: } \Delta\rho = \rho c(\delta + |q|).\end{aligned} \tag{18}$$

Using these relations and equation (9) we obtain for the change in the Madelung energy when a noble gas is dissolved:

$$\frac{(\Delta U_{\mathrm{Mad}})_{\mathrm{ng}}}{N_X} = \frac{Z^2}{2c\Omega}\sum_{k\neq 0} V_k^{\mathrm{c}}(S_{MM} - S_k) - Z\delta\rho(V_0^{\mathrm{c}} - V_0^M). \tag{19}$$

Similarly we obtain for a solution of an electronegative element, taking into account the interactions in which the negative ions are involved:

$$\frac{(\Delta U_{\mathrm{Mad}})_{\mathrm{ene}}}{N_X} = -Z\rho(|q|+\delta)(V_0^{\mathrm{c}} - V_0^M) - Z\rho|q|(V_0^{\mathrm{c}} + V_0^X)$$

$$+ \frac{1}{2c\Omega}\sum_{k\neq 0} V_k^{\mathrm{c}}[Z^2(S_{MM} - S_k) - 2S_{MX}|q|Zc^{1/2} + c|q|^2(S_{XX} - 1)]. \tag{20}$$

The change in band-structure energy upon dissolving a noble gas is easily found in the same way to be;

$$\frac{(\Delta U_{\mathrm{bs}})_{\mathrm{ng}}}{N_X} = \frac{Z^2}{2c\Omega}\sum_{k\neq 0}\frac{k^2}{4\pi e^2}\chi_k|V_k^M|^2(S_{MM} - S_k) + \frac{Z^2\delta}{2\Omega}\sum_{k\neq 0}\frac{k^2}{4\pi e^2}\psi_k|V_k^M|^2 S_k, \tag{21}$$

where we have introduced $\chi_k = 1/\epsilon_k - 1$ and $\psi_k = \Omega(\partial\chi_k/\partial\Omega)$.

Along similar lines we find for the change in band-structure energy when an electronegative element is dissolved:

$$\frac{(\Delta U_{\mathrm{bs}})_{\mathrm{ene}}}{N_X} = \frac{1}{2c\Omega} \sum_{k\neq 0} \frac{k^2}{4\pi e^2} \chi_k \; [(S_{MM} - S_k) Z^2 |V_k^M|^2 + 2c^{1/2} Z |q| S_{MX} |V_k^X V_k^M|$$

$$+ c |q|^2 S_{XX} |V_k^X|^2] + \frac{(|q| + \delta) Z^2}{2\Omega} \sum_{k\neq 0} \frac{k^2}{4\pi e^2} |V_k^M|^2 S_k \psi_k. \tag{22}$$

The quantity Q of equation (1) is obtained as the sum of the contributions (16), (19) and (21) when a noble gas is dissolved and of (17), (20) and (22) in the case of an electronegative element.

Using these same quantities the specific resistance of a dilute solution ($c \ll 1$) of X in M can be obtained from the equation of Faber and Ziman (1965):

$$\rho_{(M,X)}(c) = \frac{3\pi m^2 N_M}{4e^2 h^3 k_{\mathrm{F}}{}^6 \Omega} \int_0^{2k_{\mathrm{F}}} \frac{1}{\epsilon_k{}^2} (|V_k^M|^2 S_{MM} + 2c^{1/2} S_{MX} |V_k^M V_k^X|$$

$$+ c S_{XX} |V_k^X|^2) k^3 \, \mathrm{d}k, \tag{23}$$

where k_{F} = wave number for electrons at the Fermi surface and m = effective mass of the electrons. For the change in specific resistance when N_X atoms of X are dissolved in N_M atoms of metal we then find:

$$\Delta\rho(c) = \frac{3\pi m^2 N_M}{4e^2 h^3 k_{\mathrm{F}}{}^6 \Omega} \int_0^{2k_{\mathrm{F}}} \frac{1}{\epsilon_k{}^2} [|V_k^M|^2 (S_{MM} - S_k) + 2c^{1/2} |V_k^M V_k^X| S_{MX}$$

$$+ c |V_k^X|^2 S_{XX}] \; k^3 \, \mathrm{d}k. \tag{24}$$

4. Calculations

For the calculations, the quantities given in tables 1 and 2 have been used. The hard-sphere diameters σ_X for the halogens have been calculated on the assumption that the distance between the hard-sphere surfaces of a dissolved halogen ion and a neighbouring metal ion is equal to that between the hard-sphere surfaces of the metal ions in the pure metal. It is also assumed that the halogen ion is surrounded by six metal ions, just as in most of the solid metal halides. For this distance, d_{MX}, the sum of the ionic radii (Weast 1970) is used. The metal–metal distance is calculated as

$$d_{MM} = (4\pi/3)^{1/3} r_{\mathrm{s}}.$$

For the chosen geometry this leads to

$$\sigma_X = \sigma_M + 2d_{MX} - 2d_{MM}. \tag{25}$$

The volume coefficient δ is chosen as

$$\delta = (\sigma_X/\sigma_M)^3. \tag{26}$$

Table 1. Values of metal properties used in the calculations.

	Na	K	Rb	Cs
T_m (°C)	97·5	62·3	38·5	28·5
r_s (Å) at T_m	2·14	2·66	2·81	3·04
R_M (Å)†	0·883	1·13	1·38	1·55
σ_M (Å)‡	3·28	4·06	4·31	4·73
r_{ion} (Å)§	0·97	1·33	1·47	1·67
E_F (eV)§	2·28	2·24	2·09	1·81

† Ashcroft and Langreth (1967a); ‡ Ashcroft and Lekner (1966); § Weast (1970).

Table 2. Values of properties of electronegative elements (all data from Weast 1970).

	F_2	Cl_2	Br_2	I_2	H_2	O_2
r_{ion} (Å)	1·33	1·81	1·96	2·20	1·54	1·32
D (kJ mol^{-1})	158	242	224	213	436	499
E_A (kJ mol^{-1})	−333	−349	324	−296	−72	691

The partial structure factors were obtained using the method described by Ashcroft and Langreth (1967b). Furthermore we take the potential $V^M(r)$ of the metal ion to be an Ashcroft potential, with the Fourier transform V_k^a given by equation (7). The potential $V^X(r)$ of the negative ions is taken to be a Shaw potential with the Fourier transform V_k^b given by equation (8). Finally screening is taken into account by using the Lindhard equation (Ziman 1972) for the dielectric constant ϵ_k.

The hard-sphere diameters of the noble gases were taken to be σ_{He} = 2·61 Å, σ_{Ne} = 2·79 Å, σ_{Ar} = 3·41 Å and σ_{Xe} = 3·97 Å (Feder and Schnijders 1970).

In table 3 the results of the calculations of the partial molar enthalpy of solution and of the relative change in resistivity $\Delta\rho(c)/c\rho(c)$ are given. The only experimental values available for the relative change in resistivity are those of Arnol'dov *et al* (1974) who gave 33 for oxygen and 27 for hydrogen, and a value of 100 for oxygen given by Blake (1960). The experimental values for the partial molar enthalpy of solution are taken from Gellings *et al* (1974), except those of He and Ar which are given by Veleckis *et al* (1971), Ne as given by Shimojima and Miyaji (1975) and Xe which was taken from Veleckis *et al* (1976).

5. Discussion

The results given in table 3 show that there is a reasonable agreement between the calculated and experimental values of the enthalpy of solution. In fact the average deviation is 33 kJ mol^{-1} as compared to 56 kJ mol^{-1} obtained with our previous model (Gellings *et al* 1974). Moreover, this agreement has been reached without the use of an arbitrary empirical parameter. Also, as stated previously (Gellings *et al* 1972), deviations of less than about 40 kJ mol^{-1} are not significant due to uncertainties in the experimental data.

Table 3. Experimental and calculated partial molar enthalpies of solution in kJ mol^{-1} and calculated relative changes in resistivity.

Metal	Nonmetal	$h_X^{\ominus}$(exp)	$h_X^{\ominus}$(calc)	$h_X^{\ominus}$(calc) – $h_X^{\ominus}$(exp)	$\dfrac{\Delta\rho(c)}{c\rho(c)}$ (calc)
Na	F	–465	–399	+66	26·7
	Cl	–322	–277	–45	19·3
	Br	–263	–234	+29	18·8
	I	–200	–159	+41	19·5
	H	+ 38	+ 43	+ 5	22·2
	O	–384	–332	+52	95·4
K	F	–485	–401	+84	37·1
	Cl	–340	–302	+41	25·4
	Br	–297	–267	+30	23·2
	I	–234	–213	–21	21·0
	O	–347	–290	+57	140·8
Rb	Cl	–322	–333	–11	22·5
	Br	–288	–299	–11	20·0
	I	–238	–246	– 8	17·6
Cs	F	–485	–446	+39	29·2
	Cl	–355	–355	0	19·1
	I	–271	–260	+11	14·7
	O	–284	–309	–25	120·5
Na	He	+ 70	+ 22	–48	5·8
	Ne	+ 55	+ 25	–30	7·0
	Ar	+ 84	+ 40	–44	13·5
	Xe	+ 86	+ 58	–28	18·0

For solutions of electronegative elements, ΔU_{Mad} is by far the largest term, ΔU_{el} and ΔU_{bs} being negligible in comparison. For monovalent solutes the electron–electron energy contribution to ΔU_{Mad} dominates, but for oxygen the energy contributions in which the ions play a role dominate.

In the few cases where experimental values are available for the change in resistivity, the calculated results are of the correct order of magnitude. The number of experimental values, however, is too small to make a real test of the theory possible.

In principle it is possible to avoid some of the assumptions made in the present model by using a minimization procedure of the energy of the solution by varying r_s and the packing fraction η. The volume coefficient δ, for which we had to use the assumption of equation (26), is then eliminated. The calculations, however, become much more involved and extensive, not only because the numerical minimization necessitates repeated calculations of the energy, but also because rather large terms have to be subtracted so that the accuracy of the calculations has to be improved greatly. As long as rather rough estimates must be made for the hard-sphere radii of the dissolved atoms it does not seem to be worthwhile to use such an improved method of calculation.

Acknowledgment

The authors would like to express their gratitude to Professor Dr W J Caspers and

E G van den Broek for valuable discussions and to M Heilbron for help with the computer programs.

References

Abriata J P and Tarchitzky H M 1975 *Scr. Metall.* 9 1003
Arnol'dov M N, Ivanovskiy M N, Pleshivtsev A D and Subbotin V I 1974 *Fluid Mech. Sov. Res.* **3** 18
Ashcroft N W 1972 *Interatomic Potentials and Simulation of Lattice Defects* (New York: Plenum Press) p91
Ashcroft N W and Langreth D C 1967a *Phys. Rev.* **155** 682
—— 1967b *Phys. Rev.* **156** 685
Ashcroft N W and Lekner J 1966 *Phys. Rev.* **145** 83
Blake L R 1960 *J. Inst. Elect. Eng.* **3278M** 383
Enderby J E, North D M and Egelstaff P A 1966 *Phil. Mag.* **14** 961
Faber T E and Ziman J M 1965 *Phil. Mag.* **11** 153
Feder H M and Schnijders H M 1970 *Argonne Nat. Lab. Internal Report* ANL-7775
Gellings P J, Huiskamp G B and Van den Broek E G 1972 *J. Chem. Soc. Dalton* 151
Gellings P J, Van der Scheer A and Caspers W J 1974 *J. Chem. Soc. Faraday Trans.* 70
Greenwood G A and Ratti V K 1972 *J. Phys. F: Metal Phys.* **2** 289
Pines D and Nozières P 1966 *The Theory of Quantum Liquids* (New York: W A Benjamin) p 326
Shimojima H and Miyaji N 1975 *J. Nucl. Sci. Technol.* **12** 658
Thompson R 1972 *J. Inorg. Nucl. Chem.* **34** 2513
Veleckis E, Cafasso F A and Feder H M 1976 *J. Chem. Eng. Data* **21** 75
Veleckis E, Dhar S K, Cafasso F A and Feder H M 1971 *J. Phys. Chem.* **75** 2832
Weast R C 1970 *Handbook of Chemistry and Physics* (Cleveland, Ohio: Chemical Rubber)
Ziman J M 1972 *The Theory of the Solid State* (London: Cambridge Univ. Press) p149

The manufacture and properties of ceramic probes to measure oxygen content in liquid sodium

C C H Wheatley, F Leach, B Hudson, R Thompson, K J Claxton and R C Asher
AERE, Harwell, Didcot, Oxon OX11 0RA, UK

Abstract. In sodium-cooled fast reactors (and elsewhere) it is important to be able to measure continuously the oxygen levels in the sodium circuits. For example, a sudden increase in the oxygen level in a secondary coolant circuit of a reactor might indicate a water leak from the heat exchanger. This paper outlines some aspects of the development of an electrochemical oxygen meter for use in sodium. A ceramic probe in the form of a closed-ended tube of thoria–7½% yttria is immersed in the sodium and the EMF generated between the sodium and an internal standard reference electrode can be used to monitor the oxygen level in the sodium. The electrolyte tubes are manufactured by a procedure which gives a dense ceramic of high chemical purity and hence freedom from spurious EMFs caused by impurities. The properties of the probes and the experience obtained with them in sodium are described.

1. Introduction

In sodium-cooled fast reactors (and elsewhere) it is important to be able to measure continuously the oxygen levels in the sodium circuits. For example, a sudden increase in the oxygen level in a secondary coolant circuit of a reactor might indicate a water leak from the heat exchanger. This paper outlines some aspects of the development of an electrochemical oxygen meter for use in sodium. The meter takes the form of a galvanic cell in which one electrode is the sodium under test, the other electrode is a reference electrode at a known, constant, oxygen activity and the electrolyte separating them is a solid ceramic. The use of solid electrolytes, which conduct oxygen ions, for measuring oxygen levels in gases and liquid metals has been extensively studied over the past 20 years (Etsell and Flengas 1970) but their long-term reliable use in liquid sodium still presents problems and care must be taken in their fabrication. Generally the electrolytes are based on the tetravalent thorium, zirconium or cerium oxides containing additives of the divalent alkaline earth oxides or trivalent scandium, yttrium or rare earth oxides. When a solid solution of these oxides is formed, the presence of the divalent or trivalent cations in the lattice causes anion vacancies to be formed in order to maintain electrical neutrality. The presence of these vacancies increases the conductivity which can now be completely ionic within a particular range of temperature and oxygen pressure. However, in choosing a suitable electrolyte it must satisfy the following conditions.

(i) In order to be a specific indicator of oxygen its conductivity must be predominantly by oxygen ions under the operating conditions. For the present application, oxygen concentrations in the range 1–50 ppm have to be monitored in the temperature range 300–450 °C.

(ii) It must be chemically compatible with the environment for periods measured in terms of thousands of hours.
(iii) It must be possible to fabricate.

To meet these requirements the choice of ceramic is, in practice, more or less limited to thorium oxide doped with yttrium oxide. The techniques developed for fabricating this material in a form suitable for use in an oxygen meter and the experience gained of its operation in sodium are outlined.

2. Manufacture of the electrolyte tube

The electrolyte takes the form of a tube closed at one end. It must have a high density and low permeability and must be free from defects which may cause premature failure. The yttrium must be in solid solution and any impurities which may be present must not seriously affect the electrochemical behaviour of the electrolyte or its compatibility with the sodium.

Both thorium and yttrium oxides have refractory properties and may be fabricated by pressing and sintering techniques. The success of the fabrication is therefore dependent on the physical properties of the starting materials. The steps taken to manufacture the electrolyte are as follows.

(i) Co-precipitation of the mixed thorium and yttrium oxalates.
(ii) Filtration and drying.
(iii) Calcination at approximately 1200 °C.
(iv) Wet milling of the calcined product.
(v) Dry rotary milling followed by sieving.
(vi) Mixing with binder followed by granulation.
(vii) Isostatic pressing to form a tube.
(viii) Removal of binder from the pressing.
(ix) Sintering.

The thorium and yttrium are co-precipitated as oxalates from a mixed solution of their nitrates that has a composition chosen to give a final product containing 7½ wt.% yttria. This co-precipitation technique gives an extremely fine, well mixed powder which is dried and calcined at 1200 °C to give a solid solution of yttria in thoria. This oxide is then wet milled to densify it and make it suitable for fabrication. The product is dried, lightly milled to break up the lumps and sieved to remove any remaining hard bits which may cause defects during sintering. The sieved powder is mixed hot with the binder (molten wax) and then cooled and granulated through a coarse sieve to prepare it for pressing. The powder is loaded, with the aid of a vibrator, into a PVC tube around a mandrel and isostatically pressed at 30 000 psi to form a tube closed at one end. The wax binder is removed by heating under vacuum to 400 °C and finally the tube is sintered at 1850 °C in an atmosphere of argon with 4% hydrogen. The dimensions of the PVC tube and the mandrel used in the isostatic pressing were selected so that, after shrinkage during sintering, the final tube is about 12 mm outside diameter and 200 mm long with a wall thickness of 2 mm.

The sintered tubes made by this method are white in colour, semi-translucent and have a density of 98% of the theoretical value. To complete the electrode assembly, the reference electrode is now fitted. An air reference electrode is used for a variety of reasons, in particular simplicity and reliability. It takes the form of a platinum gauze on the inside of the closed end of the tube and is held in position using platinum paint as an 'adhesive'; this is fired at 1000 °C in the argon/4% hydrogen. The completed oxygen meter unit is shown in figure 1.

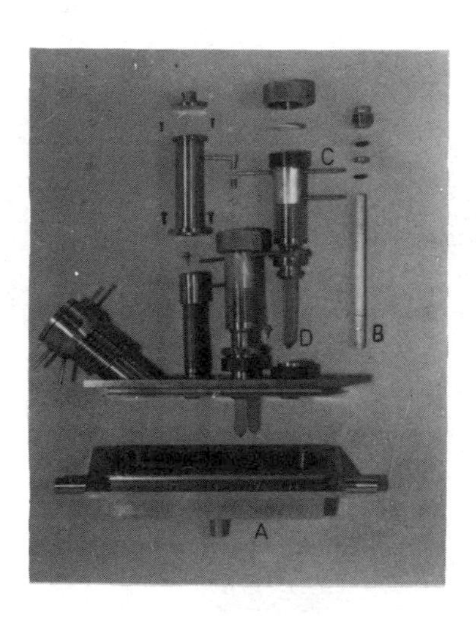

Figure 1. An exploded view of the oxygen meter unit as fitted to a dynamic sodium loop. The sodium flows through the box A. The electrolyte probe B dips into this box via the seal C enclosed in the gauze sleeve D.

3. Operating experience

At present a number of meters of this type are operating satisfactorily in sodium. This is illustrated in figure 2 which shows the recorded EMFs plotted against 'cold-trap temperature': this temperature controls the oxygen level in the sodium and can be conveniently used as a semi-quantitative indication of the oxygen level. The observations cover a period of 1000 hours and refer to two different sodium temperatures. Each point shows the EMF recorded after a change in cold-trap temperature every 24 hours.

During the early development of the thoria–yttria electrolyte there have been instances when electrolytes have had a short life (50–100 hours) and did not always respond to changes in oxygen level in the sodium. When such electrolytes were examined they showed little visible sign of attack on their outer surface but there was a change in the type of fracture from transgranular (before sodium immersion) to intergranular (after sodium immersion), showing that weakening of the electrolyte at its grain boundaries had occurred. This feature was first noticed when polished sections of the electrolyte showed an apparent marked increase in porosity extending inwards from the surface that had been exposed to the sodium. However, on closer examination it was evident that this apparent porosity was in fact a removal of the grains during polishing (see figure 3). The change in fracture characteristics is well illustrated by scanning electron microscope photographs (figures 4 and 5). Figure 6 shows the change in fracture characteristics across a tube which appears to have failed when the region of intergranular fracture reached a defect extending from the inner surface at the tip of the

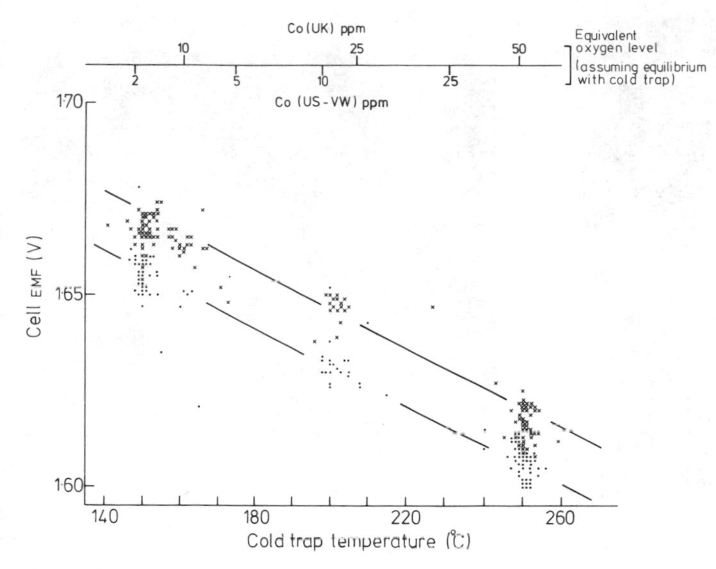

Figure 2. A graph of EMF plotted against cold-trap temperatures (oxygen concentration). Each point shows the EMF recorded after a change in cold-trap temperature over a period of approximately 1000 hours and illustrates the reliability of the electrolyte: ×, cell temperature 370 °C; •, cell temperature 300 °C.

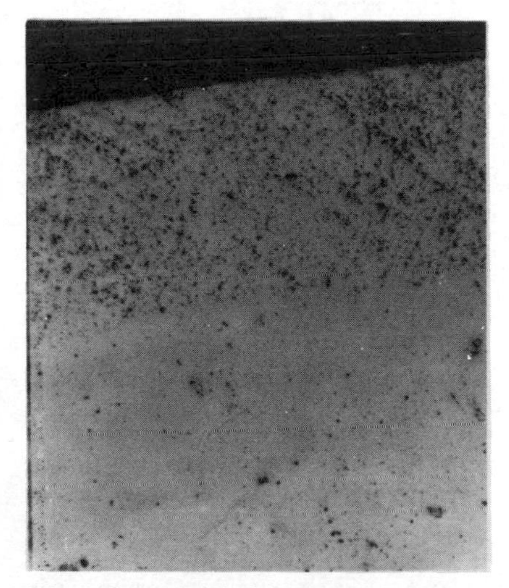

Figure 3. A polished section of electrolyte showing the apparent porosity at the surface exposed to the sodium (magnification: × 55).

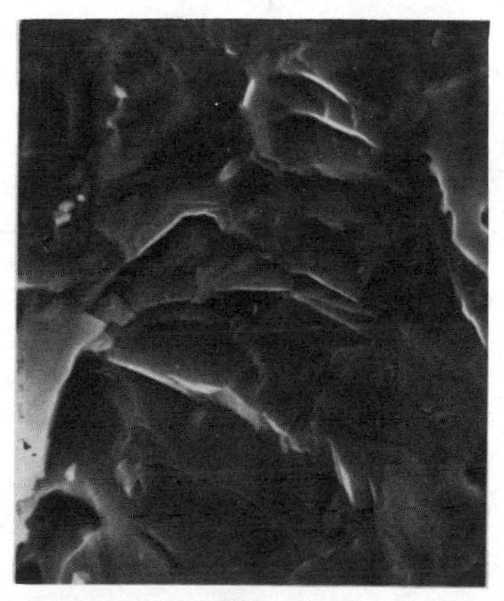

Figure 4. A fracture surface of a piece of unused electrolyte showing transgranular fracture.

tube. Freshly fractured surfaces of this tube were examined with an electron microprobe analyser and found to contain sodium on the areas of intergranular fracture. This type of failure appears to predominate in tubes which, in order to meet dimensional tolerances, have been ground to a final surface finish after firing. It is thought that the grinding process causes defects in the electrolyte surface which facilitate a rapid ingress

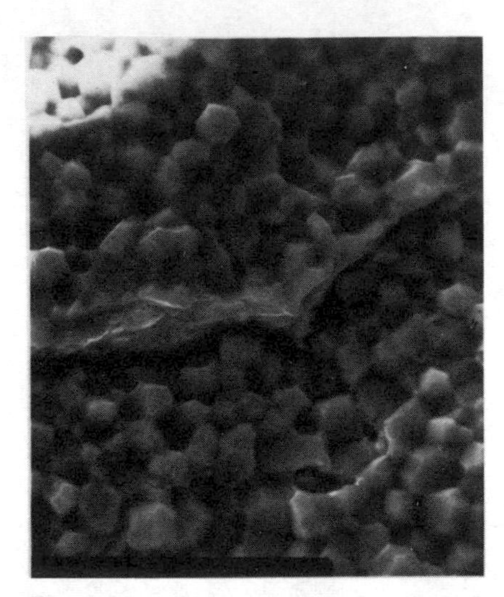

Figure 5. A fracture surface of a piece of used electrolyte showing an area of intergranular failure. (Mean grain size approximately 5 μm.)

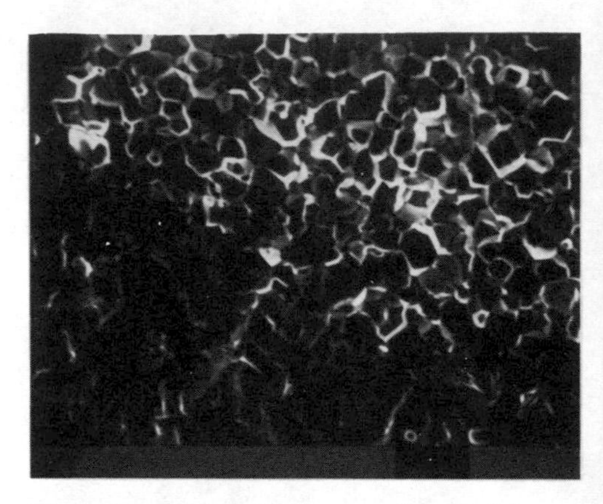

Figure 6. The changing fracture characteristics across an electrolyte tube which did not respond to changes in cold-trap temperature and failed after about 100 hours in sodium: intergranular on the sodium side and transgranular on the air side.

of sodium. This may account for the observed lack of response to changes in oxygen concentration with such tubes. It seems possible that the measured EMF is the one generated between the air electrode and the static sodium in the defects rather than the sodium at the surface of the electrolyte. The unchanging and very low oxygen activity of the trapped sodium is reflected by the high voltage observed (around 1·9 V). In recent tests, where pieces of electrolyte have been immersed in sodium containing varying concentrations of oxygen, it appears that high oxygen levels (100–500 ppm) in the sodium might also promote this effect through attack on the ceramic.

It appears therefore that the poor behaviour of the electrolyte tubes observed early in the programme was a result of incorrect fabrication technique coupled with possible high levels of oxygen in the sodium used for the tests. These problems have been overcome and good long-term behaviour of oxygen meters is now observed.

4. Conclusions

It has been established that electrolyte tubes (about 200 mm long by 12 mm diameter) are fabricated with reproducible electrical properties and the capability of operating for more than 2000 hours in sodium. However, this material is susceptible to both thermal and mechanical shock and it is therefore desirable, with a view to reliability and cost of replacement, to keep the electrolyte shape as small and simple as possible. Current development work has the objective of producing an oxygen meter in the form of a ceramic 'thimble' brazed into the end of a metal tube.

Reference

Etsell T H and Flengas S N 1970 *Chem. Rev.* **70** 3

A continuous method of determining the thermodynamic activity of carbon in sodium

R C Asher and T B A Kirstein
Applied Chemistry Division, AERE, Harwell, Didcot, Oxon., UK

Abstract. The Harwell carbon meter (HCM) provides a continuous indication of the thermodynamic carbon activity in liquid sodium. This information is important with respect to the carburization and decarburization of steel components of sodium-cooled fast reactors. A description of the HCM is given. It has been shown to respond rapidly to changes in carbon activity, to give an output which is in satisfactory agreement with spot calibrations using nickel 'tabs' and to operate reliably without failure for long periods.

1. Introduction

In sodium-cooled fast reactors, such as PFR and its commercial successors, structural metals, principally stainless steel, are brought into contact with large quantities of liquid sodium in the approximate temperature range from 400 to 650 °C. Carbon is slightly soluble in sodium (~6 ppm at 600 °C) and therefore the steel surfaces will tend to gain or lose carbon depending on whether the thermodynamic carbon activity (α_C) in the sodium is locally greater or less than the thermodynamic activity of the carbon in the steels. Therefore, transport of carbon from one steel component to another is possible and is governed by the nature and temperature of the steel as well as the carbon activity in the sodium in contact with it. Moreover there is always a possibility that additional carbon may enter the sodium accidentally, for example, as a result of failure of the cladding of shielding rods or shut-off rods, or from leakage of lubricating oil from mechanical components; even during normal operation α_C will vary round the circuit merely as a consequence of the temperature changes.

The carburizing or decarburizing effects resulting from these processes can bring about significant changes in the mechanical properties of steels. Hence, it is important to be able to predict what is happening to all the crucial reactor components during operation. This prediction will probably never be easy to carry out but nevertheless will be assisted if α_C can be measured at one or more points in the sodium circuit. In addition there is often a requirement to be able to measure α_C in out-of-reactor sodium circuits. The Harwell carbon meter provides a means of carrying out such determinations continuously.

2. Description of the Harwell carbon meter (HCM)

The principle used in the HCM has already been described (Asher *et al* 1973), but for convenience an outline is given here.

An iron 'membrane', generally in the form of a helically wound tube, is immersed in the sodium at 550–600 °C (figure 1(*a*)); the inner surface of the membrane has been previously oxidized and therefore bears a thin iron oxide film. The outer surface of the membrane will rapidly achieve the same α_C as the sodium, but at the inner surface any carbon will react with the iron oxide to form carbon monoxide, which is swept away by a flow of inert carrier gas. Thus, α_C at the inner surface of the membrane is maintained close to zero. In this way there is produced across the membrane a carbon activity gradient. The rate of diffusion of carbon through the membrane and the resulting production rate of carbon monoxide are directly proportional to α_C at the membrane outer surface and, by the same token, directly proportional to α_C in the sodium. The rate of production of carbon monoxide is measured continuously by passing the carrier gas to an analytical unit. This takes the form of a flame ionization detector (FID) which is very sensitive to compounds containing C–H bonds; for this reason the carbon monoxide is catalytically reduced to methane before passing to the FID. The analytical unit and the efficiency of the catalyst are periodically checked and calibrated by injecting carbon monoxide at a known rate into the system through a calibrated leak. The flow diagram is shown in figure 1(*b*).

A typical membrane has a surface area in the range of 50–1000 cm^2 (depending on the expected range of α_C and the temperature) and is normally made of tubing of 3 mm outside diameter and 0·2 mm thickness.

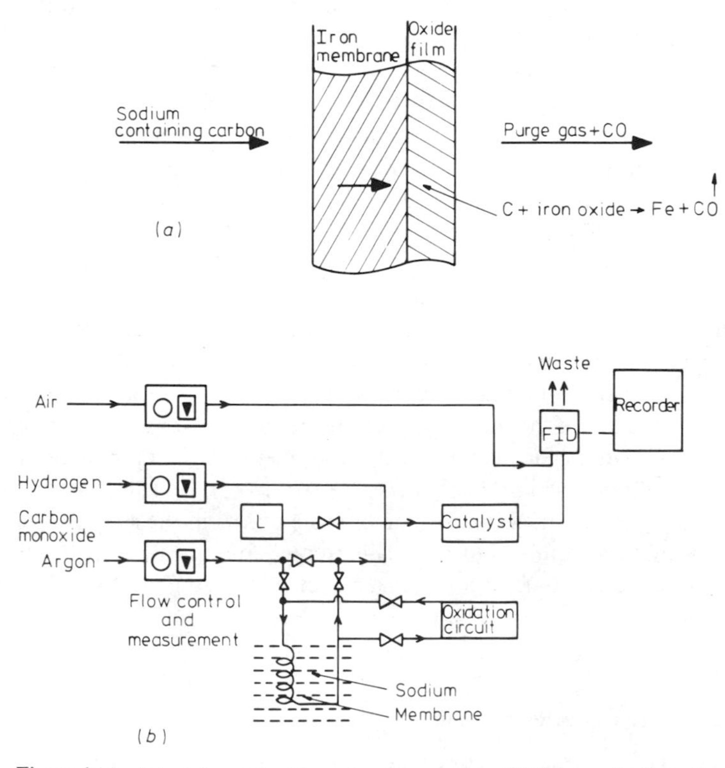

Figure 1(*a*). Principle of the Harwell carbon meter. (*b*) Schematic flow diagram of the meter. L = calibrated leak.

The carbon activity as indicated by the HCM, α_C (HCM), is calculated from the expression

$$\alpha_C(\text{HCM}) = \frac{W}{V}\,\frac{t}{A}\,\frac{1}{DS}\,\frac{L}{F(\text{L})}\,F(\text{M}) \qquad (1)$$

where W is the atomic weight of carbon, V is the gram molecular volume (22 400 cm^3), t the thickness of the membrane (cm), A the membrane area (cm^2), D the diffusion coefficient of carbon in iron (cm^2 s^{-1}), S the solubility of carbon in iron (gm cm^{-3}) and L the calibrated leak rate (cm^3 s^{-1}). F(L) is the FID reading from the calibrated leak, and F(M) is the FID reading from the HCM, both in arbitrary units.

The first term in the expression consists of universal constants, the second is dependent on the membrane geometry, the third consists of the physical constants of the membrane materials, while the fourth represents the sensitivity of the analytical instrumentation.

3. Experience with the Harwell carbon meter

The aim of the present note is to summarize operating experience with the Harwell carbon meter. A number of HCMs have been used in laboratory sodium facilities, and in larger pilot scale loops; similar units are awaiting installation in the primary circuit of the PFR. A typical membrane coil is shown in figure 2.

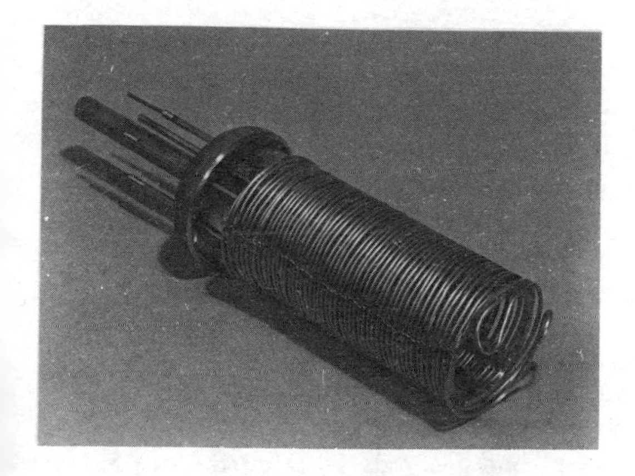

Figure 2. Typical membrane coil.

3.1. Response of the HCM to changes in α_C

Figure 3 shows the typical response of a membrane at 600 °C to changes in α_C in sodium. Firstly, the carbon activity of the sodium was increased considerably by introducing a number of carbon steel rods into the sodium and it is seen that the HCM output increased very rapidly. This part of the experiment is shown on an expanded timescale in figure 4 and it is seen that the HCM required only about 50 min to respond to the new level of α_C. The meter clearly responds rapidly to changes in α_C but, since

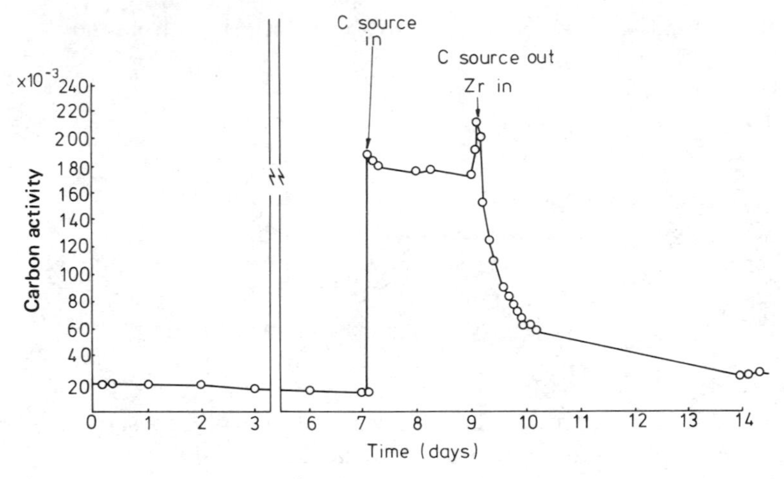

Figure 3. Response of Harwell carbon meter to changes in carbon activity.

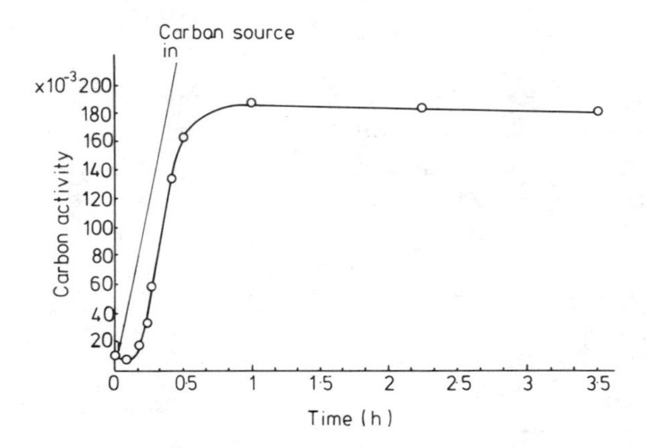

Figure 4. Response of Harwell carbon meter to a carbon source.

the net rate of addition of carbon to the sodium was not known, a response time could not be determined accurately. Even if the unrealistic assumption is made that the addition of carbon to the sodium resulted in an instantaneous change to the new carbon activity then the apparent response time is still no more than an hour or so; this would be perfectly adequate for monitoring purposes in a fast reactor sodium circuit.

Also shown on figure 3 is the effect of reducing the carbon activity. This was achieved by replacing the carbon source (the carbon steel rods) by thin strips of zirconium which acted as a carbon getter. As can be seen, there was the expected rapid decrease in the HCM output. (The slight increase in the HCM output which occurred even when the zirconium getter was introduced into the sodium is a result of the accidental introduction of small amounts of carbon-rich sodium compounds when the sealing device at the top of the sodium facility was disturbed.)

3.2. Calibration of the HCM

There is no fully developed, universally accepted method of 'absolute' determination of carbon activity in sodium. The method we have adopted to get as close as practicable to absolute calibration is the use of metal 'tabs'. These are thin foils of a suitable metal, in our case nickel, which are exposed to the sodium for long enough for α_C in the tab to come into virtual equilibrium with its surroundings. After exposure, the tab is removed from the sodium and cleaned, and its carbon content measured. Provided that the relationship between α_C and the concentration is known for the tab material then α_C in the sodium may be determined; it is usually assumed that Henry's law applies. Analysis of the tab material was by gamma activation analysis (Hislop *et al* 1975); this method gives high sensitivity and, by etching away about 40% of the tab thickness after activation, the problem of surface contamination of the tab with carbon-containing deposits can be overcome. Used in this way, nickel tabs permitted the 'calibration' of the HCM over the range of α_C from 10^{-3}–1. The results are generally expressed in terms of the calibration factor F:

$$F = \frac{\alpha_C(\mathrm{Ni})}{\alpha_C(\mathrm{HCM})}.$$

In figure 5, F is plotted against the apparent carbon activity as indicated by three different membranes in three separate experiments; the length of exposure varied from 115 to 245 days. This shows that over a considerable part of the range of interest F is

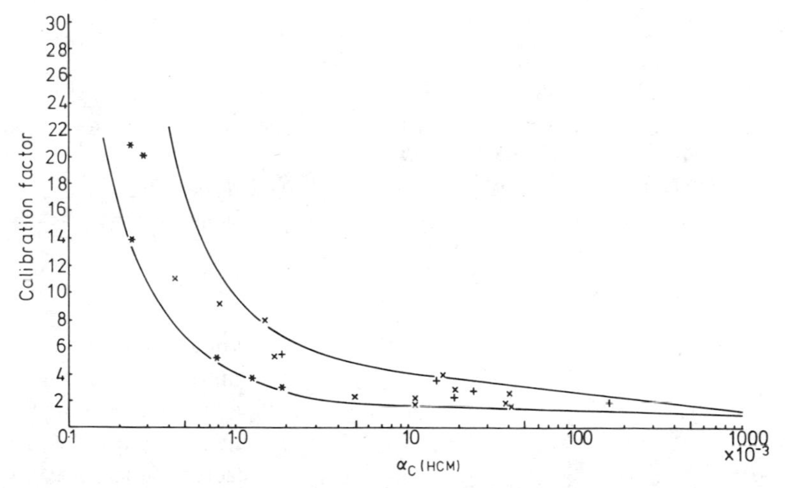

Figure 5. Calibration factor against carbon activity at 600 °C. + experiment A, x experiment B, ★ experiment C.

fairly constant and in the range 2–5 there are several possible reasons why F does not have the expected value of unity. Firstly a number of the parameters required to calculate α_C (HCM) and α_C (Ni) (in particular the diffusion coefficient of carbon in iron and the solubilities of carbon in iron and in nickel) are not known with a high degree of precision. Secondly it is not unlikely that the mechanism of operation of the HCM is not ideally simple as described above; it is possible, for example, that the carbon activity at the inner surface of the membrane (i.e. the interface between the iron and the iron oxide) is not zero, perhaps because of the difficulty with which the carbon monoxide produced can escape through the oxide film. This is a possible explanation of the large rise in F at low carbon activities indicated in figure 5.

Despite the fact that a detailed explanation of the shape of figure 5 is not yet possible, it is clear that this calibration curve enables the HCM to be used for determinations of the carburizing potential of sodium with quite good accuracy over a range of activities from 10^{-3} to about 1.

3.3. Long-term behaviour of the HCM

Several carbon meters have been tested for long periods in a pilot scale loop and satisfactory behaviour has been found. No failure of the membrane itself has been observed even after periods as long as 14 months.

4. Conclusions

It has been shown that the Harwell carbon meter provides a reliable means for determining the thermodynamic carbon activity of liquid sodium. The calibration curve provides figures for the carbon activity which are consistently in good agreement with those determined using nickel tabs. No failure of the sensing head has been observed over periods of 14 months.

References

Asher R C, Bradshaw L, Kirstein T B A, Nixon T H and Tolchard A C 1973 *Proc. BNES Conf. on Liquid Alkali Metals, Nottingham* pp 133–5

Hislop J S, Sanders T, Webber T J and Williams D R 1975 *AERE Report* 8182

Chemistry and physics at liquid alkali metal /solid metal interfaces

M G Barker

Department of Chemistry, University of Nottingham, University Park, Nottingham NG7 2RD

Abstract. This paper describes the chemistry of processes which take place at the interface between liquid alkali metals and solid metal surfaces. A brief review of wetting data for liquid sodium is given and the significance of critical wetting temperatures discussed on the basis of an oxide-film reduction mechanism. The reactions of metal oxides with liquid metals are outlined and a correlation with wetting data established. The transfer of dissolved species from the liquid metal across the interface to form solid phases on the solid metal surface is well recognized. The principal features of such processes are described and a simple thermodynamic explanation is outlined. The reverse process, the removal of solid material into solution, is also considered.

1. Introduction

Over the past decade the chemical and physical properties of the liquid alkali metals have been studied in great detail as a result of their application as heat-transfer media in fast breeder, and thermonuclear power reactors. Of main concern has been the interaction of solutions of nonmetals in the alkali metals with the materials used in the construction of the reactors. This paper will outline, in general terms, those processes which are of fundamental interest to chemists and will present the current state of the art in this area. As is the case with many areas of industrial science, only a general theoretical background exists, and specific problems cannot often be explained.

Liquid alkali metals have high thermal conductivities and low viscosities which result in a high heat-transfer coefficient, smaller energy losses on circulation and hence greater efficiency. This is illustrated by the fact that alkali metals have low Prandtl numbers (the ratio of kinematic viscosity to thermal conductivity) compared with other fluids. They also have a large liquid range and low vapour pressures at reactor operating temperatures, and being elements they do not undergo decomposition or polymerization under prolonged thermal shock and irradiation.

For use in any form of heat-transfer system, the liquid metal must be contained in metal vessels, and one important feature which determines the efficiency of heat transfer is the extent to which the solid metal is wetted by the liquid metal. Apart from a limited number of cases, an increase in the heat-transfer coefficient is observed when the solid metal is wetted by the alkali metal. Since the wetting process represents the starting point for any discussion on the transfer of species across the interface between the solid and liquid phases it is convenient to outline briefly the studies which have been carried out in this area.

2. The wetting of materials by the liquid alkali metals

The criterion of wetting is generally determined by the value of the contact angle θ between the liquid and solid surface (Bondi 1953) and the term 'complete wetting' is used to indicate that the contact angle is zero. For many systems the value of the contact angle falls between 0 and 90° and the term 'partial wetting' is applied in these cases, and 'non-wetting' is used where the contact angle is 180°. One feature of wetting, observed in liquid metal–solid metal systems, is the variation of contact angle with temperature. This variation may be a gradual change in θ from a positive value (usually 80–90°) to a value of 0° with increase in temperature. This type of variation leads to the definition of a 'wetting temperature' for the system. The other mode of variation is that in which the value of θ is drastically altered by only a small change in temperature, and a sharp transition from a non-wetting condition to an equilibrium wetting condition is observed. This transition temperature is termed the 'critical wetting temperature'.

Table 1 gives the results of several separate studies on the wetting of materials by liquid sodium. In general terms there is good agreement on the type of behaviour observed, the metals Ni, Co, Fe, Mo and W showing critical wetting-type behaviour and the refractory elements showing a different type of wetting (Addison *et al* 1968). The actual critical wetting temperatures observed by the different authors are, however, not in such good agreement. Longson and Prescott (1973) have shown that the wetting of nickel and iron is influenced by the level of dissolved oxygen in the sodium. An increase in the oxygen level was shown to decrease the wetting temperature and to increase the rate at which wetting took place. The disparity in critical wetting temperatures may

Table 1. Wetting temperatures of metals by liquid sodium.

Metal	Temperature at which contact angle changes rapidly (°C)	Method	Reference
Ni	195	vertical plate	(a)
Co	140	vertical plate	(a)
Fe	190	vertical plate	(a)
Mo	160	vertical plate	(a)
W	160	vertical plate	(a)
V	170	vertical plate	(a)
Nb	150	vertical plate	(a)
Ta	160	vertical plate	(a)
Ti	200	vertical plate	(a)
Zr	180	vertical plate	(a)
Fe	335	sessile drop	(b)
Cr	290	sessile drop	(b)
Mo	410	sessile drop	(b)
Ni	370	sessile drop	(b)
Nimonic 80A	280	sessile drop	(b)
M316L	280	sessile drop	(b)

(a) Addison *et al* (1968). (b) Hodkin *et al* (1973).

therefore be due to the different oxygen content of the sodium used in the various studies. General agreement is, however, expressed in the mechanism of wetting. Addison *et al* (1968) have suggested that three wetting processes may be considered, namely:

(i) Those materials that are wetted by chemical reduction of the oxide film present on the solid metal by the liquid sodium. The driving force for the reduction is the difference in the free energy between the metal oxide and the sodium (e.g. Fe, Co, Ni).
(ii) Materials which wet by chemical interaction involving the formation of surface compounds. In this case the formation of ternary oxides is possible and the free energies of the metal oxides are close to that of sodium oxide (e.g. V, Nb, Ta, Cr).
(iii) Materials which form oxides due to removal of oxygen from sodium. In this case the wetting process is more concerned with the wetting of the surface oxide rather than that of the metal itself (e.g. Zr).

Virtually all of the measurements of wetting behaviour tend to support the above mechanism. Of particular significance is the effect of alloying on the wetting behaviour. Hodkin *et al* (1973) have shown that Cr is more readily wetted than Ni or Fe and that alloys show increased wettability with increasing chromium content.

It is therefore apparent that the wetting process is essentially a chemical interaction between the components of the metal surface and the liquid metal with the added possibility that dissolved non-metals in the liquid metal may also play a part in the process. The mechanism proposed by Addison *et al* has been further examined by studies of the reactions of metal oxides with the liquid alkali metals and the influence of dissolved non-metals in the alkali metals on the surface of solid metals has also received much attention. It is not possible here to cover all the work published in these areas but some of the basic trends of behaviour will be explained by the use of selected examples.

3. The reactions of metal oxides with liquid alkali metals

The reactions of metal oxides with liquid alkali metals depend on the relevant free energies of formation of the oxide with that of the alkali metal monoxide. Since the free energy of formation of lithium oxide is much greater than that of any other alkali metal it follows that liquid lithium will reduce the majority of transition metal oxides directly to the metal. Liquid sodium is less reducing than lithium but will reduce the oxides of Fe, Co, Ni and Cu to the metal, with the formation of sodium oxide (Addison *et al* 1972b).

$$Co_3O_4 + 8Na \rightarrow 3Co + 4Na_2O.$$

This result clearly agrees with the wetting behaviour of these metals by sodium and would suggest a similar critical wetting-type behaviour for a large number of metals by liquid lithium. The only oxides stable towards liquid sodium are those of the scandium group together with ZrO_2 and HfO_2. The remaining transition metal oxides all exhibit ternary-oxide formation with liquid sodium, and this behaviour is closely followed by liquid potassium (table 2).

Table 2. Reactions of some alkali metals with some transition metal oxides.

Oxide	Reaction product in lithium	Reaction product in sodium	Reaction product in potassium
Fe_2O_3	$Fe + Li_2O$	$Fe + Na_2O$ (a)	$Fe + K_2O$
NiO	$Ni + Li_2O$	$Ni + Na_2O$ (a)	$Ni + K_2NiO_2$
VO_2	$V + Li_2O$ (c)	$NaVO_2$ (b)	KVO_2 (f)
V_2O_3	$V + Li_2O$	$NaVO_2 + VO$ (b)	$KVO_2 + CO$ (f)
Nb_2O_5	$Nb + Li_2O$ (d)	$Nb + Na_3NbO_4$ (e)	$Nb + K_3NbO_4$ (g)
NbO	$Nb + Li_2O$	$Nb + Na_3NbO_4$ (e)	no reaction (g)
Ta_2O_5	$Ta + Li_2O$	$Ta + Na_3TaO_4$ (e)	$Ta + K_3TaO_4$ (g)
CrO_2	$Cr + Li_2O$	$NaCrO_2$	–
Cr_2O_3	$Cr + Li_2O$	$NaCrO_2 + Cr$	–
ZrO_2	$Zr + Li_2O$ (h)	no reaction	no reaction

(a) Addison *et al* (1972b). (b) Barker and Hooper (1973). (c) Addison *et al* (1972a). (d) Barker and Bentham (1975). (e) Barker *et al* (1974). (f) Barker *et al* (1973). (g) Addison *et al* (1970). (h) Barker *et al* (1975).

The distribution of oxygen between the solid and liquid metals is of more importance when one considers the behaviour of constructional materials towards solutions of oxygen in liquid sodium. In the following section the behaviour of other dissolved metals, particularly nitrogen in liquid lithium, will be described.

4. Reactions of metals with solutions of nonmetals in liquid alkali metals

Since the partitioning of oxygen between the solid and liquid metal is thermodynamically controlled it is to be expected that a close relationship will exist between the two situations Na + M–O and Na[O] + M. Thus in the case of liquid lithium, oxygen will generally remain in solution rather than be transferred to the solid metal. In liquid sodium, oxygen may be removed from solution and will be transferred to the metal surface, forming either a ternary or a binary oxide at the solid–liquid interface. Examples of this behaviour are given in figure 1.

The high value of the free energy of formation of lithium oxide ensures that solutions of oxygen in liquid lithium will not react with transition metal surfaces such as niobium. Indeed any oxygen present on the metal surface, or in solid solution, will be removed and will go into solution in the liquid metal. Solutions of oxygen in sodium or potassium are more reactive towards metal surfaces. In the example shown in figure 1 oxygen is removed from solution and is transferred across the solid–liquid interface and diffuses into the solid metal forming initially a solid solution. As more oxygen diffuses across the interface a surface film is formed of a ternary oxide in the case of niobium or of a binary oxide in the case of zirconium. Eventually equilibrium is reached when the activities of oxygen in the alkali metal and in the surface film are balanced. This type of behaviour has been observed for a larger number of metals as shown in table 3. Some measure of the extent at which the partition of oxygen takes place may be gained from a consideration of the partition coefficient, which may be calculated

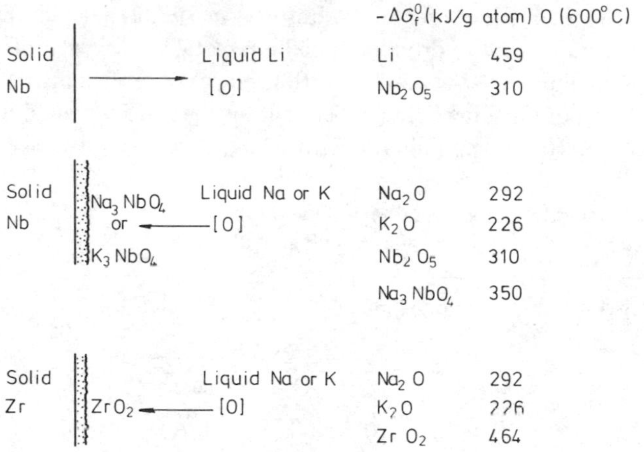

$-\Delta G_f^0$ (kJ/g atom) O (600°C)

Solid Nb | → Liquid Li [O]

Li	459
Nb_2O_5	310

Solid Nb | Na_3NbO_4 or K_3NbO_4 ← Liquid Na or K [O]

Na_2O	292
K_2O	226
Nb_2O_5	310
Na_3NbO_4	350

Solid Zr | ZrO_2 ← Liquid Na or K [O]

Na_2O	292
K_2O	226
ZrO_2	464

Figure 1. Partition of oxygen between a liquid metal and a solid metal.

Table 3. Reactions of transition metals in liquid lithium or sodium containing dissolved oxygen.

Na + M oxide	Ternary (Na–M–O) oxides				M + Na[O]				
Sc	Ti	V	Cr	Mn	Fe	Co	Ni	Cu	Zn
Y	Zr	Nb	Mo						
La	Hf	Ta	W						
Ternary oxides	M + Li[O] in liquid lithium →								

using the expression

$$K_T = \frac{C_B}{C_A} = \exp\left(\frac{\Delta G^0_{f\,(A\ oxide)} - \Delta G^0_{f\,(B\,oxide)}}{RT}\right)\frac{(C_B)_s}{(C_A)_s}$$

where C_A is the concentration of oxygen in alkali metal A and C_B the concentration of oxygen in solid metal B.

Reasonably good agreement has been obtained between calculated and experimental data for systems where binary-oxide formation takes place (Harms and Litman 1967), but the lack of accurate thermodynamic data for ternary oxides prevents universal application of the equation. For liquid lithium–oxygen–transition metal systems very low values of K_T are found which are of course observed in practice.

Liquid lithium is of particular interest in that a wide range of non-metals show reasonable solubilities in the liquid metal. It is therefore important that non-metals other than oxygen should be considered, particularly in view of the fact that the free energies of formation of Li_3N, Li_2C_2 and LiH are much lower than that of Li_2O. Thus although oxygen will remain partitioned in the liquid metal, nitrogen and carbon will possibly be transferred from solution to the solid metal surface. A study of the reactions of vanadium monoxide, with liquid lithium containing various nitrogen levels, has shown that this is experimentally observed (Addison *et al* 1972a). Vanadium monoxide reacts with liquid lithium to produce vanadium metal and oxygen dissolved in the liquid lithium. Vanadium metal in contact with lithium containing nitrogen is rapidly nitrided, the extent of nitriding being dependent on the ratio of vanadium to the nitrogen present in solution (table 4). Similar reactions have now been observed for

Table 4. The vanadium–nitrogen reaction in liquid lithium. Vanadium monoxide was added to liquid lithium at 250–600 °C.

Nitrogen content of lithium (wt %)	Solid phase	Solution
O	V	Li_2O
0·04 → 0·06	$V + V_3N$	Li_2O
0·1 → 0·8	V_3N	Li_2O
0·9 → 1·7	$V_3N + VN$	Li_2O
1·8 → 2·0	VN	Li_2O
2·6	Li_7VN_4	Li_2O

many transition metals, where a clear distinction can be made between the behaviour of oxygen and nitrogen in liquid lithium (Barker 1973). In these cases, however, it is clear that oxygen will always remain in solution in the lithium but it is possible to define a situation where both oxygen and nitrogen are capable of being partitioned across the liquid–solid interface, and are therefore in competition with each other. Such a situation exists in the reaction of dissolved oxygen and nitrogen with thorium metal (Barker *et al* 1975). Thorium metal reacts with liquid lithium containing oxygen to give thorium dioxide as a surface product

$$Th + Li[O] \rightarrow ThO_2 + Li$$

since the free energy of formation of ThO_2 is greater than that of Li_2O. Similarly, solutions of nitrogen in lithium react with thorium metal to give a surface layer of thorium nitride

$$Th + Li[N] \rightarrow ThN + Li.$$

In the competitive case, when thorium metal is in contact with both nitrogen and oxygen in solution in the liquid lithium, only ThN is produced if nitrogen is in excess.

$$Th + Li[N]_{\text{in excess}} + Li[O] \rightarrow ThN.$$

ThO_2 can only be produced if all the nitrogen in solution is removed by formation of ThN and sufficient thorium metal remains to react with the dissolved oxygen

$$Th + Li[N] + Li[O]_{\text{in excess}} \rightarrow ThN + ThO_2.$$

This order of reactivity is not unique to thorium but has also been observed in the reactions of CeO_2 with liquid lithium (Barker and Alexander 1975). In the absence of nitrogen the oxide reacts with lithium to give the ternary oxide $LiCeO_2$

$$Li + CeO_2 \rightarrow LiCeO_2.$$

Traces of nitrogen dissolved in the lithium cause decomposition of the ternary oxide with the formation of CeN and Li_2O

$$Li[N] + LiCeO_2 \rightarrow Li_2O + CeN + LiCeO_2.$$

With sufficient nitrogen in solution all the ternary oxide may be decomposed

$$Li[N] + LiCeO_2 \rightarrow Li_2O + CeN,$$

and under an excess of nitrogen the formation of a ternary nitride may take place

$$Li[N]_{\text{in excess}} + CeO_2 \rightarrow Li_2O + Li_2CeN_2.$$

An alternative view of a competition reaction exists when an alloy is exposed to solutions of oxygen in liquid sodium. Each component of the alloy may possibly interact with the oxygen in solution. An important example of this type of behaviour is the interaction of oxygen in sodium with stainless steel. The principal elements comprising stainless steel have reactivities towards oxygen in sodium which are quite different. Chromium reacts with oxygen in sodium to form a ternary oxide

$$Cr + Na[O] \rightarrow NaCrO_2.$$

This reaction takes place at virtually all levels of oxygen in sodium. Iron will only form a ternary oxide at very high oxygen levels, that is, saturation at 600 °C.

Nickel shows no interaction with oxygen in sodium but has a fairly high solubility in the liquid metal. The net result of all these reactions is that a surface layer of $NaCrO_2$ forms on stainless steel on exposure to liquid sodium. This removes chromium from the surface leaving an underlying ferritic layer. This effect is enhanced by the additional loss of nickel by direct solution in the liquid metal. With time the system will reach equilibrium and the rate of corrosion will decrease.

A final example of a selective reaction is that of molybdenum with carbon dissolved in alkali metals. Experimental tests have shown (Barker and Morris 1976) that molybdenum only removes dissolved carbon from both sodium and lithium; no interaction has been observed between molybdenum with oxygen in lithium or sodium or with nitrogen in lithium. Since sodium only contains a trace of dissolved carbon this is rapidly removed by molybdenum. In this situation carbon may be removed from other materials in contact with the sodium and transported in solution through the sodium to be deposited on the molybdenum surface. This mass-transfer phenomenon is common in liquid alkali metals and takes place whenever a difference in the chemical activity of a species exists in contact with the liquid metal.

The movement of dissolved species under the influence of an activity difference may be employed for the quantitative determination of the soluble species. Meters for the measurement of carbon, oxygen and hydrogen have been designed and since some of the papers in this conference will describe these, they need not be dealt with in this paper.

5. Conclusion

In conclusion the wetting behaviour of metals can be seen to be an extension of the known chemistry of metal oxides in alkali metals. The behaviour of dissolved non-metals in alkali metals may be adequately described by a simple thermodynamic approach, and many of the practical problems associated with the application of alkali metals, for example, corrosion and mass transfer, are due to the partitioning of non-metals in solution between the alkali metal and the solid metal.

References

Addison C C, Barker M G and Bentham J 1972a *J. Chem. Soc. Dalton Trans.* 1035
Addison C C, Barker M G and Hooper A J 1972b *J. Chem. Soc. Dalton Trans.* 1017
Addison C C, Barker M G and Lintonbon R M 1970 *J. Chem. Soc.* **A** 1465
Addison C C, Iberson E and Pulham R J 1968 *Soc. Chem. Ind. Monograph* 28 p 246
Barker M G 1973 *Proc. Int. Conf. on Liquid Alkali Metals.* (Br. Nucl. Energy Soc.) p 219
Barker M G and Alexander I C 1975 *J. Chem. Soc. Dalton Trans.* 1464
Barker M G, Alexander I C and Bentham J 1975 *J. Less-Common Metals* **42** 241
Barker M G and Bentham J 1975 *J. Less-Common Metals* **40** 1
Barker M G and Hooper A J 1963 *J. Chem. Soc. Dalton Trans.* 1520
Barker M G, Hooper A J and Lintonbon R M 1973 *J. Chem. Soc. Dalton Trans.* 2618
Barker M G, Hooper A J and Wood D J 1974 *J. Chem. Soc. Dalton Trans.* 55
Barker M G and Morris C W 1976 *J. Less-Common Metals* **44** 169
Bondi A A 1953 *Chem. Rev.* **52** 417
Harms W O and Litman A P 1967 *ASME Publication* 67 WA/AV-1
Hodkin E N, Mortimer D A and Nicholas M 1973 *Proc. Int. Conf. on Liquid Alkali Metals.* (Br. Nucl. Energy Soc.) p 167
Longson B and Prescott J 1973 *Proc. Int. Conf. on Liquid Alkali Metals.* (Br. Nucl. Energy Soc.) p 171

Individual and collective atomic motion in non-molecular liquids

J R D Copley and S W Lovesey
Institut Laue–Langevin, 156X Centre de Tri, 38042 Grenoble Cedex, France

Abstract. Recent research, dating roughly from mid-1974, into the individual and collective motion of atoms in one- and two-component non-molecular liquids, is discussed. Both inelastic neutron-scattering and computer-simulation experiments are described, as well as theoretical studies. The field is slowly expanding to include experimental and theoretical work on temperature and pressure effects, and on liquids slightly more complicated than monatomic liquids. Looking to the future it appears that really significant progress in the neutron-scattering domain will only be possible if rather specialized instruments, built on a new generation of neutron source, become available. With regard to theory, the more sophisticated theories being developed need to be compared carefully with all data available, in order to see precisely how they represent an improvement over simple theories, which already give good representations of the experimental data.

1. Prologue

Our understanding of the dynamic properties of liquids has advanced significantly over the past decade, due in large part to the advent of precise neutron-scattering and computer-simulation experiments. Many aspects of theory and experiment for monatomic liquids have been reviewed by Copley and Lovesey (1975) and by Schofield (1975), and texts by Faber (1972) and Croxton (1974) provide concise presentations of the fundamentals. In the present paper we aim to review recent experimental and theoretical activities, dating essentially from mid-1974. We shall also discuss recent work on two-component systems using neutron-scattering and computer-simulation techniques. Moreover, we shall attempt to identify areas of research which seem to us, at the time of writing, to be particularly valuable. Due to limitations on space, we shall sometimes refer to previous reviews and texts for further details. Wherever possible, we follow the notation and conventions used by Marshall and Lovesey (1971) and by Copley and Lovesey (1975). Most of the papers cited in the present review date from mid-1974, and we refer the reader to Copley and Lovesey (1975), Schofield (1975) and Larose and Vanderwal (1973) for bibliographies of earlier papers.

Section 2 contains a reasonably complete summary of the important formulae and results which relate to the scattering of neutrons from liquids. The following three sections then treat, in turn, the three main areas of active research into the dynamical behaviour of atoms in simple liquids: neutron-scattering experiments, computer-simulation studies, and theoretical advances. In the final section we attempt to bring these approaches together and to indicate the most probable directions for future research.

2. Basics

In this section we review briefly the formalism which underlies the interpretation of neutron scattering from an n-component liquid, in which the translational motion of the atoms is described by classical equations of motion.

Neutron-scattering experiments are interpreted in terms of partial differential cross sections which can, in common with the cross sections for many others types of scattering experiment, be expressed in terms of correlation functions of operators associated with the target system. If we denote the position of the jth atom by $\mathbf{R}_j$, then the correlation function which enters the neutron cross section is

$$Y_{ij}(Q,t) = \langle \exp\{-\mathrm{i}\mathbf{Q}.\mathbf{R}_i(0)\} \exp\{\mathrm{i}\mathbf{Q}.\mathbf{R}_j(t)\}\rangle. \qquad (1)$$

The angular bracket denotes a thermal average at temperature T, $\mathbf{Q}$ is the change in the neutron wavevector, and t has the dimension of time. Denoting the change in the neutron energy† by ω, the cross section of interest is simply

$$\frac{\mathrm{d}^2\sigma}{\mathrm{d}\Omega\,\mathrm{d}E'} = \frac{k'}{k}\frac{1}{2\pi}\int_{-\infty}^{\infty} \mathrm{d}t \exp(-\mathrm{i}\omega t) \sum_{ij} b_i^* b_j Y_{ij}(Q,t) \qquad (2)$$

where k and k' are the wavevectors of the incident and scattered neutrons, respectively.

The quantities b_i which enter equation (2) are the scattering lengths, which contain the details of the neutron–nucleus interaction for thermal neutron scattering. These scattering lengths vary from element to element, and from isotope to isotope. They depend also on the relative orientations of the neutron and nuclear spins. The cross section is averaged over all these possible states, and we denote the averaging procedure by a horizontal bar. Assuming that there is no correlation between values of b_i^* and b_j associated with different nuclei, then

$$\overline{b_i^* b_j} = \bar{b}_i^* \bar{b}_j + \delta_{ij}\{\overline{|b_i|^2} - |\bar{b}_i|^2\}. \qquad (3)$$

Explicit formulae for the averaged scattering lengths, in terms of isotope concentrations and spins, are given by Marshall and Lovesey (1971), for example.

Using the result (3) in equation (2), the cross section is written as the sum of a coherent and an incoherent cross section,

$$\frac{\mathrm{d}^2\sigma}{\mathrm{d}\Omega\,\mathrm{d}E'} = \left(\frac{\mathrm{d}^2\sigma}{\mathrm{d}\Omega\,\mathrm{d}E'}\right)_{\mathrm{coh}} + \left(\frac{\mathrm{d}^2\sigma}{\mathrm{d}\Omega\,\mathrm{d}E'}\right)_{\mathrm{incoh}} \qquad (4)$$

where

$$\left(\frac{\mathrm{d}^2\sigma}{\mathrm{d}\Omega\,\mathrm{d}E'}\right)_{\mathrm{coh}} = \frac{1}{N}\frac{k'}{k}\sum_{\mathrm{A}}^{n}\sum_{\mathrm{B}}^{n} \bar{b}_{\mathrm{A}}^* \bar{b}_{\mathrm{B}} (N_{\mathrm{A}}N_{\mathrm{B}})^{1/2} S_{\mathrm{AB}}(Q,\omega) \qquad (5)$$

with the scattering function,

$$S_{\mathrm{AB}}(Q,\omega) = \frac{1}{2\pi}\int_{-\infty}^{\infty} \mathrm{d}t \exp(-\mathrm{i}\omega t)(N_{\mathrm{A}}N_{\mathrm{B}})^{-1/2} \sum_{i(\mathrm{A})}^{N_{\mathrm{A}}}\sum_{j(\mathrm{B})}^{N_{\mathrm{B}}} Y_{ij}(Q,t) \qquad (6)$$

† We set $\hbar = 1$ throughout the paper.

and

$$\left(\frac{d^2\sigma}{d\Omega\, dE'}\right)_{\text{incoh}} = \frac{1}{N}\frac{k'}{k}\sum_{A}^{n} \overline{|b_A - \bar{b}_A|^2}\, N_A S_A^{(s)}(Q,\omega) \tag{7}$$

with the scattering function

$$S_A^{(s)}(Q,\omega) = \frac{1}{2\pi}\int_{-\infty}^{\infty} dt \exp(-i\omega t) N_A^{-1} \sum_{j(A)}^{N_A} Y_{jj}(Q,t). \tag{8}$$

In these formulae, N_A is the number of atoms of type A, the notation $\Sigma_{i(A)}^{N_A}$ denotes a sum over all atoms of type A, and $N = \Sigma_A^n N_A$ is the total number of atoms. The scattering functions $S_A^{(s)}(Q,\omega)$ and $S_{AB}(Q,\omega)$ are related, through Fourier transformation, to the self and total space–time correlation functions $G_A^{(s)}(r,t)$ and $G_{AB}(r,t)$, respectively (see for example Price and Copley 1975). The partial structure factors $S_{AB}(Q)$ are related to spatial Fourier transforms of appropriate radial distribution functions $g_{AB}(r)$. Explicitly,

$$S_{AB}(Q) = \delta_{AB} + (\rho_A\rho_B)^{1/2} \int d\mathbf{r} \exp(i\mathbf{Q}\cdot\mathbf{r})\{g_{AB}(r) - 1\} \tag{9}$$

where ρ_A is the mean *number* density of atoms of type A. This expression differs from that due to Faber and Ziman (1965): the definitions of $S_{AB}(Q)$ differ, but $g_{AB}(r)$ is the same in both expressions.

We write the moments of the scattering functions as

$$\langle\omega_{AB}^m\rangle = \int_{-\infty}^{\infty} d\omega\, \omega^m S_{AB}(Q,\omega) \tag{10}$$

and

$$\langle\omega_A^m\rangle^{(s)} = \int_{-\infty}^{\infty} d\omega\, \omega^m S_A^{(s)}(Q,\omega) \tag{11}$$

where $m = 0, 2, 4, \ldots$. The odd integer moments are zero for classical systems†, and

$$\langle\omega_{AB}^0\rangle = S_{AB}(Q) \tag{12}$$

$$\langle\omega_A^0\rangle^{(s)} = 1 \tag{13}$$

$$\langle\omega_{AB}^2\rangle = \delta_{AB} Q^2 kT/M_A \tag{14}$$

and

$$\langle\omega_A^2\rangle^{(s)} = Q^2 kT/M_A \tag{15}$$

where M_A is the mass of a type A atom. The $m = 4$ moments, given for example by Price and Copley (1975), involve the atomic potentials between the various types of atom. Rahman (1975) has suggested that these relations could be used to obtain the atomic potentials (at least for a monatomic system) but the accurate measurement of the fourth moments is generally fraught with difficulty. However, Woods *et al* (1976a) have reported an accurate measurement of the fourth moment for normal ^{4}He.

† For a more detailed discussion, see Copley and Lovesey (1975).

For the case of a liquid composed of one element ($n = 1$ and $N_A = N$) the formulae given above reduce to the, now standard, results given first by Van Hove (1954). We denote the coherent and incoherent scattering functions for this case by $S(Q, \omega)$ and $S_s(Q, \omega)$ respectively.

We shall now illustrate some features of the foregoing discussion for the case of a two-component system with components denoted 1 and 2. If the components are substitutional and there is no correlation between the position operators and the distribution of the two species, so that the species are randomly mixed, the coherent cross section (equation (5)) is then proportional to

$$N|c\bar{b}_1 + (1-c)\bar{b}_2|^2 S(Q, \omega) + c(1-c)|\bar{b}_1 - \bar{b}_2|^2 S_s(Q, \omega)$$

where c is the fractional concentration of type-1 atoms, whereas the incoherent cross section (equation (6)) is proportional to

$$\{c\sigma_i(1) + (1-c)\,\sigma_i(2)\}\, S_s(Q, \omega)$$

with $\sigma_i = 4\pi\overline{|b - \bar{b}|^2}$. Thus there is, in general, a contribution to the scattering cross section, proportional to $S_s(Q, \omega)$, even when the species have no incoherent cross section, due to isotope or nuclear spin disorder. This contribution is usually called the diffuse cross section.

It is sometimes useful to define different sets of scattering functions, which are related simply to the functions $S_{AB}(Q, \omega)$ (for a recent survey of the subject see Blétry 1976). In a two-component system it is instructive to consider correlations involving the total number density,

$$N(\mathbf{r}, t) = N_1(\mathbf{r}, t) + N_2(\mathbf{r}, t) \tag{16}$$

and a function which measures departures from the equilibrium concentration, namely (using $c_1 = c$ and $c_2 = 1 - c$),

$$C(\mathbf{r}, t) = c_2 N_1(\mathbf{r}, t) - c_1 N_2(\mathbf{r}, t) \tag{17}$$

where

$$N_A(\mathbf{r}, t) = \sum_{j(A)}^{N_A} \delta\{\mathbf{r} - \mathbf{R}_j(t)\}.$$

A natural choice of scattering functions (cf Bhatia and Thornton 1970) is then

$$S_{NN} = c_1 S_{11} + 2(c_1 c_2)^{1/2} S_{12} + c_2 S_{22} \tag{18}$$

$$S_{NC} = c_1 c_2 S_{11} + (c_1 c_2)^{1/2}(c_2 - c_1)\, S_{12} - c_1 c_2 S_{22} \tag{19}$$

$$S_{CC} = c_1 c_2^2 S_{11} - 2(c_1 c_2)^{3/2} S_{12} + c_1^2 c_2 S_{22}. \tag{20}$$

In terms of these various functions the coherent cross section, equation (5), becomes

$$\frac{k'}{k}\{S_{NN}(c_1\bar{b}_1 + \bar{b}_2 c_2)^2 + 2S_{NC}(\bar{b}_1 - \bar{b}_2)(c_1\bar{b}_1 + \bar{b}_2 c_2) + S_{CC}(\bar{b}_1 - \bar{b}_2)^2\}. \tag{21}$$

Clearly, S_{NN} and S_{CC} are obtained if $\bar{b}_1 = \bar{b}_2$ and $\bar{b}_1 c_1 + \bar{b}_2 c_2 = 0$ respectively. This formalism is related closely to that used by Price and Copley (1975), who consider the number density $N(\mathbf{r}, t)$ and the charge density $Q(\mathbf{r}, t) = C(\mathbf{r}, t)(Z_1 - Z_2)$, where Z_i is the valence of ion i. The present formalism has wider application since it may be used for any two-component system. The functions S_{AB} are most suitable to a treatment of the scattering at large Q, whereas the functions S_{NN}, S_{NC} and S_{CC} are the appropriate choice for small values of Q. Bhatia and Thornton (1970) have shown that, for ions 1 and 2 of the same volume, $S_{NC}(Q = 0) = 0$, and that, if the ions are also similar in shape, then S_{NC} is small for all values of Q. Furthermore, it is easily shown (see for example March *et al* 1973, and references therein) that, for an ionic system, such that $c_1Z_1 + c_2Z_2 = 0$, one has the result $S_{CC}(Q = 0) = 0$. This latter result does not hold for an alloy, where the ions carry a net positive charge.

The aim of a neutron experiment is usually to measure either the coherent or the incoherent cross section, and thereby to obtain information on collective or single-particle dynamics, respectively. However, the observed scattering is the sum of both types of scattering. Fortunately, some elements have negligible coherent cross sections, for example vanadium, or negligible incoherent cross sections, for example lead. Moreover, it is possible in some cases to obtain the two cross sections separately by using isotopically enriched samples, as was done, for example, by Sköld *et al* (1972) and Johnson *et al* (1976, 1977a,b) in their studies of liquid argon and liquid nickel, respectively. Another means of affecting the separation is to use the technique of neutron polarization analysis. Since a liquid does not possess a preferred direction, the cross section (equation (4)) is independent of the polarization state of the incident beam. However, the polarization of the scattered beam, $\mathbf{P}'$, differs, in general, from the incident polarization, $\mathbf{P}$.

A straightforward calculation shows that $\mathbf{P}$ and $\mathbf{P}'$ are related by the expression

$$\mathbf{P}'\left(\frac{d^2\sigma}{d\Omega\, dE'}\right) = \mathbf{P}\,\frac{k'}{Nk}\left[\sum_{A}^{n}\sum_{B}^{n} \bar{b}_A^* \bar{b}_B (N_A N_B)^{1/2} S_{AB}(Q,\omega) + \right.$$
$$\left. + \sum_{B}^{n}\left(\overline{|A_B|^2} - |\bar{b}_B|^2 - \frac{1}{12} I_B(I_B+1)\overline{|B_B|^2}\right) N_B S_B^{(s)}(Q,\omega)\right] \tag{22}$$

where I_B is the magnitude of the nuclear spin of type B atoms, and the quantities $\overline{|A|^2}$ and $\overline{|B|^2}$ are given explicitly by Marshall and Lovesey (1971, p 336). For a system containing a single type of atom, the single-atom coherent and total cross sections are

$$\sigma_c = 4\pi|\bar{b}|^2 \tag{23}$$

and

$$\sigma = 4\pi[\overline{|A|^2} + \tfrac{1}{4} I(I+1)\overline{|B|^2}] \tag{24}$$

and the incoherent cross section $\sigma_i = \sigma - \sigma_c$. With these definitions, the expression (22) simplifies, for a single-component liquid, to the result

$$\mathbf{P}'[\sigma_c S(Q,\omega) + \sigma_i S_s(Q,\omega)] = \mathbf{P}\left[\sigma_c S(Q,\omega) + (\sigma_i - \tfrac{4}{3}\sigma + \tfrac{16}{3}\pi\overline{|A|^2}) S_s(Q,\omega)\right]. \tag{25}$$

If there is in fact just one type of isotope, the coefficient of the incoherent scattering function on the right hand side reduces to $-\sigma_i/3$. We see from these expressions that measurements of the cross sections for non-spin-flip and spin-flip scattering, using a polarized incident beam, together with a knowledge of the various scattering lengths, enable one (at least in principle) to separate out the coherent and incoherent scattering functions.

3. Neutron-scattering studies

A fairly complete review of inelastic neutron-scattering measurements on monatomic classical liquids has appeared recently (Copley and Lovesey 1975). In this section we briefly summarize the experimental situation as it stands today, emphasizing the more significant contributions to our knowledge. We shall not discuss recent work on ^{3}He, but instead refer the interested reader to the relevant papers (Stirling *et al* 1976, Sköld *et al* 1976).

Neutron-scattering experiments on liquids† are notoriously difficult and time consuming. Perhaps the most fundamental problem is to achieve sufficient accuracy, in the measurements themselves, to be able to say something significant about the scattering functions. Since for a classical liquid the odd moments $\langle \omega^m \rangle$ vanish, and since the $m = 0$ and $m = 2$ moments are known already (assuming $S(Q)$ has been determined from a diffraction experiment), the object of an inelastic neutron-scattering experiment is to determine the *fourth* moment of $S(Q,\omega)$ and/or $S_s(Q,\omega)$. In fact, the departure from some simple-model behaviour is what one looks for in such a study. These remarks also apply to the study of solids, where the (potential term in the) fourth moment of $S(Q,\omega)$ is related very closely to the dynamical matrix. The difference is that for a liquid the scattering functions are rather smooth in general, whereas the coherent-scattering function for a solid can contain sharp peaks which arise from the creation, or annihilation, of phonons.

At small values of Q, for example $Q < 1$ Å^{-1}, the moment approach is less useful. There are extreme experimental difficulties in this Q region, largely because of lack of intensity, since $S(Q)$ is small for dense liquids far from the liquid–gas phase transition, and because of limitations on the energy transfers available for a given Q. In most careful measurements on liquids the corrections for scattering from the container and for multiple scattering are important. Finally, there are sometimes problems associated with the fact that there is unwanted incoherent (or coherent) scattering in the sample. For example, coherent-scattering studies at small Q are particularly sensitive to any appreciable incoherent-scattering cross section.

Despite these difficulties, a number of experiments on monatomic liquids have been performed, as indicated by the list of experiments in table 4 of the paper by Copley and Lovesey (1975). Since that table was compiled, a few new experimental papers have appeared. Dasannacharya *et al* (1976) and Woods *et al* (1975, 1976a, b) report on measurements of the coherent scattering in *normal helium-4* at small wavevectors ($0{\cdot}06 \leqslant Q \leqslant 0{\cdot}2$ Å^{-1}) and at small to medium wavevectors ($0{\cdot}1 \leqslant Q \leqslant 2{\cdot}6$ Å^{-1}) respectively; the extensive study of polycrystalline and liquid *rubidium* by J B Suck is now in print (Suck 1975); the observation of a metastability limit in liquid *gallium* has been described in two papers (Bosio and Windsor 1975, Bosio *et al* 1976). Furthermore, detailed measurements of both $S(Q,\omega)$ and $S_s(Q,\omega)$ for *nickel* are reported in a paper by Johnson *et al* (1976, 1977b). Table 1 summarizes some important parameters of these measurements.

† In the absence of indications to the contrary, 'neutron scattering' and 'liquids' will be taken to imply inelastic neutron scattering and simple one- or two-component liquids respectively.

Table 1. Some parameters of recently published inelastic neutron-scattering measurements on monatomic liquids. This table updates table 4 of Copley and Lovesey (1975).

Element	Reference	Spectro-meter	λ_0	N_ϕ†	Range (deg)‡	Presentation
^{4}He	Dasannacharya *et al* (1976)	TAS	$(0{\cdot}06 \leqslant Q \leqslant 0{\cdot}2$ Å$^{-1})$			Constant Q
	Woods *et al* (1975, 1976a, b); Svensson *et al* (1976)	TAS	$(0{\cdot}1 \leqslant Q \leqslant 2{\cdot}6$ Å$^{-1})$			Constant Q
Rb	Suck (1975)	TOFS	4·0	40	6–165	Constant Q, ω
Ga	Bosio and Windsor (1975); Bosio *et al* (1976)	TAS	1·90	$(Q = 2{\cdot}5, 3{\cdot}1$ Å$^{-1})$		Constant Q
Ni	Johnson *et al* (1976)	TOFS	2·05	30	38–96	Constant Q, ω

† N_ϕ is the number of scattering angles used in the experiment.
‡ The range of scattering angles.

Until recently, little was known about the incoherent-scattering function for a liquid metal. However, the measurements of Johnson *et al* (1976, 1977b) on nickel show clearly that, as one might have expected, $S_s(Q, \omega)$ is a rather featureless function, similar in shape to the incoherent-scattering function for liquid argon (Sköld *et al* 1972). On the other hand, this function clearly contains information about the dynamics of the liquid, but this information is almost irretrievably hidden in the finer details of the shape of $S_s(Q, \omega)$. The limiting behaviour of $S_s(Q, \omega)$, as Q tends to zero, is related simply to the Fourier transform $p(\omega)$ of the velocity autocorrelation function (Egelstaff 1967). Carneiro (1976) has attempted to extract information about the behaviour of $p(\omega)$ at small ω using existing neutron-scattering data for argon, hydrogen and sodium. There are indications, from this analysis, of a long-time $t^{-3/2}$ tail in the velocity autocorrelation function, but it is also clear that more neutron-scattering work must be done in order to confirm its existence.

As is true of the study of solids, the coherent function $S(Q, \omega)$ for a liquid is much richer in features than the incoherent function. The general features of $S(Q, \omega)$ are, as always, dictated by its moments. At large values of Q and ω, $S(Q, \omega)$ reflects more and more the behaviour of single particles, and therefore looks more and more like $S_s(Q, \omega)$, which itself is tending to the shape of the scattering function for an ideal gas. At intermediate values of Q (say, $1 \leqslant Q \leqslant 5$ Å^{-1}) $S(Q, \omega)$ shows characteristic features which are determined largely by the behaviour of $S(Q)$. Again the details of $S(Q, \omega)$ in this range of wavevectors contain much information, but again careful measurements are required in order to extract these features. For example, a difference in the detailed shape of $S(Q, \omega)$ is found, for $Q \sim 1{\cdot}3\, Q_0$ (where Q_0 is the position of the principal peak in $S(Q)$) when one compares results for argon (Sköld *et al* 1972) and rubidium (Copley and Rowe 1974b), as is illustrated in figure 1. The more complicated behaviour in rubidium is possibly associated with the more persistent nature of the collective excitations observed at smaller values of Q.

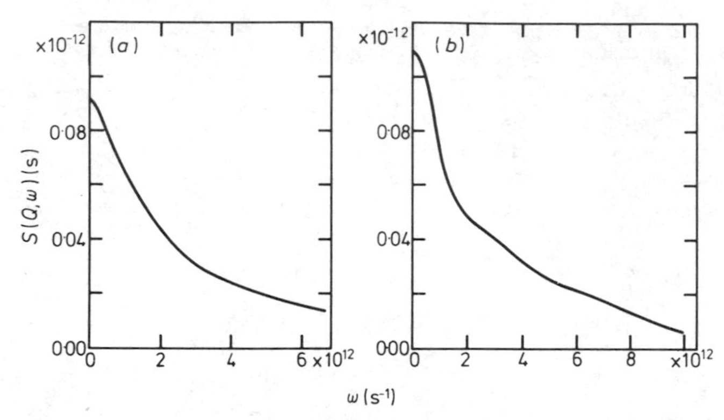

Figure 1. $S(Q, \omega)$ (*a*) for argon at $Q = 2{\cdot}6$ Å^{-1} (from table 4 of Sköld *et al* 1972), and (*b*) for rubidium at $Q = 2{\cdot}0$ Å^{-1} (Copley and Rowe 1974b). The ratio $Q/Q_0 = 1{\cdot}3$ in both cases.

For $Q < Q_0$, the integrated intensity (i.e., $S(Q)$) drops off rapidly with decreasing Q, and experiments in this region of Q are plagued with difficulties. The situation is generally worse for metals than for the inert elements, because the energy transfers of interest are normally larger, and the compressibility (which is related to $S(Q = 0)$) is usually small, and not increased easily by changing the temperature and density. On the other hand, this region of Q probably contains more information about the inter-atomic forces than the region of $Q > Q_0$. Recent experiments by the Jülich group on normal helium at 4·2 K (Dasannacharya *et al* 1976), and on fluid neon, well removed from its triple point (Bell *et al* 1973, 1975), demonstrate that $S(Q, \omega)$ has both Rayleigh and Brillouin peaks, but only for Q of order $0{\cdot}1\,Q_0$, or less. In contrast Copley and Rowe (1974a) found that side-peaks exist in liquid rubidium for $Q \leqslant 0{\cdot}65\,Q_0$. The results for neon and rubidium (see figure 2) are well supported by realistic molecular

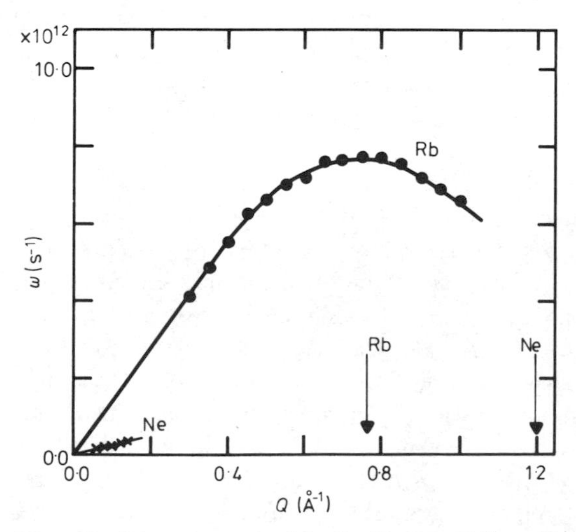

Figure 2. 'Dispersion curves' for liquid neon (Bell *et al* 1973, 1975) and liquid rubidium (Copley and Rowe 1974a). Lines through the points are drawn to guide the eye. The arrows indicate the positions defined by $Q = Q_0/2$ for each liquid.

dynamics calculations, and some progress has been made toward a theoretical understanding of these data.

A number of experiments has been performed with a view to measuring the effects of temperature and pressure on the scattering cross sections. An advantage of such experiments is that, as Egelstaff (1972) has pointed out, they often involve the subtraction of sets of data taken at different temperatures and/or pressures, and as a result the errors introduced by imperfect corrections (such as for multiple scattering) are kept to a minimum. Löffler (1973) has examined liquid gallium at 305 K and 1253 K, and Wignall and Egelstaff (1968) studied the temperature dependence of the scattering law for liquid lead. The inert elements neon and argon have been studied for several thermodynamic states, by Bell *et al* (1973, 1975) and by Hasman (1971, 1973) respectively. P Verkerk (private communication) has measured the scattering from liquid ^{36}Ar and from a liquid mixture of Ar isotopes, chosen to maximize the incoherent scattering. The data, which were taken at 120 K, at 20 and 270 atm, and for $0{\cdot}3 < Q < 2{\cdot}2$ Å^{-1}, are presently being analysed. Furthermore, C A Pelizzari and T A Postol (private communication) have measured the scattering function for ^{36}Ar gas at 300 K and 476 atm, corresponding to a density which is 1·28 times the density at a critical point. Hawkins and Egelstaff (1975) report initial experiments on dense nitrogen gas at several modest pressures. P A Egelstaff, R K Hawkins, D Litchinsky, W Gläser and J B Suck (private communication) have extended these measurements to higher pressures (of order 12 000 psi) and to many more densities; they have also done work on the pressure dependence of the scattering from krypton. In the near future they plan to extend their programme to look at liquid rubidium under pressure. The object of these pressure experiments is to extract information about the time-dependent triplet correlation function, whereas measurements of the temperature dependence of the scattering functions relate to the density–energy correlation function. This subject is extensively discussed by Egelstaff (1972, 1973a, b).

An area which has received scant attention is the study of the dynamics of the liquid–gas phase transition, though B Mozer (private communication) has done some preliminary measurements. Several experiments to investigate the solid–liquid transition have been reported, one of the most recent being that of Suck (1975), but even a rudimentary understanding of this transition is not yet available. A rather specialized study, which produced some pleasing results, was made by Bosio and Windsor (1975) on liquid gallium. These authors found that, as the limit of metastability of the liquid is approached, the width of $S(Q, \omega)$ for $Q = Q_0$ tends to zero: since $S(Q_0)$ does *not* diverge (Bizid *et al* 1974, Carlson *et al* 1974), it appears that $S(Q_0, 0)$ diverges as the temperature approaches the metastability limit.

To the best of our knowledge and belief, no alloys have as yet been studied using the inelastic neutron-scattering technique. A system of n components has in general n independent $S_s(Q, \omega)$ and $n(n+1)/2$ independent $S(Q, \omega)$, so it is clear that a separation of the various functions (using a technique such as isotopic substitution) is a formidable, if not impossible, project, even for a two-component system. In certain cases the situation is alleviated because the scattering cross sections of the elements permit some simplification. For example, the function $S_{NN}(Q, \omega)$ (see §2) is readily obtained in a two-component system if $\bar{b}_1 = \bar{b}_2$ and $\sigma_i(1)$ and $\sigma_i(2)$ are negligible. This situation obtains to a good approximation in the ionic system RbBr, which has been studied quite recently (Price and Copley 1975). As might have been expected, the function $S_{NN}(Q, \omega)$ for RbBr is reminiscent of the scattering function for a monatomic liquid. Molecular dynamics calculations on alkali halide systems (e.g., Hansen and McDonald 1975, Copley and Rahman 1976) clearly show that, for $Q \lesssim 1$ Å^{-1}, the function $S_{CC}(Q, \omega)$ has a well defined peak (plotted against ω for constant Q). The peaks lie on an 'optic' dispersion curve (Abramo *et al* 1973) which is much better defined than the 'acoustic' dispersion curve obtained from $S_{NN}(Q, \omega)$. The situation is illustrated in figure 3. These 'optic' modes of vibration, in which the two species tend to move in

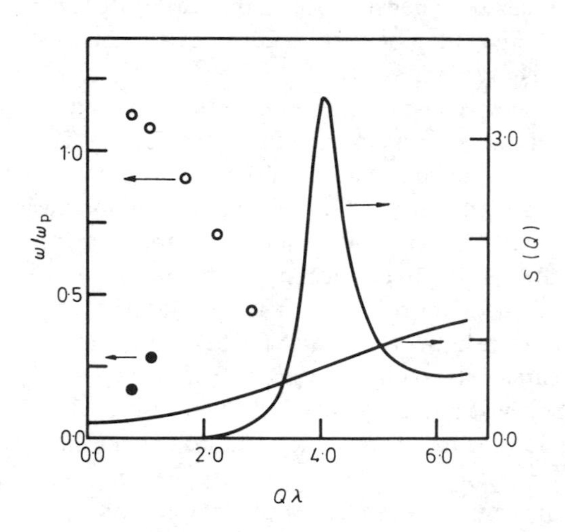

Figure 3. 'Dispersion curves' (closed and open circles), and structure factors $S(Q)$, for density and charge fluctuations in a simple molten salt (Hansen and McDonald 1975). The parameter λ is the distance at which the cation–anion potential is a minimum: ω_p is the plasma frequency of the system.

opposite directions, are not confined to ionic systems. They are likely to exist in metallic alloys, particularly in those alloys within which atoms of type 1 are more strongly attracted to atoms of type 2 and vice versa. A serious attempt has been made to observe 'optic' modes in liquid KBr (J R D Copley and G Dolling private communication). Weak evidence for such modes was obtained, for $Q \simeq 0{\cdot}8$ Å^{-1}, but alternative explanations for the observed scattering cannot be ruled out. Though these measurements are very difficult (see, for example, Copley and Rahman 1976), they are nevertheless of great interest. It is hard to judge whether it will be more difficult to see such modes in an alloy or in an ionic salt.

In principle, inelastic neutron-scattering measurements provide the most detailed experimental information about the dynamics of liquids on an atomic scale. On the other hand these measurements are hard to perform properly. In the circumstances we are extremely fortunate that we have another powerful 'experimental' tool at our disposal, the computer-simulation method. We briefly discuss it in the following section.

4. Computer-simulation experiments

A beauty of both the computer-simulation techniques, Monte Carlo and molecular dynamics, is that the input to the calculation – the potential, density and temperature in particular – is known. This makes it somewhat less difficult to identify the shortcomings of a theory used to interpret the results of the simulation. On the other hand, comparisons between neutron-scattering results and data obtained from an appropriate computer-simulation 'experiment' can shed light on the deficiencies of the input to the computer simulation.

Of late there has been considerable activity in the field of computer-simulation work, and a growing number of groups have been doing significant calculations. Rahman and co-workers (Rahman *et al* 1976, Mandell *et al* 1976) have recently explored the properties of an amorphous Lennard-Jones system at low temperature, and they have studied the phenomenon of crystal nucleation in a Lennard-Jones system, with interesting results. In the context of liquid metals, Block and Schommers (1975) investigated the static triplet correlation function for rubidium, finding 'no important qualitative difference' between their results and results obtained for the equivalent state of a Lennard-Jones system by Wang and Krumhansl (1972).

Many papers have been published describing computer 'experiments' on ionized systems, in particular on alkali halides (for a recent review see Singer 1976)†. One of the most exciting aspects of this work is the clear prediction of 'optic modes' at small Q, in both the density–density and the current–current correlation functions. For example, the results obtained by Hansen and McDonald (1975) for their simple model of a molten salt are shown in figure 3. We expect to see similar effects, but less pronounced, in an alloy. In fact, the extent to which we expect to see such modes in a two-component system is not doubt determined by the degree of difference between the various interparticle interactions at long distances.

The marked difference, both in computer-simulation and neutron-scattering experiments, between $S(Q, \omega)$ for inert gas liquids and liquid metals for $Q < Q_0$, focuses attention on the dependence of the coherent-scattering function on the interparticle potential. We shall assume in the following discussion that the interparticle potential can be represented by pair-wise additive potentials, to a good approximation.

If we denote the position of the first zero in the pair potential by σ, and the number density by ρ, then the mean free path in a liquid is of order $1/\rho\sigma^2$, and for $Q \ll \rho\sigma^2$, $S(Q, \omega)$ possesses a peak at $\omega = vQ$, where v is the velocity of sound. Computer-simulation and neutron-scattering data are obtained usually for $Q \gtrsim \rho\sigma^2$, outside the hydrodynamic regime. However, as noted in the preceding section, experiments on liquid rubidium, near its solidification line, show that $S(Q, \omega)$ possesses a peak at nonzero ω for $Q \lesssim 0{\cdot}65\, Q_0$, which we identify with the existence of a short-wavelength collective density excitation and hereafter call a viscoelastic mode. The presence of such a mode for a liquid metal, and its absence in inert gas liquids, is attributed usually to the relatively softer core of the liquid metal pair potential, since large-angle scatterings produced by a highly repulsive potential are particularly effective in destroying correlated motions between particles.

A first step toward a more quantitative understanding of the role of the potential in determining the structure of $S(Q, \omega)$ for intermediate wavevectors, $\rho\sigma^2 < Q < Q_0$, has been made by Lewis and Lovesey (1977) who have analysed computer-simulation data for a series of dense liquids interacting with a generalized Lennard-Jones potential in which the powers of the repulsive and attractive parts can be varied. They conclude that the conditions obtaining in a (12–6) Lennard-Jones liquid in the vicinity of its triple point are far removed from those required to sustain a viscoelastic mode. Such a mode exists when the spatial dependence of the potential is weakened significantly

† See also Sangster and Dixon (1976).

compared to that of a (12–6) Lennard-Jones potential: a point of particular interest is that it does not appear to be sufficient to merely soften the core of the potential. This conclusion is shown to be consistent with ideas based on an elastic continuum model, for which the Grüneisen parameter

$$\gamma_G \simeq -(r_0/6)(U'''/U'')_{r_0} \tag{26}$$

needs to be less than about two for a well defined collective mode to exist. In equation (26), U is the pair potential, $U'' \equiv \partial^2 U/\partial r^2$, and r_0 is the position of the minimum in the potential. Lewis and Lovesey (1977) also conclude that the shape of $S(Q, \omega)$ shown in figure 1(*b*) is characteristic of dense liquids in which motions of the particles are highly correlated.

5. Theory

In the following, brief survey of recent calculations of the coherent-scattering function for monatomic liquids, we shall not include progress in the study of kinetic equations for correlation functions. The reader who is interested in this field of study is referred to the paper by Jhon and Forster (1975), and the review paper by Schofield (1975). Theories of multi-component systems are reviewed by March and Tosi (1976); see also the paper of Abramo *et al* (1974) and Kerr (1976) for discussions of the scattering from molten salts.

The interpretation of data for $S(Q, \omega)$ for intermediate wavevectors has shown that a continued fraction expansion of a generalized Langevin equation for the number density autocorrelation function can generate expressions for the coherent-scattering function which give good overall agreement with available data. It appears that a good zero-order theory for $S(Q, \omega)$ is obtained with a continued fraction expansion if the first two levels are retained, and the remaining terms approximated by a simple, wavevector-dependent, termination function (Copley and Lovesey 1975). Such an approach guarantees that the resulting $S(Q, \omega)$ has the correct $m = 0$, 2 and 4 moments. In view of the success of the simple theories, it is natural to use them as the starting point for the development of more sophisticated theories as we shall describe shortly. A major stumbling block in comparing any such theory with data for $S(Q, \omega)$ is that it requires, at least, a knowledge of the three-particle distribution function. Ideally then, the theory should be compared with computer-simulation data, using as input to the theory the (exact) static correlation functions, including three-particle and, if required, higher-order correlation functions obtained in the computer simulation.

In describing the theory, it is convenient to work in terms of the intermediate scattering function $F(Q, t)$ which is the temporal Fourier transform of $S(Q, \omega)$. If we denote the Laplace transform of $F(Q, t)$ by $\tilde{F}(Q, s)$, then

$$\pi S(Q, \omega) = \mathrm{Re}\, \tilde{F}(Q, \mathrm{i}\omega)\, S(Q) \tag{27}$$

Much of the content of the preceding paragraph can be summarized with the expression

$$\tilde{F}(Q, s) = \left(s + \frac{\omega_0^2}{s + (\omega_1^2 - \omega_0^2)\tilde{K}_c(Q, s)}\right)^{-1}. \tag{28}$$

where $\omega_0^2 = \langle\omega^2\rangle/S(Q)$, and $\omega_1^2 = \langle\omega^4\rangle/\langle\omega^2\rangle$. We note that $\tilde{K}_c(Q,s)$ in equation (28) is the Laplace transform of the memory function $K_c(Q,t)$ for the current correlation function

$$C(Q,t) = -\frac{\partial^2}{\partial t^2} F(Q,t) \tag{29}$$

since,

$$\frac{\partial}{\partial t} C(Q,t) = -\omega_0^2 \int_0^t \mathrm{d}\bar{t}\, C(Q,\bar{t}) - (\omega_1^2 - \omega_0^2) \int_0^t \mathrm{d}\bar{t}\, K_c(Q,t-\bar{t})\, C(Q,\bar{t}). \tag{30}$$

The approximation

$$\tilde{K}_c(Q,s) = \left(s + \frac{1}{\tau(Q)}\right)^{-1} \tag{31}$$

with the relaxation time $\tau(Q)$ defined by Copley and Lovesey (1975) has been shown to give a good description of the computer-simulation data for a model of liquid rubidium. We note in particular that the approximation reproduces the shape of $S(Q,\omega)$ shown in figure 1(*b*), which can be viewed as a narrow peak superposed on to a broad maximum. This shape leads naturally to the concept of two relaxation times, but we would emphasize that such comments must be made in the context of a specific prescription for $S(Q,\omega)$ since, as we have noted, approximation (31), for example, reproduces the data quite well with apparently a 'single' relaxation time, $\tau(Q)$.

In setting up a more sophisticated theory it is natural to introduce a function $\tilde{M}(Q,s)$, such that

$$\tilde{K}_c(Q,s) = [s + \tilde{M}(Q,s)]^{-1}, \tag{32}$$

replaces (31), and to seek an improved approximation to $\tilde{M}(Q,s)$. Note that the instantaneous value of $M(Q,t)$ contains the sixth moment of $S(Q,\omega)$, which involves the three-particle correlation function (Bansal and Pathak 1974).

The formulation by Mori (1965a, b) of irreversible thermodynamics provides an exact expression for the function $\tilde{M}(Q,s)$ in equation (32). This can be expressed in the form of a matrix element of an operator taken between states proportional to the third time derivative of the number density, or equivalently the second derivative of the longitudinal current density. To approximate the expression for $\tilde{M}(Q,s)$ we can use a mode–mode coupling scheme, which has proved useful in a wide range of problems in irreversible thermodynamics (Pomeau and Résibois 1975). Applied to the problem in hand, the scheme suggests that we seek to represent the states used in calculating $\tilde{M}(Q,s)$ by the product of basic hydrodynamical variables, which are then assumed to be non-interacting. Götze and Lücke (1975) argue that the appropriate pair of variables is the number and longitudinal current densities, and they approximate the three-particle correlation function by the product of pair distribution functions, including hard-core correlations with a step function in the positions of the particles. A conclusion of particular interest is that longitudinal and transverse modes play an essentially equal role in determining $\tilde{M}(Q,s)$. The authors find that, for intermediate values of Q, the theory is in good agreement with computer-simulation and neutron-scattering data for liquid argon. J Bosse (private communication) has used the theory to interpret data for liquid rubidium and finds satisfactory agreement for intermediate wavevectors. Detailed tests of the more sophisticated theories are necessary if we are to learn precisely where and how they represent an improvement over simple theories, since these already give a good account of the data available, except at small frequencies.

Barker and Gaskell (1975) have attempted to elucidate the role of the potential in determining $S(Q, \omega)$ for intermediate wavevectors, and their conclusions tie in nicely with (subsequent) computer-simulation data, discussed in the preceding section. The memory function $K_c(Q, t)$ is expressed by these authors in terms of the self-correlation function, and short-range collision processes are included accurately into their prescription, so that the hard-core limit can be recovered. The authors argue the importance of obtaining this limit correctly, even at the expense of the fourth moment, as in their own formulation. Their final expression for $S(Q, \omega)$ is in good overall agreement with the computer-simulation data for liquid rubidium. The most interesting feature of the calculation is, perhaps, that $S(Q, \omega)$ is determined largely by the memory function obtained from the tail of the potential, $r > \sigma$, the effect of the core contribution to $K_c(Q, t)$ being only to reduce slightly the intensity of the collective mode peak, and to give better agreement with experiment. While this conclusion ties in nicely with conclusions drawn from computer-simulation data, as mentioned earlier, it appears to be in conflict with the authors' argument that short-range collision processes are important in determining the dynamic properties of dense liquids.

Hubbard and Beeby (1969) obtained an expression for $K_c(Q, t)$ in terms of the self-correlation function, using a quite different approach to that of Barker and Gaskell (1975). The theory of Hubbard and Beeby (1969) does not satisfy the zero-moment relation, and consequently it is not very useful for the interpretation of experimental data. Tewari and Tewari (1975) have proposed a physically motivated modification to the original theory, which permits the zero moment to be satisfied. The authors claim that the modified theory is in good agreement with the current correlation function for liquid argon. This claim is disputed by Kahol *et al* (1976), who also argue that, when the modified theory of Tewari and Tewari is evaluated correctly, the agreement between theory and experiment is inferior to that achieved with the modified theory of Pathak and Bansal (1973).

6. Discussion

In this short review we have described some recent developments in the field of liquid dynamics, confining our attention to relatively simple systems. We have said nothing about inelastic neutron-scattering work on molecular liquids such as the homonuclear diatomic fluids hydrogen and nitrogen (Carneiro *et al* 1973, Carneiro 1974, Carneiro and McTague 1975, Postol *et al* 1976, Sinclair *et al* 1975), nor have we described molecular dynamics studies of melting (e.g. Cotterill *et al* 1974). On the theory side we have also been forced, due to space limitations, to omit a number of papers from our discussion. Nevertheless, it is clear that our understanding of liquids has improved significantly in the past few years. In particular, a number of careful measurements on simple liquids, and parallel computer-simulation experiments, have been completed since the last liquid metals conference in 1972. On the theoretical side there have also been some significant advances. On the other hand it is just as clear that we still have a very long way to go to a detailed understanding of the liquid state.

In this connection the remarks made by the panel on liquids, glasses and gases, at a recent Workshop on Uses of Advanced Pulsed Neutron Sources (Carpenter and Werner 1976) are very relevant. In order to make progress toward a better understanding of the liquid state, we need to be able to do accurate neutron-scattering measurements over a wide range of values of both Q and ω. For a given Q, the interesting range in ω is dictated by the second moment, equations (14) and (15), so it is clear that the need for high incident energies, in order to obtain large ω values at modest Q values, is especially great if the temperature is high and the mass relatively small. At small values of Q the minimum incident energy is related to the sound velocity in the liquid (Copley and Lovesey 1975). Finally large ω values (which imply high incident energies) will be necessary in order to unravel the detailed dynamics of more complicated liquids. These arguments all point to the need for a high-intensity, high-energy inelastic time-of-flight spectrometer, fitted with an array of detectors at scattering angles down to (say) 5°, plus a two-dimensional multidetector to be used for small Q work. Without this type of machine it is hard to see dramatic progress being made in neutron-scattering studies of simple liquids. An exception is, perhaps, the study of effects such as temperature and pressure dependence (Egelstaff 1972): here we can look forward to considerable advances using existing instruments.

Acknowledgments

We thank J Bosse, K Carneiro, P A Egelstaff, M W Johnson, T A Postol and P Verkerk for communicating results to us prior to publication.

References

Abramo M C, Parrinello M and Tosi M P 1974 *J. Phys. C: Solid St. Phys.* **7** 4201
Abramo M C, Parrinello M, Tosi M P and Thornton D E 1973 *Phys. Lett.* **43A** 483
Bansal R and Pathak K N 1974 *Phys. Rev.* **A9** 2773
Barker M I and Gaskell T 1975 *J. Phys. C: Solid St. Phys.* **8** 3715
Bell H G, Kollmar A, Alefeld B and Springer T 1973 *Phys. Lett.* **45A** 479
Bell H G, Moeller-Wenghoffer H, Kollmar A, Stockmeyer R, Springer T and Stiller H 1975 *Phys. Rev.* **A11** 316
Bhatia A B and Thornton D E 1970 *Phys. Rev.* **B2** 3004
Bizid A, Bosio L, Curien H, Defrain A and Dupont M 1974 *Phys. Stat. Solidi* **A23** 135
Blétry J 1976 *Z. Naturf.* **31A** 960
Block R and Schommers W 1975 *J. Phys. C: Solid St. Phys.* **8** 1997
Bosio L, Schedler E and Windsor C G 1976 *J. Physique* **37** 747
Bosio L and Windsor C G 1975 *Phys. Rev. Lett.* **35** 1652
Carlson D G, Feder J and Segmuller A 1974 *Phys. Rev.* **A9** 400
Carneiro K 1974 *Danish Atomic Energy Commission Risø, Rep. No.* 308
—— 1976 *Phys. Rev.* **A14** 517
Carneiro K and McTague J P 1975 *Phys. Rev.* **A11** 1744
Carneiro K, Nielsen M and McTague J P 1973 *Phys. Rev. Lett.* **30** 481

Carpenter J M and Werner S A (eds) 1976 *Argonne National Laboratory Rep.* ANL-76-10 vol 2 p 130
Copley J R D and Lovesey S W 1975 *Rep. Prog. Phys.* **38** 461
Copley J R D and Rahman A 1976 *Phys. Rev.* **A13** 2276
Copley J R D and Rowe J M 1974a *Phys. Rev. Lett.* **32** 49
—— 1974b *Phys. Rev.* **A9** 1656
Cotterill R M J, Damgaard Kristensen W and Jensen E J 1974 *Phil. Mag.* **30** 245
Croxton C A 1974 *Liquid State Physics – A Statistical Mechanical Introduction* (London: Cambridge UP)
Dasannacharya B A, Kollmar A and Springer T 1976 *Phys. Lett.* **55A** 337
Egelstaff P A 1967 *An Introduction to the Liquid State* (New York: Academic)
—— 1972 *Neutron Inelastic Scattering 1972* (Vienna: IAEA) p383
—— 1973a *Ann. Rev. Phys. Chem.* **24** 159
—— 1973b *The Properties of Liquid Metals* ed S Takeuchi (London: Taylor and Francis) p13
Faber T E 1972 *Introduction to the Theory of Liquid Metals* (London: Cambridge UP)
Faber T E and Ziman J M 1965 *Phil. Mag.* **11** 153
Götze W and Lücke M 1975 *Phys. Rev.* **A11** 2173
Hansen J P and McDonald I R 1975 *Phys. Rev.* **A11** 2111
Hasman A 1971 *PhD Thesis* Technische Hogeschool Delft, The Netherlands
—— 1973 *Physica* **63** 499
Hawkins R K and Egelstaff P A 1975 *Molec. Phys.* **29** 1639
Hubbard J and Beeby J L 1969 *J. Phys. C: Solid St. Phys.* **2** 556
Jhon M S and Forster D 1975 *Phys. Rev.* **A12** 254
Johnson M W, McCoy B, March N H and Page D I 1976 *AERE Harwell Report* MPD/NBS/19
—— 1977a this volume
—— 1977b *Phys. Chem. Liquids* to be published
Kahol P K, Bansal R and Pathak K N 1976 *J. Phys. C: Solid St. Phys.* **9** L259
Kerr W C 1976 *J. Chem. Phys.* **64** 885
Larose A and Vanderwal J 1973 *Bibliography of Papers Relevant to the Scattering of Thermal Neutrons* (Hamilton, Ontario: McMaster University)
Lewis J W E and Lovesey S W 1977 *J. Phys. C: Solid St. Phys.* to be published
Löffler U 1973 *Thesis* University of Karlsruhe, Germany (unpublished)
Mandell M J, McTague J P and Rahman A 1976 *J. Chem. Phys.* **64** 3699
March N H and Tosi M P 1976 *Atomic Dynamics in Liquids* (London: Macmillan)
March N H, Tosi M P and Bhatia A B 1973 *J. Phys. C: Solid St. Phys.* **6** L59
Marshall W and Lovesey S W 1971 *Theory of Thermal Neutron Scattering* (London: Oxford UP)
Mori H 1965a *Prog. Theor. Phys.* **33** 423
—— 1965b *Prog. Theor. Phys.* **34** 399
Pathak K N and Bansal R 1973 *J. Phys. C: Solid St. Phys.* **6** 1989
Pomeau Y and Résibois P 1975 *Phys. Rep.* **19** 63
Postol T A, Chen S H and Sköld K 1976 *Proc. Conf. on Neutron Scattering, Gatlinburg, Tennessee* (Oak Ridge, Tennessee: Technical Information Division, ERDA)
Price D L and Copley J R D 1975 *Phys. Rev.* **A11** 2124
Rahman A 1975 *Phys. Rev.* **A11** 2191
Rahman A, Mandell M J and McTague J P 1976 *J. Chem. Phys.* **64** 1564
Sangster M J L and Dixon M 1976 *Adv. Phys.* **25** 247
Schofield P 1975 *Specialist Reports – Statistical Mechanics* vol 2, ed K Singer (London: Chemical Society)
Sinclair R N, Clarke J H and Dore J C 1975 *J. Phys. C: Solid St. Phys.* **8** L41
Singer K 1976 *SRC Atlas Symposium on Computational Physics of Liquids and Solids, Oxford* (Rutherford Laboratory: SRC)
Sköld K, Pelizzari C A, Kleb R and Ostrowski G E 1976 *Phys. Rev. Lett.* **37** 842
Sköld K, Rowe J M, Ostrowski G and Randolph P D 1972 *Phys. Rev.* **A6** 1107
Stirling W G, Scherm R, Hilton P A and Cowley R A 1976 *J. Phys. C: Solid St. Phys.* **9** 1643
Suck J B 1975 *Thesis* Kernforschungszentrum Karlsruhe, Rep. No KFK2231

Svensson E C, Stirling W G, Woods A D B and Martel P 1976 *Proc. Conf. on Neutron Scattering, Gatlinburg, Tennessee* (Oak Ridge, Tennessee: Technical Information Division, ERDA)
Tewari S P and Tewari S P 1975 *J. Phys. C: Solid St. Phys.* **8** L569
Van Hove L 1954 *Phys. Rev.* **95** 249
Wang S and Krumhansl J A 1972 *J. Chem. Phys.* **56** 4297
Wignall G D and Egelstaff P A 1968 *J. Phys. C: Solid St. Phys.* **1** 519
Woods A D B, Svensson E C and Martel P 1975 *Low Temperature Physics* (LT 14) ed M Krusius and M Vuorio (Amsterdam: North-Holland) vol 1 p187
—— 1976a *Proc. Conf. on Neutron Scattering, Gatlinburg, Tennessee* (Oak Ridge, Tennessee: Technical Information Division, ERDA)
—— 1976b *Phys. Lett.* **57A** 439

Neutron scattering and atomic dynamics in liquid nickel

M W Johnson†‡, B McCoy†, N H March† and D I Page§

† Department of Physics, The Blackett Laboratory, Imperial College, London, UK

§ Materials Physics Division, Harwell, Didcot, Oxon, UK

Abstract. The results of inelastic neutron scattering experiments on liquid nickel at a temperature of 1870 K are reported, one set of measurements being performed on natural nickel and another on an incoherently scattering mixture of ^{58}Ni and ^{62}Ni isotopes. From these data the dynamic structure factor $S(q, \omega)$ and the self function $S_s(q, \omega)$ have been extracted over a limited range of q ($2{\cdot}2 < q < 4{\cdot}4$ Å^{-1}) and ω ($0 < \omega < 80$ meV).

To interpret $S_s(q, \omega)$ a model is used which attempts, as is usual, a precise description of the short-time behaviour of the atomic motions. The theory is characterized by two quantities for which first-principle expressions are available. Because of uncertainties in (i) the interionic potential and (ii) the static triplet correlation function, these two quantities are treated as parameters in the model. The main features of the incoherent scattering data are reflected by the model.

The coherent scattering data possibly provide some preliminary evidence for the existence of collective modes in liquid nickel. Such modes, if they exist, would have to be due, at least in part, to the long-range force in the pair interaction, which has been estimated previously from the static structure factor $S(q)$.

Finally, approximations to $S(q, \omega)$ are calculated from the observed values of $S(q)$, using both measured data and the model for S_s. The usefulness and limitations of such relations are briefly assessed.

‡ Present address: Neutron Beam Research Unit, Rutherford Laboratory, Chilton, Oxon, UK.

Modified Alder–Wainwright model for the velocity autocorrelation function

T Geszti and J Kertész

Research Institute for Technical Physics of the Hungarian Academy of Sciences, H-1325 Budapest, POB 76, Hungary

Abstract. In a continuous fluid, a spherical volume of atomic size centred at $\mathbf{r} = 0$ is given an initial velocity $\mathbf{v}_0$. The corresponding velocity field is decomposed into plane waves. Subsequent evolution is described by generalized hydrodynamics linearized in the wave amplitudes. $\langle \mathbf{v}_0 \cdot \mathbf{v}(\mathbf{r} = 0, t) \rangle$ is taken as a model for the velocity autocorrelation function (VAF), which is obtained as the sum of a longitudinal and a transverse part. Metallic behaviour is simulated by a high value of the longitudinal relaxation time, which results in an oscillatory long-time tail of the VAF. Transverse viscoelasticity is found to give a large contribution to the negative part of the VAF.

1. Introduction

So far none of the existing theories of the velocity autocorrelation function (VAF) have accounted for a very characteristic difference between Lennard–Jones liquids (argon) and liquid metals: the oscillatory long-time tail of the VAF observed in molecular dynamics (MD) calculations on liquid sodium (Paskin 1967) and rubidium (Schommers 1972) but not in liquid argon. In a previous paper (Geszti 1976) we attributed the long-time oscillations to the Van Hove singularity of the velocity spectrum corresponding to the maximum of the dispersion curve $\omega_1(q)$ of weakly damped longitudinal waves, observed in liquid rubidium by means of neutron inelastic scattering by Copley and Rowe (1974) and reproduced in MD calculations by Rahman (1974).

Our aim was to make the above physical picture of VAF oscillations more explicit by a hydrodynamical model calculation. The model introduced by Alder and Wainwright (1970) has been chosen, modified by the use of generalized hydrodynamics in describing the evolution of plane wave components.

2. Description of the model

Following Alder and Wainwright (1970) we investigate the evolution of the velocity field $\mathbf{v}(\mathbf{r}t)$ of a fluid continuum, subsequent to an initial state characterized by a spatially constant density $\rho = \rho_0$ and the velocity distribution

$$\begin{aligned} \mathbf{v}(\mathbf{r}0) &= \mathbf{v}_0 \qquad \text{for } r \leqslant a, \\ &= 0 \qquad \text{for } r > a, \end{aligned} \tag{1}$$

where a is the atomic radius.

Equation (1) corresponds to a flux density $\rho_0\mathbf{v}(\mathbf{r}t)$ that can be decomposed into longitudinal and transverse plane waves:

$$\rho_0\mathbf{v}(\mathbf{r}0) = \int_0^{2\pi} \mathrm{d}v \int_0^{\pi} \mathrm{d}u \int_0^{\infty} \mathrm{d}q\,[A^{\mathrm{l}}(quv)\,\mathbf{a}_q \exp(\mathrm{i}\mathbf{q}\cdot\mathbf{r}) + A^{\mathrm{t}}(quv)\,\mathbf{a}_u \exp(\mathrm{i}\mathbf{q}\cdot\mathbf{r})] \tag{2}$$

with

$$\begin{Bmatrix} A^{\mathrm{l}}(quv) \\ A^{\mathrm{t}}(quv) \end{Bmatrix} = (2\pi^2)^{-1} \rho_0 v_0 a^2 q \mathrm{j}_1(qa) \begin{Bmatrix} \sin u \cos u \\ \sin^2 u \end{Bmatrix} \tag{3}$$

where $\mathrm{j}_1(z)$ is a spherical Bessel function; q, u and v are the spherical coordinates of $\mathbf{q}$; $\mathbf{a}_q$, $\mathbf{a}_u$ and $\mathbf{a}_v$ are the respective local unit vectors. The derivation of equations (2) and (3) as well as of equations (5) and (6) below is given elsewhere (Kertész 1976).

Our basic approximation consists of linearizing as much as possible around full equilibrium when treating the time evolution of equation (2), namely:

(i) $\rho(\mathbf{r}t)\,\mathbf{v}(\mathbf{r}t)$ is approximated by $\rho_0\mathbf{v}(\mathbf{r}t)$;
(ii) interaction between the plane wave modes in equation (2) is neglected: $\rho_0\mathbf{v}(\mathbf{r}t)$ remains a superposition of independently evolving plane waves for all $t > 0$;
(iii) the expectation value of the amplitude of a single longitudinal or transverse flux mode excited to some initial value is assumed to relax proportionally to the longitudinal current correlation function $C_{\mathrm{l}}(qt)$ or the transverse one $C_{\mathrm{t}}(qt)$ respectively.

These approximations obviously fail for very short times, when the amplitudes are large and the waves strongly coupled. Still the results will show that the scheme is quite instructive to study.

With the above approximations equations (1) and (2) give

$$\langle \mathbf{v}(00)\cdot\mathbf{v}(\mathbf{r}t)\rangle = (\rho_0)^{-1} \int_0^{2\pi} \mathrm{d}v \int_0^{\pi} \mathrm{d}u \int_0^{\infty} \mathrm{d}q$$

$$\times \left\langle \mathbf{v}_0 \cdot \left(A^{\mathrm{l}}(quv) \frac{C_{\mathrm{l}}(qt)}{C_{\mathrm{l}}(q0)} \mathbf{a}_q \exp(\mathrm{i}\mathbf{q}\cdot\mathbf{r}) + A^{\mathrm{t}}(quv) \frac{C_{\mathrm{t}}(qt)}{C_{\mathrm{t}}(q0)} \mathbf{a}_u \exp(\mathrm{i}\mathbf{q}\cdot\mathbf{r}) \right) \right\rangle. \tag{4}$$

Now comes one last step of linearization: we neglect the displacement of the particle from 0 to $\mathbf{r}$ and calculate the VAF from the $\mathbf{r} = 0$ value of equation (4). Then evaluating the angular (u and v) integrals, one obtains vector harmonics (Morse and Feshbach 1953) on which the limit $r \to 0$ is easy to take. The result takes the form of the sum of a longitudinal and a transverse term in view of the linear approximation (4):

$$z(t) = \langle \mathbf{v}(00)\cdot\mathbf{v}(0t)\rangle = z_{\mathrm{l}}(t) + z_{\mathrm{t}}(t), \tag{5}$$

where

$$\begin{Bmatrix} z_{\mathrm{l}}(t) \\ z_{\mathrm{t}}(t) \end{Bmatrix} = \langle v_0^2 \rangle \frac{2}{\pi} a^2 \int_0^{\infty} \mathrm{d}q\, q\mathrm{j}_1(qa) \begin{Bmatrix} \frac{1}{3} C_{\mathrm{l}}(qt)/C_{\mathrm{l}}(q0) \\ \frac{2}{3} C_{\mathrm{t}}(qt)/C_{\mathrm{t}}(q0) \end{Bmatrix}. \tag{6}$$

To proceed further, $C_{\mathrm{l}}(qt)$ and $C_{\mathrm{t}}(qt)$ need closer specification.

3. The current correlation functions

Within the framework of generalized hydrodynamics (Chung and Yip 1969, hereafter referred to as CY) the functions $C_l(qt)$ and $C_t(qt)$ can be given simple representations much more realistic than viscoelastic hydrodynamics.

For $C_t(qt)$ the expression given by CY is

$$C_t(qt)/C_t(q0) = \exp(-t/2\tau_t(q)) \times \{\cos[\omega_t(q)\,t] + [2\tau_t(q)\,\omega_t(q)]^{-1} \sin[\omega_t(q)\,t]\}, \quad (7)$$

where the q-dependence of the relaxation time can be approximated by CY

$$\tau_t^{-2}(q) = \tau_t^{-2}(0) + q^2 v_t^2, \quad (8)$$

v_t being the mean thermal velocity. Further,

$$\omega_t^2(q) = \hat{\omega}_t^2(q) - \tfrac{1}{4}\tau_t^{-2}(q), \quad (9)$$

where

$$\hat{\omega}_t^2(q) = -\ddot{C}_t(q0)/C_t(q0). \quad (10)$$

Values of $\tau_t(0)$ and $\hat{\omega}_t(q)$ are needed as input data.

For $C_l(qt)$ there is no explicit expression in CY. We have used a formula analogous to equation (7):

$$C_l(qt)/C_l(q0) = \exp[-t/2\tau_l(q)] \times \{\cos[\omega_l(q)\,t] - [2\tau_l(q)\,\omega_l(q)]^{-1} \sin[\omega_l(q)\,t]\}, \quad (11)$$

where the coefficient of the sine term is fixed by the requirement

$$\int_0^\infty C_l(qt)\,dt = \lim_{\omega \to 0} \frac{\omega^2}{q^2} S(q\omega) = 0. \quad (12)$$

The q-dependence of τ_l is approximated by (quoted in CY)

$$\tau_l^{-1}(q) = \tau_l^{-1}(0) + q v_t. \quad (13)$$

For $\omega_l(q)$ two different expressions suggested by CY are used according to whether $q > q_0/4$ or $q < q_0/4$, where q_0 is the wavevector of the main peak in the static structure factor:

$$\begin{aligned} \omega_l^2(q) &= \omega_0^2(q) && \text{for } q < q_0/4, \\ &= \hat{\omega}_l^2(q) - \omega_0^2(q) - \tfrac{1}{4}\tau_l^{-2}(q) && \text{for } q > q_0/4, \end{aligned} \quad (14)$$

where

$$\omega_0^2(q) = k_B T q^2/mS(q) \quad (15)$$

and

$$\hat{\omega}_l^2(q) = -\ddot{C}_l(q0)/C_l(q0). \quad (16)$$

Here the input data required are $\tau_l(0)$, $\omega_0(q)$ and $\hat{\omega}_l(q)$.

The two expressions in equation (14) do not match at $q_0/4$; we found that this inaccuracy of the calculation was not particularly significant.

The formulae (7) and (11) and all results drawn from them are invalid for very short times. Equations (10) and (16), which are the standard second moments of the current correlation spectra, must be evaluated of course from exact dynamics, not from equations (7) and (11).

4. Results of the calculation

First of all, it is easy to check that the transverse part $z_t(t)$ carries the $t^{-3/2}$ asymptotics of the VAF (see e.g. Pomeau and Résibois 1975). Indeed, for long times the q-integral for z_t in equation (6) is dominated by the slowly relaxing region around $q = 0$, where $C_t(qt)/C_t(q0) \simeq \exp(-q^2\nu_t t)$ (ν_t is the kinematic shear viscosity); further $j_1(qa) \simeq qa/3$, which substituted into equation (6) gives the time dependence proportional to $t^{-3/2}$.

On the other hand, in $z_l(t)$ the oscillatory contribution can be separated out if one approximates equation (14) by a parabola $\omega_m - \alpha(q - q_m)^2$ around the maximum of $\omega_l(q)$, then writes $q = q_m$ in all factors depending smoothly on q, finally turns from q to the integration variable $\omega_l(q) \equiv \omega$ and extends the integration from $-\infty$ to ω_m. The result is

$$z_l(t)\,|_{\rm osc} = \langle v_0^2\rangle (2/3\sqrt{\pi})\, a^2 q_m j_1(q_m a)\, \alpha^{-1/2}$$
$$\times\, t^{-1/2} \exp(-t/2\tau_l(q_m)) \left[\cos\left(\omega_m t - \frac{\pi}{4}\right) - (2\tau_l(q_m)\,\omega_m)^{-1} \sin\left(\omega_m t - \frac{\pi}{4}\right)\right]. \qquad (17)$$

The oscillation is caused by the $(\omega_m - \omega)^{-1/2}$ singularity introduced by the change of variables $q \to \omega$ (Geszti 1976)†.

Figures 1 and 2 present the results obtained by numerical evaluation of the integrals (6). In one set of calculations the input data were chosen to give a possibly close representation of liquid argon: we used $a = 1{\cdot}83 \times 10^{-8}$ cm, $\tau_t(0) = 2{\cdot}0 \times 10^{-13}$ s, $\tau_l(0) = 1{\cdot}8 \times 10^{-13}$ s (Zwanzig and Bixon 1970); $\hat{\omega}_t(q)$, $\hat{\omega}_l(q)$ and $\omega_0(q)$ have been taken from the numerical calculations of Ailawadi *et al* (1971). Figure 1 shows the results for $z(t)$ and its longitudinal and transverse parts, along with the MD results of Rahman (1964). In view of the roughness of the model the agreement can be considered quite encouraging. Two features are recommended to the reader's attention: (i) the very strongly damped oscillatory character of $z_l(t)$; (ii) the strong negative part of $z_t(t)$. This latter feature will be discussed in §5.

Figure 2 presents the same quantities evaluated for a model metal. We have tried to pin down 'metallic behaviour' of the VAF to the single quantity $\tau_l(q)$ or (in view of the approximation (13)) to the parameter $\tau_l(0)$. Accordingly, we took the same input data as in figure 1, except for $\tau_l(0) = 1{\cdot}8 \times 10^{-12}$ s, i.e. ten times longer than for argon. The

† In Geszti (1976) the reference to Randolph (1964) is misleading. The maximum longitudinal frequency ω_m for sodium can be obtained by assuming similarity of the pair-potentials of Na and Rb. Then for equal reduced temperatures the ratio ω_m/cq_0 (c is the sound velocity and q_0 the wave vector of the main peak in the structure factor) should have the same value for both liquids. Substituting known values, this gives $\omega_m(\mathrm{Na}) \simeq 19 \times 10^{12}\,\mathrm{s}^{-1}$.

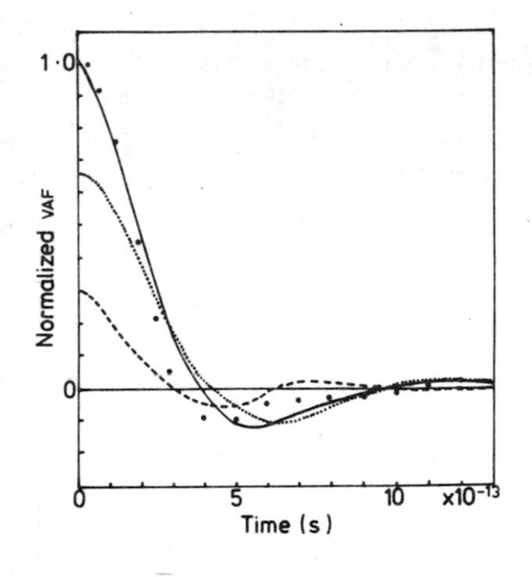

Figure 1. Calculated VAF and its longitudinal and transverse parts for liquid argon. Full curve: $z(t)/z(0)$; broken curve: $z_l(t)/z(0)$; dotted curve: $z_t(t)/z(0)$. Full circles: MD results after Rahman (1964).

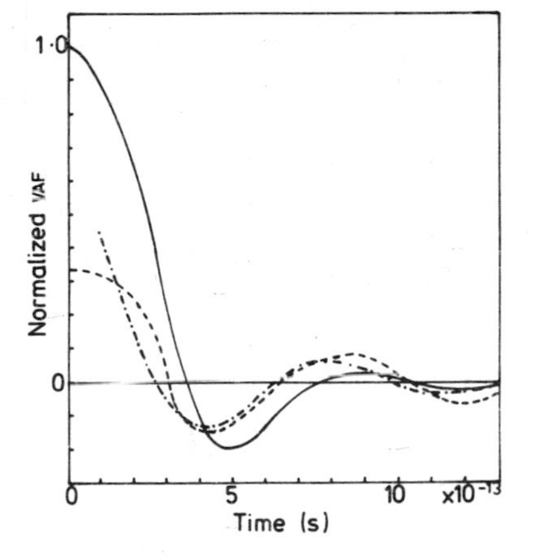

Figure 2. Calculated VAF and its longitudinal part for a model metal: $\tau_l(0)$ ten times longer than in figure 1, otherwise the same input data. Full curve: $z(t)/z(0)$; broken curve: $z_l(t)/z(0)$; chain curve: formula (17).

resulting curves show the strongly enhanced oscillatory character. The function (17) is also shown in figure 2 to illustrate that the Van Hove singularity mechanism really 'explains' the oscillations in its way.

5. Discussion

The present calculation gives good support to the qualitative picture of VAF oscillations (Geszti 1976). In that paper we exploited the connection between potential core softness and oscillatory long-time tail found by Schiff (1969). Let us mention here that Schommers (1975) arrives at the conclusion that the long-range part of the potential is very relevant in determining the oscillatory character of the VAF. This conclusion is opposite to that of Schiff. However, Schommers cuts the long-range part of the potential in such a way as to destroy also the smooth potential well around the first neighbour distance, an absolutcly essential short-range feature of the potential (cf. Schofield 1973). In our opinion this decides in favour of Schiff's conclusion.

Let us turn to the new conclusions emerging from the present work. Firstly, the approximate separation of the VAF into a longitudinal and a transverse part seems to correspond fairly well to reality. It is definitely different from, and in our view more instructive than, Rahman's (1964) separation into a 'rattling' and a 'slipping' part. An unsatisfactory feature of the longitudinal–transverse separation is that at present it is connected to the modified Alder–Wainwright model; it would be desirable to define it as a more direct approximation to exact dynamics.

Secondly, the strong negative part of $z_t(t)$ shows that the initial rebound of the moving particle is caused at least as much by shear viscoelasticity as by longitudinally elastic behaviour. These two aspects are somewhat confused in the tunnel model (Zwanzig and Bishop 1974), where 'motion along the tunnel' corresponds to shear flow, but the negative part of the VAF is caused by the row of particles ahead in the same tunnel, which must be classified longitudinal. The present treatment brings some clarification, showing that the particle is also viscoelastically pulled back by the 'walls of the tunnel'.

Finally, one remark of philosophy is in place. The Vineyard (1958) convolution approximation and its numerous refinements (e.g. Ortoleva and Nelkin 1970) tried to build up a picture of collective motion, starting from known data on individual particle behaviour. We think that there is much more physical basis for looking in the opposite direction: starting from collective behaviour which is phenomenologically quite simple (although not simple to explain), to construct a picture of the complexities of individual particle motion; in particular, the VAF. The Alder–Wainwright (1970) and Zwanzig–Bixon (1970) theories were the first steps in this direction; the present work is one more.

Acknowledgment

We are indebted to Dr T Vicsek for helpful advice on numerical work.

References

Ailawadi N K, Rahman A and Zwanzig R 1971 *Phys. Rev.* **A4** 1616–25
Alder B J and Wainwright T E 1970 *Phys. Rev.* **A1** 18–21
Chung C H and Yip S 1969 *Phys. Rev.* **182** 323–39
Copley J R D and Rowe J M 1974 *Phys. Rev. Lett.* **32** 49–52
Geszti T 1976 *J. Phys. C: Solid St. Phys.* **9** L263–5
Kertész J 1976 *Acta Phys. Hung.* in the press
Morse P M and Feshbach H 1953 *Methods of Theoretical Physics* (New York: McGraw-Hill) pp 1864–7 and 1898–901
Ortoleva P and Nelkin M 1970 *Phys. Rev.* **A2** 187–95
Paskin A 1967 *Adv. Phys.* **16** 223–40
Pomeau Y and Résibois P 1975 *Phys. Rep.* **C19** 63–139
Rahman A 1964 *Phys. Rev.* **A136** 405–11
—— 1974 *Phys. Rev. Lett.* **32** 52–4
Schiff D 1969 *Phys. Rev.* **186** 151–9
Schofield P 1973 *Comput. Phys. Commun.* **5** 17–23
Schommers W 1972 *Z. Phys.* **257** 78–91
—— 1975 *Solid St. Commun.* **16** 45–7
Vineyard G H 1958 *Phys. Rev.* **110** 999–1010
Zwanzig R and Bishop M 1974 *J. Chem. Phys.* **60** 295–6
Zwanzig R and Bixon M 1970 *Phys. Rev.* **A2** 2005–12

Self-consistent calculation of the coherent scattering function of liquid rubidium

J Bosse, W Götze and M Lücke
Max-Planck-Institut für Physik and Physik-Department der Technischen Universität, München, West Germany

Abstract. Within the framework of Mori's theory, the longitudinal and transverse current–current correlation functions of simple classical liquids are expressed in terms of restoring forces and frequency-dependent relaxation kernels. The spectra of the relaxation kernels are approximated by the mode-coupling approximation suggested by Götze and Lücke (1975). The three-particle distribution function entering the decay vertex is treated in the Kirkwood superposition approximation and, in contrast to Götze and Lücke, the full angular anisotropy of the vertex is taken into account. The resulting nonlinear integral equations are solved by iteration for liquid rubidium parameters, and the coherent scattering law $S(k, \omega)$ derived from the obtained longitudinal current correlation function is compared with neutron scattering experiments (Copley and Rowe 1974a, b) and computer simulations (Rahman 1974a, b). The theory leads, in the hydrodynamical limit, to a reasonable value for the viscosity, and the ideal gas limit is included.

References

Copley J R D and Rowe J M 1974a *Phys. Rev.* **A9** 1656
—— 1974b *Phys. Rev. Lett.* **32** 49
Götze W and Lücke M 1975 *Phys. Rev.* **A11** 2173
Rahman A 1974a *Phys. Rev. Lett.* **32** 52
—— 1974b *Phys. Rev.* **A9** 1667

The excess viscosity of liquid binary alloys

Z Morita, T Iida and M Ueda
Department of Metallurgy, Faculty of Engineering, Osaka University, Suita, Osaka 565, Japan

Abstract. An expression describing the excess viscosity of liquid binary alloys has been derived by use of some basic physical quantities, in order to estimate the viscosity of liquid alloys (considered to be one of the most important properties in metallurgical processes). Values calculated from the equation coincided qualitatively with experimental data, and for regular or nearly regular solutions in particular, an excellent agreement has been found between calculated and experimentally observed results.

1. Introduction

The basic behaviour of various elements in extraction and refining processes is dominated by their interaction strength. From the microscopic point of view, the viscosity of alloys is a physical property representing the strength of interaction among unlike components. Therefore, the viscosity of liquid metals, alloys and slags is of considerable importance in metallurgical processes. Some experimental studies have been reported on viscosities of liquid binary alloys, but it seems that viscosity measurements of transition metals and alloys are especially difficult because of their high reactivity and melting points. Thus their experimental data are scattered. On the other hand, although numerous attempts have been made to describe the viscosities of liquid binary alloys in terms of the viscosities of their components and a few other parameters attributed to interaction between them, there are none that can be regarded as satisfactory.

In this paper, an expression to describe the excess viscosity has been derived by use of some known microscopic parameters, in order to estimate viscosities of liquid binary alloys which have not yet been studied experimentally. The excess viscosity is considered to be approximately proportional to the strength of the interaction between unlike components and the equation is expressed with simple physical quantities, rigorous physical treatment being sacrificed to a certain extent.

2. Derivation of excess viscosity for liquid binary alloys

In the liquid state, the regular arrangement of ions is destroyed and the ions undergo Brownian motion according to the Langevin equation of motion. In this treatment, the motion of a given ion is assumed to be impeded by the frictional forces of its neighbours. The friction constant is related to the viscosity; for example, by Stokes's law, the friction constant ζ, for a sphere of radius a, moving in a medium of viscosity η, is $\zeta = 6\pi\eta a$. The friction constant, ζ, has two parts (Helfand 1961, Rice 1961):

$$\zeta = \zeta^{h} + \zeta^{s} \tag{1}$$

where ζ^h and ζ^s are hard and soft parts respectively. (The cross effect between the hard and soft parts is neglected.)

Now we suppose that the viscosity, η, also has two parts and similar relations to Stokes's law hold between the friction constant (ζ^h, ζ^s) and the viscosity (η^h, η^s). The assumptions are as follows:

$$\eta = \eta^h + \eta^s \tag{2}$$

and,

$$\eta^h = \zeta^h/(nr)^h, \qquad \eta^s = \zeta^s/(nr)^s \tag{3}$$

where η^h and η^s are hard and soft parts of the viscosity respectively, n is a constant and r a quantity which has dimension of length estimated to be the order of 2 to 4 Å.

2.1. *Derivation of η^h*

The contribution to the friction constant from the hard-core collisons is given by Helfand (1961):

$$\zeta^h = \frac{8}{3}\rho\sigma^2 g(\sigma)(\pi mkT)^{1/2} \tag{4}$$

where ρ is number density, σ the diameter of the sphere (radius of the exclusion sphere) and $g(\sigma)$ the contact radial distribution function. Eliminating ζ^h from equations (3) and (4), we have

$$\eta^h = \frac{8}{3}\frac{1}{(nr)^h}\rho\sigma^2 g(\sigma)(\pi mkT)^{1/2}. \tag{5}$$

Next we consider isothermal excess viscosity: in the above equation, we estimate tentatively that the effective hard-sphere diameter, σ, and atomic mass, m, may be main factors for the excess viscosity from hard-sphere interaction.

2.2. *Contribution to the excess viscosity from hard-sphere collisions*

2.2.1. Sphere size. We assume that except for σ, all the factors in equation (5) are constant regardless of the type of metal. That is,

$$\frac{8}{3}\frac{1}{(nr)^h}\rho g(\sigma)(\pi mkT)^{1/2} = \frac{4}{3}\frac{1}{(nr)^h}\rho g(\sigma)\left(\frac{mkT}{\pi}\right)^{1/2} 2\pi = 2\pi K\hat{\sigma} = \text{constant}$$

or,

$$\eta^h_\sigma = 2\pi K_\sigma \sigma^2. \tag{6}$$

We consider the excess viscosity of a liquid binary mixture of metals 1 and 2 whose viscosities are given by equation (6). Assuming a completely random state, the viscosity of a binary hard-sphere fluid, η_σ^h, would be

$$\eta_\sigma^h = 2\pi K_\sigma\left[x_1^2\sigma_1^2 + 2x_1x_2\left(\frac{\sigma_1+\sigma_2}{2}\right)^2 + x_2^2\sigma_2^2\right]. \quad (7)$$

If we denote the additive viscosity by $\eta_{\sigma\,add}^h$, the excess viscosity that is examined here is given by

$$\Delta\eta_\sigma^h = \eta_\sigma^h - \eta_{\sigma\,add}^h = -\pi K_\sigma x_1x_2(\sigma_1-\sigma_2)^2 = -\pi K_\sigma x_1x_2(\Delta\sigma)^2. \quad (8)$$

The above equation implies that $\Delta\eta_\sigma^h$ tends to become less negative as the difference between σ_1 and σ_2 increases.

From the following treatment, we have a similar expression to equation (8):

$$\Delta\eta_\sigma^h = \eta_\sigma^h - \eta_{\sigma\,add}^h = 2\pi K_\sigma[(x_1\sigma_1+x_2\sigma_2)^2 - (x_1\sigma_1^2+x_2\sigma_2^2)] = -2\pi K_\sigma x_1x_2(\Delta\sigma)^2. \quad (9)$$

To calculate the excess viscosity, it is convenient to rewrite equation (8) in the form of the product of additive viscosity, $(x_1\eta_1+x_2\eta_2)$, by a dimensionless parameter:

$$\Delta\eta_\sigma^h = -\alpha(x_1\eta_1+x_2\eta_2)\frac{x_1x_2(\Delta\sigma)^2}{x_1\sigma_1^2+x_2\sigma_2^2} \quad (10)$$

where $\alpha = \pi K_\sigma(x_1\sigma_1^2+x_2\sigma_2^2)/(x_1\eta_1+x_2\eta_2)$, x is the atomic fraction ($x_1+x_2=1$), and α can be assumed to be a function of the types of metal and the composition.

2.2.2. Atomic mass. We assume that all the factors except for m in equation (5) are constant regardless of the types of metal. That is,

$$\frac{8}{3}\frac{1}{(nr)^h}\rho\sigma^2g(\sigma)(\pi kT)^{1/2} = K_m = \text{constant}.$$

The contribution from the atomic mass to the excess viscosity of a binary hard-sphere fluid could be expressed in the form:

$$\eta_m^h = K_m(x_1m_1+x_2m_2)^{1/2}. \quad (11)$$

On the other hand, the additive viscosity, $\eta_{m\,add}^h$, of the alloy examined here, is given by

$$\eta_{m\,add}^h = K_m[x_1m_1^{1/2}+x_2m_2^{1/2}]. \quad (12)$$

As mentioned above, it is convenient to introduce a parameter defined by

$$\beta = \frac{K_m[x_1m_1^{1/2}+x_2m_2^{1/2}]}{x_1\eta_1+x_2\eta_2}. \quad (13)$$

Combining equations (11), (12) and (13), we get

$$\Delta\eta_m^h = \eta_m^h - \eta_{m\,add}^h = \beta(x_1\eta_1+x_2\eta_2)\left[\left(1+\frac{x_1x_2(\Delta m^{1/2})^2}{(x_1m_1^{1/2}+x_2m_2^{1/2})^2}\right)^{1/2} - 1\right]. \quad (14)$$

In this case, the excess viscosity, $\Delta\eta_m^h$, tends to become positive as the difference between m_1 and m_2 increases.

If the contribution to the excess viscosity from hard-sphere collisions could be split into two terms ($\Delta\eta_\sigma^h$, $\Delta\eta_m^h$), that is if the contribution to $\Delta\eta^h$ from the hard-sphere size is independent of the atomic mass, and the contribution to $\Delta\eta^h$ from the atomic mass is independent of the hard-sphere size, then the excess viscosity of a binary hard-sphere fluid, $\Delta\eta^h$, could be expressed by the sum of $\Delta\eta_\sigma^h$ and $\Delta\eta_m^h$.

2.3. *Derivation of* η^s

η^h is only the first part of the viscosity and the second part, which is not due to hard-core interaction, must be considered. The second part denoted by η^s corresponds to the soft part of the potential between unlike atoms. At present, however, the form of the pair potential is not known in detail and therefore we consider η^s from a different point of view. Moelwyn-Hughes (1964) describes the excess viscosity of a binary mixture by the equation:

$$\Delta\eta = -2(x_1\eta_1 + x_2\eta_2)\frac{x_1x_2\Delta u}{kT} = -2(x_1\eta_1 + x_2\eta_2)\frac{\Delta H}{RT} \tag{15}$$

where ΔH is the integral enthalpy of mixing ($\Delta H = x_1x_2N\Delta u$, where N is Avogadro's number). The agreement between values calculated from equation (15) and experimental data is not very satisfactory (Crawley 1972, Rialland and Perron 1974). The reason for this discrepancy is considered as follows. In many alloy systems, the excess viscosity can not be expressed in a simple equation which considers only the thermodynamic property. The hard-sphere collisions are not taken into account sufficiently by Moelwyn-Hughes (1964) and therefore the part due to hard-core interactions must be added. If we denote equation (15) by $\Delta\eta_0 = \Delta\eta^s + a\Delta\eta^h$, and take $\Delta\eta^h = a\Delta\eta^h + b\Delta\eta^h \ldots$, then, according to the law of corresponding states, we assume that $a\Delta\eta^h = c\Delta\eta_0$, where a, b, c, are constants with values between 0 and 1. From the above equations, we have

$$\Delta\eta^s = \gamma\Delta\eta_0 = -2\gamma(x_1\eta_1 + x_2\eta_2)\frac{x_1x_2\Delta u}{kT} \tag{16}$$

where $\gamma = 1 - c$.

2.4. *Evaluation of the unknown parameters* (α, β, γ)

From the above discussion, the excess viscosity of liquid binary alloys is expressed as follows:

$$\Delta\eta = \Delta\eta^h + \Delta\eta^s = \Delta\eta_\sigma^h + \Delta\eta_m^h + \Delta\eta^s = (x_1\eta_1 + x_2\eta_2)\left\{-\frac{\alpha x_1x_2(\Delta\sigma)^2}{x_1\sigma_1{}^2 + X_2\sigma_2{}^2} + \beta\left[\left(1 + \frac{x_1x_2(\Delta m^{1/2})^2}{(x_1m_1{}^{1/2} + x_2m_2{}^{1/2})^2}\right)^{1/2} - 1\right] - \frac{2\gamma x_1x_2\Delta u}{kT}\right\}. \tag{17}$$

The above equation is based on a variety of approximations and assumptions, the cumulative effect of which is uncertain. It is very difficult to evaluate the parameters strictly from a theoretical point of view. Therefore, we use a method of obtaining values of the parameters from experimental viscosity data. On the basis of experimental values of Au–Ag (1100 °C) (Gebhardt and Becker 1951), Na–K (104 °C) (Ewing *et al* 1951) and Sb–Bi (700 °C) (Iida 1970) at 50 at.%, the parameters have been evaluated by solving a set of three linear equations and we have $\alpha = 5 \times 10^{-2}$, $\beta = 2 \times 10^{-2}$, and $\gamma = 6 \times 10^{-4}$. To evaluate the parameters: for σ, we used ionic radius (after Pauling); for m, we used atomic weight divided by Avogadro's number; and for ΔH, we used values quoted from references (Hultgren *et al* 1963, Fuwa *et al* 1972). A comparison with experiment shows that the parameters are approximately constant over the complete composition range. In addition, from the law of corresponding states, the parameters are expected to be nearly constant for many liquid binary alloys. Finally, equation (17) becomes

$$\Delta\eta = (x_1\eta_1 + x_2\eta_2)\left\{-\frac{5x_1x_2(\Delta\sigma)^2}{x_1\sigma_1{}^2 + x_2\sigma_2{}^2} + 2\left[\left(1 + \frac{x_1x_2(\Delta m^{1/2})^2}{(x_1m_1{}^{1/2} + x_2m_2{}^{1/2})^2}\right)^{1/2} - 1\right] - \frac{0{\cdot}12x_1x_2\Delta u}{kT}\right\} \tag{18}$$

where $x_1x_2N_0\Delta u = \Delta H$ and $\Delta\eta$ is in cP.

For the case of a regular solution ($\Delta H = RT(x_1 \ln \gamma_1 + x_2 \ln \gamma_2)$, where γ is the activity coefficient), equation (18) can be expressed as follows:

$$\Delta\eta = (x_1\eta_1 + x_2\eta_2)\left\{-\frac{5x_1x_2(\Delta\sigma)^2}{x_1\sigma_1{}^2 + x_2\sigma_2{}^2} + 2\left[\left(1 + \frac{x_1x_2(\Delta m^{1/2})^2}{(x_1m_1{}^{1/2} + x_2m_2{}^{1/2})^2}\right)^{1/2} - 1\right] - 0{\cdot}12(x_1 \ln \gamma_1 + x_2 \ln \gamma_2)\right\}. \tag{19}$$

3. Comparison between the observed and calculated values

The results of calculations based on equations (18) and (19) are shown together with experimental data in figures 1, 2 and 3. These figures show some of the results of

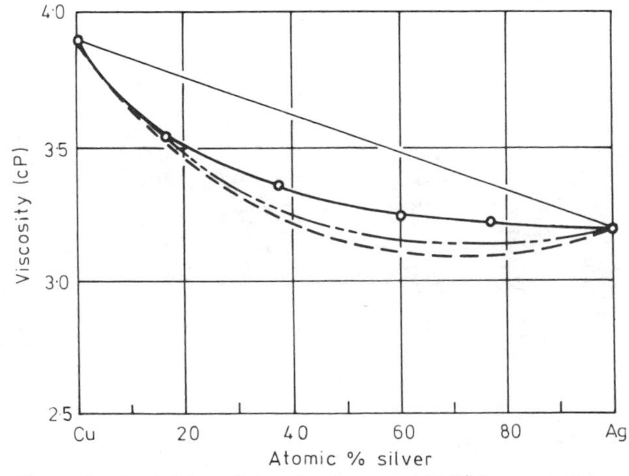

Figure 1. Viscosities of Cu–Ag alloys at 1100 °C; ○ experimental values (Gebhardt and Wörwag 1953); ------ calculated from equation (18); — -- — calculated from equation (19).

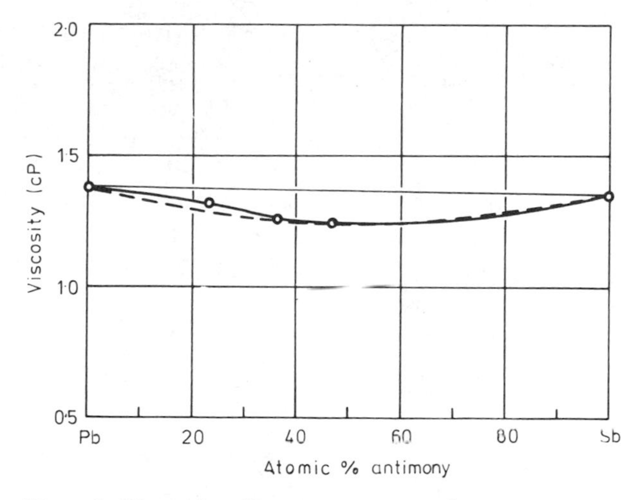

Figure 2. Viscosities of Pb–Sb alloys at 700 °C; o experimental values (Gebhardt and Köstlin 1957); ------ calculated from equation (18).

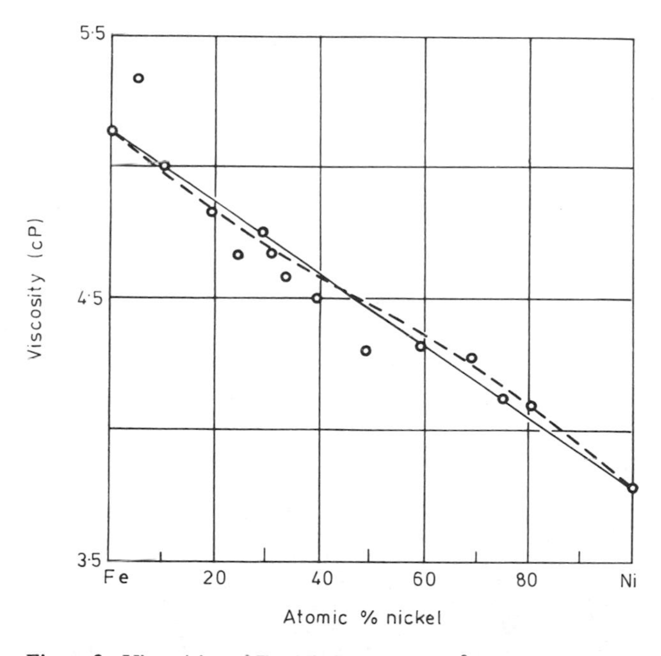

Figure 3. Viscosities of Fe–Ni alloys at 1600 °C; o experimental values (present work); ------ calculated from equation (18).

calculations for various liquid binary alloys which are regular or nearly regular solutions. As obvious from the figures, the agreement between calculation and experiment is very good. However, there are major difficulties with the theory in that it cannot describe big departures from regular solution theory.

4. Summary

In order to estimate the viscosities of liquid binary alloys from a practical standpoint, an equation describing the excess viscosity has been derived by use of some microscopic parameters. Values calculated from the equation coincided qualitatively with experimental results, and in particular, for regular or nearly regular solutions, a good agreement between observed and calculated values was obtained. By linking viscosity with thermodynamic properties, the equation is expected to be possibly useful for the purpose of evaluating the atomic or ionic interaction between components (especially unlike components), in various liquid binary alloys.

References

Crawley A F 1972 *Metall. Trans.* **3** 971
Ewing C T, Grand J A and Miller R R 1954 *J. Phys. Chem.* **73** 1086
Fuwa T, Ban-ya S, Iguchi Y, Tozaki Y and Kakizaki M 1972 *Proc. 8th Symp. Physical Properties of Molten Iron and Slags* (The Iron and Steel Institute of Japan) (in Japanese)
Gebhardt E and Becker M 1951 *Z. Metallkd.* **42** 111
Gebhardt E and Köstlin K 1957 *Z. Metallkd.* **48** 636
Gebhardt E and Wörwag G 1953 *Z. Metallkd.* **44** 385
Helfand E 1961 *Phys. Fluids* **4** 681
Hultgren R, Orr R L, Anderson P P and Kelley K K 1963 *Selected Values of Thermodynamic Properties of Metals and Alloys* (New York: Wiley)
Iida T 1970 *D.Eng. Thesis* Tohoku University (in Japanese)
Moelwyn-Hughes E A 1964 *Physical Chemistry* (Oxford: Pergamon)
Rialland J F and Perron J C 1974 *Metall. Trans.* **5** 2401
Rice S A 1961 *Molec. Phys.* **4** 305

Spin-lattice relaxation of the β-emitter ^{8}Li in liquid ^{7}Li

P Heitjans, H Ackermann, D Dubbers, M Grupp and H-J Stöckmann
Physikalisches Institut der Universität Heidelberg, Heidelberg, West Germany and
Institut Laue–Langevin, Grenoble, France

Abstract. The asymmetric β-radiation of polarized ^{8}Li ($T_{1/2}$ = 0·84 s) nuclei, produced by capture of polarized thermal neutrons, was used to measure the spin–lattice relaxation time T_1 of ^{8}Li in liquid ^{7}Li between the melting point and 1100 K in fields B_0 = 0·01–0·7 T. T_1 proved to be field independent. Contrary to the temperature dependence in solid Li, where we found T_1T = 292·7 ± 5·2 sK over a wide range of T, in liquid Li T_1T decreases continuously down to 236 sK at 1100 K. It is argued that on the one hand contributions to T_1 in the liquid additional to that due to the magnetic hyperfine interaction in pure Li can probably be excluded. On the other hand, the available experimental values of the Knight shift K seem to be practically temperature independent so that the Korringa product K^2T_1T must decrease with T. Thus, in liquid Li, there is some indication of a mechanism contributing differently to T_1 and K, although for a definite statement Knight-shift data at higher temperatures are needed.

1. Introduction

This study deals with the measurement of the nuclear spin–lattice relaxation time T_1 in liquid Li up to 1100 K by use of a nuclear radiation detection technique. Until now, measurements by nuclear magnetic resonance (NMR) methods have not been reported for temperatures substantially higher than the melting point, T_M = 454 K, due to the difficulties caused by the skin-effect of this extremely reactive metal. The present technique avoids these problems by observing T_1 on a bulk Li sample without applying any radiofrequency field.

2. Experimental method

Polarized β-active ^{8}Li ($T_{1/2}$ = 0·84 s) nuclei were produced by capture of polarized thermal neutrons in isotopically pure (99·99%) ^{7}Li (from Oak Ridge National Laboratory, metallic impurity <0·05%), enclosed in a stainless steel box. The asymmetric angular distribution of the ^{8}Li β-radiation was used to monitor the polarization of the ^{8}Li nuclear ground state. Thus, by observing transients of the β-radiation asymmetry after neutron activation pulses, the decay constant $1/T_1$ of the ^{8}Li polarization can be determined directly. As the ^{8}Li polarization is not described by a Boltzmann factor, contrary to conventional NMR techniques, the method is well suited to measurements both at high temperatures and low magnetic fields.

3. Results

The relaxation rate T_1^{-1} was measured between the melting point and 1100 K in fields

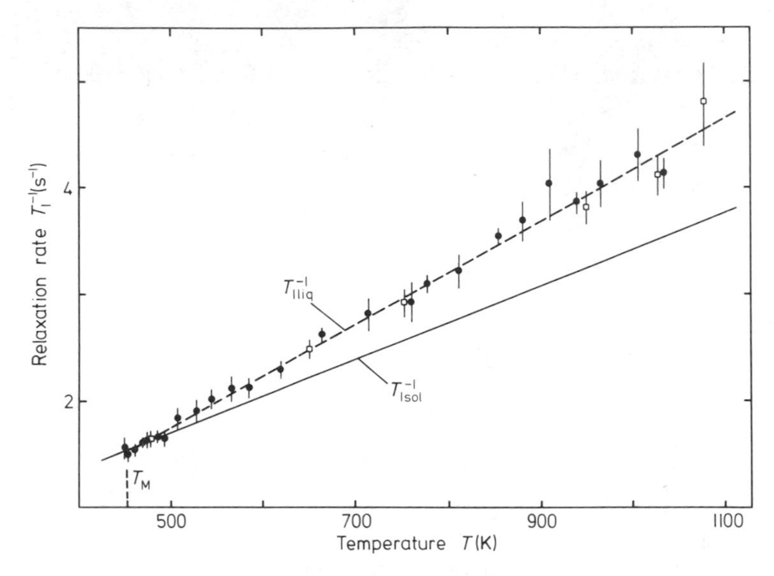

Figure 1. Spin–lattice relaxation rate of ^{8}Li in ^{7}Li as a function of temperature at $B_0 = 0{\cdot}35$ T. The broken line $T_{1\,liq}^{-1}$ represents a linear fit to all the data above the melting point. The full line $T_{1\,sol}^{-1}$ is an extrapolation of the measured relaxation rate of ^{8}Li in solid ^{7}Li due to the hyperfine interaction with the conduction electrons and corresponds to $T_1T = 292{\cdot}7$ sK. ● Li in stainless steel box; □ Li in Mo container.

$B_0 = 0{\cdot}01$–$0{\cdot}7$ T. T_1^{-1} was found to be independent of B_0, and to increase with temperature T (see figure 1). However, the slope of T_1^{-1} against T was greater than in the solid, where we found, for the relaxation of ^{8}Li due to the hyperfine interaction with the conduction electrons, $T_{1sol}^{-1} = T/C$ with $C = 292{\cdot}7 \pm 5{\cdot}2$ sK over a wide temperature range (Ackermann *et al* 1975). To make sure that the additional relaxation rate in the liquid does not originate from corrosion of the stainless steel box, we made some supplementary measurements on Li in a container of molybdenum which is known to be resistant to liquid Li up to at least 1300 K. No difference between the two samples showed up within experimental error. A fit of the function $T_{1\,liq}^{-1} = A + T/B$ to all the experimental values taken for $T > T_M$ at $B_0 = 0{\cdot}35$ T resulted in $A = -0{\cdot}66 \pm 0{\cdot}04\ s^{-1}$ and $B = 207{\cdot}0 \pm 3{\cdot}0$ sK, that is, T_1T decreases (with no discernible step at the melting point) from 293 sK to 236 sK at 1100 K.

We must now attempt to interpret the marked deviation of the temperature dependence of T_1^{-1} in the liquid from that in the solid implied by our data.

4. Discussion

Impurities forming localized moments should give rise to a temperature-independent relaxation rate (e.g., Heeger 1969). However, apart from the fact that no such contribution could be detected in the solid, a temperature-independent relaxation rate cannot explain, at least by itself, the additional rate $T_{1\,liq}^{-1} - T_{1\,sol}^{-1}$ increasing linearly with temperature. Explaining $T_{1\,liq}^{-1} - T_{1\,sol}^{-1}$ tentatively by the diffusion of impurity ions, characterized by a correlation time τ_c long compared to the ^{8}Li Larmor period $1/\omega_0$, has also to be dropped because, in this case, the relaxation rate would be expected to depend on B_0; this was not found, however. Although an impurity effect cannot completely be ruled out, it seems rather likely that the enhanced relaxation rate in the liquid is due to a mechanism intrinsic to pure Li.

As quadrupolar relaxation mechanisms are inprobable in pure Li, primarily because of its small quadrupole shielding factor (which is of the order unity) and the small quadrupole moment of ^{8}Li ($Q = 0{\cdot}032$ b), we are left with the possibility that the magnetic hyperfine interaction is modified with temperature. This could be due to the fact that the conduction electrons which are known to have a large fractional p-character in solid Li (Gaspari *et al* 1964) tend towards increased s-character for higher temperatures in the liquid when the system becomes more isotropic. This argument should apply to the Knight shift K as well. However, currently available Knight-shift data up to 673 K (Dupree and Seymour 1972, Weisman *et al* 1973, Feitsma *et al* 1975) indicate that K, to a good approximation, is independent of temperature. Combining K with our measured $T_1 T$ values, the decrease of the Korringa product $K^2 T_1 T$ with temperature seems to be significant. Nevertheless, extension of the Knight-shift measurements to higher temperatures is desirable in order to check whether there is evidence for one of the proposed mechanisms which are thought to affect T_1 and K differently (Mahanti and Das 1971, T P Das 1976 private communication).

Acknowledgment

We are very grateful to Professor T P Das for helpful comments.

References

Ackermann H, Dubbers D, Grupp M, Heitjans P, Messer R and Stöckmann H-J 1975 *Phys. Stat. Solidi* **B71** K91

Dupree R and Seymour E F W 1972 *Liquid Metals* ed S Z Beer (New York: Dekker) p461

Feitsma P D, Slagter G K and van der Lugt W 1975 *Phys. Rev.* **B9** 3589

Gaspari G D, Shyu W M and Das T P 1964 *Phys. Rev.* **134** A852

Heeger A J 1969 *Solid St. Phys.* **23** 283 (New York: Academic Press)

Mahanti S D and Das T P 1971 *Phys. Rev.* **B3** 1599

Weisman I D, Swartzendruber L J and Bennet L H 1973 *Measurements of Physical Properties* vol VI/6 ed E Passaglia (New York: Wiley) p165

Sternheimer antishielding and fluctuating electric field gradients in liquid metals

Walter Schirmacher
Fachbereich Physik, Freie Universität Berlin, Berlin, Germany

Abstract. The Sternheimer antishielding effect plays an important role in nuclear quadrupole relaxation in liquid metals. The effect is examined using pseudopotential theory and the variational procedure of Das and Bersohn. The calculations which are performed for Ga and In show that the electric field gradient due to shielding depends on the details of the expressions used for the electron screening density. The calculated effective antishielding factor which enters quadrupole relaxation theory turns out to be greater than the ionic antishielding factor $1 - \gamma_\infty$ for Ga and smaller for In.

1. Introduction

Nuclear quadrupole measurements are a useful method for studying electric field gradients (EFG) in liquid metals. This is evidenced by an increasing number of NMR and angular correlation experiments (see e.g. Sholl 1974 or Riegel 1975 for references) which show that the electric quadrupole interaction is an important part of the interaction of nuclei with their environment.

The theoretical understanding of the quadrupolar relaxation process has been appreciably improved in the course of the last ten years. The pioneering work in this field was that of Sholl (1967) who suggested that the EFG fluctuations arise from the dynamic motion of the particles in the liquid. The systematic study of the temperature dependence of quadrupolar relaxation in pure metals and of its composition dependence in alloys confirm this suggestion, although the question as to which type of molecular motions produce the EFG fluctuations has not yet finally been answered.

In any case it has become clear that quadrupole relaxation measurements can provide information about the dynamical and structural behaviour of the liquid. Later theories (Sholl 1974, Warren 1974, Gabriel 1974, Gabriel and Schirmacher 1974) have shown that the quadrupolar relaxation rate R_Q depends on approximate expressions containing the Van Hove coherent and incoherent scattering laws $S(\mathbf{k}, \omega)$ and $S_s(\mathbf{k}, \omega)$:

$$R_Q \propto \sum_{q=-2}^{+2} \int_{-\infty}^{+\infty} d^3\mathbf{k} \int_{-\infty}^{+\infty} d\omega \, |\tilde{u}_q(\mathbf{k})|^2 S(\mathbf{k}, \omega) \, S_s(\mathbf{k}, \omega). \tag{1}$$

Where $\tilde{u}_q(\mathbf{k})$ is the Fourier transform of the EFG due to the presence of a particular ion at the site $\mathbf{r}_\mu$ multiplied by the radial distribution function $g(r_\mu)$:

$$\tilde{u}_q(\mathbf{r}_\mu) = g(r_\mu) \, u_q(\mathbf{r}_\mu). \tag{2}$$

For a comparison of this theory with experiment it is of crucial importance to find reliable expressions for the EFG $u_q(\mathbf{r}_\mu)$ produced by an individual moving ion in the liquid metal. It is well known (see Sternheimer 1974 for references) that the main contribution to the EFG seen by the nucleus comes from the deformation of the ionic core caused by external fields (Sternheimer antishielding). This effect has been treated in the past by multiplying the external electric field gradients by Sternheimer's antishielding factor $1-\gamma_\infty$ which has been calculated for free atoms or ions. It has been pointed out (Sholl 1974) that this may be incorrect, but until now there has been no attempt to study the antishielding effect in liquid metals more carefully. EFG calculations in solid metals including the antishielding effect have been performed (Das 1975) and they are strongly dependent on the details of the band structure. Comparison of the measured EFG with simple pseudopotential calculations in solid metals shows (Quitman *et al* 1974) that the metallic antishielding effect should be considerable larger than the ionic one.

There are two methods of considering the Sternheimer antishielding effect and these have been shown to be equivalent (Sternheimer 1966). According to the first method the quadrupole moment ΔQ produced by the perturbation of the nuclear quadrupole moment Q_0 on the core interacts with the external EFG u^{ext}. In the second method – which is applied in the present paper – the EFG Δu caused by the perturbation of u^{ext} on the core interacts with the nuclear quadrupole moment. The ratio $\Delta Q/Q_0 = \Delta u/u^{\text{ext}} = -\gamma$ can be shown to be a constant, $-\gamma_\infty$, independent of the details of the external charge distribution provided the charges stay completely outside of the core. Therefore the atomic antishielding factor calculated for free atoms can be taken for ionic crystals as well.

The situation is different in a liquid metal, however, because the conduction electrons penetrate into the core and the ionic charges can come so close to the core that they can overlap with the edge of the core density. The effect of point charges that continuously come nearer to the core region has been calculated by Foley *et al* (1954) who replace the constant γ_∞ by a function $\gamma(r)$. The present contribution is an attempt to include also the effect of the conduction electrons. In §2 we present the expression for the perturbing 'external' EFG that has been derived by Schirmacher (1976). In §3 the variational procedure of Das and Bersohn (1956) is applied to calculate metallic antishielding functions which are discussed in §4.

2. Electric field gradients

In this section we briefly review the derivation of the non-antishielding EFG (Schirmacher 1976, hereafter referred to as I) which we call the 'external' EFG $u_q^{\text{ext}}(r_\mu)$ although the corresponding charges need not stay outside the core. The starting point is the general expression for the EFG:

$$\begin{aligned} u_q(\mathbf{r}_\mu) &= 2\left(\frac{4\pi}{5}\right)^{1/2}\int \mathrm{d}^3\mathbf{r}\,\Delta\rho(\mathbf{r},\mathbf{r}_\mu)\,r^{-3}\,Y_{2q}(\hat{\mathbf{r}}) \\ &= 2\left(\frac{4\pi}{5}\right)^{1/2}\int \mathrm{d}^3\mathbf{r}\,(\Delta\rho^{\text{ion}}+\Delta\rho^{\text{cond}}+\Delta\rho^{\text{core}})\,r^{-3}\,Y_{2q}(\hat{\mathbf{r}}) \\ &= u_q^{\text{ion}}(\mathbf{r}_\mu)+u_q^{\text{cond}}(\mathbf{r}_\mu)+\Delta u_q(\mathbf{r}_\mu) = u_q^{\text{ext}}(\mathbf{r}_\mu)+\Delta u_q(\mathbf{r}_\mu). \end{aligned} \quad (3)$$

Where $\hat{\mathbf{r}}$ is the normal vector belonging to $\mathbf{r}$ and $Y_{2q}(\hat{\mathbf{r}})$ is the second-order spherical harmonic; $\Delta\rho(\mathbf{r},\mathbf{r}_\mu)$ is the deviation of the total extra-nuclear charge density from its average value caused by the instantaneous presence of the ions at the sites $\mathbf{r}_\mu$. The ionic charges $\Delta\rho^{\mathrm{ion}}(\mathbf{r},\mathbf{r}_\mu)$ are approximated by point charges $Z\delta(\mathbf{r}-\mathbf{r}_\mu)$ and the conduction-electron charges are treated using pseudopotential theory (Harrison 1966):

$$\rho^{\mathrm{cond}}(\mathbf{r},\mathbf{r}_\mu) = \sum_{\mathbf{q}} \exp[\mathrm{i}\mathbf{q}(\mathbf{r}-\mathbf{r}_\mu)] \frac{4}{\epsilon(q)} \sum_{\mathbf{k}} \frac{\langle \mathbf{k}+\mathbf{q} \mid w_0 \mid \mathbf{k}\rangle}{|\mathbf{k}+\mathbf{q}|^2 - k^2}$$

$$= -Z \sum_{\mathbf{q}} \exp[\mathrm{i}\mathbf{q}(\mathbf{r}-\mathbf{r}_\mu)]\, \phi_{\mathrm{N}}^{\mathrm{scr}}(q). \tag{4}$$

(We use atomic units $\hbar = e = m_{\mathrm{e}} = 1$; energies are given in Rydbergs.) w_0 is a single-ion bare pseudopotential and $\phi_{\mathrm{N}}^{\mathrm{scr}}(q)$ is called the normalized screening potential. Throughout this paper we restrict ourselves to a local pseudopotential. As in I we use Ashcroft's empty-core potential (Ashcroft 1966) and have

$$\phi_{\mathrm{N}}^{\mathrm{scr}}(q) = \cos(R_{\mathrm{c}}q)\,[1-(1/\epsilon(q))], \tag{5}$$

where R_{c} is an adjustable parameter. We define $\epsilon(q)$ by

$$\frac{1}{\epsilon(q)} = 1 + \frac{4\pi}{q^2} \frac{\chi_0(q,k_{\mathrm{F}})}{1-(4\pi/q^2)\,[1-f(q,k_{\mathrm{F}})]\,\chi_0(q,k_{\mathrm{F}})} \tag{6}$$

where k_{F} is the Fermi wavenumber, $\chi_0(q,k_{\mathrm{F}})$ is the Lindhard screening function and $f(q,k_{\mathrm{F}})$ is the correction for exchange effects introduced for example by Kleinman (1967).

In equation (3) the so-called depletion hole charge density (the difference between the conduction electron density calculated by pseudopotential theory and the real one) has been omitted. This is consistent with taking a local pseudopotential. It should be noted, however, that such charges can give additional contributions to the EFG which are neglected in the present treatment. The incorporation of full non-local pseudopotential theory into the shielding calculations is straightforward but cumbersome.

If we define

$$\phi_{\mathrm{N}}(q) = 1 - \phi_{\mathrm{N}}^{\mathrm{scr}}(q) \tag{7}$$

we have

$$u_q^{\mathrm{ext}}(\mathbf{q}) = -\left(\frac{4\pi}{5}\right)^{1/2} \frac{8\pi Z}{3}\, \phi_{\mathrm{N}}(q)\, Y_{2q}(\hat{\mathbf{q}}). \tag{8}$$

It has been pointed out in I that in some of the previous papers on quadrupolar relaxation (Sholl 1973, Gabriel and Schirmacher 1974), u_q^{ext} has been erroneously associated with the interionic pair potential. Sholl (1976) has shown that outside the core the Fourier transform of equation (8) is identical with the expression for u_q^{ext} derived by him in 1967.

The explicit expression which is used in the present calculations is

$$u_q^{\mathrm{ext}}(\mathbf{r}_\mu) = \xi^{\mathrm{ext}}(r_\mu) \frac{2Z}{r_\mu^3} \left(\frac{4\pi}{5}\right)^{1/2} Y_{2q}(\hat{\mathbf{r}}_\mu) \tag{9}$$

with the normalized EFG function

$$\xi^{\text{ext}}(r_\mu) = 1 - \frac{2}{3\pi} r_\mu^3 \int_0^\infty dq q^2 j_2(qr_\mu) \cos(qR_c) \left(1 - \frac{1}{\epsilon}\right). \tag{10}$$

j_l is the lth-order spherical Bessel function.

3. Variational procedure of Das and Bersohn (1956)

To calculate the antishielding effect of the core electrons we apply stationary perturbation theory. The interaction potential of a core electron with the perturbing charge distribution $\Delta\rho^{\text{ion}}(\mathbf{r}, \mathbf{r}_\mu) + \Delta\rho^{\text{cond}}(\mathbf{r}, \mathbf{r}_\mu)$ and its multipole expansion are given by

$$\begin{aligned} \mathscr{H}_1(\mathbf{r}, \mathbf{r}_\mu) &= -2 \int d^3\mathbf{r}^2 \frac{\Delta\rho^{\text{ion}}(\mathbf{r}' - \mathbf{r}_\mu) + \Delta\rho^{\text{cond}}(\mathbf{r}' - \mathbf{r}_\mu)}{|\mathbf{r} - \mathbf{r}'|} \\ &= -\sum_{K=0}^{\infty} h_1^{(K)}(r, r_\mu) P_K(\cos\theta) = \sum_K \mathscr{H}_1^{(K)}. \end{aligned} \tag{11}$$

P_K is the Kth-order Legendre polynomial. The z-axis of the coordinate system has been taken parallel to $\mathbf{r}_\mu$. The radial part of the quadrupole term of equation (11) ($K = 2$) has the form

$$h_1^{(2)}(r, r_\mu) = \frac{20Z}{\pi} \int_0^\infty dq j_2(qr) j_2(qr_\mu) \phi_N^{(q)}. \tag{12}$$

Note that

$$\frac{10}{\pi} \int_0^\infty dq j_2(qr) j_2(qr_\mu) = r_<^2 / r_>^3 .$$

Now we assume that for the unperturbed wavefunctions $\psi_0(\mathbf{r})$ the atomic (or ionic) Hartree–Fock functions can be taken:

$$\psi_0(\mathbf{r}) = \langle \mathbf{r} | \psi_0 \rangle = \sum_{nlm} \frac{1}{r} P_{nl}^{(0)}(r) Y_{lm}(\hat{\mathbf{r}}). \tag{13}$$

The variational principle now consists in minimizing the energy functional

$$\begin{aligned} E &\simeq \frac{\langle \psi_0 + \psi_1 | \mathscr{H}_0 + \mathscr{H}_1 | \psi_0 + \psi_1 \rangle}{\langle \psi_0 + \psi_1 | \psi_0 + \psi_1 \rangle} \\ &\simeq \text{constant} + 2\langle \psi_1 | \mathscr{H}_1 | \psi_0 \rangle + \langle \psi_1 | \mathscr{H}_0 - E_0 | \psi_1 \rangle \\ &= \text{constant} + \phi_1 + \phi_2 \end{aligned} \tag{14}$$

up to second order in the perturbation with respect to the parameters α_ν of the following trial function:

$$\psi_1(\mathbf{r}) = \sum_{\nu=0}^{f} \alpha_\nu r^\nu \langle \mathbf{r} | \mathcal{H}_1^{(2)} | \psi_0 \rangle$$

$$= \sum_{\substack{nlm\lambda \\ [l-\lambda \text{ even}]}} [(2l+1)(2\lambda+1)]^{1/2} \begin{pmatrix} 2 & l & \lambda \\ 0 & -m & m \end{pmatrix} \begin{pmatrix} 2 & l & \lambda \\ 0 & 0 & 0 \end{pmatrix} \frac{1}{r} P_{nl\to\lambda}^{(1)} Y_{\lambda m}^{*}(\hat{\mathbf{r}}) \tag{15}$$

with

$$P_{nl\to\lambda}^{(1)} = \sum_{\nu=0}^{f} \alpha_\nu r^\nu h_1(r, r_\mu) P_{nl}^{(0)}(r) \tag{16}$$

where f is an integer greater than or equal to 0. The so-called radial excitations are represented by the terms $l = \lambda$, the angular ones by $l \neq \lambda$. To meet the orthogonalization condition $\langle \psi_0 | \psi_1 \rangle = 0$ in the radial case a term proportional to the unperturbed function must be subtracted:

$$P_{nl\to l}^{(1)} = \sum_{\nu=0}^{f} \alpha_\nu (r^\nu h_1 - \langle r^\nu h_1 \rangle_{nl}) P_{nl}^{(0)} \tag{17}$$

where

$$\langle X(r) \rangle_{nl} = \int_0^\infty \mathrm{d}r\, X(r)\, [P_{nl}(r)]^2. \tag{18}$$

The corresponding expressions for the functionals ϕ_1 and ϕ_2 can be derived straightforwardly and minimizing the sum $\phi_1 + \phi_2$ with respect to α_ν leads to a system of linear equations which can be solved by means of Kramer's rule.

Now we are able to calculate the EFG due to the perturbed core electrons. In the first order of the perturbation we have

$$\Delta u_0(r_\mu) = -2 \sum_{nl\lambda} \beta_{l\lambda} \sum_{\nu=0}^{f} \alpha_\nu \langle r^{-3} (r^\nu h_1^{(2)} - \langle r^\nu h_1^{(2)} \rangle_{nl} \delta_{l\lambda}) \rangle_{nl} \tag{19}$$

with

$$\beta_{l\lambda} = \frac{4}{5} (2l+1)(2\lambda+1) \begin{pmatrix} 2 & l & \lambda \\ 0 & 0 & 0 \end{pmatrix}^2.$$

By analogy with equation (9) we define the normalized shielding function

$$\Delta\xi(r_\mu) = \frac{r_\mu^3}{2Z} \Delta u_0(r_\mu). \tag{20}$$

The total EFG can now be written as

$$u_q(\mathbf{r}_\mu) = [\xi^{\text{ext}}(r_\mu) + \Delta\xi(r_\mu)] \frac{2Z}{r_\mu^3} \left(\frac{4\pi}{5}\right)^{1/2} Y_{2q}(\hat{\mathbf{r}}_\mu). \tag{21}$$

If for $h_1^{(2)}(r, r_\mu)$ the Coulomb expression $2Zr_<^2/r_>^3$ is taken and r_μ is assumed to be always greater than r, equation (19) reduces to the expression given by Das and Bersohn (1956). For the case of ionic shielding we have therefore

$$\Delta\xi^{\text{ion}} = -\gamma_\infty = -\sum_{nl\lambda\nu} \alpha_\nu\, 2\beta_{l\lambda} \langle r^{-3}[r^{\nu+2} - \langle r^{\nu+2}\rangle \delta_{l\lambda}]\rangle_{nl}. \qquad (22)$$

4. Results and discussion

To find out the optimum number of terms in the trial function ansatz (16), the ionic antishielding factors γ_∞ have been calculated according to (22) for values of f from 0 to 6. The smallest energies were obtained for $f = 2$ and $f = 4$. In table 1 the results for Ga and In are compared with the values given by Sternheimer (1963, 1967). To keep the numerical effort within reasonable limits the shielding functions were calculated with $f = 2$. (In their original paper Das and Bersohn also took $f = 2$.)

Table 1. Ionic antishielding factors calculated with two different trial functions, compared with the values given by Sternheimer (1963, 1967).

	Gallium			Indium		
	f=2	f=4	Stern-heimer	f=2	f=4	Stern-heimer
2p → p	−0·61	0·62	−0·58	−0·31	−0·33	−0·31
3p → p	−6·60	−5·58	−6·42	−3·60	−3·59	−2·13
4p → p				−26·67	−26·56	−15·77
3d → d	−3·81	−3·80	−3·29	−0·50	−0·51	−0·51
4d → d				−9·81	−9·79	−8·40
Radial	−10·02	−10·00	−10·29	−40·27	−40·78	−26·12
Angular	−0·80	−2·26	+0·79	−0·21	−0·61	+2·2
	−11·82	−12·26	−9·5	−40·48	−41·39	−24·9

We now discuss the results of the antishielding calculations for liquid Ga and In based on equation (19). The unperturbed wavefunctions have been taken from Watson and Freeman (1961) and Roetti and Clementi (1974). The calculation of the mean values $\langle X\rangle_{nl}$ have been done by numerical integration. The wavefunctions were truncated at $r = 19{\cdot}2$ a.u. For the calculation of $h_1^{(2)}$ and ξ^{ext} the fast Fourier routine of Cooley and Tukey (1965) has been used. As in the case of the ionic calculations the angular excitations turn out to be negligibly small in comparison with the radial ones. In particular, the p → p excitation of the outermost closed shell produces about 60% of the Sternheimer effect.

In figure 1 the function

$$\xi(r) = \Delta\xi(r) + \xi^{\text{ext}}(r) \qquad (23)$$

is shown together with $\xi^{\text{ext}}(r)$ (multiplied by 10) for In and Ga. For $\Delta\xi$ only the sum of the radial contributions has been taken. The values of k_{F} for Ga and In and of R_{c} for In have been taken from Cohen and Heine (1970). For Ga the value of R_{c} which best fits the liquid resistivity measured by Cusack and Kendall (1960) is 1·05 a.u. The calculated

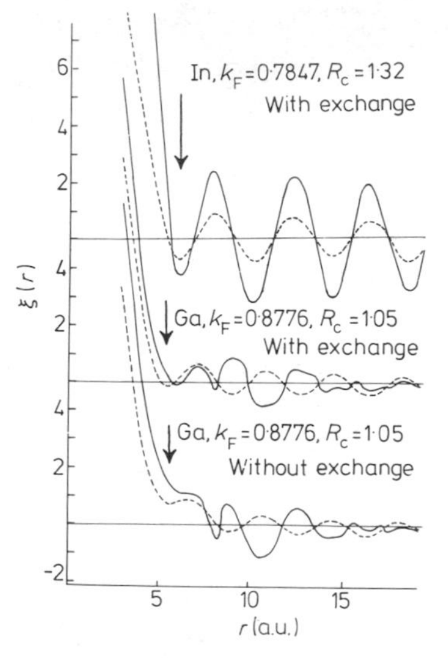

Figure 1. Normalized EFG functions including the radial part of the antishielding effect (full curve) compared with the external EFG function multiplied by 10 (broken curve). The arrow marks the maximum of the pair distribution function $g(r)$.

EFG for Ga turns out to be almost zero in the region of the peak of the pair distribution $g(r)$ (marked by an arrow). This leads to a calculated value of the relaxation rate which differs from the measured one by three orders of magnitude. (The details of the calculation of R_Q will be given in a forthcoming paper.) This discrepancy could not be removed by slight changes of R_c or k_F. If the calculation is done without Kleinman's exchange correction $f(q, k_F)$ we get very different and what appear to be more realistic results, as can be seen from figure 1. It is possible that the inclusion of the conduction-electron exchange effect is too sophisticated compared with the crudeness of the other approximations. It would be interesting to see whether keeping the exchange correction and using a non-local pseudopotential leads to a larger EFG near the peak of $g(r)$. Another interesting feature of the EFG function for Ga is that it does not follow the course of the external function and has a different sign for $r \geqslant 5{\cdot}2$ a.u. The EFG function for In does not show such a behaviour and can be represented fairly well by the external function multiplied by a constant.

We now define an effective antishielding factor which can be used for the interpretation of quadrupolar relaxation data:

$$\gamma_{\text{eff}} = (R_Q / R_Q^{\text{ext}})^{1/2}. \tag{24}$$

R_Q is the value of the quadrupolar relaxation rate calculated according to equation (1) including the shielding EFG, and R_Q^{ext} is that calculated with the external EFG only. The calculated values of γ_{eff} for Ga and In are given in table 2. For the R_Q calculation the

Table 2. Values of γ_{eff} to be used for nuclear quadrupole relaxation.

System	Angular contributions omitted	With total antishielding function
Ga with exchange	15·16	15·24
Ga without exchange	17·81	18·94
In with exchange	26·06	–

structure data of Narten (1972) for Ga and Waseda and Suzuki (1973) for In were used together with Vineyard's approximation for $S(\mathbf{k}, \omega)$ (see Sholl 1974, Gabriel and Schirmacher 1974). It is surprising that γ_{eff} for Ga is greater and for In smaller than the corresponding values $1 - \gamma_\infty$ which we calculated and which are given in table 1.

Let us now summarize the main features and deficiencies of the present treatment. Within the model of ion-like core electrons which are perturbed by ionic charges and screening conduction electrons, the antishielding EFG functions can be obtained with moderate numerical effort. The introduction of pseudopotentials that are more sophisticated than the local, empty-core potential (which has been used here) is possible and straightforward. The theory, however, does not claim to account for all shielding-like processes which can occur in a liquid metal. In particular, special effects due to p-type conduction electrons which are reported to be important in the solid (Das 1975) have been neglected as well as the effect of the depletion hole which is located immediately around the probe nucleus. Both effects may be important and should be the object of future investigations (see Sholl 1976) since present calculations, which take into account only the *pseudo*-electron density have shown that the antishielding effect is very sensitive to the various approximations which are used to calculate the electron density.

Acknowledgments

It is a pleasure to thank Professor Dr T P Das for very helpful discussions. I am indebted to Professor Dr H Gabriel for supervising the work and to Mr R Linke for giving me his Cooley–Tukey routine. This work is part of the research programme of Sonderforschungsbereich 161 of Deutsche Forschungsgemeinschaft.

References

Ashcroft N W 1966 *Phys. Lett.* **23** 48–50
Cohen M L and Heine V 1970 *Solid St. Phys.* **24** 37–248 (New York: Academic Press)
Cooley J W and Tukey J W 1965 *Math. Comput.* **19** 297–301
Cusack N and Kendall P 1960 *Proc. Phys. Soc.* **75** 309–11
Das T P 1975 *Phys. Scr.* **11** 133–9
Das T P and Bersohn R 1956 *Phys. Rev.* **102** 733–8
Foley H M, Sternheimer R M and Tycko D 1954 *Phys. Rev.* **93** 734–42
Gabriel H 1974 *Phys. Stat. Solidi* **B64** K23–6
Gabriel H and Schirmacher W 1974 *Proc. 18th Ampere Congr., University of Nottingham, September 1974* pp 331–2
Harrison W A 1966 *Pseudopotentials in the Theory of Metals* (New York: Benjamin)
Kleinman L 1967 *Phys. Rev.* **160** 585–90
Narten A H 1972 *J. Chem. Phys.* **56** 1185–9
Quitman D, Nishiyama K and Riegel D 1974 *Proc. 18th Ampere Congr., University of Nottingham, September 1974* pp 349–50
Riegel D 1975 *Phys. Scr.* **11** 228–36
Roetti C and Clementi E 1974 *J. Chem. Phys.* **61** 2062–3 suppl.
Schirmacher W 1976 *J. Phys. F: Metal Phys.* **6** L157–9
Sholl C A 1967 *Proc. Phys. Soc.* **91** 130–43
—— 1974 *J. Phys. F: Metal Phys.* **4** 1556–74
—— 1976 *J. Phys. F: Metal Phys.* **6** L161–2

Sternheimer R M 1963 *Phys. Rev.* **130** 1423–5
—— 1966 *Phys. Rev.* **146** 140–60
—— 1967 *Phys. Rev.* **159** 266–72
—— 1974 *Phys. Rev.* **A9** 1783–93
Warren W W Jr 1974 *Phys. Rev.* **A10** 657–70
Waseda Y and Suzuki K 1973 *Sci. Rep. RITU (Japan)* **A24** 139–240
Watson R E and Freeman A J 1961 *Phys. Rev.* **124** 1117–23

Atomic transport properties in liquid metals: experimental considerations

D A Rigney
Department of Metallurgical Engineering, The Ohio State University, Columbus, Ohio 43210

Abstract. Experimental aspects of diffusion, electrotransport, and thermotransport in liquid metals are discussed. The advantages, disadvantages, and special precautions associated with the various techniques are described. Several less familiar procedures are included.

1. Introduction

A complete review of the atomic transport properties of liquid metals would include the subjects of diffusion, electrotransport, thermotransport and viscosity. This article is not such a review. First, viscosity is not treated here; those interested in viscosity are referred to reviews by Thresh (1962), Wittenberg and Ofte (1970), and Beyer and Ring (1972). For the remaining three topics, experimental aspects are discussed, with convection effects particularly emphasized. Equations are included only when they serve to describe specific techniques or their limitations.

Liquid convection is not generally included together with a list of atomic transport processes. However, convection is given a prominent place in this paper, because some understanding of convection is absolutely essential for the proper design of experiments and the proper interpretation of experimental results.

2. Treatment of liquid metal diffusion data

At both the first and second conferences in this series, Nachtrieb (1967, 1972) discussed the state of our understanding of liquid metal diffusion as reflected by the appearance of new model theories and by our readiness or reluctance to accept them. Among other valuable comments, he suggested that we abandon use of the popular activation energy and pre-exponential for reporting liquid diffusion results, since their use is based on assumptions about liquid structure and diffusion mechanisms that are not well justified.

This point is worth pursuing because of its implications for experimentalists. Theories are now available which predict that the temperature dependence of D is a simple power law, $D \sim T^n$, where n may vary between $\frac{1}{2}$ and 4. Ascarelli and Paskin (1968) have suggested that $D \sim T^{3/2}/(C - T)$, and more recent linear trajectory calculations of Rao and Murthy (1975) have indicated that the temperature dependence is at least as complicated as $T/(C'T^{1/2} + C''T^{-1/2})$.

Since the variation of D with T is slow in liquid metals, almost any of these expressions can fit the experimental data. If one is to distinguish from among these many choices, it will be necessary to select data obtained over a wide temperature range with small experimental uncertainties in both D and T. Larsson *et al* (1972) have shown that this may still leave the $D(T)$ function and mechanism uncertain. They analysed $D(T)$ data for the alkali metals from near the melting point T_M to about $1{\cdot}6\,T_M$, but several functions fit the data adequately within the experimental uncertainty. In fact, three of the functions crossed near $T \simeq 1{\cdot}6\,T_M$, which would indicate that data well beyond this temperature would be needed to clarify the situation.

Liquid gallium would seem to be a prime candidate for further $D(T)$ work, since it has a low melting point (303 K) and a high boiling point (2676 K). Broome and Walls (1969) report D for Ga up to 674 K; above this temperature the Ga attacked the stainless steel shear cells. Larsson *et al* (1970) used Pyrex capillaries and they report D over a smaller temperature range. There appears to be no serious barrier to extending diffusion measurements in liquid Ga to much higher temperatures by a judicious choice of capillary or cell material.

One might ask if we should even assume that a single $D(T)$ function should be expected over a wide temperature range, particularly if structural changes are appreciable. Or one might ask whether it is reasonable to assume that one type of $D(T)$ function should hold for different liquid metals, considering the variations in structure indicated by diffraction studies. The answer to this question is particularly unclear in liquid alloys of unlike species, for which solvation effects may be important.

It appears then, that experimental diffusion data should not be reported simply as the best-fit parameters for a favourite $D(T)$ function. All such data, even those which seem to fit a given calculated function, should be reported in tabular form, together with the estimated uncertainty in both D and T. Graphical presentation and curve-fitting may be useful for liquid metal diffusion data, but they are inadequate by themselves. Without tabular presentation of the data, we lose some of the detailed information that may be needed to advance our understanding of liquid metal diffusion, and then we do not receive an optimum return on the effort involved in making the diffusion measurements.

3. Convection in liquid metals

Upon reading much of the published work on the atomic transport properties of liquid metals, one is struck by the frequent reference to convection and by the general recognition that it can influence measurements and should be carefully controlled. However, the approaches selected to minimize convection effects vary widely, and although the situation seems to be improving, a significant fraction of published papers shows evidence of inadequate understanding of convection and inadequate attention to its implications.

Perhaps the most dramatic evidence of this situation is the variety of choices made for the shape, size and orientation of the liquid metal container. For example, for capillary experiments, vertical, horizontal, tilted, zig-zag and spiral containers have been used and defended, often with the statement that the chosen arrangement would prevent convection. There are enough mysteries about other liquid metal properties without confusing the literature in this way.

The conditions for the onset of convection are reasonably well understood, especially for the simple boundary conditions that can be chosen for liquid metal experiments. This section will review those aspects of the subject which appear to be relevant to studies of liquid metal diffusion, electrotransport, and thermotransport.

One can start by examining the Navier–Stokes equation which relates the time and space variations of the fluid velocity **v** to the density ρ, kinematic viscosity ν, the acceleration due to gravity g, and the pressure. Density variations are easily included by using the thermal expansion coefficient α, and a solute expansion coefficient α'.† Verhoeven (1968) has pointed out that for $\mathbf{v} = 0$, the relation $\alpha\ \nabla T \times \mathbf{g} + \alpha'\ \nabla C \times \mathbf{g} = 0$ must hold. If convection is to be avoided, neither ∇T nor ∇C can have a component perpendicular to **g**. More bluntly, one can compose a simple rule: design liquid metal experiments so that horizontal temperature gradients and concentration gradients are both absent. Otherwise convection may be present even for small values of ∇T and ∇C. Studies related to solidification programmes have confirmed this conclusion (Cole and Winegard 1964, Hurle 1966, Carruthers 1968).

It appears easiest to avoid convection if a vertical capillary container is used. This case has been studied by Rayleigh (1916), Hales (1937), Taylor (1954), Chandrasekhar (1961), Verhoeven (1968, 1969) and others. Convection is predicted when a negative density gradient exceeds a critical magnitude. This condition can be summarized by using a dimensionless parameter called the Rayleigh number, R, which accounts for the geometry and the properties of both liquid and container. For a long capillary of radius r, the value of R for a vertical temperature gradient is given by $R_t = -g\alpha r^4 |\nabla T|/\nu K$, where K is the thermal diffusivity. The corresponding expression for a concentration gradient is $R_c = -g\alpha' r^4 |\nabla C|/\nu D$.

The critical values for R depend on the convection mode operating. Convection is easiest – that is, R(critical) is smallest – for a simple anti-symmetric mode, in which fluid moves upwards near one side and down near the other. For capillary materials that are good thermal insulators, Verhoeven (1968) and others have shown that the corresponding R(critical) = 68·0 for both R_t and R_c. When the capillary material is a good thermal conductor, higher values are appropriate for R_t, as shown by Hales (1937). The long capillary values of R(critical) apply for length/radius ratios greater than about 25; for shorter capillaries, higher values of R(critical) have been calculated (Verhoeven 1968).

Evidence is available that the expressions for R and their critical values work well, and that they may be relied upon in a nearly quantitative way. Verhoeven (1969) has shown that an oscillating temperature appears for R_t about 10% greater than the calculated value in liquid Hg. This seems to be associated with a pulsed mode of convection, and it does not preclude a slow laminar flow at values of R_t even closer to the expected value. For R_c, the data of Davis and Fryzuk (1965) on convection in Sn–Ag alloys are consistent with the calculated value.

The predicted dependence of both ∇T(critical) and ∇C(critical) on r^{-4} is consistent with the popularity of small capillaries of diameters $d \simeq 0{\cdot}5$–2 mm for transport measurements. For liquid Sn, Verhoeven (1968) has shown that critical values of $|-\nabla T|$ for d = 1, 2, and 20 mm are, respectively, about 3000, 200, and 0·02 °C/cm. Thus in simple capillary experiments, convection is ordinarily not expected for pure

† Density data for estimating α' are reviewed by Lucas (1970) and Crawley (1974).

liquid metals, except perhaps near the mouth, where boundary conditions are different. In fact a range of experimental ∇T values could be selected such that convection would not occur in a capillary, but it would occur in a larger reservoir in which the capillary is immersed. This would have practical consequences for capillary–reservoir measurement techniques.

One has to be more careful with concentration gradients. For Bi in liquid Sn, the critical value for $|-\nabla C|$ is about 350 ppm/mm if d = 1 mm, and it is as small as about 25 ppm/mm if d = 2 mm. For components which have larger differences in density, the situation is even worse. Therefore, experiments involving alloys must be even more carefully planned than those involving pure liquid metals.

Magnetic fields **H** have been used by numerous workers to either augment (Cole and Bolling 1966) or control (Utech and Flemings 1967, Hurle 1967, Uhlmann *et al* 1966) convection in liquid metals. Various geometries have been used, generally with **H** applied perpendicular to the specimen axis. Roehrig (1976) has adapted calculations of Chandrasekhar (1952) to the case of **H** parallel to a vertical capillary containing a liquid metal alloy. He predicted that the value of R(critical) could be raised by around 60% for H = 300 G, which was not enough to prevent solute-induced convection in his experiments.

It is interesting to examine containers of non-cylindrical symmetry, such as ribbons, which have been used for atomic transport measurements in liquid metals. Wooding (1960) has shown that it is much easier to establish convection in a flat capillary ribbon than in a more traditional capillary having the same cross-sectional area. Apparently, the expanded dimension of the flat capillary allows easy flow in the plane of the ribbon. Thus, ribbons are not recommended if convection is a concern.

Many authors include vibration in the list of factors which may have influenced their results, but the effects of vibration on convection in typical liquid metal studies are not at all clear. Pak *et al* (1970) have suggested that vibration could affect convection through relative motion arising from the compressibility of the liquid and the flexibility of the container. Davis (1970) has speculated that vibration causes horizontal concentration fluctuations and these lead to convection. Pak *et al* (1970) found that vibration increased the effectiveness of heat transfer at a solid–liquid interface. This could affect the critical Rayleigh number for capillaries made from good thermal conductors. In a large column of liquid, Purdy (1963) observed that acoustic vibrations initiated local convection modes at certain resonance conditions.

It is difficult to see how the kinds of complex, low-amplitude vibration spectra encountered in typical capillary experiments could cause convection, and the evidence for or against such effects is not available. Therefore, rather than continue to speculate, it would seem prudent to eliminate potential vibration problems through application of standard isolation and damping techniques (Tse *et al* 1963, Roehrig 1976). There would then be one less variable to blame for uncertain data.

4. Experimental techniques for liquid metal diffusion

The various techniques used for liquid metal diffusion measurements have been described in several recent reviews (Nachtrieb 1967, 1972, 1973, Edwards *et al* 1968, Walls 1970,

Larsson 1974). No single method emerges as a generally accepted, preferred method that is reasonably free of difficulties and clearly superior to all the others.

Nevertheless, the capillary–reservoir (CR) technique in its various forms has become a kind of popular standard. It is also commonly used for electrotransport experiments. For these reasons, it will be treated first and in greater detail than the other methods.

The technique was used as early as 1851 by Graham and in 1896 by Roberts-Austin (1896a,b, 1897). It is sometimes referred to as the method of Anderson and Saddington (1949), since they were the first to use it with radioactive tracers.

The CR technique involves use of a capillary container, 0·5–3 mm in diameter and with a length of several cm. It is sealed at one end, filled with liquid metal, and immersed in a reservoir containing a much larger volume of liquid metal having a different concentration of the species investigated. This may be a radioactive tracer for self-diffusion or a solute for interdiffusion studies. Usually the concentration is chosen to be zero in either the capillary or the reservoir at the start of an experiment. The diffusing species can be arranged to diffuse into or out of the capillary, and the open end is chosen to be up or down, depending on the relative densities of the two liquids. Capillary containers may be made from available precision bore tubing or they may be formed by drilling holes into a block of the chosen container material (Kassner *et al* 1962).

For the case of solute or tracer diffusing out of a vertical capillary at $x = l$, the solution to Fick's second law for uniform initial concentration C_0 in the capillary, zero concentration at the open end at all times, and no change at the closed end is

$$\frac{c(x,t)}{C_0} = \frac{4}{\pi} \sum_{n=0}^{\infty} \frac{(-1)^n}{(2n+1)} \exp\left(\frac{-(2n+1)^2 \pi^2 Dt}{4l^2}\right) \cos\left(\frac{(2n+1)\pi x}{2l}\right). \tag{1}$$

This solution may be combined with measured concentration profile data and used directly to determine D. The series converges rapidly if $Dt/l^2 \gtrsim 0{\cdot}2$.

Anderson and Saddington (1949) used equation (1) differently. They integrated it over the capillary length to yield an expression for the average concentration:

$$\bar{c}(t) = \frac{8}{\pi^2} \sum_{n=0}^{\infty} \frac{C_0}{(2n+1)^2} \exp\left(\frac{-\pi^2 (2n+1)^2 Dt}{4l^2}\right). \tag{2}$$

This expression converges rapidly for $Dt/l^2 \gtrsim 0{\cdot}1$, and the first term of the series is sufficient if t and l are chosen correctly. The diffusivity is then simply expressed by $D = (4l^2/\pi^2 t) \ln (\pi^2 \bar{c}/8C_0)$. The simplicity of this expression is one of the attractive features of this form of the CR technique. For $D = 5 \times 10^{-5}\ \mathrm{cm^2\ s^{-1}}$ and $l \simeq 5$ cm, appropriate times are $t \gtrsim 15$ h.

When the solute or tracer is initially in the reservoir and diffuses into the capillary, the solution to Fick's second law involves the familiar error function,

$$c(x,t) = C_0 \left[1 - \mathrm{erf}\left(\frac{x}{2(Dt)^{1/2}}\right)\right] = C_0\, \mathrm{erfc}\left(\frac{x}{2(Dt)^{1/2}}\right)$$

with x now measured from the open end of the capillary. If this solution is to remain appropriate, the concentration must not change at $x = l$. Therefore, diffusion must be terminated before $Dt/l^2 \simeq 0{\cdot}07$. For $D \simeq 5 \times 10^{-5}$ cm^{2}s^{-1} and $l \simeq 5$ cm, the corresponding time is about 10 h. A plot of erfc^{-1} (c/C_0) against x gives a straight line going through the origin. S is easily determined from the slope. The presence of end effects is indicated by a horizontal displacement of the line by an amount Δl (Davis *et al* 1965, Davis 1966).

A number of possible error sources have been widely recognized by users of the CR technique. Most of these belong to the category popularly termed 'end effects' or Δl effects, since they may be described by uncertainties in l of 0·1 to 0·5 mm. They may be caused by some combination of volume and concentration changes occurring during melting and solidification and by convection during immersion, during the extended time of the diffusion, or during removal from the reservoir. Depending on the details of design and procedure, one or more of these may dominate.

Stirring of the reservoir liquid has been commonly thought necessary to sweep away solute which would otherwise accumulate at the open end of the capillary and violate the assumed boundary condition there. The stirring may be accomplished by rotating either the crucible or the capillary assembly, by introducing a special stirring accessory, by bubbling an inert gas through the liquid, or by imposing a small inverse temperature gradient. The last method, suggested by Verhoeven (1968), depends on the r^{-4} dependence for the onset of convection in a cylinder. However, the resulting flow velocity near the open end of the capillary would be uncertain for this method, which could be a disadvantage relative to the well-characterized rotation methods.

However, Borucka *et al* (1957) and Kassner *et al* (1962) have shown that steady stirring leads directly to an end effect unless the fluid velocity is kept low. Therefore, if too much stirring leads to a large Δl effect, and if too little stirring also leads to changes in the boundary conditions, there would seem to be some intermediate choice possible. A simple approach would be to use $r \simeq (Dt)^{1/2}$ to estimate the time needed to build up a diffusion layer extending about 1 mm beyond the open end. Then, if the velocity were such that the capillary translated approximately $2r$ laterally during this time, this velocity should be sufficient. For r values in the range 0·25–1·5 mm, and $D \simeq 5 \times 10^{-5}$ cm^2 s^{-1}, the required velocity is about 10^{-3} mm s^{-1}. This corresponds to 10^{-3} rpm for capillaries positioned 10 mm away from the rotational axis, or a little larger if viscous effects are included. However, the values commonly chosen are 100–1000 times as large. This may be a case in which the usual cure has created a larger problem than the original one. It would seem that the best choice would be to select a velocity close to the minimum needed to prevent solute build-up near the open end of the capillary. It might even be better, as Davis (1970) has suggested, to eliminate stirring altogether, and to depend on long capillaries to reduce the percentage error from end effects, rather than to stir too rapidly.

Reducing or eliminating stirring raises another question. How large should the reservoir be in order to assure that the composition remains at C_0 near the mouth? Typical choices are $r \simeq 1$ mm for the capillary and $r' \simeq 10$ mm for the reservoir. If the capillary length is about 1/3 of the reservoir depth, the ratio of volumes is 1/300. This is certainly adequate if mixing in the reservoir is appreciable. Without stirring, the situation is less clear.

However, for mutual diffusion experiments, it is very likely that sufficient convection will develop near the open end as diffusion proceeds. As the solute diffuses upwards, the concentration gradient in the capillary remains stable against convection, but a region near the mouth will develop a horizontal component to the concentration gradient. In this case, a threshold value for ∇C is not needed. Thus, convection will develop, perhaps enough to limit the solute build-up at the mouth and enough to avoid a problem from reduced effective reservoir volume. This mechanism should apply for both open-end-up and open-end-down geometries unless the volume change upon alloying is close to zero.

From the preceding discussion, it would seem that some convection near the open end of the capillary is inherent in the CR technique, and that stirring is generally unnecessary when using this technique for solute diffusivities. For tracer self-diffusion, a small amount of stirring may be needed. If too much stirring is used, complex correction procedures may be needed to compensate for the resulting Δl effects (for example, see Foster and Reynik 1973).

A second source of end effects is the complex fluid motion which results when the filled capillary is lowered into the reservoir, especially when the open end breaks through the liquid metal surface. Depending on the relative surface energies, it is possible that for selected materials combinations this effect would be larger than the usual effect from stirring. Overfilling of the capillary has been used to minimize this effect (Schadler 1957), and various correction procedures have been proposed (Borucka *et al* 1957, Kassner 1960). Perhaps special immersion procedures can be developed to reduce this effect. For example, one might use a two-piece capillary, one piece of which is a short cap, closed at one end and with its open end abutting the open end of the diffusion capillary. The assembly would be immersed as a unit, and after the temperature had equilibrated, the short capillary segment and the liquid contained in it would be translated horizontally to a position safely distant from the sample capillary. The cap could then be removed from the melt or, for a large reservoir, kept immersed for the duration of the experiment. A related procedure could be followed before removing the diffusion capillary from the reservoir. That is, a suitable cover could be moved carefully into position and the new assembly lifted out together. This suggested approach would also be useful for experiments with aluminium, since it would avoid problems which might arise from this metal's tenacious oxide film.

The solidification operation following the diffusion anneal of the liquid metal sample is a crucial step; the volume changes and solute redistribution must be considered. In effect, one is faced with a miniature casting problem, and the rate of freezing, R, and the direction of freezing will influence the results. If directional solidification were used at this stage, one would have the advantages of limiting convection and assuring some reproducibility, but this would invite the disadvantages of limiting the solidification rate and of creating large changes in the solute profile, particularly at the ends. For combined radial and longitudinal heat flow there is the added danger of a large non-uniform shrinkage cavity or pipe where the last liquid freezes.

An alternative approach would be to use a rapid radial quench. The usual expression for solute redistribution during normal freezing could be used to predict the resulting radial solute profile (Chalmers 1964). This would be a flat profile except for an initial transient of approximate depth D/k_0R at the surface and a final transient of approximate depth D/R at the centre. The equilibrium distribution coefficient k_0 is defined as the ratio of the solute concentration in the solid to that in the liquid; it can be obtained from published phase diagrams. Depending on the particular alloy system, k_0 can be greater than or less than 1. For example, for Bi in Sn, $k_0 \simeq 0{\cdot}3$, and for a solidification rate of $1\ \text{mm s}^{-1}$, the depth of the initial transient is about $20\ \mu\text{m}$ and that of the final transient is about $7\ \mu\text{m}$. The horizontal gradients would thus be so limited in extent that convection should not be important. The obvious advantage of the radial freezing approach is that the longitudinal concentration profile resulting from the diffusion anneal is not changed.

Rapid freezing is favoured by using thin-wall capillaries and by using materials that are good thermal conductors. The type of capillary formed by drilling a hole into a massive block of material is not favourable for rapid quenching.

A simple way to estimate the actual freezing rates involved and the direction of solidification would be to calibrate with a eutectic system having a similar melting point. The lamellar spacing is proportional to $R^{-1/2}$, and the orientation of the lamellae is in the growth direction. Grain orientation could also provide information on freezing direction.

Edwards *et al* (1968) have summarized the surface effects that have been suggested as sources of uncertainty in D measurements, particularly for small-diameter capillaries. One might, for example, consider whether D is different near the capillary wall and in the interior of the capillary. Fixman (1958) has shown that this is a small effect, creating a 10% error within a few hundred atomic distances of the wall. This would only influence a volume fraction of around 10^{-5}, and therefore the effect is completely negligible. Rough surfaces could create a larger effect, but they would have to be very rough indeed to contribute serious error.

Careri *et al* (1958) have reported that the measured diffusivity for In depends on the capillary material. Chemical interaction is a possibility, but a detailed mechanism is not available to explain how this affects the measured diffusivity.

A different problem occurs when the liquid metal and capillary material do not react chemically. In this case, the relative surface energy is high, and it becomes increasingly difficult to fill the capillary as the diameter is decreased. Occasionally the capillary will be incompletely filled, or a void or a gap may disrupt the diffusion path. In electrotransport experiments, this can be easily detected by checking for electrical continuity *in situ*. It might be desirable to adapt this extra feature for monitoring continuity to liquid diffusion measurements as well.

For any capillary experiment, errors from uneven shape and size become more important with small-diameter capillaries. These can be controlled by selection from precision bore capillaries (available in Pyrex, quartz, and Lucalox alumina) and they can be estimated by measuring sample cross sections under a microscope.

A second major technique for measuring the liquid metal diffusivity is the diffusion couple, familiar to solid state diffusion workers (Shewmon 1963). With liquids the method is sometimes referred to as the long-capillary method. Two semi-infinite samples are placed together in the same capillary container at the start of the diffusion anneal; one contains tracer and/or solute, and the other does not.

The appropriate solution to Fick's second law involves the familiar error function:

$$2c(x, t) = C_0\left[1 - \text{erf}\left(\frac{x}{2(Dt)^{1/2}}\right)\right],$$

with $c(x, 0) = C_0$ for $x < 0$ and $C(x, 0) = 0$ for $x > 0$. If this solution is to remain appropriate, the concentration must not change at $x = \pm l$. Therefore, diffusion must be terminated before $Dt/l^2 \simeq 0{\cdot}1$. For $D \simeq 5 \times 10^{-5}\,\mathrm{cm^2\,s^{-1}}$ and $l = 5$–10 cm, the corresponding times are between 15 and 60 h. For shorter lengths or longer times, available tables may be used to determine the diffusivity (Jost 1960). When D is a function of concentration, the Matano method may be used.

Although this technique has been frequently used with a horizontal arrangement, a vertical position for the capillary would be preferable to minimize convection effects. Also, as with other capillary methods, a positive ∇T and appropriate sign for ∇C should be maintained at all times. With these precautions, convection during the diffusion anneal is not a serious problem for this method.

The most serious problem arises at the start, when the two half-samples are placed in contact. This may be accomplished by placing two solids together and then melting both, by drawing liquid of one concentration into a capillary that is already partly filled with solid of another composition, or by placing two liquid columns together. The problems encountered may involve oxide barriers, volume changes during melting or freezing, thermal expansion, different melting points, and convective mixing near the interface. The best approach appears to be the one involving careful joining of two liquid columns when they are at the desired temperature. Broome and Walls (1968) have proposed ways to estimate the errors arising from mixing at the diffusion couple junction. An adaptation of a shear-cell device (Broome and Walls 1968) or a burette-type geometry would keep this correction small. However, the design would still have to allow rapid radial solidification to assure that the concentration profile is unaffected by solute redistribution; therefore, massive shear-cell components are not suitable for this stage. There is certainly room in this area for innovative design ideas which will combine features that appear at first to be mutually exclusive.

Another popular solid state method (Shewmon 1963) that has been adapted for liquid diffusion is the use of a long sample with solute or tracer added at one end. This has been used by Eriksson *et al* (1974) and by Larsson (1974). The diffusant can be inserted at one end of the capillary as part of a small solid pellet ($\Delta l \lesssim 1$ mm) long, or as a liquid with the aid of a syringe. The remaining portion of the long capillary ($l \simeq 80$–150 mm) is subsequently filled with liquid metal. It is at this stage that some convective mixing could occur. This should not create a serious error for small-bore capillaries and $\Delta l \ll l$. However, it could be controlled still further by adding a shear-cell feature or a thin inert separating partition which could be withdrawn at the start of the diffusion anneal when the temperature had equilibrated.

As with earlier methods, a vertical arrangement is preferred, and the metals should be placed so that the heavier component is initially at the lower end.

The appropriate solution to Fick's second law is

$$c(x, t) = (\text{constant})\,(Dt)^{-1/2} \exp(-x^2/4Dt).$$

A simple plot of $\ln(c)$ against x^2 then gives a straight line with a slope of $-1/4Dt$. Larsson (1974) has shown that typical values for Δl, l, D, and t justify the use of this 'thin-film solution' for the determination of D. The experiment must be terminated before the concentration at the far end has changed; for $l \simeq 80$–150 mm, and $D \simeq 5 \times 10^{-5}\,\mathrm{cm^2\,s^{-1}}$, the maximum time is between 20 and 80 h. Again, rapid solidification is necessary to avoid profile changes after the diffusion anneal.

Larsson *et al* (1970) and Larsson (1974) have described a non-destructive method for avoiding solidification problems arising from volume and profile changes. For systems having isotopes with penetrating γ-radiation, it is possible to measure the concentration profile as a function of time during the diffusion anneal. Using this technique, Larsson found that he could improve reproducibility of his diffusion data.

It is important in all these experiments to begin with samples which have the uniform composition assumed for the diffusion calculations. It is sometimes surprising to find that, despite convective mixing, this is not a trivial matter. Some metals may dissolve slowly in others, and even induction melting may provide inadequate mixing if the alloy is made in a crucible with a large l/r ratio. Macro-segregation may also occur during the solidification step, unless the quench is sufficiently rapid. Therefore, it is good practice to analyse the composition uniformity of the reservoir and of selected capillaries before any diffusion experiments have taken place. It would also be wise to examine selected specimens metallographically. A few large inclusions in a small-bore capillary could render diffusion results meaningless.

Very few liquid metal diffusion experiments have been done under constant volume conditions at different temperatures. Two notable exceptions are the studies of Nachtrieb and Petit (1956) and of Ozelton and Swalin (1968). Constant volume data are necessary for comparison with the predictions of most liquid metal diffusion theories. In principle, any of the methods described above could be used for high pressure experiments. The capillary reservoir technique is commonly thought unsuitable for such work, but if stirring is not needed, there is no fundamental problem preventing its use.

There are many other factors which could have been included in this discussion of experimental concerns, but an effort has been made to exclude most of those which are common to diffusion experiments in solids. It is assumed, for example, that temperature will be adequately controlled and measured, that suitable chemical analysis techniques will be used, and that, for tracers, the energy spectra, half-lives, and counting statistics will be properly treated.

The preceding material has emphasized experimental aspects of the more familiar and more commonly used techniques for measuring liquid metal diffusivities. There are many others, including shear-cell techniques, neutron scattering, polarography, resistance measurements, diaphragm cell techniques, liquid electrolyte methods, and electrotransport methods. These have been reviewed by Nachtrieb (1967, 1972, 1973), Edwards *et al* (1968) and Walls (1970). Examples of dissolution studies using various geometries are those by Berg and Hall (1973), Ohno (1973), Barinov *et al* (1974) and Breitkreutz and Haedecke (1974). Figgins (1971) has described an application of inelastic light scattering to diffusion in liquids. With the advent of lasers having shorter wavelengths, it might become possible to apply this technique to the alkali metals.

Two additional techniques, although not commonly used for liquid metal diffusion measurements, will be treated next in some detail. These are the use of nuclear magnetic resonance and solid electrolytes. In each case, the systems that may be investigated are somewhat limited, but careful work has provided some interesting results.

4.1. NMR techniques

Perhaps the most sophisticated methods used for studying diffusion in liquid metals would be those employing nuclear magnetic resonance. The equipment is relatively complex compared with that required for more traditional methods, and the data analysis generally requires careful consideration of complex correction factors. Nevertheless, NMR has been applied successfully to measure self-diffusion in several interesting liquid metal systems, including Na (Murday and Cotts 1970, Garroway and Cotts 1973), natural and isotopically enriched Li (Murday and Cotts 1968, 1971, Krüger *et al* 1971, Garroway and Cotts 1973), and alloys of these alkali metals with ammonia (Garroway and Cotts 1973).

Several different NMR techniques have been described by Slichter (1963) and Drain (1967) for diffusion in solids, but these generally require some knowledge of the diffusion mechanism to extract values for D. Fortunately, the method selected for liquid metals does not require such knowledge. It is the spin-echo technique described by Carr and Purcell (1954) and Stejskal and Tanner (1965), which can be applied to fast-diffusing systems.

Two carefully timed RF pulses, separated by a time τ of the order of 5–10 ms are applied to the sample. Shortly after each RF pulse, a pulsed magnetic field gradient is imposed for about 1–2 ms. The resulting spin-echo signal at a time 2τ is only appreciable if most of the nuclear spins remain in regions of constant magnetic field. But, if diffusion is rapid, most of the atoms do not remain in regions of constant magnetic field, and the spin-echo signal is reduced in amplitude by an amount which can be related to the diffusion coefficient. NMR thus allows *in situ* measurements of liquid metal diffusion, whereas most other methods require a solidification step which can introduce changes in the concentration profiles used for analysis.

However, liquid metals present special difficulties compared with other liquids, and these limit the number of systems to which the spin-echo method may be applied. One of these is the relatively strong interaction between nuclear spins and the conduction electrons. This limits the lifetime of the nuclear spin states, characterized by a relaxation time T_1. If T_1 is too short, the spin-echo signal will be reduced, and it will not reflect diffusive motion in the magnetic field gradient. For liquid Li near its melting point, T_1 is about 80 ms, which is long compared with 2τ. However, T_1 falls off rapidly for metals with higher atomic numbers. In addition, T_1 is inversely proportional to temperature. Therefore, best results are expected for metals having low atomic numbers and low melting points. It is no surprise that Li and Na have been the first liquid metals studied. Mg, Si and K should be difficult because of unfavourable isotopic abundance and sensitivity factors. Al and Cu have short T_1 values, but these liquid metals might become candidates for liquid diffusion measurements if rapid pulses are used so that 2τ can be reduced to 1 ms or less. For example, the apparatus of I Lowe (1976 private communication) at the University of Pittsburgh might be able to cope with these liquid metals.

For the magnetic fields generally used, the radio frequencies required for resonance limit the useful size of metal samples due to the skin depth. Liquid metal droplets larger than the skin depth do not contribute fully to the NMR signal. Therefore droplets are usually chosen in the size range 50–200 μm. This in itself is not a serious problem, since the droplets can be kept from coalescing by suitable coatings and by suitable liquid or solid media inside a capsule (see, for example, Rowland 1961). However, the resulting size is comparable with the diffusion length during the time 2τ, so that diffusion is affected by the limited volume. Correction factors for this systematic error are described by Murday and Cotts (1971). The magnitude of the correction is in the range of 5–20%.

A second correction term arises from the inhomogeneous magnetization in a sample composed of small droplets. This background gradient term, which can amount to as much as 20%, can be minimized by using larger particles. In fact, Garroway and Cotts (1973) have shown that it is possible to neglect this term altogether by using a macroscopic sample having a diameter of the order of millimetres.

Faced with these complications, researchers have chosen two different approaches. Krüger *et al* (1971) elected to measure the ratio of $D(^6Li)/D(^7Li)$ using a capsule containing segregated layers of carefully prepared droplets having different isotopic concentrations of ^6Li and ^7Li. This procedure largely avoided the need to apply complex correction factors. Murday and Cotts (1968, 1970, 1971) chose to make absolute measurements, and all the correction factors were needed. The uncertainty in the resulting D values was assumed to result from contributions of ±5% from gradient pulse measurements, ±2% from bounded diffusion, ±5% from background gradients, and ±3% from overall reproducibility.

The spin-echo technique appears to give reliable and reproducible values for the self-diffusion coefficient in liquid metals. It has the advantages of being a direct *in situ* method, of being able to measure D for specific isotopes, and of being useful for studying solvation effects (Garroway and Cotts 1973). However, it appears to be limited to a small number of liquid metal systems and the range of useful temperatures is restricted, making it unlikely that NMR data will be very helpful in distinguishing from among the various proposed functions for $D(T)$.

4.2. *Solid electrolyte methods*

Certain metal/non-metal compounds having largely ionic-type bonding are good solid electrolytes at elevated temperatures. These are typically oxides or halides, often with appropriate doping to assure that the ionic conductivity is large and considerably greater than the electron or hole conductivity.

Rapp and Shores (1970) have published a list of 37 solid electrolytes in their review of solid electrolyte galvanic cells. These include compounds in which the following ions are the migrating species: F^-, Cl^-, Br^-, I^-, O^{2-}, Li^+, Na^+, K^+, Cu^+, Ag^+, Mg^{2+}, and Al^{3+}. For example, ZrO_2, with a few per cent of CaO added to partially stabilize the cubic crystal structure, is commonly used when oxygen is the diffusing species. The presence of a small amount of a second phase is tolerated because it reduces the sensitivity of the material to cracking. ZrO_2(+CaO) has been found useful in the temperature range 975–1875 K, and it has been applied for measurements of D(oxygen) in liquid Fe, Cu, Ag, Sn, Pb (see Odle and Rapp 1975) and In (Klinedinst and Stevenson 1973).

In a typical cell (see, for example, Oberg *et al* 1973), liquid metal containing oxygen is in contact with one surface of a suitable solid electrolyte. At the opposite surface, a steady supply of oxygen at a constant partial pressure is in contact with porous Pt to form the reference electrode. If a potential V_1 is applied across this oxygen concentration cell, it will fix the oxygen partial pressure $P''(O_2)$ at the liquid metal/solid electrolyte interface. At steady-state, with no ionic current flowing,

$$V_1 = (RT/4F) \ln [P'(O_2)/P''(O_2)],$$

where F is Faraday's constant and $P'(O_2)$ is the partial pressure of oxygen at the reference electrode. If V_1 is now suddenly changed to a new value V_2, $P''(O_2)$ changes rapidly to a new value, and oxygen diffuses out of or into the liquid metal, depending on the sign of ΔV. With V_2 held constant in a potentiostatic experiment, the resulting current is monitored as a function of time. The ionic part of the current can be calculated from Fick's second law, and the solution indicates that for long times, a linear plot of $\log(i_{ion})$ against t should result. The slope is proportional to D(oxygen) in the liquid metal, with the exact value depending on the geometry chosen for the experiment.

Several corrections should be considered when using these methods. The first involves the constant electronic current, i_e, which will flow for constant V_2. This can be a significant part of the measured current (more than 50% of i_{total} $(t \simeq 0)$ for liquid Fe), and it should not be neglected. Fortunately, i_e can be measured directly when i_{ion} has decayed to zero, and then $i_{ion}(t) = i_{total}(t) - i_e$ is easily obtained.

A second effect is more subtle. At short times, there is a potential drop equal to $i_{ion}(t)\,\Omega_{ion}$, where Ω_{ion} is the ionic resistance of the electrolyte. Thus, the experiment is not truly potentiostatic, and a complete solution to the diffusion equation would involve a time-dependent boundary condition. The error can be minimized by ignoring short-time data and by selecting a concentration range in which the contribution is minimized.

Still another correction (Oberg *et al* 1973) is needed for thermoelectric effects, since the lead wire attached to the Pt electrode and the one inserted in the liquid metal are generally different. The choice of the electrode wire inserted in the melt is important. Ideally, it would be completely insoluble in the liquid metal, but a certain amount of dissolution can be tolerated (as with W in Fe) if it does not affect the behaviour of the diffusing species.

Most of the experiments using these techniques have used cylindrical containers, but some investigators have chosen radial diffusion normal to the side walls of a solid electrolyte tube, while others have preferred longitudinal diffusion normal to the closed end of the liquid metal container. Either geometry is perfectly acceptable if there is no convection and if precautions are taken to eliminate local cell effects along the liquid metal/solid electrolyte interface. The radial geometry is inherently free from this effect, but the longitudinal arrangement can be affected unless the sides of the crucible are made of a non-electrolyte, as in the liquid Cu work of Osterwald and Schwarzlose (1968). They sealed a ZrO_2 electrolyte disc to an Al_2O_3 tube, and if the seal was good, their longitudinal arrangement should have yielded data comparable in quality with data obtained from radial arrangements. Horizontal concentration gradients could affect the radial method unless concentrations are kept very dilute.

Finally, as with most diffusion measurements, data obtained with these and related electrochemical techniques will be in error if convection is present. A slight positive temperature gradient should become standard practice, despite the small uncertainty in T imposed.

5. Electrotransport and thermotransport in liquid metals

Many of the experimental aspects of diffusion in liquid metals apply to electrotransport and thermotransport as well. Again, the control of convection is important, and again the experimental arrangement should involve a vertical capillary container. However, with electrotransport and thermotransport, concentration gradients develop from initially homogeneous material. Thus, unless the experiments are carefully designed, convection which was initially absent can develop as the experiment proceeds.

5.1. Electrotransport

Numerous reviews on electrotransport have been published in the last few years (Verhoeven 1963, Belashchenko 1965a, Epstein 1972, Wever 1973, Huntington 1973, 1975, Sellors and Pratt 1973, Rigney 1974). Therefore this brief section is intended only to describe some of the special features of electrotransport in liquid metals.

If one applies a direct current to a liquid alloy sample, the relative flux of each component is generally different and a concentration gradient develops. This is usually described by expressing the flux of species i as $J_i = C_i U_i E$, where C_i is the concentration of i, U_i is the mobility, and E is the applied electric field. If $U_i = qD/kT = Z_i^* |e| D/kT$, then the flux J_i is proportional to the effective valence Z^*. The force is then $\mathbf{F} = q\mathbf{E} = Z^* |e| \mathbf{E}$. For metals Z^* is usually negative, because the dominant mechanism involves momentum exchange from the conduction electrons. In general, Z^* bears no simple relationship to the normal chemical valence. Values have been reported ranging from 0 to 564. Electrotransport results have also been reported in terms of a ratio, P_i, of atom flux to electron flux, divided by the concentration (which normalizes P_i for small concentration). This ratio is typically of the order 10^{-4}, which reflects the inefficiency of the process.

The capillary–reservoir technique has been popular for electrotransport measurements. One can study the total concentration change or the concentration profile which develops in the capillary. The net flux, including back-diffusion, is

$$J_i = C_i U_i E - D_i \ (\partial C_i / \partial x).$$

If convection is present, D_i should be replaced by a larger effective diffusivity, D_{eff}. For times $t \lesssim l^2/\pi D_{\text{eff}}$ (Klemm 1946), the flux at the mouth remains constant at $C_i U_i E$, and a measurement of total concentration change is sufficient to measure J_i (and therefore U_i). This can be done by chemically analysing the entire capillary contents as a unit or by integrating a measured concentration profile (Roxbergh *et al* 1973). The latter has the advantage of testing for effects of back-diffusion on the profile, but the former can provide higher accuracy for the chemical analysis.

For long times, a steady-state profile will develop, and U_i can be determined if D_{eff} is known. However, if D_{eff} is not constant during the experiment, it is difficult to use this method. There is also a further danger involved in using the steady-state profile approach. In general, the effective valence will be concentration-dependent, sometimes strikingly so. In fact, it can even change sign, so that reversals of migration direction occur in selected systems (Blough *et al* 1972).

Various mechanisms have been proposed to explain convection effects in electrotransport experiments (Pikus and Fiks 1959, Lodding and Klemm 1962, Belashchenko 1965b, Schmidt and Verhoeven 1967). These include an electrokinetic effect involving inelastic scattering at the capillary walls, a magnetohydrodynamic effect arising from tapered or irregular capillary walls, a Joule heating effect leading to horizontal temperature gradients, and a Lorentz effect involving interaction of the charge carriers with the current-induced magnetic field. None of these has been conclusively demonstrated experimentally.

Roehrig (1976) has examined the concentration profiles resulting from field-freezing (electrotransport plus unidirectional solidification) dilute alloys of Bi in Sn. In this system, the equilibrium distribution coefficient is $k_0 \simeq 0{\cdot}3$, so that Bi accumulates ahead of the interface. With no electric current applied, the profile is as expected for normal freezing without mixing in the liquid. The same profile results if an alternating current of approximately 5 A mm^{-2} is imposed, indicating no convection for this case. The results for a direct current depend, as expected, on the polarity. If the lower end is chosen as the anode, toward which the Bi migrates, then the profile shows no evidence of convection. The reverse polarity results in a distinctly different profile that shows evidence of appreciable convection. Similar behaviour was found for Sb in Sn, but the polarities were reversed, since $k_0 > 1$ for this system. Thus, at these current densities the current does not seem to cause convection directly: the process is indirect. If the current causes a negative concentration gradient to build up, convection will appear when ∇C reaches a critical level. Obviously, if this is the principal mechanism for electrotransport-caused convection, it can be controlled by a sensible choice for migration direction.

At very high current densities, one or more of the other mechanisms may become important. For example, Belashchenko (1965b) has reported that current densities of 10–20 A mm^{-2} caused convection, whereas 1–1·5 A mm^{-2} did not. The sensitive field-freezing technique used by Roehrig would be useful for determining the maximum allowable current density if convection is to be avoided for a given alloy system.

Epstein (1972) has provided further evidence for the kind of convection found by Roehrig. He connected the anode lead to the closed end of one capillary and the cathode lead to the closed end of the other capillary and then immersed both in the reservoir with the open ends down. With this arrangement, an unfavourable ∇C must develop in one of the two capillaries, and, if α' is not too small, convection must eventually develop. Epstein's experiments confirmed this expectation with a current density of about 10 A mm^{-2} for a series of solutes in liquid Bi.

End effects with the capillary–reservoir technique for electrotransport are similar to those expected with liquid diffusion, except that immersion effects are absent, since the initial composition is generally uniform. However, two new kinds of end effect can develop. First, if the Joule heating is appreciable in the capillary, a negative ∇T will develop near the open end for the open-end-up configuration. This is unlikely to be large enough to cause convection well inside the capillary, but it could cause appreciable stirring in the reservoir near the mouth – perhaps enough to cause an end effect. The second effect involves development of an adverse concentration gradient. With the open-end-up configuration, if the lighter component migrates upwards, ∇C in the capillary itself will have the right sign, but ∇C near the open end will have a reversed sign, and convection should appear there. The unexplained results of Shaw and Verhoeven (1973) for Cd in Hg may have been influenced by these mechanisms. It may be that the capillary–reservoir technique is not a suitable choice for testing the origin of electrotransport-induced convection. A long sealed capillary arrangement may be better.

Finally, since the direction of electrotransport appears to be critical in determining the presence or absence of convection, it is important to emphasize that one cannot assume that a given component will migrate in the same direction as the composition is changed. Reversals of migration direction occur for several Hg alloys, for Na–K and Na–Rb, for Ag and Au in a number of polyvalent solvents, and perhaps for others not yet discovered (Blough *et al* 1972, Sellors and Pratt 1973, Mack 1976).

5.2. *Thermotransport*

The process in which a concentration gradient is induced by a temperature gradient in solids, liquids, or gases has been called thermotransport, thermomigration, thermal diffusion, or the Ludwig–Soret effect (Ludwig 1856, Soret 1881). The phenomenon is well studied in many materials (Grew 1969, Huntington 1973), but not in liquid metals.

In a typical experiment, a liquid with atomic fraction X_1 of component 1 is contained in a vertical capillary several cm long. It is then subjected to a temperature gradient of 1–5 °C/mm for a time $t > l^2/2D$, so that a steady-state concentration profile is developed. Early data were reported in terms of a 'thermal diffusion factor',

$$A = \log(q)/\log(T_{\text{hot}}/T_{\text{cold}}),$$

where q is the 'separation factor' given by

$$[X_1/(1-X_1)]_{\text{hot}}/[X_1/(1-X_1)]_{\text{cold}},$$

and also in terms of the Soret coefficient,

$$S = X_1(1-X_1)A = D_t/D,$$

with D_t the coefficient of thermal diffusion. For liquid metals it is now common practice to express results in terms of the net heat of transport, Q^*, which for dilute alloys is given by

$$\frac{d(\ln X_1)}{d(1/T)} = \frac{Q^*}{R}.$$

Thus, Q^* is directly obtained from a plot of $\ln X_1$ against $1/T$. For concentrated alloys one should use

$$\frac{Q^*}{R} = \left[\frac{d(\ln a_1)}{d(\ln X_1)}\right]\left[\frac{d(\ln X_1/(1-X_1))}{d(1/T)}\right]$$

(from Murarka *et al* 1974). According to the convention developed at the Marstrand Conference (Lodding and Lagerwald 1971), Q^* is positive if a component migrates to the cold end of the liquid column.

Gonzalez and Oriani (1965) have pointed out that a strong correlation exists between Q^* in solid metal systems and the effective valence measured in electrotransport experiments. Gerl (1967) has shown that this is expected if the electronic contribution to Q^* is dominant. The correlation is particularly good for dilute liquid metals, in which the component which migrates to the hot end in thermotransport is the one which goes to the anode for electrotransport. It holds also for the isotopes of Li, since the migration of ^{6}Li to the hot end is consistent with the measured Haeffner effect (Ott and Lunden 1964, Verhoeven 1963). It even appears to hold for systems having reversals of migration direction, such as Na–K (Murarka *et al* 1974, Rigney 1974), although the situation for Na–Rb is less clear (Hsieh and Swalin 1973). For concentrated alloys, the correlation of Q^* and Z^* is not as good (Winter and Drickamer 1955).

Even with the large positive temperature gradients that are generally used for these experiments, convection can develop during thermotransport if the component that migrates to the hot end has a higher density than the other component. An interesting case is that of Na–K, in which the system remains stable against convection for K in Na, but the existence of the migration reversal means that an unstable density gradient can develop for Na in K. Whenever this situation arises, it would probably be best to use a negative temperature gradient to assure that convection from concentration gradients is not a problem. Bhat and Swalin (1972) have shown that it is possible to use such a negative ∇T without causing convection inside the capillary.

Blough *et al* (1972) have used a long-time technique to measure accurately the composition of electrotransport reversals. A similar technique could be applied to thermotransport reversals. For both types of experiment, a concentration plateau should develop at one end, depending on whether the initial concentration is below or above that of the reversal.

In 1972, Bhat and Swalin stated that the study of thermotransport in liquid metals was 'in its infancy'. The same may still be said today. Experimental methods and analysis are straightforward, especially for dilute alloys, but very little work of this type has been reported. Perhaps this situation will improve before the next liquid metals conference.

6. Summary and conclusions

The goal of this review has been to provide a concise yet useful description of the experimental techniques most commonly used to measure atomic transport properties in liquid metals. Hopefully, it will help experimentalists to design their work so as to avoid the usual sources of error in their data, and perhaps also it will help others to evaluate those data. It would certainly be helpful if experimentalists would publish full experimental details, including assumptions, geometry, dimensions, materials, temperature monitoring, chemical analysis, and precautions taken to avoid convection.

Finally, the use of new techniques should be encouraged, but the resulting data may not be easy to interpret unless the same workers report data resulting from use of a 'standard' technique. Such comparative studies of techniques have been largely absent from the literature.

Acknowledgments

The author is grateful to Professors R M Cotts, A Lodding, R A Rapp and R A Swalin and Dr R J Reynik for providing helpful information and to the National Science Foundation for support under contract DMR73-02426-A01.

References

Anderson J S and Saddington K 1949 *J. Chem. Soc.* **152** 381
Ascarelli P and Paskin A 1968 *Phys. Rev.* **165** 222

Barinov G I, Krushenko G G and Shurygin P M 1974 *Izv. Vyssh. Ucheb. Zaved.* **17** 123
Belashchenko D K 1965a *Russ. Chem. Rev.* **34** 219
—— 1965b *Russ. J. Phys. Chem.* **39** 430
Berg H and Hall E L 1973 *Ann. Proc., Reliability Phys. (Symp.)* **11** 10
Beyer R T and Ring E M 1972 *Liquid Metals, Chemistry and Physics* ed S Z Beer (New York: Dekker)
Bhat B N and Swalin R A 1972 *Acta Metall.* **20** 1387
Blough J L, Olson D C and Rigney D A 1972 *J. Appl. Phys.* **43** 2476
Borucka A Z, Bockris J O'M and Kitchener J A 1957 *Proc. R. Soc.* **A241** 554
Breitkreutz K and Haedecke K 1974 *Metall* **28** (1) 31
Broome E F and Walls H A 1968 *Metall. Trans.* **242** 2177
—— 1969 *Trans. Metall. Soc. AIME* **245** 739
Careri G, Paoletti A and Vicentini A 1958 *Nuovo Cim.* **10** 1088
Carr H Y and Purcell E M 1954 *Phys. Rev.* **94** 630
Carruthers J R 1968 *J. Cryst. Growth* **2** 1
Chalmers B 1964 *Principles of Solidification* (New York: Wiley)
Chandrasekhar S 1952 *Phil. Mag.* **43** 501
—— 1961 *Hydrodynamic and Hydromagnetic Stability* (Oxford: Oxford UP)
Cole G S and Bolling G F 1966 *Trans. Metall. Soc. AIME* **236** 1366
Cole G S and Winegard W C 1964 *J. Inst. Metals* **93** 153
Crawley A F 1974 *Int. Metall. Rev. No.* 180 32
Davis K G 1966 *Can. Metall. Q.* **5** 245
—— 1970 *Can. Metall. Q.* 9 409
Davis K G and Fryzuk P 1965 *Trans. Metall. Soc. AIME* **233** 1662
Drain L W 1967 *Metall. Rev. No.* 119 p 195
Edwards J B, Hucke E E and Martin J J 1968 *Metall. Rev. No.* 120 p 13
Epstein S G 1972 *Liquid Metals, Chemistry and Physics* ed S Z Beer (New York: Dekker)
Eriksson P E, Larsson S J and Lodding A 1974 *Z. Naturf.* **29A** 893
Figgins R 1971 *Contemp. Phys.* **12** 283
Fixman M 1958 *J. Chem. Phys.* **29** 540
Foster J P and Reynik R J 1973 *Metall. Trans.* **4** 207
Garroway A N and Cotts R M 1973 *Phys. Rev.* **7A** 635
Gerl M 1967 *J. Phys. Chem. Solids* **28** 725
Gonzalez O D and Oriani R-A 1965 *Trans. Metall. Soc. AIME* **233** 1878
Graham T 1851 *Phil. Trans. R. Soc.* **A140** 805
Grew K E 1969 *Transport in Fluids* ed H J M Hanley (New York: Dekker)
Hales A L 1937 *Mon. Not. R. Astron. Soc. (Geophys. Suppl.)* **4** 122
Hsieh M Y and Swalin R A 1973 *Scr. Metall.* **7** 1195
Huntington H B 1973 *Diffusion. Papers presented at ASM Seminar 1972* (Metals Park, Ohio: Am. Soc. Metals)
—— 1975 *Diffusion in Solids, Recent Developments* ed A S Nowick and J J Burton (New York: Academic Press)
Hurle D T J 1966 *Phil. Mag.* **13** 305
—— 1967 *Crystal Growth* ed H S Peiser (New York: Pergamon)
Jost W 1960 *Diffusion in Solids, Liquids, and Gases* (New York: Academic Press)
Kassner T F 1960 *MS Thesis* Purdue University, West Lafayette, Indiana
Kassner T F, Russell R J and Grace R E 1962 *Trans. Am. Soc. Metals* **55** 858
Klemm A 1946 *Z. Naturf.* **1** 252
Klinedinst K A and Stevenson D A 1973 *J. Electrochem. Soc: Solid-state Science and Technology* **120** 304
Krüger G T, Müller-Warmuth W and Klemm A 1971 *Z. Naturf.* **26A** 94
Larsson S J 1974 *PhD Thesis* Göteborg, Sweden
Larsson S J, Broman L, Roxbergh C and Lodding A 1970 *Z. Naturf.* **25A** 1472
Larsson S J, Roxbergh C and Lodding A 1972 *Phys. Chem. Liquids* **3** 137
Lodding A and Klemm A 1962 *Z. Naturf.* **17A** 1085
Lodding A and Lagerwald T (ed) 1971 *Proc. Conf. Atomic Transport in Solids and Liquids, Marstrand, Sweden 1970* (Tübingen: Verlag der Z. Naturf.)

Lucas L D 1970 *Physicochemical Measurements in Metals Research* ed R A Rapp (New York: Wiley)
Ludwig C 1856 *Akad. Wiss.* **20** 539
Mack R 1976 *MS Thesis* Ohio State University, Columbus, Ohio
Murarka S P, Kim T Y, Hsieh M Y and Swalin R A 1974 *Acta Metall.* **22** 185
Murday J S and Cotts R M 1968 *J. Chem. Phys.* **48** 4938
—— 1970 *J. Chem. Phys.* **53** 4724
—— 1971 *Z. Naturf.* **26A** 85
Nachtrieb N H 1967 *Adv. Phys.* **16** 309
—— 1972 *Liquid Metals, Chemistry and Physics* ed S Z Beer (New York: Dekker) p 507
—— 1973 *The Properties of Liquid Metals* ed S Takeuchi (London: Taylor and Francis) p 521
Nachtrieb N H and Petit J 1956 *J. Chem. Phys.* **24** 746
Oberg K E, Friedman L M, Boorstein W M and Rapp R A 1973 *Metall. Trans.* **4** 61
Odle R R and Rapp R A 1975 *Electrochemical Society Symposium on Metal–Slag–Gas Reactions and Processes, Toronto*
Ohno R 1973 *Metall. Trans.* **4** 909
Osterwald J and Schwarzlose G 1968 *Z. Phys. Chem., NF* **62** 119
Ott A and Lunden A 1964 *Z. Naturf.* **19A** 822
Ozelton N W and Swalin R A 1968 *Phil. Mag.* **153** 441
Pak H Y, Winter E R F and Schoenhals R J 1970 *Augmentation of Convective Heat and Mass Transfer* ed A E Bergles and R L Webb (New York: Am. Soc. Mech. Eng.)
Pikus G E and Fiks V B 1959 *Sov. Phys.–Solid St.* **1** 972
Purdy K R 1963 *PhD Thesis* Georgia Institute of Technology, Atlanta, Georgia
Rao R V G and Murthy A K 1975 *Z. Naturf.* **A30** 619
Rapp R A and Shores D A 1970 *Physicochemical Measurements in Metals Research* vol 4, part 2, ed R A Rapp (New York: Wiley)
(Lord) Rayleigh 1916 *Phil. Mag.* **32** 529
Rigney D A 1974 *Charge Transfer/Electronic Structure of Alloys* ed L H Bennett and R H Willens (New York: Metall. Soc. AIME)
Roberts-Austin W C 1896a *Phil. Trans. R. Soc.* **A187** 383
—— 1896b *Proc. R. Soc.* **59** 281
—— 1897 *Nature* **55** 377
Roehrig F K 1976 *PhD Thesis* Ohio State University, Columbus, Ohio
Rowland T J 1961 *Prog. Metal Phys.* **9** 1
Roxbergh C, Persson T and Lodding A 1973 *Phys. Chem. Liquids* **4** 1
Schadler H W 1957 *PhD Thesis* Purdue University, West Lafayette, Indiana
Schmidt R L and Verhoeven J D 1967 *Trans. TMS-AIME* **239** 148
Sellors R G R and Pratt J N 1973 *Electrotransport in Metals and Alloys* (Riehen, Switzerland: Trans. Tech. SA)
Shaw R E and Verhoeven J D 1973 *Metall. Trans.* **4** 2349
Shewmon P G 1963 *Diffusion in Solids* (New York: McGraw-Hill)
Slichter C P 1963 *Principles of Magnetic Resonance* (New York: Harper and Row)
Soret C 1881 *Ann. Chim. Phys.* **22** 293
Stejskal E O and Tanner J E 1965 *J. Chem. Phys.* **42** 288
Taylor G I 1954 *Proc. Phys. Soc.* **B67** 868
Thresh H R 1962 *Trans. Am. Soc. Metals* **55** 790, 1108
Tse F S, Morse I E and Hinkle R T 1963 *Mechanical Vibrations* (Boston: Allyn and Bacon)
Uhlmann D R, Seward T P and Chalmers B 1966 *Trans. Metall. Soc. AIME* **236** 527
Utech H P and Flemings M C 1967 *Crystal Growth* ed H S Peiser (New York: Pergamon)
Verhoeven J D 1963 *Metall. Rev.* **8** 311
—— 1968 *Trans. Metall. Soc. AIME* **242** 1937
—— 1969 *Phys. Fluids* **12** 1733
Walls H A 1970 *Physicochemical Measurements in Metals Research* ed R A Rapp (New York: Wiley)
Wever H 1973 *Electro- and Thermotransport in Metals* (Leipzig: Johann Barth)
Winter F R and Drickamer H G 1955 *J. Phys. Chem.* **59** 1229
Wittenberg L and Ofte D 1970 *Physicochemical Measurements in Metals Research* ed R A Rapp (New York: Wiley)
Wooding R A 1960 *J. Fluid Mech.* **7** 501

Linear-response theory of the electromigration driving forces†

W L Schaich

Physics Department, Indiana University, Bloomington, Indiana 47401, USA

Abstract. The theory of the driving forces of electromigration has undergone considerable sophistication and improvement since Kumar and Sorbello suggested a linear-response approach. We have extended and simplified their treatment and consider a simple model in detail from a linear-response point of view. By applying many of the modern approaches to this model, we demonstrate their content and inter-relationships. We also present and discuss formulae suitable for calculations of electromigration in liquid metals.

1. Introduction

The application of an electric field to a metal causes both the electrons and the ions to move. Although the electromigration of the ions is much smaller than conductivity of the electrons, it can be detected and in some cases leads to dramatic effects (d'Heurle 1971, Jordan 1974). Extensive compilations of experimental data now exist (Wever 1973, Pratt and Sellors 1973) and the subject has been often reviewed (Verhoeven 1963, Fiks 1971, Epstein 1972, Rigney 1974, Sorbello 1975, Huntington 1975). The phenomenon is a fascinating one for theorists, in that it requires theories of both electron and ion transport, and there has recently been a resurgence of theoretical effort based on many different approaches. In our opinion a key paper in this recent work is that of Kumar and Sorbello (1975), who first suggested that one apply the formalism of linear response. This formalism allows the straightforward application to electromigration of techniques that are well known in other areas. It also provides an overview from which various specific methods and approximations may be put in common perspective. In §2 we illustrate these features by discussing a simple model – dilute impurities in jellium – from various points of view. Then in §3 we apply the formalism directly to liquid metals, presenting several formulae suitable for calculation. The advantages and limitations of these equations are discussed. Due to lack of space, detailed derivations are omitted; some of the points made here have been discussed extensively elsewhere (Sham 1975, Schaich 1976a,b, McCraw and Schaich 1976). Also, the primary concern of this paper is with the forces that drive the ion motion, although our formulae in §3 are well adapted for tackling the problem of the consequent electromigration due to these forces. To evaluate these driving forces we invoke the Born–Oppenheimer approximation so that we may consider the ion configuration as momentarily stationary while calculating the response of the electrons (Schaich 1976a,b).

2. A simple model

Here, for the sake of clarity and unification, we discuss several different approaches to a

† Supported in part by the National Science Foundation through grant DMR 74-14063.

simple model problem from the linear-response point of view. We consider a model of dilute impurities in jellium whose self-consistently screened electron–ion potentials are isotropic and spin independent. The impurities are sufficiently dilute that we may treat them independently when calculating the driving forces. Also we assume that there is a clear distinction between the core electrons associated with each impurity, which merely polarize in an applied field, and the valence electrons (Z in number) associated with each impurity, which enter the host conduction band. We shall neglect the former here (Schaich 1976a,b) and concentrate only on the electrons in the conduction band. Our formal procedure is to apply a uniform external electric field, $\mathbf{E}$, which turns out to be the same as the macroscopic internal field in steady state (Schaich 1976a,b, Sham 1975). We then ask what incremental driving force (linear in $\mathbf{E}$) develops on each impurity.

The answer is well known; the induced force $\mathbf{F}^{\mathrm{D}}$ is

$$\mathbf{F}^{\mathrm{D}} = -Ze\mathbf{E} + eN(\tau/\tau_{\mathrm{i}})\mathbf{E} \tag{1}$$

where

$$\frac{1}{\tau_{\mathrm{i}}} = \frac{1}{N_{\mathrm{i}}(\tau)}$$

$$= \frac{2\pi}{\hbar} \sum_{\mathbf{k}'} (1 - \hat{\mathbf{k}}\cdot\hat{\mathbf{k}}') \left|\langle \mathbf{k} | \mathbf{t}^{+} | \mathbf{k}' \rangle\right|^2 \delta(\epsilon_{\mathbf{k}} - \epsilon_{\mathbf{k}'}), \tag{2}$$

with $e < 0$ the charge on the electrons, $\mathbf{t}^{+}$ the t-matrix for a single impurity, $|\mathbf{k}\rangle$ and $|\mathbf{k}'\rangle$ plane wave states, $\epsilon_{\mathbf{k}}$ the free electron energy of a state with momentum $\hbar\mathbf{k}$, and τ the conductivity relaxation time for the whole system. The total number of conduction electrons N equals $N_0 + ZN_{\mathrm{i}}$ with N_0 the number of electrons associated with the host jellium and N_{i} the number of impurities. $\mathbf{F}^{\mathrm{D}}$ is the sum of the 'direct' force of the external field on the ion and an 'indirect' force due to electronic response.

In the original derivations (Fiks 1959, Huntington and Grone 1961) this indirect force, which we denote by $\mathbf{f}\cdot\mathbf{E}$, is derived as the negative, average, time rate of change of the momentum of the electrons due to scattering off the impurity in steady state. One writes

$$\mathbf{f}\cdot\mathbf{E} = -\sum_{\mathbf{k}\sigma} \hbar\mathbf{k}\,(-n_1(\mathbf{k})/\tau_{\mathrm{i}}) \tag{3}$$

where the spin sum over σ merely adds a factor of 2 and $n_1(\mathbf{k})$ is the linear (in $\mathbf{E}$) deviation of the electron distribution function. One finds $n_1(\mathbf{k})$ from the Boltzmann equation

$$e\mathbf{E}\cdot\frac{\partial n_0}{\partial(\hbar\mathbf{k})} = -n_1/\tau. \tag{4}$$

Since at zero electron temperature for this simple model $n_0(\mathbf{k}) = \theta(k_{\mathrm{F}} - k)$, with k_{F} the Fermi momentum and $\theta(x) = 1$ for $x > 0$ and zero otherwise; combining equations (3) and (4) yields the second term in equation (1).

This ballistic approach for the indirect force is only unambiguous for dilute impurities. A simple alternative approach that does allow easy generalization is to seek the induced electron charge density about each impurity (Bosvieux and Friedel 1962). One writes

$$n(\mathbf{r}) = n_0(\mathbf{r}) + n_1(\mathbf{r}) = \sum_{\mathbf{k}\sigma} [n_0(\mathbf{k}) + n_1(\mathbf{k})] \mid \psi_{\mathbf{k}}^{+}(r) \mid^2 \tag{5}$$

where $\psi_{\mathbf{k}}^{+}(r)$ is the outgoing scattered wave solution about an impurity, placed for convenience at the origin; again one determines $n_1(\mathbf{k})$ from equation (4). The indirect driving force is found by integrating $n_1(\mathbf{r})$ with the force function $\nabla_{\mathbf{r}} V(\mathbf{r})$, where V is the electron-impurity potential energy. Along with this modified mathematical approach, one changes the verbal description of the source of $\mathbf{f}$ from momentum transfer to dynamic screening, with the latter arising from the steady-state perturbed electron distribution about each ion. However, the final results are identical. We stress further that we are still using an independent particle treatment: V is a self-consistently screened equilibrium interaction. To develop the approach we expand $\psi_{\mathbf{k}}^{+}(r)$ in a series of Legendre polynomials, $\mathrm{P}_l(\hat{\mathbf{k}}\cdot\hat{\mathbf{r}})$,

$$\psi_{\mathbf{k}}^{+}(r) = \sum_l (\mathrm{i})^l (2l+1) \exp(\mathrm{i}\delta_l) R_l(k, r) \mathrm{P}_l(\hat{\mathbf{k}}\cdot\hat{\mathbf{r}}) \tag{6}$$

where R_l is the regular solution of the l-wave Schrödinger equation and δ_l is the l-wave phase shift (Messiah 1966). One then finds that $n_1(\mathbf{r})$ has the following dipolar form:

$$n_1(\mathbf{r}) = 6\bar{n}(-e\tau E/\hbar k_{\mathrm{F}})\, \hat{\mathbf{E}}\cdot\hat{\mathbf{r}} \sum_l (l+1) R_l(r) R_{l+1}(r) \sin(\delta_{l+1} - \delta_l) \tag{7}$$

with $\bar{n}$ the equilibrium electron charge density. Note that the size of $n_1(\mathbf{r})$ is controlled by the extremely small ratio $(-e\tau E/\hbar k_{\mathrm{F}})$, which is equivalent to the ratio of the electron drift speed, v_{D}, to the Fermi speed v_{F}. The asymptotic behaviour of $n_1(\mathbf{r})$ is interesting. Since $R_l(r)$ is normalized so that outside the range of V and for large kr it goes to $\sin(kr - \frac{1}{2}l\pi + \delta_l)/kr$, the basic decay of $n_1(\mathbf{r})$ is controlled by $1/r^2$ times the sum of two terms, one a constant and the other oscillatory. Using the standard formula for the scattering amplitude (Messiah 1966),

$$f(\hat{\mathbf{k}}\cdot\hat{\mathbf{r}}) = 1/k \sum_l (2l+1) \exp(\mathrm{i}\delta_l) \sin\delta_l\, \mathrm{P}_l(\hat{\mathbf{k}}\cdot\hat{\mathbf{r}}), \tag{8}$$

one finds by straightforward trigonometric reduction (Schaich 1976b)

$$n_1(\mathbf{r}) \to 3\bar{n}(-e\tau E/\hbar k_{\mathrm{F}})\, \hat{\mathbf{E}}\cdot\hat{\mathbf{r}} \left\{ \frac{1}{k_{\mathrm{F}} r^2} \operatorname{Im}[f(-1)\exp(2\mathrm{i}k_{\mathrm{F}} r)] + \frac{1}{r^2}\int \frac{\mathrm{d}\Omega}{4\pi} (1-\cos\theta) \mid f(\cos\theta) \mid^2 \right\} + \mathrm{O}(1/r^3), \tag{9}$$

where Im denotes 'imaginary part of'.

The above development differs from the usual application of this approach (Bosvieux and Friedel 1962, Gerl 1967, 1971, Sorbello 1973) in two principal ways. We have used an independent-particle model rather than one of interacting electrons and we have employed the exact $\mid \psi_{\mathbf{k}}^{+}(r) \mid^2$ rather than a first-order (in V) expansion. The simplification of the first change is evident and its justification, which is complete (Sham 1975, Schaich 1976b), will be clarified below. The extension with the second change allows one to resolve a recent controversy (Schaich 1976b). In essence, by limiting oneself to a first-order (in V) calculation of $n_1(\mathbf{r})$, the contribution in equation (9) due to the square of the scattering amplitude is omitted. This monotonic term in the asymptotic $n_1(\mathbf{r})$ is the remnant of Landauer's resistivity dipole (Landauer 1957). It was recently pointed out (Landauer and Woo 1972) that such terms in $n_1(\mathbf{r})$ yield contributions to $\mathbf{f}$ whose sign is opposite to the contributions from the term in $n_1(\mathbf{r})$ linear in V. This latter contribution yields an $\mathbf{f}$ of the form given by equation (1) but with $\mathbf{t}^+$ replaced by V. However, such an order-by-order comparison is unnecessary and misleading since one can show directly from equation (7) that the complete $\mathbf{f}$ in this approach is given by equation (1) (Schaich 1976b, Sham 1975).

Let us now turn to a third possible approach, one based on the so-called equivalence theorem (Das and Peierls 1973). This attack originates from the alternate Galilean reference frame in which the electron gas is at rest and the impurity ion is moving with velocity $-\mathbf{v}_D = -e\mathbf{E}\tau/m$, where m is the mass of an electron. The induced force on the ion now appears to be a friction force (d'Agliano *et al* 1975). This transformation in viewpoint is intuitively plausible and may be rigorously justified for the dilute impurity model (McCraw and Schaich 1976); however, it becomes ambiguous in more general situations. To extract $\mathbf{f}$ from this approach we note that the moving ion represents a time-dependent potential

$$V(\mathbf{r}, t) = V(\mathbf{r} - (-\mathbf{v}_D t)). \tag{10}$$

The linear (in $\mathbf{v}_D$) response of the electron density to this perturbation is (Landau and Lifshitz 1960, Sorbello 1973, Das 1976)

$$n_1(\mathbf{q}) = i e \operatorname{Im} \chi(\mathbf{q}, -\mathbf{q}\cdot\mathbf{v}_D)\, V(\mathbf{q}) \tag{11}$$

where $n_1(\mathbf{q})$ is the Fourier transform of $n_1(\mathbf{r})$ and $\chi(\mathbf{q}, \omega)$ is the density response function. Working to only first order in V, we find

$$n_1(\mathbf{r}) = 12\bar{n}(-e\tau E/m)\,\hat{\mathbf{E}}\cdot\hat{\mathbf{r}} \int_0^\infty \left(\frac{q}{2k_F}\right)^2 \mathrm{d}\left(\frac{q}{2k_F}\right) \mathrm{J}_1(qr)\, V(q) \lim_{\omega\to 0} \left(-\frac{q}{\omega}\chi(\mathbf{q}, \omega)\right) \tag{12}$$

where J_1 is a spherical Bessel function. From equation (12) one finds $\mathbf{f}$ by a simple integration as before, but to proceed we need to know what to use for χ.

Our approach based on an independent-electron picture sets $\chi = \chi_0$, its value for non-interacting electrons. The alternative approach (Sorbello 1973, Das 1976) is to use interacting electrons (treated within the random-phase approximation (RPA), which yields $\chi_{RPA} = \chi_0/(1 - v\chi_0)$, where v is the bare Coulomb potential) *and* to use unscreened electron–ion potentials. These methods are equivalent because, as $\omega \to 0$,

$$\operatorname{Im} \chi_{RPA} = (\operatorname{Im} \chi_0)/\epsilon_{RPA}^2 \tag{13}$$

where $\epsilon_{RPA} = 1 - v\chi_0$. The extra powers of the RPA dielectric constant in equation (13) merely act to screen the bare potentials, yielding our independent-electron final result. This equivalence is not limited to the random-phase approximation; we only miss the Fermi liquid corrections to the Boltzmann equation solution by our method here (Sham 1975, Schaich 1976b). To complete the independent-electron theory we need

$$\lim_{\omega\to 0} \left(\frac{q}{\omega}\chi_0(\mathbf{q}, \omega)\right) = (\gamma_0 N_0/v_0)\,\theta(K_0 - q) \tag{14}$$

where $\gamma_0 = \pi/2, K_0 = 2k_F, v_0 = v_F$, and $N_0 = mk_F/\pi^2\hbar^2$, the density of states. Combining equations (12) to (14) to find **f**, one finds the standard result, equation (1), with $\mathbf{t}^+$ replaced by V (Bosvieux and Friedel 1962, Sorbello 1973, Das 1976).

It is also possible with this approach to apply classical or semi-classical methods; all that is required is the appropriate χ (McCraw and Schaich 1976). For example, in two separate semi-classical calculations (Das and Peierls 1973, Rorschach 1976), one finds the quantum values for all the parameters in equation (14) except K_0, which has its classical value of infinity. In fact one must set K_0 to a finite value to avoid a logarithmic divergence in **f**, and $2k_F$ is the best choice.

The final approach to the dilute impurity model that we shall discuss is based on sum-rule arguments that derive from the steady-state conduction condition (Faber 1972, Sorbello 1975, Landauer 1975, Sham 1975, Das and Peierls 1975, Schaich 1976a, Turban *et al* 1976). Since the electrons are in a state of steady drift, the total force on them must vanish. Because these forces arise only from the external field and the ions, we have, by Newton's third law,

$$eN\mathbf{E} - \sum_i \mathbf{f}^i \cdot \mathbf{E} = 0 \tag{15}$$

In the dilute impurity model each $\mathbf{f}^i$ is identical, so we deduce

$$\mathbf{f}^i \cdot \mathbf{E} = (N/N_i)\, e\mathbf{E}, \tag{16}$$

which is implicit in equations (1) and (2). By assuming a form of Matthiessen's rule, one can extend this argument to include background phonon scattering (Landauer 1975, Das and Peierls 1975, Sham 1975, Turban *et al* 1976), but the method becomes ambiguous in the presence of band structure or in liquid alloy systems, where different driving forces act on different species.

To close this section we remark that we have not discussed one further approach, what we refer to as 'bootstrap' arguments (Kumar and Sorbello 1975, McCraw and Schaich 1976, Turban *et al* 1976). By such arguments one seeks to relate **f** to alternative, hopefully more tractable, expressions. For the simple model of this section, these methods lead again to equation (1), but their extension to more general models is subtle.

3. Application to liquid metals

We now turn to a more realistic model, though in considering a liquid metal we are in effect only relaxing the condition on the density of scattering centres. We still use a Boltzmann equation to describe the deviation in the electron distribution function, still assume independent electrons, and still concentrate on finding the induced electron density change about the ion of concern. This formal problem has already been solved (Schaich 1976b) using infinite re-summations of Green's function expansions. A notable feature of the result is that it yields the driving force on any ion without invoking an average over ion configurations.

As we have shown in §2, a perturbative treatment of the electron–ion potentials should be quite satisfactory in simple liquid metals, as long as the distinction between core and valence electrons is clear cut. One need only choose the effective screened form factor of an ion so that it closely mimics the true *t*-matrix of that ion. Taking the

z axis along **E** one finds for the indirect force on the ion α at the origin, $f_{\mu z}^{(\alpha)}E$, where (Bosvieux and Friedel 1962, Sorbello 1973)

$$f_{\mu z}^{(\alpha)} = \frac{8\pi}{3} e \left(\frac{\tau}{\hbar/E_F}\right) \sum_\beta \int_0^1 \mathrm{d}\left(\frac{q}{2k_F}\right)\left(\frac{q}{2k_F}\right)^3 \tilde{v}_\beta(q)\,\tilde{v}_\alpha(q)$$

$$\times \left\{\frac{3}{4\pi}\int \mathrm{d}\Omega\, \hat{q}_\mu\, \hat{q}_z \exp\left(\mathrm{i}\mathbf{q}\cdot\mathbf{R}_{\beta\alpha}\right)\right\} \tag{17}$$

with

$$\left\{\ \right\}_{\mathbf{R}_{\beta\alpha}\to\mathbf{R}} = \begin{cases} J_0(qR) - 2P_2^0(\cos\theta)\,J_2(qR), & \mu = z \\ -P_2^1(\cos\theta)\cos\phi\, J_2(qR), & \mu = x \\ -P_2^1(\cos\theta)\sin\phi\, J_2(qR), & \mu = y. \end{cases} \tag{18}$$

Here we have normalized the individual form factors so that

$$\tilde{v}_\alpha(q) \underset{q\to 0}{\rightarrow} -Z_\alpha.$$

We see that $\mathbf{f}^\alpha$ is a sum of contributions associated with neighbouring ions, whose relative positions are at $\mathbf{R}_{\beta\alpha}$. The J_l are spherical Bessel functions and the P_l^m are associated Legendre polynomials (Messiah 1966), depending, respectively, on the magnitudes and polar angles of the $\mathbf{R}_{\beta\alpha}$. If we were to average equation (17) over the ion configuration we obtain the formulae, suggested by Faber using sum-rule arguments, that have been recently evaluated by Stroud (Faber 1972, Stroud 1976).

However, to do so would amount to ignoring the great advantage of equation (17) which is that it is in a form that may be used as direct input for a molecular dynamic calculation of the ion motion. To the ion–ion forces that exist in the absence of electrical conduction, one merely adds the induced driving forces, $-Z_\alpha eF + \mathbf{f}^{(\alpha)}\cdot\mathbf{E}$, and then seeks the steady drift velocities of the various species present. This procedure thus retains any local correlations between driving force and ion mobility. Hence it is now possible to study the Haeffner effect (isotope separation by electromigration) as well as to do a more complete analysis on alloy separation (Rigney 1974). Note too the calculational advantage that since the driving forces and resulting electromigration are linear in E, one can greatly increase E beyond its physical value, which is limited by Joule heating. This will ensure that the electromigration occurs within a reasonable time span. Another feature worth remarking is that the sum total of all the driving forces on all the ions must vanish; there is no centre of mass acceleration. Thus one can continue to use a relatively small number of ions subject to periodic boundary conditions. The only possible restriction in this regard is the relatively slow fall-off, as $\cos(2k_FR)/(2k_FR)^2$, of the induced coupling between ions. Note that this effect is more of a problem in monovalent than polyvalent metals, where in the former case $2k_F$ lies below the first peak in the structure factor. In any case this restriction only acts as a caution rather than as a severe limitation.

In summary we believe that the linear-response approach to the theory of the driving forces of electromigration gives strong support to the useful validity, in simple liquid metals, of perturbation theory formulae such as equation (17), for the indirect force coupled with an unscreened direct force on each ion. We urge that such formulae be incorporated into a molecular dynamic calculation. For the more complex metals, alternative expressions exist which take into account the strong scattering (Schaich 1976b). Their parametrization and evaluation are, however, rather more involved.

References

d'Agliano E G, Kumar P, Schaich W and Suhl H 1975 *Phys. Rev.* **B11** 2122
Bosvieux C and Friedel J 1962 *J. Phys. Chem. Solids* **23** 123
Das A K 1976 *Preprint*
Das A K and Peierls R 1973 *J. Phys. C: Solid St. Phys.* **6** 2811
—— 1975 *J. Phys. C: Solid St. Phys.* **8** 3348
Epstein S G 1972 *Liquid Metals: Chemistry and Physics* ed S Beer (New York: Dekker)
Faber T E 1972 *Theory of Liquid Metals* (London: Cambridge UP)
Fiks V B 1959 *Sov. Phys.–Solid St.* **1** 14
—— 1971 *Proc. Conf. Atomic Transport in Solids and Liquids, Marstrand 1970* ed A Lodding and T Lagerwall (Tübingen: Verlag der Z. Naturf.)
Gerl M 1967 *J. Phys. Chem. Solids* **28** 725
—— 1971 *Z. Naturf.* **26A** 1
d'Heurle F M 1971 *Proc. IEEE* **59** 1409
Huntington H B 1975 *Diffusion in Solids, Recent Developments* (New York: Academic Press)
Huntington H B and Grone A R 1961 *J. Phys. Chem. Solids* **20** 76
Jordan R G 1974 *Contemp. Phys.* **15** 375
Kumar P and Sorbello R S 1975 *Thin Solid Films* **25** 25
Landau L D and Lifshitz E M 1960 *Electrodynamics of Continuous Media* (Oxford: Pergamon)
Landauer R 1957 *IBM J. Res. Develop.* **1** 223
—— 1975 *J. Phys. C: Solid St. Phys.* **8** L389
Landauer R and Woo J W 1972 *Phys. Rev.* **B5** 1189
McCraw R and Schaich W 1976 *J. Phys. Chem. Solids* in press
Messiah A 1966 *Quantum Mechanics* (New York: Wiley)
Pratt J N and Sellors R G R 1973 *Electrotransport in Metals and Alloys* ed Y Adda *et al* (Riehen, Switzerland: Trans. Tech. SA)
Rigney D A 1974 *Charge Transfer – Electronic Structure of Alloys* ed L H Bennett and R H Willens (New York: Metals Soc. AIME)
Rorschach H E 1976 *Ann. Phys., NY* **98** 70
Schaich W L 1976a *Phys. Rev.* **B13** 3360
—— 1976b *Phys. Rev.* **B13** 3350
Sham L J 1975 *Phys. Rev.* **B12** 3142
Sorbello R S 1973 *J. Phys. Chem. Solids* **34** 937
—— 1975 *Comm. Solid St. Phys.* **6** 117
Stroud D 1976 *Phys. Rev.* **B13** 4221
Turban L, Nozieres P and Gerl M 1976 *J. Physique* **37** 159
Verhoeven J 1963 *Metall. Rev.* **8** 311
Wever H 1973 *Electro- und Thermotransport in Metallen* (Leipzig: Barth)

Driving force for electromigration in liquid alloys: a pseudopotential calculation †

D Stroud
Department of Physics, The Ohio State University, Columbus, Ohio 43210, USA

Abstract. The average driving force for electromigration acting on ions of each species is calculated by means of a pseudopotential approach. The calculated forces are generally consistent with the experimentally observed directions of electromigration in various alloys.

In a number of liquid alloys, the constituent ions migrate in opposite directions under the influence of an applied uniform DC electric field; this effect is known as electromigration, electrotransport, or electrodiffusion (for recent reviews see Epstein 1972, Pratt and Sellors 1973, Rigney 1974). The driving force responsible for this effect is generally believed to be the sum of two terms: a direct force on the ion arising from the applied electric field itself, and an 'electron drag' component due to the scattering of conduction electrons by the ions. The former tends to carry the ions towards the cathode while the latter, as a rule, acts in the opposite direction, that is, parallel to the electronic current. The ionic species which experiences the greater drag force will thus tend to migrate towards the anode, while the other, in order not to build up a density gradient in the alloy, will drift in the opposite direction.

The drag force acting on a particular ion in a liquid alloy (or other system) can, in general, be expressed in terms of a force–current correlation function (Kumar and Sorbello 1975). If the electron–ion interaction is sufficiently weak and if the short-range order in the alloy extends over a length which is small compared to the mean free path, the correlation function can be explicitly evaluated (Schaich 1976). The result for the drag force involves the pseudopotentials and the local ionic arrangement around the ion in question. Unfortunately, information on local environments in liquid alloys is not readily available, and one must instead be content with an expression for the *average* driving force, $\mathbf{F}_\gamma$, acting on an ion of species γ. Under the assumption of a weak, local, and energy-independent interaction, $\mathbf{F}_\gamma$ takes the form (Faber 1972, Schaich 1976):

$$\mathbf{F}_\gamma = Z_\gamma |e| \mathscr{E} - Z^* \alpha_\gamma / \rho |e| \mathscr{E} \tag{1}$$

where

$$\alpha_\gamma = \frac{4\pi^3 \hbar}{e^2 k_F} Z^* \int_0^1 y^3 \, dy \sum_\beta \left(\frac{x_\beta}{x_\gamma}\right)^{1/2} S_{\gamma\beta}(y) V_\beta(y) V_\gamma(y). \tag{2}$$

† Supported in part by the US National Science Foundation under Grant No. GH-33746. A more detailed account of the calculations described here has been given in Stroud D 1976 *Phys. Rev.* **B13** 4221.

Here $\rho = \Sigma x_\gamma \alpha_\gamma$ is the alloy resistivity in the weak-scattering limit (Faber and Ziman 1965), the $S_{\gamma\beta}$ are partial ionic structure factors, $V_\gamma(y)$ is the pseudopotential form factor for an ion of species γ in units of two-thirds the Fermi energy, y is the wavevector in units of twice the Fermi wavevector k_F, and $Z^* = \Sigma x_\gamma Z_\gamma$.

The purpose of the work reported here was to see whether or not the average forces $\mathbf{F}_\gamma$ could be correlated with the observed directions of electromigration in a number of liquid alloys. Rather than study a single system with an accurate pseudopotential, it was decided to investigate a variety of systems, using simple model pseudopotentials. All the calculations were carried out using empty-core pseudopotentials (Ashcroft 1966), Hubbard screening (Hubbard 1957), and partial structure factors corresponding to a mixture of hard spheres of different diameters. The core radii characterizing the pseudopotentials were chosen so as to fit the resistivities of the pure liquid metals at melting. The hard-sphere packing fraction was chosen so as to equal 0·45, a value which reproduces the observed liquid structure factors at melting quite well (Ashcroft and Lekner 1966), and the ratio of hard-sphere diameters was (except for Na_xK_{1-x}) assumed to equal the cube root of the ratio of the valences. (The ion of larger valence is expected to have a larger hard-sphere diameter, but the specific choice of ratios is otherwise arbitrary.)

Table 1. Calculated drag coefficients and directions of migration for polyvalent solutes in liquid polyvalent hosts. Experimental results are in brackets; all are taken from the compilation by Pratt and Sellors (1973).

Alloy system	α	Direction of migration
In Sn	68 [72]	A [A]
In Pb	113 [72]	A [A]
In Bi	190 [115]	A [A]
Sn Bi	168 [75]	A [A]
Pb Bi	149 [135]	A [A]
Bi Zn	8 [40, 33, −100, −180]	C [C]
Sn Al	13 [28]	C [C]
Pb Sn	47	C [C]

Table 1 shows the computed and measured drag coefficients for a number of alloys of polyvalent solutes in polyvalent hosts. It is evident from this table that the average forces are a good indicator of the directions of electromigration in such liquid alloys. The discrepancy in magnitude is not large and should not be viewed as excessively important in view of the neglect of fluctuations mentioned above, as well as the substantial experimental uncertainties.

The system Na_xK_{1-x} is of particular interest because a reversal in electromigration direction is observed in that system as a function of composition, the Na ions migrating to the cathode for $x \gtrsim 0{\cdot}65$ and to the anode for other compositions (Epstein 1972, Rigney 1974). We have carried out calculations over the whole composition range, assuming a hard-sphere ratio $\sigma_K/\sigma_{Na} = 1{\cdot}25$, which is consistent with structural and thermodynamic data (Stroud 1973). This leads to the results of figure 1. As may be seen, the calculated average forces do not cross over with the choice of parameters used

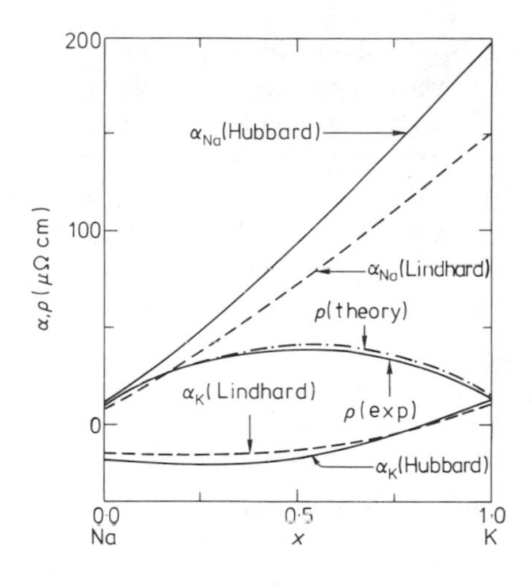

Figure 1. Electron drag coefficients α_{Na} and α_K for liquid Na_xK_{1-x} at 370 K, from equation (2), using a hard-sphere ratio σ_K/σ_{Na} = 1·25 and a packing fraction of 0·45. Calculations using both Hubbard and Lindhard screening are shown.

in this calculation. The trend of the curves is, however, clearly in the right direction, and a crossover is predicted to occur if the alloy system is placed under pressure so that the electron density is increased. A crossover is also predicted to occur if the ratio of the hard-sphere diameters is assumed equal, as is evident from figure 2, where the 'effective valence' $Z_K^{eff} = Z_K - \alpha_K/\rho$ is plotted as a function of σ_K/σ_{Na}.

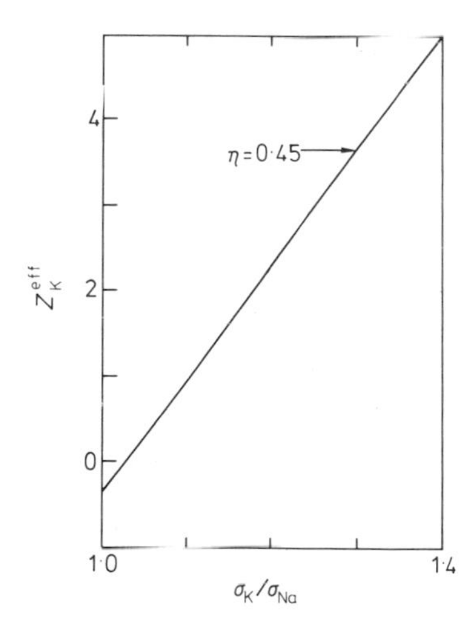

Figure 2. Computed Z_K^{eff} for liquid **Na** K, plotted as a function of σ_K/σ_{Na}.

We have also considered the driving force acting on ions of polyvalent metals dissolved in Na and K. The polyvalent ions are predicted to migrate with large effective valences to the anode, as has been observed experimentally (Pratt and Sellors 1973). Since in order to carry out the calculations we have had to extrapolate the pseudopotentials of the polyvalent ions to electron densities very different from their native environment, the calculated values for the average forces are not quantitatively reliable, but the qualitative predictions are clearly correct.

The results presented here show that the average force (equation 2) does correctly describe the trend of the experimental data. It is not adequate, however, to predict detailed behaviour such as crossover concentrations. This is almost certainly due to a neglect of fluctuations in the driving force, and the correlation of such fluctuations with fluctuations in ionic mobility. The necessity of considering fluctuations in the driving force was already pointed out several years ago (Faber 1972). They are known from explicit calculations to be very large in the solid state, the ions nearest vacancies tending to experience the largest electron drag forces (Sorbello 1973). Similar calculations in the liquid state would be possible but considerably more complicated because of the need to have an explicit model of the actual ionic configuration in the liquid.

References

Ashcroft N W 1966 *Phys. Lett.* **23** 48
Ashcroft N W and Lekner J 1966 *Phys. Rev.* **145** 83
Epstein S G 1972 *Liquid Metals: Chemistry and Physics* ed L H Bennett and R H Willens (New York: Dekker) p572
Faber T E 1972 *Theory of Liquid Metals* (Cambridge: Cambridge UP) p485
Faber T E and Ziman J M 1965 *Phil. Mag.* **11** 153
Hubbard J 1957 *Proc. R. Soc.* **A243** 336
Kumar R and Sorbello R S 1975 *Thin Solid Films* **25** 25
Pratt J N and Sellors R G R 1973 *Electrotransport in Metals and Alloys* (Reichen: Trans. Tech. SA)
Rigney D A 1974 *Charge Transfer/Electronic Structure of Alloys* ed L H Bennett and R H Willens (New York: Metallurgical Society)
Schaich W L 1976 *Phys. Rev.* **B12** 3360
Sorbello R S 1973 *J. Phys. Chem. Solids* **34** 937
Stroud D 1973 *Phys. Rev.* **B7** 4405

Reversal of electromigration direction in selected metal alloy systems

D A Rigney and R D Mack
Department of Metallurgical Engineering, The Ohio State University, Columbus, Ohio 43210, USA

Abstract. When a direct current is passed through a binary liquid metal alloy, one component will migrate to the anode and the other will concentrate at the cathode. For most binary systems reported, the same migration direction is observed for a particular component at all compositions. However, for other systems, the direction of migration reverses at a critical composition. That is, if species A migrates to the anode for a small amount of A dissolved in B, then B migrates to the anode for a small amount of B dissolved in A. At some intermediate composition, no separation is found during electromigration experiments. The Na–K and Na–Rb systems and several systems involving mercury are known to behave in this manner.

Using a simple capillary-reservoir technique, one can drive the composition at one end of the capillary toward the reversal composition, so that the composition profile features an easily measured plateau if the current is applied for a sufficiently long time. Since a literature survey indicates that additional systems are likely to feature migration reversals, we have applied our capillary technique to selected systems involving noble metals alloyed with polyvalent normal metals. The results, including critical compositions, are reported.

1. Introduction

When a direct current is passed through a liquid binary alloy, the components A and B will generally respond in such a way that a concentration gradient develops. Epstein and Paskin (1967) have suggested that one can predict the direction of motion of both components by comparing their bulk resistivities for pure liquid A and pure liquid B. They expected that this would be a reasonable way to estimate the relative size of each component's scattering cross section, S, and therefore to predict which component would receive a larger push toward the anode by receiving momentum from the conduction electrons. Their resistivity rule works well for most normal metal systems, even for concentrated alloys (Pratt and Sellors 1973). It is particularly good for alloys involving two polyvalent components, in which the scattering behaviour does not appear to change markedly with composition.

However, the simple resistivity rule is not obeyed for several alloys involving alkali metals and for a larger number of alloys involving noble metals or transition metals dissolved in polyvalent solvents (Rigney 1974). In fact, for at least five of these systems (Na–K, Na–Rb, Hg–Na, Hg–K, and Hg–Ba), it has been clearly demonstrated that a special composition exists at which the direction of migration reverses (Pratt and Sellors 1973). For example, K goes to the anode in nearly pure Na, but Na goes to the anode in nearly pure K. The data for many of the other anomalous systems are sparse, but there are definite indications that reversals of migration direction may be much more common than previously expected. This paper reports the results of a search for migration reversals in Ag–Sn and Au–Pb. These systems are the first in a series chosen to clarify electrotransport behaviour in noble metal/polyvalent metal alloys.

Kuzmenko *et al* (1962) have reported that 1 at.% Sn in liquid Ag migrates to the anode at elevated temperatures with an effective valence Z^* in the range -48 to -23. Schwarz (1940) reported that for about 17 at.% Ag, Sn also goes to the anode. These results are consistent with the resistivity rule. However, for a trace of either Ag in liquid Sn or Au in liquid Pb (Schwarz 1940, Belashchenko and Grigorev 1961, 1962), the noble metal goes to the anode. Thus reversals are indicated for small concentrations of the noble metal.

One could perform a series of experiments using different compositions to locate the reversal composition in each system, but this is not the most efficient way to proceed. Blough *et al* (1972) have described a long-time technique in which electrotransport drives the composition toward the reversal composition at one end of the liquid alloy sample. The resulting concentration plateau provides an accurate measure of the reversal composition. This method has now been applied to liquid alloys of Ag in Sn and Au in Pb.

2. Experimental details

A simple Pyrex cell which could serve for capillary-reservoir experiments in either the normal or inverted position was used (figure 1). This allowed the choice of migration direction to be made consistent with relative densities of A and B and with initial composition. Thus, a concentration plateau can develop at the closed end of the capillary without complications from convection. The capillary portion was made from Trubore Pyrex from the Ace Glass Company, 1·50 mm ID, 7·5 mm OD, and 4·0 cm long. The electrodes were 99·9% tungsten wire from the General Electric Company, with their ends electropolished in 2% NaOH solution for 30 s at a current density of 4 A cm^{-2} (AC).

Great care was taken to assure that the alloys used were initially homogeneous and free from oxides and other contaminants. Starting material was 99·999% pure. Ag and

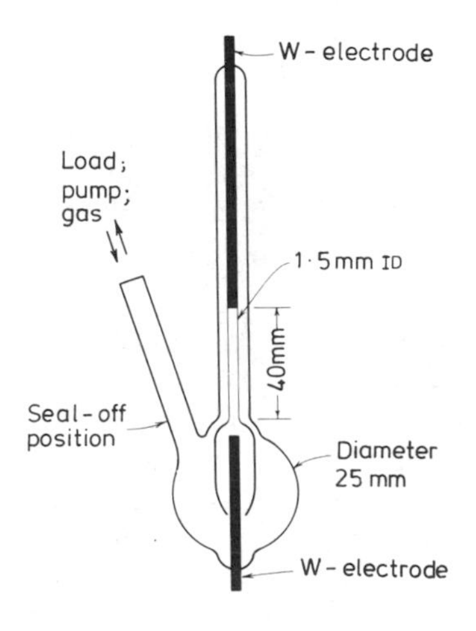

Figure 1. Schematic diagram of Pyrex cell for electrotransport experiments with dilute liquid alloys of Ag in Sn and Au in Pb.

Pb were supplied in ingot form by ASARCO, Sn in ingot form by Ventron, and Au in shot form by the United Mineral and Chemical Corporation. Initial cleaning consisted simply of removing cutting burrs with 120 grit SiC paper, and, for the Pb, further cleaning in boiling nitric acid solution (1:1 by volume with water), followed by successive rinses in distilled water and methanol.

Each alloy was initially mixed in a Pyrex test tube under argon and then poured into a second test tube held in water for rapid solidification. Most of the oxide remained on the walls of the first test tube. The alloy was then placed into the upper portion of a Vycor tube, melted under vacuum, and allowed to drip through a narrow constriction into the lower portion. The specimen was then sealed under about $\frac{1}{3}$ atm of hydrogen and thoroughly mixed by induction melting. Next, the alloy was rapidly quenched by vigorously shaking the capsule under water. It was then melted under argon again, and drawn up into 2·5 mm ID Pyrex tubing. The resulting ingot was cut into 1–1·5 cm lengths with a clean razor blade, loaded into the experimental cell, melted under vacuum, and forced into the capillary portion of the cell by back-filling with about $\frac{1}{2}$ atm of argon. Finally, the loading tube was sealed off, and, with the alloy still entirely molten, the assembly was lowered into a pre-heated tube furnace for the electrotransport experiment.

After the temperature had equilibrated for $\frac{1}{2}$ hr, a direct current of 7·0 A was passed through the liquid metal for 94·5 hr (Sn–Ag) or 74·25 hr (Pb–Au). Temperature and current were monitored for the duration of electrotransport. The two thermocouples, one close to each end of the capillary, detected no temperature increase from the current density of 4·0 A mm^{-2}. A small positive temperature gradient was used for each experiment. For Sn–Ag, T varied from 613 to 649 K along the capillary length, and for Pb–Au, the range was 635–647 K.

Since it was expected that the noble metal solute would migrate to the anode for the dilute alloys, the anode was chosen to be the lower electrode. This assured that the heavier component would migrate downward in each case. After each experiment, the alloy was rapidly frozen by forcing compressed air through the furnace. It was then removed from the capillary and cut into 20–30 pieces for chemical analysis by atomic absorption. Samples were also taken from the reservoir portion of the cell.

Atomic absorption analysis was accomplished with a Perkin-Elmer Model 360 spectrophotometer, using a HGA-2100 controller for the graphite furnace. Since the Ag and Au solutions are light sensitive, analysis was completed within hours of dissolution for both the standards and the alloy samples.

3. Results and discussion

A composition plateau developed at the anode side for both Sn–Ag and Pb–Au. These indicated reversal compositions of 1·8±0·2 at.% Ag in Sn and 2·4±0·3 at.% Au in Pb. For smaller concentrations, the noble metal migrates to the anode, and for larger concentrations it goes to the cathode.

The reversals in Na–K and Na–Rb have been explained by the crossing of the scattering cross section curves for each component as composition is varied (Epstein and Dickey 1970, Olson *et al* 1972, Stroud 1976). The reversals in liquid Hg, and the reversals in the present alloys have not previously been explained. Arguments based directly on residual resistivity are not adequate, since they may be used to predict reversals in almost any binary system, even the majority which do not have this interesting feature.

Since the reversal occurs in dilute alloys for all of the unexplained cases, it is tempting to examine a nearest-neighbour model for some insight into the reversal phenomenon. One considers the nearest-neighbour environment of each solute atom at a given instant in time. One then assumes that a solute atom that is surrounded entirely by solvent atoms behaves quite differently from one that has one or more solute atoms in nearest-neighbour sites. That is, the electronic structure of the solute is different for isolated solute atoms from the electronic structure existing when the solutes are clustered with two or more together. For the case of Au in Pb, one might imagine that the change in the electronic structures of the isolated Au atoms is such that the d-shell becomes more involved in electron scattering. Thus, when most of the Au atoms are isolated, scattering will be large, and Au will migrate to the anode in an electrotransport experiment. However, as the concentration of Au rises, the fraction of pairs and higher groupings of Au will rise, the scattering will become lower than that of the polyvalent solvent atoms, and the Au will go to the cathode.

Assuming for now that the diffusive mobilities are similar for Au in these different environments and that solute binding energies may be neglected, one would expect a reversal to occur when equal numbers of Au atoms existed as singles or clustered as pairs or larger groupings. That condition would occur when the probability of a nearest-neighbour shell having one or more Au atoms is 0·5. Or, one could say that the probability that a given nearest-neighbour shell is all Pb (solvent) is 0·5. Therefore, if C is the atomic fraction of solute, and Z is the average coordination number in the liquid, a simple equation results:

$$0{\cdot}5 = (1 - C)^Z. \tag{1}$$

For $Z = 10$, 11 and 12 for nearly close-packed liquids, $C = 6{\cdot}7$, 6·0 and 5·6 at.%, respectively.

These results should be considered upper limits to the critical concentration of solute, because solute–solute binding energies have been ignored. Such an interaction would increase the average lifetimes of the various Au-groupings above that expected for configurations based on random diffusion of isolated atoms, and therefore the reversal would occur at a lower concentration of solute. Thus, the results for Ag–Sn and Au–Pb are reasonably well described by the assumed model.

It is interesting to note an apparent connection between these results and the phenomenon of fast diffusion in solids, reviewed by Warburton and Turnbull (1975). The case of Au in Pb is a classic one, in which Au diffuses much faster than Pb. For example, at 448 K, Au diffuses about 10^5 times as fast as Pb in the Pb matrix. The diffusion has been called interstitial-like, which is difficult to reconcile with the usual size ascribed to Au atoms unless an appropriate electronic transition occurs. In polyvalent solvents, the large difference in electronic environment between Au in Au and Au in the polyvalent solvent could lead to such a transition. Rossolimo and Turnbull (1973) and Warburton (1973) have shown that a sharp change in the residual resistivity with temperature can be explained in dilute Pb–Au alloys by this assumption. The agreement is in fact quantitative if a standard Gibbs free energy difference between Au in one type of environment and Au in a different environment is assumed to be (9 ± 2) kcal mole^{-1} $- T(11{\cdot}5 \pm 4{\cdot}0)$ cal mole^{-1} K^{-1}. These results are also consistent with the slower than expected diffusion encountered in precipitation experiments with Au in Pb at low temperatures (Rossolimo and Turnbull 1973), and with the types of peaks which are found in internal friction experiments with this system (reviewed in Warburton and Turnbull 1975).

The model presented here is related to one presented by Jaccarino and Walker (1965) to explain changes in magnetic moments which occur upon alloying. In both cases, a critical number of nearest neighbours of a given type is assumed to affect the details of electronic structure of the solute. In the present case, the critical number appears to be one nearest neighbour. It should be an acceptable approach as long as a given nearest-neighbour environment is maintained for times τ longer than the time needed for a scattering event to occur. Even for the rapid diffusion encountered in liquid metals, this time τ is about 10^{-11} s, which is much longer than characteristic scattering times.

This discussion has concentrated on the noble metal/polyvalent metal systems. However, it may be that a similar picture can be used for the liquid Hg alloys. The electrotransport reversals occur for about 16 at.% Na, about 12 at.% K, and about 3·8 at.% Ba in liquid Hg (Kremann *et al* 1930, 1931). The values for Na and K are higher than expected. Unless the coordination number is very low, or unless much of the solute was lost through oxidation, it is difficult to explain these results using the nearest-neighbour model. Perhaps these experiments should be carefully repeated.

Finally, one can imagine that various properties of these liquid metal systems will be affected if the simple picture presented above is basically correct. First, resistivities of systems like Pb–Au should show anomalous behaviour over a small range of composition as the Au content in Pb is increased. Second, the magnetic behaviour of these systems may change strikingly in the same composition range. These types of data are not yet available in liquid metals.

The electrotransport work described in this paper is part of a continuing survey of reversals of migration direction in liquid metal systems. Available evidence shows that the chances are good that many more systems can be proven to have reversals at low concentrations of selected solutes.

Acknowledgments

The authors are grateful to Professor G St Pierre for helpful discussions and to the National Science Foundation for support under contract DMR73-02426-A01.

References

Belashchenko D K and Grigorev G A 1961 *Izvest. Vysshikh. Ucheb. Zaved. Chernaya Met.* **11** 116
—— 1962 *Izvest. Vysshikh. Ucheb. Zaved. Chernaya Met.* **5** 120
Blough J L, Olson D L and Rigney D A 1972 *J. Appl. Phys.* **43** 2476
Epstein S G and Dickey J M 1970 *Phys. Rev.* **B1** 2442
Epstein S G and Paskin A 1967 *Phys. Lett.* **A24** 309
Jaccarino V and Walker L R 1965 *Phys. Rev. Lett.* **15** 258
Kremann B, Bauer F, Vogrin A and Scheibel H 1930 *Monatsh. Chem.* **56** 35

Kremann B, Vogrin A and Scheibel H 1931 *Monatsh. Chem.* **57** 323
Kuzmenko P P, Ostrovsky L F and Kovalchuk V S 1962 *Phys. Metals Metallogr.* **13** 83
Olson D L, Blough J L and Rigney D A 1972 *Acta Metall.* **20** 305
Pratt J N and Sellors R G R 1973 *Electrotransport in Metals and Alloys* (Riehen, Switzerland: Trans Tech SA)
Rigney D A 1974 *Charge Transfer/Electronic Structure of Alloys* ed L H Bennett and R H Willens (New York: Met. Soc. AIME)
Rossolimo A N and Turnbull D 1973 *Acta Metall.* **21** 21
Schwarz K E 1940 *Elektrolytische Wanderung in Flüssigen und festen Metallen* (Leipzig: J A Barth)
Stroud D 1976 *Phys. Rev.* **B13** 4221
Warburton W K 1973 *J. Phys. Chem. Solids* **34** 451
Warburton W K and Turnbull D 1975 *Diffusion in Solids, Recent Developments* ed A S Nowick and J Burton (New York: Academic Press)

Determination of the diffusion coefficient in liquid metal alloys from measurements of the electrical resistivity

M Kéita†, S Steinemann†, H U Künzi‡ and H-J Güntherodt‡
† Institut de Physique Expérimentale de l'Université de Lausanne, Switzerland
‡ Institut für Physik der Universität Basel, Switzerland

Abstract. A new experimental method for measuring chemical diffusion in liquid alloys is presented. It is based on the 'capillary-reservoir technique', but instead of a chemical analysis, a continuous recording of the electrical resistance is made. These resistance–time data are fitted to the formal solution of the diffusion equation. Results for Hg–In, Al–Sn and Al–Si alloys are presented, together with the electrical resistivity of these liquid systems.

1. Introduction

The determination of diffusion coefficients in the liquid state is a difficult task. Experiments need to be conducted in such a way that transport occurs exclusively through diffusion; that is, convection currents must be absent. Simplicity of the experimental set-up (i.e., of its boundary conditions) is further required to allow easy computation of the diffusion coefficient.

The availability of precise data on diffusion is important for understanding the liquid state. The data also act as basic parameters for studying metallurgical phenomena such as nucleation and growth processes during solidification of alloys. This work describes a precise method for the measurement of the chemical diffusion coefficient in liquid binary alloys of metals.

2. Experimental methods

Walls (1970) (see also Rigney 1977) reviews the different techniques for measuring diffusion coefficients in liquids. Two types of techniques can be distinguished:

(i) The indirect methods where the diffusion coefficient is extracted from a phenomenon dependent on diffusion (such as nuclear magnetic resonance, inelastic neutron scattering, etc) and which applies especially to self-diffusion.
(ii) The direct methods like the capillary-reservoir, the long capillary or the shear-cell technique and their variations; chemical and self-diffusion (using a tracer) coefficients can be obtained.

The reservoir-capillary technique is the most commonly used method for the study of liquid metals. A capillary is in contact with an 'infinite volume reservoir' at one end. If C_r is the concentration in the reservoir, and C_0 is the concentration in the capillary at the beginning of the experiment, the solution of the diffusion equation is of the form:

$$\Psi(x,t) \equiv [C(x,t)-C_r]/(C_0-C_r) = \sum_{n=0}^{\infty} A_n(x)\, B_n(t), \tag{1}$$

where

$$A_n(x) = 4[\pi(2n+1)]^{-1} \sin(\lambda_n x), \tag{2a}$$

$$B_n(t) = \exp(-\lambda_n^2 Dt), \tag{2b}$$

$$\lambda_n = (2n+1)\,\pi/(2L), \tag{2c}$$

and L is the total length of the capillary. It is assumed that the diffusion coefficient, D, is independent of concentration; this assumption is a good approximation if $|C_r - C_0|$ is small.

Equation (1) can be integrated over x to obtain the mean concentration in the whole capillary, or between two points located at αL and βL to give:

$$\bar{\Psi}_{\alpha\beta}(t) = 8[\pi^2(\beta-\alpha)]^{-1} \sum_{n=0}^{\infty} (2n+1)^{-2}(\cos\alpha_n - \cos\beta_n)\exp(-\theta_n^2), \tag{3}$$

where

$$\alpha_n = \alpha\pi(n+\tfrac{1}{2}),$$

$$\beta_n = \beta\pi(n+\tfrac{1}{2}),$$

$$\theta_n^2 = \pi^2(2n+1)^2\, Dt/(4L^2),$$

and

$$\bar{\Psi}_{\alpha\beta}(t) = \left(\int_{\alpha L}^{\beta L} \Psi(x,t)\,\mathrm{d}x\right)\left(\int_{\alpha L}^{\beta L} \mathrm{d}x\right)^{-1}.$$

If $\alpha = 0$ and $\beta = 1$, equation (3) reduces to an expression containing only the exponential function. This applies to the case when the whole capillary is analysed.

Systematic and accidental errors can affect the capillary-reservoir technique; the most important being:

(i) Turbulence and material losses during introduction and removal of the capillary from the melt in the reservoir produce an error in the effective diffusion distance (called the 'ΔL effect'), etc.
(ii) Improper filling of the capillary (gas bubbles or oxide particles present).

This classical technique needs, in any case, a large number of experiments but, nevertheless, the accuracy of data is seldom better than 30% if severe precautions are not adopted. Some of these difficulties are absent in a continuous measurement in the liquid state such as that described in the next section.

3. Reservoir-capillary resistivity technique

The title suggests the procedure; the experimental configuration is still the capillary-reservoir arrangement, but the concentration is determined by a measurement of resistivity in the capillary, as done by Buell *et al* (1970) and Kaiser (1970). Some conditions must be fulfilled for this extension of the capillary-reservoir technique:

(i) Easy use of equation (3) is possible only if there is a linear relationship between resistivity and concentration; it is straightforward to show that in this case

$$\bar{\Psi}_{\alpha\beta}(t) = [R_{\alpha\beta}(t) - R_{\alpha\beta}(\infty)] / [R_{\alpha\beta}(0) - R_{\alpha\beta}(\infty)] . \quad (4)$$

For a nonlinear resistivity–concentration dependence, equation (1) must be used instead of (4) and the computation problem becomes slightly more complicated.
(ii) The variation of the resistivity with alloying must be sufficiently large; this is the case for liquid metals.
(iii) The resistance measurements must be AC to avoid electromigration and at a low level to avoid Joule heating. High precision is possible with phase-sensitive detection (easily 0·1%).
(iv) The diffusion coefficient and the resistivity vary with temperature, but extreme temperature stabilization is not required for the experiment; correction of the resistance and diffusion may be done knowing the temperature of the capillary.

The experimental equipment and different types of measuring cells are shown in figures 1, 2(*a*) and 2(*b*). Different orientations of the capillary (horizontal and vertical) are possible but the arrangement in the vertical case must be such as to oppose movement due to gravitational forces. The diffusion cell and electrodes must resist attack from the liquid metal; ceramics and graphite or a refractory metal are suitable. Calibration is an easy task with mercury.

A typical experiment proceeds as follows:

(i) The solvent or an alloy of given concentration is introduced in the reservoir and degassed during heat-up.
(ii) A protective gas is introduced when the measuring temperature is reached; the excess pressure pushes the melt into the capillary and the resistance is checked by

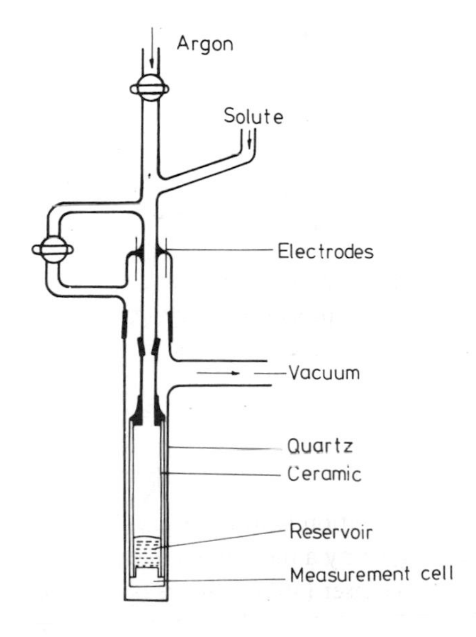

Figure 1. Sketch of the apparatus.

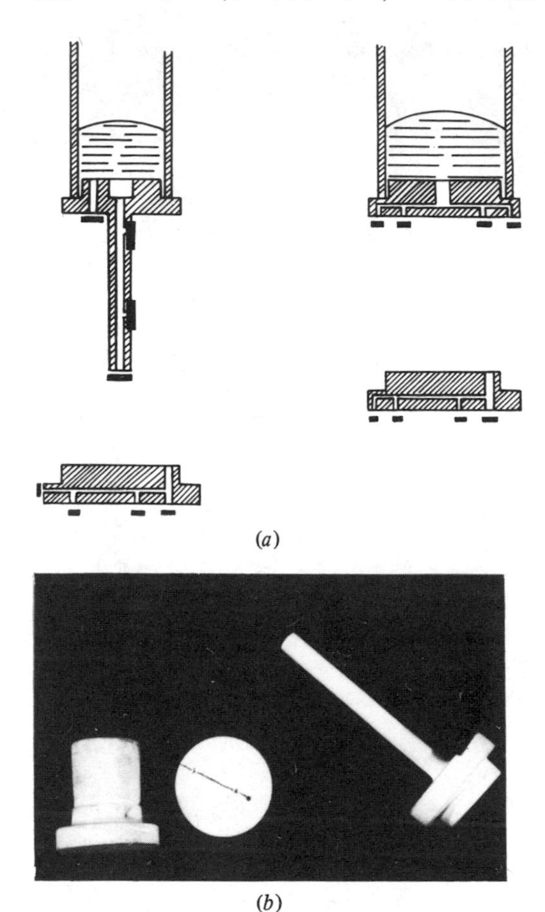

Figure 2. The different types of measuring cells: (*a*) sketch; (*b*) photograph.

comparison with a calculated value (obtained from the geometrical calibration with mercury and the known resistivity of the metal with concentration C_0).
(iii) Solute is added to the reservoir and the diffusion recorded by following the resistance.
(iv) Step (iii) is repeated for other concentrations.

Several advantages of this new method are apparent. As the whole experiment is carried out while the metal is liquid, no hydrodynamical problems related to the handling of the capillary occur. The continuous recording of resistance is equivalent to a large number of single classical experiments. It is further seen that an eventual 'ΔL effect' is corrected with ease; in fact if $\alpha \neq 0$ and/or $\beta \neq 1$, then D and L can be taken as independent parameters for fitting to a resistance–time curve.

4. Experimental results and discussion

The dependence of resistivity upon concentration and temperature of the system of interest must first be determined. This can be done in the same capillary as used for the study of diffusion. For the Hg–In system, the electronic transport properties have been fully investigated by Roll *et al* (1961) and Cusack *et al* (1964). For Sn–Al and Al–Si only few data are available and these are not very reliable (Marty 1958, Korolkov 1962). The properties for these alloys have been measured and the results are shown in figures 3 and 4.

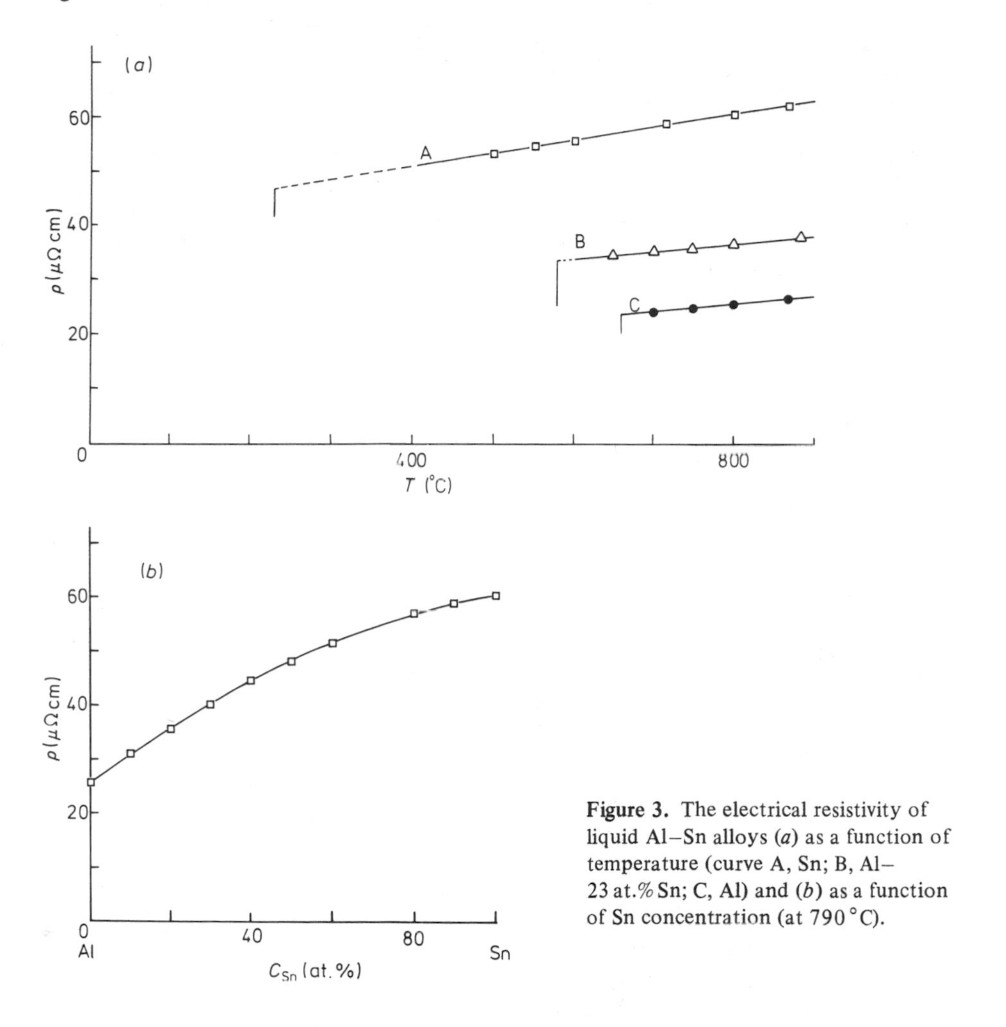

Figure 3. The electrical resistivity of liquid Al–Sn alloys (*a*) as a function of temperature (curve A, Sn; B, Al–23 at.% Sn; C, Al) and (*b*) as a function of Sn concentration (at 790 °C).

The diffusion measurement of the Hg–In system was taken in cells made of glass and with molybdenum electrodes. High-purity alumina and graphite were used for the Sn–Al and Al–Si systems. The resistivity–concentration curve of the Hg–In and Al–Sn systems are rather nonlinear. The analysis following equation (3) was made using small steps of concentration changes, $|C_r - C_0|$. The resulting errors do not exceed the uncertainties of the resistivity measurement.

The diffusion coefficients of the Hg–In and Sn–Al systems are shown in figures 5 and 6. In and Al are the lighter elements and convection mixing is readily avoided if they are the solutes for the vertical arrangement of figure 2(*a*). The Sn–Al system was also measured in the horizontal capillary and the results coincided with data taken in

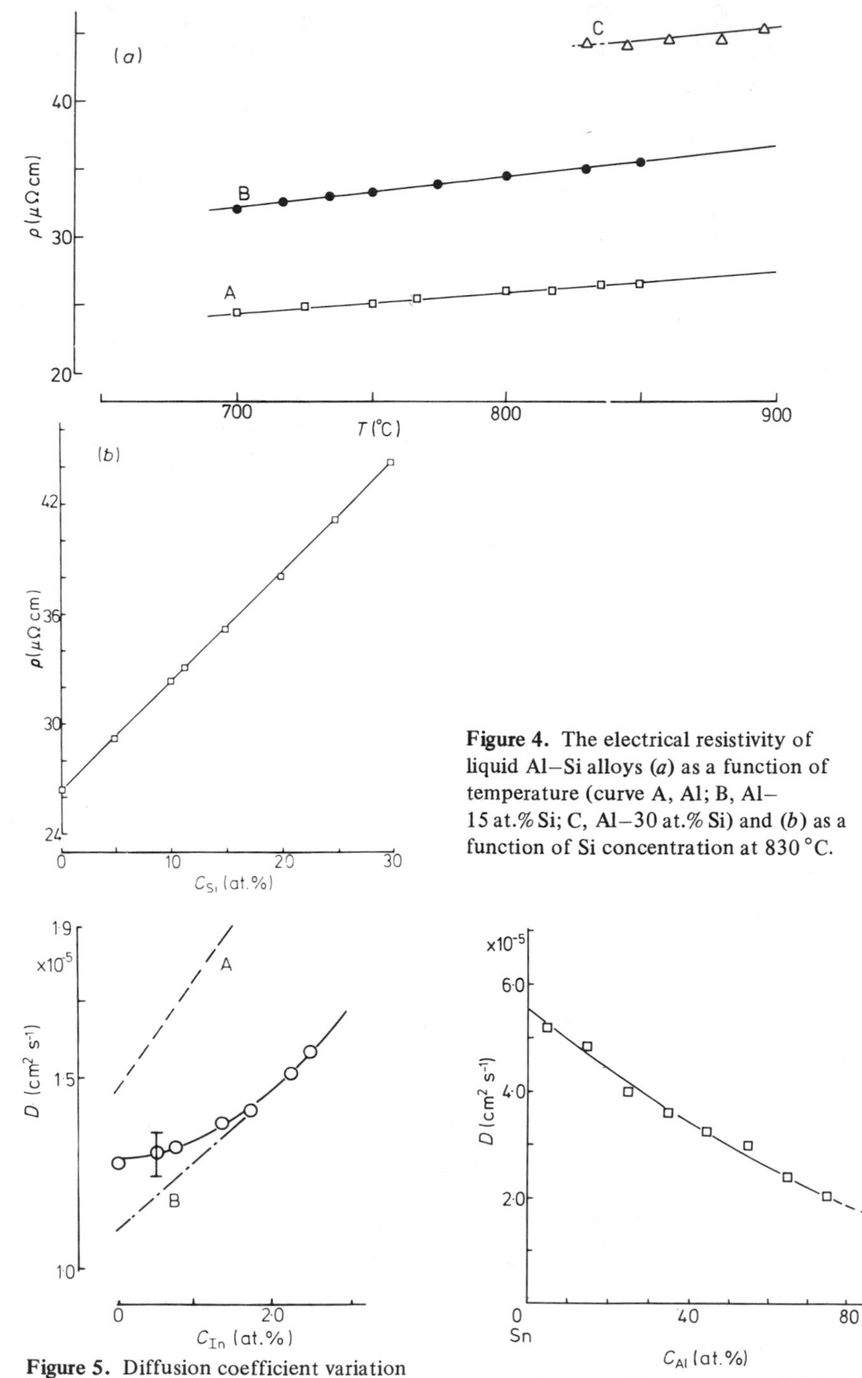

Figure 4. The electrical resistivity of liquid Al–Si alloys (*a*) as a function of temperature (curve A, Al; B, Al–15 at.% Si; C, Al–30 at.% Si) and (*b*) as a function of Si concentration at 830 °C.

Figure 5. Diffusion coefficient variation with In concentration in liquid Hg–In alloys at 25 °C: A, from Nanda *et al* (1970); B, from Kaiser *et al* (1970). The solid line represents the results of this work.

Figure 6. Diffusion coefficient variation with Al concentration in liquid Sn–Al alloys at 800 °C.

the vertical capillary, within experimental error, when small-diameter bores (1·2 mm or less) were used.

The absolute errors for the measured diffusion coefficients are estimated to be of the order of 10%. This corresponds to the 'root mean square standard deviation' calculated from fitting the experimental resistance–time curves to the formal solution (4) of the diffusion equation. When these same data are plotted in an Arrhenius plot of D versus $1/T$ a smaller error of about 5% is obtained. The reproducibility of the experiments was in any case better than 5%.

The Al–Si system was extensively studied. It was first observed that vertical capillaries always gave larger diffusion coefficients than the horizontal arrangement, even for small-diameter capillaries (1·2 mm). The data of figure 7 were obtained with the horizontal capillary of diameter 1·4 mm and total length 20 mm, these being dimensions to make easy comparison with other data. Korber *et al* (1971) review diffusion data for the Al–Si system and these are collected in figure 8. Considerable discrepancies appear but a precision of 15% is quoted. The present data (solid lines in figure 8) are the lowest

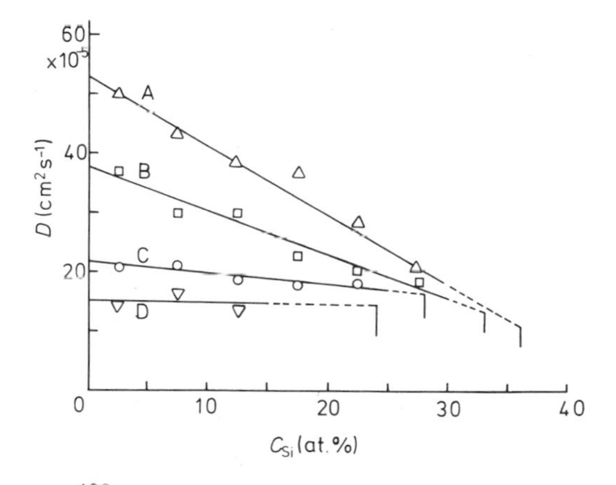

Figure 7. Diffusion coefficient variation with Si concentration in liquid Al–Si alloys: A, 903 °C; B, 863 °C; C, 801 °C; D, 755 °C.

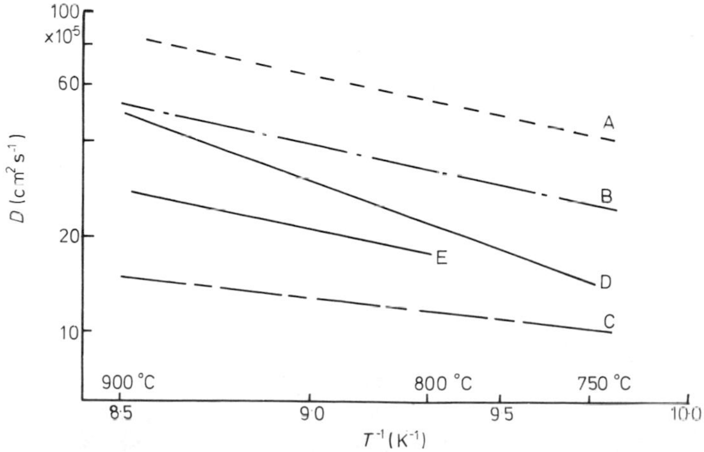

Figure 8. Arrhenius plot for comparison of some data reviewed by Korber (1971). Curve A (for 4·5–9 at.% Si) and curve B (for 7·5–14·2 at.% Si) are chemical diffusion coefficients. Curve C is the self-diffusion coefficient in eutectic Al–Si. The solid lines are the results of this work: curve D, 0–5 at.% Si and curve E, 20–25 at.% Si.

observed and it is believed that the effects of convection currents have been greatly reduced, if not eliminated.

Some general remarks concerning the influence of capillary position and geometrical factors should be added (for details see Kéita *et al* 1976). Convection is limited or suppressed only for small-bore capillaries. This requirement seems to be even more important than the orientation of the capillary as evidenced by our results. Small capillary diameters also prevent perturbation through mechanical vibrations (pumps, etc). Low values of the diffusion coefficients are likely to be a criterion for the absence of convection.

5. Conclusion

The common capillary-reservoir technique has been improved by adding a sensitive and continuous analysis of composition. The resistivity measurement assures constant control of the diffusion condition in the capillary and suppresses all errors coming from the introduction and the removal of the capillary from the melt. The resistivity can easily be measured with high precision. It should further be recalled that the technique is generally applicable to liquid alloys because their resistivities show marked dependency on concentration.

The main object of the present study was to obtain diffusion data for Al–Si alloys and emphasis was given to this system. It is part of the study of crystallization phenomena of eutectic Al–Si casting alloys.

Acknowledgments

The support of the Schweizerische Aluminium AG is gratefully acknowledged.

References

Cusack N *et al* 1964 *Phil. Mag.* **10** 871
Buell *et al* 1970 *Metall. Trans.* **1** 1875
Kaiser H 1970 *MS Thesis* Iowa State University, Ames, Iowa
Kéita M *et al* 1976 *Ber. Bunsenges. Phys. Chem.* **80** 722
Korber K *et al* 1971 *Giesserei-Forschung* **23** 169
Korolkov A M *et al* 1962 *Izv. Akad. Nauk* **1** 84
Marty W 1958 *Brown Boveri Mitt.* **45** 549
Nanda *et al* 1970 *Metall. Trans.* **1** 353
Rigney D A this volume
Roll A *et al* 1961 *Z. Metallka.* **52** 111
Walls H A 1970 *Physicochemical Measurements in Metals* **4** 459–92

Experimental studies on the diffusion and electrotransport of impurities in liquid tin

D Agnoux, M O du Fou, J M Escanyé and M Gerl
Laboratoire de Physique des Solides, Université de Nancy I, CO 140, 54037 Nancy-Cedex, France

Abstract. We describe an accurate method for measuring diffusion coefficients in liquid metals, using a shear cell device. The diffusion coefficients of ^{110m}Ag, ^{109}Cd, ^{114m}In, ^{113}Sn and ^{125}Sb, as well as the effective diffusion coefficients of In in presence of a DC current, have been measured in liquid tin.

1. Introduction

Many mechanisms have been proposed in order to interpret diffusion phenomena in liquid metals. Semi-empirical models (Escanyé and Gerl 1976) are in general transposed from diffusion theories appropriate to crystals and, as they make use of numerous parameters, their agreement with experimental data is not significant. On the other hand, statistical models evolved from the theory of dense hard-sphere fluids (Protopapas *et al* 1973) are fairly successful, although the only adjustable parameter is the hard-sphere diameter σ, the radial distribution function $g(\sigma)$ being deduced from the compressibility factor $Z = pV/NkT$. Z itself can be calculated using the Carnahan–Starling or Percus–Yevick approximations, when the packing fraction is known. The temperature dependence of the diffusion coefficient is very well reproduced when provision is made for the relative softness of the repulsive potential, leading to a temperature-dependent hard-sphere diameter $\sigma(T)$.

There is a lack of experimental data on self-diffusion and impurity-diffusion in liquid metals. In particular, it would be desirable to perform measurements of self-diffusion coefficients in a large temperature range, and a systematic study of impurity-diffusion coefficients as a function of the characteristics of the impurity (valence, mass, size etc). In the present paper we report some measurements on the diffusion and electrotransport of impurities in Sn, obtained using a shear cell device.

2. Experimental method

In the long-capillary and capillary-reservoir techniques (Nachtrieb 1972), the duration of the experiment is usually not perfectly controlled and segregation phenomena may occur when cooling the samples. The shear cell technique (Nachtrieb and Petit 1956, Potard *et al* 1972) allows perfect control of the temperature and duration of the experiment, and avoids spurious effects due to solute segregation.

In our experiments, we used a shear cell made of a stack of 20 discs of graphite or boron nitride, 4 mm thick and 42 mm diameter, adjusted on a vertical axis (figure 1).

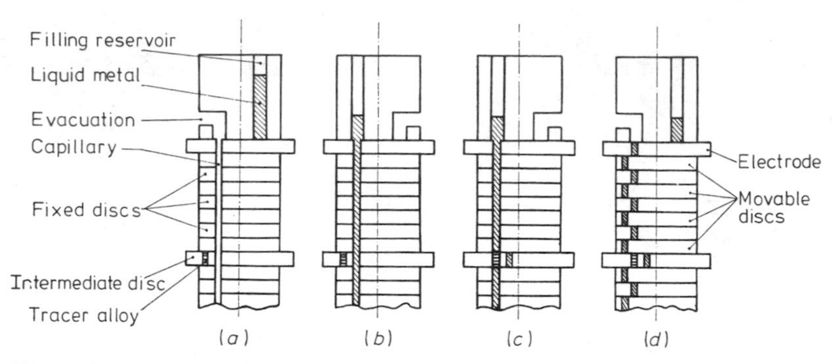

Figure 1. Four configurations of the shear cell: (*a*) evacuation of the capillaries, (*b*) filling of the capillaries and introduction of the current, (*c*) beginning of the (electro) diffusion run, (*d*) sectioning of the capillaries.

Four holes, 1·5 mm diameter, bored in each of these discs, can be aligned to form four capillaries, or sheared with respect to each other. Figure 1 shows the course of a diffusion run:

(i) In an intermediate disc, two 1·5 mm diameter holes are bored for each capillary; one of these holes is filled with an alloy of tin and the solute under investigation. The second hole allows filling of the capillary.

(ii) The filling being completed, the temperature of the experiment is carefully controlled and the desired current density is passed through the capillaries. The intermediate disc is then rotated from outside the furnace, which accurately defines the beginning of the diffusion run.

(iii) At the end of the diffusion run, the capillaries are sectioned through a rotation of the discs with respect to each other, and solidification takes place. This procedure avoids segregation phenomena when cooling the samples.

(iv) The impurity concentration in the sections of the capillary is measured using a standard (NaI–Tl) γ-spectrometer when radioactive isotopes are used, or by mass spectrometry for natural isotopes.

2.1. *Diffusion experiments*

In the diffusion experiments the thickness of the intermediate disc is so small with respect to typical mean square displacements of solute atoms that we can use the solution of Fick's equation corresponding to a thin layer in an infinite medium:

$$c(x, t) = \frac{Q}{2\sqrt{(\pi D t)}} \exp\left(-\frac{(x - x_{\mathrm{m}})^2}{4Dt}\right) \tag{1}$$

where Q is the number of solute atoms initially deposited in the thin layer, and x_{m} denotes the location of the maximum of the gaussian distribution; x_{m} is determined through a mean-square fitting to the experimental data.

There are many advantages in using this geometry:

(i) it allows an accurate determination of D from the measurement (figure 2) of the slope $-1/4Dt$ of the straight lines $\ln c = f[(x - x_{\mathrm{m}})^2]$;

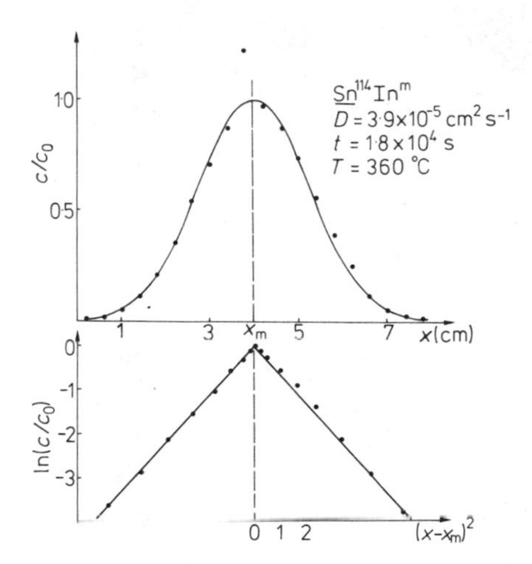

Figure 2. A typical concentration profile measured after a diffusion experiment.

(ii) any irregularity or accidental imperfection occurring during the diffusion run (convection, spurious temperature gradients, solute cluster formation etc) is clearly exhibited on the observed concentration profile (figure 2).

2.2. Electrotransport experiments

For electrotransport experiments, a shear cell made of boron nitride is used. The DC current is introduced into the capillaries by means of two stainless steel discs at each end of the shear cell. When a current density of about 600 A cm^{-2} is passed through the capillaries, the effective diffusion coefficient D_{eff} is found to be larger than D by a factor of about 50. The approximation of an infinite medium is no longer valid (figure 3) and we use the following boundary conditions:

$$\begin{cases} c(x,0) = Q\delta(x - x_0) \\ c(\infty, t) = \dfrac{\partial c}{\partial x}(\infty, t) = 0 \\ \dfrac{\partial c}{\partial x}(0,t) = \dfrac{\langle v \rangle}{D_{\text{eff}}}\, c(0,t) \end{cases} \tag{2}$$

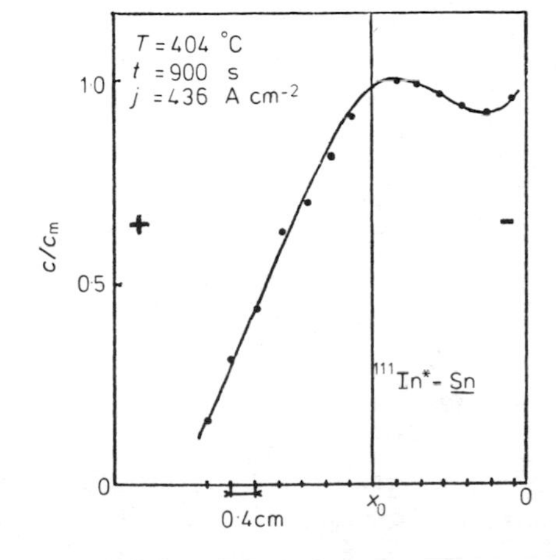

Figure 3. Concentration profile exhibiting boundary effects, after an electrotransport experiment.

where 0 is the location of the reflecting barrier and $\langle v \rangle$ is the mean drift velocity of the solute atoms under the action of the electric field. A mean-square fitting of the experimentally measured concentration to the solution of Fick's equation allows the determination of D_{eff} and $\langle v \rangle$.

3. Experimental results

3.1. Impurity diffusion

The measured diffusion coefficients of a series of radioactive isotopes in tin are given in table 1.

Table 1. Diffusion coefficients of ^{110m}Ag, ^{109}Cd, ^{114m}In, ^{113}Sn and ^{125}Sb in liquid tin; z_i is the valence of the solute relative to Sn; n is the number of different measurements of D_i, and D_s is the diffusion coefficient of ^{113}Sn.

Solute	z_i	T (°C)	t (10^3 s)	n	D_i (10^{-5} cm^2 s^{-1})	D_i/D_s
^{110m}Ag	−3	355	18·0	3	3·8 ± 0·2	0·97
^{109}Cd	−2	354	28·8	4	2·1 ± 0·1	0·52
^{114m}In	−1	360	18·0	5	3·9 ± 0·2	0·98
^{113}Sn	0	355	19·8	6	4·0 ± 0·2	1
^{125}Sb	+1	357	19·8	3	3·0 ± 0·2	0·75

These results are in fairly good agreement with those obtained in preceding investigations (du Fou and Gerl 1976).

Contrary to the investigations carried out on solid tin (Huang and Huntington 1974), the diffusion coefficients of solutes in liquid tin are rather insensitive to their relative valence z_i with respect to the solvent. This feature seems to contradict diffusion models based on hole mechanisms, as such models would predict, owing to the electrostatic interaction between the solute and the neighbouring hole, a strong dependence of D on z_i.

Our experimental results can be interpreted using the hard-sphere model of Thorne and co-workers (Chapman and Cowling 1970). According to this model, the self-diffusion (D_s) and impurity-diffusion (D_i) Enskog coefficients can be written respectively:

$$\left\{\begin{aligned} D_s &= \frac{3}{8n\sigma_s^2}\,\frac{1}{g(\sigma_s)}\left(\frac{kT}{\pi m_s}\right)^{1/2} \\ D_i &= \frac{3}{8n\sigma_{is}^2}\,\frac{1}{g_{is}(\sigma_{is})}\left(\frac{kT}{2\pi\mu}\right)^{1/2} \end{aligned}\right. \qquad (3)$$

where $\sigma_{is} = (\sigma_i + \sigma_s)/2$ is the average of the solute (σ_i) and solvent (σ_s) diameter, g_{is} is the radial distribution of unlike atoms in contact, and $\mu = m_i m_s/(m_i + m_s)$. The solute concentration being vanishingly small, we use the same number density n for the pure solvent and for the dilute alloy, and the low concentration expansion of D_i.

From the measurement of D_i/D_s the ratio g_{is}/g_s can be estimated once the diameters σ_i and σ_s are chosen. For solutes which are normally liquid at the temperature of the diffusion measurement, we use the formula given by Protopapas *et al* (1975) to determine σ_i and σ_s:

$$\sigma(T) = 1{\cdot}288 \times 10^{-8} \left(\frac{M}{\rho_m}\right)^{1/3} \left(1 - 0{\cdot}112 \left(\frac{T}{T_m}\right)^{1/2}\right) \text{(cm)}$$

where M is the molar mass and ρ_m is the mass density of the pure liquid (solvent or solute) at its melting temperature T_m. For Ag and Sb we used diameters obtained by Pauling (1960) in the solid phase. The fact that all impurity diffusion coefficients are smaller than D_s seems to indicate that the ratio g_{is}/g_s is larger than unity (table 2).

Table 2. Determination of the ratio g_{is}/g_s of the radial distribution functions at contact.

Liquid	T_m (°C)	ρ_m (g cm^{-3})	M (g)	σ (629 K) (Å)	g_{is}/g_s
Sn	232	7·00	119	2·90	1
Ag				2·88	1·038
Cd	321	8·02	112	2·75	2·066
In	156	7·03	115	2·83	1·045
Sb				3·31	1·163

In order to improve the results, account should be taken of dynamical correlations. Equations (3) have been actually written without considering these correlations which are very sensitive to the size of the solute, as shown by Alder *et al* (1974). A molecular dynamics calculation of the correction factor for Enskog's formulae should allow an accurate determination of the ratio g_{is}/g_s.

3.2. Electrotransport experiments

Table 3 and figure 4 give the effective diffusion coefficient of In in Sn when a current density j is passed through the sample.

These results can be fitted to the formula:

$$D_{eff}(j) = D_i + Bj^2$$

where D_i is the diffusion coefficient and $B = 1{\cdot}08 \times 10^{-8}\,\text{cm}^6\,\text{s}^{-1}\,\text{A}^{-2}$, showing that the electro-osmosis mechanism suggested by Pikus and Fiks (1959) should be operating.

Table 3. Effective diffusion coefficient of In as a function of the current density j.

T (°C)	403	404	408	405	400	405	400	404	407
t (s)	6300	6300	6300	6300	2100	6300	1680	900	300
j (A cm^{-2})	0	29·4	70·2	129·0	143·7	199·2	275·0	435·7	547·8
D_{eff} (cm^2 s^{-1})	5·05	6·60	15·9	29·5	38·7	52·4	95·0	216	330

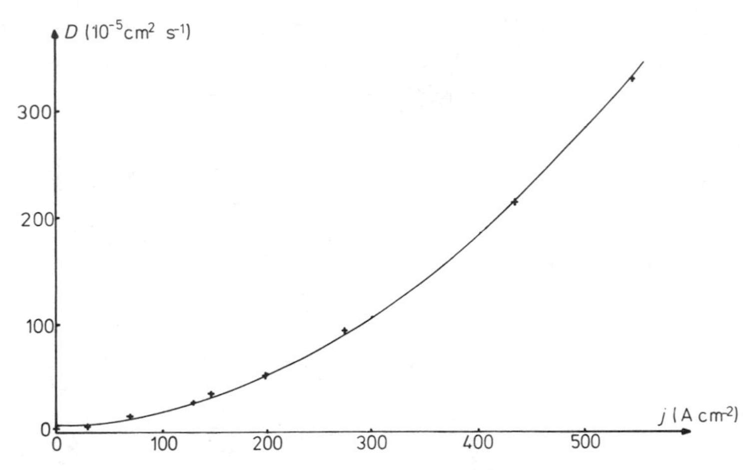

Figure 4. Variation of the effective diffusion coefficient D_{eff} with the current density j.

Possibilities of other contributions to electroconvection are, however, not to be excluded (Lodding 1967).

In order to obtain accurate values for $\langle v \rangle$, the drift velocity of the solute under the influence of an electric field, we intend to use a shear cell with a larger number of discs, which would avoid boundary effects on the diffusion profiles. The deduced effective valence will be compared to the specific resistivity of the impurity in the host metal, according to proposed theories (Turban *et al* 1976).

Acknowledgment

It is a pleasure to thank Dr A Lodding for very stimulating discussions.

References

Alder B J, Alley W E and Dymond J H 1974 *J. Chem. Phys.* **61** 1415
Chapman S and Cowling T G 1970 *The Mathematical Theory of Non-uniform Gases* 3rd edn (cam-(Cambridge: Cambridge University Press)
Escanyé J M and Gerl M 1976 *Proc. 19 ème Colloque de Métallurgie, Saclay*
du Fou M O and Gerl M 1976 *Phys. Chem. Liquids* **5** 113
Huang F H and Huntington H B 1974 *Phys. Rev.* **B9** 1479
Lodding A 1967 *J. Phys. Chem. Solids* **28** 557
Nachtrieb N H 1973 *The Properties of Liquid Metals* ed S Takeuchi (London: Taylor and Francis)
Nachtrieb N H and Petit J 1956 *J. Chem. Phys.* **24** 746
Pauling L 1960 *Nature of the Chemical Bond* 3rd edn (Ithaca, NY: Cornell University Press)
Pikus G E and Fiks V B 1959 *Sov. Phys.–Solid St.* **1** 972
Potard C, Teillier A and Dusserre P 1972 *Mater. Res. Bull.* **7** 583
Protopapas P, Andersen H C and Parlee N A D 1973 *J. Chem. Phys.* **59** 15
—— 1975 *Chem. Phys.* **8** 17
Protopapas P and Parlee N A D 1975 *J. Chem. Phys.* **11** 201
Turban L, Nozières P and Gerl M 1976 *J. Physique* **37** 159

Thermotransport of solutes in liquid tin and NFE calculation of the electronic heats of transport

M Balourdet†, Y Malmejac† and P Desre‡

† Centre d'Etudes Nucléaires de Grenoble, 85 X, 38041 Grenoble, Cedex, France

‡ Laboratoire de Thermodynamique et Physico-Chimie Métallurgiques associé au CNRS, BP44, 38401 Saint Martin d'Hères, France

Abstract. Using a capillary shear cell technique, we studied the thermomigration of trace amounts of ^{110}Ag, ^{114}In and ^{124}Sb in liquid Sn. Assuming the heats of transport to be constant in the range 250–700 °C we measured the following values:

$$Q^*_{\mathrm{Ag}} = -84 \qquad Q^*_{\mathrm{In}} = -75 \qquad Q^*_{\mathrm{Sb}} = -68 \text{ cal mole}^{-1}.$$

However, the thermotransport of Ag in liquid Sn exhibits a reversal above 600 °C: for higher temperatures Q^*_{Ag} becomes positive. We must take into account the temperature dependence of the heat of transport, due to the competition between two contributions: the interaction of a silver ion with the other ones results in an 'intrinsic' part of Q^* which can be supposed, as a first approximation, to vary as AT; the interaction with the conduction electrons results in an 'electronic' part of Q^* which we can show to vary as BT^2. Using this temperature variation of the heat of transport $Q^* = AT + BT^2$ we measured the thermotransport of ^{110}Ag in liquid Sn in the range 250–1250 °C:

$$Q^*_{\mathrm{Ag}} = -0{\cdot}6\,T + 0{\cdot}0007\,T^2 \text{ cal mole}^{-1}$$

An extension of the screening calculation of Gerl (1967) for the electronic heats of transport to the case of concentrated liquid metal alloys in the framework of Ziman's NFE theory is proposed (Faber and Ziman 1965). The electronic heat of transport Q_j of the jth constituent in a liquid alloy then reads:

$$Q^*_j = Z\frac{\pi^2 k^2 T^2}{3E_{\mathrm{F}}}\left(\xi - \frac{\rho_j}{\rho}\,\xi_j\right)$$

with

$$\xi_j = -\left(\frac{\mathrm{d}\lg\rho_j}{\mathrm{d}\lg E}\right)_{E=E_{\mathrm{F}}}$$

and

$$\rho_j = \frac{3\pi\Omega}{\hbar e^2 v_{\mathrm{F}}^2}\int_0^1\left[u_j^2 + \sum_{i=1}^{n} x_i(a_{ij}-1)\,u_i\,u_j\right]4x^3\,\mathrm{d}x$$

Here, u_i are screened pseudopotentials, a_{ij} are partial structure factors (Balourdet *et al* 1976).

The same calculation leads to the electron wind force for electrotransport:

$$F_j = -Z(\rho_j/\rho)\,|e|\epsilon$$

identical to the one deduced by Stroud (1976). Using the experimental structure factor of liquid Sn and Animalu's pseudopotentials, we computed the electronic contribution to the heat of transport of Ag in liquid Sn: $Q^*_{\mathrm{Ag\,elec}} = 0{\cdot}00068\,T^2$ cal mole^{-1} which compares favourably with the measured value.

References

Balourdet M, Malmejac Y and Desre P 1976 *Phys. Lett.* **56A** 51
Faber T E and Ziman J M 1965 *Phil. Mag.* **11** 153
Gerl M 1967 *J. Phys. Chem. Solids* **28** 725
Stroud D 1976 *Phys. Rev.* **B13** 4221

Author Index